AAPS Advances in the Pharmaceutical Sciences Series

Series Editor
Yvonne Perrie, Strathclyde Institute of Pharmacy and Biomedical Sciences, University of Strathclyde, Glasgow, UK

The AAPS Advances in the Pharmaceutical Sciences Series, published in partnership with the American Association of Pharmaceutical Scientists, is designed to deliver volumes authored by opinion leaders and authorities from around the globe, addressing innovations in drug research and development, and best practice for scientists and industry professionals in the pharma and biotech industries. Indexed in

Reaxys

SCOPUS Chemical Abstracts Service (CAS) SCImago

EMBASE

More information about this series at https://link.springer.com/bookseries/8825

Seshadri Neervannan • Uday B. Kompella
Editors

Ophthalmic Product Development

From Bench to Bedside

Springer

aaps®

Editors
Seshadri Neervannan
Chief Operating Officer
Tarsus Pharmaceuticals Inc
Irvine, CA, USA

Uday B. Kompella
Professor of Pharmaceutical Sciences
University of Colorado Anschutz
Medical Campus
Aurora, CO, USA

ISSN 2210-7371 ISSN 2210-738X (electronic)
AAPS Advances in the Pharmaceutical Sciences Series
ISBN 978-3-030-76366-4 ISBN 978-3-030-76367-1 (eBook)
https://doi.org/10.1007/978-3-030-76367-1

This Springer imprint is published by the registered company Springer Nature Switzerland AG
The registered company address is: Gewerbestrasse 11, 6330 Cham, Switzerland

Preface

Diseases of the eye, a complex sensory organ for sight, affect the quality of life of hundreds of millions of people around the world. Eye diseases can result in vision loss or in some cases blindness, if left untreated. To treat eye diseases, ocular therapies and drug products have gone through a long and extensive evolution from early mysticism to more rational approaches. A variety of ophthalmic drug products are now marketed to manage dry eye, diabetic retinopathy, macular degeneration, glaucoma, infection, inflammation, among other eye diseases.

The anatomical and physiological barriers of the eye make drug treatment enormously challenging. The ocular surface is complex with multiple tissues protecting the eye from exposure to the external environment. The cornea consists of several layers with both hydrophobic and hydrophilic properties that hinder absorption of drugs administered to the eye surface. The posterior segment of the eye is not as accessible from topical route, and treatment of diseases of this segment, unlike anterior segment, is usually invasive. Additionally, blood-tissue barriers limit drug delivery from the systemic circulation to the eye and vice versa.

Ophthalmic products have evolved steadily from approaches such as topical solutions to complex drug delivery systems, devices, and more recently to gene and cell therapy. In addition, the rapidly evolving advances in diagnostic medical technology tools such as 3D imaging with Optical Coherence Tomography and Corneal Pachymetry have greatly increased the understanding and management of eye diseases. The combination of an aging population and increasing life expectancy has brought the eye disease burden and treatments to the scientific forefront. In product development, scientists address technical complexities for each product, while applying the basic principles of formulation, dosage form development, processing, manufacturing, and the assurance of purity, safety, and efficacy of the manufactured ophthalmic product. This book attempts to piece together a comprehensive narrative of all elements of ophthalmic product development from several distinguished authors. The chapters are structured along this theme and offer a useful guide for scientists, students, regulatory experts, and other personnel working on the commercialization of ophthalmic drug products.

The book has four major parts. Part I provides a history of ophthalmic product development as well as an understanding of the anatomy and physiology of anterior and posterior segments of the eye from the perspective of product development. Part II delves into the fundamental approach to product development with a vision for high-quality products to meet regulatory and patient needs. Part III discusses various technologies and dosage forms in a product development scientist's armamentarium, addressing many products in development and in the market. Part IV takes a deeper look at some specialty approaches including artificial tears and devices that are an essential part of the treatment continuum. This part also presents the more recent advances such as gene and cell therapy that are shaping the future of ophthalmic medicine.

The distinguished multidisciplinary author panel represents several disciplines involved in understanding the anatomy, physiology, and pharmacology as well as development, regulatory oversight, and commercialization of ophthalmic drug products. The individual chapters provide new knowledge, discuss the challenges, and summarize innovative solutions with expert opinions to address emerging challenges for ophthalmic drug and specialty product development. The diverse expertise that the authors bring through this book should stimulate and encourage the reader, especially the pharmaceutical product development professional, to go beyond one's own area of experience and engage in interdisciplinary discussions.

We would like to convey our appreciation to all the authors for their outstanding work and their dedication in contributing their expertise to this book. We are thankful to Springer and the American Association of Pharmaceutical Sciences (AAPS) for the opportunity to prepare this book and for their constant encouragement and support in completing this volume. We are indebted to Rachel Hartman at the University of Colorado for her help with chapter reviews, support in communicating with the authors, and compilation of various components of the book. We are also grateful to our mentors, peers, colleagues, students, and trainees who helped shape our knowledge, understanding, skills, and expertise in developing ophthalmic drug products for treating eye diseases.

Irvine, CA, USA Seshadri Neervannan
Aurora, CO, USA Uday B. Kompella

Contents

Part I
Introduction and Background

Introduction and History of Ophthalmic Product Development

Seshadri Neervannan

Abstract Ocular applications have a fascinating history from the time of various ancient civilizations. From the first known written record of medical treatment of eye diseases dating back to 1500 BCE to the current amazing array of products including gene and cell therapy treatments, ophthalmic products have reached a point where no segment of the eye is ignored and curing blindness is in the horizon. An increase in the aging population has increased the prevalence of age-related ocular conditions that are difficult to treat, and the treatments are still evolving. Regulations and science are converging around the world, and with the increasing access to information, many new innovative products are in the works that will significantly alter the treatment of ophthalmic conditions in the years and decades to come.

Keywords Ophthalmic products · Topical administration · Retinal delivery · Device products · Gene therapy · Generic product development · Preservatives · Packaging development · Nanoparticulates · Biologics products

Ophthalmic product development has long been considered a niche expertise area practiced by select experts. The main reason is that the eye is a specialized organ and is unique in its anatomy, physiology, and etiology of various diseases that afflict the eye. From ancient times, the eye has been a subject of various debates in terms of understanding its purpose, how it functioned, as well as the various remedies that were prescribed. Medical treatments for eye diseases have been traced back to ancient Egypt (1500 BCE) and to Ayurvedic Indian healing in which over 60 ocular diseases and over 50 surgical treatments were noted (Hirschberg 1982; Andersen 1997; Velpandian and Gupta 2016). The key advances were then perpetuated by ancient Greeks (Aristotle documented three layers in the eye), and in the second century AD, the Greek physician, Galen, identified the curvature of the lens and

S. Neervannan (✉)
Chief Operating Officer, Tarsus Pharmaceuticals Inc, Irvine, CA, USA
e-mail: seshaneer11@gmail.com

S. Neervannan, U. B. Kompella (eds.), *Ophthalmic Product Development*, AAPS Advances in the Pharmaceutical Sciences Series 37,
https://doi.org/10.1007/978-3-030-76367-1_1

cornea as well as the structure of the optic nerve and tear ducts. His model of the eye is closest to the contemporary models and served as a springboard for more innovations to come.

We have come a long way when it comes to understanding treatments for the eye. A prescription for dryness of the eye from the Assyro-Babylonian ophthalmology (Krauss 1934) describes it as "…he shall rub an onion, drink it in beer, apply oil to his eyes…." A German Egyptologist put together a collection called *Ebers Papyrus* that describes eye paste with basic elements; Celtic treatments for dimness of the eye contained pound fennel roots and honey (Fig. 1).

During medieval times, inventions such as microscope and lenses advanced the understanding of the ocular anatomy and physiology greatly. Hermann von Helmholtz invented the ophthalmoscope that revolutionized the practice of medicine. This led to various surgical treatments and use of devices to correct various limitations of the eye (Fig. 2). Ernst Abbe, a German scientist, contributed to significant advances to corrective lenses and is considered the father of modern optics.

Systematic understanding of pharmacological treatments for eye conditions started with the discovery of atropine in the early 1800s as well as with advances on a clear understanding of the differences between cataracts and glaucoma. Even before that, Greek women used extracts of *Atropa belladonna* as a cosmetic to enlarge their pupils. Atropine is still being used to dilate the iris and for other conditions. Subsequently, the late nineteenth century and twentieth century saw

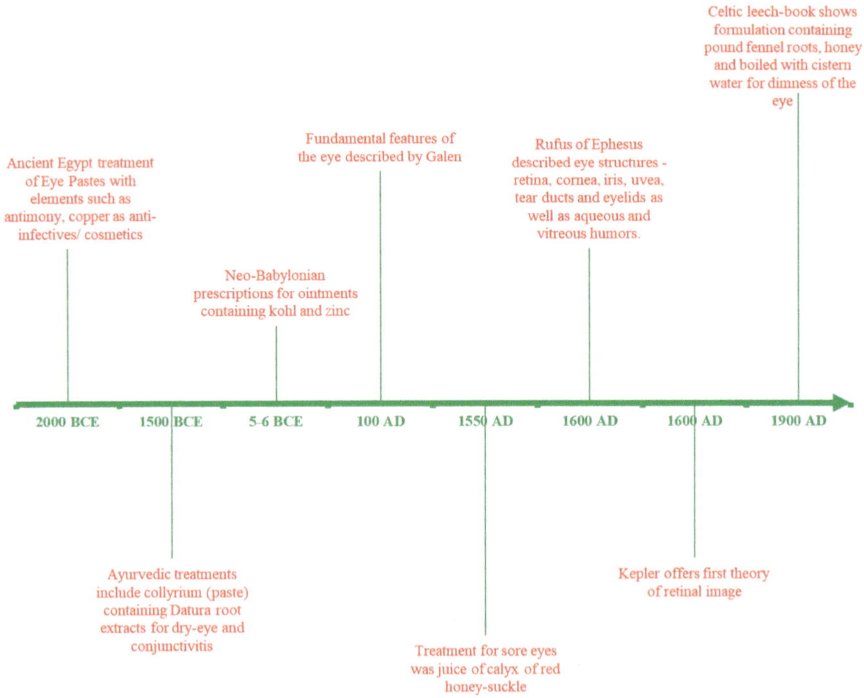

Fig. 1 Ancient history of ophthalmology

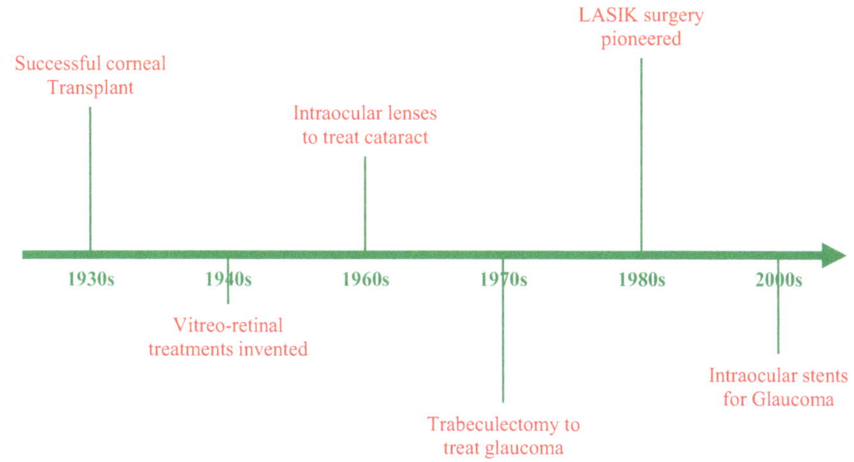

Fig. 2 Twentieth-century advances in ophthalmic treatments (surgery and devices)

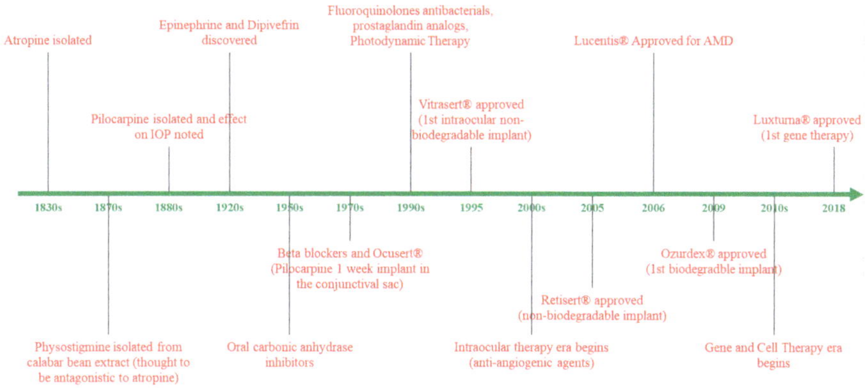

Fig. 3 Modern pharmacological and drug delivery advances

more advances with discovery of pilocarpine and other nonselective sympathomimetics that led to more sophisticated treatments for glaucoma (Velpandian and Gupta 2016) (Fig. 3). A significant shift in innovation occurred in the early twenty-first century when scientists successfully started using antiangiogenic agents for retinal indications that were caused by vascular leakage. Thus, the age of intraocular treatments for the retina began that also coincided with the increase in aging population. These conditions had very poor options for treatment of retinal conditions prior to anti-VEGF treatments and usually led to progressive blindness. These advances in intraocular treatments also necessitated a big shift in ophthalmic product development as conventional systems for topical ocular surface delivery were no longer effective; invasive treatments had to be invented and this charted another inflection point in drug delivery innovation (Fig. 3). The advances in treatments with gene and cell therapy technologies in the last decade is shifting the paradigm.

Ophthalmic disease disproportionately affects the aging population. It is estimated that, globally, there are >150 million blind and >300 million visually impaired individuals, discounting uncorrected refractive error (Access Economics 2010). There are over 100,000 blind from glaucoma in the USA alone representing about 10% of the blinded population (Quigley and Vitale 1997). The percentage of people over age 65 is growing rapidly (Population Division, DESA 2002). Hence, ocular impairment will be an even greater burden on society in the future. In addition, as most drugs are applied directly to the eye, patient compliance, especially with the elderly, poses additional challenges to ophthalmic product development. There is a great need for medications to mitigate ophthalmic disease and reduce the overall healthcare burden.

Challenges and Uniqueness for Ocular Product Development

The eye is a very sensitive and complex organ and is highly impervious to external agents. It has naturally built defense mechanisms that provide significant anatomical and physiological barriers for drug treatments (Cholkar et al. 2013) (Fig. 4).

For most ophthalmic conditions, topical administration is preferred over systemic therapy. There are significant advantages to topical treatment over systemic therapy for treating ophthalmic conditions. Because of direct delivery to the target tissues, the treatment is highly localized, and this minimizes systemic exposure to drugs that may cause other unwanted side effects. However, delivering drugs to the topical targets is beset with many challenges. This topic is addressed extensively in other chapters, but suffice to say, a few key barriers for low ocular bioavailability include nasolacrimal drainage, reflex blinking, corneal barriers (both physical and metabolic), and low residence time on the ocular surface (Lee and Robinson 1986). Even for ocular surface conditions such as dry eye, the barriers are compounded because multiple target tissues are involved in the etiology of the disease and targeted delivery of drugs becomes highly challenging (Stern et al. 2013).

Advances in understanding of retinal diseases have created opportunities but also have its unique challenges. Topical treatments for retinal and other posterior chamber conditions have not been successful. Blood-retinal barriers are a major impediment to systemic treatment. The blood-retinal barrier is anatomically separated into inner and outer blood barriers, the endothelial cells of the retinal vasculature, and the retinal pigmented epithelium, respectively (Cunha-Vaz 1979). Very lipophilic compounds may penetrate the blood-retinal barriers but they may not be very efficient. This approach requires very large doses and hence significantly increases the risk of systemic safety. The default approach for most drugs to treat posterior segment of the eye has been through local drug administration such as intravitreal or periocular injections.

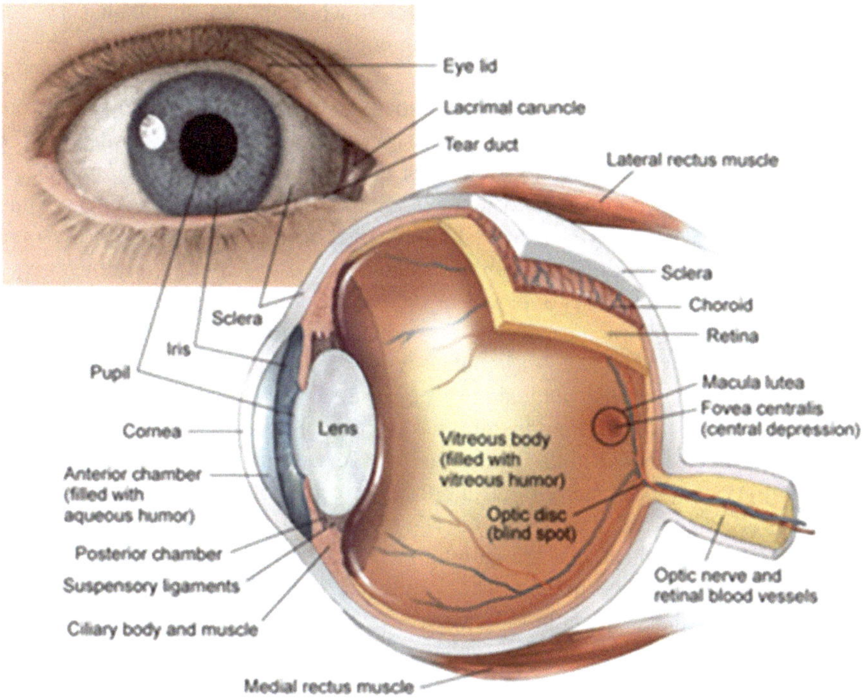

Fig. 4 Structure and features of the eye

Ophthalmic Products and Their Progressive Complexity

Conventional Topical Products

Conventional ophthalmic preparations primarily consisted of topical approach because of its noninvasive nature and the ability to directly access the organ of interest. Key requirements for such preparations were that it be a sterile liquid, with minimal foreign particulate matter to not cause any eye irritation. Also, other key goals are to maximize drug delivery to target tissues while minimizing toxic effects to local tissues. Patient acceptance and compliance is key especially with respect to stinging, blurring, redness, and any general discomfort (Richardson et al. 2013; Gooch et al. 2012). In addition, the active component needs to be stable for the intended storage duration of at least 2 years, and more importantly the formulation must prevent contamination from pathogens as infections of the eye can be a major cause of blindness. A survey of available ophthalmic drug formulations indicated that over 60% of them are solutions followed by ointments and suspensions (Lang 1995). The reasons for the overwhelming majority for solutions as a preference include blurring from ointments and discomfort and irritation from suspensions.

Developing a topical ophthalmic preparation, while appears simple, has a lot of constraints. A solution requires that the drug not only be soluble but also be chemically stable. Many conventional solubilizers used in other dosage forms are restricted as many excipients are not compatible with ocular surface tissues. Other key factors for topical formulations include compatibility of pH and being iso-osmotic with the tear fluid. One of the biggest constraints for drug bioavailability is the rapid clearance of the formulation from the ocular surface that minimized drug residence time for absorption. Over the years, several approaches have been employed to resolve the issues including addition of polymers and other viscosity-modifying agents to reduce clearance time. Other approaches such as prodrugs that improved drug penetration also found limited success. Many other innovative approaches to improve upon drug residence time such as contact lenses coated with drugs, ocular inserts such as Ocusert®, soluble ophthalmic drug inserts (SODI) which are small oval wafers produced with acrylamide, minidisks, artificial tear inserts (Lacrisert®), collagen shield, etc. proved highly limiting and had various issues including ocular compatibility, unnoticed excretion from the eye, feeling of foreign body in the eye, and more (Baranowski et al. 2014).

More recent advances that are gaining significant foothold both with sustained drug delivery to the anterior segment tissues and minimizing patient compliance include biodegradable and nonbiodegradable ocular implants. Durysta® is an investigational stage product that is designed to deliver drugs to the anterior chamber to effectively control IOP. Many other approaches are in the works as depicted in Fig. 5 (Verma 2018).

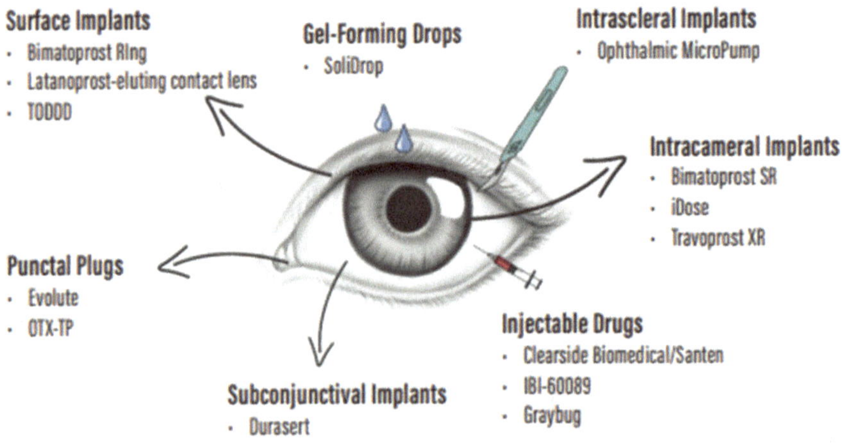

Fig. 5 Various sustained-release ocular products in development

Posterior Segment Drug Products

Delivery to the posterior segment of the eye is more challenging. As mentioned earlier, neither the topical route nor systemic route has been effective, and the only way to access the tissues has been with invasive direct intravitreal injections. Such an approach is currently being used for the administration of drugs such as Macugen® (pegaptanib), Avastin® (bevacizumab injection), Lucentis® (ranibizumab injection), and Eylea® (aflibercept). In addition, the clearance from intravitreal space is high (even though macromolecules tend to hang around a bit longer). Sustained-release approach is not only desired but also a necessity to minimize frequency of administration.

Non-erodible implants for treating posterior segment disease were developed initially to enable sustained release. These implants provide near-zero order release without significant burst. The disadvantage with these systems is that they may require surgical removal after their drug payload is exhausted. Retisert™ and Vitrasert™ are approved for the delivery of fluocinolone acetonide and ganciclovir, respectively. They both have to be surgically removed. Iluvien™ is a fluocinolone acetonide intravitreal insert. It is a smaller size than Retisert® and is designed to release drug over 18–30 months as well as expected to stay in the eye permanently, eliminating complications associated with implant extrusion.

On the other hand, biodegradable implants erode or dissolve eliminating the need for removal. This minimizes the risks associated with surgery as the implant erodes on its own without having to retrieve it. They are biocompatible and are eliminated safely from the body. They also can be administered by injection rather than surgery. The polymer matrix comprising these systems degrades into nontoxic metabolites as drug is delivered. Drug release is generally first order. Ozurdex® is a biodegradable polymeric implant containing PLGA copolymers delivering dexamethasone and approved for retinal vein occlusion, noninfectious posterior uveitis, and diabetic macular edema (Chang-Lin et al. 2011).

The twenty-first century is bringing even more innovation in ocular therapy. Inherited ocular retinal diseases have become a perfect laboratory for AAV-based gene therapy. More than 20 products are in development today, with Luxturna® being the first approved to treat the effects of biallelic RPE65 mutation-associated retinal dystrophy, a rare genetic eye disease. It is administered as a subretinal injection (Rodrigues et al. 2018). All viral vectors are susceptible to degradation—both physical and chemical. The current formulation approach is primarily a frozen solution (−70 °C), which carries significant constraints in logistics, cost, and flexibility.

Other Key Components of an Ophthalmic Product

Ophthalmic products are more than just active drug and "inert" additives. As stated earlier, excipients play a key role as they have to be compatible with ocular tissues to minimize local tolerability (Abelson et al. 2017; McCann 2011). In some cases, they also elicit placebo response (e.g., dry eye conditions). Preservatives have often been used to keep the product sterile for multidose use; however, preservative-free approaches are gaining in popularity, especially with the advent of new ophthalmic dispensers that are designed to prevent microbial contamination and obviate the need for a preservative.

In addition, primary and secondary packaging are highly critical as many topical ophthalmic products are in semipermeable containers and are susceptible to oxygen and water transport. Many components of the container closures, including the volatile parts in the glue contained in secondary packaging, are known to leach into the products causing adverse events (Lynch 2011).

Artificial Tears Are a Key Segment in Ophthalmic World

The total artificial tears market is valued at over $2 billion, with a projected growth rate of 8% (MarketResearch.biz 2019). The market is growing due to the aging population as well as from other factors such as growing dry eye conditions from environmental effects. As aging results in weaker eyesight, people are more at risk of ophthalmic disorders. The rising use of contact lenses and digital devices not only in aging population but also in younger ones creates future opportunities for the growth of the market. Many artificial tears developed recently mimic real human tears and provide a great tool in the arsenal to treat dry eye conditions. Many of the innovations came out of a formulator's toolkit from years of understanding the interplay of excipients. Naturally occurring polymers such as methylcellulose derivatives to synthetic polymers such as polyvinyl alcohol, povidone, and Carbopol have been integrated into artificial tears. More recently hyaluronic acid-containing tears as well as omega-3 oils have added to increasing the comfort and compatibility of the tears products (Abelson 2014). The regulatory landscape is also changing, and many different pathways are now regulating these products (Table 1).

Repurposing Has Been a Key Stratagem in Ophthalmology

Ophthalmology has had the unique place in using old drugs for new applications. Drugs such as atropine and pilocarpine are still in use in clinical practice and finding new applications. Some drugs designed for ophthalmic use have been repurposed for other conditions such as bimatoprost (originally developed for glaucoma) for

Table 1 Regulatory pathways for artificial tears products around the globe

	Pharmaceutical	OTC drug	Medical device
USA		No submission required: if monograph requirement met	Contact lens rewetter indication: 510(k) clearance required class II device
Canada		CTD format (CMC data only) US OTC monograph requirements	
EU	Optional: requires clinical development and scientific advice		CE marking
Asia Pacific	CTD: most countries Korea: stricter CMC requirements Japan: more stringent requirements than most		NZ: No submission AUS: EU Technical file Singapore and others: EU Technical file and CE mark
China	CTD: Lengthy approval time. Requires local clinical trial		Contact lens rewetter indication only—prerequisite is 510(k) or CE marking
Latin America	Argentina, Brazil, and Columbia: CTD		Chile and Mexico
Other markets	CTD: most countries Middle East: unique stability requirements Russia: unique clinical requirements		Some countries: individual requirements vary, but mostly based on EU Technical file

eyelash growth (Law 2010) or expanded use of Botox® (which originally was developed for ophthalmic use) (Basar and Arici 2016). Even some of the placebo vehicles have been repurposed as artificial tears products. More recent examples include expanded use of antiangiogenic drugs that were originally developed for oncology conditions and now playing a major role in AMD (age-related macular degeneration) and DME (diabetic macular edema) among other retinal conditions.

Regulations Are Still Evolving, Especially for Complex Products

As ophthalmic treatments grow in complexity, so does the type of products. Regulating approval of new products and post-approval changes as well as bringing generic products that are bioequivalent with comparable safety and efficacy profile to the originator product requires a keen understanding of the interplay between ocular anatomy, physiology, disease state, and the product type. Since most ocular administration is local, directly to the organ and sometimes to the tissue of relevance, many of the common bioequivalence tools are rendered ineffective. New tools are needed. Many regulatory agencies have employed a case-by-case approach with product-specific guidances or through other guidance documents such as the recently issued draft guidance for "Drug Products, Including Biological Products, that Contain Nanomaterials—Guidance for Industry" (https://www.fda.gov/media/109910/download). FDA defines a complex product as (www.fda.gov/drugs/guidances-drugs/):

- A product with:
  - A complex active ingredient(s) (e.g., peptides, polymeric compounds, complex mixtures of APIs, naturally sourced ingredients)
  - A complex formulation (e.g., liposomes, colloids)
  - A complex route of delivery (e.g., locally acting drugs such as dermatological products and complex ophthalmological products and otic dosage forms that are formulated as suspensions, emulsions, or gels)
  - A complex dosage form (e.g., transdermals, metered-dose inhalers, extended-release injectables)
- Complex drug-device combination products (e.g., autoinjectors, metered-dose inhalers)
- Other products where complexity or uncertainty concerning the approval pathway or possible alternative approaches would benefit from early scientific engagement

In addition, call for new research to improve the tools to assess bioequivalence as well as characterizing these complex products has also been commissioned by the FDA as outlined in the report entitled "FY2018 GDUFA Science and Research Report: Ophthalmic Products" (https://www.fda.gov/media/130617/download). As the scientific understanding of complex ophthalmic products continues to improve, high-quality products that are safe and effective for patients will continue to grow and thrive.

Conclusion

We are in the midst of an explosive growth of ophthalmic products and treatments, especially with the aging population increasing the need for such products. As science and technology evolves, there is a greater need to further understand the complex organ that we all cherish and would like to preserve. The following chapters have attempted to capture a comprehensive view of the knowledge today, especially from the product development standpoint, covering the structure and function of the eye segments to the various product types, drugs, devices, and combination products alike.

References

Abelson MB. 3500 Years of artificial tears. Rev Ophthalmol. 2014.
Abelson MB, Rimmer D, Hollander DA. The foundation of a good formulation. Rev Ophthalmol. 2017.
Access Economics, prepared for AMD Alliance International. The global economic cost of visual impairment, March 2010. www.amdalliance.org/en/.
Andersen SR. The eye and its disease in ancient Egypt. Acta Ophthalmol Scand. 1997;75:338.
Baranowski P, Karolewicz B, Gajda M, Pluta J. Ophthalmic dosage forms: characterisation and research methods. Sci World J. 2014:861904.
Basar E, Arici C. Use of botulinum neurotoxin in ophthalmology. Turk J Ophthalmol. 2016;46(6):282–90.
Chang-Lin JE, Attar M, Acheampong AA, Robinson MR, Whitcup SM, Kuppermann BD, et al. Pharmacokinetics and pharmacodynamics of a sustained-release dexamethasone intravitreal implant. Invest Ophthalmol Vis Sci. 2011;52:80–6.
Cholkar K, Dasari SR, Pal D, Mitra AK. Eye: anatomy, physiology and barriers to drug delivery. In: Mitra AK, editor. Ocular transporters and their role in drug delivery. Elsevier; 2013. p. 1–36.
Cunha-Vaz J. The blood-ocular barriers. Surv Ophthalmol. 1979;5:279–96.
Gooch N, Molokia S, Condie R, et al. Ocular drug delivery for glaucoma management. Pharmaceutics. 2012;4:197–211.
Hirschberg J. The history of ophthalmology, vol. I: antiquity. Translated by F.C. Blodi. Bonn: Verlag J.P. Wayenborgh; 1982.
Krauss A. Assyro-Babylonian ophthalmology. Ann Med Hist. 1934;6:42–55.
Lang JC. Ocular drug delivery conventional ocular formulations. Adv Drug Deliv Rev. 1995;16:39–43.
Law SK. Bimatoprost in the treatment of eyelash hypotrichosis. Clin Ophthalmol. 2010;4:349–58.
Lee VHL, Robinson JR. Topical ocular drug delivery: recent developments and future challenges. J Ocul Pharmacol. 1986;2:67–108.
Lynch MP. Extractables and leachables: quality concerns and considerations for ophthalmic and injectable products. Am Pharm Rev. 2011.
MarketResearch.biz. Report generated by MarketResearch.biz. Global artificial tears market analysis trends, applications, analysis, growth, and forecast to 2028. 2019. https://marketresearch.biz/report/artificial-tears-market/.
McCann J. Advances and challenges in topical ocular medications. Glaucoma Today. 2011.
Population Division, DESA. United Nations, World population ageing 1950–2050;2002.
Quigley HA, Vitale S. Models of open-angle glaucoma prevalence and incidence in the United States. Inv Ophthalmol Vis Sci. 1997;38:83–91.
Richardson C, Brunton L, Olleveant N, et al. A study to assess the feasibility of undertaking a randomized controlled trial of adherence with eye drops in glaucoma patients. Patient Prefer Adherence. 2013;7:1025–39.
Rodrigues GA, Shalaev E, Karami TK, Cunningham J, Slater NKH, Rivers HM. Pharmaceutical development of AAV-based gene therapy products for the eye. Pharm Res. 2018;36(2):29.
Stern ME, Schaumburg CS, Pflugfelder SC. Dry eye as a mucosal autoimmune disease. Int Rev Immunol. 2013;32(1):19–41.
Velpandian T, Gupta SK. Ocular pharmacology and therapeutics: origin, principle, challenges and practices. In: Velpandian T, editor. Pharmacology of ocular therapeutics. Cham: Springer; 2016. p. 1–11.
Verma D. Sustained release drug delivery: closer than you think. Glaucoma Today. 2018.

Ocular Surface Anatomy and Physiology: Impact on Product Development

Patrick M. Hughes and Jie Shen

Abstract Topical administration of drops is by far the most common and convenient way to administer drug to the eye. However, the efficiency of this route of delivery remains poor. Much of the failure of topical drug delivery is caused by poor compliance and difficulties with administration; however very low bioavailability also limits the effectiveness of topical ophthalmic medications. The bioavailability ranges from 1 to 5%. To improve therapeutic efficacy, the ophthalmic formulation scientist must take into consideration the unique anatomy and physiology of the eye.

This chapter will review the ocular surface anatomy and physiology and discuss their impact on ophthalmic product development. The structure and function of the tear film, eyelids, and nasolacrimal system will be overviewed. Further, the impact that conjunctival and corneal permeability and absorption have on bioavailability will be discussed. The goal of this chapter is to provide an overview of the ocular surface anatomy and physiology relevant to drug delivery to enable engineering of optimized ophthalmic delivery systems.

Keywords Drug delivery · Tear film breakup · Tear turnover · Dry eye · Ocular anatomy · Benzalkonium chloride

Abbreviations

BAK	Benzalkonium chloride
BCRP	Breast cancer resistance protein
Cps	Centipoise
DED	Dry eye disease
IL-1β	Interleukin 1 beta

P. M. Hughes (✉)
Pharmaceutical Development, Visus Therapeutics, Irvine, CA, USA

J. Shen
Clinical Pharmacology, Nonclinical and Translational Sciences, AbbVie, Irvine, CA, USA

S. Neervannan, U. B. Kompella (eds.), *Ophthalmic Product Development*,
AAPS Advances in the Pharmaceutical Sciences Series 37,
https://doi.org/10.1007/978-3-030-76367-1_2

IOP	Intraocular pressure
MGD	Meibomian gland dysfunction
MMP-9	Matrix metallopeptidase 9
mOsm	Milliosmole
MRP	Multidrug resistance protein
OAG	Open-angle glaucoma
OHT	Ocular hypertension
OSD	Ocular surface disease
P-gp	p-Glycoprotein
ppm	Parts per million
TBU	Tear film breakup
TBUT	Tear film breakup time
TFLL	Tear film lipid layer
TNF-α	Tumor necrosis factor

Introduction

The ocular surface is continuously exposed to environmental insults. As a result, evolution has engineered the eye to possess exquisitely protective barriers including the cornea, conjunctiva, eyelids, and tear film to endure these exogenous threats. These same barriers however render topical drug delivery to the ocular tissues challenging. Additionally, it is quite difficult for patients, particularly in the elderly population, to remember and accurately self-administer topical ocular medications (Newman-Casey et al. 2015). As a consequence, improper dosing or poor compliance with the topical administration of eye drops is commonplace (Kulkarni et al. 2008; Richardson et al. 2013). To render ocular drug delivery effective, these constraints must be taken into account in the development of ophthalmic drug products. This requires optimization of ophthalmic drug substances and dosage forms for the unique challenges associated with drug administration in the eye.

The eye can be anatomically separated into the anterior and posterior segments (Fig. 1). The anterior segment is comprised of tissues in front of the vitreous humor. These include the lids, conjunctiva, cornea, aqueous humors, trabecular meshwork, iris ciliary body, and lens. Additionally, Schlemm's canal, the collector channels, and the aqueous and episcleral veins play a critical role in controlling intraocular pressure. The anterior segment of the eye makes up approximately 1/3 of the globe. For the purpose of this review, we will define the periocular structures, the lids, the Meibomian and lacrimal glands, and the nasolacrimal duct as part of the anterior segment.

By far the most common means to deliver drugs to the eye is by topical ocular administration to the anterior surface of the eye. Numerous researchers and reviews have identified the barriers to productive topical absorption over the years (Hughes et al. 2005; Kompella et al. 2010; Lee and Robinson 1986). These include precorneal, corneal, and intraocular variables. Estimates of ocular bioavailability from topical administration range from 1 to 5% (Gaudana et al. 2009). Precorneal factors including lacrimation and tear dilution, blinking, cul-de-sac volume limitations,

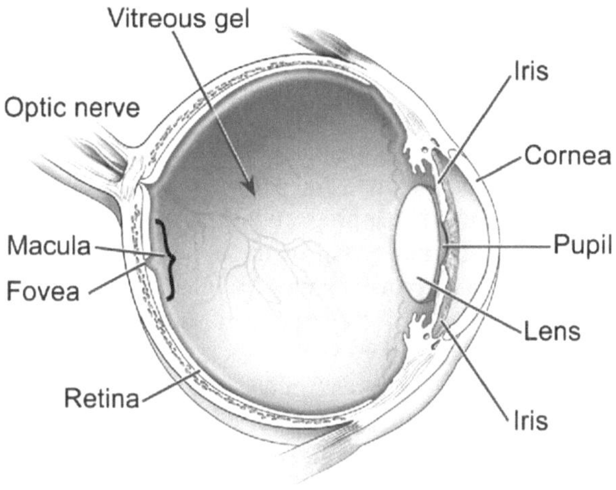

Fig. 1 Cross section of the eye. National Eye Institute, National Institutes of Health

conjunctival absorption, and nasolacrimal drainage lead to rapid clearance of topically administered drugs. Most of the administered dose is typically cleared from the precorneal space within a matter of minutes (Chrai et al. 1974; Ghate and Edelhauser 2008; Sieg and Robinson 1976). The cornea is a tri-laminate structure, with both highly lipophilic and hydrophilic layers, and also poses a significant barrier to topical absorption. Maintenance of effective intraocular drug concentrations is further limited by the rapid clearance of drug from the intracameral space due to the aqueous humor turnover. Clearly, the unique structure and function of the eye present significant barriers to topical drug delivery.

The overall goal of this chapter is to provide a brief introduction into the anatomy and physiology of the ocular surface in the context of drug delivery and to assess their impact on product development. More specific formulation approaches are provided in subsequent chapters. A cross section of the human eye with the major anatomical features labeled is shown in Fig. 1.

Lids and Blinking

The eyes' first line of defense from external insults is the eyelids. The eyelids are skin folds over the cornea that open and close under voluntary and involuntary (spontaneous and reflex) control. Spontaneous blinking serves to form and spread the tear film across the cornea, remove debris from the ocular surface, and facilitate tear drainage. The outer surface of the eyelid is a thin layer of skin with minimal squamous epithelium and a thin epidermis (Cochran et al. 2018). There is no subcutaneous fat in the eyelid. These factors render the eyelid a potential dosing site for ophthalmic drugs (See et al. 2017). The remainder of the lid is comprised largely of the orbicularis oculi muscle, the palpebral fascia, the tarsal plates, and the palpebral conjunctiva. The lid margin is lined with eyelashes that protect the eye from dust and foreign particulates. Meibomian glands in the lids secrete the lipids forming the tear film lipid layer (TFLL).

Several researchers have exploited the permeable nature of the eyelid to evaluate drug delivery from the skin surface of the lid. This route may be useful in targeting the Meibomian, lacrimal glands, and conjunctival goblet cells. In US patent US9034830B2, "Methods and compositions for sustained delivery of drugs," the inventors demonstrate increased tear production in two human subjects from a pilocarpine lotion applied to the outer skin of the upper eyelid (Nandurit and Dyer 2009). They administered a lotion containing 1% pilocarpine and 0.16% caffeine to the upper eyelid of two severe dry eye patients. Increased tear production was immediately observed as measured by Schirmer's test. Further examples were given with timolol, physostigmine, latanoprost, and brimonidine. In US patent application US20070053964 A1, "Ophthalmic percutaneously absorbed preparation containing muscarinic receptor agonist," the inventors demonstrate the efficacy of pilocarpine administered in an ointment to the outer surface of rabbit lids on lacrimal secretion as measured with Schirmer's strips (Isowaki and Ohtori 2004). The inventors shaved and taped stripped the lids of Japanese white male rabbits. The impact of tape stripping and shaving the skin was not discussed. Then either 7.4 mg of a 1% pilocarpine hydrochloride ointment was applied to the outer eyelid skin or a 50 μL drop of a 10% pilocarpine solution was instilled into the eye. The lacrimation was greater and more prolonged for the cohort receiving the application to the outer lid than was achieved with topical dosing to the cul-de-sac. In US patent application US20090209632A1, "Percutaneously absorptive ophthalmic preparation comprising olopatadine," the same inventors demonstrated better efficacy of applying a 1% topical olopatadine adhesive to the lid at preventing histamine-induced conjunctival chemosis in guinea pigs than a 0.1% instilled drop.

Blinking occurs with the relaxation of the levator palpebrae superioris muscle and contraction of the orbicularis oculi muscle. This initiates a downward movement of the upper lid representing the major portion of the blink. The lower lid elevates to a lesser extent causing lid closure from the temporal to the nasal side in a zipper-like movement (Doane 1980). Two nasolacrimal ducts drain the tears from the precorneal space. These ducts have openings of about 0.3 mm in diameter and drain through canaliculi into the lacrimal sac (Wilson et al. 2007). The closing of the orbicularis oculi muscle also opens the lacrimal sac. This creates a negative pressure relative to the canaliculi and drains the tears into the lacrimal sac (Lemp and Weiler 1983). Tears then flow from the lacrimal sac into the nasolacrimal duct and on into the nasal cavity. The upper lid motion also serves to remove debris from the ocular surface on closing and to redistribute tears and lipids from the lower tear meniscus across the ocular surface upon opening. The tear film stabilizes in about 0.3–1 s on the ocular surface in normal individuals as measured by lipid layer interferometry. This can take up to 3 s in the aqueous deficient state (Goto and Tseng 2003a, b). Blinking occurs in normal individuals at a rate of approximately 15 times per minute on average and lasts from 0.3 to 0.4 s (Doughty 2001). Reading or computer work has been shown to decrease the blinking frequency by up to fivefold, and blinking frequency is also decreased in dry eye disease (Nakamori et al. 1997; Patel et al. 1991; Tsubota and Nakamori 1995).

The reflex blinking caused by topical ocular drop instillation results in a rapid removal of an applied dose by the above mechanisms as well as spreading of the applied dose across ocular surfaces. It has been estimated that the drug deposited onto the precorneal tear film is spread within the first couple of blinks after topical instillation. Any intervention that can reduce precorneal clearance of an administered dose may improve the bioavailability of topically applied drugs.

Punctual plugs are approved as medical devices in the treatment of dry eye to reduce tear drainage and are placed by an ophthalmologist or optometrist into the ophthalmic punctum. Punctal plugs have also been utilized for drug delivery. Drug-eluting punctual plugs have the dual advantage of reducing precorneal tear film drainage and sustaining the release of drugs to the surface of the cornea. Several companies have been developing drug-eluting punctual plugs for ocular delivery, most notably Ocular Therapeutix and Mati Therapeutics. Ocular Therapeutix is evaluating moxifloxacin-, dexamethasone-, and travoprost-eluting punctual plugs. They have recently received FDA approval for their dexamethasone punctual plug, Dextenza® (dexamethasone ophthalmic insert) 0.4 mg, for the treatment of postoperative ophthalmic pain. Mati Therapeutics is developing latanoprost plugs for glaucoma and olopatadine plugs for allergy relief. Plug retention was one of the biggest hurdles to overcome for these technologies.

Tears and Nasolacrimal Drainage

Covering the cornea is the tear film, a clear thin aqueous layer. The tear film functions to protect the surface of the eye from evaporative effects, pollution, and microbes and contributes to clarity and refraction. Tears and the cornea also contribute significant refractive power to the eye, up to 60% of total focusing power (Mishima 1965). Hence, transient alterations of the tear film affect visual clarity. The tear film is ~7 μm thick and 7–10 μL in volume, with an estimated turnover rate of approximately 0.5–2.2 μL/min or about 15% per minute (Tomlinson and Khanal 2005). Tear health relies on a balance between tear production, drainage, and evaporation. Tear loss in normal individuals to evaporation is about 10–15% of the tear turnover with most of the remaining due to nasolacrimal drainage (Tomlinson and Khanal 2005). The pH of the tear film ranges from 7.3 to 7.7, and the corneal surface temperature is approximately 34–35 °C (Stjernschantz and Astin 1993).

The classical model of the tear film is as a tri-laminate structure: an outer tear film lipid layer (TFLL), an intermediate aqueous layer, and an inner mucin (glycocalyx) layer. The tear film and ocular structures involved in tear production are depicted in Fig. 2. The lipid layer ranges from 20 to 160 nm thick and is comprised of an outer nonpolar lipid layer and an inner polar lipid layer (Butovich et al. 2007). The lipid constituents of the TFLL are secreted by the Meibomian glands which are located inside the tarsal plates of the lids. Components of the lipid layer include

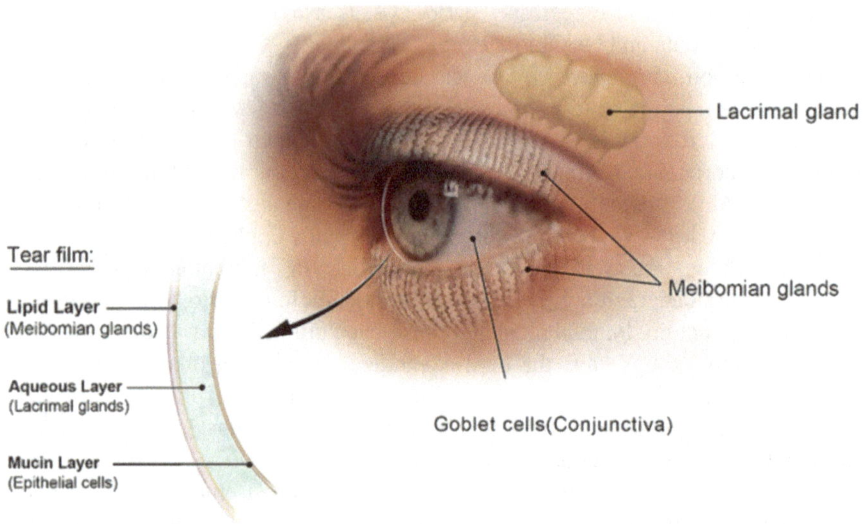

Fig. 2 The tear film and ocular structures involved in tear production. National Eye Institute, National Institutes of Health

waxes, cholesterol and cholesterol esters, fatty acids, and phospholipids (Chen et al. 2010; King-Smith et al. 2010). Recently it has been suggested that proteins are intercalated between the lipids in the TFLL (Green-Church et al. 2011; Saaren-Seppala et al. 2005). The lipid components of the TFLL change with Meibomian gland dysfunction (MGD) (Joffre et al. 2008; Mathers and Lane 1998). The inner sub-layer of the TFLL is composed of the polar lipids. This layer stabilizes the non-polar lipids of the outer sub-layer of the TFLL and communicates with the aqueous layer. Estimates of the TFLL lipid composition are 80% polar and 20% nonpolar (Butovich 2009). The TFLL prevents evaporation and reduces surface tension of the tear film facilitating tear spreading across the cornea. Studies have shown that the lipid composition of the tear film is different in MGD patients than normal patients. Meibomian gland dysfunction and its consequential effects on the TFLL is the major cause of dry eye disease (DED) (Craig and Tomlinson 1997).

The intermediate aqueous layer is composed of water, salts, proteins, and mucin. Lipocalin and lysozyme represent the most abundant protein components of the tear film (Dartt 2011; Wiesner and Vilcinskas 2010). This layer is 4–8 μm thick and produced by the secretory lacrimal gland and accessory glands at a rate of about 1–2 μL per minute (Mishima et al. 1966). The rate of tear production is highly dependent on emotional state, environmental stimulus, disease, and age. There is some thought that the tear lipocalins also help anchor the TFLL to the underlying glycocalyx.

The inner mucin layer is 0.02–0.5 μm in thickness (Wilson et al. 2007). This layer is viscous and composed of membrane-bound and free mucin. It facilitates the adherence of the tear film and wetting of the corneal surface. Mucin is a glycoprotein that also gives the tear film a non-Newtonian rheologic behavior. This allows for shear thinning to facilitate spread, but higher viscosities at rest for improved residence time of the tears. Mucin is produced by the conjunctival goblet cells, and inflammation and DED can damage these cells leading to low mucin production (Gilbard et al. 1988; International Dry Eye WorkShop 2007).

As previously discussed, blinking serves to spread the tear film across the cornea. The tear film is spread over the cornea and stabilizes itself within about 1 s in normal individuals (Owens and Phillips 2001). After a period of time, the tear film thins and destabilizes causing it to break down. This is tear film breakup (TBU). With each downward blink, lipids in the tear film are squeezed into a thick layer at the lower tear meniscus and then redistributed into a monolayer across the ocular surface upon opening of the lids. Blinking further serves to distribute instilled drug across the surface of the cornea post instillation. Tear film breakup time (TBUT) is a clinical measure of tear film stability. The tear film breakup time is greater than 10 s in normal individuals. Tear film breakup times of less than 5–10 s are correlated with evaporative dry eye disease. Clinically TBUT is assessed by the administration of fluorescein and determining the time it takes for the first dry spot to appear from the time of the last blink. The TBUT in relationship to blinking frequency becomes critical as the tear film lipids are replaced with each blink.

A healthy tear film requires the appropriate balance between tear production, drainage, and evaporation. Reduced aqueous production and increased evaporation lead to hyperosmolarity. The tear hyperosmolarity is the primary cause of discomfort and tissue damage in DED (Farris et al. 1983; Gilbard et al. 1978). Normal tear film osmolarity is 296 + 9.8 mOsm/L with hyperosmolarity being defined as 308–320 mOsm/L depending on reference and measurement technique (Gilbard et al. 1978; International Dry Eye WorkShop 2007; Keech et al. 2013; McCann et al. 2012; Messmer et al. 2010). Considerable overlap exists in the measured tear osmotic pressure of normal and DED patients. Tear film hyperosmolarity is an important pathologic mechanism leading to the inflammation that causes DED. Hyperosmolarity can damage epithelial cell membranes and the mucin-producing goblet cells and alter the interactions of lipids and proteins in the tear film (Gilbard et al. 1988). Epithelial damage leads to release of interleukin 1 beta (IL-1β), matrix metallopeptidase 9 (MMP-9), tumor necrosis factor (TNF-α), and stimulating inflammatory cytokines leading to DED (Li et al. 2004, 2006).

The potential exists for preservatives such as benzalkonium chloride (BAK) to further disrupt the tear film leading to a lipid instability and hyperosmolarity. The use of preservatives in eye drops remains controversial, especially the toxicity of benzalkonium chloride (BAK). Benzalkonium chloride is typically used in ophthalmic formulations at concentrations of 40–200 parts per million (ppm). Benzalkonium chloride is known to be cytotoxic to corneal and conjunctival epithelial cells in vitro and ex vivo. Studies have shown the BAK induces apoptosis in conjunctival epithelial cells and damages corneal epithelial cells in cell culture and explants (Pauly et al. 2009; Pisella et al. 2004; Steven et al. 2018). Altered morphology and pro-apoptotic activity in corneal and conjunctival epithelium has also been observed ex vivo. Based on SEM and microscopic analysis of BAK-treated explants, it has been shown that BAK can damage the corneal and conjunctival epithelium causing morphologic changes and sloughing of the epithelium (Smith et al. 1991).

However, the BAK exposure in these studies is usually the result of bathing the cells or tissues in a BAK solution, whereas topical eye drops would result in very short, transient exposure to BAK. Alterations of TBUT with increased corneal staining, goblet cell density, and mucin production have been noted due to BAK exposure in animal models (Pisella et al. 2000). Increased corneal staining indicative of keratitis has also been observed with BAK-preserved formulations clinically (Jaenen et al. 2007; Steven et al. 2018).

Clinical data from the use of BAK is conflicting. Many studies have shown that switching from BAK-preserved formulations to non-preserved formulations results in a reduction in adverse symptoms including signs of ocular surface disease (OSD) (Uusitalo et al. 2016). Other studies have concluded that no significant differences in corneal staining or toxicity occur from BAK formulations (Trocme et al. 2010). However, these studies are often confounded by switching to entirely different medications, e.g., one prostaglandin for another.

Very few actual randomized double-blinded clinical studies have compared BAK-preserved formulations to their non-preserved counterparts. The results of the few well-controlled clinical studies examining BAK-preserved and non-preserved formulations indicate that there is no significant difference in adverse events (Katz et al. 2010; Steven et al. 2018). Specifically, no differences in self-reported adverse events or objective clinical measures such as corneal staining, hyperemia, or foreign body sensation were observed in these studies. However, given the correlation in noncontrolled clinical studies with increased signs and symptoms of ocular surface disease with BAK formulations in dry eye patients, it was deemed prudent to use non-preserved formulations in this population when possible. The risk is associated with the decreased tear production and tear turnover in OSD patients. The toxicity of BAK is a function of its concentration in the formulation, the frequency of dosing, the tear health, and existing ocular surface disease.

The Management and Therapy Subcommittee of the International Dry Eye Workshop (2007) reviewed the existing literature as well as the Dry Eye Preferred Practice Patterns of the American Academy of Ophthalmology and the International Task Force (ITF) Delphi Panel on Dry Eye (International Dry Eye WorkShop 2007). The subcommittee made several conclusions and recommendations. For patients with mild to no OSD who take four to six drops per day or less, there is no need to switch to preservative-free formulations. Those taking multiple topical medications or with existing OSD would benefit from preservative-free formulations.

The impact of precorneal tear film dynamics is a rapid loss of an instilled topical dose from the cul-de-sac. Most drugs enter the eye by passive diffusion across the cornea. Hence, it is critical to establish a concentration gradient from the tear film across the cornea to drive this diffusion. The area under the tear film concentration time profile as well as the maximal concentrations achieved in the precorneal space must be sufficient to provide productive and therapeutic disposition into the eye. Precorneal clearance of drugs is the most significant factor in reducing this and limiting bioavailability (Maurice and Mishima 1986; Maurice 2002). Precorneal

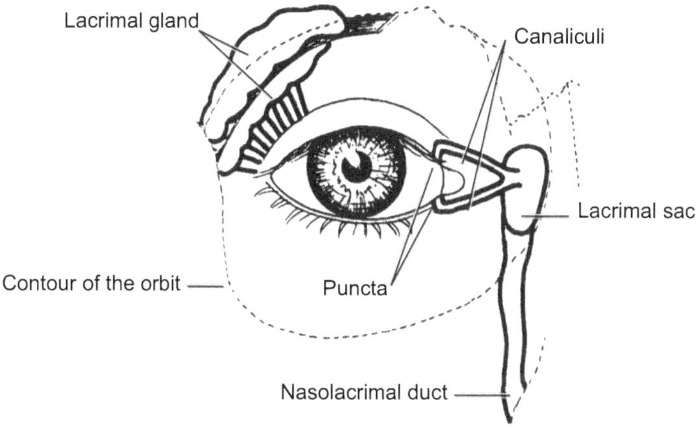

Fig. 3 Secretory and drainage apparatus of the eye. (Adapted from Grant-Kels and Kels (1992))

clearance is a result of spilling of the instilled dose over the lid margins and down the cheek, blinking, nasolacrimal drainage, lacrimation and subsequent dilution, tear turnover, and nonproductive conjunctival absorption. The cul-de-sac has a resting volume of about 7–9 μL (Ahmed and Patton 1985; Maurice and Mishima 1986; Maurice 2002; Sieg and Robinson 1976). This capacity can transiently expand to 30 μL but rapidly normalizes through nasolacrimal drainage. The typical drop is 35–50 μL in volume. Hence, much of the initial instilled volume is immediately lost through spillage over the lid margin or nasolacrimal drainage. This loss can lead to systemic absorption across the nasal mucosa and result in systemic side effects. The transcorneal diffusion of drug is further impacted by dilution in the tears and tear turnover. Various factors are known to cause lacrimation. These include formulation factors such as the tonicity and pH of the drug product, the active compound, irritation, and even physiological state. Furthermore, administration of a second drop to the same eye within 30 s, 2 min, and 5 min of the first will result in a loss of 45%, 17%, and 0% of the first drug administered, respectively (Chrai et al. 1974). This is why if two drugs are to be administered to the same eye, it is suggested to wait at least 5 min between drops (Fig. 3).

Conjunctiva

The conjunctiva is a thin vascularized mucous membrane that covers the anterior sclera and inner eye lids. The conjunctiva is separated into the bulbar conjunctiva and the palpebral conjunctiva joining at the fornix. The bulbar conjunctiva covers the anterior sclera and connects with the globe at the corneal scleral limbus. The palpebral conjunctiva lines the inner portion of the eyelid with attachment at the lid margin. The conjunctiva is composed of two layers: an outer layer of stratified columnar epithelial cells and an inner stromal layer. The epithelium possesses tight junctions limiting paracellular transport. Goblet cells in the bulbar conjunctiva secrete mucin. The conjunctiva serves important barrier functions as well as participates in immunity.

Instilled topical ocular drugs can achieve concentrations in the aqueous humor and iris by the conjunctival/scleral route of absorption. This is termed the noncorneal route of absorption. This route is largely considered nonproductive for drug absorption owing to the static and dynamic barriers offered by the conjunctiva and surrounding vasculature. Nonproductive absorption involves absorption across the conjunctiva and clearance to the systemic circulation. The conjunctiva contributes to approximately 20% of the systemic absorption from topically instilled drugs with nasolacrimal drainage and absorption by the nasal mucosa making up the remaining 80% (Wilson et al. 2007). However, this route of absorption into the ocular tissues will predominate for compounds that are poorly permeable to the cornea. Studies have shown that macromolecules such as inulin as well as small molecules such as bimatoprost access the intraocular structures primarily by this route. The conjunctiva surface area is four times that of the cornea, thus facilitating absorption (Ahmed and Patton 1985; Bito and Baroody 1981; Olsen et al. 1998; Schoenwald et al. 1997).

There are both static and dynamic barriers to conjunctival absorption of molecules with the dynamic barriers being more significant (Lee et al. 2010; Robinson et al. 2006). Static barriers include a permeability resistance offered by the conjunctival epithelium and conjunctival transporters. The conjunctiva is more permeable to hydrophilic compounds and macromolecules than the cornea. Macromolecules with estimates ranging from 40 kDa to full-size antibodies of 150 kDa can penetrate the conjunctiva (Ahmed 2003). Using cassette dosing in an Ussing chamber with excised porcine bulbar conjunctiva, Urtti and fellow investigators demonstrated that permeation through the conjunctiva is inversely related to a molecule's polar surface area and hydrogen bond donors and directly related to halogen content (Ramsay et al. 2017).

Passive and active membrane transporters exist in the conjunctival epithelium that can also impact drug delivery (Kadam et al. 2013). These transporters should be taken into consideration or perhaps exploited when developing ophthalmic formulations. Transporters in the eye include both efflux and influx systems. The efflux systems tend to reduce bioavailability and eliminate drug from the eye. Several efflux systems are known to exist in the conjunctiva and include ATP-dependent drug efflux pumps p-glycoprotein (P-gp), multidrug resistance protein (MRP), and breast cancer resistance protein (BCRP). P-gp actively transports neutral or positively charged lipophilic compounds. MRPs are organic anion transporters with hydrophobic substrates. BCRP transports hydrophilic compounds and conjugated organic anions. Influx transporters are also found in the conjunctiva and include transporters for amino acids and peptides, but also nucleosides and glucose among others. Prodrug strategies can exploit these systems by transient alterations of molecular structure to either facilitate absorption by uptake transporters or avoid efflux pumps (Vooturi et al. 2012).

The dynamic barriers are comprised of the conjunctival blood and lymphatic vessels and the episcleral blood flow. The dynamic barriers have been shown to be a much greater contributor to reducing conjunctival delivery than the static barriers (Lee et al. 2010; Robinson et al. 2006). The conjunctival blood and lymphatic supply rapidly clear drugs from the subconjunctival space rendering conjunctival absorption largely nonproductive. The episcleral veins further act to reduce the concentration of any drug permeating the conjunctiva. However, the subconjunctival space is an attractive area for depot delivery to the eye. The space is easily accessible by simple injection and can expand to accommodate drug and drug delivery system depots. From this space drug can theoretically diffuse into the anterior chamber through the limbus and cornea or iris route and the posterior segment by transscleral diffusion. Steroids, cyclosporine, and glaucoma agents have been successfully delivered in clinical and preclinical models via this route (Cheng et al. 2014; Clayton et al. 2009; Hadayer and Schaal 2016; Natarajan et al. 2012).

Cornea

The cornea contributes significantly as a barrier to the ocular penetration of topically applied ophthalmic compounds. The cornea is about 545 μm at its center, increasing to approximately 600–700 μm at the limbus (Doughty and Zaman 2000; Reinstein et al. 2008). The cornea can be thought of as a tri-laminate structure relative to drug delivery sandwiching two thin membranes. However, the epithelium and stromal layers have the major impact on transcorneal diffusion of drugs. Microvilli are present at the surface squamous epithelial cells.

As shown in Fig. 4, the outer corneal epithelial layer is about 52 μm thick and approximately five to seven cell layers (Reinstein et al. 2008). It is composed of lipophilic epithelial cells with closely packed tight junctions. The corneal epithelium has about 100× the lipid content of the stroma (Wilson et al. 2007). Because of the tight junctions and the lipid content of the epithelium, diffusion across the

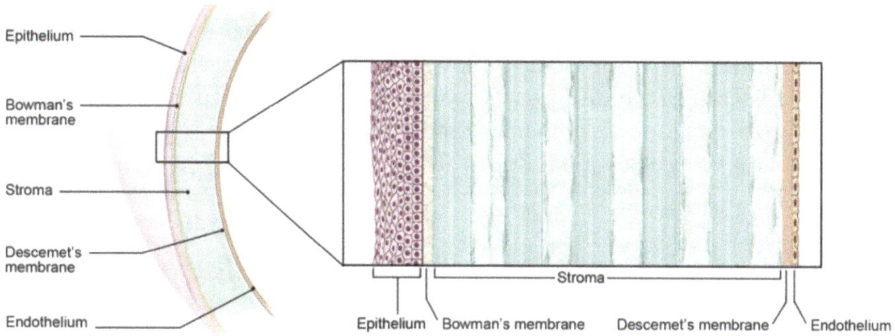

Fig. 4 Structure of the cornea. National Eye Institute, National Institutes of Health

corneal epithelium is limited to paracellular transport of lipophilic small molecules. The middle layer, the stroma, has a mean thickness that can range from about 450 to 500 μm thick (Reinstein et al. 2008). The stroma is composed of 78% water, 15% collagen, and 5% non-collagen proteins with 1% glycosaminoglycans and salts (Fatt and Weissman 1992). The high water content of the stroma offers more resistance to lyophilic drug penetration. The inner most layer is a single layer of polygonal cells, the corneal endothelium, and is responsible for corneal deturgescence. The endothelial cells possess a Na/K pump that balances the passive diffusion of water into the stroma with active transport out of the stroma, maintaining hydration of the cornea (Fatt and Weissman 1992). Loss of endothelial function will lead to clouding of the cornea. Endothelial cells are not regenerated. Hence, it is important that minimal impact to the endothelium occurs via any drug delivery. The endothelium is 200× more permeable to solutes than the epithelium (Maurice and Polgar 1977). Because of the lipophilic epithelium and hydrophilic stroma, the transcorneal flux of small molecules is optimized at a lipophilicity corresponding to an octanol/water partition coefficient of about 2–4 logP (Schoenwald and Huang 1983).

Two membranes separate the stroma from the epithelium and endothelium: Bowman's and Descemet's membranes, respectively. Bowman's membrane is approximately 8–14 μm thick (Stjernschantz and Astin 1993). Descemet's membrane is the basement membrane for the endothelial cells and is a single cell layer of epithelial cells approximately 10–15 μm thick (Dua et al. 2013; Morrison and Khutoryanskiy 2014). Neither of these two membranes contributes significant resistance to the transcorneal permeation of compounds.

The focus of this chapter is the anterior ocular anatomy and physiology; hence it will only briefly focus on the sclera. Transscleral diffusion has been suggested as a mechanism of topically applied drugs getting to the posterior segment. The sclera is the tough fibrous shell of the eye. It makes up 80% of the total surface area of the eye, the remaining 20% being the cornea. There are three layers to the sclera, the episclera, the stroma, and the lamina fusca, and it is composed of connective tissue of collagen and elastin fibers arranged in a crisscrossing fashion. Proteoglycans make up an amorphous interfibrillar space. This tight bundle of overlaying fibers gives the sclera its opacity and strength. The sclera serves to protect the eye from external forces and maintain the eyes shape. The sclera is thickest at the posterior pole at 1 mm and thins toward the equator (0.3 mm) again thickening to about 0.5 mm at the limbus (Olsen et al. 1998). The interfibrillar space is aqueous providing low resistance to the diffusion of hydrophilic drugs. The permeability of the sclera is proportional to a compound's molecular radius and consistent with free diffusion (Cruysberg et al. 2005; Kao et al. 2005).

Aqueous Humor

The anterior chamber is the space formed from the iris, lens, and cornea. The anterior chamber is filled with the aqueous humor that serves to supply nutrients to the cornea, maintain intraocular pressure, and supply antioxidants to the anterior chamber. The aqueous humor normally maintains the intraocular pressure at around 15 mmHg in humans. The aqueous humor is transparent with a composition and osmolality that is similar to plasma with the exception that it has about 1/2 the protein content of the plasma. The aqueous humor is produced by the ciliary body at a rate of approximately 2 μL per minute maintaining a volume of about 150–250 μL. In an adult human, aqueous humor is estimated to have a turnover rate of 60–90% per hour depending on time of the day (Goel et al. 2010). The actual clearance of drugs from direct administration to the intracameral space has been measured for only a few drugs with aqueous humor half-lives ranging from 0.4 to 2.2 h (Durairaj 2017).

Several technologies are in development to directly deliver drug to the aqueous humor, bypassing the precorneal barriers to drug absorption. Injectable intracameral depots and implants have shown success in delivering therapeutic drug concentrations for several months. This route of delivery inherently results in a high bioavailability and can also minimize some of the side effects from topical administration such as hyperemia and chemosis. Because of the invasive nature of direct intracameral injection, sustained-release implants have been developed for this route. Bimatoprost SR is an injectable bioerodible implant of bimatoprost developed by Allergan plc and is currently in Phase 3 clinical studies for the treatment of ocular hypertension (OHT) and open-angle glaucoma (OAG). In a Phase 1/2 clinical trial, 91% and 71% of the patients treated with the Bimatoprost SR had the intraocular pressure (IOP) controlled for 16 weeks and 6 months, respectively (Lewis et al. 2017). Glaukos Corporation is developing iDose, an implantable sustained-release delivery system for travoprost. The implant continuously elutes travoprost to lower IOP in OHT and OAG patients. iDose achieved sustained IOP reduction with minimal adverse events in an interim analysis of a Phase 2 study and is currently in Phase 3 clinical trials in the USA (Glaukos Corp 2018).

An injectable bioerodible suspension depot of dexamethasone has been developed for the treatment of post-cataract surgery inflammation. DEXYCU™ is a 9% suspension of dexamethasone in a bioerodible depot utilizing Icon Bioscience's Verisome technology. DEXYCU™ is administered as a single intracameral injection at the conclusion of surgery. Compared to placebo DEXYCU™ demonstrated significant anterior chamber cell clearing lasting for 30 days after a single injection at the end of cataract surgery compared to placebo (Donnenfeld and Holland 2018).

The posterior chamber is the compartment created by the lens, iris-ciliary body, and anterior vitreous and is also filled with aqueous humor. The aqueous humor formed by the ciliary processes in the posterior chamber flows into the anterior chamber. Clearance of the aqueous humor occurs through the trabecular meshwork into Schlemm's canal, on into the collector channels and the aqueous veins, and finally out the episcleral veins. This is the conventional outflow pathway and is

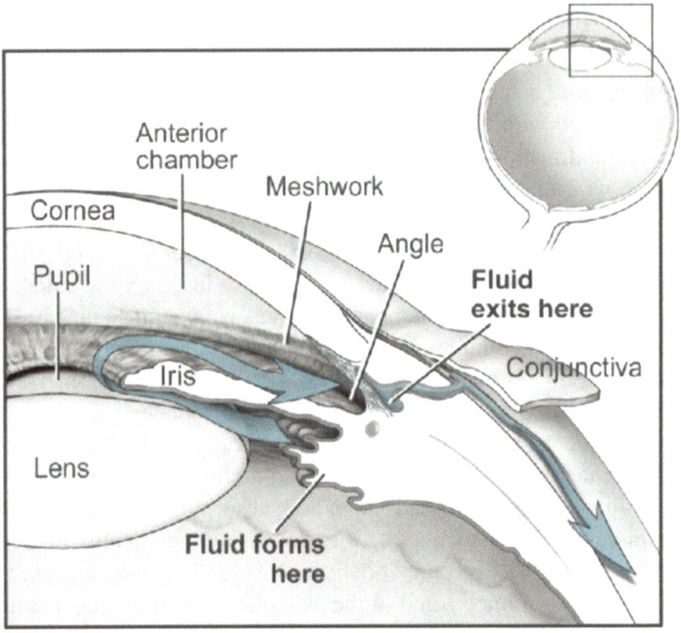

Fig. 5 The formation and outflow of aqueous humor. National Eye Institute, National Institutes of Health

sensitive to episcleral venous pressure, preventing the intraocular pressure from going too low (hypotony). Clearance also takes place by uveoscleral outflow, the nonconventional clearance pathway. The ratio of the conventional clearance to nonconventional is 3:1. The formation and flow of aqueous humor are depicted in Fig. 5.

Formulation Approaches

Clearly the anatomy and physiology of the ocular surface poses considerable challenges to the formulator of ophthalmic drops. Key barriers that must be overcome to achieve optimal therapy include precorneal factors such as rapid clearance of instilled drops (nasolacrimal drainage, blinking, tear turnover, and spilling out the cul-de-sac), lacrimation and dilution, and nonproductive conjunctival absorption. The lipophilic tight epithelium and aqueous stroma of the cornea as well as the rapid aqueous humor turnover further impede achieving prolonged therapeutic levels of drugs in the intraocular tissues. Formulation approaches to improve the ocular performance of topically applied drugs must take into consideration these anatomical and physiologic eccentricities of the anterior segment (Dua et al. 2013).

Other chapters in this book will deal with specific formulation strategies to mitigate these constraints. However, a brief overview is introduced here. Improving corneal permeability can be directly accomplished through optimizing the drug's lipophilicity and solubility. Flux across the cornea will be a function of both the hydrophilic/lipophilic balance (permeability) and solubility (concentration gradient). The lipophilicity of a drug can be addressed directly for new chemical entities intended for ophthalmic use at the time of development. Unfortunately, this is most often not the case. However, the ophthalmic formulator can utilize formulation approaches to optimize the compound's lipophilicity for repurposed compounds. The pH of a formulation can be adjusted within a limited range around the compound's pKa to optimize the fraction of drug ionized/unionized. This approach has been utilized with some success in the formulation of brimonidine tartrate and ketorolac (Acheampong et al. 2002; Attar et al. 2010). The goal is to increase the fraction of drug unionized and its permeability to a greater extent than the concomitant decrease in solubility. The prodrug approach has also been used in ophthalmology to optimize a drug's lipophilicity. Examples include propine, the di-pivalyl ester of epinephrine, as well as the esters and amides of the prostaglandin analogs latanoprost, travoprost, and bimatoprost. Lipophilic modification improves the corneal permeability of these compounds. Upon or during diffusion across the cornea, esterases and other enzymes cleave the drugs back into their parent moieties.

Surfactants, solubilizers, and complexation can be used to enhance a compound's solubility. Surfactants are commonly used to increase the solubility of compounds in ophthalmic drops and may also have the ancillary benefit of increasing corneal permeability. Lumigan 0.01% formulation is formulated with 200 ppm of BAK as opposed to the 50 ppm of BAK used in the Lumigan 0.03% formulation (European Medicines Agency 2010). Both formulations show similar efficacy, and this is thought to be the results of the increased BAK concentration in the 0.01% formulation. Common surfactants used in ophthalmic drops include polysorbate 20, polysorbate 60, polysorbate 80, poloxamer 188, poloxamer 407, polyoxyl 40 hydrogenated castor oil, polyoxyl 35 castor oil, polyoxyl 40 stearate (polyoxyethylene 40 stearate), nonoxynol-9, and tyloxapol (Shen et al. 2018). Surfactants have been shown to increase the bioavailability from topical administration depending on the compound's physicochemical properties and the nature of the surfactant used. The impact is generally the greatest on more hydrophilic drugs suggesting a permeation enhancement. Caution should be used when using surfactants on the surface of the eye as the tear film can be adversely affected. In general, neutral surfactants are better tolerated after topical administration followed by cationic and then anionic surfactants (Sahoo et al. 2014). Complexing agents that have met with success in clinical and preclinical studies include polyvinylpyrrolidone and cyclodextrins (Gamache 2014; Loftsson et al. 1994; Loftsson and Stefansson 2002; Lorenzo-Veiga et al. 2019; Soliman et al. 2016). Complexation increases the solubility of the drugs and thus the driving force for transcorneal diffusion. Additionally, for suspensions particles size reduction has been used to enhance the bioavailability of topically applied medications (Ali et al. 2011; Kassem et al. 2007).

Decreasing precorneal clearance and increasing the area under the tear film drug concentration time profile can also enhance bioavailability. Increasing formulation viscosity has been shown to reduce precorneal clearance of topically applied medications and improve bioavailability. Polymers commonly used in ophthalmology to increase viscosity of topical drops include hydroxypropyl cellulose, hydroxypropyl methylcellulose, methylcellulose, sodium hyaluronate, polyvinyl alcohol, and polyvinylpyrrolidone (Shen et al. 2018). Improvements in precorneal retention time are typically seen with increasing viscosity up to about 15–30 cps after which there is a diminishing benefit (Patton and Robinson 1975). However, newer systems with more complex rheology utilizing polymers such as carbomer 947P, carbomer 980 NF, carbomer 1342, polycarbophil, xanthan gum, and gellan gum have shown dramatic improvement at even higher or more complex viscosities (Shen et al. 2018).

Precorneal inserts have also been contemplated and developed to improve residence time. Ocusert was the first precorneal insert approved for ophthalmic drug delivery. Ocusert was a reservoir-based drug delivery insert that eluted 20 or 40 μg of pilocarpine per hour. The insert was placed in the lower cul-de-sac and delivered pilocarpine to the tear film for 1 week. Ocusert demonstrated moderate IOP reduction of 20% in open-angle glaucoma (OAG) patients, but was withdrawn from the market in 2007 (Lee et al. 1975). Unfortunately, the device experienced dose dumping, insert expulsion, and movement on the cornea as well as a pronounced foreign body sensation (Armaly and Rao 1973; Pollack et al. 1976).

Newer precorneal drug delivery devices have been developed to address these issues and improve upon the performance of Ocusert. Punctal plugs have received a lot of attention as potential drug delivery systems and have been discussed previously in this chapter. Numerous researchers have attempted to deliver drugs from contact lenses. However, clinical and commercial success has not been achieved by this approach. Amorphex Therapeutics' TODDD device is a modified contact lens that is designed as a precorneal insert that rests on the sclera. The elastomeric implant is curved to the radius of the sclera and is placed the precorneal space to deliver drug to the eye. A canine study has demonstrated the potential for sustained IOP reduction from a latanoprost TODD eluting device. The device achieved a reduction in IOP of 7–10 mmHg from baseline at 16 days postimplantation in the dog (Crawford et al. 2013).

Another precorneal insert that has found success in the clinic is the bimatoprost ocular ring insert. This device is a flexible ring-shaped insert that delivers bimatoprost over a 6-month period. The insert is placed under the lids. It's left in place to release clinically effective concentrations of the agent over 6 months. A Phase 2 study has demonstrated IOP lowering of 3.2–6.4 mmHg over 6 months with acceptable side effects. Approximately 88.5% of the patients retained the ring for 6 months (Brandt et al. 2016).

Species Difference and Impact on Product Development

Many of the product development decisions are based on preclinical animal studies. It is critical to understand species differences relative to humans and how they may impact developmental decisions. Much of the animal literature relative to ocular drug delivery, whether to assess pharmacokinetics or tolerability, have been conducted in rabbits. Anatomically and physiologically rabbit eyes are more like human eyes than rodents. Rabbits share comparable ocular size, vitreous volume, and intraocular structure with humans. However, other animal models may be more translatable depending on the study objective (Rodrigues et al. 2018). For example, ocular drug exposure can be confounded by blinking rate differences between species, reported to be 0.05–0.3 blink/min in rabbits (Maurice 1995), 1–2 blinks/min in dogs, and 4–22 blinks/min in primates and human (Stevens and Livermore Jr 1978). It has also been established that rabbit may not be the most sensitive species to chemicals causing ocular irritation (Bito 1984), raising the question if it should be the default toxicology species for ocular toxicology studies for product development. Rabbits and dogs both have a nictitating membrane (a third eyelid) that is absent in primates and humans. The membrane may affect drug distribution following dosing or interfere with placement of drug delivery devices placed on the ocular surface.

Therefore, the choice of animal species for testing ocular drug delivery should take into account similarities and differences in anatomy, physiology, ocular pharmacokinetic properties (e.g., enzyme and transporter expression), and targeted drug tissue in comparison to humans.

A cross-species comparison of ocular anatomy and aqueous humor physiology relevant for drug delivery is shown in Tables 1 and 2.

Table 1 Cross-species comparative physical dimensions of the eye

	Rat	Rabbit	Dog	Monkey	Human
Corneal thickness (mm)	0.17	0.40	0.585–0.670	0.44	0.52
Anterior chamber volume (mL)	0.015	0.25–0.3	0.4–0.77	0.072–0.220	0.1–0.25
Posterior chamber volume (mL)	Unknown	0.06	0.2	Unknown	0.06
Lens volume (mL)	Unknown	0.2	0.5	Unknown	0.2
Vitreous volume (mL)	<0.02	1.4–1.7	3.2	1.5–4.0	3.9–5

Data from Attar et al. (2013), Gum and MacKay (2013), Shen et al. (2018)

Table 2 Estimates of aqueous humor dynamics between species

	Human	Dog	Rabbit	Nonhuman primate
Estimated normal IOP (mmHg)	13–18	15–18	15–20	13–15
"C" outflow (μL/mmHg/min by tonography)	Unknown	0.24–0.30	0.22–0.28	0.24–0.28
Uveoscleral outflow (μL/min)	40–50%	15%	13–25%	5–40%
Episcleral venous pressure (mmHg)	8–10	10–12	9	10–11
Aqueous formation (μL/min)	2.4	5.22	1.84	2.75

Data from Goel et al. (2010) and Gum and MacKay (2013)

Conclusion

In this chapter we endeavored to deliver a working knowledge of the ocular surface anatomy and physiology and the impact it has on product development for ophthalmic drugs. Successful formulation development demands an integration of the unique challenges offered by the eye and advanced pharmaceutical technologies. As noted these challenges include precorneal elimination of instilled drugs, nonproductive losses, poor corneal permeability, and rapid elimination for the internal ocular tissues. Preclinical in vivo evaluation of formulations can guide development and gauge the success in overcoming these delivery constraints. Additionally, animal studies not only direct our development decisions, but they are a requisite part of new drug registration. However, it is imperative to understand the anatomical and physiologic differences between species to accurately interpret preclinical results. Keeping the anatomy and physiology in mind when developing drugs, the pharmaceutical scientist will be able to maximize the therapeutic benefits to the patient.

References

Acheampong AA, Small D, Baumgarten V, Welty D, Tang-Liu D. Formulation effects on ocular absorption of brimonidine in rabbit eyes. J Ocul Pharmacol Ther. 2002;18:325–37. https://doi.org/10.1089/10807680260218498.

Ahmed I. The noncorneal route in ocular drug delivery. In: Mitra AK, editor. Ophthalmic drug delivery systems. New York: Marcel Dekker; 2003. p. 335–63.

Ahmed I, Patton TF. Importance of the noncorneal absorption route in topical ophthalmic drug delivery. Invest Ophthalmol Vis Sci. 1985;26:584–7.

Ali HS, York P, Ali AM, Blagden N. Hydrocortisone nanosuspensions for ophthalmic delivery: a comparative study between microfluidic nanoprecipitation and wet milling. J Control Release. 2011;149:175–81. https://doi.org/10.1016/j.jconrel.2010.10.007.

Armaly MF, Rao KR. The effect of pilocarpine Ocusert with different release rates on ocular pressure. Invest Ophthalmol Vis Sci. 1973;12:491–6.

Attar M, Schiffman R, Borbridge L, Farnes Q, Welty D. Ocular pharmacokinetics of 0.45% ketorolac tromethamine. Clin Ophthalmol. 2010;4:1403–8. https://doi.org/10.2147/OPTH.S15146.

Attar M, Brassard JA, Kim A, Matsumoto S, Ramos M, Vangyi C. Safety evaluation of ocular drugs. In: Faqi AS, editor. A comprehensive guide to toxicology in preclinical drug development. London: Elsevier Inc.; 2013. p. 570.

Bito LZ. Species differences in the responses of the eye to irritation and trauma: a hypothesis of divergence in ocular defense mechanisms, and the choice of experimental animals for eye research. Exp Eye Res. 1984;39:807–29.

Bito LZ, Baroody RA. The penetration of exogenous prostaglandin and arachidonic acid into, and their distribution within, the mammalian eye. Curr Eye Res. 1981;1:659–69.

Brandt JD, Sall K, DuBiner H, Benza R, Alster Y, Walker G, Semba CP. Six-month intraocular pressure reduction with a topical bimatoprost ocular insert: results of a phase II randomized controlled study. Ophthalmology. 2016;123:1685–94. https://doi.org/10.1016/j.ophtha.2016.04.026.

Butovich IA. The Meibomian puzzle: combining pieces together. Prog Retin Eye Res. 2009;28:483–98. https://doi.org/10.1016/j.preteyeres.2009.07.002.

Butovich IA, Uchiyama E, Di Pascuale MA, McCulley JP. Liquid chromatography-mass spectrometric analysis of lipids present in human meibomian gland secretions. Lipids. 2007;42:765–76. https://doi.org/10.1007/s11745-007-3080-2.

Chen J, Green-Church KB, Nichols KK. Shotgun lipidomic analysis of human meibomian gland secretions with electrospray ionization tandem mass spectrometry. Invest Ophthalmol Vis Sci. 2010;51:6220–31. https://doi.org/10.1167/iovs.10-5687.

Cheng YH, et al. Sustained delivery of latanoprost by thermosensitive chitosan-gelatin-based hydrogel for controlling ocular hypertension. Acta Biomater. 2014;10:4360–6. https://doi.org/10.1016/j.actbio.2014.05.031.

Chrai SS, Makoid MC, Eriksen SP, Robinson JR. Drop size and initial dosing frequency problems of topically applied ophthalmic drugs. J Pharm Sci. 1974;63:333–8.

Clayton JA, Perry C, Robinson M. A Phase 1 clinical trial to study the safety of a sustained-release subconjunctival cyclosporine (CsA) implant for ocular graft-vs-host disease (GVHD). Paper presented at the Association for Research in Vision and Ophthalmology Fort Lauderdale, FL; 2009.

Cochran ML, Lopez MJ, Czyz CN. Anatomy, head and neck, eyelid. StatPearls Publishing; 2018. https://www.ncbi.nlm.nih.gov/books/NBK482304/

Craig JP, Tomlinson A. Importance of the lipid layer in human tear film stability and evaporation. Optom Vis Sci. 1997;74:8–13.

Crawford K, Ellis J, Rulander J, Johnston S, Lai F, Ellis E, Leahy C. Sustained delivery of prostaglandin from drug-containing depots using ocular rings in beagles. Paper presented at the 2013 ARVO Annual Meeting, Seattle, Washington, 05–09 May 2013; 2013.

Cruysberg LP, Nuijts RM, Geroski DH, Gilbert JA, Hendrikse F, Edelhauser HF. The influence of intraocular pressure on the transscleral diffusion of high-molecular-weight compounds. Invest Ophthalmol Vis Sci. 2005;46:3790–4. https://doi.org/10.1167/iovs.04-1414.

Dartt DA. Tear lipocalin: structure and function. Ocul Surf. 2011;9:126–38.

Doane MG. Interactions of eyelids and tears in corneal wetting and the dynamics of the normal human eyeblink. Am J Ophthalmol. 1980;89:507–16.

Donnenfeld E, Holland E. Dexamethasone intracameral drug-delivery suspension for inflammation associated with cataract surgery: a randomized, placebo-controlled, phase III trial. Ophthalmology. 2018;125:799–806. https://doi.org/10.1016/j.ophtha.2017.12.029.

Doughty MJ. Consideration of three types of spontaneous eyeblink activity in normal humans: during reading and video display terminal use, in primary gaze, and while in conversation. Optom Vis Sci. 2001;78:712–25.

Doughty MJ, Zaman ML. Human corneal thickness and its impact on intraocular pressure measures: a review and meta-analysis approach. Surv Ophthalmol. 2000;44:367–408.

Dua HS, Faraj LA, Said DG, Gray T, Lowe J. Human corneal anatomy redefined: a novel pre-Descemet's layer (Dua's layer). Ophthalmology. 2013;120:1778–85. https://doi.org/10.1016/j.ophtha.2013.01.018.

Durairaj C. Ocular pharmacokinetics. Handb Exp Pharmacol. 2017;242:31–55. https://doi.org/10.1007/164_2016_32.

European Medicines Agency. Assessment report for Lumigan. 2010. http://www.ema.europa.eu/docs/en_GB/document_library/EPAR_-_Assessment_Report_-_Variation/human/000391/WC500074096.pdf. Accessed 19 May 2019.

Farris RL, Gilbard JP, Stuchell RN, Mandel ID. Diagnostic tests in keratoconjunctivitis sicca. CLAO J. 1983;9:23–8.

Fatt I, Weissman BA. Physiology of the eye: an introduction to vegetative functions. 2nd ed. London: Butterworth-Heinemann; 1992.

Gamache DA. High concentration olopatadine ophthalmic composition. US Patent US8791154B2, July 29, 2014; 2014.

Gaudana R, Jwala J, Boddu SH, Mitra AK. Recent perspectives in ocular drug delivery. Pharm Res. 2009;26:1197–216. https://doi.org/10.1007/s11095-008-9694-0.

Ghate D, Edelhauser HF. Barriers to glaucoma drug delivery. J Glaucoma. 2008;17:147–56. https://doi.org/10.1097/IJG.0b013e31814b990d.
Gilbard JP, Farris RL, Santamaria J 2nd. Osmolarity of tear microvolumes in keratoconjunctivitis sicca. Arch Ophthalmol. 1978;96:677–81.
Gilbard JP, Rossi SR, Gray KL, Hanninen LA, Kenyon KR. Tear film osmolarity and ocular surface disease in two rabbit models for keratoconjunctivitis sicca. Invest Ophthalmol Vis Sci. 1988;29:374–8.
Glaukos Corp. Glaukos Corporations iDose travoprost achieves sustained IOP reduction and favorable safety profile in 12-month interim cohort; 2018.
Goel M, Picciani RG, Lee RK, Bhattacharya SK. Aqueous humor dynamics: a review. Open Ophthalmol J. 2010;4:52–9. https://doi.org/10.2174/1874364101004010052.
Goto E, Tseng SC. Differentiation of lipid tear deficiency dry eye by kinetic analysis of tear interference images. Arch Ophthalmol. 2003a;121:173–80.
Goto E, Tseng SC. Kinetic analysis of tear interference images in aqueous tear deficiency dry eye before and after punctal occlusion. Invest Ophthalmol Vis Sci. 2003b;44:1897–905.
Grant-Kels JM, Kels BD. Human ocular anatomy. Dermatol Clin. 1992;10:473–82.
Green-Church KB, Butovich I, Willcox M, Borchman D, Paulsen F, Barabino S, Glasgow BJ. The international workshop on meibomian gland dysfunction: report of the subcommittee on tear film lipids and lipid-protein interactions in health and disease. Invest Ophthalmol Vis Sci. 2011;52:1979–93. https://doi.org/10.1167/iovs.10-6997d.
Gum GG, MacKay EO. Physiology of the eye. In: Gelatt KN, Gilger BC, Kern TJ, editors. Veterinary ophthalmology. 5th ed. Ames, IA: Wiley; 2013. p. 192.
Hadayer A, Schaal S. Delivery of steroids into the eye for the treatment of macular edema. Expert Opin Drug Deliv. 2016;13:1083–91. https://doi.org/10.1080/17425247.2016.1182488.
Hughes PM, Olejnik O, Chang-Lin JE, Wilson CG. Topical and systemic drug delivery to the posterior segments. Adv Drug Deliv Rev. 2005;57:2010–32. https://doi.org/10.1016/j.addr.2005.09.004.
International Dry Eye WorkShop. Research in dry eye: report of the Research Subcommittee of the International Dry Eye WorkShop (2007). Ocul Surf. 2007;5:179–93.
Isowaki A, Ohtori A. Ophthalmic percutaneously absorbed preparation containing muscarinic receptor agonist. US Patent; 2004.
Jaenen N, Baudouin C, Pouliquen P, Manni G, Figueiredo A, Zeyen T. Ocular symptoms and signs with preserved and preservative-free glaucoma medications. Eur J Ophthalmol. 2007;17:341–9.
Joffre C, et al. Differences in meibomian fatty acid composition in patients with meibomian gland dysfunction and aqueous-deficient dry eye. Br J Ophthalmol. 2008;92:116–9. https://doi.org/10.1136/bjo.2007.126144.
Kadam RS, Vooturi SK, Kompella UB. Immunohistochemical and functional characterization of peptide, organic cation, neutral and basic amino acid, and monocarboxylate drug transporters in human ocular tissues. Drug Metab Dispos. 2013;41:466–74. https://doi.org/10.1124/dmd.112.045674.
Kao JC, Geroski DH, Edelhauser HF. Transscleral permeability of fluorescent-labeled antibiotics. J Ocul Pharmacol Ther. 2005;21:1–10. https://doi.org/10.1089/jop.2005.21.1.
Kassem MA, Abdel Rahman AA, Ghorab MM, Ahmed MB, Khalil RM. Nanosuspension as an ophthalmic delivery system for certain glucocorticoid drugs. Int J Pharm. 2007;340:126–33. https://doi.org/10.1016/j.ijpharm.2007.03.011.
Katz G, Springs CL, Craven ER, Montecchi-Palmer M. Ocular surface disease in patients with glaucoma or ocular hypertension treated with either BAK-preserved latanoprost or BAK-free travoprost. Clin Ophthalmol. 2010;4:1253–61. https://doi.org/10.2147/opth.S14113.
Keech A, Senchyna M, Jones L. Impact of time between collection and collection method on human tear fluid osmolarity. Curr Eye Res. 2013;38:428–36. https://doi.org/10.3109/02713683.2013.763987.

King-Smith PE, Hinel EA, Nichols JJ. Application of a novel interferometric method to investigate the relation between lipid layer thickness and tear film thinning. Invest Ophthalmol Vis Sci. 2010;51:2418–23. https://doi.org/10.1167/iovs.09-4387.
Kompella UB, Kadam RS, Lee VH. Recent advances in ophthalmic drug delivery. Ther Deliv. 2010;1:435–56. https://doi.org/10.4155/TDE.10.40.
Kulkarni SV, Damji KF, Buys YM. Medical management of primary open-angle glaucoma: best practices associated with enhanced patient compliance and persistency. Patient Prefer Adherence. 2008;2:303–14.
Lee VH, Robinson JR. Topical ocular drug delivery: recent developments and future challenges. J Ocul Pharmacol. 1986;2:67–108.
Lee P, Shen Y, Eberle M. The long-acting Ocusert-pilocarpine system in the management of glaucoma. Invest Ophthalmol Vis Sci. 1975;14:43–6.
Lee SJ, He W, Robinson SB, Robinson MR, Csaky KG, Kim H. Evaluation of clearance mechanisms with transscleral drug delivery. Invest Ophthalmol Vis Sci. 2010;51:5205–12. https://doi.org/10.1167/iovs.10-5337.
Lemp MA, Weiler HH. How do tears exit? Invest Ophthalmol Vis Sci. 1983;24:619–22.
Lewis RA, et al. Bimatoprost sustained-release implants for glaucoma therapy: 6-month results from a phase I/II clinical trial. Am J Ophthalmol. 2017;175:137–47. https://doi.org/10.1016/j.ajo.2016.11.020.
Li DQ, Chen Z, Song XJ, Luo L, Pflugfelder SC. Stimulation of matrix metalloproteinases by hyperosmolarity via a JNK pathway in human corneal epithelial cells. Invest Ophthalmol Vis Sci. 2004;45:4302–11. https://doi.org/10.1167/iovs.04-0299.
Li DQ, Luo L, Chen Z, Kim HS, Song XJ, Pflugfelder SC. JNK and ERK MAP kinases mediate induction of IL-1beta, TNF-alpha and IL-8 following hyperosmolar stress in human limbal epithelial cells. Exp Eye Res. 2006;82:588–96. https://doi.org/10.1016/j.exer.2005.08.019.
Loftsson T, Stefansson E. Cyclodextrins in eye drop formulations: enhanced topical delivery of corticosteroids to the eye. Acta Ophthalmol Scand. 2002;80:144–50.
Loftsson T, Frithriksdottir H, Stefansson E, Thorisdottir S, Guthmundsson O, Sigthorsson T. Topically effective ocular hypotensive acetazolamide and ethoxyzolamide formulations in rabbits. J Pharm Pharmacol. 1994;46:503–4.
Lorenzo-Veiga B, Sigurdsson HH, Loftsson T. Nepafenac-loaded cyclodextrin/polymer nanoaggregates: a new approach to eye drop formulation. Materials (Basel). 2019;12:229. https://doi.org/10.3390/ma12020229.
Mathers WD, Lane JA. Meibomian gland lipids, evaporation, and tear film stability. Adv Exp Med Biol. 1998;438:349–60.
Maurice D. The effect of the low blink rate in rabbits on topical drug penetration. J Ocul Pharmacol Ther. 1995;11:297–304. https://doi.org/10.1089/jop.1995.11.297.
Maurice DM. Drug delivery to the posterior segment from drops. Surv Ophthalmol. 2002;47(Suppl 1):S41–52.
Maurice DM, Mishima S. Ocular pharmacokinetics. Handbook of experimental pharmacology: pharmacology of the eye. Springer, Berlin; 1986.
Maurice DM, Polgar J. Diffusion across the sclera. Exp Eye Res. 1977;25:577–82.
McCann LC, Tomlinson A, Pearce EI, Papa V. Effectiveness of artificial tears in the management of evaporative dry eye. Cornea. 2012;31:1–5. https://doi.org/10.1097/ICO.0b013e31821b71e6.
Messmer EM, Bulgen M, Kampik A. Hyperosmolarity of the tear film in dry eye syndrome. Dev Ophthalmol. 2010;45:129–38. https://doi.org/10.1159/000315026.
Mishima S. Some physiological aspects of the precorneal tear film. Arch Ophthalmol. 1965;73:233–41.
Mishima S, Gasset A, Klyce SD Jr, Baum JL. Determination of tear volume and tear flow. Invest Ophthalmol. 1966;5:264–76.
Morrison PW, Khutoryanskiy VV. Advances in ophthalmic drug delivery. Ther Deliv. 2014;5:1297–315. https://doi.org/10.4155/tde.14.75.

Nakamori K, Odawara M, Nakajima T, Mizutani T, Tsubota K. Blinking is controlled primarily by ocular surface conditions. Am J Ophthalmol. 1997;124:24–30.

Nandurit P, Dyer A. Methods and compositions for sustained delivery of drugs. US Patent; 2009.

Natarajan JV, Ang M, Darwitan A, Chattopadhyay S, Wong TT, Venkatraman SS. Nanomedicine for glaucoma: liposomes provide sustained release of latanoprost in the eye. Int J Nanomedicine. 2012;7:123–31. https://doi.org/10.2147/IJN.S25468.

Newman-Casey PA, Robin AL, Blachley T, Farris K, Heisler M, Resnicow K, Lee PP. The most common barriers to glaucoma medication adherence: a cross-sectional survey. Ophthalmology. 2015;122:1308–16. https://doi.org/10.1016/j.ophtha.2015.03.026.

Olsen TW, Aaberg SY, Geroski DH, Edelhauser HF. Human sclera: thickness and surface area. Am J Ophthalmol. 1998;125:237–41.

Owens H, Phillips J. Spreading of the tears after a blink: velocity and stabilization time in healthy eyes. Cornea. 2001;20:484–7.

Patel S, Henderson R, Bradley L, Galloway B, Hunter L. Effect of visual display unit use on blink rate and tear stability. Optom Vis Sci. 1991;68:888–92.

Patton TF, Robinson JR. Ocular evaluation of polyvinyl alcohol vehicle in rabbits. J Pharm Sci. 1975;64:1312–6.

Pauly A, Meloni M, Brignole-Baudouin F, Warnet J-M, Baudouin C. Multiple endpoint analysis of the 3D-reconstituted corneal epithelium after treatment with benzalkonium chloride: early detection of toxic damage. Invest Ophthalmol Vis Sci. 2009;50:1644–52. https://doi.org/10.1167/iovs.08-2992.

Pisella PJ, Fillacier K, Elena PP, Debbasch C, Baudouin C. Comparison of the effects of preserved and unpreserved formulations of timolol on the ocular surface of albino rabbits. Ophthalmic Res. 2000;32:3–8. https://doi.org/10.1159/000055579.

Pisella PJ, Debbasch C, Hamard P, Creuzot-Garcher C, Rat P, Brignole F, Baudouin C. Conjunctival proinflammatory and proapoptotic effects of latanoprost and preserved and unpreserved timolol: an ex vivo and in vitro study. Invest Ophthalmol Vis Sci. 2004;45:1360–8.

Pollack IP, Quigley HA, Harbin TS. The Ocusert pilocarpine system: advantages and disadvantages. South Med J. 1976;69:1296–8.

Ramsay E, et al. Impact of chemical structure on conjunctival drug permeability: adopting porcine conjunctiva and cassette dosing for construction of in silico model. J Pharm Sci. 2017;106:2463–71. https://doi.org/10.1016/j.xphs.2017.04.061.

Reinstein DZ, Archer TJ, Gobbe M, Silverman RH, Coleman DJ. Epithelial thickness in the normal cornea: three-dimensional display with Artemis very high-frequency digital ultrasound. J Refract Surg. 2008;24:571–81.

Richardson C, et al. A study to assess the feasibility of undertaking a randomized controlled trial of adherence with eye drops in glaucoma patients. Patient Prefer Adherence. 2013;7:1025–39. https://doi.org/10.2147/PPA.S47785.

Robinson MR, et al. A rabbit model for assessing the ocular barriers to the transscleral delivery of triamcinolone acetonide. Exp Eye Res. 2006;82:479–87. https://doi.org/10.1016/j.exer.2005.08.007.

Rodrigues GA, Lutz D, Shen J, Yuan X, Shen H, Cunningham J, Rivers HM. Topical drug delivery to the posterior segment of the eye: addressing the challenge of preclinical to clinical translation. Pharm Res. 2018;35:245. https://doi.org/10.1007/s11095-018-2519-x.

Saaren-Seppala H, Jauhiainen M, Tervo TM, Redl B, Kinnunen PK, Holopainen JM. Interaction of purified tear lipocalin with lipid membranes. Invest Ophthalmol Vis Sci. 2005;46:3649–56. https://doi.org/10.1167/iovs.05-0176.

Sahoo RK, Biswas N, Guha A, Sahoo N, Kuotsu K. Nonionic surfactant vesicles in ocular delivery: innovative approaches and perspectives. Biomed Res Int. 2014;2014:263604. https://doi.org/10.1155/2014/263604.

Schoenwald RD, Huang HS. Corneal penetration behavior of beta-blocking agents I: physiochemical factors. J Pharm Sci. 1983;72:1266–72.

Schoenwald RD, Deshpande GS, Rethwisch DG, Barfknecht CF. Penetration into the anterior chamber via the conjunctival/scleral pathway. J Ocul Pharmacol Ther. 1997;13:41–59. https://doi.org/10.1089/jop.1997.13.41.

See GL, Sagesaka A, Sugasawa S, Todo H, Sugibayashi K. Eyelid skin as a potential site for drug delivery to conjunctiva and ocular tissues. Int J Pharm. 2017;533:198–205. https://doi.org/10.1016/j.ijpharm.2017.09.070.

Shen J, Lu GW, Hughes P. Targeted ocular drug delivery with pharmacokinetic/pharmacodynamic considerations. Pharm Res. 2018;35:217. https://doi.org/10.1007/s11095-018-2498-y.

Sieg JW, Robinson JR. Mechanistic studies on transcorneal permeation of pilocarpine. J Pharm Sci. 1976;65:1816–22.

Smith LGM, Berdy G, Abelson M. Investigative Ophthalmology and Visual Science, annual meeting. Abstract issue. April 28–May 3, 1991. Sarasota, Florida. Invest Ophthalmol Vis Sci. 1991;32:733.

Soliman KA, Ibrahim HK, Ghorab MM. Effect of different polymers on avanafil-beta-cyclodextrin inclusion complex: in vitro and in vivo evaluation. Int J Pharm. 2016;512:168–77. https://doi.org/10.1016/j.ijpharm.2016.08.044.

Steven DW, Alaghband P, Lim KS. Preservatives in glaucoma medication. Br J Ophthalmol. 2018;102:1497. https://doi.org/10.1136/bjophthalmol-2017-311544.

Stevens JR, Livermore A Jr. Eye blinking and rapid eye movement: pulsed photic stimulation of the brain. Exp Neurol. 1978;60:541–56.

Stjernschantz J, Astin M. Anatomy and physiology of the eye—physiological aspects of ocular drug therapy. In: Edman P, editor. Biopharmaceutics of ocular drug delivery. Boca Raton, FL: CRC Press; 1993. p. 1–25.

Tomlinson A, Khanal S. Assessment of tear film dynamics: quantification approach. Ocul Surf. 2005;3:81–95.

Trocme S, Hwang LJ, Bean GW, Sultan MB. The role of benzalkonium chloride in the occurrence of punctate keratitis: a meta-analysis of randomized, controlled clinical trials. Ann Pharmacother. 2010;44:1914–21. https://doi.org/10.1345/aph.1P268.

Tsubota K, Nakamori K. Effects of ocular surface area and blink rate on tear dynamics. Arch Ophthalmol. 1995;113:155–8.

Uusitalo H, Egorov E, Kaarniranta K, Astakhov Y, Ropo A. Benefits of switching from latanoprost to preservative-free tafluprost eye drops: a meta-analysis of two Phase IIIb clinical trials. Clin Ophthalmol. 2016;10:445–54. https://doi.org/10.2147/opth.S91402.

Vooturi SK, Kadam RS, Kompella UB. Transporter targeted gatifloxacin prodrugs: synthesis, permeability, and topical ocular delivery. Mol Pharm. 2012;9:3136–46. https://doi.org/10.1021/mp300245r.

Wiesner J, Vilcinskas A. Antimicrobial peptides: the ancient arm of the human immune system. Virulence. 2010;1:440–64. https://doi.org/10.4161/viru.1.5.12983.

Wilson CG, Semenova EM, Hughes P, Olejnik O. Eye structure and physiological functions. In: Touitou E, Barry BW, editors. Enhancement in drug delivery. Boca Raton, FL: Taylor & Francis; 2007. p. 473–87.

Anatomy and Physiology of the Anterior Chamber: Impact on Product Development

Sangly P. Srinivas, Giovanna Guidoboni, Anirudh Burli, Bhavya Harjai, and Uday B. Kompella

Abstract Ophthalmic drugs are administered as drops on the ocular surface because of simplicity, although they are lost to the nasolacrimal duct rapidly (half-life at the ocular surface is small ~4 min). They access the anterior chamber predominantly by penetration across the cornea, which resembles an oil-water-oil matrix. Hence, in addition to short half-life at the ocular surface, the tight junctions of the epithelium oppose the penetration into the anterior chamber, especially for hydrophilic drugs. While efficiently partitioning into the epithelium, lipophilic drugs do not partition well into the stroma, and hence their bioavailability in the anterior chamber is limited. Following transcorneal transport, the drugs undergo mixing in the anterior chamber by convection currents. Concomitantly, they are cleared from the anterior chamber by transport into iris and lens but mainly by entrainment with the bulk flow of aqueous humor exiting the eye. The trabecular meshwork (>70% of the outflow) and the uveoscleral routes (15–30% of the outflow; decreasing with age) constitute the two outflow pathways of aqueous humor. With the volume of the aqueous humor at ~310 μL and the average secretion rate at

S. P. Srinivas (✉)
School of Optometry, Indiana University, Bloomington, IN, USA
e-mail: srinivas@indiana.edu

G. Guidoboni
Electrical Eng., Computer Science, and Mathematics, University of Missouri – Columbia, Columbia, MO, USA

A. Burli
Indiana University, Bloomington, IN, USA

B. Harjai
Computer Science, DSCE, Bangalore, India

U. B. Kompella
Professor of Pharmaceutical Sciences, University of Colorado Anschutz Medical Campus, Aurora, CO, USA

S. Neervannan, U. B. Kompella (eds.), *Ophthalmic Product Development*, AAPS Advances in the Pharmaceutical Sciences Series 37,
https://doi.org/10.1007/978-3-030-76367-1_3

~3 μL/min, the half-life of a drug in the anterior chamber would be ~72 min. The aqueous secretion, which depends on the HCO_3^--mediated active ion transport mechanisms, is modulated by antiglaucoma drugs, including α_2 agonists, β_1/β_2 antagonists, and carbonic anhydrase inhibitors. The aqueous secretion also shows a circadian peak in the morning hours. While the new antiglaucoma drugs such as ROCK inhibitors affect the trabecular outflow, the established prostaglandin analogs enhance the uveoscleral outflow. Overall, the aqueous humor clearance of topical drugs is mainly determined by aqueous secretion and outflow, which may be influenced by many topical pharmacological agents and the circadian rhythm associated with aqueous secretion.

Keywords Ciliary epithelium · Trabecular meshwork · Uveoscleral pathway · Convection · Intraocular pressure

Abbreviations

Δm	Change in the mass of fluorescein in the anterior eye (m) during Δt (μg)
Δt	Observation time interval (min)
π_p	Oncotic pressure (mmHg)
π_s	Osmotic pressure (mmHg)
σ_p	Reflection coefficient for proteins
σ_s	Reflection coefficients for small solutes
C_a	The average concentration of fluorescein in the anterior chamber
cBP	Blood pressure in the capillaries (mmHg)
C_c	The concentration of fluorescein in the cornea
C_{tm}	Outflow facility across the trabecular meshwork ((μL/min)/mmHg)
EVP	Episcleral vein pressure (mmHg)
F_S	Rate of aqueous secretion (μL/min)
IOP	Intraocular pressure (mmHg)
JCT	Juxtacanalicular tissue
L_{in}	The inflow facility ((μL/min)/mmHg)
NPE	Nonpigmented epithelial cells
PE	Pigmented epithelial cells
PKPD	Pharmacokinetics and pharmacodynamics
RPE	Retinal pigment epithelium
SC	Schlemm's canal
TM	Trabecular meshwork
U	Outflow rate through the uveoscleral pathway (μL/min)
V_a	The volume of the anterior chamber (μL)
V_c	The volume of the cornea (μL)

Introduction

Ophthalmic drugs are administered mainly as drops (~30 μL) into the lower conjunctival cul-de-sac (Yavuz and Kompella 2017). Intraocular and periocular injections or drug-loaded implants in the anterior chamber or the vitreous cavity are also employed, but less frequently (Yavuz and Kompella 2017; Bourges et al. 2006; del Amo et al. 2017; Haghjou et al. 2011; Patel et al. 2013; Subrizi et al. 2019; Zhang et al. 2017). Once instilled, the blink dynamics induces clearance of the topical drug with tears (>95%) into the nasal cavity via the nasolacrimal duct (Yavuz and Kompella 2017; Patel et al. 2013). The residence time on the ocular surface, which depends on drop viscosity, blink activity, and tear secretion, is characterized by a half-life of 3–4 min (Yavuz and Kompella 2017). Transcorneal penetration is the main route of entry of the drug into the anterior chamber (Gupta et al. 2010, 2012; Srinivas and Maurice 1992). The drug penetrates the cornea through successive partitioning and diffusion across the multilayered cornea, eventually into the anterior chamber (Gupta et al. 2010, 2012; Srinivas and Maurice 1992). The drug undergoes convective mixing in the aqueous humor (Missel 2012; Missel and Sarangapani 2019; Wyatt 2004; Bhandari et al. 2020; Chen et al. 2015) and is simultaneously distributed into the ciliary body, iris, and lens with the potential for prolonged sequestration in each of the tissues. The drug undergoes clearance from the anterior chamber by entrainment with the aqueous humor outflow. Since the turnover of aqueous humor in the anterior chamber is about 1% per min, drugs typically show a half-life of ~72 min in the anterior chamber. Thus, the pharmacokinetics and pharmacodynamics (PKPD) of topical drugs are dependent on mechanisms that affect ocular surface dynamics, transcorneal penetration, aqueous humor dynamics, mixing in the anterior chamber, and partitioning of the drug into the surrounding tissues. These mechanisms also affect the rate of drug release and PKPD of drug implants positioned in the anterior chamber (Bhandari et al. 2020; Chen et al. 2015). This chapter presents an overview of anterior segment anatomy and aqueous humor dynamics that can impact the PKPD of topical drugs to assist pharmaceutical scientists in developing strategies for ocular drug delivery.

Relevant Anatomy of the Anterior Segment

As an extension of the brain, the eye maintains strict cellular barriers against the entry of drugs into intraocular compartments from the systemic circulation (Yavuz and Kompella 2017; Agrahari et al. 2016). Although the nerves and blood vessels penetrate the eye, it is protected by three distinct anatomical layers. The innermost layer is the retina, which is comprised of nervous tissue and photoreceptors. The outer layer is a fibrous layer, which includes the cornea and sclera. The middle layer is the vascular layer, referred to as the uveal tract, and it consists of the iris, ciliary body, and choroid. These layers form the globe of the eye with a total volume of

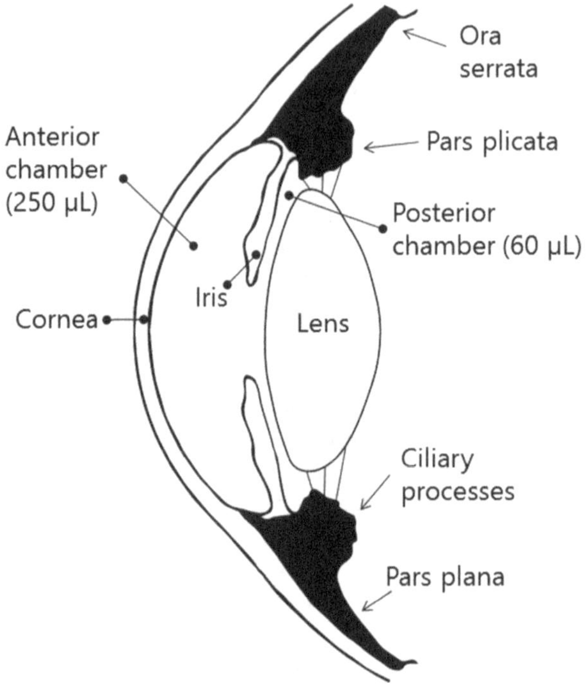

Fig. 1 Anterior segment of the eye. It is comprised of the cornea, ciliary body and aqueous humor outflow pathways, lens, and the two-fluid compartments, viz., anterior and posterior chambers. The ciliary body houses the mechanisms for aqueous humor secretion and accommodation of the crystalline lens

about 7000 μL. The intraocular structures and the fluid compartments within the globe are classified as part of either the anterior or posterior segments. The anterior segment consists of structures in front of the vitreous humor, including the crystalline lens, ciliary body, iris, and cornea (Fig. 1). In contrast, the posterior segment consists of components posterior to the lens, including the vitreous humor, optic disk, retina, and choroid. This chapter focuses on the structural and functional aspects of the anterior segment and their potential impact on topical drug delivery. The ciliary body, iris, and aqueous humor dynamics play a significant role in the PKPD.

Ciliary Body

The ciliary body, summarized in Figs. 2 and 3, houses the mechanisms for aqueous humor secretion (Kiel et al. 2011). Its cross section is an isosceles triangle with its base forward and apex tapering back to the ora serrata (anterior limit of the retina). A portion of the anterior ciliary body contributes to the formation of the anterior

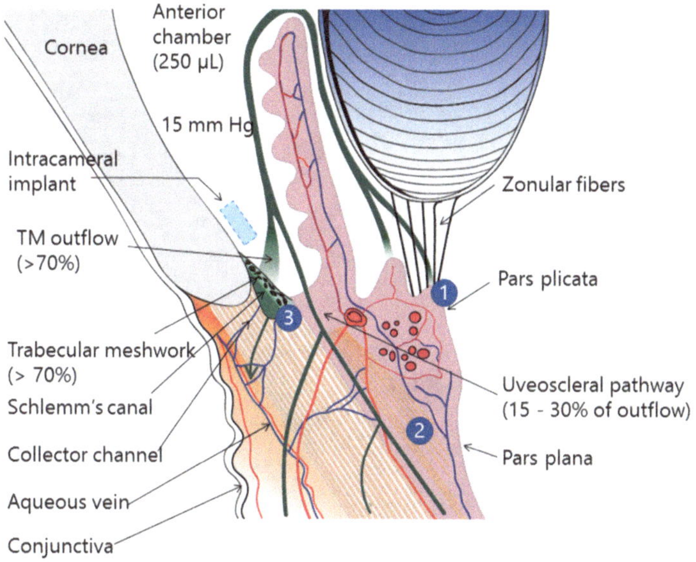

Fig. 2 Overview of aqueous humor, secretion, and outflow. The aqueous humor (total volume ~310 μL) is secreted at ~2.5 μL/min into the posterior chamber by the ciliary epithelium. It escapes into the anterior chamber through the pupil and then undergoes mixing by convection currents induced by the temperature difference between the corneal surface (~33 °C) and the vascularized iris (37 °C). The aqueous humor exits the eye via the trabecular meshwork (a.k.a. conventional pathway; via Schlemm's canal, collector channels, and episcleral veins) and uveoscleral (a.k.a. unconventional pathway) routes into the systemic circulation and suprachoroidal space, respectively. The intracameral implants positioned at the angle are useful to sustain release of drugs to treat glaucoma. Key: *TM* trabecular meshwork; 1, ciliary epithelial cells are the targets of carbonic anhydrase inhibitors, α2 agonists, and β1/β2 blockers; 2, ciliary muscle cells are targets of prostaglandin analogs (via FP receptors) and muscarinic agonists (e.g., pilocarpine, outdated drug); 3, TM cells are targets of Rho kinase (ROCK) inhibitors and β2 agonists (e.g., dipivefrin)

chamber angle. The outer side of the ciliary body lies adjacent to the sclera with suprachoroidal space in between. The ciliary body is divided anatomically into two regions: pars plicata (anterior 1/3; ~2 mm wide) and pars plana (posterior 2/3; ~4 mm wide) (Figs. 2 and 3). The latter is relatively avascular; hence it is penetrated for intravitreal injections and posterior segment procedures. Internally, the ciliary body consists of ciliary muscles, ciliary stroma, and ciliary epithelia. The ciliary stroma is a loose connective tissue of collagen and fibroblasts surrounded by blood vessels, nerve fibers, melanocytes, and mast cells. The capillaries in the stroma of the ciliary processes are partly fenestrated but do not show fenestrations deeper in the ciliary muscle. Posteriorly, the most interior layer of the ciliary body is continuous with the neural retina, while the ciliary stroma continues as the choroid.

The choroid is a dark brown pigmented vascular layer that provides nutrients to all the layers of the eye. Its pigments absorb light crossing the retina, thereby preventing reflection and scattering of light within the eye. It extends from the ora serrata up to the aperture of the optic nerve in the sclera.

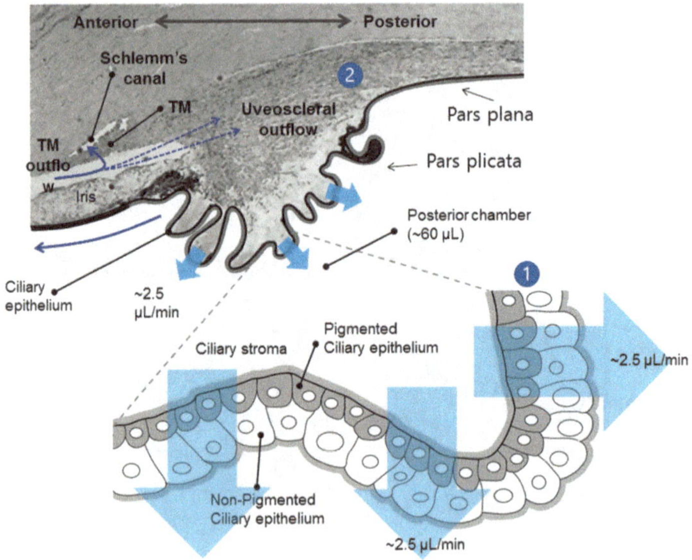

Fig. 3 Schematic of the ciliary epithelium. The ciliary epithelium lines the ciliary body. It consists of outer pigmented ciliary epithelium (PE) and inner nonpigmented ciliary epithelium (NPE). These cells together secrete aqueous humor into the posterior chamber. Apart from PE and NPE cells, the ciliary body is comprised of ciliary stroma, capillaries, and ciliary muscles. The ciliary body also receives both sympathetic and parasympathetic innervations for autonomic control of aqueous humor secretion and accommodation of the lens, respectively. The space between the ciliary muscles forms the route for the uveoscleral pathway. Key: *TM* trabecular meshwork; 1, ciliary epithelial cells are the targets of carbonic anhydrase inhibitors, α2 agonists, and β1/β2 blockers; 2, ciliary muscle cells are targets of prostaglandin analogs (via FP receptors) and muscarinic agonists (e.g., pilocarpine, outdated drug)

The ciliary muscles, which are smooth muscle cells, occupy the bulk of the ciliary body. There are three distinct groups of ciliary muscles, viz., radial, circular, and longitudinal ciliary muscles. The longitudinal ciliary muscles are coupled to trabecular meshwork via the scleral spur. The radial and inner circular ciliary muscles, along with longitudinal ciliary muscle, contribute to accommodation. The accommodation is aided by the suspensory ligaments (zonules of Zinn), which emanate from the ciliary body and hold the lens in place while the ciliary muscles extend the ligaments to change the thickness of the lens.

Iris

The iris controls the light entry into the retina by adjusting the pupil size. It is attached to the middle of the base at the apex of the anterior ciliary body (Fig. 2). It is a colored circular diaphragm with an opening in the center, and it forms the boundary between the anterior chamber (~250 μL) and the posterior chamber (60 μL). The iris is characteristically pigmented with melanocytes dispersed in the iris stroma, fibroblasts, blood vessels, and nerves. In contrast to the ciliary stroma, the blood vessels in the iris stroma are lined by non-fenestrated endothelium. Thus, the tight junctions between the vascular endothelia in the iris partly contribute to the blood-aqueous barrier. The pupil diameter depends on the contraction of its smooth muscles, including the circular sphincter muscle and radial dilatory papillae. While the sphincter smooth muscle at the pupillary border receives parasympathetic innervation, the radial muscle extending from the root of the iris receives sympathetic innervation.

Ciliary Epithelium

The ciliary epithelium (Fig. 3), consisting of two layers, forms the interior surface of the ciliary body which houses the mechanisms responsible for the active secretion of aqueous humor (Kiel et al. 2011). The outer layer is pigmented, while the inner layer is nonpigmented. Posteriorly, the pigmented ciliary epithelium (PE) is continuous with the retinal pigment epithelium (RPE). Anteriorly, PE extends over the ciliary body up to the root of the iris and continues as the anterior pigmented epithelium of the iris. The nonpigmented ciliary epithelium (NPE), which lines the inner surface of the ciliary body (i.e., adjacent to the posterior chamber), is continuous with the posterior pigmented epithelial layer of the iris. In the pars plicata, the ciliary body shows ~70 fingerlike projections into the posterior chamber, referred to as ciliary processes (Fig. 3). These projections constitute the site of aqueous humor production. In the pars plana, NPE cells are columnar; they become cuboidal in the pars plicata. NPE cells are firmly attached to PE cells by a series of desmosomal junctions that are also present between cells in each layer. The basal side of NPE cells faces the posterior chamber, while the basal side of the PE cells faces the ciliary stroma. Thus, the apical surfaces of PE and NPE cells are opposed to one another. Most importantly, the apical domain of NPE cells shows tight junctions that confer the main blood-aqueous barrier (Fig. 4).

Aqueous Humor

Aqueous humor is a clear fluid with a refractive index of ~1.33 (Table 1). Its secretion and outflow are essential for the maintenance of intraocular pressure (IOP), which in turn confers distention of the cornea. Aqueous humor is also responsible for delivering plasma-derived O_2 and nutrients to the lens, the corneal endothelium, and the posterior limbus. The aqueous outflow helps in the clearance of cellular debris and inflammatory products from the anterior chamber. Aqueous humor is

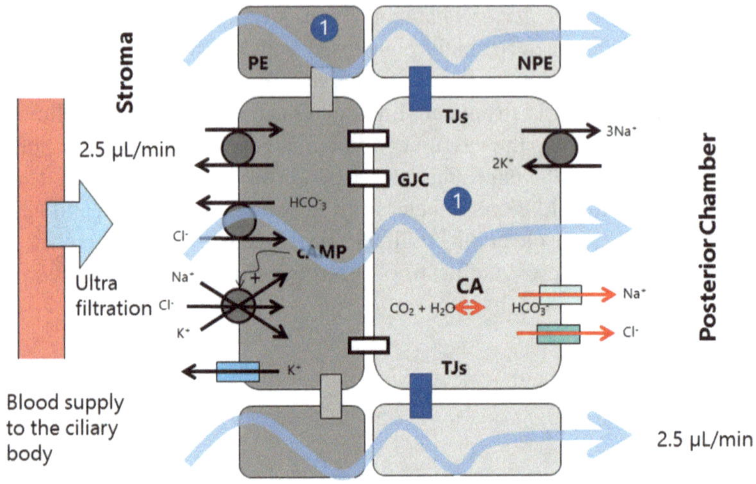

Fig. 4 Mechanisms of aqueous humor secretion. Ultrafiltration of plasma from ciliary capillaries accumulates in the ciliary stroma, which is subsequently exported into the posterior chamber as aqueous humor. The ion transport mechanisms that are expressed in the NPE and PE cells together bring about aqueous humor secretion. The activity of the Na^+-K^+-ATPase and carbonic anhydrases (membrane-bound and cytosolic) present in the PE and NPE function in concert with other mechanisms of ion transport to produce a net ionic movement into the posterior chamber, which, in turn, elicits aqueous secretion by osmotic coupling. Key: *CA* carbonic anhydrase, *GJC* gap junctional complex, *NPE* nonpigmented ciliary epithelium, *PE* pigmented ciliary epithelium, *TJs* tight junctions. Key: 1, ciliary epithelial cells are the targets of carbonic anhydrase inhibitors, α2 agonists, and β1/β2 blockers

Table 1 Characteristics of aqueous humor in humans

Volume	310 μL (posterior chamber ~60 μL; anterior chamber ~250 μL)
Secretion rate	3.0 μL/min morning hours (turnover, 1% per min)
Viscosity	1.0
Refractive index	1.33332
Electrolyte composition	$[Na^+]$ ~ 142 $[K^+]$ ~ 3 $[Ca^{2+}]$ ~ 2 $[Cl^-]$ ~ 131 HCO_3^- ~ 20 Osmolarity ~ 304
Nonelectrolytes	Ascorbic acid, immunoglobulins, glutathione, amino acids
Dissolved O_2	25 mmHg
Proteins	0.013 g/100 mL (albumin ~0.010 g/100 mL and globulin ~0.003 g/100 mL)
pO_2	13–80 mmHg

secreted at a rate of ~2–4 μL/min but slows with age by 4% per decade of life. The secretion is continuous with a characteristic circadian rhythm with peak secretion occurring during the morning hours (Nau et al. 2013). The steady-state volume of aqueous humor is ~310 μL; of this, the volume in the posterior chamber is ~60 μL, while the remainder circulates in the anterior chamber. The anterior chamber volume, however, decreases with age by 14–24 μL per decade. Aqueous humor is slightly acidic compared to plasma, with a pH of 7.2 in the anterior chamber. Although a fluid derived from the plasma, aqueous humor contains unusually high ascorbate (15×) compared to plasma (Table 1).

Secretion by Active Solute Transport

Aqueous humor is derived from the plasma, which is deposited into the ciliary stroma by ultrafiltration from the capillaries of the ciliary body (Fig. 4). Nearly 4% of the capillary flow gets into the ciliary stroma as the plasma ultrafiltrate across the fenestrated capillaries in the ciliary body. The ultrafiltrate is then transported by the ciliary epithelium that lines the pars plicata. Active ion transport mechanisms associated with NPE and PE cells bring about the fluid movement (Civan et al. 1997; Jacob and Civan 1996; Macknight et al. 2000; McLaughlin et al. 2001) (Fig. 4). The aqueous humor then escapes into the anterior chamber, where it is continuously mixed by convection currents caused by the temperature difference between the cornea, which is exposed to the external ambient conditions, and the vascularized iris, which is at body temperature (Purslow and Wolffsohn 2005).

The secretion of aqueous humor is dependent on several HCO_3^--dependent ion transport mechanisms (Civan et al. 1997; Jacob and Civan 1996; Macknight et al. 2000; McLaughlin et al. 2001; Do and To 2000; Coca-Prados and Escribano 2007; Civan and Macknight 2004). These are expressed in the apical and basolateral membranes of NPE and PE cells (Fig. 4). The basolateral Na^+/K^+-ATPase on NPE and PE set up the Na^+ and K^+ gradients across their respective apical/basolateral membranes. Carbonic anhydrase (intracellular/membrane-bound), Na^+/H^+ exchanger, and Cl^-/HCO_3^- exchanger collectively support the HCO_3^--dependent mechanisms. These mechanisms, in coordination with $Na^+/K^+/Cl^-$ cotransport and Cl^- channels, drive net Na^+ and Cl^- flux into the posterior chamber. Coupled to the net ionic movement is the water efflux into the posterior chamber as the aqueous humor. In particular, water and isosmotic NaCl with HCO_3^- constitute the aqueous humor. Ascorbic acid and glucose, along with several amino acids, enter the posterior chamber through secondary active transport mechanisms based on Na^+ gradients. While a detailed analysis of aqueous humor secretion in relation to ciliary blood flow is reviewed by Kiel et al. (2011), the mechanisms of ion transport are reviewed by Do and To (2000) as well as Coca-Prados and Escribano (2007).

Table 2 Pharmacology of aqueous secretion and outflow

Tissue	Receptor	G protein/ signaling	IOP	Hormone/nT*	Clinical significance
Ciliary epithelial cells	β_1	Gs ↑ cAMP	↑	Epinephrine Norepinephrine	β_1 and β_2 antagonists are used to reduce IOP in the treatment of glaucoma For example, timolol (β_1 and β_2 antagonist) Betaxolol (β_1 antagonist)
Ciliary epithelial cells	β_2	Gs ↑ cAMP	↑	Epinephrine Norepinephrine	
Ciliary epithelial cells	α_2	Gi ↓ cAMP	↓	Epinephrine Norepinephrine	α_2 agonists such as apraclonidine and brimonidine reduce IOP. Brimonidine is used widely to reduce IOP in the treatment of glaucoma
Ciliary smooth muscle	M_3	Gq ↑ Ca^{2+}	↓	ACh	Pilocarpine has been used to treat glaucoma. It results in a reduction of IOP by an increase in the trabecular outflow. The contraction of the ciliary muscles induces traction of the corneoscleral meshwork, leading to increased porosity in the corneoscleral domain of the trabecular meshwork
Ciliary smooth muscle	FP	Gq ↑ Ca^{2+}	↓	PGF2α	Bimatoprost, latanoprost, and travoprost are PGF2α analogs. They likely activate FP receptors, leading to transcriptional activation of MMPs along the uveoscleral pathways
Trabecular meshwork endothelial cells	β_2	Gs ↑ cAMP	↓	Epinephrine	Dipivefrine is a prodrug that activates β_2 receptors on trabecular meshwork endothelial cells. The increased cAMP remodels the trabecular meshwork increasing the outflow

Pharmacology of Secretion

The α_2, β_1, and β_2 adrenergic receptors (GPCRs, G protein-coupled receptors) expressed on the ciliary epithelia modulate the rate of secretion of aqueous humor (Table 2) (Coca-Prados and Wax 1986; Cooper et al. 1990; Krupin et al. 1991; Wax and Barrett 1993; Wax and Molinoff 1987). In addition to parasympathetic innervation of the ciliary muscles, the sympathetic nerve fibers innervate the ciliary body. Although glucocorticoids elevate IOP, their effect via modulation of aqueous humor secretion is minimal. In terms of the autonomic receptors, α_2, β_1, and β_2 receptors are expressed by the ciliary epithelium, but their expression levels may vary across the pars plana and pars plicata regions. The activation of β_1 and β_2 receptors increases the rate of aqueous humor secretion because of elevated cAMP (Table 2). In contrast, activation of the α_2 receptors reduces the rate of aqueous humor secretion by the reduction of cAMP. Although muscarinic agonists reduce IOP, the aqueous formation is unaffected. Thus, β_1 and β_2 blockers (i.e., antagonists) and α_2 agonists are commonly employed agents to reduce IOP in the treatment of glaucoma (Table 2).

Measurement of Secretion Rate

The rate of secretion of aqueous humor is measured by anterior segment fluorometry (Nau et al. 2013; Johnson et al. 2017; McLaren 2009; McLaren et al. 1990; Radenbaugh et al. 2006; Topper et al. 1984). The stroma is loaded by topical fluorescein (~10%). After 5–6 h, fluorescein levels in the cornea and anterior chamber are measured at 2-h intervals for several hours. The average aqueous flow rate (F_S) for each interval is estimated from the rate of loss of fluorescein from the cornea and anterior chamber given by

$$\Delta m = F_S * \Delta t * C_a$$

where Δm corresponds to the change in mass of fluorescein in the cornea and anterior chamber over Δt and C_a is the average concentration of fluorescein in the anterior chamber during the interval. Δm can be calculated using the equation below:

$$m = V_c * C_c + V_a * C_a$$

where V_c is the volume of the cornea and V_a is the volume of the anterior chamber. C_c and C_a represent the concentration of fluorescein in the cornea and anterior chamber, respectively. Both concentrations can be measured by fluorometry. Thus, we can estimate Δm and therefore F_S.

Circadian Rhythm in Aqueous Secretion

In normal subjects, the mean secretion is normally distributed with a mean of 3.0 ± 0.8 μL/min in the morning hours (Nau et al. 2013; Radenbaugh et al. 2006). Although the secretion rate remains steady for most of the day, it is reduced by half during sleep (Radenbaugh et al. 2006). This rhythm is attributed mainly due to β-adrenergic stimulation during the day and its absence during sleep (Nau et al. 2013). Timolol, which blocks the β-adrenergic receptors, suppresses flow during the day to rates similar to those during sleep (Nau et al. 2013). Although some critical efforts were made in understanding the influence of circadian rhythm on ocular drug delivery (Lee et al. 1996), additional studies are required to understand the influence of circadian rhythm and drug effects on aqueous secretion/outflow and, hence, drug delivery.

Convective Mixing and Turnover in the Anterior Chamber

For an aqueous volume of 310 μL, the aqueous production represents ~1% per min turnover of its total volume per minute (3 μL/min × 100/310 μL) during wakeful hours. At this rate, a drug in the anterior chamber shows a half-life of ~72 min (= ln (2)/0.0096) (assuming mixing with the posterior chamber) (Missel 2012; Missel and Sarangapani 2019). Though fluorescein clearance measurements of aqueous humor flow rate of the anterior chamber demonstrate a 25% decrease as an effect of aging in the period of 20–80 years, there is also a coupling of an increase of aqueous turnover rate of 20% during this time; this is due to a ~40% decrease in the anterior chamber volume. In addition to the bulk motion induced by its production and drainage, the flow of aqueous humor within the anterior chamber exhibits convective currents induced by the temperature difference between the corneal surface that is exposed to the temperature of the external ambient and the vascularized iris, which is at body temperature (Purslow and Wolffsohn 2005). The techniques to measure the flow and the vector field of the convection currents include particle image velocimetry (Chung and Kim 2008; Lindken et al. 2009) and particle-tracking velocimetry (Kaji et al. 2012; Feng et al. 2011; Khalighi and Lee 1989; Peterson et al. 2012).

The strength and the direction of the convective currents also depend on the posture since it changes the relationship between buoyancy and gravity (Wyatt 2004; Chen et al. 2015; Heys and Barocas 2002; Villamarin et al. 2012; Kumar et al. 2007). Computational models have also been utilized to unravel and quantify the relative influence of external temperature and posture on the convection currents within the anterior chamber (Heys and Barocas 2002; Ooi and Ng 2008; Boushehrian et al. 2016; Abdelhafid et al. 2019; Dvoriashyna et al. 2020). Interestingly, even if the external temperature is kept constant, the temperature distributions and the flow streamlines characterizing the aqueous motion within the anterior chamber are markedly different depending on whether the individual is in a supine, prone, or standing position. Shown in Fig. 5 is the case of an external temperature of 25 °C (Abdelhafid et al. 2019). The computational domain was a two-dimensional cross section of the anterior chamber delimited by the cornea, the iris, and the pupil. Changes in the posture were simulated by changing the direction of the gravitational acceleration **g**. Thus, in the supine, prone, and standing positions, **g** is oriented downward (Fig. 5a, b), upward (Fig. 5c, d), and rightward (Fig. 5e, f), respectively. Temperature profiles in the supine (Fig. 5a) and prone (Fig. 5c) positions are similar, whereas the symmetry is broken when standing (Fig. 5e). The effect of posture on aqueous humor flow is more complex. When standing, a single major vortex is dominating the aqueous flow (Fig. 5f), while in the supine and prone positions, two

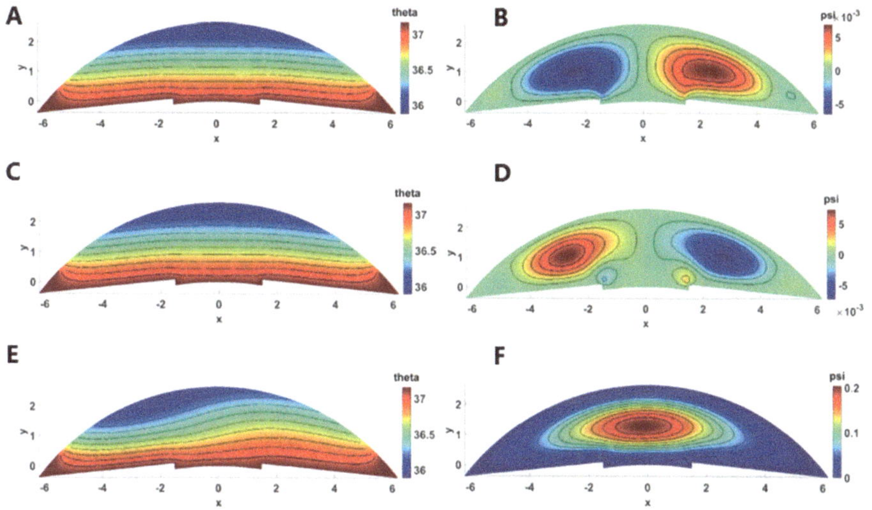

Fig. 5 Simulated temperature profiles (**a**, **c**, **e**) and flow streamlines (**b**, **d**, **f**) in the supine, prone, and standing positions of a human subject for an external temperature of 25 °C. Theta is the temperature in °C and psi is the vorticity in 1/s. The direction of the gravitational acceleration is downward in the supine position (**a**, **b**), upward in the prone position (**c**, **d**), and rightward in the standing position (**e**, **f**)

vortices appear in the middle of the anterior chamber. Interestingly, in the supine and prone postures, the aqueous fluid vortices rotate in opposite directions (Fig. 5b, d). Given a certain posture, changes in the external temperature can modulate the intensity of the convective currents by influencing the internal temperature difference. For example, Fig. 6 shows how a drop of external temperature from 45 °C (Fig. 6a, c, e) to 15 °C (Fig. 6b, d, f) may change the intensity of the internal vortices by two orders of magnitude. In summary, results show how computer simulations could be used as virtual laboratories to identify factors influencing internal convection currents. This is particularly relevant in the context of pharmacology since convection currents potentially influence both the residence time and distribution kinetics of topical drugs and drug release from implants positioned in the anterior chamber (Bourges et al. 2006; Missel and Sarangapani 2019; Wyatt 2004; Villamarin et al. 2012; Ooi and Ng 2008; Abdelhafid et al. 2019). When integrated with pharmacokinetic models, the model described above may inform the influence of patient orientation on drug delivery. Furthermore, convection affects the settlement of cells on the endothelial surface, like in the case of allograft rejection. The characteristic circulation of aqueous humor underlies the formation of the Krukenberg spindle: a vertical band of pigment on the corneal endothelium in certain pathologies (Heys and Barocas 2002; Kumar et al. 2007).

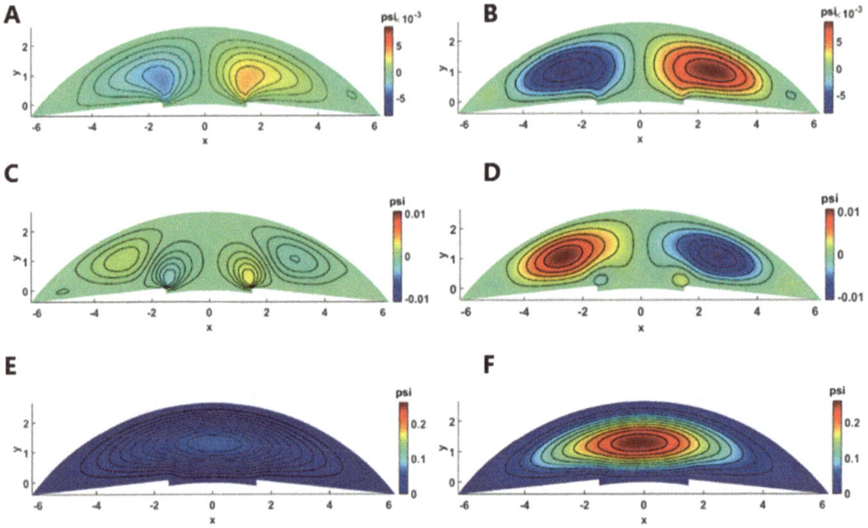

Fig. 6 Simulated flow streamlines in the supine, prone, and standing positions for the external temperatures of 45 °C (**a**, **c**, **e**) and 15 °C (**b**, **d**, **f**). The direction of the gravitational acceleration is downward in the supine position (**a**, **b**), upward in the prone position (**c**, **d**), and rightward in the standing position (**e**, **f**). The units of the vorticity psi are 1/s

Aqueous Flare

A breakdown of the blood-aqueous barrier during trauma, intraocular surgery, chemical irritation, vasodilation, intraocular inflammation, prostaglandin use, uveitis, and anterior segment ischemia results in the entry of proteins and cells into the anterior chamber (Agarwal et al. 2019; Sawa 2017; Sudhir et al. 2018; Tugal-Tutkun and Herbort 2010). Consequently, an increase in light scatter from the anterior chamber, called the aqueous flare, can be observed (i.e., Tyndall effect). In the absence of intraocular inflammation, aqueous humor has no cells, and protein content is much less than in the plasma (0.02% in aqueous vs. 7% in plasma). Uveitis is graded clinically based on the number of cells and quantifying the flare from the anterior chamber (Agarwal et al. 2019; Tugal-Tutkun and Herbort 2010). Assessment of aqueous flare through a measurement of light scatter can be used for enhanced clinical management of uveitis and pharmacological evaluation of new or existing drugs (Agarwal et al. 2019; Tugal-Tutkun and Herbort 2010). Breakdown of the blood-aqueous barrier in the above conditions may influence drug pharmacokinetics.

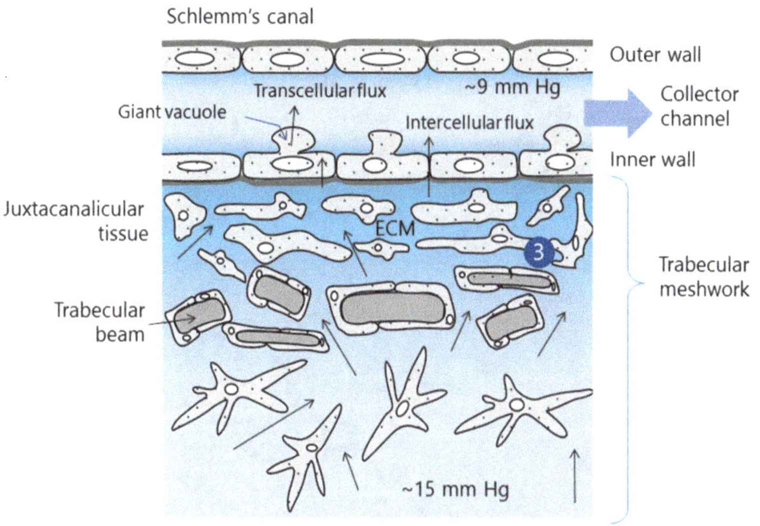

Fig. 7 Anatomy of the trabecular meshwork. The trabecular meshwork is divided into three regions. The innermost is the uveal meshwork of high porosity. The middle layer is the corneoscleral meshwork with reduced porosity and yet with negligible resistance to the passage of aqueous humor. The juxtacanalicular tissue (a.k.a. cribriform layer) is the outermost layer of the trabecular meshwork. Juxtacanalicular tissue shows small porosity and hence is expected to offer the most resistance to aqueous outflow. However, the specialized inner wall of Schlemm's canal may also contribute to the resistance. Thus, a pressure drop of 6 mmHg between the anterior chamber and Schlemm's canal drops across the combined resistance of juxtacanalicular tissue and the inner wall of Schlemm's canal. Schlemm's canal cells lining the inner wall are specialized for the transport of aqueous through the formation of giant vacuoles. Key: *ECM* extracellular matrix; 3, TM cells are targets of Rho kinase (ROCK) inhibitors and β2 agonists (e.g., dipivefrin)

Aqueous Humor Outflow Pathways

There are two distinct routes of aqueous humor outflow in humans (Carreon et al. 2017) (Figs. 7, 8 and 9). Nearly 70–85% of outflow occurs through the trabecular meshwork route, although this is subject to a significant change in the elderly and during glaucoma. The remainder of the aqueous exits the eye through the uveoscleral pathway (Carreon et al. 2017). The resistance to aqueous outflow gives rise to IOP (15.5 ± 2 mmHg; ranges ~10–21 mmHg). Since the aqueous humor secretion exhibits a circadian rhythm, IOP also shows a parallel diurnal variation of 3–4 mmHg. IOP is a risk factor for primary open-angle glaucoma (POAG). However, many with open angle show glaucomatous damage at IOP <15 mmHg (Investigators 2000). Lowering of IOP is the only pharmacological approach to treat glaucoma, even in normotensive glaucoma (Investigators 2000). Angles between the iris and surface of the trabecular meshwork falling in the range of 20–45° are deemed wide angles, while angles less than 20° are deemed narrow angles as per Shaffer's classification (Chan et al. 1981; Kashiwagi et al. 2005).

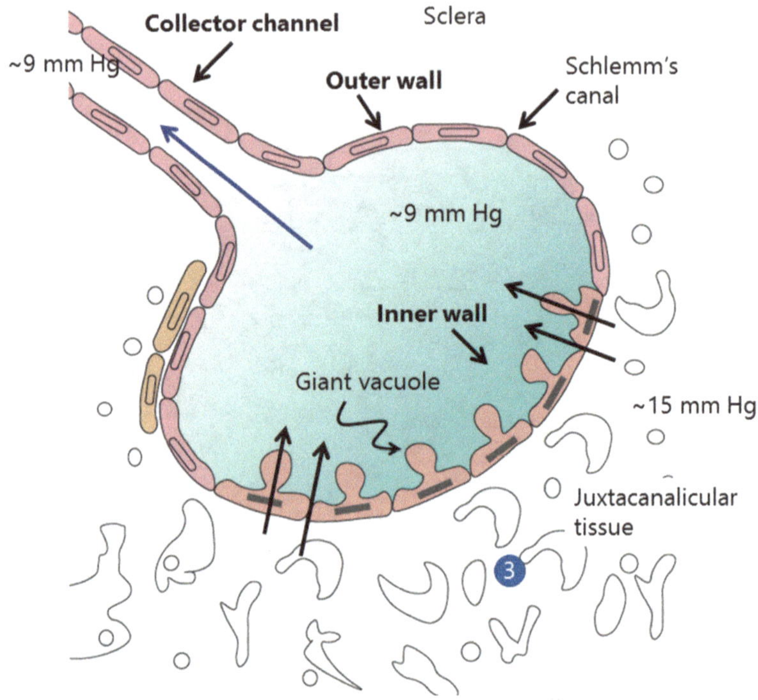

Fig. 8 Inner wall of Schlemm's canal. Schlemm's canal is a lymphatic-like flaccid vessel. The wall next to juxtacanalicular tissue is referred to as the inner wall of Schlemm's canal. The endothelial cells of the inner wall show characteristic giant vacuoles, which enable aqueous humor movement from juxtacanalicular tissue. The endothelial cells lining the outer wall of Schlemm's canal do not possess giant vacuoles. Schlemm's canal does not offer any resistance to aqueous humor from reaching the episcleral veins through collector channels. Hence, the fluid pressure in Schlemm's canal lumen is close to the episcleral vein pressure, which is ~9 mmHg. Key: 3, TM cells are targets of Rho kinase (ROCK) inhibitors and β2 agonists (e.g., dipivefrin)

Trabecular Meshwork Outflow

The trabecular meshwork (TM) is a sieve-like triangular matrix of collagen beams close to the root of the iris, leading the aqueous to Schlemm's canal (SC) (Figs. 2 and 3). It is located in the scleral sulcus and bounded by the periphery of the cornea and anterior surface of the ciliary body, including the scleral spur. The trabecular meshwork can be divided into three regions based on its microscopic organization. The uveal meshwork forms the innermost region of the trabecular meshwork and consists of a thick band of connective tissue that originates from the iris root and extends up to Schwalbe's line. The uveal meshwork has large pores; hence, it offers negligible resistance to aqueous outflow. Next is the corneoscleral (CS) meshwork: the middle and largest region of the trabecular meshwork. Although the porosity of CS is significantly smaller than the uveal meshwork, the resistance to aqueous flow is negligible. The longitudinal ciliary muscle is mechanically coupled to the CS via

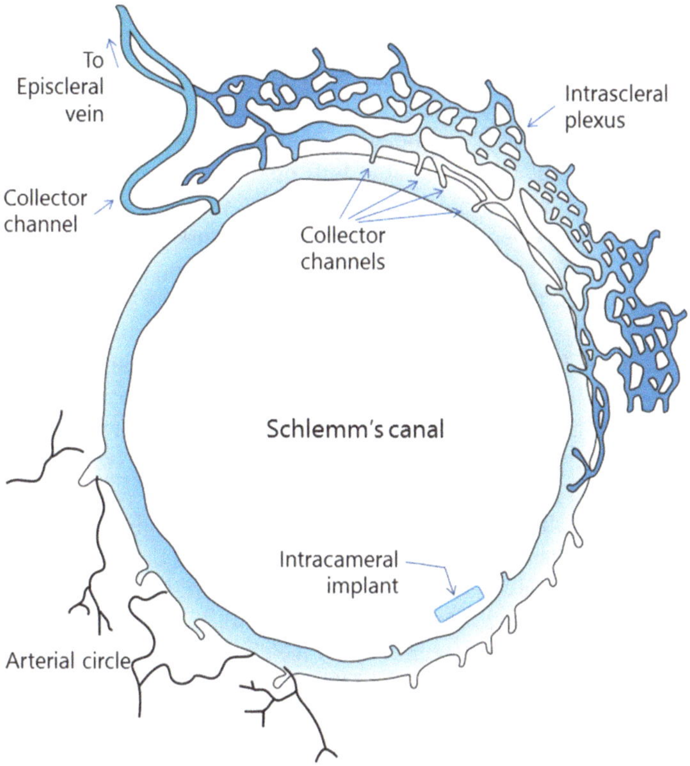

Fig. 9 Distribution of collector channels: The aqueous humor reaches the episcleral veins via collector channels (~25–30 channels around the circumference) that insert into Schlemm's canal. The flow across the trabecular meshwork is predominant around these collector channels and thus creates a segmented outflow of aqueous humor

the scleral spur. Thus, the ciliary muscle contraction in response to muscarinic stimulation induces traction on the CS, leading to transient changes in porosity of the trabecular meshwork. The most anterior region of the trabecular meshwork next to the CS is the juxtacanalicular meshwork (JCT; a.k.a. cribriform layer), which is characterized by minimal porosity and the presence of endothelial cells on the trabecular beams (called trabecular meshwork cells) at a higher density. As the name implies, JCT borders the SC, which may augment resistance to aqueous humor outflow (Fig. 7).

The SC is a circular (oval-shaped in cross section) vessel lined by a monolayer of the endothelium, which shows characteristics of the lymphatic endothelium (Fig. 8). In particular, the cells show expression of PROX1, VEGFR3, CCL21, and FOXC2 (Aspelund et al. 2014; Karpinich and Caron 2014; Kizhatil et al. 2014; Park et al. 2014; Truong et al. 2014). However, since the cells do not express LYVE1 and PDPN, SC is considered as only a lymphatic-like vessel (Aspelund et al. 2014; Karpinich and Caron 2014; Kizhatil et al. 2014; Park et al. 2014; Truong et al. 2014).

The transport of aqueous across the JCT-SC interface is still poorly defined. The current understanding is that SC cells transport the aqueous from JCT to the lumen of SC in "packages" called giant vacuoles. The paracellular pathway between SC cells is thought to be a minor second route. The presence of giant vacuoles in the normal outflow process has been demonstrated by electron microscopy (Lai et al. 2019; Parc et al. 2000). The endothelial cells on the outer wall of the SC, which is not adjacent to JCT, are devoid of giant vacuoles and appear flat in the electron microscopic images. Overall, the trabecular meshwork outflow consists of the movement of the fluid from the anterior chamber to the lumen of the SC by crossing uveal meshwork, CS, JCT, and the inner wall of the SC.

From the SC, the aqueous eventually reach general circulation after passage through collector channels, aqueous veins, and the episcleral vein, respectively (Figs. 8 and 9). Being a flaccid vessel, the SC offers no resistance to fluid flow. The aqueous in the SC travel through collector channels which cross the limbal sclera to reach the episcleral veins (Fig. 9). The SC is perforated by ~25–30 aqueous collector channels on the scleral side. The episcleral veins drain into the ophthalmic veins which empty into the general circulation. The episcleral venous pressure is ~9 mmHg, close to the pressure in the SC (Sit et al. 2011; Sit and McLaren 2011).

The aqueous drainage across the trabecular meshwork is segmented (Loke et al. 2018; Swaminathan et al. 2014; Vranka et al. 2020). In other words, only certain portions of the trabecular meshwork around the circumference are utilized for outflow, seemingly showing preference near the collector channels. Aqueous angiography in glaucoma patients demonstrated segmental aqueous humor outflow (Loke et al. 2018; Swaminathan et al. 2014; Vranka et al. 2020).

Uveoscleral Outflow

The aqueous outflow through the uveoscleral pathway is considered to be the minor and unconventional route in humans, accounting for 15–30%. However, it is high in young adults and decreases with age. Interestingly, the uveoscleral outflow is 5% in cats and rabbits, while it is 50% in monkeys.

In the uveoscleral pathway, the aqueous flows from the angle of the anterior chamber into the connective tissue spaces in the ciliary muscle through the iris root and anterior side of the ciliary body (Fig. 2). Both the anterior ciliary body and the iris root lack endothelial linings; therefore there is an absence of cellular barriers for passage of aqueous into the ciliary body. After passage through the spaces between ciliary muscles, aqueous percolates into the suprachoroidal space and leaves the eye through scleral perforations or the vortex veins. Prostaglandins, a major category of drugs used to treat glaucoma, act on the uveoscleral pathway (Weinreb et al. 2020; Gaton et al. 2001). Uveoscleral flow is independent of IOP at levels greater than 7–10 mmHg (Johnson et al. 2017). Decline of uveoscleral outflow due to aging is induced by a thickening of elastic fibers in the ciliary muscles. The extracellular matrix along the uveoscleral pathway, which forms the main resistance to outflow, is synthesized by ciliary muscle cells. Collagens I and III form the structural interstitial matrix, while collagen IV, laminin, and fibronectin form the basement membrane. The matrix metalloproteinases (MMPs; expressed as an inactive proenzyme) regulate the amount of extracellular matrix and hence affect the uveoscleral outflow rate. MMP-1 (a.k.a. interstitial collagenase) breaks down collagens I and III, while MMP-9 (gelatinase-B) breaks down collagen IV and laminin (Schachtschabel et al. 2000). MMPs, in conjunction with tissue inhibitors of metalloproteases, control the extracellular remodeling and thereby regulate the uveoscleral flow. Indeed, the intraocular pressure-lowering effect of prostaglandins may in part be due to the elevated expression of MMPs (Weinreb et al. 2020; Gaton et al. 2001).

Outflow Facility

The outflow across the trabecular meshwork is driven by $\Delta P = \text{IOP} - P_{\text{EVP}}$, where P_{EVP} is the episcleral vein pressure (~9 mmHg). SC and collector channels' resistance to aqueous flow is assumed to be negligible. The Goldman equation describes the outflow across the trabecular meshwork by

$$\left[F_{\text{S}} - U\right] = \left(\text{IOP} - P_{\text{EVP}}\right) * C_{\text{tm}}$$

Here, F_{S} is the aqueous secretion ($F = 2.5$ μL/min), and U is the outflow rate through the uveoscleral pathways (at 10%, i.e., 0.25 μL/min). C_{tm} is the outflow facility through the trabecular meshwork (~0.35 μL/min per mmHg). A C_{tm} value of less than 0.2 μL/min per mmHg is indicative of the glaucomatous range. Ocular fluorometers can be used to assess the outflow facility. First, F_{S} is determined by measuring the disappearance of fluorescein from the anterior chamber. Next, topical dorzolamide and/or timolol is administered to reduce the IOP (i.e., IOP2 − IOP1; measured by tonometry) and aqueous secretion (i.e., $F_{\text{S2}} - F_{\text{S1}}$; measured by fluorometry). The outflow facility is then calculated using the equation $C_{\text{tm}} = (F_{\text{S2}} - F_{\text{S1}})/(\text{IOP2} - \text{IOP1})$.

Rho kinase inhibitors (i.e., ROCK inhibitors) that have been recently been introduced to reduce IOP are thought to act by way of increasing the outflow facility through the TM (Prasanna et al. 2016; Ramachandran et al. 2011; Nakajima et al. 2005; Honjo et al. 2001; Komizo et al. 2019; Tanihara et al. 2008; Andrew et al. 2020; Jayanetti et al. 2020). In principle, the prostaglandin analogs and α2 agonists reduce IOP by increasing the uveoscleral outflow facility and decreasing aqueous humor secretion, respectively (Jayanetti et al. 2020). They may also affect the TM outflow facility, although their mechanisms of action are not described.

Further Considerations on the Balance Between Aqueous Inflow and Outflow

IOP is determined by the balance between aqueous inflow and outflow. Numerous factors influence this balance, including blood pressure and osmotic pressure, whose variations are extremely difficult to identify and isolate in clinical and experimental studies. To this end, mathematical models have been proposed to test a variety of clinical scenarios (Kiel et al. 2011; Mauri et al. 2016; Sacco et al. 2020; Siggers and Ethier 2012; Szopos et al. 2016). As an example, we describe the model previously developed by our team members (Szopos et al. 2016) to predict the efficacy of IOP-lowering medications in different individuals. We denote J_{in} and J_{out} as the total flows of aqueous humor in and out of the anterior chamber. The inflow is due to a combination of ultrafiltration, a passive mechanism, and ion transport, an active mechanism; thus, we can write

$$J_{\text{in}} = L_{\text{in}}\left[\left(\text{cBP} - \text{IOP}\right) - \sigma_{\text{p}}\Delta\pi_{\text{p}} - \sigma_{\text{s}}\Delta\pi_{\text{s}}\right]$$

where L_{in} denotes the inflow facility, cBP is blood pressure in the capillaries of the ciliary body, σ_{p} and σ_{s} are the reflection coefficients for proteins and small solutes, and $\Delta\pi_{\text{p}}$ and $\Delta\pi_{\text{s}}$ are the differences in osmotic and oncotic pressures across the ciliary body, respectively. The outflow occurs through the trabecular and uveoscleral pathways; thus, we can write

$$J_{\text{out}} = C_{\text{tm}} * \left(\text{IOP} - P_{\text{EVP}}\right) + C_{\text{uv}} * \text{IOP}$$

where C_{tm} and C_{uv} denote the outflow facility characterizing the trabecular and uveoscleral routes and P_{EVP} is the episcleral vein pressure. Finally, the balance between inflow and outflow can be written as

$$J_{\text{in}} = J_{\text{out}}$$

which leads to the following equation, which is useful to predict IOP in different clinical circumstances:

$$\text{IOP} = L_{\text{in}}\left(\text{cBP} - \sigma_{\text{p}}\Delta\pi_{\text{p}} - \sigma_{\text{s}}\Delta\pi_{\text{s}}\right) / \left(L_{\text{in}} + C_{\text{tm}} + C_{\text{uv}} * P_{\text{EVP}}\right)$$

At a healthy baseline, the values of the parameters in the formula above are reported by Szopos et al. (2016). At baseline, IOP is ~15 mmHg. A reduction of C_{tm}, which implies obstruction within the trabecular meshwork, elevates IOP. For example, when C_{tm} is reduced by 25%, 50%, and 75%, the model predicts IOP levels of 16 mmHg, 18 mmHg, and 20 mmHg, respectively. If the medications are prescribed to lower the IOP by reducing aqueous humor secretion on three individuals with different degrees of obstruction in the trabecular meshwork, say by 25%, 50%, and

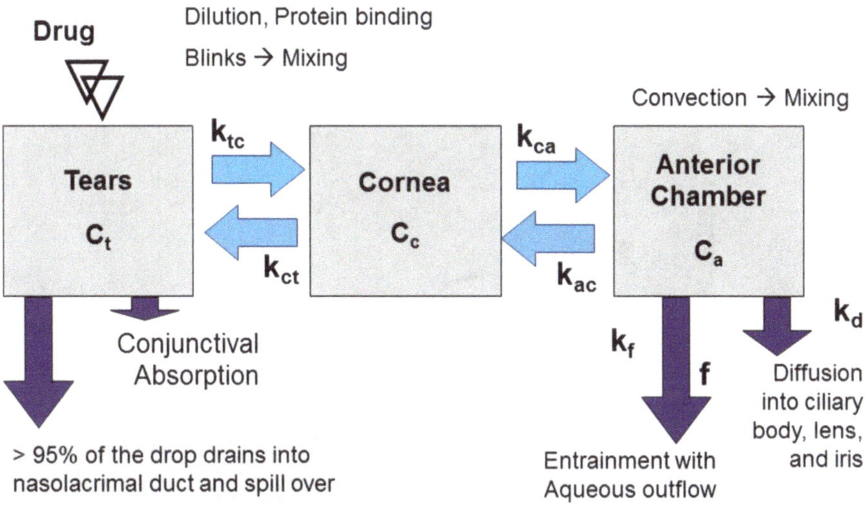

Fig. 10 Generalized compartmental model highlighting drug clearance from the anterior chamber. Mixing in the anterior chamber is affected by convection. The volume of distribution of a drug in the anterior chamber could be affected by sequestration into the iris, lens, and ciliary body. Drug clearance from the anterior chamber could be affected by loss into iris vasculature but occurs largely by entrainments with the aqueous humor outflow through the trabecular meshwork. Key: *C* concentration, *k* first-order rate constant, *f* flow rate

75%, we can reduce the osmotic pressure difference $\Delta\pi_s$ by, say, 10% and estimate IOP reductions of 6.7%, 7.4%, and 8.2%, respectively. Thus, the model is suitable for predicting changes in IOP under different pathological and clinical settings. Indeed, a number of physiological, pharmacological, and pathological factors are known to affect IOP. For instance, sitting position, progesterone/estrogen, heroin, marijuana, alcohol, HIV infection, hyperthyroidism, myotonic dystrophy, and anterior uveitis may reduce IOP. On the other hand, supine position especially in glaucomatous eyes, blinking especially in Graves' infiltrative ophthalmopathy, adrenocorticotrophic hormone, glucocorticoids, growth hormones, lysergic acid diethylamide, caffeine, systemic hypertension, diabetes, and hypothyroidism may increase IOP. However, since the estimation of various parameters in the experimental and clinical settings is difficult, more simplified mathematical models are warranted. Such models in conjunction with pharmacokinetic models (Kompella et al. 2020) may be useful in predicting the influence of various factors that influence IOP on antiglaucoma drug delivery. Figure 10 shows a general pharmacokinetic model for topically administered drugs, and Fig. 11 shows distribution of rhodamine B, a fluorescent lipophilic dye after topical application in the rabbit model. Drug levels are high in the solid tissues, consistent with a high affinity of the drug to these tissues. Low levels of the dye in the aqueous humor are consistent with the low solubility and affinity of this molecule to aqueous humor.

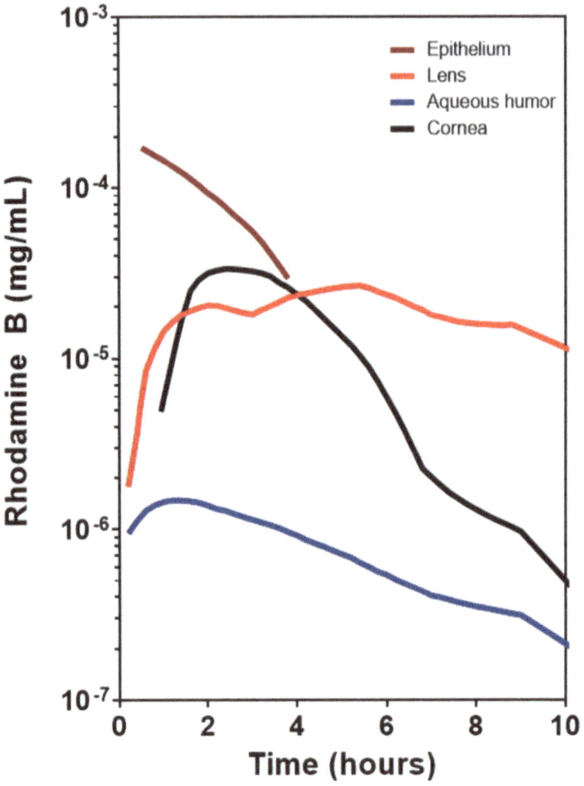

Fig. 11 Partitioning of drugs into the crystalline lens: The lens acts as a reservoir for lipophilic drugs. The anterior lens facing the anterior chamber is covered with an epithelial layer with tight junctions. The potential for accumulation of lipophilic drug is being demonstrated in rabbit eyes after topical application of the lipophilic fluorescent molecule, rhodamine B (MW: Log $P = 2.3$). The evolution of fluorescence, which is the proportional concentration of rhodamine B, in the tears, cornea, anterior chamber, and lens, is shown

Summary

The ciliary epithelium secretes aqueous humor at 2–4 μL/min into the posterior chamber with secretion being dependent on the HCO^-_3-mediated active ion transport mechanisms. It is modulated by the autonomic nervous system via α_2, β_1, and β_2 adrenergic receptors. The aqueous humor escapes into the anterior chamber and then exits the eye via two outflow pathways, viz., trabecular meshwork route and uveoscleral route. A hydraulic pressure gradient drives the trabecular meshwork route ($\Delta P = \text{IOP} - \text{EVP}$). The resistance to outflow lies mainly in the juxtacanalicular tissue of the trabecular meshwork, and it is possibly augmented by the inner wall of Schlemm's canal. The outflow is then directed to episcleral veins via collector channels, which are distributed around the trabecular meshwork. The outflow occurs predominantly at regions proximal to collector channels, therefore segmenting trabecular meshwork outflow. The resistance to flow in juxtacanalicular tissue is sensitive to agents that modulate the contractility of trabecular meshwork cells. Agents that reduce the contractility (ROCK inhibitors) increase the outflow facility. On the other hand, matrix metalloproteases modulate the resistance to outflow via the uveal route, which break down the extracellular matrix around ciliary muscles. Convection drives the aqueous to be mixed in the anterior chamber. The resistance to aqueous humor outflow sets up intraocular pressure. Intraocular pressure shows a circadian rhythm, peaking in the morning hours. Elevated intraocular pressure is a major risk factor for the onset of glaucoma. Accordingly, the decreasing intraocular pressure is the principal approach in the treatment of glaucoma. The rate of aqueous turnover is 1% per min, which corresponds to a half-life of ~72 min for drugs and solutes in the anterior chamber if their removal is solely dependent on aqueous humor turnover. Therefore, factors including circadian rhythm, therapeutic agents, and pathological conditions that influence aqueous humor turnover can potentially influence drug clearance from the aqueous humor.

References

Abdelhafid F, et al. Operator splitting for the simulation of aqueous humor thermo-fluid-dynamics in the anterior chamber. In: Proceedings of the V International Conference in Applied Mathematics, Modeling and Computational Science (AMMCS 2019). 2019 (In Press): AMMCS.

Agarwal A, et al. Application of laser flare photometry in the quantification of blood-aqueous barrier breakdown after micro-incision vitrectomy. Ocul Immunol Inflamm. 2019;27(4):651–7.

Agrahari V, et al. A comprehensive insight on ocular pharmacokinetics. Drug Deliv Transl Res. 2016;6(6):735–54.

Andrew NH, Akkach S, Casson RJ. A review of aqueous outflow resistance and its relevance to microinvasive glaucoma surgery. Surv Ophthalmol. 2020;65(1):18–31.

Aspelund A, et al. The Schlemm's canal is a VEGF-C/VEGFR-3-responsive lymphatic-like vessel. J Clin Invest. 2014;124(9):3975–86.

Bhandari A, Bansal A, Sinha N. Effect of aging on heat transfer, fluid flow and drug transport in anterior human eye: a computational study. J Control Release. 2020;328:286–303.

Bourges JL, et al. Intraocular implants for extended drug delivery: therapeutic applications. Adv Drug Deliv Rev. 2006;58(11):1182–202.

Boushehrian H, et al. Relationship between saccadic eye movements and formation of the Krukenberg's spindle—a CFD study. Mat Med Biol. 2016;34(3):293–312.

Carreon T, et al. Aqueous outflow—a continuum from trabecular meshwork to episcleral veins. Prog Retin Eye Res. 2017;57:108–33.

Chan RY, Smith JA, Richardson KT. Anterior segment configuration correlated with Shaffer's grading of anterior chamber angle. Arch Ophthalmol. 1981;99(1):104–7.

Chen H, et al. Numerical investigation of topical drug transport in the anterior human eye. Int J Heat Mass Transfer. 2015;85:356–66.

Chung SK, Kim SK. Digital particle image velocimetry studies of nasal airflow. Respir Physiol Neurobiol. 2008;163(1–3):111–20.

Civan MM, Macknight AD. The ins and outs of aqueous humour secretion. Exp Eye Res. 2004;78(3):625–31.

Civan MM, et al. Potential contribution of epithelial Na+ channel to net secretion of aqueous humor. J Exp Zool. 1997;279(5):498–503.
Coca-Prados M, Escribano J. New perspectives in aqueous humor secretion and in glaucoma: the ciliary body as a multifunctional neuroendocrine gland. Prog Retin Eye Res. 2007;26(3):239–62.
Coca-Prados M, Wax MB. Transformation of human ciliary epithelial cells by simian virus 40: induction of cell proliferation and retention of beta 2-adrenergic receptors. Proc Natl Acad Sci U S A. 1986;83(22):8754–8.
Cooper HS, Manning DR, Wax MB. G protein complement of SV40-transformed ciliary epithelial cells. Curr Eye Res. 1990;9(5):493–9.
del Amo EM, et al. Pharmacokinetic aspects of retinal drug delivery. Prog Retinal Eye Res. 2017;57:134–85.
Do CW, To CH. Chloride secretion by bovine ciliary epithelium: a model of aqueous humor formation. Invest Ophthalmol Vis Sci. 2000;41(7):1853–60.
Dvoriashyna M, et al. Mathematical models of aqueous humor production, flow and drainage. In: Guidoboni G, Harris A, Sacco R, editors. Ocular fluid dynamics. anatomy, physiology, imaging techniques, and mathematical modeling. New York: Springer; 2020.
Feng Y, Goree J, Liu B. Errors in particle tracking velocimetry with high-speed cameras. Rev Sci Instrum. 2011;82(5):053707.
Gaton DD, et al. Increased matrix metalloproteinases 1, 2, and 3 in the monkey uveoscleral outflow pathway after topical prostaglandin F(2 alpha)-isopropyl ester treatment. Arch Ophthalmol. 2001;119(8):1165–70.
Gupta C, et al. Measurement and modeling of diffusion kinetics of a lipophilic molecule across rabbit cornea. Pharm Res. 2010;27(4):699–711.
Gupta C, Chauhan A, Srinivas SP. Penetration of fluorescein across the rabbit cornea from the endothelial surface. Pharm Res. 2012;29(12):3325–34.
Haghjou N, Soheilian M, Abdekhodaie MJ. Sustained release intraocular drug delivery devices for treatment of uveitis. J Ophthalmic Vis Res. 2011;6(4):317–29.
Heys JJ, Barocas VH. A boussinesq model of natural convection in the human eye and the formation of Krukenberg's spindle. Ann Biomed Eng. 2002;30(3):392–401.
Honjo M, et al. Effects of rho-associated protein kinase inhibitor Y-27632 on intraocular pressure and outflow facility. Invest Ophthalmol Vis Sci. 2001;42(1):137–44.
Investigators A. The Advanced Glaucoma Intervention Study (AGIS): 7. The relationship between control of intraocular pressure and visual field deterioration. The AGIS Investigators. Am J Ophthalmol. 2000;130(4):429–40.
Jacob TJ, Civan MM. Role of ion channels in aqueous humor formation. Am J Physiol. 1996;271(3 Pt 1):C703–20.
Jayanetti V, Sandhu S, Lusthaus JA. The latest drugs in development that reduce intraocular pressure in ocular hypertension and glaucoma. J Exp Pharmacol. 2020;12:539–48.
Johnson M, McLaren JW, Overby DR. Unconventional aqueous humor outflow: a review. Exp Eye Res. 2017;158:94–111.
Kaji Y, et al. Visualization of irrigation fluid flow and calculation of its velocity distribution in the anterior chamber by particle image velocimetry. Graefes Arch Clin Exp Ophthalmol. 2012;250(7):1023–7.
Karpinich NO, Caron KM. Schlemm's canal: more than meets the eye, lymphatics in disguise. J Clin Invest. 2014;124(9):3701–3.
Kashiwagi K, et al. Agreement between peripheral anterior chamber depth evaluation using the van Herick technique and angle width evaluation using the Shaffer system in Japanese. Jpn J Ophthalmol. 2005;49(2):134–6.
Khalighi B, Lee YH. Particle tracking velocimetry: an automatic image processing algorithm. Appl Opt. 1989;28(20):4328–32.
Kiel JW, et al. Ciliary blood flow and aqueous humor production. Prog Retin Eye Res. 2011;30(1):1–17.

Kizhatil K, et al. Schlemm's canal is a unique vessel with a combination of blood vascular and lymphatic phenotypes that forms by a novel developmental process. PLoS Biol. 2014;12(7):e1001912.
Komizo T, et al. Additive intraocular pressure-lowering effects of the Rho kinase inhibitor ripasudil in Japanese patients with various subtypes of glaucoma. Jpn J Ophthalmol. 2019;63(1):40–5.
Kompella UB, Hartman RR, Patil MA. Extraocular, periocular, and intraocular routes for sustained drug delivery for glaucoma. Prog Retin Eye Res. 2020;82:100901.
Krupin T, et al. Effects of adrenergic agents on transepithelial electrical measurements across the isolated iris-ciliary body. Exp Eye Res. 1991;53(6):709–16.
Kumar S, Acharya S, Beuerman R. Deposition of particles on ocular tissues and formation of Krukenberg spindle, hyphema, and hypopyon. J Biomech Eng. 2007;129(2):174–86.
Lai J, et al. The role of Schlemm's canal endothelium cellular connectivity in giant vacuole formation: a 3D electron microscopy study. Invest Ophthalmol Vis Sci. 2019;60(5):1630–43.
Lee VH, et al. Basis for dosing time-dependent changes in the ocular and systemic absorption of topically applied timolol. J Ocul Pharmacol Ther. 1996;12(2):103–13.
Lindken R, et al. Micro-Particle Image Velocimetry (microPIV): recent developments, applications, and guidelines. Lab Chip. 2009;9(17):2551–67.
Loke CY, et al. Segmental aqueous humour outflow and eye orientation have strong influence on ocular drug delivery. Appl Math Model. 2018;57:474–91.
Macknight AD, et al. Formation of the aqueous humor. Clin Exp Pharmacol Physiol. 2000;27(1–2):100–6.
Mauri AG, Sala L, Airoldi P, Novielli G, Sacco R, Cassani S, Guidoboni G, Siesky BA, Harris A. Electro-fluid dynamics of aqueous humor production: simulations and new directions. J Model Ophthalmol. 2016;1(2):48–58.
McLaren JW. Measurement of aqueous humor flow. Exp Eye Res. 2009;88(4):641–7.
McLaren JW, et al. Rate of flow of aqueous humor determined from measurements of aqueous flare. Invest Ophthalmol Vis Sci. 1990;31(2):339–46.
McLaughlin CW, et al. Timolol may inhibit aqueous humor secretion by cAMP-independent action on ciliary epithelial cells. Am J Physiol Cell Physiol. 2001;281(3):C865–75.
Missel PJ. Simulating intravitreal injections in anatomically accurate models for rabbit, monkey, and human eyes. Pharm Res. 2012;29(12):3251–72.
Missel PJ, Sarangapani R. Physiologically based ocular pharmacokinetic modeling using computational methods. Drug Discov Today. 2019;24(8):1551–63.
Nakajima E, et al. Contribution of ROCK in contraction of trabecular meshwork: proposed mechanism for regulating aqueous outflow in monkey and human eyes. J Pharm Sci. 2005;94(4):701–8.
Nau CB, et al. Circadian variation of aqueous humor dynamics in older healthy adults. Invest Ophthalmol Vis Sci. 2013;54(12):7623–9.
Ooi EH, Ng EY. Simulation of aqueous humor hydrodynamics in human eye heat transfer. Comput Biol Med. 2008;38(2):252–62.
Parc CE, Johnson DH, Brilakis HS. Giant vacuoles are found preferentially near collector channels. Invest Ophthalmol Vis Sci. 2000;41(10):2984–90.
Park DY, et al. Lymphatic regulator PROX1 determines Schlemm's canal integrity and identity. J Clin Invest. 2014;124(9):3960–74.
Patel A, et al. Ocular drug delivery systems: an overview. World J Pharmacol. 2013;2(2):47–64.
Peterson K, et al. Single-camera, three-dimensional particle tracking velocimetry. Opt Express. 2012;20(8):9031–7.
Prasanna G, et al. Pharmacology of novel intraocular pressure-lowering targets that enhance conventional outflow facility: Pitfalls, promises and what lies ahead? Eur J Pharmacol. 2016;787:47–56.
Purslow C, Wolffsohn JS. Ocular surface temperature: a review. Eye Contact Lens. 2005;31(3):117–23.
Radenbaugh PA, et al. Concordance of aqueous humor flow in the morning and at night in normal humans. Invest Ophthalmol Vis Sci. 2006;47(11):4860–4.

Ramachandran C, et al. Rho-Rho kinase pathway in the actomyosin contraction and cell-matrix adhesion in immortalized human trabecular meshwork cells. Mol Vis. 2011;17:1877–90.
Sacco R, et al. A theoretical approach for the electrochemical characterization of ciliary epithelium. Life (Basel). 2020;10(2):8.
Sawa M. Laser flare-cell photometer: principle and significance in clinical and basic ophthalmology. Jpn J Ophthalmol. 2017;61(1):21–42.
Schachtschabel U, Lindsey JD, Weinreb RN. The mechanism of action of prostaglandins on uveoscleral outflow. Curr Opin Ophthalmol. 2000;11(2):112–5.
Siggers JH, Ethier CR. Fluid mechanics of the eye. Annu Rev Fluid Mech. 2012;44(1):347–72.
Sit AJ, McLaren JW. Measurement of episcleral venous pressure. Exp Eye Res. 2011;93(3):291–8.
Sit AJ, et al. A novel method for computerized measurement of episcleral venous pressure in humans. Exp Eye Res. 2011;92(6):537–44.
Srinivas SP, Maurice DM. A microfluorometer for measuring diffusion of fluorophores across the cornea. IEEE Trans Biomed Eng. 1992;39(12):1283–91.
Subrizi A, et al. Design principles of ocular drug delivery systems: importance of drug payload, release rate, and material properties. Drug Discov Today. 2019;24(8):1446–57.
Sudhir RR, et al. Ocular spot fluorometer equipped with a lock-in amplifier for measurement of aqueous flare. Transl Vis Sci Technol. 2018;7(6):32.
Swaminathan SS, et al. Aqueous outflow: segmental and distal flow. J Cataract Refract Surg. 2014;40(8):1263–72.
Szopos M, et al. Mathematical modeling of aqueous humor flow and intraocular pressure under uncertainty: towards individualized glaucoma management. J Model Ophthalmol. 2016;1(2):29–39.
Tanihara H, et al. Intraocular pressure-lowering effects and safety of topical administration of a selective ROCK inhibitor, SNJ-1656, in healthy volunteers. Arch Ophthalmol. 2008;126(3):309–15.
Topper JE, McLaren J, Brubaker RF. Measurement of aqueous humor flow with scanning ocular fluorophotometers. Curr Eye Res. 1984;3(12):1391–5.
Truong TN, et al. Novel characterization and live imaging of Schlemm's canal expressing Prox-1. PLoS One. 2014;9(5):e98245.
Tugal-Tutkun I, Herbort CP. Laser flare photometry: a noninvasive, objective, and quantitative method to measure intraocular inflammation. Int Ophthalmol. 2010;30(5):453–64.
Villamarin A, et al. 3D simulation of the aqueous flow in the human eye. Med Eng Phys. 2012;34(10):1462–70.
Vranka JA, et al. Elevated pressure influences relative distribution of segmental regions of the trabecular meshwork. Exp Eye Res. 2020;190:107888.
Wax MB, Barrett DA. Regulation of adenylyl cyclase in rabbit iris ciliary body. Curr Eye Res. 1993;12(6):507–20.
Wax MB, Molinoff PB. Distribution and properties of beta-adrenergic receptors in human iris-ciliary body. Invest Ophthalmol Vis Sci. 1987;28(3):420–30.
Weinreb RN, et al. Matrix metalloproteinases and glaucoma treatment. J Ocul Pharmacol Ther. 2020;36(4):208–28.
Wyatt HJ. Modelling transport in the anterior segment of the eye. Optom Vis Sci. 2004;81(4):272–82.
Yavuz B, Kompella UB. Ocular drug delivery. Handb Exp Pharmacol. 2017;242:57–93.
Zhang F, Chen H, Huang Y. Computer modeling of drug delivery in the anterior human eye after subconjunctival and episcleral implantation. Comput Biol Med. 2017;89:162–9.

Part II
Fundamental Approach to Ophthalmic Product Development

Back of the Eye Anatomy and Physiology: Impact on Product Development

Clive G. Wilson

Abstract Treatment of ocular tissues is attempted to remedy infection, inflammatory disease and the sequelae of pathological systemic changes such as diabetes and degenerative changes associated with ageing. The anatomical features of the orbit and physiological mechanisms are important in understanding the possible limitations in drug delivery objectives and are reviewed in this short chapter with the aim of illustrating the behaviour of potential drug delivery spaces which have been reported by researchers in ocular physiology and in pharmaceutics. As such, this article is not a comprehensive primer but highlights prominent physiological processes and anatomical features that are useful to know at the outset of ocular product design.

Keywords Posterior segment · Anatomy · Retina · Drug delivery · Vitreous flow · Suprachoroidal space

Introduction

The eye is an extremely fragile structure partially protected by the forehead and the bony orbit. Along the eye axis, there is no protection, and sharp stereoscopic vision enables us to attempt to avoid trauma by closing the eyes and turning the head. Material that lands on the eye benignly is cleared by lacrimation and outflow of the tears down the cheek. Moreover, the blood-retinal barrier prevents accumulation of material from the systemic circulation isolating the tissues of the delicate tissues of the eye as much as possible. From the facial aspect, evolution has provided an efficient clearance which provides barriers to noxious agents and to organisms which would otherwise colonize the surface tissue. In addition, tear secretion ensures that

C. G. Wilson (✉)
Strathclyde Institute of Pharmacy and Biomedical Sciences, Glasgow, Scotland, UK
e-mail: c.g.wilson@strath.ac.uk

S. Neervannan, U. B. Kompella (eds.), *Ophthalmic Product Development*, AAPS Advances in the Pharmaceutical Sciences Series 37,
https://doi.org/10.1007/978-3-030-76367-1_4

the cornea is kept as a good optical surface and, with the lens, allows focusing of the image on the retina.

Treatment of diseases in the posterior eye is always difficult. Solutions placed on the cornea will be cleared quickly by tear drainage and venous and lymphatic outflow, and loosely attached external structures may be dislodged unless the patient becomes deliberately tolerant. Any temporary degradation of the image provokes tearing, forced blinking, movement of facial muscles and the urge to rub the eyes. Delivery beyond the lens requires invasive procedures and technological approaches to sustaining levels of active treatment. Flow will dilute and distribute drug released from devices and only that fraction that reaches the retina allows treatment. Around the eye, broaching the sclera will allow closer access of a drug from an exterior reservoir, but deeper penetration will be strongly influenced by the choroidal circulation.

In respect of drug delivery as shown in Fig. 1, the anterior eye can be considered as two zones comprising the frontal periocular tissues and cornea externally and interiorly the aqueous anterior chamber, iris and ciliary body. Drug delivery to the lens is generally approached with anterior eye delivery technologies, i.e. drops and suspensions (Fig. 1). To treat the retina, we have to utilize posterior eye delivery technologies, which are usually invasive, and these aspects are the main topic of the review. For completeness, reference is made to flow processes which connect the posterior media to uveoscleral clearance and aqueous flow as this drives advection. The most important of these processes is the generation of intraocular pressure by the formation of aqueous humour.

General Anatomy and Major Compartments

The eye is situated in the anterior portion of the orbit closer to the lateral than the medial wall and nearer to the roof than to the floor of the skull. The eye is held in place by tendons attached to the sclera and operated by the attached muscles to allow tracking of objects of interest. Around the eye, the gaps between muscle and tendon allow the placement of drug reservoirs, and at the back of the eye, the long optic nerve 47–55 mm in length connects the optic disc of the retina to the optic chiasma terminating on the opposite side of the brain. The nerve runs through the orbital socket and the optic canal in the splenoid bone and is cushioned by the intraconal fat (the central surgical space) which is enclosed by the intramuscular septum attached to the rectus muscles. This in turn is surrounded by the extraconal fat and is clearly discriminated by MRI (see Fig. 2). The injection into the intraconal zone (retrobulbar block) is usually attempted to anaesthetize the ciliary nerves and the II, III and VI cranial nerves which run through this space allowing intraocular surgical procedures. The volume injected is typically 3–4 mL.

The human eye has a spherical shape with an approximate average diameter of 22–24 mm and a volume of ~6 cm^3. The range of the axial length of the globe varies between 21 and 25 mm. Eyes with an axial length <24 mm are typically considered

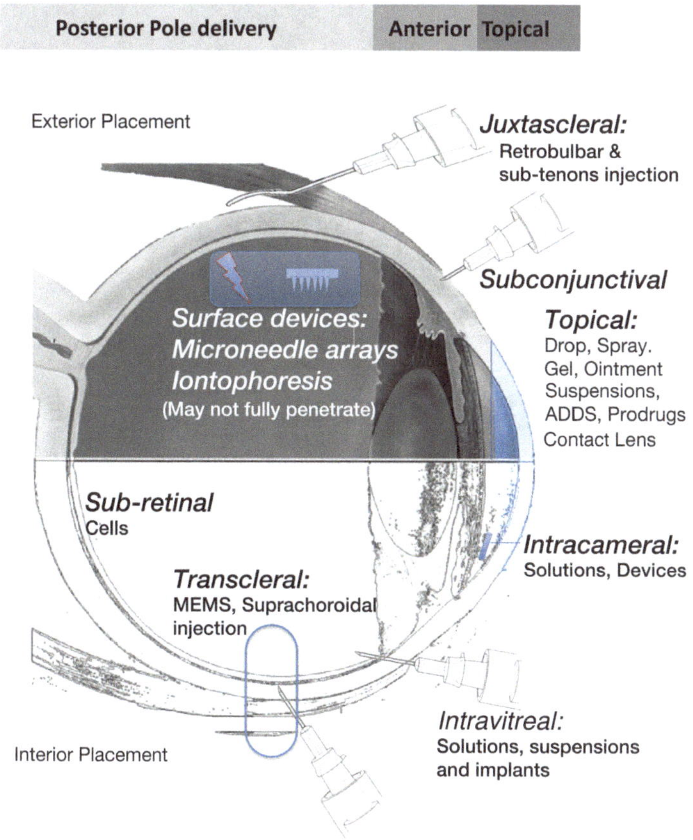

Fig. 1 Common routes of ocular drug delivery. (Adapted from Wilson et al. (2016))

hyperopic (or hypermetropic), whilst eyes with an axial length of >25 mm are typically considered myopic. Thus, for an average of +3 dioptres hyperopia (±2.0), the average axial length was estimated to be in the range 22.62 ± 0.76 mm, whilst for an average of −3.3 dioptres myopia (±2.0), the axial length was 25.16 ± 1.23 mm (Llorente et al. 2004).

The eye is filled with liquid, and the frontal anterior chamber holds a volume of approximately 200 μL in man (Zhu et al. 2018), with a turnover rate of between 1 and 1.5% per minute (Gabelt and Kaufman 2003). Approximately 50 μL is present at the site of formation in the posterior chamber (PC), the narrow space between the lens and iris, but this is affected by the extent of pupil dilation (Barocas and Huang 2006) which traps fluid in the PC (Civan 2008). Toris and colleagues measured a significant difference in anterior chamber volume in healthy normotensive volunteers according to age. The younger group had a volume of 247 ± 39 μL versus 160 ± 39 μL in the >60 years group (Toris et al. 1999).

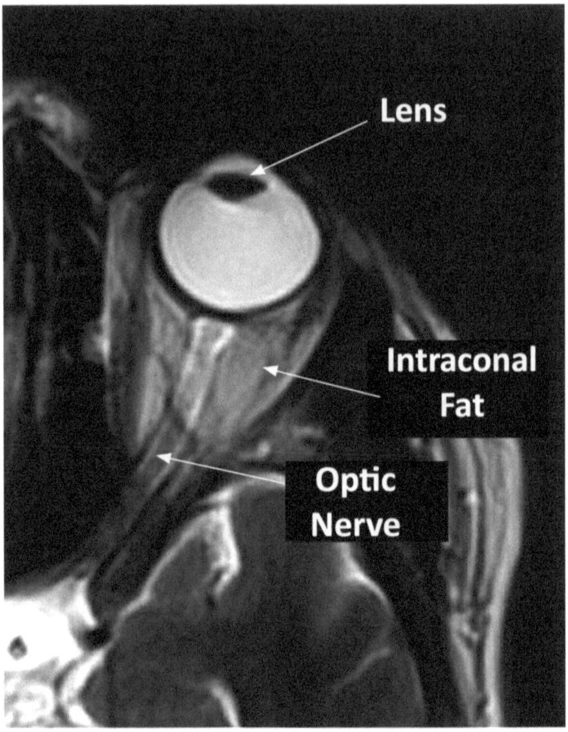

Fig. 2 MRI illustrating intraconal fat at the back of the eye

The intraocular pressure generated by secretion of aqueous humour by the ciliary processes produces the geometry needed for focusing on the image of the retina. The aqueous humour provides nutrients and oxygen for the internal avascular structures – the cornea and the lens – and flows from the site of production in the posterior chamber around the lens and through the pupil to enter the anterior chamber exiting through the canal of Schlemm. The baseline flow rate is very rapid, resulting in the total replacement of ciliary epithelial intracellular fluid in approximately 4 min (Civan 2008). The majority of the fluid escapes though anterior chamber drainage (70%), but unconventional drainage routes are important particularly uveoscleral outflow. In monkeys, the flow pathway from the ciliary muscle through the uveoscleral tract contributes between 40 and 60%, whereas in man, the figure is nearer 25%.

The Lymphatic Pathways

The lymphatic pathway plays an important part in the removal from depot injections placed around the eye, and on death, the barrier functions cease. Li and colleagues used gadolinium-DTPA as a medium-weight hydrophilic MRI marker and showed insignificant amounts of the marker entered the eye by this route following subconjunctival injection into a rabbit eye, although when injected into a dead animal, the material rapidly penetrated into the vitreous humour (Li et al. 2012). The eye is supported in the socket by the extraorbital muscles within the conical space of the intraconal orbital compartment and cushioned behind by the intraconal fat. The MRI scans in the live animal showed that a portion of the dose was delivered to the surrounding tissues outside the globe and exited through the fat to the vasculature and the lymphatics.

Quantum dots (QD), nano-sized CdSe-ZnS core-shell particles with a broad excitation and a specific size-related high-intensity emission spectrum are useful tools to track flow although issues of toxicity generally limit investigations to animal studies rather than clinical exploitation. Tam and colleagues have used QD655 preparations coated with carboxylic acids injected into the anterior chamber of a mouse using 33-gauge needle (Tam et al. 2011). Two hours after injection, a strong QD signal was visible in the neck region, and postmortem examination of the tissues suggested that drainage occurred through lymphatic channels in the ciliary body. The same group showed that intracameral latanoprost into the left eye increased lymphatic drainage of QD655 with accumulation of the tracer in the left submandibular node, suggesting manipulation of the uveo-lymphatic axis as a possible future target in glaucoma (Tam et al. 2013).

Convective and Advective Flow

In household physics, we refer to convection as the process by which heat is transferred from boilers to radiators by the bulk movement of liquids, which heats our living space by radiation and air movement. The differential heat content of molecules causes physical movement at a much faster rate than diffusion and results in redistribution from the highest concentration of hot molecules by diffusive processes, which are poor drivers of mixing. We can think of many analogies with regard to the movement of drugs where fluid flows generated by the heart; muscular contraction and compression causes mixing and flow along the path of least resistance. Since liquids are not compressible, changes in the pressure pathway result in the translocation of molecules and the physical deformation of tissues. Tuma and colleagues have described a computational model that examined the influence of scleral wall movement on the flow within a viscoelastic medium, i.e. the vitreous humour (Tuma et al. 2018). They show that the major effect is on the stress distribution exerted by the vitreous humour on the tissues, but the overall flow effects are relatively insensitive to the rheological characteristics of the medium. In this simulation, the influence of a superficial fluid zone is not considered, but displacement of the retina following the onset of PVD is probably strong affected by syneresis, and the fluid matrix is complex.

In ocular drug delivery, advective flow occurs on the surface of the cornea, within the aqueous humour, in the uveal tract and in the vitreous humour, especially following syneresis. The dissolved or suspended material moves with the bulk flow of the fluids, and when flow stops, diffusion moves the final distances under Brownian motion. Other processes become important as the rate of bulk flow slows – for example, surface binding which may either stop or assist more translocation. This is one of the roles of blood in tissue clearance: the binding of drugs to plasma proteins facilitates transport by supporting the concentration difference across a membrane and removing the diffusing species. In contrast, melanin binding may simply adsorb drug resulting in diminution of effect without transfer. Advection is often a neglected function in drug permeability measurements since the preparation may be static, the animal anaesthetized or elements of a model deficient in this respect. At the other extreme, the receptor and donor compartments may be so well stirred as to bear no resemblance to partially stopped or periodically stopped flow in a living system. These processes occur on the outer surfaces of the eye, within the blood stream supplying the tissues and within the tissue itself. Kim and colleagues commented that calculations based on permeability values or diffusion coefficients per se cannot be used to describe the transport of solutes in ocular tissue, as disposition is strongly affected by metabolic and dynamic factors (Kim et al. 2007).

Temperature gradients can contribute to convective mixing and are important in small eyes. The temperature of the anterior chamber is affected by three factors: the external temperature, the internal core temperature and blinking, since the eyelids will help to equilibrate corneal temperature. The difference in the temperatures between the cornea, exposed to the external environment, and the iris receiving a central blood supply results in buoyancy-driven currents. The stirring of the compartment can be explored pharmacologically as in the experiments of Wyatt and colleagues (1996). When a mydriatic drug was applied to the eye, it was noted that the initial response was highly asymmetric, with a larger response on the inferior nasal side than superior temporal, which in contrast hardly moved. The authors comment that initial response appears to be driven by diffusion, whereas latter distribution occurs by convective movement and trans-pupillary flow (Chen and Wyatt 2004). Vogel and colleagues have shown the application of precision thermal imaging to measure ocular surface temperature (OST) and compared rat, rabbit and human (Vogel et al. 2016). Unlike in man, OST was significantly higher in rats and rabbits by almost 1 °C, compared to rectal temperature, and it was concluded that this was attributed to the lack of fur that insulates the eye from the surrounding tissue in man. Fitt and Gonzalez applied a fluid mechanics treatment of flows generated by temperature-dependent buoyancy, aqueous production, flow generated by lens tremor (phakodonesis) and rapid eye movements (REM) (Fitt and Gonzalez 2006). They concluded that buoyancy-driven flow could produce flows as large as 0.1 mm s^{-1}, which was by far the highest magnitude effect. REM sleep has often been thought of as reprogramming event whilst the brain has less stimulation although finding a clear role is difficult. A controversial theory was proposed by David Maurice (1996), who suggested the role of REM was to assist in stirring the anterior chamber although this explanation is not supported by the Fitt and Gonzalez analysis (Fitt and Gonzalez 2006) as the contribution is probably much smaller than buoyant flow effects.

The direct use of point-source laser-generated heating in the eye to assist convection-assisted mixing has been explored in a two-dimensional computational model (Narasimhan and Sundarraj 2015). The authors suggest that heating at various points in the vitreous increases the mass transport of drug from a central depot by a factor of more than 5. They envisage the use of heating to direct drug to the posterior segment and suggest that even with high molecular weight drugs with a lower diffusivity, this approach could be useful.

The posterior eye is occupied by the vitreous humour, which makes up to approximately 80% of the eye volume. In the young and in many laboratory animals, the vitreous humour is a relatively stiff gel that, in man, liquifies with age. The volume is approximately 4 mL. Unlike the front of the eye, which has a flow process turning over the volume at around 1% total per minute, the vitreous is relatively static and viscous, and whilst well-formed, free diffusion of small molecules through the gel occurs. Bulk flow is generated in part through retro-zonular generation and flow through the retina (Lund-Andersen 2003). As will be discussed later, because the structure of VH is inhomogeneous and imperfectly viscoelastic, advective and oscillatory forces can be generated by eye motion (Bonfiglio et al. 2013).

The Sclera

The sclera is the protective outer layer of the eye containing collagen and elastic fibres and forms the posterior five-sixths of the connective tissue coat of the eyeball. On the rear side of the eyeball, the sclera is continuous with dura mater (as the outermost layer enveloping the optic nerve), whilst on the front side, it connects with the cornea through the limbus. The total thickness of the human sclera is estimated at approximately 670 μm, the maximal thickness occurring at the posterior pole (~1000 μm) and the minimal at the equator (~490 μm) (Norman et al. 2010). The sclera is relatively inert metabolically and the primary role is to protect the intraocular contents. It is an elastic tissue which allows a more even distribution of blood to the choroid and so to the retina whilst having enough rigidity to achieve stability of the optical axis.

Histological Structure The sclera has four layers: episclera, scleral stroma, lamina fusca and endothelium. Running parallel to the scleral surface, Tenon's capsule gloves the whole eye as the *fascia bulbi*. Composed of hypocellular collagen bundles, it is attached anteriorly to the conjunctiva. Human beings are distinct from other primates by having a white sclera. From the front conjunctiva, it extends backwards becoming more loosely attached beyond the limbus and extends backwards as a condensed bundle of fibres. Merging with the dural sheath of the optic nerve and with fibrous connections linking the orbit to the globe, it provides the socket in which the eye moves. The separation between sclera and sheath forms part of the periscleral lymph space which is continuous with the subdural and subarachnoid cavities. The episclera is a relatively well-vascularized outer layer of the sclera, and

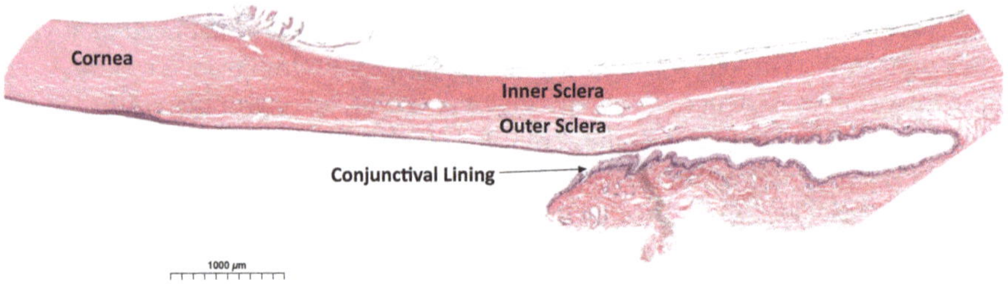

Fig. 3 The transition from the cornea to the more hydrophobic inner sclera; because the corneal fibres are hydrated, they appear more loose on fixing and staining. The outer epithelial layers stain darkly because of the higher protein content of these cells. Note that the inner retinal structures are not shown

the bundles of collagen are arranged in a circumferential pattern, anchoring the blood vessels tightly (see Fig. 3). It attaches to Tenon's capsule near to the limbus.

The scleral stroma forms a strong matrix with bundles of parallel aligned collagen fibres which form a lamellar structure becoming more interwoven in the deep sclera. In Fig. 3, this can be seen as the more dense pink-stained inner sclera in whole eye section. At the limbus, more superficial fibres merge with the episcleral fibres to form the scleral spur, an annular structure that protrudes into the anterior chamber and provides the attachment zone for the longitudinal fibres of the iris. It is attached anteriorly to the trabecular meshwork. A short spur is recognized as a risk factor in development of primary open-angle glaucoma (Swain et al. 2015).

The thickness of the sclera has been of great interest in drug delivery, since thickness varies across the globe and that might indicate the most favourable site of application for drug delivery. Olsen and colleagues examined fixed bank eyes and made cross-sectional measurements across the surfaces (Olsen et al. 1998). The authors characterized the mean scleral thickness at the corneoscleral limbus (0.53 ± 0.14 mm) and the equator (0.39 ± 0.17 mm) and saw that it increased near to the optic nerve (0.9–1 mm). They therefore determined that the equatorial sclera would be the appropriate position to attempt to cross the sclera from the outside surface. Similar changes in dimensions have been confirmed using MRI to measure cross-sectional thickness in enucleated eyes (Norman et al. 2010). Approaching the limbus, the fibrillar components of the matrix become less and some interlace with a circumcorneal annulus of fibres which are thought to contribute to the curvature of the cornea. This area of the sclera contains a high proportion of the antigen-presenting cells which contribute to inflammatory signalling in the cornea and sclera (Newton and Meek 1998).

In the posterior sclera round the optic nerve head, a sieve-like conical plug, the *lamina cribrosa*, separates the intra- and extraocular spaces. Through the holes of the *lamina cribrosa*, the nerve axons and blood vessels project into interior. Jonas and colleagues characterize the structure as a pressure barrier, serving to balance fluid pressures between CSF and IOP which may be damaged by sustained elevated IOP (Jonas et al. 1991, 2003). As humans get older, intraocular pressure gradually rises, and the prevalence of open-angle glaucoma similarly increases probably due to anatomical changes due to sustained changes on the optic nerve head and remaining nerve fibres (Morrison et al. 2005). Mechanically, the elastic modulus of the posterior sclera alters by stiffening (Geraghty et al. 2012), with a change in the collagen and elastin fibres. These fibres were found to strengthen in the circumferential direction suggesting mechanical reinforcement was occurring in response to sustained pressure rises which might even be responsible for optic nerve hypoplasia (ONH) changes (Coudrillier et al. 2012).

The microscopic structure of the sclera shows that it is bound together with parallel collagen fibres which run circularly around the eyeball in distinct layers around 50 μm thick which crisscross (Watson and Young 2004). Scanning electron microscopy (SEM) reveals fibres running in a bundle across the field of view, sandwiched by fibre bundles in cross section. This has been described as resembling a rhombic lattice. At the surface, the fibres are finer and are interwoven to form a dense netted mat, with occasional blood vessels visible. Drugs applied to the surface diffuse through this layer to the deeper scleral planes, and in a preparation where the eye is perfused, lateral movement of material can be seen (Fig. 4). For example, in the experiments conducted in our laboratory using the artificially perfused bovine eye preparation, a tazarotenic acid-loaded polyvinyl alcohol-loaded thin slab was placed at the equator of the sclera (Kek et al. 2010). Fluorescence in the fibres generated by two-photon excitation at 745 nm showed solvent drag occurring in the mid-sclera, moving the drug faster laterally than it penetrated. Previous workers have measured human scleral hydraulic conductivity in an Ussing chamber apparatus (Jackson et al. 2006), but this measurement technique does not account for lateral travel.

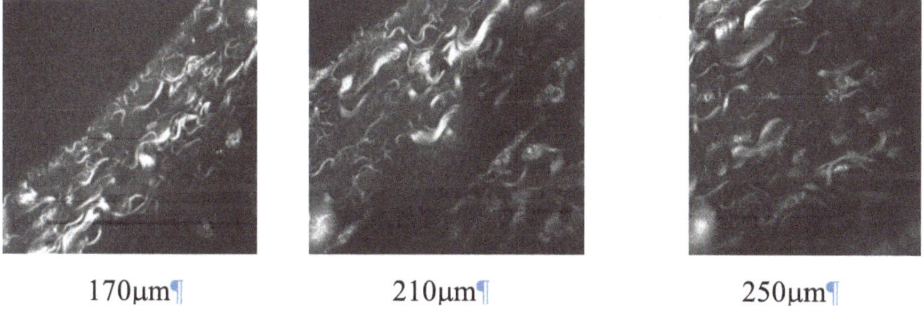

Fig. 4 Two photon images of sclera, showing tazarotenic acid moving laterally in the mid-sclera after application of the drug on the surface as a PVA-loaded patch. (From Kek et al. (2010))

The Suprachoroidal Space (SCS)

The suprachoroidal space (SCS) is often described as a virtual anatomical space, since it is visible on injection of a fluid with a short needle through the sclera. Moisseiev and colleagues describe the SCS as the potential space between the relatively well-defined inner border of the choroid, Bruch's membrane, and looser lamellar tissue (Moisseiev et al. 2016). However, the SCS can be seen on an enhanced depth imaging OCT scan and is spontaneously evident in 45% of healthy subjects (Yiu 2014). It appears as a 150 μm zone running parallel to the choriocapillaris when liquid is injected through a short needle, or the eye is perfused through a needle patch (Chiang et al. 2017). After injections of liquids of different viscosities, it was found that viscous CMC solutions were retained near the point of injection, expanding the SCS as water entered the hydrogel. Particles administered with viscous gel co-locate, whereas less viscous solutions result in separation of fluid and particles (size range 0.02–2 μm) (Chiang et al. 2016a). In the 3D model constructed by Zhang and co-workers, 90% of an injection of bevacizumab into the SCS would cleared in 12 h (Zhang et al. 2018). Krohn and Bertelsen prepared corrosion casts of the human (Krohn and Bertelsen 1997a) and pig eye (Krohn and Bertelsen 1997b) to examine the drainage into the uveoscleral tissue and also made Indian ink-stained gelatin injections in the human cadaver eye (Krohn and Bertelsen 1998). The authors described the function of the SCS as allowing outflow of the aqueous humour through the ciliary muscle and outwards through the scleral veins. They comment that the SCS might be a result of channels formed early in development that grow too far posteriorly, forming a beginning cluster of collector vessels close behind the scleral spur. Seiler and colleagues showed that the anterior SCS of porcine and canine cadaver eyes can be distended and postulated on the basis of latex casts and ultrasound contrast agents that about 50% of the posterior space could be reached (Seiler et al. 2011). The circumferential spread of fluorescent nanoparticles, 200 nm diameter, injected through a 750 μm 33 G hollow needle, was examined in rabbit and human cadaver eyes (Chiang et al. 2016b). In the rabbit, the anterior spur of the SCS limited the spread anteriorly, and spread was restricted posteriorly by the long posterior ciliary artery (LPCA). In the human eyes, the short posterior ciliary restricted flow in a ring preventing spread to the optic nerve. In contrast to the rabbit, the LPCA in humans enters the sclera further back and does not adhere the sclera to the choroid. It therefore does not interfere with the dispersion of particles in the SCS.

One of the attractions of SCS is that the intraocular space is not breached and the sclera can be bypassed. This has led to considerable interest in this route, and further explorations, including the use of iontophoresis to increase posterior segment targeting, have been explored by Mark Prausnitz's group (Jung et al. 2018).

Effects of Raised Intraocular Pressure

An elevated intraocular pressure is a risk factor of several ocular diseases, notably glaucoma, and results in reduction of aqueous humour flow. Increasing uveoscleral outflow remains an important additional target for IOP lowering with prostaglandins. Liquid flow through these pathways will contribute to advection and to the distribution and access of drugs in solution which should be increased in patients receiving therapy with prostaglandins. Generally, prostaglandins act exclusively on this route without an effect on aqueous humour kinetics even when no lateral flow is applied. Aihara and colleagues describe changes in human scleral permeability which occur after application of up to 200 nM latanoprost which causes a twofold increase in the permeation of basic fibroblast nerve factor (FGF-2) in a time-dependent manner (Aihara et al. 2001).

Vitreous Humour

As mentioned previously, the vitreous space is a large hydrogel compartment filling the space between the lens and the retina. In the young adult, the volume is approximately 4 mL with a wet weight of 4 g. The gel consists of 98–99% water and is structured from collagen type II fibres, coated with type IX fibres. The structuring of the collagen bridged by glycosaminoglycans gives the gel a viscoelastic nature, which is nonuniform. The central core is enclosed in a *cisterna,* first described by Jongebloed and Worst, who delineated the structure with red and white ink (Jongebloed and Worst 1987). The central zone is surrounded by a stiffer outer cortical zone, but under gravity, the whole structure collapses, especially when obtained as an elderly human donor eye. Thus to examine it, the tissue must be supported in situ, which also allows perfusion of the ciliary artery to study flows within the vitreous. This is commonly achieved using a modified Miyake-Apple preparation (Davis et al. 2003). This is shown in Fig. 5, a photograph taken after removal of the heating, perfusion and monitoring lines placed in and around a sheep eye. The spread of injected fluorescent particles can be seen behind lens by illumination through the cornea, which becomes opaque when perfusion is terminated. Removal of the vitreous from the globe still allows the exploration of local viscosity differences and the in vitro visualization of cisternal structures with adherent beads (Watts et al. 2014).

Flow Processes in the Vitreous

The movement of drug directly introduced into the intravitreal compartment must occur by three principal processes: diffusion (Moseley 1981), hydrostatic pressure (Fatt 1977) and convective flow (Stay et al. 2003). The contribution of these three

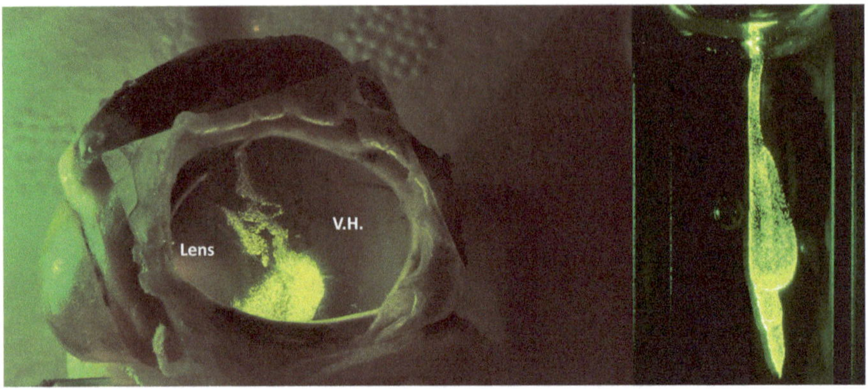

Fig. 5 The useful Miyake-Apple preparation, which shows spread of the injected particles (left). Decanting fresh vitreous into a cuvette and adding particles to the top as shown on the right reveals the cisternae of the vitreous. (Adapted from Wilson et al. (2011a))

parameters alters with the nature of the experimental design, and the simplest model assumes molecular diffusion through a static medium. Flux is then driven by the outward direction from a reservoir by Brownian motion, the motive drive being driven until the establishment of equilibration through the tissue space. If flow is principally in one direction or clearance is occurring at a boundary layer, a concentration gradient would be maintained. The balance between convective and diffusive processes is described by the Péclet number (Laude et al. 2010). The flow process was first described by Duke-Elder, who observed that in the rabbit, fluid originated from the ciliary body and pars planar region and exited near the optic nerve head (Duke-Elder 1930).

Stay argued that transport of a small solute within the vitreous depends on both convection and diffusion, but this argument was challenged by Missel who argued that convective flow through the vitreous is only significant for small molecules not efficiently cleared by the choroid (Missel 2002). Finite element modelling showed that flow was greatest near the hyaloid inlet at the edge of the lens and is rapidly lost progressing towards the posterior pole. It is important to remember that in most simulations, the vitreous is treated as a single compartment and for small molecules, this is not significant; however, as will be seen, the issues for large particles and macromolecules are different. Moreover, primate-laboratory animal differences become important. Stay does make the extremely important observation that convection in the vitreous of small eyes is not an important process for drug distribution since the distances are small. In addition, the lens occupies a significantly larger part of the globe compared to primates.

Missel's analysis suggested that useful flow conditions might be created near to the edge boundary, and an old paper by Fowlkes mentioned an intriguing finding (Fowlks et al. 1963). When blue tetrazolium chloride was injected into the eyes of anaesthetized rabbits, which were killed and the eyes frozen and removed, a thin stained area no more than about 3.5 mm thick developed along the edge of the

VH. This stained zone was quite symmetrical, and Fowlkes suggested that this was evidence of meridional flow, originating from behind the ciliary body and flowing largely posteriorly. He noted that it only occurred when the head was alive (and therefore blood flow was necessary). His thoughts were that the orientation of the collagen fibrils in the pararetinal region which run parallel to the surface of the inner limiting membrane as later noted by Le Goff and Bishop was important (Le Goff and Bishop 2008). He argued that this radial orientation was responsible for a higher concentration of hyaluronate nearer to the retina as had been noted by Balazs (1961). In a Miyake-Apple perfused eye preparation which has received an injection of fluorescent-tagged beads, ciliary artery perfusion-infusion causes detectable movement of fluorescent particles near to the surface layers (Wilson et al. 2011b). It has already been noted that the retina contains many aquaporin channels, and so the question of whether Fowlkes observation reflects the establishment of a superficial thin, more liquid zone of the VH is worth considering. Micro-rheological investigations using laser-trapped beads conducted by our group (Watts et al. 2013) support the concept that the very outermost layer of the vitreous humour is more fluid, which may be a mechanism how a proportion of a partially refluxed injection may spread laterally in agreement with Fowlkes.

To scan an object of interest, keeping it within the highest resolving zone of vision, the eyes make fast simultaneous sweeps or saccadic movements. This jerky movement is characterized by sudden starts and stops and exerts a torsional force on the viscoelastic vitreous gel. Womersley described a mathematical treatment of viscous flow movement in arteries, related to their calibre (Womersley 1955). David and colleagues used the Womersley model to describe these torsional forces applied to the retina (David et al. 1998). Such forces are important in propagating retinal tears, and understanding through modelling reveals how fluid motion affects the onset of posterior vitreous detachment and subsequent formation of retinal tears. The treatment of saccadic movement suggests that small flows begin to develop in a toroidal shape across the equatorial plane. This is commonly referred to as Darcy flow. Darcy's law describes the flow behaviour within and along a porous substrate as a function of resistance to flow. Although these movements of the eyeball are small, they exert an influence over a long period of time and allow complete recirculation of the dissolved materials within the vitreous. Zhang and colleagues observe that injected solution, presumably of higher viscosity, is mixed in a portion of the vitreous and sinks to the lowest surface as a denser sub-fraction (Zhang et al. 2018). The group at Johns Hopkins simulated the injection of various volumes and concluded that ranibizumab equilibrated within the vitreous at 2 days and 90% was cleared at 2 weeks postinjection.

Formulations can bind within the vitreous by virtue of strong ionic charges, and for cationic particles, the binding within the vitreous can greatly diminish the exposure of the retina from a nanoparticulate depot. This has been particularly important for polyethyleneimine-based systems used for the delivery of DNA (Pitkanen et al. 2003). For a further discussion, the reader is referred to the review by Mains and Wilson (2013a).

Pressure Effect

The injection of even small volumes of liquid into the eye will cause a transient increase in intraocular pressure (commonly called a 'spike'), with rises in measured hydrostatic pressure up to 30 mmHg (Lemos et al. 2015). Mortlett and Young reported rises of 38 mmHg associated with a 0.1 mL injection and described the use of previous ocular decompression to reduce the rise (Mortlet and Young 1993). The magnitude of the pressure increase is related to the volume injected and the speed of injection, since the sclera must deform to accommodate the extra liquid (Grzybowski et al. 2018). The rise in pressure restricts retinal blood flow and may be sustained in patients with glaucoma. Although the healthy eye can withstand these transient rises in pressure without lasting damage, those with retinal central vein occlusion or anterior ophthalmic optic neuropathies are at risk. Needle gauge is also a factor. A narrower gauge needle allows less reflux and therefore is associated with a higher postinjection spike (Pang et al. 2015; Alagöz et al. 2016). Although the injected additional volume is small, the drug itself may contribute to the rise in intraocular pressure as noted with bevacizumab injections (Hollands et al. 2007). Finally, in a meta-analysis of major anti-VEGF trials conducted by Bracha and colleagues (Bracha et al. 2018) and also Bakri and colleagues (Bakri et al. 2014), the authors noted that there was a subset of patients who appear to develop a sustained and maintained rise in IOP due to chronic therapy.

Removal of the Lens or Vitreous

Surgical removal of the lens is a common procedure in cataract surgery, and vitrectomies may be performed in the treatment of retinal detachment and haemorrhage. If small molecules are removed by an anterior pathway through the trabecular meshwork, then the effect of lensectomy should be predominately to increase the apparent volume of vitreal distribution. Studies on rabbits that had been made surgically aphakic or had both lens and vitreous removed (aphakic/vitrectomized) have been carried out by several authors. Ficker and colleagues compared the clearance of cefazolin in rabbits that had undergone these procedures, including the establishment of an inflammatory state. Inflammation caused by the injection of heat-treated *S. epidermidis* decreased rates of clearance, and in aphakic/vitrectomized eyes, clearance was faster (Ficker et al. 1990). In phakic, non-inflamed eyes, levels at the beginning were lower, but thereafter clearance rates were similar to control. Pearson and colleagues conducting a similar protocol with ciprofloxacin noted that in normal eyes, the elimination half-life was 2.2 h with a distribution volume of 1.2 mL. In aphakic vitrectomized eyes, the half-life was 1 h and the distribution volume was 1.4 mL (Pearson et al. 1993). Stern and colleagues injected [^{125}I]-labelled liposomes into rabbits which had been lensectomied and vitrectomized and noted that the clearance occurred with a half-life of 3.02 ± 0.4 h (Stern et al. 1987). The disposition of the negatively charged, gold colloid-labelled liposomes was seen in fibroblasts in different layers of the epiretinal fibroblasts but not in the retina.

A faster clearance of VEGF was noted in rabbits post-vitrectomy (Lee et al. 2010). The authors noted that measured half-life of $hVEGF_{165}$ (<3 h) was much shorter than would be expected with a molecule of 42 kDa and commented that a retinal adaptation to clear VEGF might be induced by the treatment. Vitrectomy in the cynomolgus macaque decreases the VEGF concentrations measured in the aqueous humour and significantly increased the elimination rate of bevacizumab injected 12 weeks after surgery (Kakinoki et al. 2012). The decrease in VEGF and erythropoietin to produce an antiangiogenic shift post-vitrectomy in patients with proliferative diabetic retinopathies has also been reported (Yoshida et al. 2012). The outcome of central core vitrectomy coupled with ranibizumab was investigated by Schramm and colleagues who reported that the combination produced the same outcome in terms of increased visual acuity but the procedure resulted in less reinjections (Schramm et al. 2014). The creation of the central pathway increasing convective flow of VEGF away from the macula may therefore be a beneficial surgical procedure in therapy. Finally, vitrectomy of rabbit eyes has little effect on the rate of dexamethasone clearance from a drug-loaded implant (Chang-Lin et al. 2011). This would be expected in view of the control of rate of delivery by such a device which will reside in the vitreous cavity until degradation.

The Ageing Eye

The remnant vitreous pulls away from the retina as the consequence of two processes: the liquefaction of the vitreous gel and the dehiscence of the membranes, associated with a progressive weakening in the adhesion between the ILM and the vitreous cortex (Sebag 1987, 2004). The changes in internal structure are illustrated in Fig. 6. Studies in our group have described a model of partial vitreous liquefaction in the rabbit by treatment with ovine testicular hyaluronidase (Tan et al. 2011). The degree of vitreous liquefaction generated was similar to that reported in humans between 55 and 60 years of age, with approximately 40% gel phase remaining. In partially liquefied vitreous, high molecular weight fluorescein isothiocyanate-dextran, average molecular weight 150 kDa was cleared faster than in controls, indicating greater convective movement in the liquefied vitreous. The ocular fluorimetry data showed that the gradient patterns of the fluorescent probes in this early synergetic model along the central axis were similar. This suggests that although the rate of elimination increased, the pathway remained unchanged. In an advanced state, the complete detachment results in the formation of two chambers, the remnants of the vitreous remaining attached at the back of the lens as a collapsed bag. The precise impact of this is not known, as by this stage the patient may have undergone operations to remove cataract which must also disturb the outflow pathway.

Loch and colleagues have recently described an approach to measure dose deposition within a spherical reservoir of polyacrylamide gel in a round-ended plastic

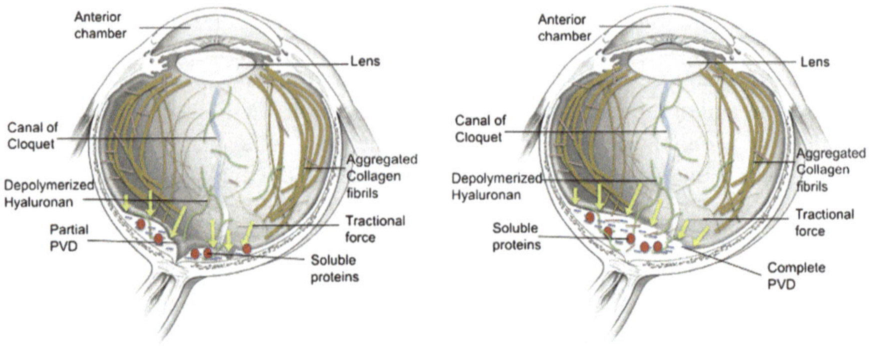

Fig. 6 Fine fibre collagen motifs of 'sticky' type II collagen covered with less adhesive type IX collagen in the juvenile eye are gradually lost on ageing, together with the bridging hyaluronates. (From Laude et al. (2010))

vessel (Loch et al. 2014). The container containing the gel is subjected to saccadic oscillatory forces by servos programmed from an Arduino microcontroller to mimic slow and fast pursuit of an object. The objective was to simulate forces applied to the vitreous body under conditions in which the environment is a stiff or a partially liquefied gel. This is because the ageing eye begins to undergo a process of liquefaction, eventually resulting in posterior inner limiting membrane detachment and the formation of two chambers within the vitreous space, with the remnants of the collapsed gel attached to the posterior surface of the lens. Tuma and colleagues have modelled the effects of elastic deformation of the eye and conclude that the stresses exerted by the vitreous are strongly correlated with its rheological properties (Tuma et al. 2018). Del Amo and Urtii have criticized authors (Del Amo and Urtti 2015), including my own group, in pointing out the shortcomings of rabbit data when attempting to extrapolate drug kinetics to man. Whilst the rabbit does provide a useful reference point, a significant issue that concerned our group is that a partially syneretic VH as found in aged patients receiving therapy for wet AMD is likely to deviate in behaviour from a juvenile animal model when considering nano and macro drug delivery constructs for reasons as stated above.

Access to the Retina

The retina extends from the back of the eye forward to form a wine glass-shaped layer inside the sclera extending to the frontal pars planar, which is devoid of neuronal cells and avascular. This translucent zone is used as the access point for injection. The retina can be stripped from the basement membrane and dissected out to a Maltese cross-shape object just over 40 mm in diameter and approximately 0.5 mm thickness.

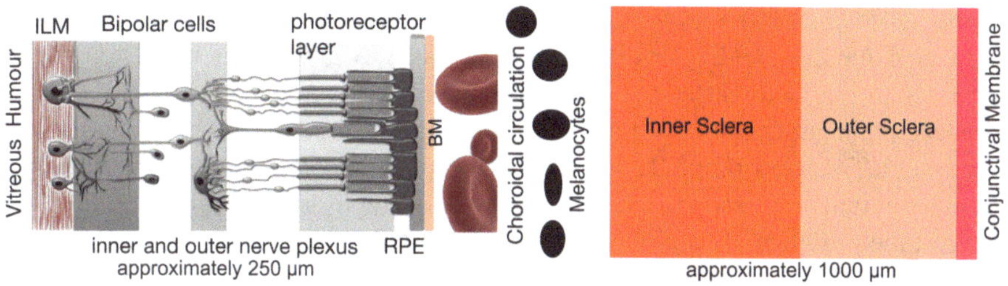

Fig. 7 The retina and the inner and outer barriers. *ILM* inner limiting membrane, *RPE* retinal pigmented epithelium, *BM* Bruch's membrane. The sclera and retina are not shown to relative scale

The inner retina is composed of the inner limiting membrane, the retinal ganglions and neuronal fibres through to about mid-thickness retina—the external limiting membrane. The general arrangement as shown in Fig. 7 is a three-component chain of the ganglion cell joined to bipolar cells and then the photoreceptors. This is associated with the accompanying cells (horizontal, amacrine and Mueller cells) (Wilson et al. 2006). The outer retina contains the photoreceptors through to the choroid. The structure is morphologically complex, the cells interdigitated and the whole retina can be discriminated into an assembly of 13 or so layers, but a barrier function in the posterior eye is only evident for some of the more distinct anatomical features, namely, the inner limiting membrane, the choroid and the retinal pigment epithelium. The last two features are often grouped as the outer blood-retinal barrier. In the anterior eye, the blood-aqueous layer limits the flow out of the iris capillaries. In these tissues the tight junctions of the endothelium restrict movement of substances towards the retina. This is referred to as the blood-aqueous barrier (see review by Lau and colleagues) (Lau et al. 2018).

The Inner Limiting Membrane

The boundary between the neutral retina and the basal lamina of the vitreous humour is formed from neural and connective components of the retina and is known as the inner limiting membrane (ILM). We have proposed that the ILM is an important barrier to drug delivery of large molecules since nanoparticulates and large proteins can be seen to accumulate at the ILM prior to translocation through the retina (Heiduschka et al. 2007; Mains and Wilson 2013b). Dalkara and others showed that the ILM formed a barrier to the entry of adenovectors expressing GFP. Fluorescently labelled AAV serotype capsids of types AAV2, AAV8 and AAV9 were seen to accumulate at the vitreoretinal junction, suggesting the presence of appropriate receptors for these serotypes at the vitreoretinal junction, whereas AAV1 and AAV5 showed no image enrichment at the ILM (Dalkara et al. 2009). They showed that protease digestion using Pronase, a cocktail of protease injected with the AAV, increased retinal expression of the GFP. Modern strategies have involved modification of residues of the AAV capsids to improve transduction or farming from mixtures of randomly generated mutants yielding novel mutants such as AAV2-7m8 which transduced the outer retina following intravitreal injection into mice (Trapani and Auricchio 2018). However, in a larger eye, either a greater dilution or a thicker ILM caused transfection to fail. This view of the ILM as a barrier to macromolecular drug constructs has been supported by Chawla and colleagues who assemble similar arguments and also suggest the presence of an intact ILM can prevent potential spread of infection and observe that in vitreous haemorrhage, the ILM prevents the spread of blood towards the photoreceptors (Chawla et al. 2017).

The ILM is a scaffolding tissue, supporting glial cells, fibrocytes and the retinal pigment epithelium. Of particular importance in the ILM cuticular layer are the Mueller cells, which perform many important functions including regulation of ion balance, secretion of glutathione as an antioxidant and other functions. The outer surface of the ILM facing the retina is uneven, whereas the inner vitreous-facing surface is smooth. The outer irregularity is formed from Mueller cell extensions (Mueller cell footplates) which are neural processes covered with glycoprotein. The Mueller cell foot is metabolically active, particularly with regard to acid-base adjustment through ion channels. The Mueller cell extensions also confer mechanical strength to the retina and anchor between the ILM and the external limiting membrane (Wollensak et al. 2006).

When the eye is observed with an ophthalmoscope, the ILM can be seen as a partially reflective, glistening layer, especially in a young individual. The ILM is approximately 10 μm thick and is continuous with the inner limiting membrane of the ciliary body at the front of the eye (Semeraro et al. 2015). At the optic disc, the processes from astrocytes replace those of the Mueller cells. The inner limiting membrane (ILM) is stiffer than the underlying neutral cells, and any distortion of the vitreomacular surface is transmitted to the retina. The Mueller cells respond to this mechanical and inflammatory stimulus by increase in activity and proliferation. The ILM is thickest at the macula, becoming thinner at the fovea. Generally, at points where the ILM is thinnest, the anchoring to the retina is firm (Morescalchi et al. 2017). The traction exerted on the posterior pole may cause the formation of a macular hole. As tractional forces increase, the Mueller cell layer forms a cone shape tissue mass that can become separated and suspended on the posterior cortex, leading to a foveal cyst that precedes the formation of a macular hole. From a surgical point of view, removal of the ILM (ILM peeling) might increase the elasticity of the underlying macula and therefore not uncommonly performed when the underlying pathology is related to the vitreomacular surface such as macular hole or continuing traction. In the past surgeons stained the ILM with the dye indocyanine green; however, alternatives such as violet 17 as less toxic alternatives are being explored (Tura et al. 2016). In any case, the operation approached with caution as the outcome is not always favourable (Morescalchi et al. 2017).

Blood Flow

The eye is circled by the choroidal flow under the sclera. The retina has a high oxygen demand to support a significant glycolysis. In the cat, the ocular blood flow measured from pulsing [^{85}Kr] in saline into the carotid artery was 1.2 L min^{-1} per 100 g tissue in the choriocapillaris (Friedman et al. 1964). The extensive flow in the choriocapillaris restricts the access of drug to the retina by vascular clearance, rather like a fast-flowing river. The isolated perfused pig eye has been used to study the influence of choroidal flow, and it was found that the uptake of both a hydrophilic probe, sodium fluorescein, and a lipophilic dye 1,1′-dioctadecyl-3,3,3′,3′-tetramethylindocarbocyanine perchlorate (DiI) was higher in the non-perfused samples, again supporting the widely held view that choroidal flow is a limiting factor in suprachoroidal injection (Abarca et al. 2013). Tsukahara and Maurice described experiments in which fluorescein or carboxyfluorescein were injected subconjunctivally into rabbit eyes. Autoradiography suggests that a drug injected into the subconjunctival space passes directly into the vitreous through the underlying tissues. When pressure was applied at the site of topical application, there were a 30-fold increase in penetration of fluorescein and about a sevenfold increase in CF which persisted for some time after the manoeuvre (Tsukahara and Maurice 1995).

As the retina becomes hypoxic through transient ischemia, the reduction in oxygen and nutrients and accumulation of waste products decrease the supply of substrates through the retinal cell layers. Re-establishment of flow results in increased oxidative stress and inflammation and if withheld for sufficiently long results in an infarct (tissue death) (Osborne et al. 2004). Retinal ganglion cell death occurs by apoptosis or necrosis. The hypoxia induces a cascade of processes including elevation of hypoxia-inducible factor-1α which in turn increases VEGF and nitric oxide synthase (Kaur et al. 2008). Mueller cells may attempt to protect the endothelial cells, and new vasculature originates from the choroid and penetrates through the retinal pigment epithelium, modulating the barrier function. The leaky, immature vessels, under increased signalling from VEGF and nitric oxide synthase, allow the formation of oedema under the retina.

Outer Blood-Brain Barrier

The retinal pigment epithelium forms an important barrier for drug delivery to the retina. The monolayer is associated with Bruch's membrane on the basolateral side and faces the outer photoreceptor segments on the apical side. Phagocytosis of the outer photoreceptor fragments occurs, the visual pigments being recycled and returned to help rebuild the receptor.

The cholesterol associated with the shed photoreceptor fragments is a waste product that must be eliminated. It has been shown that the RPE cells use the ABCA1 transporter to efflux the lipid to extracellular receptors on both sides of the monolayer (Storti et al. 2017). Failure to mobilize the cholesterol away from the membrane contributes to drusen on both sides of the membrane (Caceres and Rodriguez-Boulan 2020). The role in processing the shed photoreceptors is a reason for high degradative enzyme activity, especially around the macula (Strauss 2005). In addition, the RPE also secretes immunosuppressive factors. The neuronal and photoreceptor activity generates ion fluxes which are accompanied by water transport, and intraocular pressure generates a movement of water from the vitreous to the retina. To eliminate the water load in the choroid, water must flow by specific mechanisms as the RPE is a 'tight' epithelium. Calcium-dependent chloride and the potassium channels drive water flux. Chloride and HCO_3 transport are linked, providing a mechanism to increase intracellular pH.

Bruch's membrane, on the basolateral side of the RPE, has a composition which is influenced by the composition of the extracellular matrix (ECM). The membrane is composed of elastins, laminin and collagens, and during ageing, the components change which can lead to decreased water permeability and detachment from the RPE. Bruch's membrane is acellular and functions as a semipermeable membrane. Diffusion of nutrients and drugs depends on hydrostatic pressure. All such molecules bind to the BM or are taken up by the RPE through the BM (Booij et al. 2010).

Melanin

The melanins are a group of natural pigments widely distributed throughout the bacterial, plant and animal kingdoms. The precursors vary across species but generally are hydrophobic, aromatic flat-sheet polymers aggregated into particulates with a high surface area. In mammals, the melanins are a mix derived from dihydroxyindoles (eumelanin) and benzothiazinylalanines (pheomelanin). Melanin is distributed unevenly in the eye, suggesting a specific physiological role which appears to be concerned with avoidance of reflection and attenuation of the effects of harmful ultraviolet radiation. High concentrations are found in the pigmented epithelial layer of the retina and the uveal tract, and differences in drug response between pigmented and nonpigmented irides are well documented. In the early 1990s, it was noted that chloroquine persisted in the melanized irides for a long period which was probably an explanation for retinal toxicity. Dursch and colleagues showed clenbuterol, a beta-adrenoreceptor agonist, accumulates in pigmented tissues—hair and the eyes of rats at a ratio of 50:1 compared to albino animals after 1 week of treatment (Dürsch et al. 1995). Melanin forms complexes with drugs such as chlortetracycline, and if UV irradiated, melanin confers less protection, and cells lose viability due to induction of oxidative stress (Rok et al. 2019). This runs counter to an old proposition by Leblanc and colleagues that ocular melanin binding per se was not predictive of toxicity and was more related to the intrinsic activity of the molecule (Leblanc et al. 1998). Urtti's group recently presented a significant overview on the effects of ocular melanin on pharmacokinetics, supported by modelling (Rimpelä et al. 2018). The histological age-related depletion of melanin in the retinal pigmented epithelium has suggested that the consistent changes seen at the chorio-retinal junction might be associated with age-related changes in vision, being seen in a more pronounced form in patients with AMD (Gupta et al. 2017).

Concluding Remarks

The relatively small size of the eye and the proportion of blood flow that it receives from the systemic circulation, coupled with the high metabolic activity of the retina, puts the tissue at risk of anoxia as ageing proceeds. The resistance to aqueous humour outflow decreases the efficiency of perfusion which in turn further limits access of drug to the retina. Thus, any surgical or pharmacological manoeuvre that decreases intraocular pressure to the low teens results in greater fluxes of drug through and along the retina. It is probable that a more efficient delivery system could be made by addressing these factors which would be of complimentary benefit in treating glaucoma; however access to the posterior pole remains problematic.

The pattern of world disease is changing: more complications of diabetes in Asia and a worldwide expansion in extreme myopia are noticed in children. This is not just attributable to genetics and habits: reading too many books or playing too many video games but by not being outside. Wandering in the environment causes us to adjust our vision to use short and far focus and change accommodation to focus on objects at the side of the field of view. In a review alerting the public to the rather alarming new trends in increasing myopia in the population, the journal *Nature* reminded the audience of the observations of a noted British ophthalmologist, Henry Edward Juler, who suggested that when myopia had become stationary, a change of air—a sea voyage—should be prescribed (Dolgin 2015). After the strike of the COVID pandemic, it is doubtful whether many will be rushing to follow this advice. Myopia is a risk factor for maculopathy, and so, cognizant of the risk, perhaps you should now put down this article and follow the '20/20/20 rule'. Every 20 min, stare at something 20 ft away from you for 20 s. An interesting idea, worthy of debate, as to understanding how our eyes work might lead to reduced compromised vision!

References

Abarca EM, Salmon JH, Gilger BC. Effect of choroidal perfusion on ocular tissue distribution after intravitreal or suprachoroidal injection in an arterially perfused ex vivo pig eye model. J Ocular Pharmacol Ther. 2013;29(8):715–22.

Aihara M, Lindsey JD, Weinreb RN. Enhanced FGF-2 movement through human sclera after exposure to latanoprost. Investig Ophthalmol Vis Sci. 2001;42(11):2554–9.

Alagöz N, Alagöz C, Yılmaz I, Yıldırım Y, Baz Ö, Altan Ç, et al. Immediate intraocular pressure changes following intravitreal dexamethasone implant. J Ocul Pharmacol Ther. 2016;32:44–9.

Bakri SJ, Moshfeghi DM, Francom S, et al. Intraocular pressure in eyes receiving monthly ranibizumab in 2 pivotal age-related macular degeneration clinical trials. Ophthalmology. 2014;121(5):1102–8.
Balazs EA. Molecular morphology of the vitreous body. In: Smelser GK, editor. Structure of the eye. New York: Academic Press; 1961. p. 293–310.
Barocas VH, Huang EC. Computer modeling of pupil dilation: is aqueous humor trapped in the posterior chamber? Investig Ophthalmol Vis Sci. 2006;47(13):223.
Bonfiglio A, Repetto R, Siggers JH, Stocchino A. Investigation of the motion of a viscous fluid in the vitreous cavity induced by eye rotations and implications for drug delivery. Phys Med Biol. 2013;58(6):1969.
Booij JC, Baas DC, Beisekeeva J, Gorgels TG, Bergen AA. The dynamic nature of Bruch's membrane. Progr Retinal Eye Res. 2010;29(1):1–18.
Bracha P, Moore NA, Ciulla TA, WuDunn D, Cantor LB. The acute and chronic effects of intravitreal anti-vascular endothelial growth factor injections on intraocular pressure: a review. Survey Ophthalmol. 2018;63(3):281–95.
Caceres PS, Rodriguez-Boulan E. Retinal pigment epithelium polarity in health and blinding diseases. Curr Opin Cell Biol. 2020;62:37–45.
Chang-Lin JE, Burke JA, Peng Q, Lin T, Orilla WC, Ghosn CR, Zhang KM, Kuppermann BD, Robinson MR, Whitcup SM, Welty DF. Pharmacokinetics of a sustained-release dexamethasone intravitreal implant in vitrectomized and nonvitrectomized eyes. Investig Ophthalmol Vis Sci. 2011;52(7):4605–9.
Chawla R, Tripathy K, Temkar S, Kumar V. Internal limiting membrane: the innermost retinal barrier. Med Hypotheses. 2017;98:60–2.
Chen Y, Wyatt HJ. Effects of changing the direction of gravity on sector pupil dilations. Optometry Vis Sci. 2004;81(8):629–35.
Chiang B, Venugopal N, Edelhauser HF, Prausnitz MR. Distribution of particles, small molecules and polymeric formulation excipients in the suprachoroidal space after microneedle injection. Exp Eye Res. 2016a;153:101–9.
Chiang B, Kim YC, Edelhauser HF, Prausnitz MR. Circumferential flow of particles in the suprachoroidal space is impeded by the posterior ciliary arteries. Exp Eye Res. 2016b;145:424–31.
Chiang B, Venugopal N, Grossniklaus HE, Jung JH, Edelhauser HF, Prausnitz MR. Thickness and closure kinetics of the suprachoroidal space following microneedle injection of liquid formulations. Investig Ophthalmol Vis Sci. 2017;58(1):555–64.
Civan MM. Formation of the aqueous humor: transport components and their integration. Curr Top Membr. 2008;62:1–45.
Coudrillier B, Tian J, Alexander S, Myers KM, Quigley HA, Nguyen TD. Biomechanics of the human posterior sclera: age-and glaucoma-related changes measured using inflation testing. Investig Ophthalmol Vis Sci. 2012;53(4):1714–28.
Dalkara D, Kolstad KD, Caporale N, Visel M, Klimczak RR, Schaffer DV, Flannery JG. Inner limiting membrane barriers to AAV-mediated retinal transduction from the vitreous. Mol Ther. 2009;17(12):2096–102.
David T, Smye S, Dabbs T, James T. A model for the fluid motion of vitreous humor of the human eye during saccadic movement. Phys Med Biol. 1998;43(6):1385.
Davis BL, Nilson CD, Mamalis N. Revised Miyake-Apple technique for postmortem eye preparation. J Cataract Refract Surg. 2003;30:546–9.
Del Amo EM, Urtti A. Rabbit as an animal model for intravitreal pharmacokinetics: clinical predictability and quality of the published data. Exp Eye Res. 2015;137:111–24.
Dolgin E. The myopia boom. Nature. 2015;519:276–8. https://doi.org/10.1038/519276a.
Duke-Elder WS. The nature of the vitreous body. Monograph supplement IV. Br J Ophthalmol Monogr Suppl. 1930;V:44.
Dürsch I, Meyer HHD, Karg H. Accumulation of the β-agonist clenbuterol by pigmented tissues in rat eye and hair of veal calves. J Anim Sci. 1995;73(7):2050–3.

Fatt I. Hydraulic flow conductivity of the vitreous gel. Investig Ophthalmol Vis Sci. 1977;16(6):565–8.

Ficker L, Meredith TA, Gardner S, Wilson LA. Cefazolin levels after intravitreal injection. Effects of inflammation and surgery. Investig Ophthalmol Vis Sci. 1990;31(3):502–5.

Fitt AD, Gonzalez G. Fluid mechanics of the human eye: aqueous humour flow in the anterior chamber. Bull Math Biol. 2006;68(1):53.

Fowlks WL, Havener VR, Good JS. Meridional flow from the corona ciliaris through the pararetinal zone of the rabbit vitreous. Investig Ophthalmol Vis Sci. 1963;2(1):63–71.

Friedman E, Kopald HH, Smith TR, Mimura S. Retinal and choroidal blood flow determined with krypton-85 anesthetized animals. Investig Ophthalmol Vis Sci. 1964;3(5):539–47.

Gabelt BT, Kaufman PL. Aqueous humor hydrodynamics. In: Hart WM, editor. Adler's physiology of the eye. 9th ed. St. Louis, MO: Mosby; 2003.

Geraghty B, Jones SW, Rama P, Akhtar R, Elsheikh A. Age-related variations in the biomechanical properties of human sclera. J Mech Behav Biomed Mater. 2012;16:181–91.

Grzybowski A, Told R, Sacu S, Bandello F, Moisseiev E, Loewenstein A, Schmidt-Erfurth U. 2018 update on intravitreal injections: euretina expert consensus recommendations. Ophthalmologica. 2018;239:181–93.

Gupta T, Saini N, Arora J, Sahni D. Age-related changes in the chorioretinal junction: an immunohistochemical study. J Histochem Cytochem. 2017;65(10):567–77.

Heiduschka P, Fietz H, Hofmeister S, Schultheiss S, Mack AF, Peters S, Ziemssen F, Niggemann B, Julien S, Bartz-Schmidt KU, Schraermeyer U. Penetration of bevacizumab through the retina after intravitreal injection in the monkey. Investig Ophthalmol Vis Sci. 2007;48(6):2814–23.

Hollands H, Wong J, Bruen R, Campbell RJ, Sharma S, Gale J. Short-term intraocular pressure changes after intravitreal injection of bevacizumab. Can J Ophthalmol. 2007;42:807–11.

Jackson TL, Hussain A, Hodgetts A, Morley AM, Hillenkamp J, Sullivan PM, Marshall J. Human scleral hydraulic conductivity: age-related changes, topographical variation, and potential scleral outflow facility. Investig Ophthalmol Vis Sci. 2006;47(11):4942–6.

Jonas JB, Mardin CY, Schlötzer-Schrehardt U, Naumann GO. Morphometry of the human lamina cribrosa surface. Investig Ophthalmol Vis Sci. 1991;32(2):401–5.

Jonas JB, Berenshtein E, Holbach L. Anatomic relationship between lamina cribrosa, intraocular space, and cerebrospinal fluid space. Investig Ophthalmol Vis Sci. 2003;44(12):5189–95.

Jongebloed WL, Worst JGF. The cisternal anatomy of the vitreous body. Doc Ophthalmol. 1987;67:183–96.

Jung JH, Chiang B, Grossniklaus HE, Prausnitz MR. Ocular drug delivery targeted by iontophoresis in the suprachoroidal space using a microneedle. J Controll Release. 2018;277:14–22.

Kakinoki M, Sawada O, Sawada T, Saishin Y, Kawamura H, Ohji M. Effect of vitrectomy on aqueous VEGF concentration and pharmacokinetics of bevacizumab in macaque monkeys. Investig Ophthalmol Vis Sci. 2012;53(9):5877–80.

Kaur C, Foulds WS, Ling EA. Hypoxia-ischemia and retinal ganglion cell damage. Clin Ophthalmol. 2008;2(4):879.

Kek WK, Foulds WS, McConnell G, Wright AJ, Girkin JM, Wilson CG. Two-photon fluorescence excitation microscopy to assess transscleral diffusional pathways in an isolated perfused bovine eye model. Investig Ophthalmol Vis Sci. 2010;51(10):5182–9.

Kim SH, Lutz RJ, Wang NS, Robinson MR. Transport barriers in transscleral drug delivery for retinal diseases. Ophthal Res. 2007;39(5):244–54.

Krohn J, Bertelsen T. Corrosion casts of the suprachoroidal space and uveoscleral drainage routes in the human eye. Acta Ophthalmol Scand. 1997a;75:32–5.

Krohn J, Bertelsen T. Corrosion casts of the suprachoroidal space and uveoscleral drainage routes in the pig eye. Acta Ophthalmol Scand. 1997b;75:28–31.

Krohn J, Bertelsen T. Light microscopy of uveoscleral drainage routes after gelatine injections into the suprachoroidal space. Acta Ophthalmol Scand. 1998;76:521–7.

Lau CML, Yu Y, Jahanmir G, Chau Y. Controlled release technology for anti-angiogenesis treatment of posterior eye diseases: current status and challenges. Adv Drug Deliv Rev. 2018;126:145–61.

Laude A, Tan LE, Wilson CG, Lascaratos G, Elashry M, Aslam T, Patton N, Dhillon B. Intravitreal therapy for neovascular age-related macular degeneration and inter-individual variations in vitreous pharmacokinetics. Progr Retinal Eye Res. 2010;29(6):466–75.

Le Goff MM, Bishop PN. Adult vitreous structure and postnatal changes. Eye. 2008;22(10):1214.

Leblanc B, Jezequel S, Davies T, Hanton G, Taradach C. Binding of drugs to eye melanin is not predictive of ocular toxicity. Regul Toxicol Pharmacol. 1998;28(2):124–32.

Lee SS, Ghosn C, Yu Z, Zacharias LC, Kao H, Lanni C, Abdelfattah N, Kuppermann B, Csaky KG, D'Argenio DZ, Burke JA. Vitreous VEGF clearance is increased after vitrectomy. Investig Ophthalmol Vis Sci. 2010;51(4):2135–8.

Lemos V, Cabugueira A, Noronha M, Abegão Pinto L, Reina M, Branco J, et al. Intraocular pressure in eyes receiving intravitreal antivascular endothelial growth factor injections. Ophthalmologica. 2015;233:162–8.

Li SK, Hao J, Liu H, Lee JH. MRI study of subconjunctival and intravitreal injections. J Pharm Sci. 2012;101(7):2353–63.

Llorente L, Barbero S, Cano D, Dorronsoro C, Marcos S. Myopic versus hyperopic eyes: axial length, corneal shape and optical aberrations. J Vis. 2004;4(4):288–98.

Loch C, Bogdahn M, Stein S, Nagel S, Guthoff R, Weitschies W, Seidlitz A. Simulation of drug distribution in the vitreous body after local drug application into intact vitreous body and in progress of posterior vitreous detachment. J Pharm Sci. 2014;103(2):517–26.

Lund-Andersen H. Adler's physiology of the eye (vitreous) (Chap. 7). In: Kaufman PL, Alm A, editors. St Louis, MO: Mosby; 2003. p. 291–316.

Mains J, Wilson CG. The vitreous humor as a barrier to nanoparticle distribution. J Ocular Pharmacol Therap. 2013a;29(2):143–50.

Mains J, Wilson CG. The vitreous humor as a barrier to nanoparticle distribution. J Ocular Pharmacol Ther. 2013b;29(2):143–50.

Maurice DM. The Von Sallmann Lecture 1996: an ophthalmological explanation of REM sleep. Exp Eye Res. 1996;66:139–45.

Missel PJ. Hydraulic flow and vascular clearance influences on intravitreal drug delivery. Pharm Res. 2002;19(11):1636–47.

Moisseiev E, Loewenstein A, Yiu G. The suprachoroidal space: from potential space to a space with potential. Clin Ophthalmol. 2016;10:173.

Morescalchi F, Costagliola C, Gambicorti E, Duse S, Romano MR, Semeraro F. Controversies over the role of internal limiting membrane peeling during vitrectomy in macular hole surgery. Survey Ophthalmol. 2017;62(1):58–69.

Morrison JC, Johnson EC, Cepurna W, Jia L. Understanding mechanisms of pressure-induced optic nerve damage. Prog Retinal Eye Res. 2005;24(2):217–40.

Mortlet N, Young SH. Prevention of intraocular pressure rise following intravitreal injection. Br J Ophthalmol. 1993;77:572–3.

Moseley H. Mathematical model of diffusion in the vitreous humor of the eye. Clin Phys Physiol Measure. 1981;2(3):175.

Narasimhan A, Sundarraj C. Convection-enhanced intravitreous drug delivery in human eye. J Heat Transf. 2015;137(12):121003.

Newton RH, Meek KM. Circumcorneal annulus of collagen fibrils in the human limbus. Investig Ophthalmol Vis Sci. 1998;39(7):1125–34.

Norman RE, Flanagan JG, Rausch SM, Sigal IA, Tertinegg I, Eilaghi A, Portnoy S, Sled JG, Ethier CR. Dimensions of the human sclera: Thickness measurement and regional changes with axial length. Exp Eye Res. 2010;90(2):277–84.

Olsen TW, Aaberg SY, Geroski DH, Edelhauser HF. Human sclera: thickness and surface area. Am J Ophthalmol. 1998;125(2):237–41.

Osborne NN, Casson RJ, Wood JP, Chidlow G, Graham M, Melena J. Retinal ischemia: mechanisms of damage and potential therapeutic strategies. Progr Retinal Eye Res. 2004;23(1):91–147.

Pang CE, Mrejen S, Hoang QV, Sorenson JA, Freund KB. Association between needle size, postinjection reflux and intraocular pressure spikes after intravitreal injections. Retina. 2015;35:1401–6.

Pearson PA, Hainsworth DP, Ashton P. Clearance and distribution of ciprofloxacin after intravitreal injection. Retina. 1993;13(4):326–30.

Pitkanen L, Ruponen M, Nieminen J, Urtti A. Vitreous is a barrier in nonviral gene transfer by cationic lipids and polymers. Pharm Res. 2003;20:576–83.

Rimpelä AK, Reinisalo M, Hellinen L, Grazhdankin E, Kidron H, Urtti A, del Amo EM. Implications of melanin binding in ocular drug delivery. Adv Drug Deliv Rev. 2018;126:23–43.

Rok J, Rzepka Z, Respondek M, Beberok A, Wrześniok D. Chlortetracycline and melanin biopolymer—the risk of accumulation and implications for phototoxicity: an in vitro study on normal human melanocytes. Chem Biol Interact. 2019;303:27–34.

Schramm K, Mueller M, Koch FH, Singh P, Kohnen T, Koss MJ. Effects of core vitrectomy in the treatment of age-related macular degeneration. Acta Ophthalmol. 2014;92(5):465–72.

Sebag J. Age-related changes in human vitreous structure. Graefe's Arch Clin Exp Ophthalmol. 1987;225(2):89–93.

Sebag J. Seeing the invisible: the challenge of imaging vitreous. J Biomed Opt. 2004;9(1):38–47.

Seiler GS, Salmon JH, Mantuo R, Feingold S, Dayton PA, Gilger BC. Effect and distribution of contrast medium after injection into the anterior suprachoroidal space in ex vivo eyes. Investig Ophthalmol Vis Sci. 2011;52(8):5730–6.

Semeraro F, Morescalchi F, Duse S, Gambicorti E, Russo A, Costagliola C. Current trends about inner limiting membrane peeling in surgery for epiretinal membranes. J Ophthalmol. 2015;2015:671905.

Stay MS, Xu J, Randolph TW, Barocas VH. Computer simulation of convective and diffusive transport of controlled-release drugs in the vitreous humor. Pharm Res. 2003;20(1):96–102.

Stern WH, Heath TD, Lewis GP, Guerin CJ, Erickson PA, Lopez NG, Hong KL. Clearance and localization of intravitreal liposomes in the aphakic vitrectomized eye. Investig Ophthalmol Vis Sci. 1987;28(5):907–11.

Storti F, Raphael G, Griesser V, Klee K, Drawnel F, Willburger C, Scholz R, Langmann T, von Eckardstein A, Fingerle J, Grimm C. Regulated efflux of photoreceptor outer segment-derived cholesterol by human RPE cells. Exp Eye Res. 2017;165:65–77.

Strauss O. The retinal pigment epithelium in visual function. Physiol Rev. 2005;85(3):845–81.

Swain DL, Ho J, Lai J, Gong H. Shorter scleral spur in eyes with primary open-angle glaucoma. Investig Ophthalmol Vis Sci. 2015;56(3):1638–48.

Tam AL, Gupta N, Zhang Z, Yücel YH. Quantum dots trace lymphatic drainage from the mouse eye. Nanotechnology. 2011;22(42):425101.

Tam AL, Gupta N, Zhang Z, Yücel YH. Latanoprost stimulates ocular lymphatic drainage: an in vivo nanotracer study. Transl Vis Sci Technol. 2013;2(5):3. https://doi.org/10.1167/tvst.2.5.3.

Tan LE, Orilla W, Hughes PM, Tsai S, Burke JA, Wilson CG. Effects of vitreous liquefaction on the intravitreal distribution of sodium fluorescein, fluorescein dextran, and fluorescent microparticles. Investig Ophthalmol Vis Sci. 2011;52(2):1111–8.

Toris CC, Yablonski ME, Wang YL, Camras CB. Aqueous humor dynamics in the aging human eye. Am J Ophthalmol. 1999;127:407–12.

Trapani I, Auricchio A. Seeing the light after 25 years of retinal gene therapy. Trends Mol Med. 2018;24(8):669–81.

Tsukahara Y, Maurice DM. Local pressure effects on vitreous kinetics. Exp Eye Res. 1995;60(5):563–73.

Tuma K, Stein J, Průša V, Friedmann E. Motion of the vitreous humour in a deforming eye–fluid-structure interaction between a nonlinear elastic solid and a nonlinear viscoelastic fluid. Appl Math Comput. 2018;335:50–64.

Tura A, Alt A, Lüke J, Grisanti S, Haritoglou C, Meyer CH, Nassar K, Lüke M. Biocompatibility of the vital dye Acid Violet-17 on retinal pigment epithelial cells. Clin Ophthalmol. 2016;10:1435.

Vogel B, Wagner H, Gmoser J, Wörner A, Löschberger A, Peters L, Frey A, Hofmann U, Frantz S. Touch-free measurement of body temperature using close-up thermography of the ocular surface. MethodsX. 2016;3:407–16.
Watson PG, Young RD. Scleral structure, organisation and diseasee. A review. Exp Eye Res. 2004;78(3):609–23.
Watts F, Tan LE, Wilson CG, Girkin JM, Tassieri M, Wright AJ. Investigating the micro-rheology of the vitreous humor using an optically trapped local probe. J Opt. 2013;16(1):015301.
Watts F, Tan LE, Wilson CG, Girkin JM, Tassieri M, Wright AJ. Investigating the micro-rheology of the vitreous humor using an optically trapped local probe. J Opt. 2014;16:1–7.
Wilson CG, Semenova EM, Hughes PM, Olejnik O. Eye structure and physiological functions. In: Touitou EW, Barry BW, editors. Enhancement in drug delivery. CRC Press; 2006. p. 473–87.
Wilson CG, Tan LA, Mains J. Principles of retinal drug delivery from within the vitreous. In: Kompella UB, Edelhauser HF, editors. Drug product development for the back of the eye. AAPS advances in the pharmaceutical sciences, vol. 2. AAPS Press Series; 2011a. p. 125–58.
Wilson CG, Tan LE, Mains J. Principles of retinal drug delivery from within the vitreous. In: Drug product development for the back of the eye. Boston, MA: Springer; 2011b. p. 125–58.
Wilson CG, Badawi M, Hilliery AM, Borooah S, Megaw R, Dhillon B. Ophthalmic drug delivery. In: Hillery A, Park K, editors. Drug delivery and targeting. New York: Taylor & Francis; 2016. p. 306–30.
Wollensak G, Spoerl E, Grosse G, Wirbelauer C. Biomechanical significance of the human internal limiting lamina. Retina. 2006;26(8):965–8.
Womersley JR. Method for the calculation of velocity, rate of flow and viscous drag in arteries when the pressure gradient is known. J Physiol. 1955;127(3):553–63.
Wyatt HJ. Ocular pharmacokinetics and convectional flow: evidence from spatio-temporal analysis of mydriasis. J Ocular Pharmacol Ther. 1996;12(4):441–59.
Yiu G. Advances in choroidal imaging with EDI-OCT. Retina Today. 2014;2014:82–4.
Yoshida S, Nakama T, Ishikawa K, Arima M, Tachibana T, Nakao S, Sassa Y, Yasuda M, Enaida H, Oshima Y, Kono T. Antiangiogenic shift in vitreous after vitrectomy in patients with proliferative diabetic retinopathy. Investig Ophthalmol Vis Sci. 2012;53(11):6997–7003.
Zhang Y, Bazzazi H, e Silva RL, Pandey NB, Green JJ, Campochiaro PA, Popel AS. Three-dimensional transport model for intravitreal and suprachoroidal drug injection. Investig Ophthalmol Vis Sci. 2018;59(12):5266–76.
Zhu Y, Zhu H, Jia Y, Zhou J. Changes in anterior chamber volume after implantation of posterior chamber phakic intraocular lens in high myopia. BMC Ophthalmol. 2018;18(1):185.

Physicochemical and Biological Fundamentals for Drug Delivery to the Eye

Imran Ahmed

Abstract The eye is an intricate and complex organ comprising numerous, intimately connected tissues serving specialized functions. The barriers to ocular drug penetration include static barriers, dynamic barriers, and metabolic barriers. Together, they severely restrict the entry and movement of drugs within the eye resulting in very low ocular bioavailability and rapid elimination of drugs from the eye. Some of these barriers may be mitigated through judicious drug and drug delivery system design. With the recent emphasis on the treatment of back of the eye disease, alternate modes of administration, such as biodegradable implants, intraocular and periocular injection modalities, and bioavailability-enhancing formulation technologies, have gained increasing focus of current research. This chapter outlines the general approach for a successful ophthalmic drug development program that is based on an integrated approach considering physicochemical drug properties, drug delivery system design, and ocular biology to overcome ocular barriers while ensuring patient safety, treatment efficacy, and compliance.

Keywords Ocular barriers · Physicochemical attributes · Developability assessment · Ocular delivery systems · Ocular bioavailability

Introduction

Ophthalmic medications allow us to treat and manage a variety of ocular disorders. A fundamental requirement in pharmacotherapy is for the drug to reach its site of action in enough concentration and duration to elicit the desired pharmacological effect. A successful ophthalmic drug development program must be based on an integrated approach that considers physicochemical drug properties, drug delivery system design, and ocular biology to overcome ocular barriers while ensuring patient safety, treatment efficacy, and compliance.

I. Ahmed (✉)
American Regent, New Albany, OH, USA

S. Neervannan, U. B. Kompella (eds.), *Ophthalmic Product Development*,
AAPS Advances in the Pharmaceutical Sciences Series 37,
https://doi.org/10.1007/978-3-030-76367-1_5

An understanding of the movement of a drug molecule from its site of application to its intraocular target can be based on a knowledge of the physicochemical properties of the drug molecule along with the barrier properties of the various ocular structures. Due to its critical functionality, a myriad of anatomical, physiological, biochemical, and immunological barriers has uniquely evolved to protect the eye against external insults and to prevent the intraocular entry of non-endogenous substances including drugs. Description of the structures and function of the eye and ocular tissues has been provided in the medical and scientific literature. Excellent reviews have been written on the ocular drug delivery barriers and challenges in drug delivery in ocular disease treatment.

In ophthalmic drug development, therapeutic and commercial success is predicated by the design and judicious selection of the drug delivery system (DDS). The DDS comprises the drug molecule, the dosage form type, the formulation technology, and the method of administration. In this context, a successful ophthalmic drug must simultaneously possess physicochemical attributes that provide (a) druggability, target engagement to elicit the desired therapeutic effect, and (b) developability—functional fit in the delivery system that enables the drug to reach its site of action in the optimal spatial and temporal pattern needed for bioavailability and efficacy. In addition, the final drug product must be stable and manufacturable with the requisite quality and controls for regulatory approval and patient safety and efficacy.

The pathologies of the eye are usually divided into different categories based on the affected region, namely, ocular surface, anterior segment (front of the eye), and posterior segment (back of the eye) diseases. This chapter presents a critical review of the biological and physicochemical fundamentals of ocular drug delivery and provides guidance for the design and development of ocular drugs and delivery systems for safe and effective treatment of eye diseases.

Drug Delivery to the Eye: Biological Fundamentals

Relevant Ocular Anatomy and Physiology

The eye is a complex sensory organ comprising numerous highly differentiated tissues with specialized functions, most notably the capture, control, and conversion of light from the outside world into electrical nerve impulses. These impulses are then carried to the part of the brain responsible for vision (the visual cortex) where they are processed and interpreted as a visual image. The aim of this section is to familiarize the reader with the anatomy and physiology of the eye relevant to ocular drug disposition and drug delivery.

The human eye is among the smallest body organs, and both eyes together account for only about 0.02% of body mass (Bekerman et al. 2014). The eye consists of two globe-shaped structures comprising the orbit and the eyeball (Fig. 1). The eyeball and its adjoining muscles are surrounded by a layer of orbital fat that

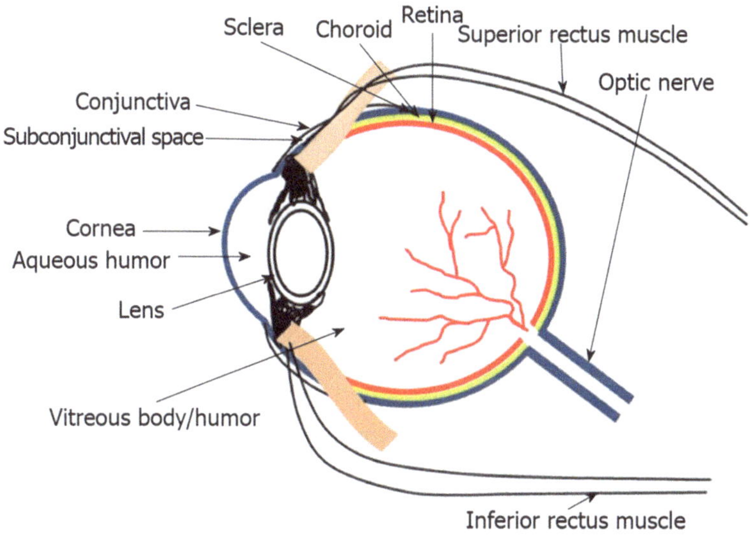

Fig. 1 Anatomy of the human eye. (Ashaben P, Cholkar K, Agrahari V, Mitra AK (2013) Ocular drug delivery: An overview. World J Pharmacol 2: 47–64 (Permission pending))

cushions the eyeball and permits smooth rotation of the eyeball about a virtually fixed point, the center of rotation. The tissues comprising the eyeball may be divided into tissues of the ocular coat and tissues of the inner eye.

Orbit

The eyeball is protected anteriorly by the structures and physiological processes associated with the lacrimal apparatus. The tear fluid present in the orbit is derived from the secretory glands. The eyelids act like a windshield wiper to sweep the secretions of the lacrimal apparatus and other glands over the surface of the eye at regular intervals to keep the cornea moist, prevent excessive evaporation of tears, and control the blink reflexes to remove foreign bodies. The tear film is a triple-layered structure. The outermost layer of the tears is oily in nature and originates from the secretions of the Meibomian glands. This layer is approximately 0.1 μm thick and retards the evaporation of the tear film. In some dry eye disease conditions, this oily layer is malformed or absent, resulting in an increase in tear evaporation rate (Vicario-de-la-Toree et al. 2007).

The middle tear layer is aqueous in nature and is secreted by the lacrimal gland and the accessory glands of Krause and Wolfring. This aqueous layer is 6.5–7.5 μm thick and accounts for more than 90% of the total tear film thickness. The innermost layer of the tear film adjacent to the surface of the eyeball is mucoidal on nature. The mucin or glycocalyx elements in the inner layer originate from the glands of Manz, crypts of Henle, and the conjunctival goblet cells. Together, the total thickness of the tear film covering the ocular surface is 6–8 μm, and the resident volume of tears in the precorneal area is 7.5–8 μL (Shastri et al. 2010). The basal rate of tear production or turnover in humans is approximately 16% per minute or about 1.5 μL per minute (Mishima et al. 1966; Glasson et al. 2006).

The excretory components of the lacrimal apparatus are responsible for the removal of tears and any excess fluid (e.g., instilled eye drops) from the precorneal. The fluid is drained from the precorneal area through small openings called puncta. There is a punctum located on the inner edge of both the upper and lower eyelid; hence, there are four puncta in humans. The puncta comprise openings of small channels known as canaliculi, which join, forming the ampulla leading into the lacrimal sac. The lacrimal sacs lead into the nasolacrimal ducts, which eventually open to the nose (Schoenwald 1997). When the volume in the precorneal area exceeds the resident value, an increased rate of drainage is observed (Chrai et al. 1973). In addition, any physiological alteration in the tear fluid, such as pH, osmolarity, or drug exposure causes induced lacrimation. As a result, an eye drop is rapidly removed from the conjunctival sac upon instillation, resulting in a very short absorption window, rapid depletion of drug concentration in the tear film, and systemic drug absorption from the nasolacrimal area.

The tears contain numerous components: proteins, lipids, electrolytes, water, and organic solutes (Gillan 2010). Approximately 500 different proteins have been identified in the tears (Gachon et al. 1979; Janssen et al. 1983; Coyle and Sibony 1989). The three principal proteins found in the tear are lysozyme, lactoferrin, and tear-specific prealbumin (lipocalin). These proteins serve useful physiological and homeostasis functions, such as infection prevention, tear film integrity, and maintenance of corneal health. Tear proteins can also bind drugs reducing ocular drug bioavailability. Precorneal drug clearance and tear biochemistry present a major barrier to topical drug absorption.

Tissues of the Ocular Coat

The coat of the eyeball is made of three layers. The outer coat or fibrous tunic that make up the surface tissues of the eye are the conjunctivae, the sclera, and the cornea. The outer covering of the eyeball is made of a relatively tough, dense connective tissue called the sclera. The sclera protects the eyeball and maintains its shape. The sclera is a dense, opaque structure constituting the posterior 5/6th or nearly 80% of the total external surface area of the eye globe. It contains few blood vessels and its anterior portion is visible, constituting the "white" of the eye. The sclera is composed primarily of collagen and mucopolysaccharides with a high degree of hydration estimated at 68% water by weight (Newell 1986). Although the sclera is poorly vascularized, it is perforated by numerous channels through which blood vessels enter the back of the ocular eye tissues, namely, the uvea and the retina. Anteriorly, adjacent to the limbus, the sclera is approximately 0.5 mm thick thinning to as little as 0.2 mm in some areas. The sclera is relatively permeable to drugs affording a route of entry into the eye with appropriate methods of administration. The scleral route of intraocular drug entry has garnered considerable interest in recent years and is sometimes referred to as the "non-corneal" route for intraocular entry (Ahmed and Patton 1985, 1987).

Near the front of the eye, in the area protected by the eyelids, the sclera is covered by a thin, transparent mucous membrane, the conjunctiva, which runs to the edge of the cornea (bulbar conjunctiva). The conjunctiva also covers the moist back surface of the eyelids and eyeballs and attaches the eyelids to the globe (palpebral conjunctiva). The conjunctival membrane has an epithelial covering with an underlying stroma that is richly supplied with both lymphatic and blood vessels. These blood vessels serve to remove permeants diffusing across the conjunctival epithelium, effectively reducing drug entry into the eye. This loss of drug to the conjunctival vasculature is often termed as "nonproductive absorption" and comprises a major barrier to intraocular drug entry (Thombre and Himmelstein 1984). The bulbar conjunctiva has microvillous processes like those seen on the surface of the cornea. The conjunctival surface area is 18 cm^2—which is about 17 times greater than the surface of the cornea in humans. The thickness of the conjunctiva varies slightly in different areas with the palpebral and bulbar conjunctiva averaging about 40 and 30 μm, respectively (Watsky and Jablomski 1988).

The most anterior structure of the globe is a transparent membrane known as the cornea. The cornea and the sclera are contiguous. The cornea-scleral junction or limbus is an important region of the eye because it encompasses the trabecular meshwork and the canal of Schlemm, which together form the drainage system of the anterior chamber in humans. This system plays an important role in maintenance of intraocular pressure and is the target of many antiglaucoma drugs. Light enters the eye through the cornea, a transparent, multilayered membrane that occupies the front center part of the external tunic. The corneal diameter is about 11.7 mm with a radius of curvature of about 7.8 mm (Watsky and Jablomski 1988). The overall thickness of the cornea in man is approximately 0.5–0.7 mm and is composed of three main layers (Reinsten et al. 1994; Hitzenberger et al. 1994). The outermost layer consists of five to six layers of squamous epithelial cells. The epithelium is microvillous and capable of rapidly regenerating. These cells are joined together on the tear side by tight junctions or zonulae occludentes. Due to these tight junctions and its lipophilic nature, the corneal epithelium offers considerable resistance to the permeation of drugs (Agrahi et al. 2016). The stroma or middle layer is largely structural and is composed of collagen fibers. It accounts for over 90% of the total corneal thickness and is highly hydrated. The stroma is essentially an aqueous solution of collagen and mucopolysaccharides and is compositionally very similar to the sclera. The aminoglycans in the corneal stroma result in a net negative charge (Schoenwald 1997). The thinnest part of the cornea is a single layer of flat, hexagonal cells called the endothelium. The endothelium has loose intracellular junctions. Unlike the corneal epithelium, it does not rapidly regenerate and plays an important role in the normal hydration and optical clarity of the cornea (Clinic 2003).

The tissues of the middle coat or the vascular tunic, also known as the uvea, are choroid, the ciliary body, and the iris. Although anatomically distinct, these structures are all continuous with each other. The most posterior section of the vascular coat is the choroid. The choroid is mainly composed of blood vessels and supplies oxygen and nutrition for the eye. A dark pigment, called melanin, occurs throughout the choroid. A primary purpose of melanin is to control light reflection within the eye (Kadam and Kompella 2010).

The anterior part of the choroid passes into the ciliary body. The ciliary body and the ciliary muscle anchor the lens in place and modulate the lens shape to provide visual accommodation or refractive power. The ciliary body is a highly specialized structure which functions to produce, secrete, and control the outflow of the aqueous humor through Schlemm's canal. The inner surface of the ciliary body has fingerlike projections known as the ciliary processes. Two layers of cuboidal epithelial cells line the inner surface. The layer facing the aqueous humor is characterized by tight junctions (zonulae occludentes). The tight junctions restrict molecular movement through the intercellular spaces and constitute part of the blood-aqueous barrier in the ciliary body (Ghate and Edelhauser 2006).

The iris is a circular, thin structure which forms the most anterior structure of the uveal tract. The iris is connected to the front part of the ciliary body and covers the top of the lens. It functions to regulate the size and diameter of the pupil and controls the amount of light that enters the eye. The iris also contains pigments, the amount of which determines a person's eye color. The front surface of the iris is covered with a porous lining of cells which resemble the corneal endothelium (Kryczka et al. 2014). This permits facile communication between the constituents of the iris stroma and the aqueous humor. The iris stroma consists of loose connective tissue with a rich vascular supply. The capillary walls of these vessels have a thick basement membrane and no fenestrae. These tight capillary junctions restrict molecular movement and account for the existence of the blood-aqueous barrier in the iris. The posterior surface of the iris is covered with two layers of epithelial cells which are extensions of the epithelium lining of the ciliary processes.

The main tissue comprising the inner coat of the eye is the retina. The retina is responsible for the perception of image and vision and is derived from the optic cup, which is a part of the brain, embryologically arising as a hollow outgrowth of the forebrain (Benjamin 2011). The retina contains photoreceptors—cells that sense light—and blood vessels that nourish them. A small area within the retina containing the highest density of photoreceptors is known as the macula. There are two types of photoreceptors—cones and rods. The cones are responsible for the central vision and color perception and localized primarily in the macula. The rods—more numerous than cones—are responsible for night vision and peripheral vision and are located mainly in the peripheral areas of the retina. The photoreceptors in the retina convert the visual image into electrical signals, which are carried to the visual cortex of the brain by the optic nerve. Each photoreceptor is linked to a nerve fiber, which are bundled together to form the optic nerve. Located at the back of the eye at the point where the optic nerve fibers depart from the eyeball is the optic disk. There are no photosensitive cells in the optic dick; it is, thus, insensitive to light and termed the "blind spot." The retina is essential to vision and is the target tissue in treating back of the eye diseases, such as age-related macular degeneration, diabetic retinopathy, and a myriad of other eye disorders.

Tissues of the Inner Eye

The interior eyeball is divided into two sections, each of which is filled with fluid. The pressure generated by these fluids fills out the eyeball and helps maintain its shape. The front or anterior segment extends from the inside surface of the cornea to the front surface of the lens. The anterior segment tissues consist of the cornea, conjunctiva, aqueous humor, iris, ciliary body, and lens. The chamber comprising the anterior segment is filled with a fluid—the aqueous humor—that nourishes the internal structures. The anterior segment is in turn divided into two fluid chambers of unequal volume: the anterior chamber and the posterior chamber separated by the iris. The front anterior chamber extends from the cornea to the iris. The back or posterior chamber extends from the iris to the lens.

The aqueous humor is produced at the ciliary processes in the posterior chamber and flows in a forward direction through the pupil flow into the anterior chamber. It primarily drains out of the eyeball through a porous network of tissues comprising outflow channels located where the iris meets the cornea (iridocorneal angle), eventually emptying into an extraocular venous plexus (Kaufman and Alm 2003). Another route, called uveoscleral drainage, may also be important in the overall turnover of aqueous humor. This route involves the bulk flow of aqueous humor through the anterior uvea into episcleral tissues. In humans, the uveoscleral drainage has been estimated to account for 4–14% of the total aqueous outflow. The overall turnover rate of the aqueous humor is approximately 1–1.5% of the volume per minute or about 2 μL per minute (Järvinen et al. 1995). The aqueous humor functions primarily as a source of nutrients for the surrounding ocular tissues.

Residing directly behind the iris is the lens. The lens is an avascular, olive-shaped tissue that focuses light rays passing through the eyeball to the back of the eye, that is, to the retina. The lens is encased in an elastic capsule which, along with the lens fibers, anchors the lens to the ciliary muscles enabling the lens to change the focus or accommodation. The lens is a dense structure wherein much of the lens is composed of tightly packed fibers with the density increasing toward the core. The extracellular space of the lens is approximated at about 3% of the total volume (Durairaj et al. 2009). Due to its dense, stratified structure, drug permeation across the lens from the anterior segment to the posterior segment of the eye is believed to be extremely unlikely. Abnormalities of the lens are implicated in various age-related eye diseases, such as cataract and presbyopia.

The back section of the eye referred to as the posterior segment extends from the back of the lens to the retina. The posterior segment mainly consists of the vitreous humor, retina, choroid, and optic nerve. The chamber comprising the posterior segment is filled with a jellylike fluid called the vitreous humor. It is situated between the lens and the retina and comprises almost two-thirds of the entire eyeball. By pushing the retina toward the choroid, the vitreous keeps the retina in place and serves as a source of nutrients. The vitreous humor is largely aqueous (98% water) but also contains substantial amounts of hyaluronic acid and collagen (Durairaj et al. 2009). The vitreous chamber is the preferred site of administration of drugs targeting tissues of the back of the eye. The distribution of substances injected into the vitreous humor suggests an anterior-posterior fluid flow. It is estimated that the vitreal water is turned over every 15 min.

In summary, despite its small size, the eye is an intricate and complex organ comprising numerous, intimately connected tissues serving specialized functions. A myriad of processes within the eye tissues and fluids act in an integrated manner to keep out drugs or non-endogenous substances.

Biological Barriers to Ocular Drug Penetration

The barriers to ocular drug penetration are divided into three categories: static barriers, dynamic barriers, and metabolic barriers.

Static Barriers

Static barriers comprise tissues of the ocular surface, the anterior segment, and the posterior segment that present a tortuous path to drug diffusion and uptake. These include the ocular surface tissues, namely, the cornea, conjunctiva, and sclera. The cornea is a complex, multilayered tissue—each layer with different polarity and cellular microstructure presenting differential barrier properties to drug permeation. The corneal epithelium is the main barrier of drug absorption into the eye. For optimal permeation across the cornea, a molecule requires the right balance between lipophilicity and hydrophilicity (partition coefficient), molecular size, charge, and degree of ionization. The corneal transport of an ionizable drug can be further influenced by the charged state in the tear fluid. For example, with pilocarpine a linear relationship was found between the observed corneal permeability and the fraction of unionized drug, with the efflux of unionized drug twofold higher than the flux of ionized (charged) drug (Mitra and Mikkelson 1988). Charged drugs are also capable of penetrating the cornea depending on its overall lipophilicity and size (Prausnitz 1998; Ghosn et al. 2007; Li et al. 2005).

The conjunctiva is a thin, transparent, highly vascularized mucous membrane that covers the sclera and the inner surface of the eyelids. The outer epithelium of the conjunctiva plays a protective role by the tight junctional barrier at the apical surface restricting the permeation of hydrophilic drugs and large molecules. The conjunctiva in humans is over seven times the corneal surface area (14.5 cm^2 for the conjunctiva versus 1.5–2.0 cm^2 for the cornea), and the conjunctival epithelium is more permeable than the corneal epithelium. However, due to its rich vasculature and lymphatics, the major portion of the drug that crosses the conjunctiva is carried away. Hence, conjunctival drug absorption is generally regarded as "nonproductive" as a portal of entry of drugs into the eye (Ahmed and Patton 1987).

The sclera is structurally like the corneal stroma, containing numerous channels and consisting mainly of mucopolysaccharides (Kim et al. 2007a, b). The portion of the drug which crosses the conjunctiva and is not lost to the local vasculature can diffuse into the sclera. Once in the sclera, the drug would have access to various parts of the eye, including the anterior chamber and the uveal tract (Ranta et al. 2010).

The internal solid tissues of the eye can present barrier to drug penetration and distribution within the eye. These internal eye tissues comprising the static barrier include the iris, the ciliary body, and the lens. The porous nature of the anterior surface of the iris allows free molecular communication between the aqueous humor and the iris stroma. The iris stroma is composed of vascularized, connective tissues. The capillary walls of the blood vessels in the iris stroma have a thick basement membrane without fenestrae (Ranta et al. 2010; Yavuz and Kompella 2017; Occhiutto et al. 2012). These tight capillary junctions restrict molecular movement and account for the blood-aqueous barrier in the iris. Permanently charged or highly water-soluble drugs are likely to encounter the most diffusional resistance in the iridal capillary walls. The clearance of such molecules from the iris is dictated by the capillary permeability and the bulk aqueous flow. In contrast, the clearance of lipophilic molecules from the iris is dictated by the iridal blood flow rate.

The iris root connects the iris with the ciliary body. The inner surface of the ciliary body has fingerlike projections known as the ciliary processes lined with two layers of epithelial cells. The layer facing the aqueous humor is characterized by tight junctions (zonulae occludentes) that connect the cells. These tight junctions greatly limit the molecular movement through the intercellular spaces and form the blood-aqueous barrier in the ciliary body (Ghate and Edelhauser 2006).

Situated directly behind the iris is the lens. The lens is a round, avascular structure comprising of elongated fibers that stretch in arcuate fashion going from the anterior to posterior poles. The bulk of the lens is composed of these fibers as is referred to as the crystalline lens. The lens is completely surrounded by a collagenous capsule. The lens is bathed on the anterior side by the aqueous humor and on the posterior side by the vitreous humor. The lens depends on these fluids for nutrients as well as for the removal of its metabolic waste products. On the aqueous humor side, the capsule encloses a single layer of cuboidal epithelium. However, the portion of the crystalline lens facing the vitreous humor has no epithelium and is covered with the thinnest segment of the capsule. The lens presents a tortuous path to the diffusion of drugs from the front of the eye to the back of the eye. The lens can also act as a reservoir for drugs that penetrate the lens (Babizhayev et al. 2014; Ahmed et al. 1989). Material taken up by the lens is not readily washed out, but slowly diffuses the entire mass (Ahmed et al. 1989).

Dynamic Barriers

Dynamic barriers include various ocular fluid systems that dilute and remove drugs from the site of application and diminish the concentration gradient for diffusion into disease-relevant ocular tissues. The fluid systems comprising the ocular dynamic barriers include the tears, the blood, aqueous humor, and vitreous humor. The fluid systems in the eye play an important role in the delivery of drugs into the eye and dictate the concentration gradient for drug diffusion into disease-relevant ocular tissue. The fluid systems that variously provide dynamic barriers to intraocular drug entry are the tear film, blood flow, aqueous humor, vitreous humor, and, to a lesser extent, lymphatic flow.

The primary reason for the exceedingly low ocular bioavailability (<5%) of topically applied drugs is the precorneal factors that rapidly diminish the drug concentration and residence time in the tear film. These factors are reflex blinking, induced lacrimation, tear dilution, nasolacrimal drainage, and mucin turnover (Chrai et al. 1973; Agrahi et al. 2016; Järvinen et al. 1995; Le Bourlais et al. 1998). The normal resident tear volume is 7–9 μL. After instillation of an eye drop, typically around 30 μL, the excess volume is rapidly drained away through the puncta into the nasolacrimal ducts resulting in rapid restoration of the resident tear volume within 2–3 min. Most of the applied eye drop is washed away within 15–30 s after instillation. The initial first-order drainage rate of eye drops from the ocular surface is 1.2 μL/min in humans (Mishima et al. 1966). In addition, lacrimation induced by pharmacological or physiological stimuli accelerates the dilution and removal of drugs from the conjunctival sac. Binding of drugs to the proteins in the tear fluid may influence transport into the eye (Janssen et al. 1983). Metabolism of drugs in the precorneal area may also diminish drug availability (Shell 1982). Together, these precorneal factors remove drug from the absorption site and deplete the concentration gradient for drug absorption into the eye after instillation of an eye drop. Consequently, less than 5% and more typically 1–2% of the applied dose penetrates the eye and is available to access disease-relevant intraocular tissues (Patel 2013; Thombre and Himmelstein 1984; Kesavan et al. 2011).

Drug permeating across the cornea enters the anterior chamber that is filled with aqueous humor. Once a drug crosses the cornea, it is distributed in the aqueous humor. The aqueous humor, like the tears, is constantly turned over or replaced. Therefore, both the aqueous humor dynamics and the kinetics of transfer from the aqueous to other ocular structures in the anterior segment can influence the overall disposition of a drug in the eye. Aqueous humor is produced by the ciliary processes and flows from the posterior chamber out through the iridocorneal angle in the anterior chamber. Drugs entering the aqueous humor are mixed by convection currents associated with this bulk flow. The rate of loss of aqueous humor is about 1% per minute (Järvinen et al. 1995). This represents the maximum rate at which a drug is lost from the aqueous humor. Some drugs, such as pilocarpine, may affect the normal rate of aqueous turnover and thus their own elimination (Thombre and Himmelstein 1984). In addition to this loss by bulk flow, a drug can be absorbed into the ocular structures or lost into the blood circulating through them. The proportion of a topically administered drug lost by these routes varies greatly based on the drug's pharmacology and physicochemical properties. The aqueous humor has substantially less protein than the tears or plasma (Chen 2009). Nonetheless, the possibility of protein binding cannot be overlooked. The aqueous humor like the tears is constantly renewed providing a continual source of fresh proteins. Secondly, many disease states, particularly inflammatory disorders, result in elevated protein levels in the aqueous humor. These factors could provide the circumstances which protein binding in the aqueous humor would be of clinical significance.

The main fluid compartment of the back of the eye is the vitreous humor. The vitreous humor is a hydrogel which is largely aqueous (98% water) but also contains significant amounts of hyaluronic acid and collagen (Swindle and Ravi 2007). The distribution of drugs in the vitreous body is subjected to anteroposterior fluid flow (Durairaj et al. 2009; Tojo et al. 1999).

The eye is protected from xenobiotics in the bloodstream by blood-ocular barriers. In addition, drug loss to blood flow or systemic clearance is also a major barrier to the ocular entry of topically periocular and intraocularly administered drugs in the eye. These barriers have two parts: blood-aqueous and blood-retinal barrier. The blood-aqueous barrier is located anteriorly and resides primarily in the uvea. This barrier prevents the access of plasma albumin as well as hydrophilic drugs from the plasma into the aqueous humor. Disease conditions may disrupt this barrier. Posteriorly, the barrier between the bloodstream and eye is comprised of retinal pigment epithelium (RPE) and the tight walls of the retinal capillaries (Jiang et al. 2006; Myles et al. 2005).

Metabolic Barriers

The existence of various enzyme systems and transporters in the eye has been reported (Gaudana et al. 2010; Barar et al. 2008). Enzymes and transporters can present metabolic barriers to the intraocular entry and access of drugs to target tissues impacting the pharmacokinetics and pharmacodynamics of therapeutic agents and disease pathophysiology [64]. Ocular enzyme systems and transporters may open possibilities for enhancing ocular drug delivery and site-specific bio-activation via prudent drug design strategies (Reddy and Bodor 1993; Shell 1984; Hughes et al. 1993).

Both phase I and phase II metabolic activity has been reported in ocular structures (Al-Ghananeem and Crooks 2007). Cytochrome P450 (CYPs), namely, CYP1, CYP2, and CYP3 families, collectively known as the "drug-metabolizing P450s" typically oxidize drugs to generate more polar products as a first step in their elimination and detoxification. These modified compounds are then conjugated to polar compounds in phase II reactions, such as reactions catalyzed by transferase enzymes such as glutathione S-transferase. In general, phase I metabolic activities are highest in those ocular structures adjacent to regions of highest blood flow, namely, the iris-ciliary body, the retina, and the choroid (Kishida et al. 1986; Schwartzman et al. 1985). P450-dependent metabolism has also been shown to occur in the corneal epithelium (Asakura 1992). Ocular tissues display substrate-dependent differences in activities in various ocular regions indicative of the presence of multiple forms of cytochrome P450 distributed heterogeneously. Figure 2 shows the distribution of cytochrome P450-dependent activities reported in the eye and their locations.

Ocular esterase, a major class of phase I enzymes, has been intensively investigated since the 1960s. Localization of ocular cytochrome P450 and esterase enzymes in ocular tissues is excerpted from the review by Nakano et al. (2014). Ocular tissues are also capable of phase II conjugation activity. Watkins and Wirthwein (1991)

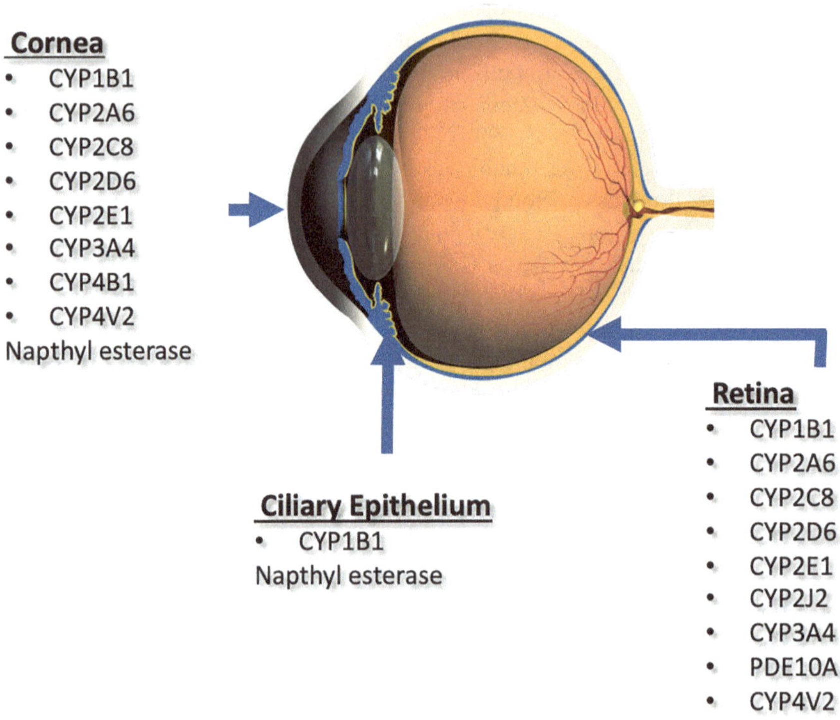

Fig. 2 Location of ocular cytochrome P450 and esterase enzymes in ocular tissues. (Nakano M, Lockhart CM, Kelly EJ, Rettie AE (2014) Ocular cytochrome P450s and transporters: roles in disease and endobiotic and xenobiotic disposition. Drug Metab Rev. 1–14 (Figure re-drawn)

conducted a comprehensive study to show evidence of activities for *N*-acetyltransferase, glutathione transferase, UDP-glucuronosyltransferase, and 2-naphthol sulfotransferase in the all tested ocular tissues—choroid, iris, retina, and cornea—except in the lens.

Researchers have applied prudent drug design to create prodrugs to exploit metabolic pathways in the eye to improve ocular bioavailability and reduce systemic toxicity and for site-specific tissue delivery (Loftsson 2015; Shell 1985; Heikkinen et al. 2018; Majumdar and Sloan 2007). Chemistry intervention has involved adding functional groups to increase drug solubility and lipophilicity resulting in improved membrane permeability, distribution, and tissue penetration (Rautio et al. 2008). Prodrugs have been investigated for improved ocular delivery of pilocarpine (Bundgaard et al. 1985, 1986a, b), epinephrine (Hussain 1976), β-blockers (Sasaki et al. 1993), and prostaglandins (Hellberg et al. 2003) with several successes entering market. Codrugs are bio-convertible similar to prodrugs wherein two molecules are covalently linked with a metabolically labile linker that is chemically or enzymatically cleaved to release the active ingredients in the optimal spatial and temporal modality (Al-Ghananeem and Crooks 2007; Hamad et al. 2006).

Recently, there has been considerable interest in transporters in the eye. Transporters are membrane-bound proteins that play an important role in active transport of nutrients across biological membranes. The presence of transporters has been reported, specifically transporters localized in the epithelia of the cornea, conjunctiva, and retina (Yavuz and Kompella 2017; Novack and Robin 2016). These transporters are of two types: efflux transporters and influx transporters. Ocular efflux transporters which lower bioavailability by pumping solutes out include P-gp, multidrug resistance protein (MRP), and BCRP (Mitra and Hughes 2013). Influx transporters facilitate the translocation of nutrients and xenobiotics across biological membrane and may be potentially useful for intraocular drug delivery as shown for valacyclovir. Influx transporters are also of various types, namely, amino acid and peptide transporters, organic cation/anion (SLC22), monocarboxylate (SLC16), nucleoside transporters (SLC28 and 29), and vitamin transporters (Novack and Robin 2016; Mannermaa et al. 2006). Discovery efforts to rationally design drug molecules and formulation technologies to exploit ocular transporters are emerging as a future opportunity in ocular drug delivery.

Drug Delivery to the Eye: Physicochemical Fundamentals

Candidate Selection

Safe and efficacious ocular delivery of drugs to treat eye disease depends on the interplay of physicochemical and biological factors. Physicochemical drug properties and characteristics of the drug delivery system affect a compound's ability to reach its site of action and interact with the disease target. Although many molecular targets may be associated with disease etiology, not all targets may be druggable. A target is considered druggable, that is, amenable to drug treatment, only if it can be modulated in vivo by a drug-like molecule. To understand what makes a target druggable, one must first understand what makes a compound a drug (Templeton et al. 2015; Lipinski 2000).

A compound must possess essential "drug-like" attributes to be considered a viable drug candidate. The three fundamental "druggability" requirements for compound to be a potentially useful therapeutic agent for treating eye disease are (1) the ability to get to the site of action in safe and effective concentrations, (2) persistence at the site of action for an optimal duration of time, and (3) to interact with one or more molecular targets to express the desired pharmacology. Additionally, to be a successful drug, a compound must be "developable" from a chemistry-manufacturing-control (CMC) perspective (Ghosh and Ahmed 2013).

Druggability Assessment

Drugs applied topically to the eye must be absorbed across the outer ocular membranes to penetrate the eye and access intraocular targets. Conceptually this is like oral drugs that must be absorbed across the intestine to enter the bloodstream and eventually access target tissues. Lipinsky et al. noted that the majority of oral drugs shared a few specific physicochemical parameters relating to size and lipophilicity, collectively referred to as the rule of five (Lipinski 2000). The rule of five (Ro5) states that poor absorption or permeation is more likely when drugs exceed the limits against the four following physicochemical parameters: the molecular weight (MW) is over 500; there are more than five H-bond donors (NHDs—calculated as the sum of -OHs and NH bonds); there are more than ten H-bond acceptors (NHA—calculated as the sum of nitrogen and oxygen atoms); and the cLogP (calculated octanol: water partition coefficient) is over 5. Whereas the Ro5 may be conceptually applicable to non-oral drugs, it may not be optimal. Choy and Prausnitz (2011) proposed modification to the Ro5 based on statistics arguing new thresholds for current ophthalmic drugs are NHD ≤3, NHA ≤8, and log P ≤ 3.4. They evaluated 59 topically applied ophthalmic drugs approved by the FDA and found good adherence with Ro5. Only 5 out of the 59 ophthalmic drugs evaluated violated Ro5 with only 1 drug (cyclosporin; MW = 1203; NHA = 12) that received an "alert" status for violating 2 combination parameters (MW and NHA). Although the permeability barrier of the cornea is greater than the intestine due to higher epithelial layer resistance (Ghate and Edelhauser 2006), qualitatively the barrier properties of the outer ocular membranes that a drug must pass through to gain intraocular access to disease targets and ocular biological fluids are similar to the intestinal barrier that oral drugs must pass through.

Topically or periocular administered drugs enter the eye and surrounding tissues mainly by passive diffusion from the eye surface or the dosing site according to Fick's law. That is, the drug is driven into the eye by virtue of the concentration gradient of the dissolved drug molecules:

$$J = D_{\mathrm{m}} \mathrm{d}C_{\mathrm{m}} / \mathrm{d}x = D_{\mathrm{m}} \left[C_{\mathrm{m}}^{\mathrm{o}} ? C_{\mathrm{m}}^{h} \right) / h \tag{1}$$

where J is the flux, in units of mol cm^{-3} s^{-1}; where $C_{\mathrm{m}}^{\mathrm{o}}$ and C_{m}^{h} are the concentrations, in mol cm^{-3} units, of uncharged form of the solute within the membrane at the two water-membrane boundaries (at positions $X = 0$ and $X = h$); where h is the thickness of the membrane in cm units; and where D_{m} is the diffusivity of the solute within the membrane, in units of cm^2 s^{-1}.

Considering the distribution coefficient of the drug molecule between bulk water and the membrane, log K_{d} (the pH-dependent apparent partition coefficient), Eq. 1 can be expressed as:

$$J = D_{\mathrm{m}} K_{\mathrm{d}} \left(C_{\mathrm{D}} ? C_{\mathrm{A}} \right) / h \tag{2}$$

where the C_D and C_A are the drug concentration in the donor and acceptor compartments, respectively.

Further simplification of Eq. 2 by lumping D_m, K_d and membrane thickness into a composite parameter called "effective permeability" P_e, and further reduction assuming "sink" conditions wherein $C_D \gg C_A$, results in an easy-to-conceptualize representation as follows:

$$J = P_e C_D \tag{3}$$

According to Eq. 3, flux, which reflects how quickly molecules cross through a membrane, depends on the product of effective permeability of the membrane and the total drug concentration in the donor compartment. For example, in ocular drug delivery, determinants of the rate of ocular penetration would be corneal permeability of the drug and the concentration of the drug in the precorneal area.

Another useful equation borrowed from oral drug delivery which estimates the maximum amount of drug that can be absorbed across a membrane from a donor compartment is commonly referred to as the maximum absorbable dose or MAD:

$$\text{MAD} = SK_a Vt \tag{4}$$

where S is the solubility of the drug in the donor compartment (mg/mL), K_a is the absorption rate constant (min^{-1}), V is the volume of fluid in the donor compartment (mL), and t is the residence time in the donor compartment or absorption site.

Taken together, Eqs. 3 and 4 provide a simple framework to conceptualize and understand the interplay of physicochemical, physiological, and anatomical factors affecting drug delivery to the eye and approaches to overcome barriers to ocular drug delivery. For example, in topical ocular drug delivery, the relevant parameter for P_e is corneal permeability, C_D is the drug concentration in the eye drop or the drug solubility (S) in the tear film (i.e., if suspension or if dose is greater than the drug solubility in the tear film), K_a is the absorption rate into the anterior chamber, V is the volume in the precorneal area, and t is the residence time of the drug in the precorneal area. The complexity associated with ocular drug delivery is that in the eye each of these parameters varies as a function of time and is highly impacted by the dosing method, dosage form, and the formulation characteristics. This renders accurate prediction of the ocular bioavailability and pharmacokinetics difficult and elusive. Due primarily to this reason, there is no regulatory accepted bioequivalence bridging strategy in ophthalmology. Hence, even small formulation changes often require repeat of clinical studies which may be cost prohibitive and risky. Therefore, in ophthalmic drug development, a right-first-time strategy is imperative, wherein all effort should be made to lock in the final, market image formulation at early clinical stage avoiding formulation changes going from proof of concept to pivotal studies.

Drug permeability across eye barriers is a critical step in the assessment of eye-related bioavailability. Quantitative structure-property relationships (QSPR) approaches have been developed to correlate the biological activity of a molecule with its physicochemical properties through a variety of molecular descriptors (Menichetti et al. 2019; Velpandian et al. 2011). Yoshida and Topliss (1996) proposed an algorithm to predict the corneal permeability (PC) for no congeneric compounds using two molecular descriptors, namely, $\Delta\log P$ and $\log D$:

$$\log \mathrm{PC} = ?0.404(?0.114)?\log P(?0.090)\log D?3.862(?0.451) \tag{5}$$

wherein PC denoted the permeability coefficient, $\Delta\log P$ is the difference between the octanol-water and the alkane-water partition coefficients, and D is the dissociation constant. Another QSPR model developed by Fu and Liang (2002) was based upon charge and molecular volume as the molecular descriptors to predict the corneal partition coefficient (PC) of no congeneric compounds:

$$\log \mathrm{PC} = ?5.566Q_{\mathrm{H}}^{2} + 3.027Q_{\mathrm{H}} ?0.155Q_{\mathrm{O,N}} ?9.413?10^{?4}V?4.278 \tag{6}$$

wherein Q_{H} is the sum of absolute values of net atomic charge of hydrogen atoms, $Q_{\mathrm{O,N}}$ is the sum of the absolute values of the net atomic charges of oxygen and nitrogen atoms, and V is the molecular volume.

The limitation with the QSPR approach is that the predictive accuracy is often predicated and constrained by the relevance and quality of the data set. Hence, the utility as a predictive tool diminishes with changing chemotypes. Therefore, new drug development commonly invokes experimental measurement of permeability via in vivo pharmacokinetics (Urtti 2006; Shell 1982; Robinson 1997), in situ isolated membranes (Mitra and Mikkelson 1988; Francoeur and Patton 1979), or, more recently, use of cell cultures techniques (Barar et al. 2014; Toropainen et al. 2001).

From a "druggability" perspective, a successful drug candidate must have the requisite physicochemical attributes to overcome the barriers to ocular entry. Drug entry into the eye is hampered by three major obstacles (Gan et al. 2013). The first major obstacle is the aqueous drug solubility and associated physicochemical parameters that affect drug solubility (S) (Avdeef 2001). Only dissolved drug molecules and "free" drug molecules that are not protein-bound or otherwise associated with a carrier can permeate through biological membranes (Petty 1993). Ophthalmic drugs must possess enough solubility in the aqueous tear fluid or the interstitial fluid at the periocular injection site to enter the eye. Most drugs are typically weak acids or bases. The proportion of drugs with an ionizable group has been estimated at 95% (Wells 1998) in which 6% of that total collection were ionizable between a pH of 2 and 12 (JT 2001). The charged state of the molecule is important, as the flux of the unionized molecule is substantially higher than that of the ionized or charged molecule. Mitra and Mikkelson (1988) showed that the permeability of ionized pilocarpine as well as the quaternary form was shown to be one-half of that for the unionized pilocarpine species. Not only the degree of ionization but also the charge of the molecule affects their corneal penetration (Rojanasakul et al. 1992; Liaw and Rojanasakul 1992). The corneal epithelium is negatively charged above its isoelectric point (pI 3.2) (Rojanasakul 1989). Consequently, hydrophilic charged cationic compounds permeate more easily through the cornea than anionic species. At the physiological pH of 7.4, the transport number for positive over negative ions was 1.63 (Rojanasakul et al. 1992). It is also preferable to design away from an ionization constant in the proximity of the pH of the tear fluid since this will create a large change in the unionized fraction of drug at near-physiological pH (Ahmed and Chaudhuri 1988). To maximize solubility amine drugs are frequently formulated at an acidic pH wherein they exist predominantly in the ionized state. Moreover, the formulations may be buffered to minimize pH drift. The buffer may, in turn, retard pH re-equilibration post-instillation exacerbating lacrimation and drug dilution, as well as reducing the concentration of unionized drugs (Chrai et al. 1973; Conrad et al. 1978; Ahmed and Patton 1984). Zwitterionic drugs can be particularly challenging from a solubility perspective if the isoelectric point at a near-physiological pH. Some drugs are amphipathic, that is, possess surface active properties and a high tendency to self-associate. This can reduce the amount of free drug available for diffusion depending on the association constant. The overall solubility target required for efficacy is difficult to predict a priori as it depends on the potency of the molecule and the drug levels required in the target tissue. This has to be determined experimentally.

Ocular Bioavailability

The major obstacle for intraocular entry from the topical route is the rapid turnover rate of the tear fluid and the consequent decrease in concentration of dissolved drug molecules. This obstacle and impact thereof are reflected in the parameters: CD, V, and t in Eqs. 3 and 4. Drug concentration decreases due to dilution by the tear turnover, as well as corneal and non-corneal absorption. Therefore, intraocular entry of topically applied drugs is constrained by a very short absorption window which diminishes the absorbable dose. The instilled volume of a typical eye drop is approximately 30 μL (Lederer Jr 1986). The physiological response of the eye is to rapidly drain away the excess instilled volume that is not lost to spillage until the resident tear volume of approximately 7 μL is reestablished. This translates to a volume-dependent drainage rate constant of about 1.5 μL per minute in humans (Lee 1986) resulting in a resident tear volume in approximately 5 min. Another route of drug loss in the precorneal area is the normal tear turnover, which is about 15–16% per minute or 1.2 μL in humans. Tear turnover serves to dilute the drug and, hence, reduce the gradient for transport through the cornea. An additional, associated problem is induced lacrimation resulting in reflex tearing in response to a physiological insult to the eye, such as an acidic pH or osmotic shift in the tear film invoked by the administered formulation. An important route of drug loss from the precorneal area is systemic absorption through the conjunctiva of the eye. Conjunctival permeability coefficients of most drugs are higher than their corneal permeabilities (Chien et al. 1991). In addition, the surface area of the conjunctiva (16–18 cm^2) is larger than that of the cornea (1 cm^2) (Watsky and Jablomski 1988). Thombre and Himmelstein developed a mathematical model to quantify the initial disposition of topically applied drugs and their availability for systemic and local absorption (Thombre and Himmelstein 1984):

$$\mathrm{d}C_{\mathrm{t}} / \mathrm{d}t = ?Q_{\mathrm{t}}C_{\mathrm{t}} ? p_{\mathrm{n}}C_{\mathrm{t}} ?0.0299 / \left[V_{\mathrm{d}} \exp\left(?K_{\mathrm{t}}\right) + V_{\mathrm{o}}\right] \tag{7}$$

$$1 / \left(Q_{\mathrm{t}} + p_{\mathrm{n}}\right) \ln(C_{\mathrm{t}} + 0.0299 / \left[Q_{\mathrm{t}} + p_{\mathrm{n}}\right] = t / V_{\mathrm{o}} + 1 / V_{\mathrm{o}}K \ln[V_{\mathrm{o}} + V_{\mathrm{d}} \exp\left(?K_{\mathrm{t}}\right] + C \tag{8}$$

where C_{t} is the drug concentration in the tear film, Q_{t} is the tear production rate, p_{n} is the rate constant of conjunctival drug loss, 0.0299 μg/min is the estimated rate of drug loss into the aqueous humor (i.e., corneal absorption), V_{d} is the instilled drop volume, V_0 is the resident tear film volume, and K is the proportionality constant that is a function of the instilled drop size, V_{d}:

$$V_{\mathrm{t}} = V_{\mathrm{d}} \exp\left(?K_{t}\right) + V_{\mathrm{o}}; \quad \text{where, } V_{t} \text{ is the total tear film volume} \tag{9}$$

All parameters except p_{n}, the rate constant for conjunctival absorption, were known from experimental measurement. The mathematical model simulated a tear drug concentration profile that was in good agreement with measured concentration. Although drainage is responsible for ≈43% of drug loss in terms of mass, conjunctival absorption is a primary factor responsible for the dramatic fall in tear film drug concentration. Overcoming precorneal drug loss, concentration gradient depletion, and short precorneal residence time requires formulation technology intervention or alternate route of administration, such as semi-invasive periocular injection.

The drug molecules must partition from the tear film into the corneal or conjunctival membrane before they can passively permeate the membrane barrier and enter the eye. Physicochemical determinants of membrane permeability are molecule size and lipophilicity. Large molecules (≫500 Da) do not easily cross the corneal membrane due to tight junctions. Corneal absorption of topically applied drugs generally exhibits a parabolic relationship with lipophilicity with an optimum partition coefficient of 2–3 (Prausnitz 1998).

The physicochemical requirements in terms of candidate attributes for periocular and intraocularly administered drugs are like topically applied drugs. The advantage of periocular drug administration is the avoidance of precorneal drug loss due to drainage and lacrimation and the localization of high drug concentration and dose at the site of absorption. The disadvantage for many of the periocular routes is the high fraction of nonproductive drug loss to the systemic circulation. Selective regional targeting may be feasible based on the site and method of drug administration. The suprachoroidal route has been reported to be an option for effectively delivering some drugs, such as triamcinolone acetonide, to the retina and back of eye tissues.

Intravitreal (IVT) injection is commonly used to treat back of the eye diseases wherein the drug is placed in closed proximity of the target tissues in the back of the eye. Besides the obvious pharmacology and potency considerations, drug solubility and stability in interstitial or intraocular fluid is important in intraocular and IVT injection. Furthermore, it is important to take into consideration the components in the vitreous, notably hyaluronic acid – which can bind the drug and is available in enough concentration to affect the aqueous fraction available for drug absorption (Haghjou et al. 2011). IVT formulations are most commonly solutions although there is considerable ongoing research toward the development of intravitreal implants. Introducing suspensions or high viscosity formulation to the IVT space is not recommended due to "snow globe" effect and intraocular pressure spike. Notable disadvantages of the IVT route, besides its obvious invasiveness and associated infection risk, are the rapid drug clearance and the convective and diffusive barrier to access to the target tissue.

Developability Assessment (DAS)

Developability assessment (DAS) of new molecular entities (NMEs) includes physicochemical and biopharmaceutical characterization; development of suitable formulations for pharmacokinetic (PK), efficacy, and toxicity studies; and technology selection for clinical proof of concept and eventual market image (Saxena et al. 2009). A framework for developability assessment for ophthalmic drug candidate is presented in Table 1 Compound Personality Assessment (COPA).

Table 1 Developability assessment framework (DAS) for ophthalmic drugs

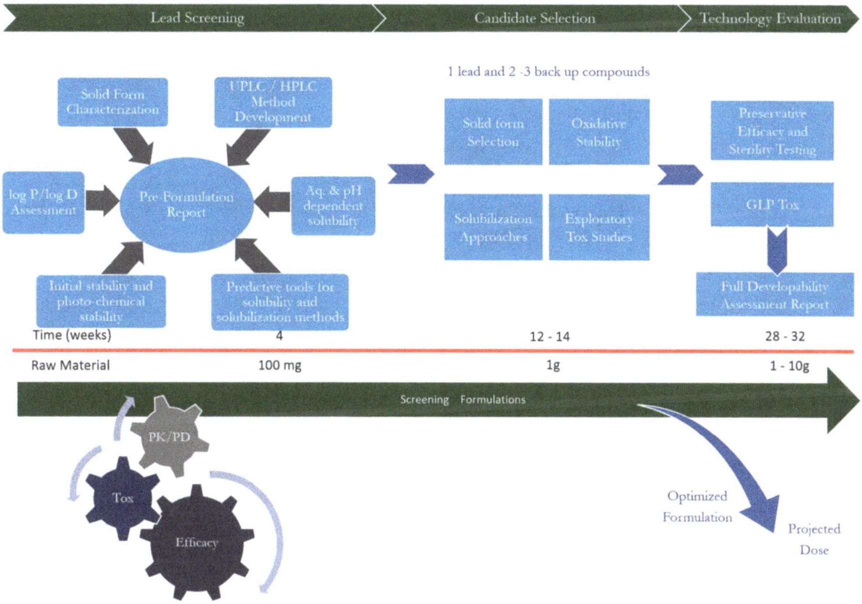

The physicochemical requirements for ophthalmic drug candidates were reviewed in the previous section. Whenever possible, the requisite solubility and stability attributes should be built into the molecule in the design and lead seeking phase of drug discovery. However, the reality is that ophthalmic drug candidates are often selected with an emphasis on optimal pharmacology and target engagement considerations. The challenge then falls upon the formulation scientists to rapidly ascertain whether the selected candidate has the requisite solubility and stability for the route of administration, method of delivery, and indication of use. Often, it is necessary to invoke formulation interventions to overcome intrinsic solubility or stability deficits with the candidate molecule. Compound Personality Assessment (COPA) provides an assessment of risks associated with early compound development concerning the following studies: (1) selection of suitable physical form; (2) solubility evaluation; (3) physical and chemical stability; (4) biopharmaceutical studies to identify suitable formulations for safety, PK, and efficacy studies; and (5) formulation technology options.

The selection of excipients suitable and qualified for use in ophthalmic products is very limited. A list of commonly used excipients with precedent for use in clinical or marketed drug products is provided in Table 2. Solubilizers and stabilizers commonly used in oral and injectable products may not be suitable for use in ophthalmic products. The acceptable levels in ophthalmic products are typically lower than in oral or injectable products. Ophthalmic formulations are predominantly aqueous based except for devices, such as ocular inserts and intraocular implants.

Technology Selection

Technology selection, vis-à-vis dosage form, formulation, and manufacturing process, is key to a successful chemistry, manufacturing, and control (CMC) strategy to ensure clinical development, product registration, and launch. The technology selection decision tree for topical ophthalmic drug delivery systems presented by Ghosh and Ahmed (2013) is shown in Table 3. A solution dosage form is most preferred for topical ophthalmic drug delivery. For ionizable drugs, pH adjustment is the first choice for solubility enhancement. The acceptable pH range for ophthalmic products is narrow, typically between 4 and 8, and preferably between 6.5 and 7.5. Nonphysiological pH can cause pain, irritation, and lacrimation. If a nonphysiological pH adjustment is necessary for solubility or stability purposes, the formulation should be unbuffered or very lightly buffered. The buffer selection also needs to be done prudently to minimize return of the tear film pH to its resident physiological pH. Other strategies used for the solubilization of ophthalmic drugs include complexation with cyclodextrins or other ligands, micelle formation with surfactants, co-solvency in mixed solvent systems, encapsulation in liposomes, and emulsification with oils. The solubilizer choice depends on the physicochemical properties of the drug molecule and the level of solubilization required.

Alternate, delivery technologies including suspensions, gels, ointments, and emulsions may be considered if a solution approach is not feasible (Manish and

Table 2 Technology selection decision tree

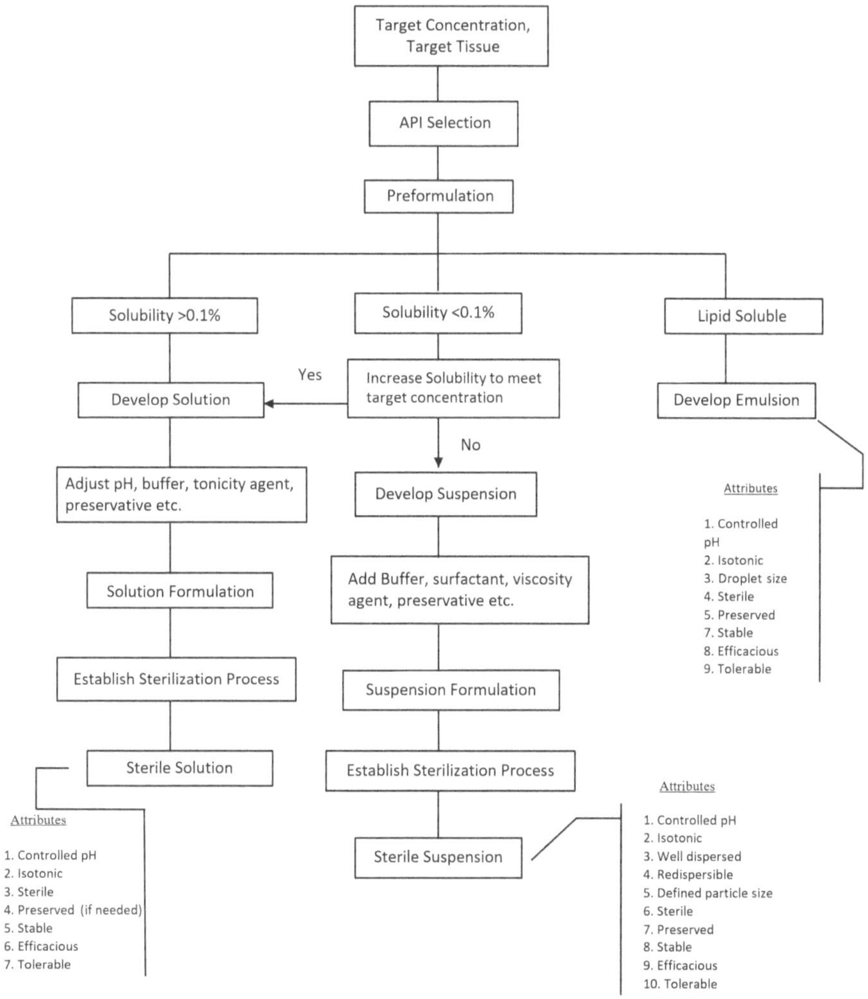

Adapted from Ghosh and Ahmed (2013)

Kulkarni 2012). There are many ophthalmic suspensions in the market. Particle size and distribution is very important for ophthalmic suspensions because particles >10 μm can cause ocular discomfort and blurriness.

Recent research and development focus have shifted from the front of the eye to the posterior pole targeting therapies and technologies to treat a myriad of back of the eye diseases (Hughes et al. 2005; Kompella et al. 2010; Jiang et al. 2018; Patel et al. 2011; Kim et al. 2007a, b). Various devices and controlled-release drug delivery systems have also been evaluated for intraocular and back of the eye drug delivery (Kim et al. 2004; Thakur and Kashiv 2011). For intraocularly applied sustained-release devices, such as intraocular implants and inserts, drug must also

Table 3 Ophthalmology excipient list

Functionality	Material	Typical levels used in formulation (%w/v)
Buffering agent	Acetic acid, citric acid, sorbic acid, tromethamine, sodium carbonate, boric acid	0.2–4% Note: Boric and sorbic acid may serve as preservative/preservative aid
Preservative agent	Benzalkonium chloride, benzododecinium chloride, benzethonium chloride, polyquaternium-1, sorbic acid, zinc chloride	0.0005–0.2% (higher levels up to 2% of benzalkonium chloride are reported as a permeation enhancer)
Surfactants/ solubilizers	Tyloxapol, polysorbate 80, polysorbate 60, polysorbate 20, polyoxy35 castor oil, polyoxyl40 hydrogenated castor oil, poloxamer 188, poloxamer 407	0.05–15% (most commonly between 0.05 and 0.5%)
Tonicity agent	Propylene glycol 400, glycerin, mannitol, sorbitol, sodium chloride, calcium chloride, sodium nitrate, sodium sulfate	0.03–5% (high-level mannitol and sodium chloride are reported for therapeutic use)
Viscosity agents	Sodium carboxymethyl cellulose, hydroxyethyl cellulose, hydroxypropyl methylcellulose, xanthan gum, gellan gum, guar gum, polyethylene glycol 8000, povidone K30 and K70, polyvinyl alcohol	0.5–2%
Oils for emulsification/ ointment base	Light mineral oil, lanolin alcohol, lanolin, mineral oil, petrolatum, castor oil	2–89%
Stabilizers/ antioxidants	Sodium metabisulfite, sodium thiosulfate, potassium metabisulfite, sodium EDTA, vitamin E TPGS	0.2–2.2%

be stable in the intraocular delivery system and also the intraocular environment in which the delivery system will reside. This requirement can be particularly challenging, particularly for large molecules, such as monoclonal antibodies and proteins that are prone to aggregations and loss of bioactivity. Polylactic polyglycolic acid copolymers are often used to fabricate intraocular drug delivery systems. Degradation and erosion of these polymers can result in a highly acidic microenvironment that can be detrimental to drug stability. Furthermore, the drug fabricated in a long-acting drug delivery system, such as an insert, implant, or microparticles, exposed to high humidity and body temperature existing inside the eye. Therefore, drug stability and release must be characterized under physiologically relevant conditions.

A special challenge for many ophthalmic products (except for injectables) is packaging. Most eye drops are packaged in plastic dropper bottles which are very difficult to protect from environmental factors, such as light or oxygen. For example, oxygen exclusion commonly used in injectable products to minimize oxidation is typically not feasible in plastic eye dropper bottles constructed out of low-density polyethylene (LDPE). Packaging innovation is an unmet need in ophthalmology product development.

Concluding Remarks

In recent years there have been major advances in ophthalmic formulation technology and drug delivery system design. The majority of the marketed ophthalmic drug products and new approvals are predominantly conventional eye drops or intraocular injections. Research and development continue with a traditional mindset focused on molecules with the best on-target pharmacology rather than differentiation based on drug delivery optimization. Integration of novel formulation technology and new drug delivery systems may become critical path adding significant cost, time, risk, and complexity to the development program. With increasing regulatory and payer expectation of improved outcomes and the advent of precision medicine away from a "one size fits all" approach, the scenario is likely to change in the near future.

Acknowledgments The author wish to thank Ms. Kalpita Nadkar for her assistance on preparing the manuscript.

References

Agrahi V, Mandal A, et al. A comprehensive insight on ocular pharmacokinetics. Drug Deliv Transl Res. 2016;6:735–54.

Ahmed I, Chaudhuri B. Evaluation of buffer systems in ophthalmic product development. Int J Pharm. 1988;44:97–105.

Ahmed I, Patton TF. Effect of pH and buffer on the precorneal disposition and ocular penetration of pilocarpine in rabbits. Int J Pharm. 1984;19:215–27.

Ahmed I, Patton TF. Importance of the non-corneal absorption route in topical ophthalmic drug delivery. Investig Ophthalmol Vis Sci. 1985;26:584–7.

Ahmed I, Patton TF. Disposition of timolol and inulin in the rabbit eye following corneal versus non-corneal absorption. Int J Pharm. 1987;38:9–21.

Ahmed I, Francouer ML, et al. The kinetics of timolol in the rabbit lens: implications for ocular drug delivery. Pharm Res. 1989;6:772–8.

Al-Ghananeem AM, Crooks PA. Phase I and phase II ocular metabolic activities and the role of metabolism in ophthalmic prodrug and codrug design and delivery. Molecules. 2007;12:373–88.

Asakura TSH. Cytochrome P450-mediated prostaglandin mega/omega-1 hydroxylase activities in porcine ciliary body epithelial cells. Exp Eye Res. 1992;55:377–84.

Avdeef A. Physicochemical profiling (solubility, permeability and charged state). Curr Top Med Chem. 2001;1:277–351.

Babizhayev MA, Li DW-C, Kasus-Jacobi A, et al. Studies on the cornea and lens. New Jersey: Springer; 2014.

Barar J, Javadzadeh AR, Omidi Y. Ocular novel drug delivery: impacts of membranes and barriers. Expert Opin Drug Deliv. 2008;5:567–81.
Barar J, Asadi M, Mortazavi-Tabatabaei SA, Omidi Y. Ocular drug delivery; impact of in vitro cell culture models. J Ophthalmic Vis Res. 2014;4:238–52.
Bekerman I, Gottlieb P, Vaiman M. Variations in eyeball diameters of the healthy adults. J Ophthalmol. 2014;2015:1–5.
Benjamin ER. Development of the retina and optic pathways. Vis Res. 2011;51:613–32.
Bundgaard H, Falch E, Larsen C, Mosher GLMT. Pilocarpic acid esters as novel sequentially labile pilocarpine prodrugs for improved ocular delivery. J Med Chem. 1985;28:79–81.
Bundgaard H, Falch E, Larsen CMT. Pilocarpine prodrugs I. Synthesis, physicochemical properties and kinetics of lactonization of pilocarpine acid esters. J Pharm Sci. 1986a;75:36–43.
Bundgaard H, Falch E, Larsen C, Mosher GLMT. Pilocarpine prodrugs II. Synthesis, stability, bioconversion, and physicochemical properties of sequentially labile pilocarpine acid diesters. J Pharm Sci. 1986b;75:755–83.
Chen L. Ocular lymphatics: state-of-the art review. Lymphology. 2009;42:66–76.
Chien D-S, Sasaki HBH, et al. Role of enzymatic lability in corneal and conjunctival penetration of timolol ester prodrugs in the pigmented rabbit. Pharm Res. 1991;8:728–33.
Choy YB, Prausnitz MR. The rule of five for non-oral routes of drug delivery: ophthalmic, inhalation and transdermal. Pharm Res. 2011;28:943–8.
Chrai SS, Patton TF, et al. Lacrimal and instilled fluid dynamics in rabbit eyes. J Pharm Sci. 1973;62:1112–21.
Clinic M. Biology of the corneal endothelium in health and disease. Eye (Lond). 2003;17(8):912–8.
Conrad JM, Reay WA, et al. Influence of tonicity and pH on lacrimation and ocular drug availability. J Parent Drug Assoc. 1978;32:150–61.
Coyle PK, Sibony PAJC. Electrophoresis combined with immunologic identification of human tear proteins. Invest Ophthalmol Vis Sci. 1989;30:1872–8.
Durairaj C, Kim SJ, Edelhauser HF, et al. Influence of dosage form on the intravitreal pharmacokinetics of diclofenac. Investig Ophthalmol Vis Sci. 2009;50:4887–97.
Francoeur ML, Patton TF. Kinetics of corneal drug uptake by corneal perfusion in situ 1. Evaluation of system and uptake of ethyl p-aminobenzoate in rabbits. Int J Pharm. 1979;2:337–42.
Fu XC, Liang W. A simple model for the prediction of corneal permeability. Int J Pharm. 2002;232:192–7.
Gachon AM, et al. Immunological and electrophoretic studies of human tear proteins. Exp Eye Res. 1979;29:539–53.
Gan L, Wang J, et al. Recent advances in topical ophthalmic drug delivery with lipid-based nanocarriers. Drug Disc Today. 2013;18:290–7.
Gaudana R, Ananthula HK, Parenky A, Mitra AK. Ocular drug delivery. AAPS J. 2010;12:348–60.
Ghate D, Edelhauser HF. Ocular drug delivery. Expert Opin Drug Deliv. 2006;3:275–87.
Ghosh M, Ahmed I. Chemistry, manufacturing and control of ophthalmic formulations. In: Methods in pharmacol toxicol. New York: Springer; 2013. p. 1–27.
Ghosn MG, Tuchin VV, Larin KV. Nondestructive quantification of analyte diffusion in cornea and sclera using optical coherence tomography. Investig Ophthalmol Vis Sci. 2007;48:2726–33.
Gillan WDH. Tear biochemistry: a review. African Vis Eye Heal. 2010;69:100–6.
Glasson MJ, Stapleton F, et al. The effect of short term contact lens wear on tear film and ocular surface characteristics of tolerant and intolerant wearers. Contact Lens Anterior Eye. 2006;29:41–7.
Haghjou N, Abdekhodaie MJ, Cheng YL, Saadatmand M. Computer modeling of drug distribution after intravitreal administration. World Acad Sci Eng Technol. 2011;77:706–16.
Hamad MO, Kiptoo PK, Stinchcomb AL. Synthesis and hydrolytic behavior of two novel tripartate codrugs of naltrexone and 6b-naltrexol with hydroxybupropion as alcohol cessation agents. Bioorg Med Chem. 2006;14:7051–61.
Heikkinen EM, del Amo EM, Ranta VP, et al. Esterase activity in porcine and albino rabbit ocular tissues. Eur J Pharm Sci. 2018;123:106–10.

Hellberg MR, Ke TL, et al. The hydrolysis of the prostaglandin analog prodrug bimatoprost to 17-phenyltrinor PGF2a by human and rabbit ocular tissues. J Ocul Pharm Ther. 2003;19:97–103.
Hitzenberger CK, Baumgartner A, Drexler WFA. Interferometric measurement of corneal thickness with micrometer precision. Am J Ophthalmol. 1994;118:468–76.
Hughes PM, Krishnamoorthy R, Mitra AK. Effect of acylation on the ocular disposition of acyclovir I: synthesis, physicochemical properties, and antiviral activity of 2′-esters. J Ocul Pharmacol. 1993;9:287–97.
Hughes PM, Olejnik O, Chang-Lin JE, Wilson CG. Topical and systemic drug delivery to the posterior segments. Adv Drug Deliv Rev. 2005;57:2010–32.
Hussain ATJ. Prodrug approaches to enhancement of physicochemical properties of drugs IV. Novel epinephrine prodrugs. J Pharm Sci. 1976;65:1510–2.
Janssen PT, et al. Origin and biosynthesis of human tear proteins. Invest Ophthal Vis Sci. 1983;24:623–30.
Järvinen K, Järvinen T, Urtti A. Ocular absorption following topical delivery. Adv Drug Deliv Rev. 1995;16:3–19.
Jiang J, Geroski DH, Edelhauser HF, Prausnitz MR. Measurement and prediction of lateral diffusion within human sclera. Investig Ophthalmol Vis Sci. 2006;47:3011–6.
Jiang S, Franco YL, Zhou Y, Chen J. Nanotechnology in retinal drug delivery. Int J Ophthalmol. 2018;11:1038–44.
JT C. Lipophilicity profiles. In: Optimization in drug research: biological, physicochemical and computational strategies. Zurich: Wiley; 2001.
Kadam RS, Kompella UB. Influence of lipophilicity on drug partitioning into sclera, choroid-retinal pigment epithelium, retina, trabecular meshwork, and optic nerve. J Pharmacol Exp Ther. 2010;332:1107–20.
Kaufman PL, Alm AKA. Alder's physiology of the eye. St. Louis: Mosby; 2003.
Kesavan K, Balasubramaniam J, Kant S, et al. Newer approaches for optimal bioavailability of ocularly delivered drugs: review. Curr Drug Deliv. 2011;8:172–93.
Kim H, Robinson MR, Lizak MJ, et al. Controlled drug release from an ocular implant: an evaluation using dynamic three-dimensional magnetic resonance imaging. Investig Ophthalmol Vis Sci. 2004;45:2722–31.
Kim SH, Galbán CJ, Lutz RJ, et al. Assessment of subconjunctival and intrascleral drug delivery to the posterior segment using dynamic contrast-enhanced magnetic resonance imaging. Investig Ophthalmol Vis Sci. 2007a;48:808–14.
Kim SH, Lutz RJ, Wang NS, Robinson MR. Transport barriers in transscleral drug delivery for retinal diseases. Ophthalmic Res. 2007b;39:244–54.
Kishida K, Matsumoto K, Manabe RST. Cytochrome P450 and related components of the microsomal electron transport system in the bovine ciliary body. Curr Eye Res. 1986;5:529–33.
Kompella UB, Kadam RS, Lee VHL. Recent advances in ophthalmic drug delivery. Ther Deliv. 2010;1:435–56.
Kryczka T, Wylęgała E, Dobrowolski D, Midelfart A. NMR spectroscopy of human eye tissues: a new insight into ocular biochemistry. Sci World J. 2014;2014:1–9.
Le Bourlais C, Acar L, Zia H, et al. Ophthalmic drug delivery systems—recent advances. Prog Retin Eye Res. 1998;17:33–58.
Lederer CMHR Jr. Drop size of commercial glaucoma medications. Am J Ophthal. 1986;101:691–4.
Lee VHLRJ. Review: topical ocular drug delivery: recent development and future challenges. J Ocul Pharmacol. 1986;2:67–108.
Li Y, Liu J, Pan D, Hopfinger AJ. A study of the relationship between cornea permeability and eye irritation using membrane-interaction QSAR analysis. Toxicol Sci. 2005;88:434–46.
Liaw J, Rojanasakul YRJ. The effect of drug charge type and density on corneal transport. Int J Pharm. 1992;88:111–24.
Lipinski CA. Drug-like properties and the causes of poor solubility and poor permeability. J Pharmacol Toxicol Methods. 2000;44:235–49.

Loftsson T. Physicochemical properties and pharmacokinetics. Essent Pharmacokinet. 2015;21:85–104.
Majumdar S, Sloan KB. Enhancing topical delivery using prodrugs. Pharm Technol Eur. 2007;19:33–5.
Manish K, Kulkarni GT. Recent advances in ophthalmic drug delivery system. Int J Pharm Pharm Sci. 2012;4:387–94.
Mannermaa E, Vellonen KS, Urtti A. Drug transport in corneal epithelium and blood-retina barrier: emerging role of transporters in ocular pharmacokinetics. Adv Drug Deliv Rev. 2006;58:1136–63.
Menichetti R, Kanekal KH, Bereau T. Drug-membrane permeability across chemical space. ACS Cent Sci. 2019;5:290–8.
Mishima S, Gasset A, et al. Determination of tear volume and tear flow. Invest Ophthalmol Vis Sci. 1966;5:264–76.
Mitra AK, Hughes P, editors. Treatise on ocular drug delivery. Sharjah: Bentham Books; 2013.
Mitra AK, Mikkelson TJ. Mechanism of transcorneal permeation of pilocarpine. J Pharm Sci. 1988;77:771–5.
Myles ME, Neumann DM, Hill JM. Recent progress in ocular drug delivery for posterior segment disease: emphasis on transscleral iontophoresis. Adv Drug Deliv Rev. 2005;57:2063–79.
Nakano M, Lockhart CM, Kelly EJ, Rettie AE. Ocular cytochrome P450s and transporters: roles in disease and endobiotic and xenobiotic disposition. Drug Metab Rev. 2014;46:247–60.
Newell FW. Ophthalmology. Principles and concepts. St. Louis: CV Mosby Co.; 1986.
Novack GD, Robin AL. Ocular pharmacology. J Clin Pharmacol. 2016;56:517–27.
Occhiutto ML, Freitas FR, Maranhao RC, Costa VP. Breakdown of the blood-ocular barrier as a strategy for the systemic use of nanosystems. Pharmaceutics. 2012;4:252–75.
Patel A. Ocular drug delivery systems: an overview. World J Pharmacol. 2013;2:47.
Patel SR, Lin ASP, Edelhauser HF, Prausnitz MR. Suprachoroidal drug delivery to the back of the eye using hollow microneedles. Pharm Res. 2011;28:166–76.
Petty HR. Transport across membranes. In: Molecular biology of membranes. New York: Springer; 1993.
Prausnitz MR. Permeability of cornea, sclera, and conjunctiva: a literature analysis for drug delivery to the eye. J Pharm Sci. 1998;87:1479–88.
Ranta VP, Mannermaa E, Lummepuro K, et al. Barrier analysis of periocular drug delivery to the posterior segment. J Control Release. 2010;148:42–8.
Rautio J, Kumpulainen H, Heimbach T, et al. Prodrugs: design and clinical applications. Nat Rev Drug Discov. 2008;7:255–70.
Reddy IK, Bodor N. Design and delivery of ophthalmic drugs: chemical approaches. Indian J Pharm Sci. 1993;55:163–7.
Reinsten DZ, Silverman RH, Rondeau MJCD. Epithelium and corneal thickness measurements by high frequency by high frequency ultrasound digital signal processing. Ophthalmology. 1994:140–6.
Robinson R. Ocular pharmacokinetics/pharmacodynamics. Eur J Pharm Biopharm. 1997;44:71–83.
Rojanasakul YRJ. Transport mechanism of the cornea: characterization of barrier permselectivity. Int J Pharm. 1989;55:237–46.
Rojanasakul Y, Wang L-Y, et al. The transport barrier of epithelia: a comparative study on membrane permeability and charge selectivity in the rabbit. Pharm Res. 1992;9:1029–34.
Sasaki H, Igarashi Y, Nishida KNJ. Ocular delivery of beta blocker, tilisolol through the prodrug approach. Int J Pharm. 1993;93:49–60.
Saxena V, Panicucci R, Joshi YGS. Developability assessment in pharmaceutical industry: an integrated group approach for selecting developable candidates. J Pharm Sci. 2009;98:1962–79.
Schoenwald RD. Ocular pharmacokinetics. In: Zimmerman TJ, et al., editors. Textbook of ocular pharmacology. Pennsylvania: Lippincott-Raven; 1997. p. 119–38.

Schwartzman ML, Abraham NG, Masferrer J, Dunn WMJ. Cytochrome P450 dependent metabolism of arachidonic acid in bovine corneal epithelium. Biochem Biophys Res Comm. 1985;132:343–51.

Shastri D, Shelat P, Shukla A, Patel P. Ophthalmic drug delivery system: challenges and approaches. Syst Rev Pharm. 2010;1:113.

Shell JW. Pharmacokinetics of topically applied ophthalmic drugs. Surv Ophthalmol. 1982;26:207–18.

Shell JW. Ophthalmic drug delivery systems. Surv Ophthalmol. 1984;29:117–28.

Shell JW. Ophthalmic drug delivery systems. Drug Dev Res. 1985;6:245–61.

Swindle KE, Ravi N. Recent advances in polymeric vitreous substitutes. Expert Rev Ophthalmol. 2007;2:255–65.

Templeton AC, Byrn SR, Haskell RJ, Prisinzano TE, editors. Discovering and developing molecules with optimal drug-like properties. New York: Springer; 2015.

Thakur RR, Kashiv M. Modern delivery systems for ocular drug formulations: a comparative overview W.R.T conventional dosage form. Int J Res Pharm Biomed Sci. 2011;2:8–18.

Thombre AG, Himmelstein KJ. Quantitative evaluation of topically applied pilocarpine in the precorneal area. J Pharm Sci. 1984;73:219–22.

Tojo K, Nakagawa K, Morita Y, Ohtori A. A pharmacokinetic model of intravitreal delivery of ganciclovir. Eur J Pharm Biopharm. 1999;47:99–104.

Toropainen E, Ranta VP, Talvitte A, et al. Culture model of human corneal epithelium for prediction of ocular drug absorption. Investig Ophthalmol Vis Sci. 2001;42:2942–8.

Urtti A. Challenges and obstacles of ocular pharmacokinetics and drug delivery. Adv Drug Deliv Rev. 2006;58:1131–5.

Velpandian T, Sharma C, Biswas NR, et al. Development of novel in silico model to predict corneal permeability for congeneric drugs: a QSPR approach. J Biomed Biotechnol. 2011;2011:1–11.

Vicario-de-la-Toree MH-VR, et al. New formulations in dry eye treatment. Arch Soc Esp Ophthal. 2007;82:395–6.

Watkins J, Wirthwein DP SR (1991) Comparative study of phase II biotransformation in rabbit ocular tissues. Drug Metab Dispos 19:708–713\.

Watsky MA, Jablomski MMEH. Comparison of the conjunctival and corneal surface areas in rabbit and humans. Curr Eye Res. 1988:483–6.

Wells H. Pharmaceutical preformulation. London: Eills Hoowood Ltd; 1998.

Yavuz B, Kompella UB. Ocular drug delivery. Handbook Exp Pharmacol. 2017;242:57–93.

Yoshida F, Topliss J. Unified model for the corneal permeability of related and diverse compounds with respect to their physicochemical properties. J Pharm Sci. 1996;85:810–23.

Transcorneal Kinetics of Topical Drugs and Nanoparticles

Sangly P. Srinivas, Wanachat Chaiyasan, Anirudh Burli, Giovanna Guidoboni, Riccardo Sacco, Abhishek Anand, Amreen H. Siraj, Hemant Kumar Daima, and Uday B. Kompella

Abstract Topical drugs to the eye, common among ophthalmic drugs, access the anterior chamber via corneal or noncorneal pathways with penetration across the cornea, an oil:water:oil matrix, preferred by low-molecular-weight lipophilic drugs. The kinetics of their transcorneal penetration also depends on the barrier integrity of the corneal epithelium. Over the years, we have measured time-dependent ocular surface dynamics and transcorneal kinetics of several fluorescent molecules (employed as drug surrogates) and nanoparticles using custom-built fluorometers: spot fluorometer and confocal scanning microfluorometer (CSMF), respectively. The spot fluorometer has enabled novel approaches to assess the efficacy of different vehicles for enhanced bioavailability of topical drugs. It has also quantified the variability in the ocular surface dynamics of topical drops. On the other hand, the

S. P. Srinivas (✉)
Optometry, Indiana University, Bloomington, IN, USA
e-mail: Srinivas@indiana.edu

W. Chaiyasan
Optometry, Naresuan University, Phitsanulok, Thailand

A. Burli
Indiana University, Bloomington, IN, USA

G. Guidoboni
Electrical Engineering, Computer Science, and Mathematics, University of Missouri, Columbia, MO, USA

R. Sacco
Dipartimento di Matematica, Politecnico di Milano, Milan, Italy

A. Anand · A. H. Siraj
Computer Science, DSCE, Bangalore, Karnataka, India

H. K. Daima
Amity Center for Nanobiotechnology and Nanomedicine, Amity Institute of Biotechnology, Amity University, Jaipur, India

U. B. Kompella
Professor of Pharmaceutical Sciences, University of Colorado Anschutz Medical Campus, Aurora, CO, USA

S. Neervannan, U. B. Kompella (eds.), *Ophthalmic Product Development*, AAPS Advances in the Pharmaceutical Sciences Series 37,
https://doi.org/10.1007/978-3-030-76367-1_6

CSMF has led to a microscopic view of the transport of fluorescent dyes across the cornea, highlighting the cellular barriers and partitioning of dyes. The data accrued from these observations have enabled rational modeling of the spatiotemporal corneal pharmacokinetics of topical drugs. Unlike the compartmental models, our models incorporate the physicochemical properties of the drugs to explain the kinetics of their penetration. In this chapter, we review our unique experimental observations and describe the framework of our pharmacokinetic model for topical drugs.

Keywords Topical delivery · Cornea · Lipophilicity · Partition coefficient · Diffusion · Modeling · Fluorescent drug surrogates

Abbreviations

CSMF	Confocal scanning microfluorometer
FITC	Fluorescein isothiocyanate
Log *P*	Logarithm (base 10) of octanol-water partition coefficient at pH 7.4
RhB	Rhodamine B
SPR	Structure permeability relationships
SRB	Sulforhodamine B

Introduction

A variety of eye diseases and conditions, including dry eye, glaucoma, allergies, inflammatory conditions, infectious diseases, and injuries of the anterior segment of the eye, employ topical ophthalmic formulations including solutions, suspensions, emulsions, and ointments for drug therapy, with solutions comprising the majority of drug products (Fig. 1) (Kompella et al. 2020; Alvarez-Trabado et al. 2017; Barar et al. 2016; Boddu et al. 2014; Kompella et al. 2010; Gaudana et al. 2010). While diseases such as dry eye and glaucoma require chronic treatment (Ratna and Rina 2011), others such as allergies typically require short-term therapy (Mashige 2017). The specific pharmacological intervention depends on the site, severity, and treatment duration for a given disease. The majority of ophthalmic formulations are intended to access the aqueous humor or anterior chamber of the eye (Agrahari et al. 2016; Yavuz and Kompella 2017). Drug access to this compartment can occur via corneal and noncorneal pathways, with the corneal pathway preferred by low-molecular-weight lipophilic drugs (Kompella et al. 2020; Agrahari et al. 2016; Yavuz and Kompella 2017). The bioavailability of topically applied drugs is usually very small (<5%) and 0% at the anterior chamber and retina, respectively, as per one estimate (Maurice 2002), making drug delivery a unique challenge in the eye. To overcome these limitations, a variety of approaches, including nanoparticles (Almeida et al. 2015; Battaglia et al. 2016; Janagam et al. 2017; Kompella et al.

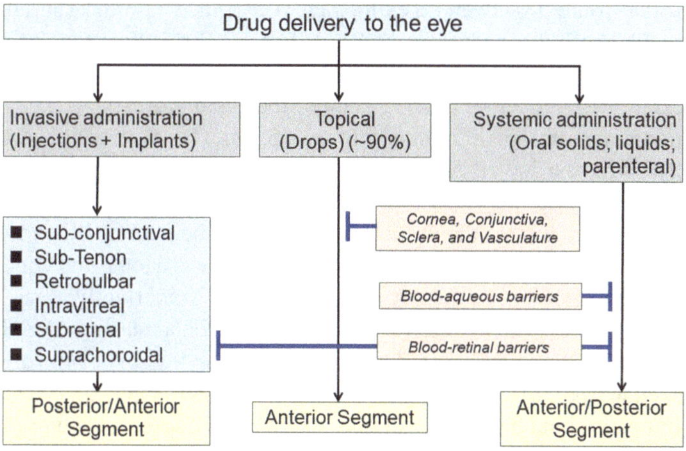

Fig. 1 Routes of drug administration to the eye. The drugs administered by the systemic route (oral or parenteral) encounter multiple barriers restricting access to the intraocular structures, leading to lower, negligible bioavailability compared to topical drugs. The blood-aqueous barrier, inner blood-retinal barrier, and outer blood-retinal barrier comprise the three barriers of entry for systemic drugs. Tight junctions of the ciliary epithelium lining the posterior chamber and the vascular endothelium of the iris vasculature confer the blood-aqueous barriers—tight junctions associated with the vascular endothelium lining the retinal capillaries from the inner blood-retinal barrier. Tight junctions of the retinal pigment epithelium confer the outer blood-retinal barrier. Due to the combination of the blood-ocular barriers and the need for local administration, the posterior segment drug delivery is typically achieved by extraocular local injections and intraocular/extraocular implants

2013; Meza-Rios et al. 2020), have been assessed with limited success. The focus of this chapter is drug delivery across the cornea, specifically drug movement across the corneal epithelium, stroma, and endothelium at a microscopic level for therapeutic benefit in the anterior chamber and surrounding tissues. This chapter describes transcorneal penetration kinetics for fluorescent drug surrogates and nanoparticles.

Ophthalmic Drug Products and Precorneal Drug Dynamics

Drug products are dosed to the eye using topical, intraocular, or systemic routes of administration (Fig. 1). Of these products, the majority are dosed topically as eye drops. The drug constituted in an eye drop has to overcome multiple static and dynamic barriers before it enters the anterior segment of the eye, a target suitable for therapeutic benefits with eye drops (Yavuz and Kompella 2017). Following systemic administration, the drug has to overcome blood-tissue barriers to enter the eye tissues (Yavuz and Kompella 2017). Invasive administration to the eye overcomes one or more barriers relative to topical administration. The subconjunctival route overcomes the conjunctiva barrier and, to an extent, rapid drainage by tears. The suprachoroidal route also overcomes the sclera barrier, while the subretinal route overcomes the choroid barrier as well (Rai Udo et al. 2015; Yiu et al. 2020). These and other invasive routes are used to treat back of the eye diseases primarily since topical dosing is inefficient in treating back of the eye diseases.

The most common topical ophthalmic dosage form, a 30 to 50 μL eye drop, is administered into the lower cul-de-sac of the eye. Upon installation, blinking in humans promotes drug mixing with tears. Tears are continuously secreted onto the eye surface and drained into the nasolacrimal duct via the puncta. Mixing with tears reduces drug concentration on the eye surface, thereby reducing the concentration gradient for transcorneal transport gradient. More importantly, tear drainage removes the drug and limits its residence time on the ocular surface (Agrahari et al. 2016; Yavuz and Kompella 2017). Tear volume and the volume of the drop can be used to estimate the dilution by blinking. In humans, without the dry eye disease, the tear volume is about 8 μL and is distributed over the cornea and conjunctiva with a tear film thickness of 2–5 μm (Agrahari et al. 2016; Yavuz and Kompella 2017; Hosaka et al. 2011; Azartash et al. 2011; King-Smith et al. 2004; Werkmeister et al. 2013). Since blinking is highly variable, the precorneal residence time of drugs shows variabilities within and between subjects, with a half-life of <4 min based on some studies (Lee et al. 1993), it is estimated that >95% of the instilled drop reaches the nasolacrimal surface, with the drug eventually entering into the systemic circulation (Lee et al. 1993). While the topical drug administration is easy and permits local delivery to the anterior segment, it is limited by pulsatile drug delivery and low intraocular bioavailability (Maurice 1980, 1993, 2002).

Transcorneal Penetration

Topically applied lipophilic drugs are believed to access the anterior chamber predominantly by penetration across the cornea (which behaves as an oil:water:oil matrix) (Figs. 2, 3 and 4). The corneal epithelium is stratified, with lipid-rich plasma membranes (Fig. 3). Moreover, the anterior squamous layers (i.e., the top two layers), referred to as the superficial epithelium, also express tight junctions (Fig. 3) (Sasaki et al. 1999). As such, the superficial epithelium is expected to be a critical permeability barrier for several topically applied hydrophilic drugs, including macromolecules and other charged species. However, a lipophilic drug can penetrate the epithelium by partitioning into the membrane lipids (Gupta et al. 2010). Therefore, lipophilicity is in part responsible for determining the bioavailability of topical drugs. After traversing the epithelium, the drug has to partition into the stroma, which is 90% of the corneal volume, and 80% hydrated (making the stroma the watery layer of the cornea). While the collagen fibrils pose a steric hindrance to the movement of macromolecules, low-molecular-weight soluble compounds diffuse freely (Edwards and Prausnitz 1998). Thus, the stroma acts as a depot and a barrier for hydrophilic and lipophilic drugs, respectively. Following the stroma, the drug has to cross the monolayer of the endothelium, which can be modeled as a thin-oily layer. The endothelium is leaky with a fewer number of tight junctional strands,

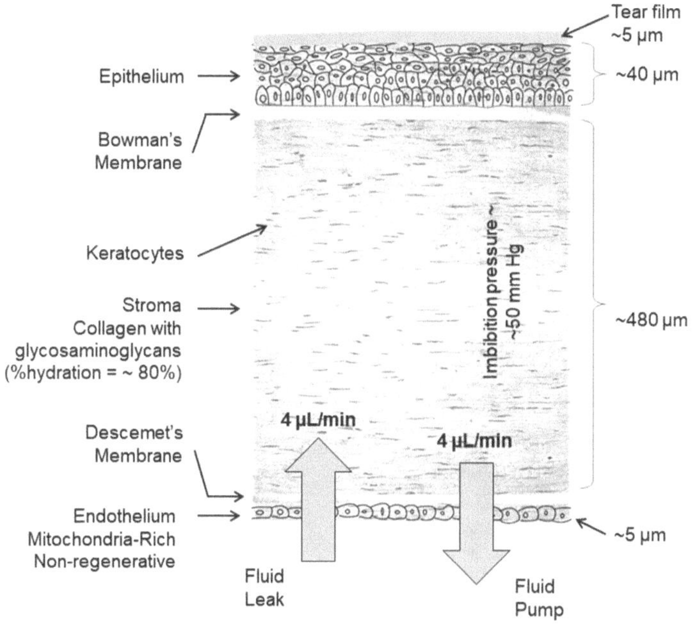

Fig. 2 Schematic of barriers for topical drugs. Histology of the cornea highlighting the potential physiological barriers for penetration of topical nanoparticles. The superficial layers (top two layers) of the corneal epithelium possess tight junctions. As such, when the corneal surface is inflamed or mechanically damaged, the nanoparticles can penetrate the epithelium only by endocytosis. The Bowman's membrane, which contains condensed collagen, next to the corneal epithelium, can offer diffusional resistance. Similar to Bowman's membrane, the collagenous matrix of the stroma can offer diffusion resistance. The potential fluid flow in the stroma may permit bulk movement in the direction of the flow. Being yet another condensed collagenous matrix, resistance to transport across the Descemet's layer would be similar to that of the Bowman's membrane. The endothelial layer, which is a monolayer and is leaky, may offer the least resistance to the transport of particles into the anterior chamber. Known for phagocytosis, the particles could be transported across the endothelium by endocytosis and exocytosis into the anterior chamber

unlike the superficial epithelium. In fact, large MW solutes have been demonstrated to cross the endothelium through the paracellular route (Maurice and Srinivas 1994). Once the drug crosses the cornea, its clearance by aqueous humor outflow is a primary determinant of its residence time in the anterior chamber (Missel 2012; Missel and Sarangapani 2019). The lens and iris may form a depot for some lipophilic drugs (Heikkinen et al. 2019; Guss et al. 1984; Kaiser and Maurice 1964). The overall kinetics of a typical drug is shown schematically in Fig. 5.

Ocular Fluorometry for Assessment of Topical Drug Kinetics

Since the cornea and intraocular structures are transparent to visible light, fluorescence spectroscopy can be used both in vivo and ex vivo to investigate drug transport (Gupta et al. 2010, 2012; Maurice and Srinivas 1992, 1994; Srinivas and

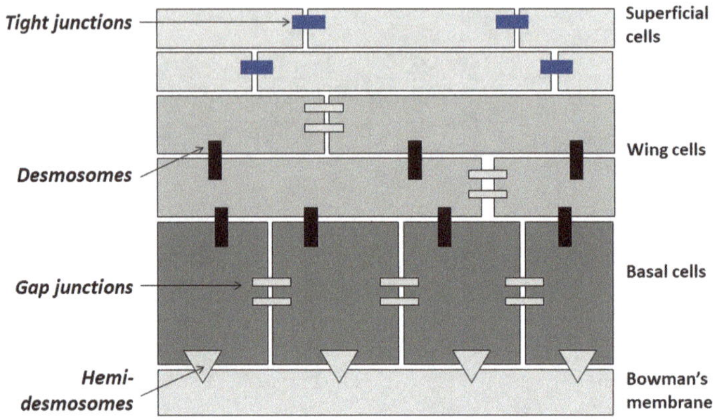

Fig. 3 Location of tight junctions in the corneal epithelium: The tight junctions are located in the squamous layers (superficial two layers in humans)

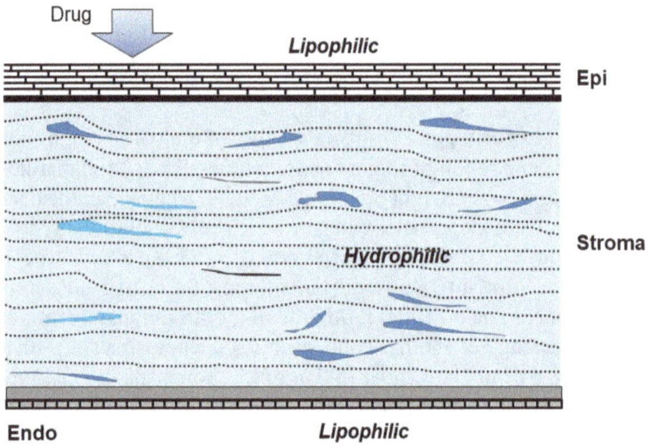

Fig. 4 Cornea as an oil:water:oil matrix. Epithelium, which is packed with cells bounded by the lipid bilayer, behaves as an oil phase. The stroma, which is 80% hydrated, forms the water phase. The lipid bilayer of the endothelium makes the monolayer behave as the second but thin oily layer. An increase in the partition coefficient of the topical drug typically improves the transcorneal permeability. However, increasing beyond an optimal value (indicated by the down arrow in the plot shown in the inset), the transcorneal permeability would decrease as the drug is unlikely to partition into the stroma or dissolved in tears. Thus, topical drug design requires optimization of the partition coefficient of the drug

Maurice 1992; Srinivas et al. 2018a, b; Niamprem et al. 2019a; McNamara et al. 1997; Chaiyasan et al. 2017, 2018). Accordingly, several types of fluorometers have been advanced for studying the eye, including ocular pharmacokinetics using fluorescent molecules as tracers or drug surrogates. Since drugs themselves are typically non-fluorescent, fluorescent dyes have been useful in examining ocular pharmacokinetics as drug surrogates (Fig. 6). Although only the dye fluorescein is approved for diagnostic use in humans, other surrogate dyes can be used in animal

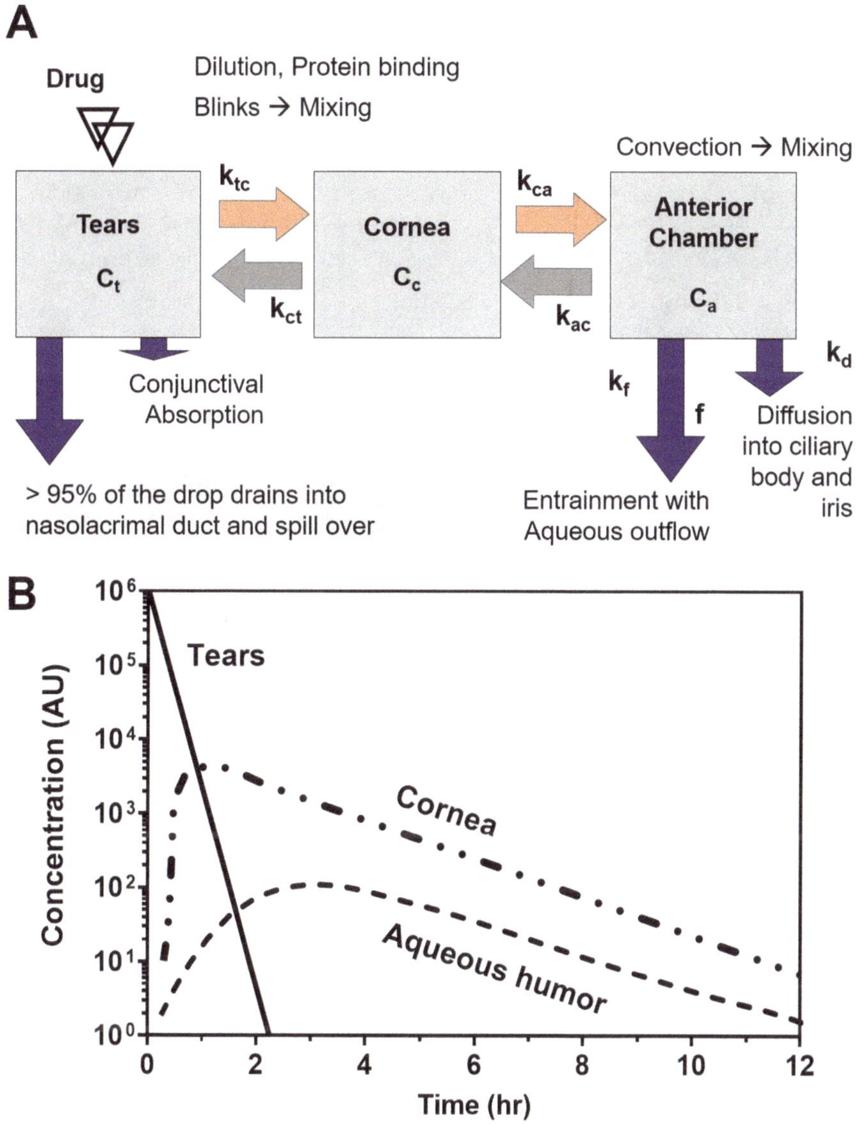

Fig. 5 Pharmacokinetics of topical drugs. (**a**) Compartmental modeling of topical drugs. Tears, cornea, and anterior chamber constitute the principle compartments. Since transcorneal measurements are typically not known for drugs, the heterogeneity of the tissue is disregarded. (**b**) Typical transient profiles of drug concentration in the tears, cornea, and aqueous humor

Carboxyfluorescein	Fluorescein*	Rhodamine B
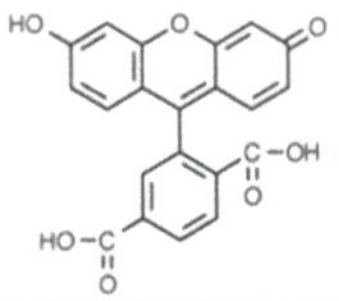	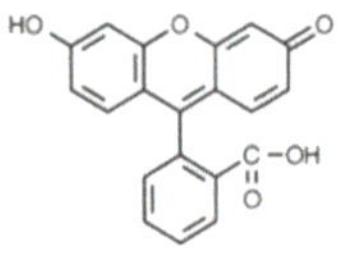	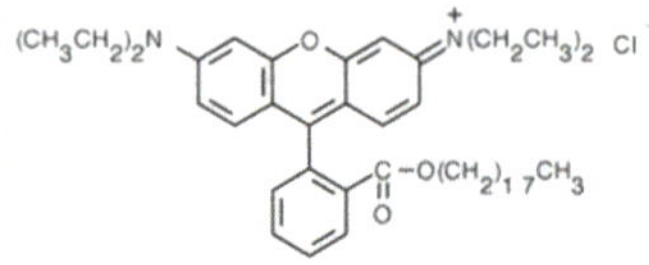
$\lambda_{ex} = 488$ nm $\lambda_{em} = 525$ nm	$\lambda_{ex} = 488$ nm $\lambda_{em} = 525$ nm	$\lambda_{ex} = 560$ nm $\lambda_{em} = 585$ nm
PC =0.0007	PC =0.61	PC = 274
MW = 520 Da	MW =332 Da	MW = 479 Da

Fig. 6 Some fluorescent dyes employed as drug surrogates

models. Over the years, we have advanced two instruments, viz., spot fluorometer (Srinivas et al. 2018a) and confocal scanning microfluorometer (CSMF) for in vivo and ex vivo applications (Gupta et al. 2010, 2012; Srinivas and Maurice 1992; Srinivas et al. 2018a, b; Niamprem et al. 2019a, b; Chaiyasan et al. 2013, 2015, 2017, 2018), respectively.

Spot Fluorometer

The spot fluorometer has been used to assess the efficacy of different vehicles used in topical drug delivery and to characterize the permeability of cellular barriers of the cornea. For example, Maurice and Srinivas employed fluorescein to assess the efficacy of a cation-sensitive polysaccharide (Gelrite™) (Maurice and Srinivas 1992), which gels on the ocular surface and thereby enhances topical drug bioavailability (e.g., timolol) (Laurence et al. 1993; Nelson et al. 1996). A small quantity of fluorescein was mixed with the polysaccharide and administered as drops in the lower cul-de-sac. The fluorescence measurements in the tear film started immediately after the drop from the ocular surface using a spot fluorometer that has been described in Brubaker et al. (Brubaker et al. 1990). Next, we estimated the area under the curve (AUC) of the resulting fluorescence decay as an index of total corneal drug exposure over time. A similar assessment with saline vehicle led us to calculate the enhancement ratio (ER), which can be defined as ER = $[AUC_{Gel}/AUC_{Saline}]$, where AUC_{Gel} is AUC obtained with fluorescein in Gelrite while AUC_{Saline} is AUC obtained with fluorescein in saline. We validated the significance of ER to use as an index of the efficacy of Gelrite in enhancing drug delivery to the anterior chamber by determining the ratio of fluorescein in the anterior chamber with and without Gelrite™, each of which was measured by the fluorometer 2 h after

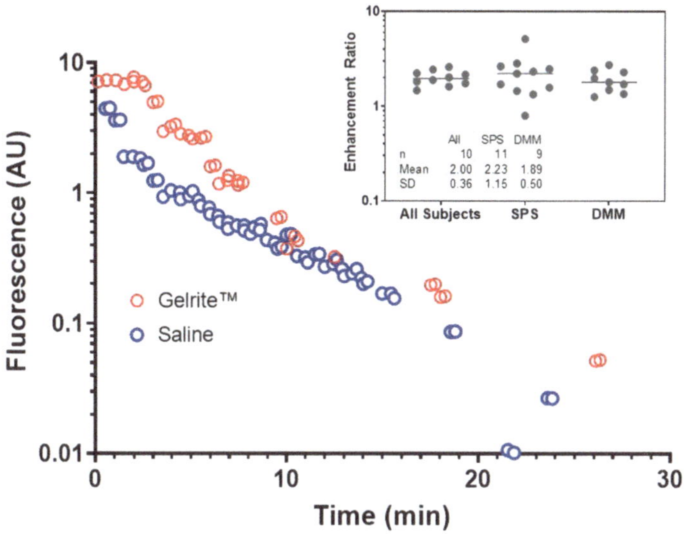

Fig. 7 Enhanced drug delivery with Gelrite™: Typical long-term time course of fluorescence in the precorneal tear film after instillation of 25 μL of the buffer. The eyes were washed after the recording, and the background values were subtracted from the measurements. Inset shows a plot of enhancement ratio for two subjects (SPS and DMM; multiple trials) and other subjects (single trials). Each point represents the ratio of $AUC_{Gelrite}$ to AUC_{Saline}; AUCs were estimated from fluorescence decay curves defined by red (Gelrite) and blue circles (saline). (Redrawn from Maurice and Srinivas (1992))

administration of drops (Fig. 7). In another application, we used the same fluorometer to study the increase in bioavailability of fluorescein administered with biodegradable collagen shields (Srinivas 1994). In particular, we compared fluorescein levels in the anterior chamber after wearing fluorescein-loaded shields with dye administered as multiple drops. The results demonstrated the superior efficacy of collagen shield to deliver hydrophilic fluorescein to the ocular surface and enhanced delivery to the anterior chamber by fivefold (Fig. 8).

Recently, we modified the spot fluorometer for enhanced fluorescence measurements from any focal spot of interest on the ocular surface or the cornea and anterior chamber (Srinivas et al. 2018a). A schematic of the modified fluorometer is given in Fig. 9a. The illumination assembly of the slit lamp constitutes the excitation arm of the fluorometer. To enable synchronous detection, we replaced the halogen lamp originally in the slit lamp with a high-power LED (e.g., Cree Inc.; XML-P2; 10 Watts). The output intensity of the LED was modulated as a sine wave (200 Hz–1 MHz), and a collimated beam was directed to the eye via the slit assembly of the slit lamp. A linear power amplifier (DC −1 MHz; 20 Watts), receiving sine wave input from a function generator (SD345; Stanford Research Systems), was used to modulate the LED output. For fluorescein excitation, the LED output was passed through a blue filter (490 ± 10 nm). The emission through the collection slit at the camera port (positioned at the conjugate plane of the illumination slit) was

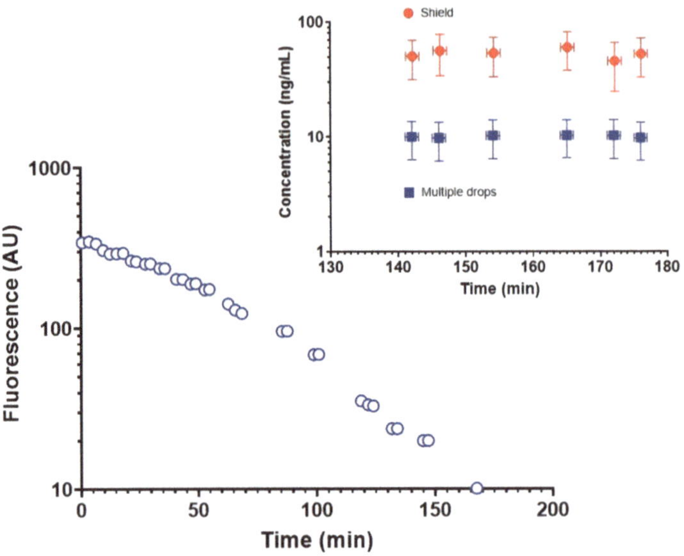

Fig. 8 Enhanced drug delivery with collagen shields: Collagen shields (120 μm thickness) soaked in 0.25% for 20 min were placed on the ocular surface like soft contact lenses, and fluorescence in the shield was measured. In 2 h, the fluorescence dropped by ~2 decades, much slower than topical drops. The decay was typically biphasic, with the first phase being slower. Inset: Concentration of fluorescein in the anterior chamber measured in one subject ~2 h after administration of fluorescein on the ocular surface via shield or 4 drops (0.25%). The levels remained steady for 40–60 min. Thus, fluorescein bioavailability in this trial was 5× that obtained with 4 drops of fluorescein. The increase in bioavailability was 3.8× (SD = 1.5; $n = 9$). (Redrawn from Srinivas (1994))

passed through an interference filter (530 ± 10 nm) and led to a photomultiplier tube (R928, Hamamatsu). The outputs of the photomultiplier and photodetector along the excitation beam were fed to a lock-in amplifier (MFLI; Zurich Instruments) to the "input signal" and "reference signal," respectively. The fluorescence from a focused spot of interest was acquired by a trigger generated through a hand-held push-button switch. The new modifications led to enhanced depth resolution of ~200 μm while measuring fluorescein at 10 nM and a signal-to-noise ratio > 40. The application of lock-in amplification blocked corruption of the fluorescence signal of interest by the ambient light and electronic noise.

Overall, the new spot fluorometer is easy to use, as it is built around a typical slit lamp. In addition, the instrument offers high sensitivity, fast sampling, and increased depth resolution. We have employed the instrument already for clinical measurements in the anterior segment, including endothelial and epithelial barrier integrity and aqueous flare measurements in uveitis patients (Srinivas et al. 2018a; Sudhir et al. 2018). In Figs. 9b and 10, we illustrate the use of the spot fluorometer in highlighting the variability in ocular surface residence time following topical drops in healthy volunteers (Srinivas et al. 2018a). The measurements were required to assess epithelial permeability to fluorescein, which is often used as an index of

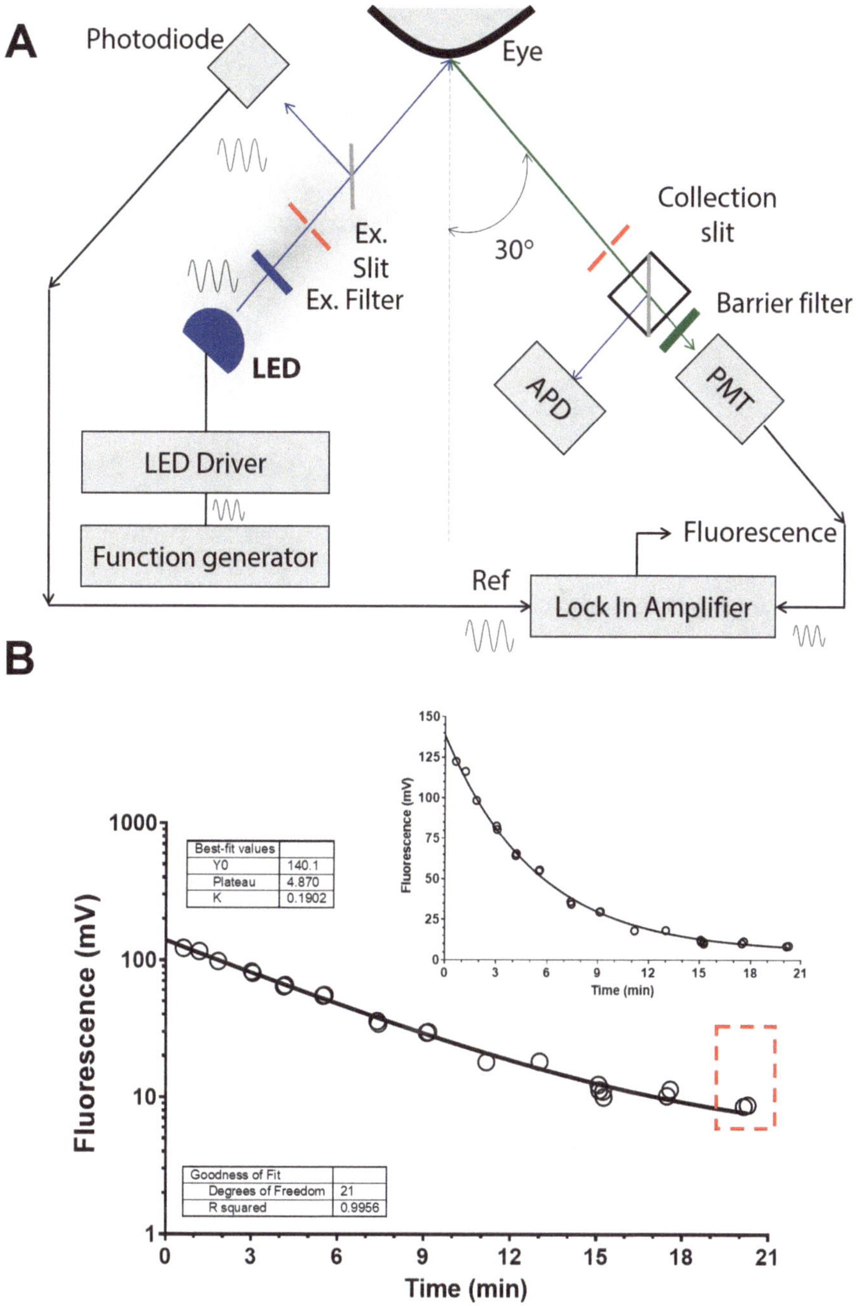

Fig. 9 New spot fluorometer. (**a**) Schematic of the optical and electronic assembly of the spot fluorometer. (**b**) Typical fluorescence decay after topical instillation of a micro drop of fluorescein into the cul-de-sac in a healthy subject. Inset shows the mono-exponential decay after administration of the drop. The decay rate constant, given by the slope of the semi-log plot, is given by $k_d = 0.19$/min

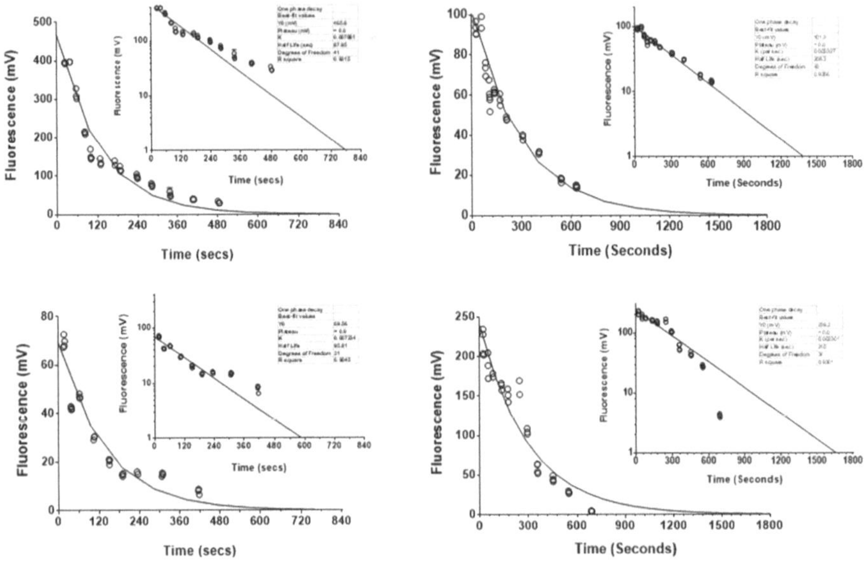

Fig. 10 Variability in the elimination rate constant (k_d). (Redrawn from Srinivas et al. (2018a))

inflammation/damage to the ocular surface. In these studies, we instilled a 0.5% fluorescein drop (2 μL in Fig. 9b) in the lower cul-de-sac, followed by measurement of fluorescence from the corneal surface at periodic intervals. Exponential decay of the tear fluorescence is plotted on a semi-log scale in Fig. 9b (inset of Fig. 9b). The decay rate constant was estimated by non-linear least square analysis (GraphPad™ software). Figure 10 shows within- and between-subject variability in the decay rate constant, highlighting the potential for variability in drug bioavailability of topical drugs in the anterior chamber.

While the spot fluorometer is useful at a macroscopic level in vivo, we developed a microscopic equivalent of the spot fluorometer, referred to as the confocal scanning microfluorometer (CSMF), to assess the kinetics of diffusion of fluorescent molecular across the cornea (Srinivas and Maurice 1992).

Confocal Scanning Microfluorometer

Topical drug pharmacokinetics has been frequently described by compartmental models (Amrite et al. 2008; Friedrich et al. 1993; McLaren et al. 1993; Ranta et al. 2003). These models assume that the drug is uniformly distributed throughout the cornea after topical administration. In other words, the compartmental models do not account for the heterogeneity of the cornea and the diffusive nature of transport in each of its three main layers. Accordingly, several attempts have been advanced to depict the multi-laminate structure in the models for pharmacokinetics of topical drugs (Avtar and Tandon 2008; Yamamura et al. 1999; Zhang et al. 2004). However, these studies are based on theoretical considerations and were not derived out of measurements of transcorneal concentration profiles of topical drugs. In order to establish more precise pharmacokinetic relationships based on physiologically relevant models, Srinivas and Maurice (Srinivas and Maurice 1992) developed a confocal scanning microfluorometer (CSMF) to characterize the time-dependent kinetics of transport of fluorescent molecules. The instrument has been described previously (Srinivas and Maurice 1992), but a schematic is redrawn in Fig. 11. Briefly, CSMF was designed exclusively for determining depth-resolved fluorescence measurements across the cornea. The output of an LED (10 W) filtered through appropriate interference filters was used to register transcorneal fluorescence of fluorescein, carboxyfluorescein, rhodamine B (RhB), and FITC dextrans. The emission collected through a collection slit, which was in the conjugate plane of the excitation slit, was filtered and led to a photomultiplier. The output of the photomultiplier was amplified by a lock-in amplifier. The synchronous detection was made possible by modulating the LED output intensity by a power amplifier as with the spot fluorometer described above. Transcorneal fluorescence profiles were obtained by depth scanning (~ 20 μm/s). The instrument has been applied so far to assess the diffusion of fluorescein (Srinivas and Maurice 1992; Gupta et al. 2012), RhB (Gupta et al. 2010), sulforhodamine B, FITC dextrans (Maurice and Srinivas 1994), and fluorescent nanoparticles across the cornea (Niamprem et al. 2019a; Chaiyasan et al. 2013, 2015, 2017). As noted earlier, the fluorescent dyes of different lipophilicities served as drug surrogates. The instrument's depth resolution is ~7 μm with a 40× objective (working distance = 1.2 mm; NA = 0.75; water immersion; Zeiss) and the LED light source. In the following, we provide transient transcorneal profiles of the dyes and nanoparticles, with a view to highlighting transcorneal kinetics of topical drugs.

Penetration of Rhodamine B (as a Lipophilic Fluorescent Drug Surrogate)

We envision the overall transcorneal penetration of drugs to consist of several serial and parallel processes involving drug binding, drug partitioning, and diffusion in tissues/cells. To demonstrate these principles at a microscopic scale, we established concentration vs. depth profiles of the fluorescent dye rhodamine B (RhB) as a surrogate of lipophilic drugs (Gupta et al. 2010). Figure 12 shows a family of transient profiles of RhB after exposure at corneal surface over an extended period. The transcorneal profiles point to the transport of RhB with characteristic jumps at the interfacial boundaries between epithelium-stroma and stroma-endothelium. In addition, we can observe the non-uniformity of fluorescence across the cornea, with elevated fluorescence in the lipophilic epithelium and endothelium compared to the hydrophilic stroma. Taken together, it is evident that RhB sequesters preferentially into

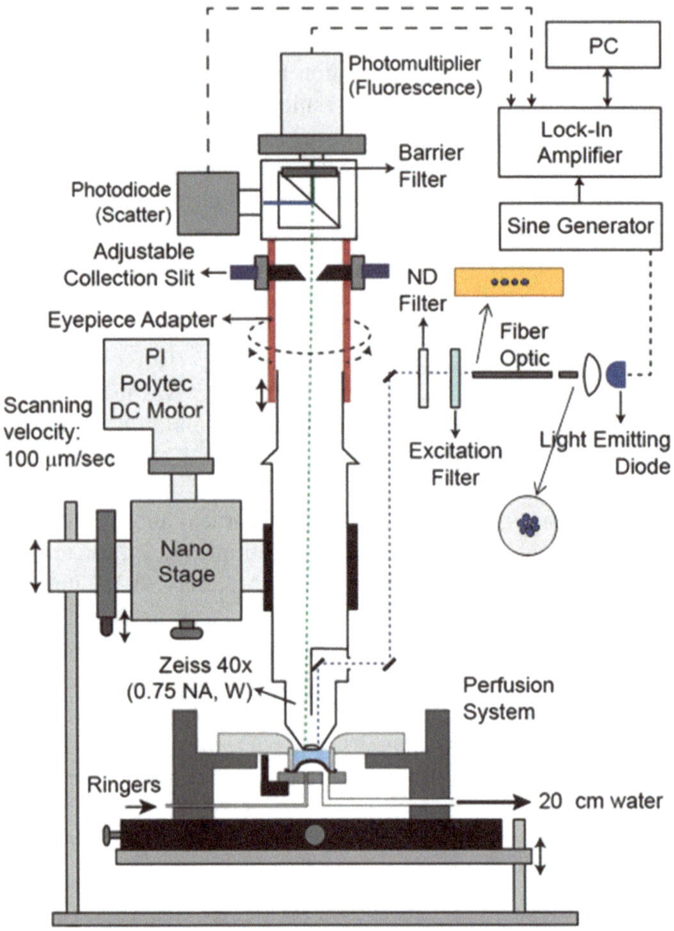

Fig. 11 Optical, mechanical, and electronic organization of the confocal scanning microfluorometer (CSMF). Excitation light was derived from a white LED (10 W; XHP-70 Cree). The output of the LED was focused on the circular end of a fiber optic consisting of 16 fibers (250 μm diameter). The linear end of the fiber optic was employed as the excitation slit. The emission collected through a confocal slit is split by a dichroic mirror to detect the fluorescence and scattered using a photomultiplier tube (PMTs; R928HA) and a photodiode, respectively. The corresponding outputs are amplified by two independent lock-in amplifiers (SR830; Stanford Research Systems, CA, USA; only one lock-in is shown for simplicity). The reference inputs of the lock-ins were coupled to the sync signal of the sine generator that was employed for the modulation of the LED (10 kHz). We carried out depth scans across the cornea held under the objective (40×, 0.75 NA; Zeiss; Water immersion type). The cornea was maintained at a steady thickness for 3–4 h by continuous perfusion of the tissue with a bicarbonate-rich medium at 37 °C. (Redrawn from Srinivas and Maurice (1992))

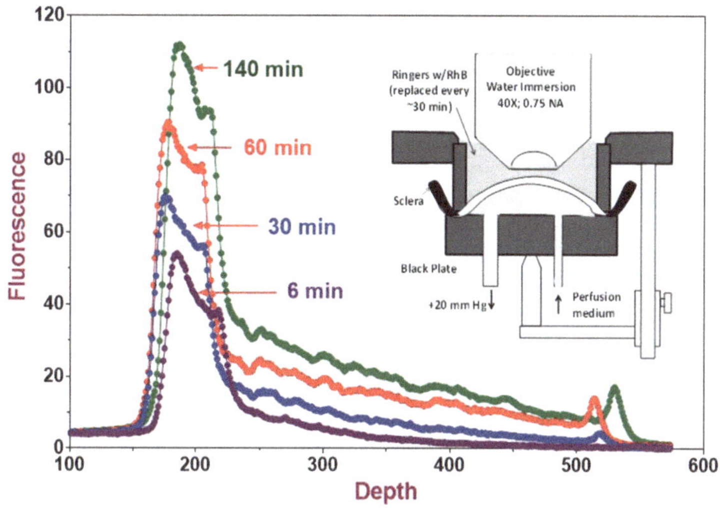

Fig. 12 Transcorneal kinetics of rhodamine B (RhB). The profiles show the kinetics of lipophilic RhB (MW = 479 Da, octanol/water partition coefficient = 274) across rabbit cornea (mounted ex vivo) in response to a constant tear side concentration. The dye was excited at 530 ± 10 nm, while the emission was collected at 585 ± 10 nm. Additional details are provided in Gupta et al. (2010). (Redrawn from Gupta et al. (2010))

lipophilic structures across the cornea. Spatial and temporal fluorescence gradients, apparent in the epithelium and stroma, indicate diffusional resistance for RhB transport. The non-uniform RhB concentrations in the epithelium and stroma support the idea that neither is a well-mixed homogenous compartment, highlighting the difficulty in describing RhB concentration gradients with conventional compartmental models. Further, the fluorescence profiles in Fig. 12 also show that fluorescence of RhB at the epithelial surface and in the endothelium continue to increase over time, with fluorescence at the endothelium beginning to increase after 30 min. These findings suggest a significant accumulation of RhB in the epithelium and endothelium compared to that in the stroma, and moreover, the RhB in the epithelium and stroma has no uniform concentrations across their thickness as assumed in the compartmental modeling. We have modeled the transient RhB profiles in Fig. 12 phenomenologically using a diffusive transport model (Gupta et al. 2010). It demonstrates a microscopic approach to correlate RhB's physicochemical properties to its transport properties across the cornea. We believe that similar analysis with other dyes of varying partition coefficients, for example, will characterize the effect of lipophilicity on transcorneal kinetics. Thus, lipophilicity, which can be modified using rational drug design, can be parameterized into transcorneal kinetics.

Penetration of Sulforhodamine B and Fluorescein (as a Hydrophilic Fluorescent Drug Surrogates)

In contrast to RhB, sulforhodamine B (SRB) and fluorescein are relatively hydrophilic (Fig. 13), and their permeability can be a marker of the breakdown of tight junctions (Chaiyasan et al. 2018). Thus, mouse corneas exposed to potentially toxic substances showed increased accumulation of SRB (Maurice and Singh 1986). Moreover, unlike that of fluorescein and carboxyfluorescein, the fluorescence of SRB is pH insensitive (Schulz et al. 2009); therefore, it serves as a superior dye for the assessment of barrier integrity of epithelial or endothelial layers (Araie 1986). In a series of experiments, we employed CSMF to assess the properties of SRB (Figs. 13 and 14a) (Chaiyasan et al. 2018). As expected, SRB fluorescence was negligible in the corneal epithelium and stroma even after 12 hrs. of exposure to SRB at the ocular surface (Fig. 13a). Conversely, when the epithelium was removed, SRB rapidly partitioned into the stroma within 30 min (Chaiyasan et al. 2018). Likewise, when epithelium disrupted by exposure to 0.5% Tween 20, a nonionic detergent, topical SRB led to significant accumulation in the stroma (Fig. 13b). Similar to Tween 20, penetration of SRB also increased after microneedle injury to the corneal epithelium (Chaiyasan et al. 2018). The findings from these experiments were reconfirmed by exposure to SRB to the endothelial surface. As shown in Fig. 13a, we injected SRB into the anterior chamber and followed its transport across the corneal for up to 100 min. The SRB fluorescence across the stroma increased in a time-dependent manner indicating significant permeability of the dye across the corneal endothelium and consequent accumulation in the stroma. Despite the accumulation of the dye in the stroma to very high levels, it did not subsequently penetrate into the epithelium. We show similar experiments with fluorescein in Fig. 14b (Chaiyasan et al. 2018). Only high levels of fluorescein in the anterior stroma over a long duration led to a slight accumulation of the dye in the epithelium (Srinivas and Maurice 1992; Gupta et al. 2012). Overall, drugs/solutes of lipophilicity close to those of SRB or fluorescein can be expected to show negligible bioavailability in the anterior chamber. We have also employed strategies to overcome the challenges of delivering hydrophilic solutes by the topical route. For example, we could employ CSMF to clearly demonstrate iontophoretic delivery of hydrophilic dyes such as fluorescein and riboflavin. Example data for delivery of fluorescein through topical anionic iontophoresis is shown in Fig. 15. Clearly, very large amounts of dye could be delivered by short-term iontophoresis.

Penetration of Nanoparticles

A set of physiological barriers (i.e., tight junctions and fluid flows) oppose the penetration of nanoparticles across the cornea (Diebold and Calonge 2010). Tear and aqueous humor dynamics on the ocular surface and anterior chamber, respectively,

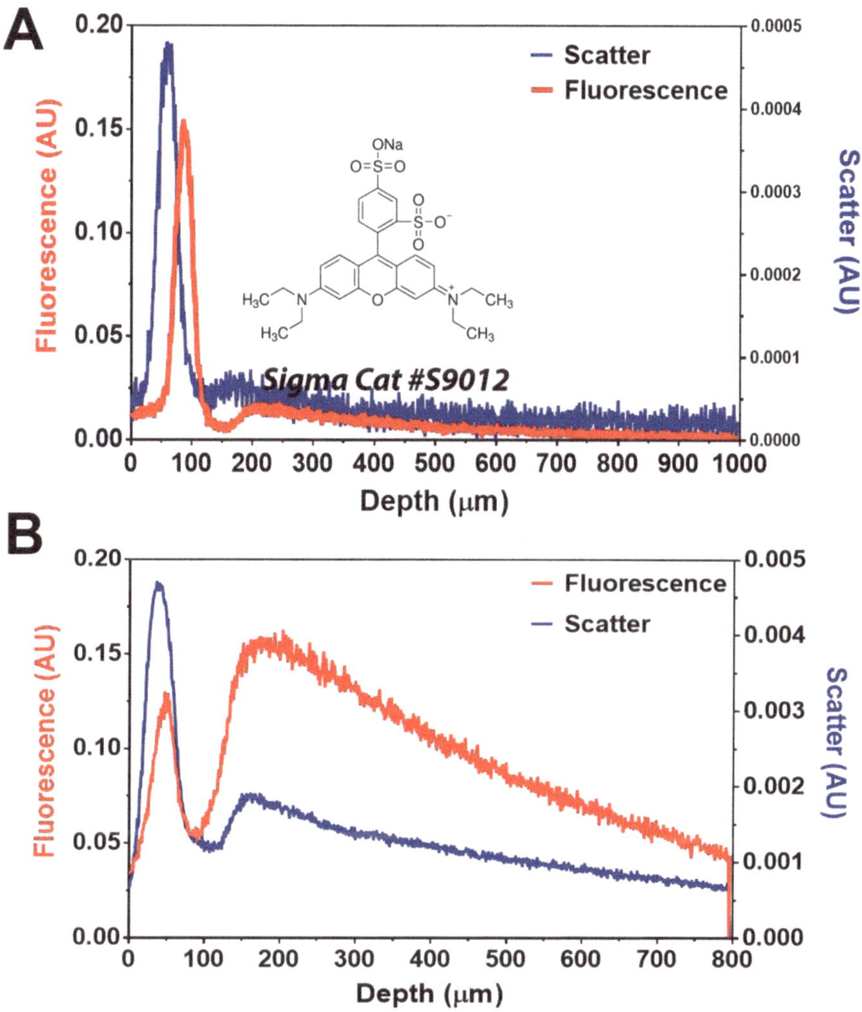

Fig. 13 Transcorneal kinetics of topical sulforhodamine B (SRB). (**a**) Prolonged exposure of SRB (a Na^+ Salt; 12 h) did not result in any significant accumulation of the dye in the epithelium/stroma. The fluorescence peak is thus attributed to dye adherent on the epithelial surface. (**b**) When SRB dissolved in PBS containing Tween 20 (0.5%) was exposed to the corneal surface, a significant accumulation of the dye in the stroma is noticed after 12 h. Similar observations were noticed with fluorescein (data not shown). The experiments were performed with the porcine cornea, ex vivo. Since the working distance of the objective was 1 mm, the cornea was scanned only for the top 800 μm. The results shown are similar to 6 other independent experiments. (Redrawn from Chaiyasan et al. (2018))

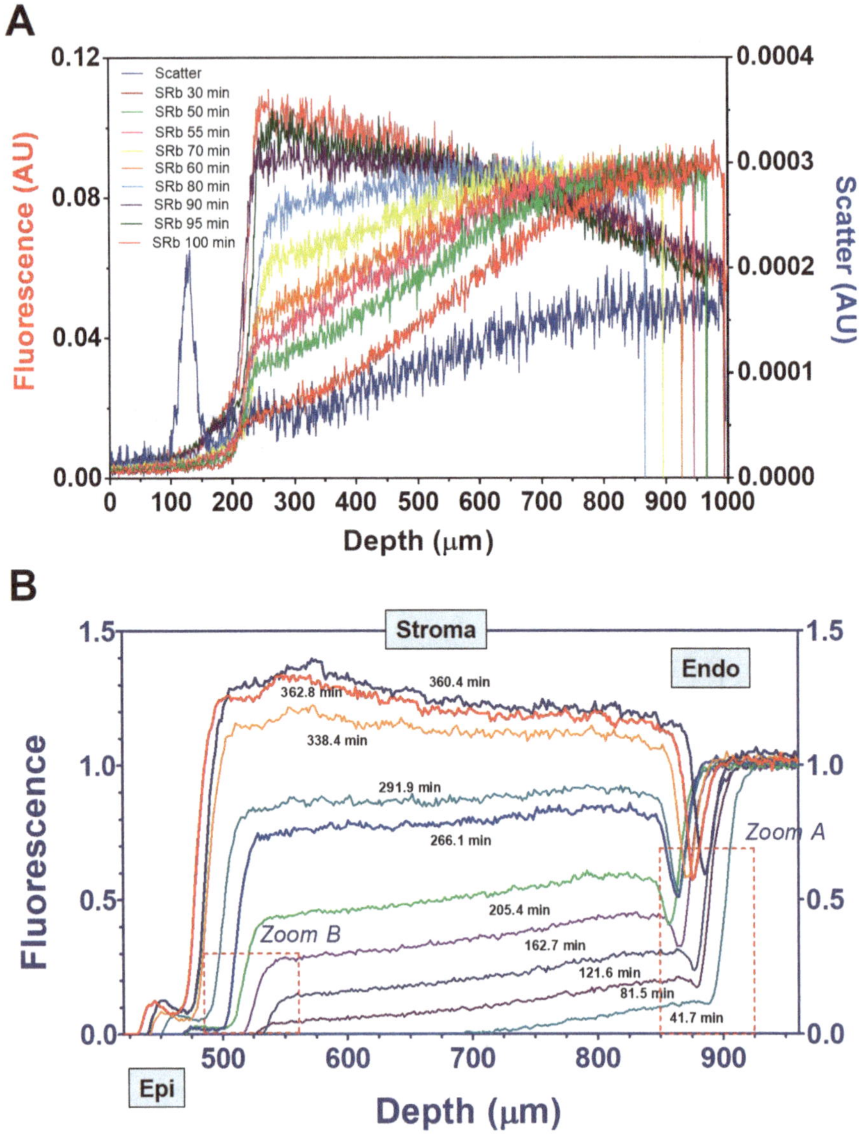

Fig. 14 Penetration of SRB and fluorescein from the anterior chamber. (**a**) Transient fluorescence profiles of SRB. This experiment was performed with the porcine cornea, ex vivo. SRB (0.1%) was injected into the anterior chamber. Subsequently, transcorneal fluorescence and scatter profiles were obtained periodically for 100 min. SRB, being highly hydrophilic, does not show any accumulation in the epithelium (corresponding to blue scatter peak). The results shown are similar to 6 other independent experiments. Redrawn from Chaiyasan et al. (2018). (**b**) Transient fluorescence profiles of fluorescein across the rabbit cornea after the endothelial side was exposed to a fixed concentration of the dye. SRB, being highly hydrophilic, does not show any accumulation in the epithelium (corresponding to blue scatter peak). Fluorescein, which is hydrophilic, shows very little accumulation in the epithelium after 6 h. (Redrawn from Gupta et al. (2012))

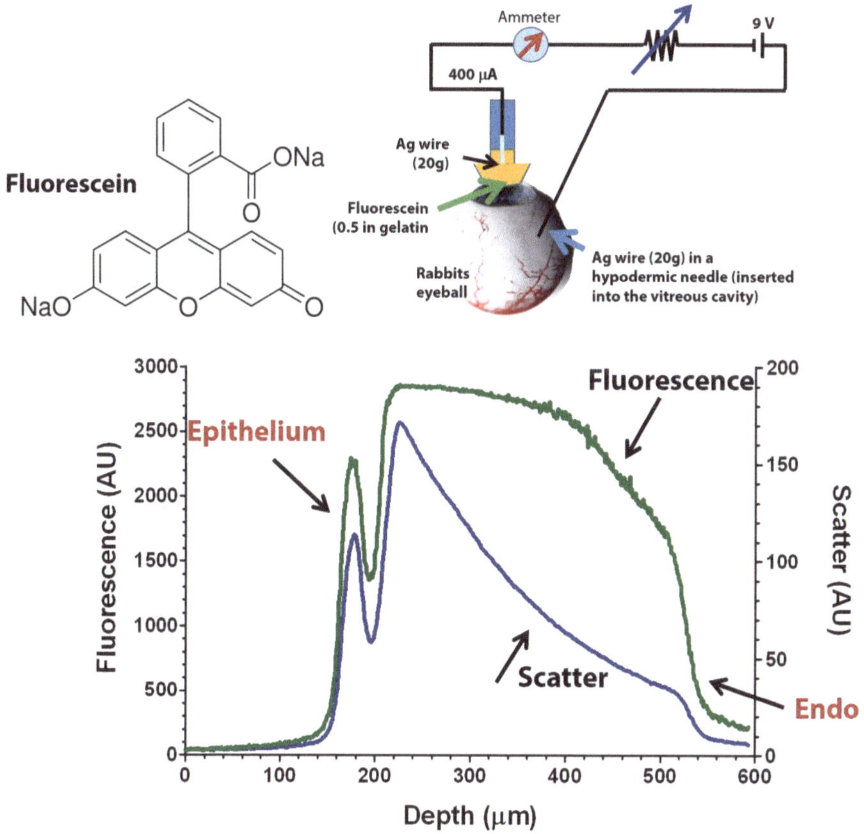

Fig. 15 Enhanced transcorneal fluorescein delivery by iontophoresis. Na-fluorescein iontophoresis across rabbit cornea with intact epithelium is shown. The fluorescence and scatter scans were carried out after 5–10 min after iontophoresis. Insets of the hydrophilic structure of fluorescein (left) and electrical circuit employed to induce anionic iontophoresis. The resistance was adjusted to obtain a current flow of 400 μA. The iontophoresis was carried out for 20 s. The dye was dissolved in gelatin gel, which was prepared with deionized water and held in a micropipette tip

create conditions for short residence time for nanoparticles, akin to topical drugs. To highlight these characteristics, we started investigating the nanoparticle drug delivery to the cornea and anterior chamber at a microscopic level (Srinivas et al. 2018b; Niamprem et al. 2019a, b; Chaiyasan et al. 2013, 2015, 2017). We employed fluorescent-dyed nanoparticles for real-time observations after topical exposure of the particles in suspension. We first assessed the penetration dynamics of monodispersed silica nanoparticles stained with RhB or labeled with FITC using the CSMF (Srinivas et al. 2018b).

Data in Fig. 16 show typical fluorescence profiles regarding penetration of RhB-stained silica nanoparticles. We used nanoparticles soaked in RhB for 24 h. The particles were separated by centrifugation and repeatedly washed with PBS. After

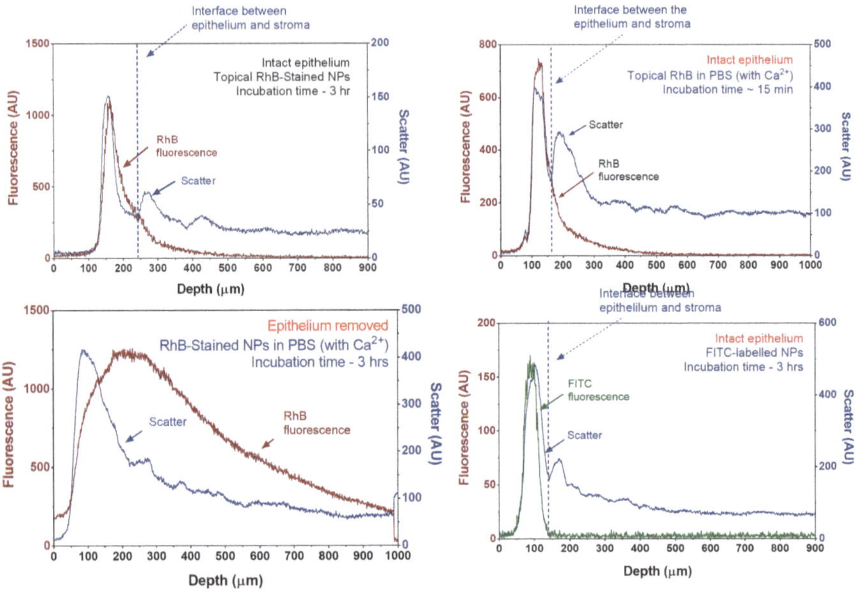

Fig. 16 Topical administration of silica nanoparticles. (**a**) The epithelial surface was exposed to a suspension of RhB-stained silica nanoparticles (0.1 mg/mL) at RT for 3 h and then scanned for fluorescence and scattered profiles. Accumulation of RhB in the epithelium and in the anterior stroma is evident. Vertical dashed line demarcates the interfacial region between the epithelium and the stroma, as noted by sharp changes in the scatter intensity. (**b**) Fluorescence and scatter profiles after exposure of the epithelial surface to a solution of RhB (0.1 mg/mL) for 15 min. Rapid accumulation of the dye in the epithelium and in the anterior stroma is comparable to that in Panel **a**. (**c**) RhB-stained silica nanoparticles in suspension were administered on the bare stroma at RT for 3 h. The significant fluorescence across the stroma is evident in the penetration of the RhB nanoparticles into the stroma. (**d**) Experiment similar to Panel C but with FITC-labeled silica nanoparticles exposed to the corneal surface for 3 h at RT. The absence of FITC fluorescence in the stroma demonstrates that nanoparticles did not cross the epithelium into the stroma. (Redrawn from Srinivas et al. (2018b))

the final wash, supernatant RhB fluorescence was negligible, but the particles remained fluorescent. The corneal surface was exposed to these RhB-stained nanoparticles at room temperature for 3 h before a series of fluorescence/scatter scans. We observed that RhB fluorescence from the epithelium and the anterior stroma significantly increased after 3 h (Fig. 16a). The increase could be due to penetration of RhB-stained silica nanoparticles and/or penetration of RhB released from the nanoparticles. As discussed earlier, RhB is a lipophilic dye (log $P = 2.43$) capable of penetrating and sequestering in the epithelium (Fig. 16b) following topical application (Fig. 12) (Srinivas et al. 2018b).

To validate findings in Fig. 16a, repeat the experiments with silica nanoparticles that were covalently linked to FITC (Srinivas et al. 2018b). Free FITC was washed off repeatedly with PBS until the eluent contained negligible dye. As shown in Fig. 16d, topical FITC-silica nanoparticles for 3 h led to accumulation in the epithelium but not in the stroma. Also, irrigation of PBS on the epithelial surface did not decrease the measured fluorescence suggesting that the particles were possibly endocytosed. However, the sudden decline in fluorescence at the epithelial and stromal interface confirms findings in Fig. 16b. In particular, the particles did not cross the epithelium. The nanoparticles may have been hindered by the Bowman's membrane since there is fluorescence in the stroma that could be expected by exocytosis from the basal layer of the epithelium. When RhB-stained and FITC-labeled silica nanoparticles were instilled on the bare stroma, we observed significant penetration after 3 h (Fig. 16c for RhB-stained nanoparticles), partly by entrainment with potential water influx into the stroma. In short, we have demonstrated that the silica nanoparticles of ~7 nm did not escape the corneal epithelium into the corneal stroma even after prolonged exposure. However, the endocytosis of the nanoparticles across the different layers of the epithelium is evident (Srinivas et al. 2018b). Prior studies have suggested endocytosis-based corneal epithelial entry of nanoparticles conjugated with peptide ligands for cell surface receptors (Kompella et al. 2006).

Modeling of Pharmacokinetics of Topical Lipophilic Drugs

Conventional compartmental models treat the cornea as a single, well-stirred compartment based on the average drug concentration in the entire cornea. At the next level of refinement, such models can be extended to represent the cornea as a multilayered structure while treating each layer as homogeneous with unique diffusivity and partition coefficients. Additional enhancements can be obtained by assigning heterogeneity to the layers (i.e., each layer is not assumed to be a well-stirred compartment). Importantly, model refinement is limited by the type of experimental data that are available. Robust identification of model parameters requires transcorneal concentration profiles. The data in Fig. 12 describe the transcorneal concentration profiles (with a resolution <8 μm) of rhodamine B (RhB), which we consider here as a small MW lipophilic drug surrogate. The data, for the first time, permitted the inclusion of the basic mechanisms of drug transport into transient pharmacokinetics of topical drugs (Gupta et al. 2010). The previous structure permeability type of analyses attempted to correlate the overall drug corneal permeability to key physicochemical properties such as partition coefficient (Friedrich et al. 1993; Ranta et al. 2003; Yoshida and Topliss 1996; Wu et al. 1993; Worth and Cronin 2000; Schoenwald and Ward 1978; Schoenwald and Huang 1983; Mitra and Mikkelson 1988; Kidron et al. 2010; Shirasaki 2008; Edward and Prausnitz 2001; Conroy and Maren 1999; Ashton et al. 1992). For example, Worth and Kronin and Kidron et al. have summarized datasets (summarized in Fig. 17) that highlight the partition coefficient's role as the major determinant of transcorneal permeability (Worth and Cronin 2000; Kidron et al. 2010). In the following, we describe a general unsteady-state model for a lipophilic solute administered on the ocular surface as one or more drops separated by a finite time (5 min or 6 h).

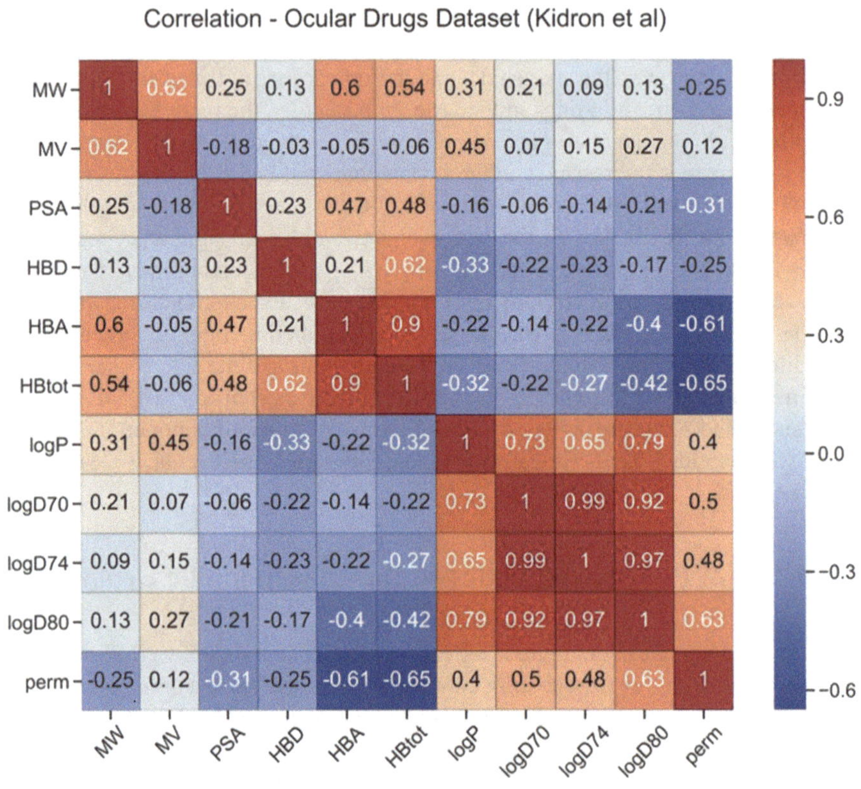

Fig. 17 Correlation matrix heatmap showing interdependence of physicochemical properties of topical drugs to their respective transcorneal permeability. Positive correlations are depicted in red (warm colors), while negative correlations are depicted in blue (cool colors). Partition coefficient (Log *P*) and distribution coefficient (log *D*) (of different pH values) have a positive correlation with permeability. Molecular weight (MW), polar surface area (PSA), and the number of putative hydrogen bonds (HBtot) are negatively correlated with permeability. Moreover, the molecular volume (MV) does not have any correlation with permeability. Abbreviations: *MW* molecular weight, *MV* molecular volume, *PSA* polar surface area, *HBA* no. of hydrogen bond acceptors, *HBD* no. of hydrogen bond donors, *HBtot* HBD + HBA, *log P* partition coefficient; log *D* 7.0 = distribution coefficient (at pH = 7); log *D* 7.4 = distribution coefficient (at pH = 7.4); log *D* 8.0 = distribution coefficient (at pH = 8.0); perm = permeability coefficient. All the plots are based on a dataset of 58 ocular drugs described by Kidron et al. (2010)

In the model, we consider the cornea to be composed of three layers, namely, the epithelium (layer 1), the stroma (layer 2), and the endothelium (layer 3). In each layer, the drug of interest is described by its concentration C. In layers 1 and 3, the drug can be found in unbounded and bounded forms, which will be denoted as C_1 and C_1^b in layer 1 and C_3 and C_3^b in layer 3. In layer 2, the drug is assumed to be found only in the unbounded form C_2. Denoting by y the direction along with the

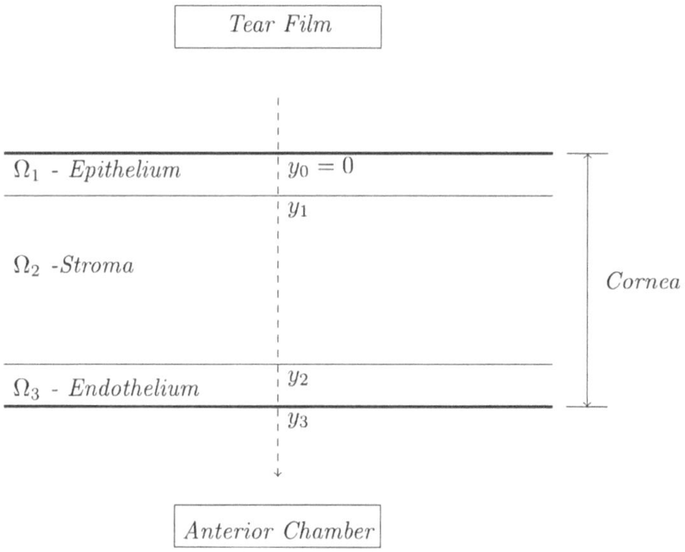

Fig. 18 Schematic of the one-dimensional domain for transcorneal transport of the drug

thickness of the cornea and by L_1, L_2, and L_3 the thickness of each layer, we can describe the three layers as the intervals $\Omega_1 = (y_0, y_1)$, $\Omega_2 = (y_1, y_2)$, and $\Omega_3 = (y_2, y_3)$, with $y_0 = 0$, $y_1 = L_1$, $y_2 = L_1 + L_2$, and $y_3 = L_1 + L_2 + L_3$ as indicated in Fig. 18.

As noted earlier, immediately after topical administration, all drugs undergo clearance from the ocular surface into the nasolacrimal duct along with tears. During its presence on the ocular surface, the drug penetration across the cornea can be envisioned as follows (Gupta et al. 2010, 2012). The lipophilic drug separates into the lipid bilayers of the superficial corneal epithelial plasma membranes in contact with tears (Step 1, Fig. 19). Then, a fraction of the drug in the epithelial membrane partitions into the hydrophilic cytoplasm, determined by the partition coefficient (Step 2). From the cytoplasm, the drug may partition into lipid membranes of the intracellular organelles (e.g., endoplasmic reticulum) (Step 3). On the other hand, small MW hydrophilic drugs might pass through paracellular pathways. The paracellular permeability of drugs would be relatively independent of their partition coefficient and degree of ionization (dotted arrow).

We express the rate of drug transport from the epithelial membrane to the intracellular lipophilic domains as the product of a rate constant (k_1), and a net driving force is given by $\left(C_1 - C_1^b / K_1\right)$, where K_1 is the ratio of C_1^b and C_1 at equilibrium (Table 1). In addition, the drug can diffuse within the lipid bilayer with an effective diffusion coefficient denoted by D_1. As a result, the drug concentration's rate of change in its unbound and bound forms within the corneal epithelium is expressed

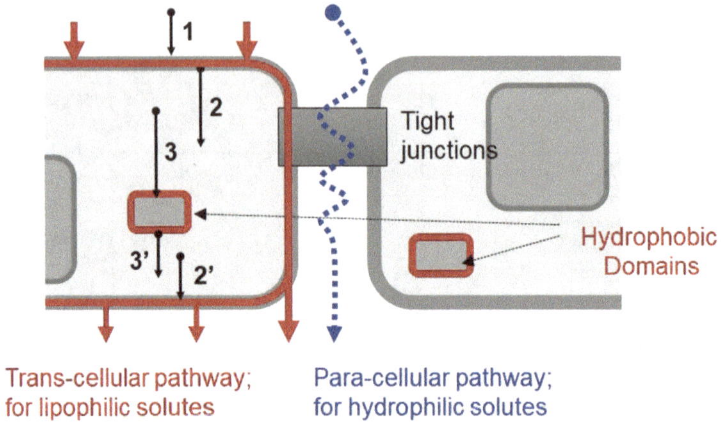

Fig. 19 Transport across the cellular layers: Two modes of transport can be envisioned across the epithelial layers. For lipophilic drugs, the main mechanism would through the lipid bilayers of the plasma membrane highlighted in red. The lipophilic drug may also partition into the cytoplasm and then accumulate in the intracellular hydrophobic domains (e.g., membrane-associated with endoplasmic reticulum; mechanisms numbered 2 and 3 as well as 2′ and 3′). On the other hand, hydrophilic drugs could pass through paracellular pathways. (Redrawn from Gupta et al. (2010))

Table 1 Model parameters estimated based on transcorneal data in Fig. 11 (Gupta et al. 2010)

Parameter	Units	Definition
ϕ_{10}	–	Partition ratio between epithelium and tears
ϕ_{21}	–	Partition ratio between stroma and epithelium
ϕ_{32}	–	Partition ratio between endothelium and stroma
D_1	m^2/s	Diffusion coefficient of RhB in lipid bilayers in the epithelium
D_2	m^2/s	Diffusion coefficient of RhB in stroma
D_3	m^2/s	Diffusion coefficient of RhB in lipid bilayers of the endothelium
K_1	–	The ratio of concentration in epithelial bilayers to that in the intracellular hydrophobic regions
k_1	s^{-1}	Permeability of cytoplasmic medium separating lipid bilayers and internal hydrophobic regions in the epithelium
k_{perm}	m/s	The permeability coefficient of the epithelium-stroma interface
K_3	–	The ratio of average concentration in endothelium bilayers (based on total cell volume) to that in internal hydrophobic regions (based on total cell volume) at equilibrium
k_3	s^{-1}	Permeability of cytoplasmic medium separating lipid bilayers and internal hydrophobic regions in the endothelium

by Eqs. (1) and (2) in Fig. 20. These are partial differential equations, where the space variable y varies in the interval Ω_1 representing layer 1, whereas the time variable t varies in the interval $(0, T)$.

We remark that Eq. (1) expresses the mass balance of the free drug. The first term on the right side represents the diffusion of the drug through the lipid bilayers, which is the dominant transport mechanism. The slow accumulation of the drug in

$$\frac{\partial C_1}{\partial t} = D_1 \frac{\partial^2 C_1}{\partial y^2} - k_1 \left(C_1 - \frac{C_1^b}{K_1} \right) \qquad \text{in } \Omega_1 \times (0,T) \qquad (1)$$

$$\frac{\partial C_1^b}{\partial t} = k_1 \left(C_1 - \frac{C_1^b}{K_1} \right) \qquad \text{in } \Omega_1 \times (0,T) \qquad (2)$$

$$\frac{\partial C_2}{\partial t} = D_2 \frac{\partial^2 C_2}{\partial y^2} \qquad \text{in } \Omega_2 \times (0,T) \qquad (3)$$

$$\frac{\partial C_3}{\partial t} = D_3 \frac{\partial^2 C_3}{\partial y^2} - k_3 \left(C_3 - \frac{C_3^b}{K_3} \right) \qquad \text{in } \Omega_3 \times (0,T) \qquad (4)$$

$$\frac{\partial C_3^b}{\partial t} = k_3 \left(C_3 - \frac{C_3^b}{K_3} \right) \qquad \text{in } \Omega_3 \times (0,T) \qquad (5)$$

Fig. 20 Transport model across the cornea. Summary of differential equations expressing the mass balance of the drug within the corneal layers (also see Gupta et al. (2010))

the epithelium (Fig. 12), however, is modeled by the second term on the right side of Eq. (1). This term is akin to modeling the accumulation by a first-order and reversible binding of the drug. Snow accumulation can result from either slow binding or transport through the cytoplasm to internal organelles.

The transport of the drug in the endothelial layer, represented by Ω_3, is governed by similar processes as those just discussed for the epithelial layer, represented by Ω_1. Thus, the mass balance of the drug in its unbounded and bounded form is described mathematically by similar equations, namely, Eqs. (4) and (5) in Fig. 20. Here, the parameters k_3, K_3, and D_3 have the same physical meaning as k_1, K_1, and D_1 for layer 1.

Finally, let us consider the stroma represented by layer 2. The corneal stroma is composed of collagen lamellae held together by glycosaminoglycans (GAGs) (Komai and Ushiki 1991). We assume no binding of the drug in the stroma consistent with the reports on the diffusion of small molecules in artificial collagen networks, suggesting rapid binding-unbinding events, with the transport of small drugs governed primarily by diffusion (Wallace and Rosenblatt 2003). Therefore, mass balance for the drug in the stroma is expressed by Eq. (3) (Fig. 20), with D_2 representing the effective diffusion coefficient in layer 2.

The mathematical model for drug transport across the cornea must be completed with suitable conditions describing the phenomena at the interfaces between layers (Fig. 21). At the interface between the tear film and the epithelium (i.e., $y = y_0$), we suppose that C_0 is the concentration of the topical drug in tears at a time instant t. The partitioning of the drug into the epithelium results in a concentration C_1 at $y = y_0$ at its outer boundary given by $\phi_{10}C_0$, where ϕ_{10} is the partition coefficient between tears (equivalent to a buffer) and lipid-rich epithelial membrane (equivalent to octanol). As a result, we have the drug dynamics at the tear-epithelium interface described mathematically by Eq. (6) (Fig. 22). Partitioning is also occurring at the epithelium-stroma interface located at $y = y_1$, as expressed by Eq. (7) (Fig. 22). In Eq. (7), ϕ_{21} is the partition coefficient between the stroma and the epithelium (Table 1, Fig. 21) and hence can be given by $\phi_{21} = \phi_{20}\phi_{10}$, where ϕ_{20} and ϕ_{10} are the partition coefficients of layers 2 and 1, respectively, with respect to buffer/tears.

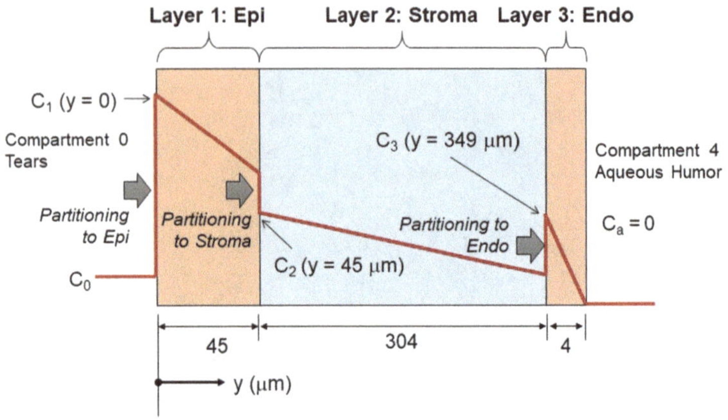

Fig. 21 Schematic of transport of a lipophilic drug surrogate based on partition and diffusion. C_0 is the concentration of the drug surrogate in tears at an instant t. The surrogate then undergoes equilibrium partitioning into the epithelium. This increases the concentration of the surrogate at $y = 0$ to C_1 (given by PC × C_0, where PC is the partition coefficient of the surrogate between tears (~ buffer) and the epithelium (~ octanol)). The partitioned surrogate then diffuses along its concentration gradient in the epithelium. Abbreviations: C_1 concentration in the epithelium, C_2 concentration in the stroma, C_3 concentration in endothelium, C_a concentration in the anterior chamber, y depth across the cornea

$$C_1 = \phi_{10}\, C_0 \qquad \text{for } y = y_0,\ t \in (0,T) \tag{6}$$

$$C_1 = \phi_{21}\, C_2 \qquad \text{for } y = y_1,\ t \in (0,T) \tag{7}$$

$$D_1 \frac{\partial C_1}{\partial y} = D_2 \frac{\partial C_2}{\partial y} \qquad \text{for } y = y_1,\ t \in (0,T) \tag{8}$$

$$C_3 = \phi_{32}\, C_2 \qquad \text{for } y = y_2,\ t \in (0,T) \tag{9}$$

$$D_2 \frac{\partial C_2}{\partial y} = D_3 \frac{\partial C_3}{\partial y} \qquad \text{for } y = y_2,\ t \in (0,T) \tag{10}$$

$$C_3 = 0 \qquad \text{for } y = y_3,\ t \in (0,T) \tag{11}$$

$$C_{total}(t,y) = \begin{cases} C_1(t,y) + C_1^b(t,y) & \text{for } y \in (y_0, y_1)\ t \in \times(0,T) \\ C_2(t,y) & \text{for } y \in (y_1, y_2)\ t \in \times(0,T) \\ C_3(t,y) + C_3^b(t,y) & \text{for } y \in (y_2, y_3)\ t \in \times(0,T) \end{cases} \tag{12}$$

$$C_{model}(t,y') = \int_0^{\infty} C_{total}(t,y)\mathrm{IRF}(y-y')\,dy \tag{13}$$

Fig. 22 Summary of equations expressing drug partitioning and flux continuity across the corneal layers (also see Gupta et al. (2010))

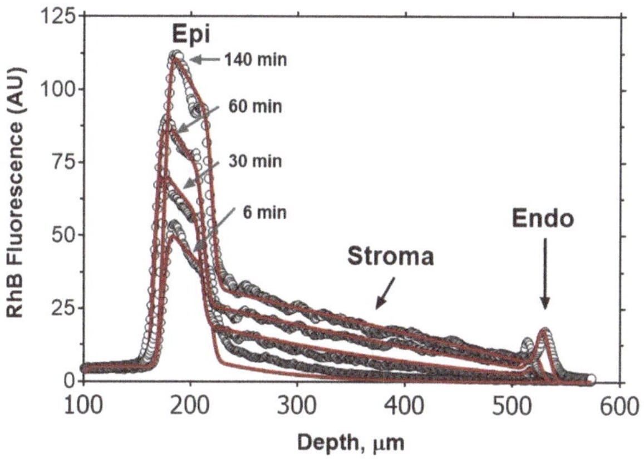

Fig. 23 Summary of equations expressing boundary conditions, convolution with the impulse response function of CSMF, and dynamics in the anterior chamber (also see Gupta et al. (2010))

Furthermore, we need to ensure flux continuity at the interface, as described by Eq. (8) (Fig. 22). Similar conditions are imposed at the interface between stroma and endothelium, located at $y = y_2$, leading to Eqs. (9) and (10) (Fig. 22). Finally, at the endothelium-aqueous humor interface located at $y = y_3$, the diffusing drug is swept away rapidly from the interface, and thus a reasonable condition is to set the concentration equal to zero (i.e., sink condition), as described by Eq. (11) (Fig. 22). This perfect sink condition is a reasonable assumption due to the high volume (~300 μL) of the receiver chamber and the rapid perfusion rate of ~3 μL/min (Agrahari et al. 2016; Gupta et al. 2010). Finally, the mathematical model summarized in Figs. 20, 21 and 22 is completed by initial conditions corresponding to zero drug concentration across the entire cornea and a given initial concentration of the drug in tears. The model identification with unsteady-state concentration profiles of RhB in Fig. 12 resulted in parameters as shown in Table 1. Figure 23 highlights the comparison of model predictions (solid lines) and experimental measurements (circles) for the transient fluorescence profiles across the cornea at $t = 6$, 30, 60, and 140 min. To compare the model prediction C_{Model} with measured fluorescence values, we defined the total drug concentration C_{total} in each corneal layer as detailed in Eq. (12), and we performed the convolution of C_{total} with IRF (impulse response function or point spread function) of the CSMF as described in Eq. (13) (Fig. 22).

In summary, we have established kinetic models for characterizing the transport of lipophilic drugs across the cornea. This is an advancement over the method of compartmental modeling (Amrite et al. 2008; Friedrich et al. 1993; McLaren et al. 1993). Previously, few studies characterized the diffusive transport across each of the corneal layers but without experimental transcorneal concentration profiles (Avtar and Tandon 2008; Yamamura et al. 1999; Zhang et al. 2004). Thus, the models that we have developed account not only for the multi-laminate structure of the cornea (Figs. 2 and 4) but are also based on the spatiotemporal experimental transcorneal concentration profiles of fluorescent drug surrogates.

Summary

In this chapter, we have provided an overview of the topical drug kinetics to the eye. In particular, we have emphasized applications of ocular fluorometry to quantitatively investigate the topical pharmacokinetics making use of fluorescein and other fluorescent surrogates. The spot fluorometer that we have developed is suitable for studies with humans and large-animal models, while CSMF is appropriate for detailed investigative work with ex vivo corneas. All major determinants of topical drug bioavailability can be investigated using the two fluorometers. Thus, we have demonstrated the application of the spot fluorometer for characterizing the impact of drug delivery modalities (e.g., viscosity-enhancing agents such as an in situ forming gel or collagen shield) on the half-life of topical drugs (Srinivas et al. 2018a; McNamara et al. 1997; Maurice and Srinivas 1992; Srinivas 1994). Secondly, the spot fluorometer can also be used to assess the barrier integrity of the corneal epithelium. Finally, we have reviewed the use of CSMF to reveal the drug and nanoparticle transport across the cornea at a microscopic scale. The concentration jumps at the interfaces between layers highlight the relevance of the partition coefficient of the drug on the transcorneal transport. The knowledge being accrued with fluorescent dyes is enabling the development of physiologically based pharmacokinetic modeling of topical drugs, which is vital to rational ophthalmic product development.

References

Agrahari V, et al. A comprehensive insight on ocular pharmacokinetics. Drug Deliv Transl Res. 2016;6(6):735–54.

Almeida H, et al. Nanoparticles in ocular drug delivery systems for topical administration: promises and challenges. Curr Pharm Des. 2015;21(36):5212–24.

Alvarez-Trabado J, Diebold Y, Sanchez A. Designing lipid nanoparticles for topical ocular drug delivery. Int J Pharm. 2017;532(1):204–17.

Amrite AC, Edelhauser HF, Kompella UB. Modeling of corneal and retinal pharmacokinetics after periocular drug administration. Invest Ophthalmol Vis Sci. 2008;49(1):320–32.

Araie M. Carboxyfluorescein. A dye for evaluating the corneal endothelial barrier function in vivo. Exp Eye Res. 1986;42(2):141–50.

Ashton P, Clark DS, Lee VH. A mechanistic study on the enhancement of corneal penetration of phenylephrine by flurbiprofen in the rabbit. Curr Eye Res. 1992;11(1):85–90.

Avtar R, Tandon D. Modeling the drug transport in the anterior segment of the eye. Eur J Pharm Sci. 2008;35(3):175–82.

Azartash K, et al. Pre-corneal tear film thickness in humans measured with a novel technique. Mol Vis. 2011;17:756–67.

Barar J, et al. Advanced drug delivery and targeting technologies for the ocular diseases. Bioimpacts. 2016;6(1):49–67.

Battaglia L, et al. Application of lipid nanoparticles to ocular drug delivery. Expert Opin Drug Deliv. 2016;13(12):1743–57.

Boddu SH, Gupta H, Patel S. Drug delivery to the back of the eye following topical administration: an update on research and patenting activity. Recent Pat Drug Deliv Formul. 2014;8(1):27–36.

Brubaker RF, Maurice D, McLaren JW. Fluorometry of the anterior segment. In: Masters BR, editor. Noninvasive diagnostic techniques in ophthalmology. New York: Springer Verlag; 1990. p. 248–80.
Chaiyasan W, Srinivas SP, Tiyaboonchai W. Mucoadhesive chitosan-dextran sulfate nanoparticles for sustained drug delivery to the ocular surface. J Ocul Pharmacol Ther. 2013;29(2):200–7.
Chaiyasan W, Srinivas SP, Tiyaboonchai W. Crosslinked chitosan-dextran sulfate nanoparticle for improved topical ocular drug delivery. Mol Vis. 2015;21:1224–34.
Chaiyasan W, et al. Penetration of mucoadhesive chitosan-dextran sulfate nanoparticles into the porcine cornea. Colloids Surf B Biointerfaces. 2017;149:288–96.
Chaiyasan W, et al. Penetration of hydrophilic sulforhodamine b across the porcine cornea ex vivo. Int J Appl Pharmaceut. 2018;10(6):94–106.
Conroy CW, Maren TH. Corneal sequestration and delivery to ciliary process of six topically applied sulfonamides. J Ocul Pharmacol Ther. 1999;15(2):179–87.
Diebold Y, Calonge M. Applications of nanoparticles in ophthalmology. Prog Retin Eye Res. 2010;29(6):596–609.
Edward A, Prausnitz MR. Predicted permeability of the cornea to topical drugs. Pharm Res. 2001;18(11):1497–508.
Edwards A, Prausnitz MR. Fiber matrix model of sclera and corneal stroma for drug delivery to the eye. AICHE J. 1998;44(1):214–25.
Friedrich SW, Cheng YL, Saville BA. Theoretical corneal permeation model for ionizable drugs. J Ocul Pharmacol. 1993;9(3):229–49.
Gaudana R, et al. Ocular drug delivery. AAPS J. 2010;12(3):348–60.
Gupta C, et al. Measurement and modeling of diffusion kinetics of a lipophilic molecule across rabbit cornea. Pharm Res. 2010;27(4):699–711.
Gupta C, Chauhan A, Srinivas SP. Penetration of fluorescein across the rabbit cornea from the endothelial surface. Pharm Res. 2012;29(12):3325–34.
Guss R, Johnson F, Maurice D. Rhodamine B as a test molecule in intraocular dynamics. Invest Ophthalmol Vis Sci. 1984;25(6):758–62.
Heikkinen EM, et al. Distribution of small molecular weight drugs into the porcine lens: studies on imaging mass spectrometry, partition coefficients, and implications in ocular pharmacokinetics. Mol Pharm. 2019;16(9):3968–76.
Hosaka E, et al. Interferometry in the evaluation of precorneal tear film thickness in dry eye. Am J Ophthalmol. 2011;151(1):18–23 e1.
Janagam DR, Wu L, Lowe TL. Nanoparticles for drug delivery to the anterior segment of the eye. Adv Drug Deliv Rev. 2017;122:31–64.
Kaiser RJ, Maurice DM. The diffusion of fluorescein in the Lens. Exp Eye Res. 1964;3:156–65.
Kidron H, et al. Prediction of the corneal permeability of drug-like compounds. Pharm Res. 2010;27(7):1398–407.
King-Smith PE, et al. The thickness of the tear film. Curr Eye Res. 2004;29(4–5):357–68.
Komai Y, Ushiki T. The three-dimensional organization of collagen fibrils in the human cornea and sclera. Invest Ophthalmol Vis Sci. 1991;32(8):2244–58.
Kompella UB, et al. Luteinizing hormone-releasing hormone agonist and transferrin functionalizations enhance nanoparticle delivery in a novel bovine ex vivo eye model. Mol Vis. 2006;12(134–135):1185–98.
Kompella UB, Kadam RS, Lee VHL. Recent advances in ophthalmic drug delivery. Ther Deliv. 2010;1(3):435–56.
Kompella UB, et al. Nanomedicines for back of the eye drug delivery, gene delivery, and imaging. Prog Retin Eye Res. 2013;36:172–98.
Kompella UB, Hartman RR, Patil MA. Extraocular, periocular, and intraocular routes for sustained drug delivery for glaucoma. Prog Retin Eye Res. 2020:100901.
Laurence J, et al. A double-masked, placebo-controlled evaluation of timolol in a gel vehicle. J Glaucoma. 1993;2(3):177–82.

Lee YH, Kompella UB, Lee VH. Systemic absorption pathways of topically applied beta adrenergic antagonists in the pigmented rabbit. Exp Eye Res. 1993;57(3):341–9.
Mashige KP. Ocular allergy. Health SA Gesondheid. 2017;22:112–22.
Maurice DM. Factors influencing the penetration of topically applied drugs. Int Ophthalmol Clin. 1980;20(3):21–32.
Maurice DM. Prolonged-action drops. Int Ophthalmol Clin. 1993;33(4):81–91.
Maurice DM. Drug delivery to the posterior segment from drops. Surv Ophthalmol. 2002;47(Suppl 1):S41–52.
Maurice D, Singh T. A permeability test for acute corneal toxicity. Toxicol Lett. 1986;31(2):125–30.
Maurice DM, Srinivas SP. Use of fluorometry in assessing the efficacy of a cation-sensitive gel as an ophthalmic vehicle: comparison with scintigraphy. J Pharm Sci. 1992;81(7):615–9.
Maurice DM, Srinivas SP. Fluorometric measurement of light absorption by the rabbit cornea. Exp Eye Res. 1994;58(4):409–13.
McLaren JW, Ziai N, Brubaker RF. A simple three-compartment model of anterior segment kinetics. Exp Eye Res. 1993;56(3):355–66.
McNamara NA, et al. Measurement of corneal epithelial permeability to fluorescein. A repeatability study. Invest Ophthalmol Vis Sci. 1997;38(9):1830–9.
Meza-Rios A, et al. Therapies based on nanoparticles for eye drug delivery. Ophthalmol Therapy. 2020;9(3):1–14.
Missel PJ. Simulating intravitreal injections in anatomically accurate models for rabbit, monkey, and human eyes. Pharm Res. 2012;29(12):3251–72.
Missel PJ, Sarangapani R. Physiologically based ocular pharmacokinetic modeling using computational methods. Drug Discov Today. 2019;24(8):1551–63.
Mitra AK, Mikkelson TJ. Mechanism of transcorneal permeation of pilocarpine. J Pharm Sci. 1988;77(9):771–5.
Nelson MD, et al. Ocular tolerability of timolol in Gelrite in young glaucoma patients. J Am Optom Assoc. 1996;67(11):659–63.
Niamprem P, Srinivas SP, Tiyaboonchai W. Penetration of Nile red-loaded nanostructured lipid carriers (NLCs) across the porcine cornea. Colloids Surf B Biointerfaces. 2019a;176:371–8.
Niamprem P, et al. Impact of nanostructured lipid carriers as an artificial tear film in a rabbit evaporative dry eye model. Cornea. 2019b;38(4):485–91.
Rai Udo J, et al. The suprachoroidal pathway: a new drug delivery route to the back of the eye. Drug Discov Today. 2015;20(4):491–5.
Ranta VP, et al. Ocular pharmacokinetic modeling using corneal absorption and desorption rates from in vitro permeation experiments with cultured corneal epithelial cells. Pharm Res. 2003;20(9):1409–16.
Ratna S, Rina LDN. Glaucoma and dry eye disease: the role of preservatives in glaucoma medications. Med J Indones. 2011;20(4):302–5.
Sasaki H, et al. Enhancement of ocular drug penetration. Crit Rev Ther Drug Carrier Syst. 1999;16(1):85–146.
Schoenwald RD, Huang HS. Corneal penetration behavior of beta-blocking agents I: physiochemical factors. J Pharm Sci. 1983;72(11):1266–72.
Schoenwald RD, Ward RL. Relationship between steroid permeability across excised rabbit cornea and octanol-water partition coefficients. J Pharm Sci. 1978;67(6):786–8.
Schulz A, et al. Evaluation of fluorescent polysaccharide nanoparticles for pH-sensing. Org Biomol Chem. 2009;7(9):1884–9.
Shirasaki Y. Molecular design for enhancement of ocular penetration. J Pharm Sci. 2008;97(7):2462–96.
Srinivas SP. In situ measurement of fluorescein release by collagen shields in human eyes. Curr Eye Res. 1994;13(4):281–8.
Srinivas SP, Maurice DM. A microfluorometer for measuring diffusion of fluorophores across the cornea. IEEE Trans Biomed Eng. 1992;39(12):1283–91.

Srinivas SP, et al. Corneal epithelial permeability to fluorescein in humans by a multi-drop method. PLoS One. 2018a;13(6):e0198831.
Srinivas SP, et al. Penetration of fluorescent silica nanoparticles into the cornea. Mater Today Proc. 2018b;5:11072–9.
Sudhir RR, et al. Ocular spot Fluorometer equipped with a lock-in amplifier for measurement of aqueous flare. Transl Vis Sci Technol. 2018;7(6):32.
Wallace DG, Rosenblatt J. Collagen gel systems for sustained delivery and tissue engineering. Adv Drug Deliv Rev. 2003;55(12):1631–49.
Werkmeister RM, et al. Measurement of tear film thickness using ultrahigh-resolution optical coherence tomography. Invest Ophthalmol Vis Sci. 2013;54(8):5578–83.
Worth AP, Cronin MT. Structure-permeability relationships for transcorneal penetration. Altern Lab Anim. 2000;28(3):403–13.
Wu NC, Chiang CH, Lee AR. Studies of carbonic anhydrase inhibitors: physicochemical properties and bioactivities of new thiadiazole derivatives. J Ocul Pharmacol. 1993;9(2):97–108.
Yamamura K, et al. Characterization of ocular pharmacokinetics of beta-blockers using a diffusion model after instillation. Pharm Res. 1999;16(10):1596–601.
Yavuz B, Kompella UB. Ocular drug delivery. Handb Exp Pharmacol. 2017;242:57–93.
Yiu G, et al. Suprachoroidal and subretinal injections of AAV using transscleral microneedles for retinal gene delivery in nonhuman Primates. Mol Ther Methods Clin Dev. 2020;16:179–91.
Yoshida F, Topliss JG. Unified model for the corneal permeability of related and diverse compounds with respect to their physicochemical properties. J Pharm Sci. 1996;85(8):819–23.
Zhang W, Prausnitz MR, Edwards A. Model of transient drug diffusion across cornea. J Control Release. 2004;99(2):241–58.

Topical Ophthalmic Dosage Form Development: Key Components and Critical Quality Attributes

Anuradha Gore and Chetan Pujara

Abstract General principles for the development of a topical ophthalmic dosage form intended for application to ocular structures are discussed. Topical ophthalmic route presents unique challenges for drug delivery due to anatomical and physiological barriers such as rapid clearance through nasolacrimal drainage, low permeability through corneal epithelium, and limitations to formulation ingredients that can be tolerated by the ocular structures. The specific quality and performance requirements and general considerations in designing these into the product attributes during the development of these products are discussed. Formulation development strategy based on a quality by design (QbD) approach is used as a tool to help formulation scientist develop dosage forms. The first step for any new product should be to define the product appropriately by establishing a quality target product profile which includes all aspects of the desired quality attributes that the product should possess. The seven-step systematic approach to product development based on QbD principles is illustrated using a case example of a sterile ophthalmic solution product.

Keywords Topical ophthalmic products · Ocular structure · Preservatives · Tonicity agents · Eye drops · Quality by design (QbD) · Target product profile (TPP) · Quality target product profile (QTPP) · Critical quality attributes (CQAs) · Critical process parameters (CPPs)

Abbreviations

APET	Antimicrobial Preservative Efficacy Test
BAK	Benzalkonium chloride

A. Gore (✉)
Asset Strategy Lead, Eye Care, Abbive, Irvine, CA, USA
e-mail: anu.gore@abbvie.com

C. Pujara
Pharmaceutical Sciences, AbbVie, Irvine, CA, USA
e-mail: chetan.pujara@abbvie.com

S. Neervannan, U. B. Kompella (eds.), *Ophthalmic Product Development*, AAPS Advances in the Pharmaceutical Sciences Series 37,
https://doi.org/10.1007/978-3-030-76367-1_7

BID	*Bis in die* (twice a day)
CCS	Container closure system
CPP	Critical process parameter
CQAs	Critical quality attributes
EMA	European Medicines Agency
FDA	Food and Drug Administration
GLP	Good laboratory practices
HDPE	High-density polyethylene
ICH	International Council for Harmonisation of Technical Requirements for Pharmaceuticals for Human Use
LDPE	Low-density polyethylene
PhEur	European Pharmacopoeia
QbD	Quality by design
QD	*Quaque die* (once a day)
QTPP	Quality Target Product Profile
SAL	Sterility assurance level
TPP	Target product profile
USP	United States Pharmacopeia

Introduction

Ophthalmic products are sterile products that are intended for application to any ocular structure, including any space adjacent to an ocular structure and its immediate surrounding spaces. The routes of administration of ophthalmic products fall into three general categories: topical, intraocular injections, and extraocular injections. Topical drug products are intended to be administered to an ocular surface component, such as the eyelid, conjunctiva, or cornea, and can produce local or systemic effects (United States Pharmacopeia and National Formulary (USP 41-NF 36) 2018a). Topical ophthalmic products are administered to the eye in a wide variety of dosage forms, including but not restricted to: solutions, suspensions, ointments, gels, emulsions, strips, and inserts (United States Pharmacopeia and National Formulary (USP 41-NF 36) 2018b).

Ophthalmic Drug Delivery Route Considerations

Structure of the Eye

The eye is a specialized sensory organ with a unique anatomy and physiology. In general, it is relatively secluded from systemic access by the blood-retinal, blood-aqueous, and blood-vitreous barriers and thus presents some unique challenges as well as opportunities for drug delivery (Henderer and Rapuano 2017). Drug delivery, pharmacokinetics, and pharmacology for the eye have been reviewed extensively previously (Henderer and Rapuano 2017; Urtti 2006; Maurice and Mishima 1984; Schoenwald 1997), and for the current review we will focus on an overview of these in the context of considerations for dosage form development.

The eye is generally referred to as a spherical structure and can be considered as made of two major segments—anterior and posterior. The anterior segment tissues include the cornea, conjunctiva, iris, ciliary body, trabecular meshwork, aqueous humor, and lens structure (Chen et al. 2018; Cholkar et al. 2013); the posterior segment tissues include vitreous, retina, macular, optic nerve, choroid and sclera, and other structures (Gaudana et al. 2009). The eye is protected by the eyelids and by the orbit, a bony cavity of the skull that has multiple fissures and foramina that conduct nerves, muscles, and vessels (Henderer and Rapuano 2017). The eyelids serve several functions. Foremost, their dense sensory innervation and eyelashes protect the eye from mechanical and chemical injuries. Blinking, a coordinated movement of the orbicularis oculi, levator palpebrae, and Müller's muscles, serves to distribute tears over the cornea and conjunctiva. In humans, the average blink rate is 15–20 times/minute. The external surface of the eyelids is covered by a thin layer of skin; the internal surface is lined with the palpebral portion of the conjunctiva, which is a vascularized mucous membrane continuous with the bulbar conjunctiva. At the reflection of the palpebral and bulbar conjunctivae is a space called the fornix, located superiorly and inferiorly behind the upper and lower lids, respectively. Topical medications usually are placed in the inferior fornix, also known as the inferior cul-de-sac (Henderer and Rapuano 2017; Maurice and Mishima 1984).

Typically, topical ocular medications commonly administered to the surface of the eye do not reach the posterior segment, and most diseases of this segment cannot be effectively treated by topical ophthalmic dosage forms (Urtti 2006). Thus, in clinical practice, ophthalmic formulations are typically developed to treat diseases and conditions affecting the anterior segment—namely, cornea, conjunctiva, anterior chamber, and iris-ciliary body as well as other periocular structures such as lachrymal glands and meibomian glands (Fig. 1).

Routes of Drug Delivery to Ocular Tissues

A schematic of the various routes of drug delivery to ocular tissues is depicted in Fig. 2. To achieve an effective and safe rate and extent of absorption, the drug product must accurately and differentially interact with each relevant tissue and the tear composition of a diseased eye that may change over the time course of the disease. In addition, the topical ophthalmic formulation must deliver and release its active ingredient to the relevant ocular tissues in a timeframe of mere minutes. Because topical ocular drug availability is extremely low, there is little or no margin for error (Gore et al. 2017).

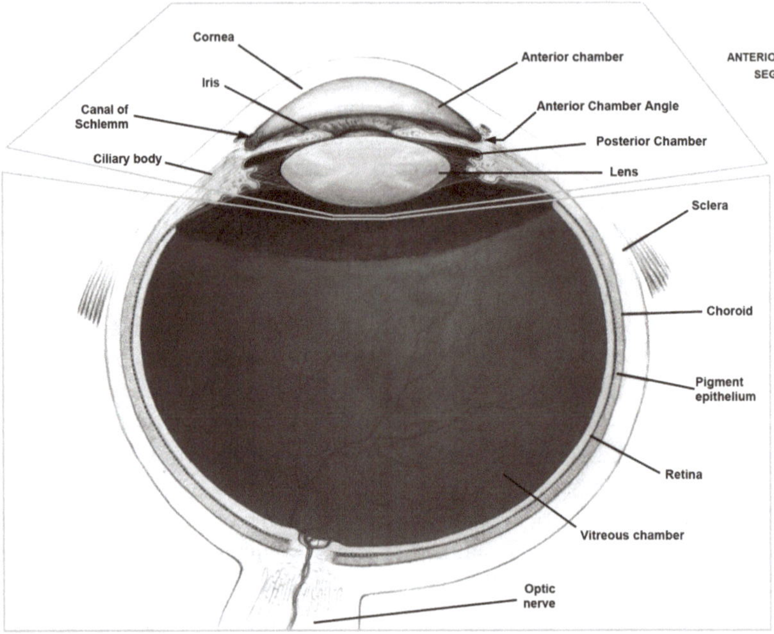

Fig. 1 Structure of the eye (Attar et al. 2013)

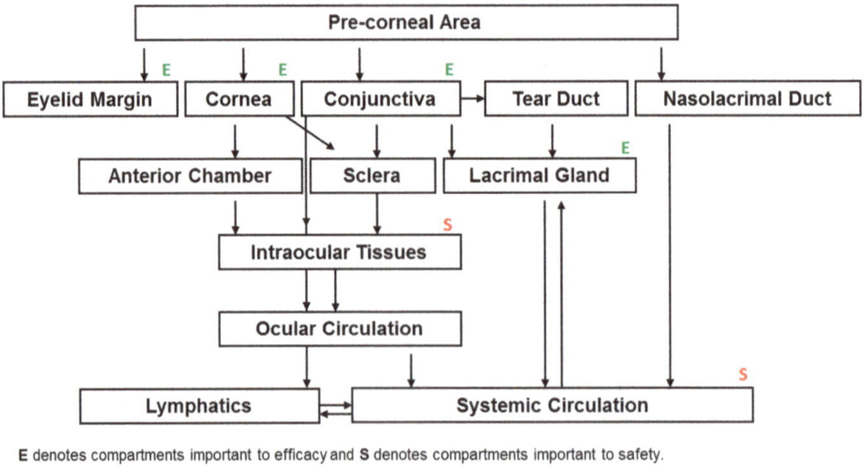

Fig. 2 Schematic of routes of ocular drug delivery (Gore et al. 2017). Note: Direct entry of drug to the blood and lymphatics is possible in several regions. For example, the conjunctiva, episclera, etc. have extensive network of blood vessels, while lymphatic drainage may be present in tissues such as lower eyelid and conjunctiva

Complexity in Drug Delivery to Target Tissues

Unlike other drug delivery routes, a topical ophthalmic formulation usually delivers drug to the ocular tissues in a relatively short timeframe of a few minutes. An eye drop, irrespective of the instilled volume, often eliminates rapidly within 5 min after administration with majority of the applied topical ophthalmic formulation being lost via nasolacrimal drainage. As a result, topical ocular drug availability is very low, and only a small fraction (1–7%) of the drug substance is delivered to the tear film and/or is absorbed and becomes bioavailable in ocular tissues (Henderer and Rapuano 2017; Durairaj 2017).

Normal human tear turnover is approximately 16% per minute, and this turnover acts to remove drug solution from the conjunctival cul-de-sac. Turnover may also be stimulated by many other factors including ocular irritation, which renders topical application of ophthalmic solutions to the cul-de-sac imprecise and extremely inefficient (Schoenwald 1997).

Formulation excipients and excipient quality can stimulate tear production and dilution, which may further enhance drug elimination. Simple dilution of instilled drug in the tears acts to reduce the transcellular availability and flux of drug that remains in the conjunctival cul-de-sac.

Bioavailability of a topically applied drug is a result of complex differential rate processes and precorneal film dynamics that adjust continually toward the equilibrium:

1. Precorneal clearance of the applied dose (e.g., due to blinking and lacrimation)
2. Tear film drug concentration time curve (i.e., amount of drug in the tears)
3. Tissue permeability
4. Post-tissue clearance

General Considerations for Ophthalmic Dosage Form Development

The goal of any dosage form is to deliver the drug/active to its target tissue so that the intended therapeutic effect can be achieved. The topical ophthalmic delivery is intended to be administered to the membrane surfaces of the eye, which are generally categorized as mucosal membranes (United States Pharmacopeia and National Formulary (USP 41-NF 36) 2018c). The target tissues intended for action using these dosage forms are generally the local tissues in the area proximate to application—systemic absorption is not typically desired and is considered unnecessary for therapeutic effect. In fact in many cases for ophthalmic dosage form development, the goal is to develop a dosage form that can localize the drug delivery to intended target tissue and minimize systemic absorption to avoid adverse events related to systemic exposure. Topical ophthalmic products can be developed in a wide variety of dosage forms including solutions, suspensions, ointments, gels, emulsions, strips, and inserts to meet these goals. In addition, newer dosage forms such as the ocular ring inserts are under development or in research.

The starting point of any new formulation development project requires collection of information that is required to define the product as it relates to its intended use. At this stage, the formulator starts compiling the list of the potential quality attributes for the product that are required to be designed into the product. Topical ophthalmic products have the same or similar requirements as injectables and implants. Some of these requirements are common to all ophthalmic formulations, while others may be specific to the dosage form selected. The criteria for topical ophthalmic products can be divided into two categories: those that assess general product quality attributes and those that assess product performance. Quality tests assess the integrity of the dosage form, whereas the performance tests assess drug release and other attributes that relate to in vivo drug performance. Taken together, quality and performance tests ensure the identity, strength, quality, purity, and efficacy of the ophthalmic drug product. In the case of topical ophthalmic products having a localized and immediate response when applied to the eye (e.g., topically applied dosage forms, including dispersed systems, having very short residence time for absorption), a dissolution/drug release test may have no practical value (United States Pharmacopeia and National Formulary (USP 41-NF 36) 2018d). Ophthalmic products have the same or similar requirements as injectables and implants. In addition, ophthalmic route of delivery falls under the category of mucosal drug products, and there are specific quality and performance requirements to be considered (United States Pharmacopeia and National Formulary (USP 41-NF 36) 2018a, c, d, e). Table 1 consolidates typical quality requirements for ophthalmic products and general considerations in designing these into the product attributes. It is important that sufficient consideration and rigor are given to selection of these parameters to ensure that the desired quality and performance are built into the final product.

Dosage Form Selection

The types of formulations that can be developed for dosing via the topical ophthalmic route can range from simple aqueous solutions to highly complex systems such as emulsions or gels. The complexity in the development of these dosage forms arises mainly from the sensitive nature of the organ that is being treated and the impact on its function, which can limit the range and type of formulations that can be developed. Although the wide ranges for quality attributes may be listed in pharmaceutical compendia, often it is observed that combining variables at the limits of their ranges which may be acceptable for other parenteral routes such as injections is generally not suitable for ophthalmic dosing. Examples of this include extremes of pH with high buffer strength, extremes of osmolality with high viscosity or high concentrations of excipients, or vehicles with high lipid-based content. As a result, most commercial ophthalmic formulations are aqueous based, within a narrow pH

Table 1 Typical quality requirements and general considerations (United States Pharmacopeia and National Formulary (USP 41-NF 36) 2018b, c, d, e, f)

Requirement	General consideration
Common to all topical ophthalmic dosage forms[a]	
Appearance	A qualitative description of the drug product, which includes a description of color, clarity, and dosage form
Identification	Tests that can determine the identity of the drug or drugs present in the dosage form
Assay	Specific and stability indicating requirement to ensure that the drug content in the dosage form is appropriate throughout its shelf-life
Impurities	These typically include organic impurities arising from the degradation of the drug substance in the drug product and those impurities arising during the manufacturing process of the drug product. Generally impurities in drug substance are not included unless they are also degradation products in the drug product
pH	Normal tears have a pH of about 7.4. The eye can tolerate products over a range of pH values from about 3.0 to about 8.6, depending on the buffering capacity of the formulation. The pH value of the formulation should be the one where the drug product is the most stable. Formulations that target the extremes of the acceptable pH range will have better patient acceptability if the formulations have a low buffering capacity
Osmolarity	Ophthalmic products may be tolerated over a fairly wide range of tonicity (0.5–5% sodium chloride, equivalent to about 171–1711 mOsm/kg, e.g., Muro 128®) (Bausch and Lomb n.d.). Hypotonic solutions are better tolerated than hypertonic solutions. Precautions should be taken to ensure that the product maintains its osmolarity during shelf-life
Particulate and foreign matter	All ophthalmic products should be inspected for package integrity and, to the extent possible, for the presence of observable foreign and particulate matter (visible particulates). For sub-visible particulates, USP guidance is followed (United States Pharmacopeia and National Formulary (USP 41-NF 36) 2018f)
Sterility	All ophthalmic products are required to be sterile and must meet sterility testing requirements described in the compendia. Methods of sterilization are further described in section "Sterilization Method Selection"
Antimicrobial preservative	These agents are generally included in dosage forms that are packed in containers that allow for the withdrawal or administration of multiple doses. These products must meet the requirements of both the content and effectiveness of the antimicrobial agent
Uniformity of dosage units	Only applicable to single-unit containers and can be demonstrated in either content uniformity or weight variation
Container contents	Ensure that the requirements for minimum fill volume of each container are met
Leachables and extractables	The packaging system should not interact with the product to alter the strength, quality, or purity of the drug product. Since most ophthalmic products are packaged in plastic containers which are semi-permeable in nature, care must be taken to select the secondary packaging systems carefully to avoid migration of leachables into the product during its shelf-life

(continued)

Table 1 (continued)

Requirement	General consideration
Container closure integrity	The package selected should be capable of being closed or sealed in such a manner as to prevent contamination or loss of contents and should provide evidence of being tamper proof. Validation of container integrity must demonstrate no penetration of microbial contamination or of chemical or physical impurities
Specific to certain topical ophthalmic dosage forms	
Viscosity	An increase in viscosity increases the residence time in the eye. However, drug diffusion out of the formulation into the eye may be inhibited due to high product viscosity and highly viscous products may cause blurred vision. These should be considered during the design of the product for its intended purpose
Antioxidant content	If antioxidants are used, the rationale for the selection and levels included in the product should be established
Resuspendability/ redispersibility	Specific to suspension and certain emulsion products to ensure re-dispersibility and uniformity of dosage
Particle size and particle size distribution	Specific to suspension and emulsion products to ensure physical stability of the product throughout its shelf-life
Drop size	For ophthalmic drug products dispensed as drops, drop sizes may typically range from 20 to 70 μL. Drop size can be controlled by weight or by volume, and it is typically evaluated during product development
Added substances	The use of ingredients solely to impart a color, odor, or flavor is prohibited for topical ophthalmic products. Certain substances may be added if required for stability provided they are safe and compatible and do not interfere with product efficacy
Performance tests specific to certain topical ophthalmic dosage forms	
Dissolution/drug release	Applicable to products that have an extended-release mechanism (beyond 1 day); the dissolution/drug release rate is rate limiting for absorption and is expected to provide a controlled therapeutic response. Test can be developed using any apparatus described in the pharmacopoeias, while novel dosage forms may require the use of non-compendial equipment and/or conditions. The test conditions should reasonably mimic the method of administration of the product and in vivo conditions to establish, if possible, an in vivo-in vitro correlation that can be used to predict in vivo performance of the product (United States Pharmacopeia and National Formulary (USP 41-NF 36) 2018d)

[a]Typically bacterial endotoxin test is not required for topically applied ophthalmic products

range, with osmolality close to that of tears, and limited number and level of additives. A few nonaqueous formulations which are typically petrolatum or mineral oil based are also available, but these are not widely accepted especially for daytime use due to their effect on vision. In recent years, a few other nonaqueous formulations are being evaluated in clinical studies, which are based on semi-fluorinated alkanes as the vehicle carrier (Dutescu et al. 2014; Wirta et al. 2019), but these are not yet approved for use in US market or other broad markets. Some of the available options are listed in Table 2.

Table 2 Dosage form options for ophthalmic delivery

Dosage form	Description	Advantages	Disadvantages
Aqueous solutions	Drug is in dissolved state in an aqueous vehicle. May contain solubilizers for solubility enhancement and preservatives for multi-dose systems	Most convenient and commonly used. Drug is immediately available for absorption and effect. Typically well-tolerated	Short duration on ocular surface; higher risk of stability of drug substance; preservatives may increase incidence of local irritation
Viscous aqueous solutions (includes gels)	Similar to solutions above with inclusion of viscosity polymers. These may include viscoelastic polymers, muoadhesives, shear thinning polymers, or in-situ gelling systems	Increase ocular residence to increase bioavailability	No clear relationship between increased viscosity and enhanced bioavailability. Increased blinking reflex may clear the viscous solution almost as rapidly as non-viscous Blurring of vision due to high viscosity. Longer retention of drug on ocular surface may increase incidences of adverse events
Emulsions	Typically O/W emulsions with drug dissolved in oil phase	Suitable for delivery of oil-soluble compounds. Presence of oil can provide evaporative barrier and reduce symptoms of dry eye	Oily material can cause temporary blurring of vision; high levels of surfactants may increase incidence of local tolerability-related adverse events; complex processes for sterile emulsion manufacture
Suspensions	Solid drug particles are suspended in a liquid phase	Suitable for delivery of poorly soluble drugs; may be suitable for drugs with poor solution stability	Low volume of tears on ocular surface may not provide sufficient dilution for dissolution leading to poor ocular bioavailability. High solid content and viscosity may cause blurred vision
Ointments	Ointments are semisolid preparations usually containing less than 20% water and volatiles and more than 50% hydrocarbons, waxes, or polyols as the vehicle	Provide occlusive barrier for tear evaporation and can reduce symptoms of dry eye disease. Suitable carrier for drugs unstable or insoluble in aqueous vehicles	Generally produce blurred vision and are limited to nighttime use. The oily nature of vehicles may not be esthetically suitable for daytime use

(continued)

Table 2 (continued)

Dosage form	Description	Advantages	Disadvantages
Nonaqueous solutions	Hydrocarbon-based vehicles such as semi-fluorinated alkanes (SFAs) proposed as nonaqueous vehicles. Currently, no approved products	Appear to be well tolerated; low surface tension allows for small drop size; suitable for drugs unstable in aqueous conditions	Long-term effects of SFAs for chronic dosing are unknown. High cost and regulatory challenges. Possible environmental hazard considerations for SFAs
Solid inserts	Polymer inserts intended for placement into the cul-de-sac area for slow release of medication	Longer duration and ocular retention	May cause patient discomfort. Care and training required for proper administration and removal
Other specialized dosage forms Ocular ring	Polymer ring loaded with drug that fits around the eyeball to release drug over prolonged periods. Currently in development—No approved products	Long duration (6 months) of release, improved patient compliance	Possibility of ring getting dislodged and removal by patient

At the beginning of any new formulation development project, the necessary and desired product attributes should be captured as a part of the target product profile (TPP) for the new development project. The portion of the TPP that summarizes the attributes related to quality, such as dosage form and strength, description, storage, and handling and other relevant attributes from Table 1 is summarized as the quality target product profile (QTPP). Although there may be an initial selection of the desired dosage form included in the QTPP, the final product developed is often determined by the stage/phase of product development, properties of the drug substance, and the dosage strength required.

Phase/Stage of Development

At early stages of a project, when the drug is being evaluated in animal studies through Phase 1 and Phase 2 studies, the dosage form selection is often based on "fit-for-use" formulations with the intent to develop commercializable formulations later in development. For topical ophthalmic formulation development, this strategy poses some unique challenges, mainly due to the complexity of the ocular route of delivery and difficulties in demonstrating bioequivalence of dosage forms. For other delivery routes, e.g., oral delivery, changes in formulation at late stages of development can be often managed by demonstrating bioequivalence of the new dosage form with the one utilized in earlier stage utilizing systemic pharmacokinetic studies. In the case of ophthalmic delivery, the target tissue for efficacy of drug is often localized within the anterior segment of the eye, and sampling of these tissues for PK determination is not feasible or ethical. Comparative clinical endpoint studies may be required if formulation changes occur late in development, i.e., after efficacy and safety has been demonstrated. To avoid the need for these expensive and time-intensive studies, it is essential that the dosage form selection and formulation composition is finalized prior to initiation of Phase 3 studies. In the case of complex dosage forms, such as emulsions and suspensions, it is essential to ensure that the manufacturing process utilized for Phase 3 product is representative of the process used for commercial product manufacture.

Physicochemical Properties of the Drug

The physicochemical properties of the drug are the primary factors that dictate the selection of a dosage form for ophthalmic delivery. Thus, the starting point of any product development activities is to collect as much information as available from the pre-formulation assessment of the drug substance. Important pre-formulation attributes that need to be considered in selecting a formulation approach are listed in Table 3.

Dose Strength

Selection of appropriate dose strength is critical in ophthalmic formulations mainly due to the poor bioavailability in ocular tissues, limitations on dosing volume, and ocular residence time. As mentioned before, the bioavailability for most drugs after administration as an eye drop is low, so it is important to ensure that the dose

Table 3 Physicochemical property information required for drug substances for dosage form development

Dosage form	Physicochemical property information required
Common to all dosage forms	• Molecular weight • log P • pKa • pH-solubility profile • Solubility in co-solvent systems, solubilizers • Stability at different pH • Degradation kinetics/pathways • Permeability
Suspension dosage form—Additional requirement	• Particle size • Crystallinity • Polymorphism • Thermal properties (e.g., melting point, glass transition, solid-state transitions)
Emulsion dosage form—Additional requirement	• Oil solubility • Stability in oil-surfactant mixtures
Nonaqueous dosage form—Additional requirement	• Solubility in nonaqueous vehicle • Stability in nonaqueous vehicle

strength is sufficient to deliver efficacious levels of drugs to the tissues. Further, it is not feasible to increase dose volume beyond the typical ranges listed previously in Table 1—the additional volume would simply drain out through the nasolacrimal drainage system without being absorbed in ocular tissues. Selecting a dose strength too high (or dosing volume) can lead to gastrointestinal or systemic adverse effects as the excess drug travels down the nasolacrimal drainage system.

During the non-clinical evaluation of a new drug in studies such as the GLP-toxicology studies, it is desirable to dose much higher than the expected therapeutic dose to obtain safety margins prior to the clinical studies. Further, a broad range of dose strengths may be required to allow for dose escalation during early stages of the project while the target dose is being identified. These situations pose unique challenges to the formulators to select an appropriate vehicle matrix, especially if the drug substance has poor solubility properties. Higher levels of solubilizers are required to accommodate the higher dose strengths, but maintaining these at the lower dose strength can negatively affect bioavailability. In the case of the latter, the drug may remain complexed/bound with the solubilizer and the dilution in tears combined with short residence time may not be sufficient to release the drug on the ocular surface (Gore et al. 2017; Graham et al. 2008). Emulsion formulations with the poorly soluble drug dissolved in a lipid phase are often considered for lipophilic drugs, but can bring its own set of challenges. The drug can partition into the various phases of the complex emulsion formulation at different concentrations depending on its physicochemical properties. This drug localized in different phases may preferentially target partitioning into different ocular tissues depending on their lipophilic or hydrophilic characteristics. For example, the drug in the aqueous phase shows greater affinity toward tissues such as the cornea or conjunctiva, and the oil compartment shows greater affinity toward lipid tissues such as the eyelid margin containing the meibomian glands. Thus, the rate and extent of distribution in these tissues may be affected by the amount of drug in the different phases of the emulsion (Gore et al. 2017). Although it may not be possible to avoid these issues completely, it is important to recognize these as potential risks to the selection of the efficacious dose strength in the selected dosage form.

Selection of Components

Drug

In early stages of a discovery research, formulators can influence the selection of the drug that is escalated into clinical phase conducting developability assessment. This is typically done by pre-formulation characterization and assessment of risks and liabilities intrinsic to the molecule. If it is determined that the liabilities of the compound cannot be overcome by formulation approaches, then it is important to halt the molecule from progressing into development. Once the molecule has moved into GLP-toxicology and clinical development stage, the formulators can still influence the selection of drug substance form or properties that are important for formulation and manufacturing processes—e.g., salt form or free acid/base, polymorph, and particle size. Emulsion dosage forms typically require free base or free acid forms of drug substance, while salt forms, polymorphs, and particle size are important considerations for solution and suspension dosage forms. These properties constitute the critical quality attributes (CQAs) of the drug substance, and it is important to identify these early during development to ensure that these are built into the drug substance specifications/requirements.

Preservative

Microbial contamination of the product may occur during use—i.e., during their instillation in the patient's eye. The main source of contamination of the dropper tips or even the solutions inside the bottle results from physical contact with microbe-harboring surfaces like fingers, eyelashes, etc. This contamination may cause a physicochemical deterioration of the ophthalmic solutions or a risk of (additional) infection for the patient's eye (Furrer et al. 2002). To minimize this risk, the formulator has two options—include a preservative in the formulation or select a preservative-free formulation system.

In ophthalmic preparations, preservatives are chemical ingredients that are added to prevent microbial growth—either by destroying microorganisms (bactericidal effect) or at least preventing their growth (bacteriostatic effect). Ideally, a preservative should provide numerous qualities like broad antimicrobial activity, chemical/thermal stability, compatibility with the container and other compounds present, as well as innocuousness toward ocular tissues (Furrer et al. 2002). Unfortunately none of the preservatives has all the required qualities so as to be used universally for any ophthalmic preparation. Indeed preservatives that kill or damage growing microbial cells may also be toxic to growing cells of the ocular tissues (Furrer et al. 2002). When selecting a preservative for an ophthalmic formulation, the following considerations must be taken into account (Gangrade et al. 1996): (a) irritation potential, (b) pH range for maximal antimicrobial activity, (c) compatibility with other ingredients, (d) synergism or antagonism in antimicrobial activity, and (e) processing conditions such as heat or packaging.

Detailed reviews on commonly used preservatives in ophthalmic formulations have been published previously (Furrer et al. 2002; Gangrade et al. 1996; von Deylen et al. 2018; Freeman and Kahook 2009). An overview of the commonly used ones is shown in Table 4.

Excipients

Selection of excipients and optimization of the amount of each excipient is formulation dependent. Ophthalmic formulations are typically comprised of the following ingredient classifications listed in Table 5.

Table 4 Preservatives commonly used in ophthalmic formulations (compiled from Furrer et al. (2002), von Deylen et al. (2018), and Freeman and Kahook (2009))

Preservative	Usual conc. (%)	Description
Benzalkonium chloride (BAK)	0.004–0.02	Most commonly used preservative for ophthalmic formulations mainly due to its high antimicrobial efficacy It is a mixture of alkyl-benzyl-dimethyl-ammonium chlorides with *n*-alkyl chains between C8 and C18 (Furrer et al. 2002). Excellent antimicrobial efficacy and well-established familiarity in industry. Side effects may include breakdown of corneal epithelium, apoptosis of ocular surface cells, accumulation in surface tissues, and tear-film instability. Disruption of corneal cell-cell junctions may allow BAK to act as a permeation enhancer, for poorly permeable hydrophilic compounds (Freeman and Kahook 2009)
Chlorobutanol	0.5	Chlorobutanol is an alcohol and exhibits an unspecific but broad effect on microorganism. It has some side effects like keratitis and irritation to ocular surface and is not thermostable and volatile (Freeman and Kahook 2009)
Chlorhexidine	(0.01)	Chlorhexidine offers good activities against bacteria. Possible side effects are tear film instabilities, irritation reactions, and corneal edema (Furrer et al. 2002)
Cetrimide	0.005	Cetrimide belongs to the group of quaternary ammonium cations. In in vitro studies it caused toxic side effects such as necrosis of human conjunctival cells (von Deylen et al. 2018)
Sodium perborate		Sodium perborate acts as oxidative preservative, releasing hydrogen peroxide. In contact with the precorneal tear film it dissociates. It is considered to be less toxic than BAK (Freeman and Kahook 2009)
Purite®	0.005–0.01	Purite® is a stabilized oxychloro complex and acts as oxidative preservative. Like sodium perborate, it is inactivated by the precorneal tear film. The toxicity is believed to be lower compared to BAK (Freeman and Kahook 2009)
SofZia®		SofZia® is an ionic buffer system containing boric acid, zinc chloride, sorbitol, and propylene glycol. Some studies suggested a better ocular tolerance compared to BAK (Freeman and Kahook 2009)
Thiomersal	(0.01–0.02)	Thiomersal is an organomercury compound that offers a broad antimicrobial spectrum. Nevertheless, its use is decreased because of its toxic side effects and its allergenic potential (Furrer et al. 2002)

Container Closure System

Every container closure system (CCS) should be shown to be suitable for its intended use: it should adequately protect the dosage form, it should be compatible with the dosage form, and it should be composed of materials that are considered safe for use with the dosage form and the route of administration. If the packaging system has a performance feature in addition to containing the product, the assembled container closure system should be shown to function properly (U.S. Food and Drug Administration 2004). The selected CCS for ophthalmic products should provide adequate protection from solvent loss and chemical contamination as well as

Table 5 Typical ingredient classifications and examples for ophthalmic delivery

Excipient classification	Role	Examples
Vehicle—Aqueous	Carrier for drug and other excipients for dosing	Typically water for injection utilized. Alternatively purified water, USP with endotoxin controls may be used
Vehicle—Nonaqueous	Carrier for drug and other excipients	Petrolatum based such as mineral oil, white petrolatum. Recent product in EU with semi-fluorinated alkanes
Preservative	Antimicrobial agent—Maybe bactericidal or bacteriostatic to prevent microbial growth in case of accidental contamination	Examples in Table 4
Tonicity agent	Adjust tonicity of drug product to match that of tears	Salts such as sodium chloride and potassium chloride; co-solvents such as glycerin, propylene glycol, or PEG
Buffer	Maintain the pH of the product	Phosphate, phosphate-citrate, lactate, borate, TRIS
Viscosity agent	Adjust viscosity to the desired range	Polymers such a HPMC, CMC, or carbomers; natural gums such as guar gum
Solubilizer	Solubilize poorly soluble drugs in vehicle	Surfactants (polysorbate, cremophor), complexing agents (cyclodextrin), co-solvents (PEG, propylene glycol), oils/lipids (castor oil, MCT)
Stabilizer	Enhance chemical stability of drugs	Chelating agents (EDTA); antioxidants (BHT, sodium metabisulfite)
Other specialized ingredients	Ingredients for specific functions, e.g., penetration enhancer for improved absorption; emollients for comfort	Penetration enhancers—Benzalokonium chloride, cyclodextrins; emollients such as glycerin, propylene glycol, or PEG

protect from microbial contamination. Since ophthalmic products are required to be sterile, they represent a high-risk category for causing damage to the eye if any contamination is introduced by contact with the packaging material or due to failure of the CCS in providing adequate protection. Further, since these products are usually liquid based with direct contact with the CCS, the potential effects of packaging component/dosage form interactions are numerous. These include higher risk to compatibility with the materials of the CCS which may result in impurities, decreased potency of drug or preservative due to adsorption or absorption, possibility of extractables/leachables ingress into the drug product during its shelf-life, and loss of solvent due to semi-permeable nature of most CCSs used for ophthalmic products. Further, as most of the packaging materials for ophthalmic products are typically made of low-density polyethylene (LDPE), these may allow permeation of volatiles; as a result, extractable/leachables from labels, adhesives, inks, and secondary containers may enter the product resulting in impurities. Lastly, the CCSs for these products include a performance feature—typically these serve as the

Table 6 Ophthalmic formulation container closure system examples

Container closure	Compatible dosage form	Considerations
Multi-dose bottle tip cap	Solution, suspension, or emulsion formulations containing anti-microbial preservatives	Typically made of LDPE polymer. Tip size selected determines the drop size delivered. Components manufactured and sterilized separately. Product filling into bottles under aseptic conditions for sterility
Unit dose vials	Preservative-free solutions, suspensions, emulsions	Typically made of LDPE, manufactured using blow-fill-seal process in aseptic conditions. Single use systems
Tubes	Ointments, gels	Metal tubes or laminated polypropylene tubes. Dispenser nozzle to allow for dosing of thin ribbon of product
Multi-dose preservative-free systems	Preservative-free solutions or emulsions	Typically utilize LDPE bottles with specialized tip assembly utilizing one-way valves or filters to prevent microbial contamination after multiple-dose administration and ensure sterility of product

dosing device allowing administration of a single drop or aliquot of the product. Functionality testing for this performance feature needs to be considered during development. Some of the commonly used CCSs for ophthalmic products are described in Table 6.

Sterilization Method Selection

Sterility is a critical quality attribute for all ophthalmic products. Further, it is not possible to assure sterility by testing, so it needs to be assured by the use of a suitable and validated manufacturing process. While developing manufacturing processes for ophthalmic product, it is essential to consider the sterilization method that is appropriate for the product and build it into the process. Generally, sterilization methods can be classified into two broad categories:

1. Terminal sterilization: Product is sterilized in its final container closure system.
2. Aseptic processing: Process performed maintaining the sterility of a product that is assembled from components, each of which has been using appropriate processes. Aseptic processing is not considered to be a sterilization process as it does not reduce any microbiological contamination but describes techniques to process sterile components without adding any microbiological contamination.

Products intended to be sterile should be terminally sterilized in their final container whenever possible, as clearly stated in the PhEur, general section "Phase/Stage of Development" (European Medicines Agency 2016). Terminal sterilization is preferred to sterilization by filtration and/or aseptic processing because it provides a sterility assurance level (SAL) that is possible to calculate, validate, and control, and thus incorporates a safety margin. For aseptic processes, a SAL is not applicable as accidental contamination caused by inadequate technique cannot be reliably eliminated by monitoring, control, or validation. Therefore, terminal sterilization provides the highest assurance of sterility and should be used whenever possible (European Medicines Agency 2016). The preferred method of sterilization is application of heat, as this method provides the maximum assurance of sterility. When terminal sterilization by heat is not possible, the application of an alternative method of terminal sterilization, sterilizing filtration and/or aseptic processing, may be considered.

For most topical ophthalmic products, terminal sterilization in the final packaging container is typically not feasible. Most container closures used for these products are comprised of HDPE, LDPE, or other heat-labile materials and are essential to product performance. Aseptic processing is the most commonly utilized—where drug product is manufactured by selection of appropriate methods of sterilization and then assembled in the final product and filled into the container under aseptic conditions. This requires that appropriate sterilization methods are required to selected processes developed and validated to ensure sterility of individual components. Further, it has to be demonstrated and validated that the aseptic process for the final product assembly to incorporate the pre-sterilized components and then the filling of the product into its container closure is capable of maintaining the sterility of the product.

Details of the sterilization methods typically utilized in the manufacture of sterile dosage forms are covered in elaborate details with guidances from the FDA (U.S. Food and Drug Administration 2004), EMA (European Medicines Agency 2000, 2016), and USP (United States Pharmacopeia and National Formulary (USP 41-NF 36) 2018g). Broadly, the sterilization methods used for topical ophthalmic products are listed in Table 7.

Other methods of sterilization, such as liquid phase sterilization and vapor phase sterilization, are also available which utilize either liquid phase or gaseous phases of chemical agents, respectively. The liquid chemical agents may include aldehydes, acids, bases, and strong oxidants in solution, while the vapor phase chemical agents include hydrogen peroxide, peracetic acid, formaldehyde, and glutaraldehyde in aqueous solution. However, these are not currently used in ophthalmic products due to the challenges in ensuring that the chemical agent is completely removed and any residual agent will not impact product quality or safety.

The EMA guidance (European Medicines Agency 2000, 2016) also provides decision trees that are intended to assist in the selection of the optimal sterilization method taking into account the various issues to be considered. When moving down the decision trees, the methods generally show decreasing levels of sterility assurance and therefore the first possible option should normally be chosen. This would ensure that the highest level of sterility assurance is achieved in conjunction with the lowest level of pre-sterilization bioburden appropriate. For topical ophthalmic products which are generally manufactured utilizing aseptic compounding, these decision trees would apply to the selection of sterilization methodology for sterilization of sub-components and process intermediate sub-parts.

Table 7 Sterilization methods applicable for topical ophthalmic products (for more details in each method, refer to European Medicines Agency (2016) and United States Pharmacopeia and National Formulary (USP 41-NF 36) (2018g))

Sterilization method	Description
Steam sterilization	The steam in the chamber directly contacts the surface of load items to effect sterilization. Generally used for sterilization of parts, hard goods, or porous items which are capable of withstanding these conditions
Moist heat sterilization	Typically applicable to aqueous products or sub-parts primarily in closed containers. Accomplished by application of heat to the container, heating of the container wall, and ultimately heating of the internal liquid volume
Dry heat sterilization	Utilized for heat-stable items (glass, stainless steel, nonaqueous liquids, powders, etc.) that are unsuited for steam sterilization. The process relies on air for the transfer of heat to and from the load items and takes longer than a steam process for a comparable size item or load. Lengthy heating and cooling periods require that the load items be unaffected by heat over a long period of time and also require the use of the overkill method for cycle development and validation
Sterilizing filtration of liquids	Utilizes an appropriate filter for the physical removal of microorganisms depending on the bioburden of the solution to be filtered, the properties of the solution, the filtration conditions, and the filter itself
Gas sterilization	Utilizes sterilizing gases (most common—Ethylene oxide) for the preparation of materials and equipment. Many polymeric materials, especially medical devices, are surface sterilized in this manner, as is non-pressure-rated process equipment. Limited to surface sterilization as gases do not penetrate into solid surfaces
Ionizing radiation sterilization	Utilizes the lethal effect of various forms of radiation as a means of microbial destruction. Ionizing radiation (gamma, x-ray, or beam) sterilization is used extensively for the sterilization of medical devices and for a variety of other materials such as packaging components. Non-ionizing radiation (microwave, UV rays) is not commonly used for sterilization

Formulation Development Strategy

Dosage form development for any route of administration should align with the overall project strategy for the stage of development, company priorities, and as described in previous section, drug substance properties. The following sections outline a quality by design approach to dosage form development. Following this approach can help a product development scientist to develop dosage forms.

QbD Approach to Formulation Development

Quality by design (QbD) is a systematic approach to development that begins with predefined objectives and emphasizes product and process understanding and process control, based on sound science and quality risk management. The overall principle of this approach is that quality should be built into products by design and that quality cannot be tested into products.

The following is a seven-step approach that can help a product development scientist to systematically develop products in an integrated manner from concept to commercialization (Pujara 2012).

Step 1: Define the desired dosage form and performance attributes through the quality target product profile (QTPP) as it relates to quality, safety, and efficacy.
Step 2: Identify approach to drug product formulation and manufacturing process development using domain expertise/prior knowledge and drug substance information.
Step 3: Identify potential critical quality attributes (CQAs) of drug substance, excipients, process intermediates, and drug product.
Step 4: Identify potential critical process parameters (CPPs), i.e., process parameters that may impact CQA, of all unit operations in the manufacturing process.
Step 5: Using risk assessment and experimental approaches, determine the functional relationships that link raw material CQAs and unit operation CPPs to drug product CQAs.
Step 6: Refine formulation and manufacturing process, if necessary, and repeat steps 3–5 to meet the QTPP defined in Step 1.
Step 7: Use the enhanced product and process understanding in combination with quality risk management to establish design space (includes raw material properties and process variables) and control strategy.

Quality Target Product Profile

A failure to define the product before development begins can be a major cause of both new-product failure and serious delays in time to market. In the pharmaceutical industry, until recently, most companies did not include a formal definition of commercial product quality attributes in drug development plans.

The FDA guidance on target product profile (TPP) is a summary of the drug development program described in terms of labeling concepts, mainly focusing on the safety and efficacy. This guidance does not address quality aspects of the product profile.

The International Council for Harmonisation (ICH) introduced the concept of quality target product profile (QTPP) (International Council for Harmonisation (ICH) 2009). The QTPP consists of quality characteristics (attributes) that the drug product should possess in order to reproducibly deliver the therapeutic benefit promised in the label. Considerations for the quality target product profile could include:

- Intended use in clinical setting, route of administration, dosage form, delivery systems, storage, and shipping conditions
- Dosage strength(s)

- Container closure system
- Therapeutic moiety release or delivery and attributes affecting pharmacokinetic characteristics (e.g., dissolution, aerodynamic performance) appropriate to the drug product dosage form being developed
- Drug product quality criteria (e.g., sterility, purity, stability, and drug release) appropriate for the intended marketed product

A QTPP should be established at the beginning of a project prior to the start of product development and included in integrated product development plans that include all aspects of product development (e.g., clinical, marketing). This should ensure a common understanding of the desired product attributes and align the teams toward a common goal.

A sample QTPP for a sterile ophthalmic product is shown below in Table 8.

During early development stages, a simple QTPP as shown in Table 9 may suffice to meet the intended purpose of early clinical trials.

Table 8 Example of a QTPP for a sterile ophthalmic product

Quality attribute of drug product	Target	Optimum	Minimum
Dosage form	Solution	Solution	Suspension
Dose strengths	0.01–0.1% w/v	0.01–0.15% w/v	0.01–0.1% w/v
pH	6.5–7.5	6.8–7.4	5.0–8.0
Preservative	BAK, meets PhEur A criteria	BAK, meets PhEur A criteria	BAK, meets USP
Assay	95–105%	95–105%	90–110%
Osmolality	280–330 mOsm	280–330 mOsm	280–330 mOsm
Viscosity	<100 cps	<50 cps	<300 cps
Dosing	QD	QD	BID
Storage	18 months at room temp	36 months at room temp	24 months refrigerated until dispensed
Packaging	Multi-dose bottles	Multi-dose bottles	Multi-dose bottles

Table 9 Example of a simplified QTPP for a sterile ophthalmic product during early development stages

Quality attribute	Target
Dosage form	Solution
Dose strength range	0.025–0.1%
Number of dose strengths	2 strengths for phase 1
pH	5–8
Preservative	Preservative-free
Dosing	QD
Storage	12–18 months at room temperature
Fill volume and container/closure	Unit of use in plastic bottles

Critical Quality Attributes (CQAs) for Ophthalmic Dosage Forms

A CQA of a material is a physical, chemical, biological, or microbiological property or characteristic that should be within an appropriate limit, range, or distribution to ensure the desired product quality. CQAs must be identified for drug product, drug substance, key excipients, in-process materials, and packaging components.

For example, purity, strength, and stability are CQAs that would be generally applicable to all dosage forms. Sterility is a CQA for ophthalmic dosage forms. Particle size would be a CQA of drug substances or excipients used in ophthalmic suspensions. Peroxide level could be a CQA for polysorbate 80 used in ophthalmic dosage forms. Tip size could be a CQA for a dropper bottle.

A table format could be used to identify critical quality attributes for further evaluation. The list of potential CQAs can be modified when the formulation and manufacturing process are selected and as product knowledge and process understanding increase. The following table (Table 10) is presented as an example and not to specifically categorize the listed quality attributes.

Critical Process Parameters (CPPs)

A critical process parameter, e.g., temp, time, speed, is a process parameter when a variable can affect the CQA of a product or process. Critical process parameters (CPPs) can be identified and investigated using risk analysis and DOE.

For example, in the drug substance table in section "Routes of Drug Delivery to Ocular Tissues," crystal form is considered a CQA of the drug substance. A cross-functional team could work together to develop an Ishikawa or fishbone diagram (example shown below in Fig. 3) to identify potential variables which can have an impact on drug substance crystal form.

The process of ranking the variables through a systematic approach has been described in ICH Q9 (Quality Risk Management). Using failure mode and effects analysis (FMEA) or similar tools based on prior knowledge and initial experimental

Table 10 Example of a list of potential CQAs during development

Quality attribute of drug substance	Target	Critical to drug product formulation and/or manufacturing process
Appearance	White to off-white	Not critical
Particle size	D90 < 25 μm	**Critical**
Impurities	Meets ICH and qualified limits	**Critical**
Optical rotation	+33–+39	Not critical
Assay	98–102%	**Critical**
Solid state form	Form I	**Critical**

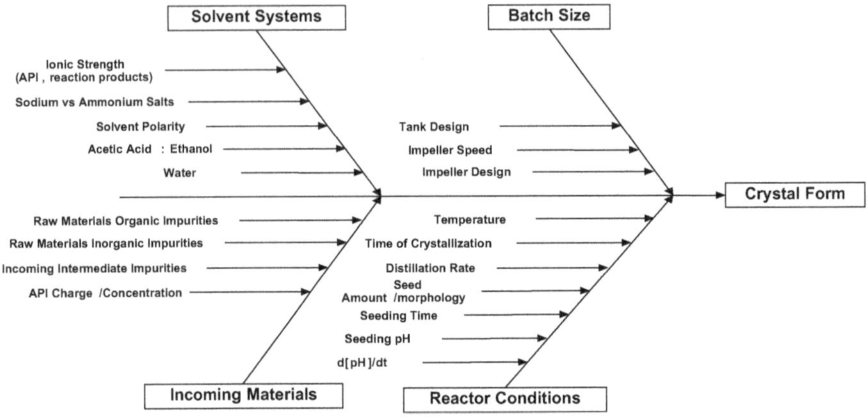

Fig. 3 Example of an Ishikawa or fishbone diagram to identify potential variables which may impact drug substance crystal form

data, the variables in the fishbone diagram can be ranked based on probability, severity, and detectability. Design of experiments or other experimental approaches could then be used to evaluate the impact of the higher ranked variables, to gain greater understanding of the process and identify critical process parameters.

A process parameter that is identified as low risk which leads to low probability of product failure is called a noncritical process parameter.

Relating Component CQAs and Manufacturing Process CPPs with CQAs of Dosage Form

The schematic below (Fig. 4) illustrates the concept of design space and control space (Adapted from MacGregor and Bruwer (2008)).

Design space is defined in ICHQ8 as "the multidimensional combination and interaction of input variables (e.g., material CQAs) and process parameters (unit operation CPPs) that have been demonstrated to provide assurance of quality." Therefore, the relationship between material CQAs, unit operation CPPs, and the product CQAs is described in the design space. A design space is useful in setting drug product specifications and in performing continuous improvement throughout the product life cycle.

The control space is a subset of the design space and is the internal target within the design space that is used to manufacture product of acceptable quality. A control strategy in the context of manufacturing process is a methodology to mitigate risks associated with the batch when the critical and noncritical process parameters fall outside the control space but within the design space.

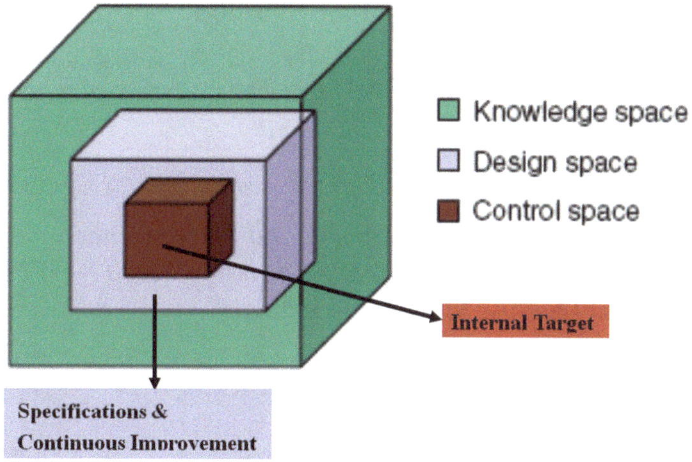

Fig. 4 Illustration of the concept of design space and control space (Adapted from MacGregor and Bruwer (2008)

Control Strategy

From ICH Q10 (International Council for Harmonisation (ICH) 2008), a control strategy is a planned set of controls, derived from current product and process understanding that ensures process performance and product quality. The controls can include parameters and attributes related to drug substance and drug product materials and components, facility and equipment operating conditions, in-process controls, finished product specifications, and the associated methods and frequency of monitoring and control. Therefore, reducing uncertainty and controlling variability are important to components as a good control strategy.

A control strategy can include, but is not limited to, the following:

- Control of input material attributes (e.g., drug substance, excipients, primary packaging materials) and intermediates based on an understanding of their impact on processability and product quality
- Product specification(s)
- Controls for unit operations that have an impact on downstream processing or product quality (e.g., the impact of drying on degradation or the impact of particle size distribution of the granulate on dissolution)

In-process or real-time release testing in lieu of end-product testing (e.g., measurement and control of CQAs during processing)A monitoring program (e.g., full product testing at regular intervals) for verifying multivariate prediction models.

Case Example: Sterile Ophthalmic Solution

The following is an example to illustrate the concepts presented in the previous section.

Step 1: Define Desired Dosage Form and Performance Attributes through the Quality Target Product Profile (QTPP) as it Relates to Quality, Safety, and Efficacy

For this exercise, a simple QTPP is used Table 11.

Step 2: Identify Approach to Formulation and Manufacturing Process Development of the Drug Product Using Domain Expertise/Prior Knowledge and Drug Substance Information

From prior knowledge and existing technology, a sterile preserved solution will be developed using conventional ophthalmic excipients and manufacturing process. Drug substance physicochemical property information is available.

Step 3: Identify Potential Critical Quality Attributes (CQAs) of Drug Substance, Excipients, Process Intermediates, and Drug Product

Critical quality attributes of the drug product are presented in Table 12.

It is also important to identify potential CQAs for drug substance and for key excipients.

Table 11 Case study: example of a QTPP

Quality attribute	Minimum	Target	Optimum
Dosage form	Emulsion with low oil content	Solution	Solution
Dose strengths	0.05–0.1% w/v	0.03–0.04% w/v	0.01–0.02% w/v
pH	5.0–8.0	6.8–7.4	6.5–7.8
Preservative	Meets USP	Meets PhEur A	Meets PhEur A
Dosing	BID	QD	QD
Storage	18 months refrigerated until dispensed	18 months at room temperature	24 months at room temperature
Packaging	Multi-dose bottles	Multi-dose bottles	Multi-dose bottles

Table 12 Critical quality attributes of case study drug product

Critical quality attribute	Target
Dosage form	Solution
Potency (strength)	0.05–0.1%
Appearance	Clear and colorless, no visible particulates
Identity	Positive for compound X
Assay	95–105% at release90–110% at end of shelf
Impurities	No single impurity greater than 0.9%
pH	Physiological pH
Isotonicity	280–330 mOsm
APET	Meet USP and PhEur A criteria
Sterility	Meet USP and PhEur
Particulate matter	Meet USP and PhEur
Leachables	Below safety threshold

Step 4: Identify Potential Critical Process Parameters (CPPs), i.e., Process Parameters that May Impact CQA, of All Unit Operations in the Manufacturing Process

As an example, for dosage forms such as solutions, the CPPs for mixing unit operation may be identified as mixing speed, mixing time, and temperature.

Step 5: Using Risk Assessment and Experimental Approaches, Determine the Functional Relationships That Link Raw Material CQAs and Unit Operation CPPs to Drug Product CQA

Continuing with the mixing unit operation, tables can be generated to represent acceptable ranges for each of these parameters (mixing time, mixing speed, and temperature). When each unit operation has been evaluated, a risk assessment can be performed to identify critical unit operations that have impact on drug product critical quality attributes.

For example, Table 13 shows unit operations that may present a high risk for a particular drug product CQA.

Table 13 Linking unit operation CPPs to drug product CQA for the case study

DP CQA Unit operation	Order of addition	Mixing	pH adjust	Filtration	Filling	Packaging
Appearance	Med	Med	Low	Low	Low	Low
Assay	Low	High	High	High	Low	Low
Impurity	Low	Med	High	High	Low	High
pH	Low	Low	Med	Low	Low	Low
Tonicity	Low	Med	High	Med	Low	Low
BAK	Low	Low	Low	High	Low	Low
Sterility	Low	Low	Low	High	High	High

Table 14 Design space for formulation and a mixing manufacturing process for the case study

Formulation ingredients or unit operation CPP	Design space
Solubilizer concentration	0.5–1.0% w/v
Viscosity agent concentration	1–2% w/v
Drug substance concentration	98–102%
Buffer concentration	+/− 10% of target
Tonicity agent	+/− 10% of target
Mixing time	30–120 min
Mixing speed	100–200 rpm
Temperature	NMT 50 °C

Step 6: Refine Formulation and Manufacturing Process, if Necessary, and Repeat Steps 3–5 to Meet the QTPP Defined in Step 1

After analyzing information from steps 3 through 5, any residual risk that remains after formulation and process development can be addressed by refining the formulation and/or manufacturing process to meet the QTPP.

For example, increase solubilizer to ensure drug substance is completely dissolved throughout shelf-life or add a pre-mixing step to ensure preservative concentration is in the acceptable range in the final drug product.

Step 7: Use the Enhanced Product and Process Understanding in Combination with Quality Risk Management to Establish Design Space (Includes Raw Material Properties and Process Variables) and Control Strategy

Table 14 illustrates sample design space for formulation and a mixing manufacturing process. A complete design space would have all the input variables (raw material/formulation CQAs and unit operation CPPs) included in the design space.

Conclusion

Ocular delivery presents unique challenges for a formulation development scientist in designing appropriate delivery systems that consistently and reproducibly deliver the therapeutic benefit to target tissues. A wide variety of dosage forms are feasible to deliver therapeutic benefit while meeting the rigorous safety and tolerability requirements for ocular tissues. Prior to initiating dosage form development, it is important that sufficient consideration and rigor are given to select and compile a list of the potential quality and performance attributes that are required to be designed into the product. A quality by design (QbD) approach can help product development scientists to systematically develop patient-centric products in an integrated manner from concept to commercialization.

References

Attar M, et al. Safety evaluation of ocular drugs. In: A comprehensive guide to toxicology in preclinical drug development; 2013. p. 567–617.

Bausch & Lomb. Muro 128® 5% (sodium chloride hypertonicity ophthalmic solution, 5%; NDC 24208-277-15) and Muro 128® (sodium chloride hypertonicity ophthalmic ointment, 5%; NDC 24208-385-55). Bausch & Lomb; n.d.

Chen H, et al. Recent developments in ophthalmic drug delivery systems for therapy of both anterior and posterior segment diseases. Coll Interface Sci Commun. 2018;24:54–61.

Cholkar K, et al. Eye: anatomy, physiology and barriers to drug delivery. In: Ocular transporters and receptors; 2013. p. 1–36.

Durairaj C. Ocular pharmacokinetics. Handb Exp Pharmacol. 2017;242:31–55.

Dutescu RM, et al. Semifluorinated alkanes as a liquid drug carrier system for topical ocular drug delivery. Eur J Pharm Biopharm. 2014;88(1):123–8.

European Medicines Agency. Decision trees for the selection of sterilization methods (annex to note for guidance on development pharmaceutics) CPMP/QWP/054/98 Corr; 2000.

European Medicines Agency. Guideline on the sterilization of the medicinal product, active substance, excipient and primary container—draft. EMA/CHMP/CVMP/QWP/BWP/850374/2015; 2016.

Freeman PD, Kahook MY. Preservatives in topical ophthalmic medications: historical and clinical perspectives. Expert Rev Ophthalmol. 2009;4(1):59–64.

Furrer P, Mayer JM, Gurny R. Ocular tolerance of preservatives and alternatives. Eur J Pharm Biopharm. 2002;53(3):263–80.

Gangrade NKGNB, Ganesan MG, Redy IK. Topical ophthalmic formulations: basic considerations. In: Reddy IK, editor. Ocular therapeutics and drug delivery: a multidisciplinary approach. Lancaster, PA: Technomic Publishing Co., Inc.; 1996. p. 371–85.

Gaudana R, et al. Recent perspectives in ocular drug delivery. Pharm Res. 2009;26(5):1197–216.

Gore A, et al. Ocular emulsions and dry eye: a case study of a non-biological complex drug product delivered to a complex organ to treat a complex disease. Generics and Biosimilars Initiative J. 2017;6(1):13–23.

Graham RS, Tien WL, Attar M, Schiffman R, Morgan A, Hollander DA. Cyclosporin compositions. USA: USPTO; 2008.

Henderer JD, Rapuano CJ. Ocular pharmacology. In: Brunton LL, Hilal-Dandan R, Knollmann BC, editors. Goodman & Gilman's: the pharmacological basis of therapeutics. 13th ed. New York: McGraw-Hill Education; 2017.

International Council for Harmonisation (ICH). Pharmaceutical quality system. 2008.
International Council for Harmonisation (ICH). Q8 (R2) Pharmaceutical development. 2009.
MacGregor JF, Bruwer M-J. A framework for the development of design and control spaces. J Pharm Innov. 2008;3(1):15–22.
Maurice DM, Mishima S. Ocular pharmacokinetics. In: Sears ML, editor. Pharmacology of the eye. Berlin: Springer; 1984. p. 19–116.
Pujara C. QbD and QRM strategies in pharmaceutical product development. Arlington, VA: Association of Pharmaceutical Scientists, Southern California Discussion Group; 2012.
Schoenwald RD. Ocular pharmacokinetics. In: Zimmerman TJ, editor. Textbook of ocular pharmacology. Philadelphia, PA: Lippincott-Raven; 1997. p. 119–38.
U.S. Food and Drug Administration. Guidance for industry—sterile drug products produced by aseptic processing. Current good manufacturing practice. Silver Spring, US: FDA; 2004.
United States Pharmacopeia and National Formulary (USP 41-NF 36). <771> Ophthalmic product—quality tests. Rockville, MD: United States Pharmacopeial Convention; 2018a.
United States Pharmacopeia and National Formulary (USP 41-NF 36). <1151> Pharmaceutical dosage forms. Rockville, MD: United States Pharmacopeial Convention; 2018b.
United States Pharmacopeia and National Formulary (USP 41-NF 36). <4> Mucosal drug products—product quality tests. Rockville, MD: United States Pharmacopeial Convention; 2018c.
United States Pharmacopeia and National Formulary (USP 41-NF 36). <1771> Ophthalmic products—performance tests. Rockville, MD: United States Pharmacopeial Convention; 2018d.
United States Pharmacopeia and National Formulary (USP 41-NF 36). <1> Injections and implanted drug products (parenterals)—product quality tests. Rockville, MD: United States Pharmacopeial Convention; 2018e.
United States Pharmacopeia and National Formulary (USP 41-NF 36). <789> Particulate matter in ophthalmic solutions. Rockville, MD: United States Pharmacopeial Convention; 2018f.
United States Pharmacopeia and National Formulary (USP 41-NF 36). <1229> Sterilization of compendial articles. Rockville, MD: United States Pharmacopeial Convention; 2018g.
Urtti A. Challenges and obstacles of ocular pharmacokinetics and drug delivery. Adv Drug Deliv Rev. 2006;58(11):1131–5.
von Deylen D, et al. New classes of polycationic compounds as preservatives for ophthalmic formulations. Pharm Res. 2018;36(1):11.
Wirta DL, et al. A clinical phase II study to assess efficacy, safety, and tolerability of waterfree cyclosporine formulation for treatment of dry eye disease. Ophthalmology. 2019;126(6):792–800.

Intraocular Injectable Dosage Form Development: Key Components and Critical Quality Attributes

Ken Reed

Abstract This chapter focuses on products that are administered in an intraocular manner (occurs within the eye) and the numerous product characteristics which must be identified and successfully incorporated to ensure the medicinal products are effective and safe. These key product characteristics known as Critical Quality Attributes (CQAs) are physical, chemical, biological, or microbiological characteristics that define whether a product has been manufactured with sufficient quality.

Commonly recognized CQAs (e.g., drug purity, sterility for injectable products) are mostly determined by the product's dosage form. Advanced delivery systems, including intravitreal injections, injections of small implants, and the surgical placement of implants, demand special considerations in the definition of Critical Quality Attributes (CQAs). The products final compilation of CQAs is a combination of standard accepted CQAs, and the products unique CQAs determined experimentally during the development cycle.

A development plan delineates the experimental studies needed to define a product's CQAs and its quality design space. Relevant product attributes are used to select the final product formulation and packaging. Manufacturing process CQAs should be utilized in the definition of the manufacturing process as used in the production of pivotal clinical and primary stability batches. The knowledge and implementation of a drug product's CQAs enables the elucidation of quality standards and results in the manufacture and distribution of quality drug product.

Keywords Product attributes · Quality assurance · Quality design · Product profile · Development · Regulatory · Manufacturing · Stability · Specifications · Intraocular route · Critical quality attributes · Product development · Biodegradable · Quality target product profile

---

K. Reed (✉)
Platform Ophthalmic Innovations, Brentwood, TN, USA
e-mail: ken.reed@belmont.edu

S. Neervannan, U. B. Kompella (eds.), *Ophthalmic Product Development*, AAPS Advances in the Pharmaceutical Sciences Series 37,
https://doi.org/10.1007/978-3-030-76367-1_8

Introduction and Background

Overview

The major objective of the development path of a new drug product is to identify the attributes (characteristics) that are critical to the quality of that drug product. It is generally understood that the development of a new product should occur according to a plan that focuses on defining the essential product characteristics. The resultant complete scientific understanding of essential product characteristics enables the definition of relevant product design space by the development scientist. The design space of a medicinal product is defined as "The multidimensional combination and interaction of input variables (e.g., material attributes) and process parameters that have been demonstrated to provide assurance of quality" (ICH 2009). The definition of a critical quality attribute (CQA) is given as "a physical, chemical, biological, or microbiological property or characteristic that should be within an appropriate limit, range, or distribution to ensure the desired product quality" (ICH 2009). A good approach to defining the design space of the product is to identify product CQAs as based upon possible detrimental effects upon efficacy and safety (Maguire and Peng 2015). Examples of CQAs include the potency of the drug, the amount and type of drug substance-related impurities, and the amount and type of drug product-related impurities resulting from shelf-life time and/or the manufacturing process.

Good practice entails generating a quality target product profile (QTPP) at the beginning of the development process for a new chemical entity (NCE) (ICH 2009). Background information should include what the intended use is, the proposed route of administration, strength, dosage form(s), why the type of dosage form(s) was selected, the definition of acceptable safety, the definition of acceptable therapeutic effectiveness, and additional relevant information. It is helpful to design an internal drug product profile (IDPP) that includes plans to address company goals that are in addition to the product performance goals contained in the QTPP.

Intraocular Administration

The definition of the intraocular route of administration as given by the FDA is simply "administration within the eye" (FDA CDER Data Element Number 2019). Similarly, the FDA definition of an intravitreal route of administration is "administration within the vitreous body of the eye" (FDA CDER Data Element Number 2019). That is, an intravitreal injection occurs when the medication is injected directly into the posterior section of the eye. This mode of administration ensures the delivery of a drug, independent of its molecular size, to the back of the eye (Choonara et al. 2010; Thrimawithana et al. 2011). An intravitreal injection is an invasive procedure, but the use of this injection technique dramatically reduces side effects resulting from systemic distribution of the drug when administered via alternate pathways (Kim et al. 2014). Intravitreal injections are conducted using a syringe equipped with either narrow diameter 27- or 30-gauge needles. The injection volume of these medications varies from 20 to 100 μL. The FDA does not allow the intravitreal injection of large volumes of medication (Mandal et al. 2018). Alternate injection sites, such as subretinal and suprachoroidal, have been investigated to improve clinical outcomes (Hartman and Kompella 2018). For the purpose of this chapter, they are also considered as being administered intraocularly as well as sub-conjunctival, intracameral, and intra-scleral administrations. It is assumed in this chapter that quality attributes described for products designed for intravitreal injection are, for the most part, generally applicable to medications injected at the subretinal and suprachoroidal sites. However, it should be kept in mind that each product has its own particular quality performance characteristics that will determine the final defined CQAs.

Intravitreal Administration of VEGF Inhibitors

The VEGF inhibitor macromolecules have proven clinically effective for inhibiting neovascularization. Anti-VEFG macromolecules marketed for treating age-related macular degeneration (AMD) include pegaptanib (Macugen®), ranibizumab (Lucentis®), and aflibercept (Eylea®) (Kim et al. 2014). They are considered protein- and peptide-based biopharmaceutical agents (Mandal et al. 2018). Pegaptanib is a pegylated aptamer, ranibizumab (~48 kDa) is a monoclonal IgG1 antibody fragment, and aflibercept (~97 kDa) is a recombinant fusion protein (Mandal et al. 2018). In order to reach their site of action in sufficient quantities, these agents are given as intravitreal injections (Kim et al. 2014; Mandal et al. 2018; Radhakrishnan et al. 2017). In general, an intravitreal injection procedure can be performed safely by qualified health professionals (Kim et al. 2014). However; serious side effects are reported with the intravitreally injected VEGF inhibitors that include endophthalmitis, retinal hemorrhage, retinal detachment, vitreous hemorrhage, blindness, retinal tear, vitreous detachment, and increased intraocular pressure (EMEA 2018a, b, c). Complications that occur more frequently but are less serious include anterior chamber inflammation, blepharitis, conjunctival hemorrhage, dry eye, eye irritation, eye pain, eye pruritus, foreign body, visual disturbance, and vitreous floaters (EMEA 2018a, b, c). It is stated by the manufacturers that most of the adverse reactions are due to the intravitreal injection procedure itself (EMEA 2018a, b, c).

Alternate Modes of Administration

The seriousness and frequency of side effects attributable to intravitreal injection warrant both the exploration of alternate injection sites and the development of novel drug delivery technologies designed to decrease the frequency of injection (Choonara et al. 2010; Thrimawithana et al. 2011; Kim et al. 2014). The injection of small implants (rods) or the surgical placement of implants into the posterior ocular space has been designed to enable the delivery of both small and large molecular weight drug molecules for long periods of time. The major advantage of the implants is the avoidance of side effects attributable to the intravitreal injection process itself by decreasing the frequency of injections or surgical procedures (Choonara et al. 2010; Delplace et al. 2015). The marketed products Vitrasert® and Retisert® represent nonbiodegradable systems that are surgically anchored to the sclera. Drug molecules can be dispersed in a reservoir of a nonbiodegradable implant that is surrounded by a polymer membrane that controls the rate of drug release. The design of these implants results in a rate of drug release that is fairly constant and approaches zero order (Yasin et al. 2014).

Polymers that undergo biodegradation in the eye can be used in the fabrication of drug-loaded implants. A design can be used that involves the drug being dispersed throughout the biodegradable polymer that is often referred to as a matrix design. An initial burst of released drug results when the matrix system is used that is followed by a more constant rate of drug release (Yasin et al. 2014). The major advantages to using inserts fabricated from biodegradable polymers include the following: (1) the implants do not have to be surgically removed, (2) more efficient drug loading is possible, and (3) it is possible to have drug release occurring up to several months (Kim et al. 2014). Disadvantages to using biodegradable implants include a significant drug release burst effect and an overall release rate that is less constant than observed with nonbiodegradable implants (Kim et al. 2014). The biodegradable polymer polylactic-co-glycolic acid (PLGA) is utilized in the composition of the insert Ozurdex®. The Ozurdex® product is a biodegradable rod which is injected into the vitreous (Kim et al. 2014; Yasin et al. 2014). It has demonstrated sustained drug release and high drug-loading capacity (Kim et al. 2014).

Biodegradable polymers have also been used in the manufacture of both microspheres and nanospheres for periocular, suprachoroidal, and intravitreal injections (Thrimawithana et al. 2011; Kim et al. 2014). It has been proposed that the intravitreal injection of drug-loaded microspheres or nanospheres will result in prolonged ocular delivery and reduction in frequency of administration. PLGAs are commonly used in the fabrication of micro- and nano-particles (Mandal et al. 2018). The rate of drug release from these systems is complex and involves several product attributes (Choonara et al. 2010). These advanced delivery systems thus demand special considerations in the definition of CQAs. For example, the in vitro rate of release must be well characterized and should be related, if possible, to in vivo activity.

Drug Development Process

The drug development process has an end point of submission and then approval of the proposed new product by the health authorities. The International Council for Harmonisation (ICH) Guideline "Pharmaceutical Development Q8" describes the information requested for the pharmaceutical development section of the common technical document (CTD) which in turn is used for submission to health authorities (ICH 2009). It gives an overview of drug development information required for a complete characterization of a new drug product and is an excellent reference to use, along with the QTPP and IDPP, in the generation of a development plan. The development plan should delineate future experimental studies that will define the product's CQAs and design space. In turn, the product development experience itself should consist of the conduct of experiments which accumulate data probing the parameters that are critical to the performance of the drug product. The experience gained from previous product development studies conducted on products with similar target profiles is invaluable in the generation of projected CQAs. Historical and relevant product development data should include data from commercially manufactured products, literature sources, and information from toxicology, pharmacology, and clinical studies of previous similar drug compounds. Formal and supportive stability data that includes both long-term and accelerated studies are particularly helpful. Previous manufacturing experience from scale-up studies and process validation studies gives valuable insight into possible critical manufacturing process parameters and any needed in-process testing.

Linkage of Product Characteristics Throughout Development

The defining characteristics of the drug substance should be monitored and documented throughout the development process. Any changes to the characteristics of the active compound (e.g., changes in the structure, polymorphic form, solvents, etc.) are to be thoroughly examined and justification provided as to how preclinical and clinical study results are affected (or not affected) by these changes. New chemical entities are often given code names during the early stages of development. It is advisable that the naming system is well documented and understandable. Any code name changes during the development process must be clearly identified and tracked as to which code names were used with the various development studies. At the time of submission, it should be clearly communicated as to whether any changes to the drugs structure, salt form, polymorphic form, etc. occurred during the development process of the new drug product. The product development scientis should be aware of which scale up lots of active substance are being used during the various stanges of product development. The formulation composition(s) used during the different phases of development should be documented and easily reported both in summary and in detail. Formulation identification codes are often used to facilitate documentation and tracking of the composition. Documentation and tracking is still needed even if the formulation composition is constant during the conduct of all clinical trials. The differences between candidate formulations and the progression to the final defined formulation(s) should be clearly documented. Ongoing communication and coordination between biochemists, pharmacologists, toxicologists, analysts, formulators, and process development scientists are incredibly important and should be well documented. Changes often occur during the development cycle in the means by which the product is manufactured. These changes and any possible failures present an important opportunity to gather information regarding the design space of the drug product. Unexpected experimental results can be especially helpful in this regard.

It is important to consider final submission requirements during the preclinical and early clinical stages. It must be clear how the lead compound (and related compounds such as impurities) studied in early preclinical toxicological and pharmacological studies is linked structurally to the drug molecule as used in clinical studies. Early preclinical studies are used to assist in the definition of the apparent safety and efficacy of the NCE. It is expected that these pharmacology and toxicology studies will assist in determining target drug concentration, expected duration of activity, nature of toxic effects, concentrations at which toxic effects appear, etc. It is understood that additional characterization data generated during the development process may be utilized in product improvement. However, any improvements must be done in a manner so as to maintain product consistency. That is, the results from preclinical toxicology and pharmacology studies must remain relevant for all clinical studies (Finn 2016).

Submission Components

The following submission components are to be available at the end of the drug development process: (1) the identification and characterization of the drug substance, (2) the definition of pertinent analytical methods with their validations, (3) a description of the drug product composition (the description is to include an identification of the drug product inactive ingredients (excipients) with their concentrations and a rationale is to be given as to the selection of the excipients and their quality requirements), (4) the selection and rationale of the container closure system and any associated dosing device (if needed), (5) a defined and well-controlled manufacturing process, (6) stability data supporting the conduct of GLP preclinical and GMP clinical studies, and (7) stability data used to establish an expiration date for commercial material.

Identification and Characterization of Drug Substance

The drug development process for low molecular weight drug substance is initiated by obtaining the drug's structural formula, the stereochemistry of the drug, and the drug's molecular mass. Additional physicochemical properties of the drug substance that should be determined include solubility, water content, and particle size. Characterization of the solid state form of the drug substance should include crystal properties (including polymorphic forms) and any changes that occur in crystal structure upon storage. For small proteins, the overall amino acid composition may determine the desired biological effect and must be measured. However, the overall amino acid composition probably does not determine the desired biological effect by itself for large proteins and is hence less useful. In general, large protein biopharmaceuticals have complex structural requirements for biological activity. Information required for biological protein drug substances includes primary, secondary, and higher-order structural information that is in addition to the physicochemical properties that have already been listed.

The biological and permeability properties of a biopharmaceutical drug substance(s) need to be consistently controlled throughout the development cycle. Possible changes to the structure and purity of the biopharmaceutical molecule are to be monitored during the development process and assessments made as to any possible affects upon efficacy and safety. Proteins normally display some structural heterogeneity and the desired macromolecule often will contain post-translationally modified forms (e.g., glycoforms). That is, the biopharmaceutical drug substance can include several different molecular forms or variants (ICH 1999a). Elucidation of any of a biopharmaceutical's post-translational forms and any immunochemical properties (e.g., schematic amino acid sequence indicating glycosylation sites) must be determined (ICH 1999a). The pattern of heterogeneity needs to be well defined and monitored during the development process in order to assure consistency. This is especially pertinent when comparing drug substance lots used in the manufacture of supplies used for preclinical studies versus lots used in the manufacture of clinical studies (ICH 1999a). The overall conformation of biopharmaceuticals can change throughout the development of manufacturing procedures and can be due to many possible causes such as oxidation and deamination (Eon-Duval et al. 2012). Additionally, process-related impurities and contaminants may arise during manufacturing procedures (Eon-Duval et al. 2012). "Process-related impurities encompass those that are derived from the manufacturing process, i.e., cell substrates (e.g., host cell proteins, host cell DNA), cell culture (e.g., inducers, antibiotics, or media components), or downstream processing" (ICH 1999a). Raw materials (e.g., components of the cell culture media) may contaminate the biopharmaceutical by co-eluting with the drug biomolecule (Eon-Duval et al. 2012). It should be emphasized that known contaminants must be controlled. It is imperative to monitor, document, and communicate to the development team any changes to (1) the manufacturing steps intended to remove or inactivate viral contaminants, (2) the controls for viral contamination, (3) the source of the cell substrate, (4) the expression construct used to genetically modify cells, (5) the initial cell clone used to develop the master cell bank, and (6) changes in cell line stability during production and storage, especially in regard to possible viral contamination of the cells and cell banks (ICH 2002).

Analytical Methods

All pertinent analytical methods must be properly validated at the time of submission as given in the International Council for Harmonisation (ICH) validation guidelines for analytical procedures (ICH 1994). In many cases, analytical methods will evolve during the drug product development process. During the initial stages of development, an analytical method may be used in which the analyst has confidence that the method meets the criteria of accuracy and precision and can be fully validated once the formulation is defined. It is understood that any changes to the formulation will result in the analytical method being revalidated. It is a good development strategy that the potency assay is one of the first methods to be developed (Finn 2016). This important analytical method should then be quickly followed by the development of analytical methods for impurities, especially impurities that possess pharmacological and/or toxic properties. An evaluation should be made as to whether the stereochemistry of the drug substance and its relationship to pharmacological activity justify specific identification testing and/or performance of a chiral assay. It is important to be able to track the state of development of the analytical method (e.g., drug assay) used to generate data during development. An ongoing objective overview of the evolution of the development of analytical methods should be performed to assign a level of confidence in analytical results generated at the early stages of drug product development. It is critical to link the identification and measured amounts of drug impurities that were obtained for early pharmacology, toxicology, and stability studies to impurity levels seen in later studies. This is especially true for later clinical studies and long-term stability studies. Drug substance and drug product quality attributes must be monitored using appropriate analytical methods during scale-up from the bench, preclinical studies, clinical studies, and commercial manufacture.

Identification and Rationale of Product Composition

It is prudent to conduct characterization studies of the new chemical entity (NCE) prior to defining the product composition and packaging. The experimental generation of this physiochemical data is often referred to as generating a preformulation profile. A potency assay is needed at this point of development in which there is confidence by the analyst that it can be validated. It is prudent that the assay is robust and can be properly used (minor or no method variations) with newly developed formulations. The solubility of the NCE is a physical property that likely has an impact upon product performance. The solubility of the NCE should be determined at a range of pH values that includes physiological pH. It is advantageous to conduct the solubility study in the presence of possible tonicity agents that are both ionic and nonionic in nature. It is expected that stability studies will be conducted which indicate that pH where the NCE is most stable. The accelerated stability conditions of elevated temperature, added peroxide (or high oxygen levels), and intense light should be conducted in order to provide insight as to the NCE's shelf-life and whether the NCE is sensitive to oxidation and/or light. Potential antioxidants should be examined if it has been determined that the drug molecule is likely subject to oxidative decomposition. If analytical methods have progressed sufficiently, additional information into possible degradation products may be generated. Studies should be conducted to determine the compatibility of the drug substance with possible excipients and includes testing for any possible specific buffer catalyzed instability. Conformational transformations of proteins may occur due to high salt concentration and other environmental factors (Mandal et al. 2018; Baid et al. 2011). Analytical methods (including any possible biological assays) should be selected which can accurately and precisely measure loss of biological activity. The generation of pH stability profiles is especially important for protein drugs which are often found to be unstable at physiological pH. Biologics are often unstable at room temperature and accelerated studies may be designed with refrigerated conditions (4–8 °C) considered as the storage condition for the future product label. Relevant higher-ordered structure of biologics should be monitored during the conduct of preformulation stability studies. Instability of biologics may manifest itself in denaturation, adsorption, aggregation, and precipitation.

The formulation development scientist gives a scientific defense as to why the composition (formulation) of the drug product was selected in the final submission to the health authorities. Important attributes of the formulation should become clear during experimentation performed to define the final formulation. The rationale for formulation selection should lead to identification of critical formulation attributes once the final formulation is defined. A rationale and appraisal are needed for any identified differences between preclinical formulation(s) and the formulations used in clinical studies. Additionally, a rationale needs to be developed to justify any changes between the formulations used in pivotal clinical batches versus formulation(s) used in primary stability batches. The selection of excipients in the drug product formulation should be based upon their ability to assist in meeting the desired drug product characteristics. A special consideration in the product design of medications intended for intravitreal injection is the desire for transparency of the product in order to minimize interference with vision (Kim et al. 2014). It should be demonstrated that excipients at their selected concentrations are safe and meet their purpose (functionality). Allowable ranges should be justified on the basis of development and manufacturing experience. The selected concentration should be at the lowest amount that meets the intended purpose.

Possible excipients include tonicity agents, buffers, antioxidants, and antimicrobial agents. The inclusion of a buffer may contribute to pain upon injection and the minimal amount should be used (Bontempo 1997a). An antimicrobial preservative is needed if the drug product is packaged as a multidose. It is unlikely that a product intended to be administered intraocularly would be developed as a multidose. This is because of additional risks as compared to a single-dose presentation. Nevertheless, if a multidose presentation is to be developed, then the stability profiles of possible antimicrobial agents should be determined. The effectiveness of the antimicrobial in the full formulation should be assessed using an accepted pharmacopeia antimicrobial preservative effectiveness test. Antimicrobial preservative effectiveness should be demonstrated during the entire development process, including scale-up, clinical, and primary stability batches. A metal-chelating agent such as EDTA may be added to both small MW drugs and large MW biologics in order to reduce oxidation and/or other adverse effects upon product quality due to heavy metal contamination. This may be especially important for protein drug substances as they may pick up heavy metals during their manufacturing process (Bontempo 1997a). Surfactants may be added to biologics to interfere with the formation of protein aggregates (Mandal et al. 2018; Bontempo 1997a) and/or to prevent protein unfolding when exposed to air (Baid et al. 2011). Polyols (e.g., mannitol, sorbitol, and glycerol), small sugars (e.g., trehalose), and polysaccharides (e.g., dextrans) have been shown to stabilize proteins (Mandal et al. 2018; Jain and Roy 2009). The polyol mannitol can also serve as a nonionic tonicity agent. Excipients should be used in the product composition that are generally recognized as safe (GRAS) if possible (Baid et al. 2011). The quality standards for inactive ingredient quality standards should be based (if possible) upon compendia monograph(s), the function(s) of the inactive ingredient, and the need to control for detrimental contaminants (especially for biological raw materials) (ICH 1999a). It is preferred that (1) less expensive excipients be utilized that fully meet quality needs, (2) three or more vendor lots of critical excipients should be evaluated, and (3) one or two backup vendors are evaluated for critical excipients.

It may be decided that the drug product should be separated into two or more components that include a liquid diluent. This approach adds complexity to the dosage form that will usually result in increased risks (e.g., decreased assurance of sterility) and costs. Therefore, a clear rationale must be established for this approach such as an increase in stability. The following must be addressed in the rationale: (1) the possible precipitation of drug substance in solution, (2) any possible sorption to packaging components, and (3) any possible negative impact upon the product's sterility as presented to the patient.

The release profile from both biodegradable polymeric inserts and particle systems is affected by the amount of loaded drug, the drug's solubility in the polymer matrix, the polymer composition (molecular weight if PLGA), the ability of the drug to diffuse through the polymer matrix, and the surface area (geometric shape) of the dosage form (Choonara et al. 2010; Kim et al. 2014). The total surface area of particulates is determined by the particle size distribution. If the drug is presented as a microsphere or nanosphere, an acceptable particle size distribution must be defined. Particle size data is usually given as the mean particle size and the percentage of total particles in upper and lower ranges. The particle size range is also related to the ability of the product to be resuspended and the correct particle size distribution will assist in the patient receiving a homogeneous dose. The IDPP should include how vigorously and for how long the medication must be agitated by the user prior to administration. These properties should be assessed during product development along with time sedimentation profiles and the effect of particle size upon the ability to inject the product. A major positive attribute for both implants and particles is the possibility of continuous presentation of medication without frequent injections. Assuming that an insert or particulate dosage form was selected in order to prolong the presentation of drug to the retina, it is likely that dissolution data will be required. It is desired that a strong correlation is achieved between in vitro release profiles and in vivo clinical performance, so that acceptance criteria are more easily established for release profile results. However, it is very challenging to achieve such a correlation. The in vitro release profile is still often considered a CQA even if the correlation between in vitro and in vivo results has not been achieved. It is believed that the in vitro release profiles from both polymer implants and particulate products are important measures of the rate at which drug will be presented to the tissue.

Selection and Rationale of Container Closure System

The preferred packaging configuration and its dimensions are usually part of the IDPP. However, the selection of the final container closure system(s) should be based upon experimental outcomes that assess drug product stability, the ability to manufacture, and any other pertinent product performance characteristics. The type of material (e.g., glass, polypropylene, etc.) used for the primary packaging should be selected on the basis of (1) whether sorption of the drug to the container occurs, (2) whether leaching occurs from the container into the drug formulation, (3) if needed, whether the material can protect the medication from water, and (4) if needed, whether the material can protect the medication from light. USP Type I glass is often favored for protein drug products (Bontempo 1997b). The material used in any elastomeric components of the packaging design must be examined in regard to absorption of drug into the elastomeric material and/or leaching of elastomeric components into the medication. Quality designs of sterile product container closure systems have one and preferably more than one microbial barrier point of contact. Studies must be performed to demonstrate that the integrity of the container is such that microbial contamination is prevented. The total drug product includes the drug formulation, the primary packaging (e.g., syringe), the label that is in immediate contact with the primary packaging, and any secondary packaging (e.g., box). Labels and adhesives are to be selected which have minimal potential interactions with the medication. Experimental data must be generated demonstrating the lack of significant interactions between the pharmaceutical formulation and the label (and label adhesive) attached to the primary packaging. Historical data from the packaging and label vendors along with previous development history assist the drug development scientist in the initial selection of packaging. If a diluent is provided, then the diluent's container and closure must be evaluated in the same manner as the packaging containing the drug molecule. An evaluation is made as to whether secondary packaging materials can interact with the primary packaging material. In the case of light-sensitive products, the secondary packaging may give additional light blockage and should be optimized. Any negative effects (e.g., generation of particulates) in the use of secondary packaging are to be minimized. An evaluation should be made as to the suitability of the complete product for transportation and experimental temperature cycling and/or vibration testing conducted in the lab if there are concerns, especially for sensitive protein products.

Usage instructions are evaluated as to the rationale in the design of the instructions, their complexity, and whether the instructions are in keeping with both drug product profiles (QTPP and IDPP). The product with the final formulation and candidate packaging should be tested under simulated use conditions. An important consideration in the use of intraocular injectable dosage forms (e.g., VEGF inhibitors packaged in prefilled syringes) is how high the injection back pressure is when the product is administered. Solutions that have a high content of proteins can be very viscous. High viscosity products require a large amount of force to deliver the solution with needles that are appropriately small in diameter. Products which require a large amount of force are considered to be poorly syringeable by the administrator. An approach to lowering the viscosity of these high concentration protein solutions is to add inorganic salts or amino acids (e.g., lysine and arginine) (Mandal et al. 2018).

Definition of Well-Controlled Manufacturing Process

It is advantageous to reference previous manufacturing experience for similar pharmaceutical products when selecting the manufacturing process. This includes identifying which type of equipment best supports the manufacturing process (e.g., an in-line homogenizer). Product characteristics and critical formulation attributes should be considered in the choice of manufacturing equipment. Intraocular injectable dosage forms must be sterile and an appropriate means of sterilization must be selected and justified. The use of terminal sterilization is preferred. If an aseptic process is necessary, then the simplest process that results in the best sterility assurance is to be utilized. Manufacturing filters (e.g., sterilizing filters) should be tested for possible sorption of the NCE and/or extractables from the filter. Biomolecules are especially sensitive during the manufacturing process and product impurities may arise from the process. "Product-related impurities (e.g., precursors, certain degradation products) are molecular variants arising during manufacture and/or storage, which do not have properties comparable to those of the desired product with respect to activity, efficacy, and safety" (ICH 1999a). Preliminary studies should be conducted for possible drug sorption to encountered materials such as steel and/or polymers. The sensitivity of the biomolecule to shear should also be evaluated prior to designing the manufacturing process. It is prudent to assess the stability of biological products after being subjected to temperature, shear, and other manufacturing conditions that are anticipated as being used in manufacturing processes (Baid et al. 2011). The manufacturing process is envisioned as occurring in a sequential manner that will eventually be documented as such. As the batch scale is increased during the development process, any anticipated changes in the manufacturing process are to be planned in advance. Previous drug product manufacturing history is helpful in this regard. It is understood that there will be significant scale-up experimentation performed from preclinical to the final manufacturing process. Designing scale-up studies to focus on the definition of major critical manufacturing processes such as degree and time of agitation, temperature, etc. is very helpful in support of the final manufacturing process. The scale-up study design should include an appraisal whether any in-process testing is to be conducted in order to better understand and/or control the manufacturing process. Differences in the manufacturing process often occur during the development of new products, and some of the differences may be significant. The reasons for the manufacturing process changes should be documented along with a quality appraisal. If novel processes, technologies, and/or packaging operations are utilized, it should be documented how these unique manufacturing steps possibly affect product quality. The following are key batches that usually signal increases in batch size and/or manufacturing knowledge: (1) preclinical (e.g., toxicology supplies), (2) phase I clinical supplies, (3) scale-up manufacturing studies, (4) phase II clinical supplies, (5) pilot manufacturing studies, (6) phase III clinical supplies, and (7) production scale. It is prudent to make clinical manufacturing processes as close as possible to the envisioned final manufacturing process. It is also advantageous to use product development data generated during the manufacture of clinical batches to make any *minor* adjustments to product CQAs prior to production-scale manufacture. Any changes in the nature of the manufacturing equipment (e.g., the addition of an in line homogenizer) should be evaluated for the potential effect upon product performance characteristics, manufacturability, and quality. In process and final batch testing, results should be evaluated to identify critical process parameters and to verify that consistent product quality has been achieved from lot to lot. A determination must be made whether any manufacturing steps are to be monitored and/or controlled for quality purposes. It is especially important that the manufacturing process is well defined and under control for those batches used for pivotal clinical trials and primary stability studies.

Stability Studies

Stability studies consist of placing the drug substance or drug product under defined conditions of temperature, time, and humidity. Stability studies are conducted in order to determine the profile of product performance during storage. Numerous product characteristics are monitored during the conduct of the study. Depending upon the nature of the product and the state of development, tested product characteristics include potency, appearance of decomposition products, and additional tests such as pH, particle size, etc. The identified product characteristics are monitored and the generated data indicates the degree to which product attributes are maintained. Initial stability evaluations are conducted on the drug substance during the preformulation stage of development. These preliminary studies are often centered on the stability of the drug molecule itself under various conditions. It is expected that informal stability studies and studies examining packaging interactions are included as part of formulation development. Abbreviated stability studies are conducted in support of preclinical studies to ensure product quality during the conduct of toxicology and pharmacology GLP studies. These studies are limited in scope but give important information in regard to product performance. At this point in development, it is important to have an assay that can monitor impurities (including degradation products) in the tested drug product. Any generated impurities should be evaluated for biological activity and possible toxicity according to the ICH guidance "Impurities in New Drug Products" (ICH 2006). It is very important that the identification and concentrations of impurities as tested in toxicology and pharmacology studies are clearly linked to later clinical studies. That is, the nature and level of impurities tested in preclinical safety studies need to demonstrate that the product can be safely used in human clinical trials. Clinical supplies are to be manufactured at GMP quality and formal stability studies are conducted in support of the clinical studies using validated analytical methods. Primary stability studies are conducted in the final product presentation and produced using the finalized commercial manufacturing procedure. Primary stability studies are utilized in the establishment of the product's storage conditions and expiration dating. More than one lot (preferably three or more) of drug substance should be evaluated in primary stability batches. The use of different vendor lots of each of the inactive ingredients is also encouraged. It is often helpful to conduct a shipping study to ensure that product quality is maintained during shipment. For sensitive drug products, such as biomolecules, it may be important to include the finished product as packaged in the bulk shipping container. Aggregation of proteins can occur for various reasons and can occur during manufacture, storage, or handling (including shipping) (Eon-Duval et al. 2012). The formation of aggregates can reduce efficacy, but more importantly they may lead to immune responses (Eon-Duval et al. 2012). Thus, the formation of protein aggregates must be monitored throughout the development process and a strong rationale generated for setting the value for the maximum allowable limit that still maintains safety and efficacy.

Critical Quality Attributes and Specifications

It is well accepted by various health authorities and the pharmaceutical industry that numerous product characteristics must be successfully satisfied in order to ensure that medicinal products are effective and safe. These characteristics include that the correct drug molecule is included in the medication and that the amount of drug is the same as indicated on the label. Additionally, the drug product has ingredients at the listed amounts with contaminants at acceptable low amounts. It is generally accepted that the phrase "achieving quality" means that the manufactured drug product possesses the necessary product characteristics (CQAs) to achieve safety and efficacy. There are a few universally recognized CQAs such as drug purity, and some CQAs are common for a class of drug products such as sterility for injectable products. However, many of a product's CQAs are determined experimentally during the development cycle for that specific product. The knowledge of a drug product's CQAs enables the definition of quality standards that in turn are achieved by having drug substance and drug product specifications in place. "A specification is defined as a list of tests, references to analytical procedures, and appropriate acceptance criteria, which are numerical limits, ranges, or other criteria for the tests described" (ICH 1999b). Furthermore, specifications are described as "critical quality standards that are proposed and justified by the manufacturer and approved by regulatory authorities as conditions of approval" (ICH 1999a). The concept and use of specifications are utilized to assure that the patient receives a drug product that is manufactured in a manner that CQAs are consistently achieved. The lists of specifications could be looked at as a contract between the manufacturer of the drug product and the patient. The vendor provides a drug product of sufficient quality in return for monetary recompense. The health authorities represent the interest of the patients and their physicians to assure that clinical outcomes will be met and safety assured when the drug product is used properly. A list of specifications is generated for both the drug substance and the drug product. A common approach is to have a list of specifications for the release of the drug product and another list of specification for meeting stability criteria. The release specifications have acceptance criteria that are usually more restrictive than the acceptance criteria for stability specifications. That is, the acceptance criteria for the release specifications are "nested" within the stability specifications. This allows for the release of product that has a reasonable expiration date. This practice is especially pertinent for the control of drug potency and drug degradation products. Some health authorities stipulate that release specifications are to be considered as in-house limits and are not considered regulatory limits. Under these regulatory requirements, release specifications are relevant for the drug manufacturer, but the stability specifications are most relevant for product quality assessment. The stability specifications give assurance to the health practitioner and patient that the product meets the quality standards throughout the life of the product.

The generation (setting) of specifications and manufacturing controls involves extensive and transparent coordination of several R&D groups. The process of providing justification for each specification is critical and should take into account any pharmacopoeia standards. Specifications should be linked to results from preclinical/clinical studies and stability results for both drug substance and drug product. In-process and end batch testing analytical results from scale-up batches and validation batches should be utilized in the setting of specifications with an emphasis on the primary stability batches (ICH 1999b). The following are common issues with the selection of specifications: (1) the specifications are not based on CQAs, (2) the acceptance criteria do not match the results from manufacturing experience, (3) the specification does not have sufficient supportive data, (4) the rationale for the specification is not adequate, (5) the specification(s) does not address cellular impurities that pose efficacy and/or safety issues, (6) the acceptance criteria are too broad and do not match product attributes for batches used in the clinical trials, and (7) product development data is misinterpreted in the definition of the specification (Finn 2016).

The following tests (along with definitions) have been given by the ICH as generally applicable to new drug substance: (1) description, (2) identification, (3) assay, and (4) impurities (organic and inorganic) (ICH 1999b). The following tests are product specific and may apply to both the drug substance and drug product: (1) physicochemical properties (e.g., pH), (2) particle size, (3) polymorphic forms, (4) chiral new drug substances, (5) water content, (6) inorganic impurities, and (7) microbial limits (ICH 1999b). Relevant pharmacopeia tests for new drug products should be included in the list of specifications. In addition to whatever pharmacopeia tests are listed as specifications, the following general tests are applicable on a case-by-case basis for new parenteral drug products: (1) uniformity of dosage units (mass or content of the active substance), (2) pH range, (3) sterility, (3) endotoxins and pyrogens, (4) particulate matter, (5) water content for nonaqueous products or if the product consists of a powder to be reconstituted, (6) antimicrobial preservative content, (7) antioxidant content, (8) extractables, (9) functionality testing of delivery systems (such as syringeability and seal integrity), (10) osmolarity, (11) particle size distribution for injectable suspensions, (12) redispersibility for injectable suspensions, and (13) reconstitution time, if applicable (ICH 1999b). The specifications for the container closure system(s) are based upon relevant compendia monograph(s), controls established by the manufacturer, the rationale for any special design features, and how well the package performs under usage conditions.

A guidance document has been generated concerning the generation of specifications specifically for biological drug molecules and applies to proteins (polypeptides) or their derivatives that are produced from both recombinant and nonrecombinant cell-culture expression systems. This is due to protein characteristics that are different from those encountered for small molecular weight drug molecules. In particular, there is a need to establish biological activity and immunochemical specifications in addition to traditional purity testing. Pharmacopoeia monographs for biotechnological and biological products usually include standards for ".....sterility, endotoxins, microbial limits, volume in container, uniformity of dosage units and particulate matter" (ICH 1999a). The specifications for biologics address the need for additional controls for the drug substance purification process in order to assure lot to lot consistency (Bontempo 1997b).

The analytical test results for drug substance and product indicate that the drug substance and drug product conform to the specifications if the results lie within the appropriate acceptance criteria. That is, confirmation of product quality is achieved when test results for the drug product meet the acceptance criteria of the listed specifications. It must be pointed out that product quality cannot be tested into the finished product. In addition to specifications, key components of the drug development process contribute to total quality assurance to the patient. Total quality control involves an extensive characterization of the product. Therefore, specifications are used to confirm product quality, but the entire development process should be designed to thoroughly understand the drug product to an extent that critical quality attributes can be accurately identified. This is accomplished through product knowledge accumulated through pharmaceutical development studies.

References

Baid R, Tyagi P, Durazo SA, Kompella UB. Protein drug delivery and formulation development. In: Drug product development for the back of the eye. Boston, MA: Springer; 2011. p. 409–48.

Bontempo JA. Preformulation development of parenteral biopharmaceuticals. Drugs Pharm Sci. 1997a;85:91–108.

Bontempo JA. Formulation development. Drugs Pharm Sci. 1997b;85:109–42.

Choonara YE, Pillay V, Danckwerts MP, Carmichael TR, Du Toit LC. A review of implantable intravitreal drug delivery technologies for the treatment of posterior segment eye diseases. J Pharm Sci. 2010;99(5):2219–39.

Delplace V, Payne S, Shoichet M. Delivery strategies for treatment of age-related ocular diseases: from a biological understanding to biomaterial solutions. J Control Release. 2015;219:652–68.

EMEA. Annex I: summary of product characteristics, Macugen 0.3 mg solution for injection. https://www.ema.europa.eu/documents/product-information/macugen-epar-product-information_en.pdf. Accessed 5 Dec 2018a.

EMEA. Annex I: summary of product characteristics, Lucentis 10 mg/ml solution for injection. https://www.ema.europa.eu/documents/product-information/lucentis-epar-product-information_en.pdf. Accessed 14 Nov 2018b.

EMEA. Annex I: summary of product characteristics, Eylea 40 mg/ml solution for injection in pre-filled syringe. https://www.ema.europa.eu/documents/product-information/eylea-epar-product-information_en.pdf. Accessed 5 Dec 2018c.

Eon-Duval A, Broly H, Gleixner R. Quality attributes of recombinant therapeutic proteins: an assessment of impact on safety and efficacy as part of a quality by design development approach. Biotechnol Prog. 2012;28(3):608–22.

FDA CDER Data Element Number: C-DRG-00301, Data Element Name. Route of Administration. https://www.fda.gov/Drugs/DevelopmentApprovalProcess/FormsSubmissionRequirements/ElectronicSubmissions/DataStandardsManualmonographs/ucm071667.htm. Accessed 2 Jan 2019.

Finn T. Establishing and applying critical quality attributes during the product development lifecycle ISCT Liaison Meeting October 19, 2016. Product Reviewer Office of Tissues and Advanced Therapies FDA/CBER. http://www.aabb.org/advocacy/government/fdaliaison/ct/Documents/Presentation%201b_Critical%20Quality%20Attributes%20During%20PD%20Lifecycle_T.%20Finn.pdf. Accessed 14 Dec 2018.

Hartman RR, Kompella UB. Intravitreal, subretinal, and suprachoroidal injections: evolution of microneedles for drug delivery. J Ocul Pharmacol Ther. 2018;34(1–2):141–53.

ICH. The international conference on harmonization of technical requirements for registration of pharmaceuticals for human use, quality guideline Q2(R1) validation of analytical procedures: text and methodology. 1994. https://www.ich.org/fileadmin/Public_Web_Site/ICH_Products/Guidelines/Quality/Q2_R1/Step4/Q2_R1__Guideline.pdf. Accessed 16 Feb 2019.

ICH. The international conference on harmonization of technical requirements for registration of pharmaceuticals for human use, quality guideline Q6B test procedures and acceptance criteria for biotechnological/biological products. 1999a. https://www.ich.org/fileadmin/Public_Web_Site/ICH_Products/Guidelines/Quality/Q6B/Step4/Q6B_Guideline.pdf. Accessed 16 Feb 2019.

ICH. The international conference on harmonization of technical requirements for registration of pharmaceuticals for human use, quality guideline Q6A test procedures and acceptance criteria for new drug substances and new drug products: chemical substances. 1999b. https://www.ich.org/fileadmin/Public_Web_Site/ICH_Products/Guidelines/Quality/Q6A/Step4/Q6Astep4.pdf. Accessed 16 Feb 2019.

ICH. The international conference on harmonization of technical requirements for registration of pharmaceuticals for human use, M4Q(R1) quality overall summary of module 2. Module 3: quality current step 4 the common technical document for the registration of pharmaceuticals for human use. 2002. https://www.ich.org/fileadmin/Public_Web_Site/ICH_Products/CTD/M4_R1_Quality/M4Q__R1_.pdf. Accessed 16 Feb 2019.

ICH. The international conference on harmonization of technical requirements for registration of pharmaceuticals for human use, quality guideline Q3B(R2) impurities in new drug products. 2006. https://www.ich.org/fileadmin/Public_Web_Site/ICH_Products/Guidelines/Quality/Q3B_R2/Step4/Q3B_R2__Guideline.pdf. Accessed 16 Feb 2019.

ICH. The international conference on harmonization of technical requirements for registration of pharmaceuticals for human use, quality guideline Q8(R2) pharmaceutical development. 2009. http://www.ich.org/fileadmin/Public_Web_Site/ICH_Products/Guidelines/Quality/Q8_R1/Step4/Q8_R2_Guideline.pdf. Accessed 16 Feb 2019.

Jain NK, Roy I. Effect of trehalose on protein structure. Protein Sci. 2009;18(1):24–36.

Kim YC, Chiang B, Wu X, Prausnitz MR. Ocular delivery of macromolecules. J Control Release. 2014;190:172–81.

Maguire J, Peng D. How to identify critical quality attributes and critical process parameters. In: FD6/PQRI 2nd conference, 6 Oct 2015.

Mandal A, Pal D, Agrahari V, Trinh HM, Joseph M, Mitra AK. Ocular delivery of proteins and peptides: challenges and novel formulation approaches. Adv Drug Deliv Rev. 2018;126:67–95.

Radhakrishnan K, Sonali N, Moreno M, Nirmal J, Fernandez AA, Venkatraman S, Agrawal R. Protein delivery to the back of the eye: barriers, carriers and stability of anti-VEGF proteins. Drug Discov Today. 2017;22(2):416–23.

Thrimawithana TR, Young S, Bunt CR, Green C, Alany RG. Drug delivery to the posterior segment of the eye. Drug Discov Today. 2011;16(5–6):270–7.

Yasin MN, Svirskis D, Seyfoddin A, Rupenthal ID. Implants for drug delivery to the posterior segment of the eye: a focus on stimuli-responsive and tunable release systems. J Control Release. 2014;196:208–21.

Microbiological Considerations for Ophthalmic Products: Sterility, Endotoxin Limits, and Preservatives

Ayako Hasegawa, Melissa Gulmezian-Sefer, Ying Cheng, and Ramakrishnan Srikumar

Abstract All ophthalmic products must meet regulatory standards for microbiological quality and safety for their intended use. This chapter outlines holistic control strategies for sterility, endotoxin, and preservative efficacy to achieve ophthalmic product safety. While finished product testing for sterility or pyrogens confirms the microbiological quality of the products, there are certain challenges in developing a sensitive microbiological test that assures product quality. Product matrices/composition may interfere with the test. Additionally, testing only a small portion of the batch poses a statistical challenge to the confirmation of product quality of the entire batch. Therefore, stringent microbiological control cannot solely rely on end-product testing. The risk of microbial and/or pyrogenic contamination can be minimized only when proper microbiological control strategies, employed throughout the manufacturing process, are combined with finished product testing. While proper manufacturing process control provides assurance to the finished product microbiological quality, preservatives in a formulation are added to provide adequate protection from microbial contamination or proliferation that may arise during the use of the product. Preservative concentration must be safe to the patient but must robustly meet the requirements for preservative effectiveness of killing microorganisms for ophthalmic products.

Keywords Ophthalmic products · Preservatives · Endotoxins · Sterility · Sterility assurance · LER

A. Hasegawa (✉) · M. Gulmezian-Sefer · Y. Cheng · R. Srikumar
Abbvie, Irvine, CA, USA
e-mail: ayako.hasegawa@abbvie.com

S. Neervannan, U. B. Kompella (eds.), *Ophthalmic Product Development*,
AAPS Advances in the Pharmaceutical Sciences Series 37,
https://doi.org/10.1007/978-3-030-76367-1_9

Abbreviations

AAMI	Association for the Advancement of Medical Instrumentation
AC	Anterior chamber
ANSI	American National Standards Institute
Anti-VEGF	Anti-vascular endothelial growth factor
APET	Antimicrobial preservative efficacy test
BAK	Benzalkonium chloride
BET	Bacterial endotoxins test
BFS	Blow-fill-seal
CCIT	Container closure integrity test
CDRH	Center for Devices and Radiological Health
CE	Conformité Européenne
CFR	Code of Federal Regulations
CFU	Colony-forming unit
CMC	Critical micelle concentration
D values	Decimal reduction times
DPBS	Dulbecco's phosphate-buffered saline
E. coli	*Escherichia coli*
EMA	European Medicines Agency
EP	European Pharmacopoeia
EU	Endotoxin unit
FDA	Food and Drug Administration
HEPA	High-efficiency particulate air
h	Hour
ICH	International Council for Harmonisation of Technical Requirements for Pharmaceuticals for Human Use
ISO	International Organization for Standardization
IVT	Intravitreal
JP	Japanese Pharmacopoeia
kg	Kilogram
LAL	Limulus amebocyte lysate
LER	Low endotoxin recovery
MAT	Monocyte activation test
MIC	Minimum inhibitory concentrations
mL	Milliliter
μm	Micrometer
NMT	Not more than
OSD	Ophthalmic squeeze dispenser
OVD	Ophthalmic viscosurgical device
PDA	Parenteral Drug Association
Post-IVT IOI	Post-intravitreal intraocular inflammation
PS	Polysorbate
PTP	Proactive TASS Program
Q&A	Questions and answers

RABS	Restricted access barrier system
rFC	Recombinant factor C
RPT	Rabbit Pyrogen Test
SOC	Stabilized oxychloro complex
TASS	Toxic anterior segment syndrome
TGA	Australian government Department of Health Therapeutic Goods Administration
US	United States
USP	United States Pharmacopeia
WFI	Water for injection

Introduction

Microbial contamination of ophthalmic products has been linked to ophthalmic conditions such as conjunctivitis, keratitis, and infectious endophthalmitis leading, in the worst case, to even blindness (Jokl et al. 2007; Snyder and Glasser 1994; Mayo et al. 1987; Goldberg et al. 2012; U.S. Food and Drug Administration 2013). Vision loss may also result from noninfectious endophthalmitis caused by bacterial endotoxin contamination of ophthalmic pharmaceuticals and devices that are injected or implanted into the eye (Agrawal et al. 2013; Rajendran et al. 2017). Based on these cause-and-effect relationships, all ophthalmic pharmaceuticals and devices are required to be manufactured sterile, while ophthalmic injection and implant products must also be controlled for endotoxins. Additionally, aqueous multi-dose products must contain antimicrobial preservatives to prevent microbial contamination during use by the patient unless other measures can ensure in-use microbiological quality of the product. This chapter outlines holistic microbial control and testing strategies that meet global regulatory standards and ensure microbiological quality and safety of the ophthalmic products for their intended use.

Sterility Requirement for Ophthalmic Products

Serious ocular infections, several even resulting in blindness, were caused by the use of topical ophthalmic solution contaminated with viable microorganisms (Theodore and Feinstein 1953). To address the concern, the US Food and Drug Administration (FDA) in 1953 led a survey of medical opinion which concluded that liquid preparations intended for ophthalmic use need to be sterile (Federal Register 1953). Fast-forward to now, all ophthalmic preparations must be sterile to be suitable for safe use (Code of Federal Regulations 2018a). This applies to ophthalmic preparations that are regulated as drugs including biologics (Code of Federal Regulations 2018b) and over-the-counter ophthalmic drug products (Code of Federal Regulations 2018c) and medical devices (Code of Federal Regulations 2018d). Three general categories of ophthalmic product routes of administrations are intraocular injection, extraocular injection, and topical application on the ocular surface. In addition to the obvious risk to injections, the 1953 survey highlighted the microbial risk of topical ophthalmic preparations to patients. Based on the high likelihood of product entering the eye, even products that are administered in the vicinity of the eye such as eyelids are required to be sterile (United States Pharmacopeia and National Formulary (USP 41-NF 36) 2018a). Applicators such as eye cups, eye droppers, and other dispensers intended for ophthalmic use must also be sterile (Code of Federal Regulations 2018a; European Pharmacopeia 2019a).

Sterility Test Consideration

An ophthalmic product needs to comply with a validated sterility test (European Pharmacopeia 2019a; United States Pharmacopeia and National Formulary (USP 41-NF 36) 2018b; Japanese Pharmacopeia 2016a). Harmonized compendial chapters, namely, United States Pharmacopeia (USP) <71>, European Pharmacopoeia (EP) 2.6.1, and Japanese Pharmacopoeia (JP) 4.06, prescribe test requirements for ophthalmic pharmaceuticals including test procedures and media (European Pharmacopeia 2019a; United States Pharmacopeia and National Formulary (USP 41-NF 36) 2018b; Japanese Pharmacopeia 2016a). Ophthalmic medical device that directly contacts the eye and an applicator of ophthalmic product supplied separately also must comply with the validated sterility test methods (European Pharmacopeia 2019a; United States Pharmacopeia and National Formulary (USP 41-NF 36) 2018b).

In order to detect microbial contamination which may be subvisible, the standard compendial method is growth based. The potential microbial contamination in the product is amplified in a growth media for 14 days for detection by visual inspection. Products with propensity to inhibit microbial growth are neutralized to remove the antimicrobial effect prior to the incubation to allow growth of potential microbial contamination in the sample for detection. There are two main techniques of neutralization provided in the compendia: membrane filtration and direct inoculation of the culture medium. For membrane filtration approach, the product is filtered through a membrane with a nominal pore size not greater than 0.45 μm, which effectively retains microbial contamination in the product on the membrane (Carter 1996). The antimicrobial substance in the product is filtered through the membrane. Any residual antimicrobial effect is neutralized through subsequent rinsing of the membrane. The microbial contaminant retained on the membrane is transferred into the growth media to allow for amplification and visual detection. Filterable products such as topical ophthalmic solutions and injection products are subjected to filtration method. Soluble solids and ointments that can be solubilized with a non-microbicidal solvent such as USP rinse fluids or isopropanol myristate to facilitate the use of the filtration method. Products that cannot be rendered filterable through manipulation such as emulsion, suspension, and solid device are subjected to the direct inoculation method. Neutralization of such product is achieved through dilution directly in the growth media or treated with a neutralizing buffer preceding dilution in growth media. For example, benzalkonium chloride, a quaternary ammonium compound that is typically used as a preservative in ophthalmic products, can be neutralized with lecithin. Product-specific suitability of test method needs to be verified through recovery of a standard compendial panel of microorganisms representing Gram-negative, Gram-positive, and anaerobic bacteria, in addition to yeast and mold.

Products that don't exhibit enough solubility in microorganism-compatible solvents face a challenge in sterility testing as the entire product may not be testable without complete dissolution. These products may dissolve only in organic solvents, but such solvents can affect viability of potential microbial contaminants, thus compromising the integrity of the test. In the absence of suitable solvent that can be microbiologically validated, sterility test by direct inoculation method of these poorly soluble products primarily addresses sterility of exterior surface of the products. To minimize the potential for release of nonsterile product, a risk assessment may be performed to identify effective mitigation plans. The risk to product sterility assurance, as it pertains to controlling the pre-sterilization bioburden, can be determined based on parameters such as raw material propensity for microbial contamination, manufacturing environmental control, equipment setup, and process capability in bioburden reduction such as high temperature mixing or suspension in microbicidal solution. The risk to product sterility can be assessed based on the choice of sterilization process such as terminal, overkill processes versus aseptic manufacturing processes. For example, terminally sterilized product carries lower risk of product contamination and may not require additional mitigation. On the other hand, products that are subjected to aseptic manufacturing processes carry higher risk for contamination due to product handling post-sterilization. Poorly soluble products that are manufactured using aseptic process may require additional testing to ensure the finished product quality. For instance, in-process sterility test may be performed on the sample collected downstream of bulk sterilization and prior to formation of the poorly soluble state of the product. The in-process sample represents what will become the interior of the finished product which cannot be evaluated in a sterility test method testing only the surface. The in-process sterility test may be implemented as supplement to the release sterility test to corroborate sterility of both interior and exterior of the finished product (Toguchi 1999). Furthermore, a modified sterility test verified using non-compendial microorganisms resistant to manufacturing and test conditions may be performed in addition to the standard sterility test. Through a risk assessment, the types of microorganisms anticipated in the finished product may be narrowed down based on their resistance to product-specific inhibition of microbial growth or sterilization process. In the development of a modified sterility test, product solubilization conditions using an organic solvent can be optimized to recover those organisms resistant to processing conditions as well as testing conditions. For instance, spore-forming microbes tend to withstand various factors such as heat, irradiation, desiccation, and chemicals including organic solvents. A modified sterility test using organic solvent may not recover a full panel of compendial standard microorganisms but may recover the organisms deemed higher risk to the manufacturing process such as spores. Such a modified sterility test may also be performed as a supplemental release test (Toguchi 1999).

Challenges with Sterility Tests

While sterility tests are required by regulation for sterile products (Code of Federal Regulations 2018e; U.S. Food and Drug Administration 2004), there are several challenges with sterility tests.

One of the challenges with sterility testing is lab contamination leading to a false-positive test. To prevent false positives, analysts need to be well trained and qualified for gowning and aseptic handling. To further minimize the risk, the sterility test environment including facility and controls need to be comparable to that of aseptic filling operations (U.S. Food and Drug Administration 2004). Products that are terminally sterilized may be released parametrically and eliminate the need for sterility test at release. Parametric release is defined as a sterility assurance release program where demonstrated control of the sterilization process enables a firm to use defined critical process controls in lieu of the sterility test. Parametric release fulfills the intent of Code of Federal Regulations Title 21 (21 CFR) 211.165(a) and 211.167(a), and is also accepted by the European Community (European Medicines Agency 2001a; European Commission, EudraLex 2018). Sterility test at shelf life in stability studies can also be replaced with a validated container and closure system integrity test (CCIT) (U.S. Food and Drug Administration 2008; European Medicines Agency 2018a; International Council for Harmonisation (ICH) 1995). The premise of this approach is that a product manufactured sterile will remain sterile over the shelf life unless the sterile barrier provided by the primary container closure is breached. The acceptability of these alternative approaches to sterility test is subject to review and approval by respective Regional Regulatory Agencies.

Another challenge with sterility test is that it is statistically impossible to demonstrate the sterility of the entire batch. For example, if 20 units from a 10,000-unit lot with a 0.1% contamination level was sterility tested, there is a 98% chance that the batch will pass as sterile (U.S. Food and Drug Administration 2004). In fact, USP<71> discloses that "[t]hese pharmacopoeial procedures are NOT by themselves designed to ensure that a batch of product is sterile or has been sterilized." Manufacturing of sterile product is accomplished primarily by routine qualification and validation of the sterilization processes or of the aseptic processing procedures.

Terminal Sterilization Versus Aseptic Manufacturing Process

Whereas terminal sterilization uses lethal treatment at the end of the manufacturing process to kill microorganisms in the product, aseptic processing relies on the removal of microorganisms from the product. To enhance the sterility assurance of the product, terminal sterilization is preferred over filtration and/or aseptic processing (U.S. Food and Drug Administration 2004; European Medicines Agency 2000; European Pharmacopeia 2019b). In terminal sterilization, the sterilization process applied to the drug product in the final package mitigates potential contamination of the product post-sterilization. For example, sterile products that are compatible with heat or irradiation can be terminally sterilized by heat or irradiation which provides a great assurance of sterility. However, many ophthalmic products are incompatible with terminal sterilization due to formulation instability and/or use of thermolabile plastic container closure. Therefore, alternative sterilization, filtration, and aseptic processing are applied in a decreasing order of sterility assurance. For instance, biologics like anti-vascular endothelial growth factor (anti-VEGF) agents administered through intravitreal injection have made a huge impact in the treatment of neovascular eye diseases. However, biologic products are often heat labile and therefore may not be suitable for terminal sterilization. Another example is ophthalmic products that come in plastic eye dropper. Incompatibility of the container closure system with terminal sterilization process alone is usually not enough to justify the selection of aseptic manufacturing process. Eye drops, however, are a unique class of products where the benefit of drop administration using plastic bottles is used to rationalize the use of aseptic manufacturing processing for these products (European Medicines Agency 2016). Consequently, many ophthalmic products are manufactured using filter sterilization and aseptic manufacturing processes. Given a lower power of sterilization provided by filter sterilization process compared to heat or irradiation sterilization, sterile product manufacturing through filter sterilization combined with aseptic processing calls for a sound microbiological control strategy. The process control should (i) minimize in-coming bioburden prior to sterilization to maximize the sterility assurance level and (ii) minimize contamination risk of sterilized components and bulk formulation during aseptic handling and filling.

Control Strategy for Assurance of Ophthalmic Product Sterility

In order to holistically approach sterile product manufacturing, the microbiological control strategy should include the following: careful design and maintenance of the plant, facilities, and equipment; sound design and validation of manufacturing process; proper cleaning and disinfection; robust control of utilities, raw materials, and container closure in-coming testing; comprehensive training of personnel; and monitoring and trending of environment including in-process sample and personnel (U.S. Food and Drug Administration 2004; Code of Federal Regulations 2018f; European Commission, EudraLex 2008). Due to the lower sterilization power, filter sterilization process requires tighter control over in-coming bioburden. For example, the European Medicines Agency (EMA) recommends tighter pre-sterilization in-process bioburden limits of not more than (NMT) 10 colony-forming unit (CFU)/100 mL for filter sterilization process compared to NMT 100 CFU/100 mL for bulk heat sterilization (European Medicines Agency 2016). One of the most critical factors for controlling in-coming bioburden is raw materials. Drug substance and excipients are recommended to be tested routinely against appropriate specification limits. USP <1111> provides acceptance criteria for raw materials used for manufacturing of nonsterile products, which may also be applied for sterile product raw material control. However, for filter-sterilized product manufacturing, tighter acceptance criteria may be considered based on risk assessment such as weighted contribution, historical data, and propensity for microbial growth during manufacturing. For example, water is the major ingredient in ophthalmic solution formulation. If there is no bioburden reduction step prior to filter sterilization, it requires water for injection quality (NMT 10 CFU/100 mL) to meet the in-process limit of NMT 10 CFU/100 mL. Microbiological risk assessment also helps to establish skip-lot or reduced testing of raw materials without compromising the quality of the product (International Council for Harmonisation (ICH) 1999). If released based on supplier certificate of analysis, the manufacturer should validate supplier's test results initially and periodically to establish reliability of the supplier's analysis (Code of Federal Regulations 2018g).

Once bulk product is sterilized, aseptic processing is applied to ensure manufacturing of sterile product. All product contact parts downstream of product sterilization need to be sterilized with validated processes. Primary container closure also must be sterile at the time of filling (Code of Federal Regulations 2018a). The environment where the sterilized product or any sterile product contact part is exposed needs to be maintained at Class 100 (International Organization for Standardization: ISO 5) condition. One of the most critical activities performed in the ISO 5 condition is filling and closing operations. Any activities conducted by personnel present a high risk of microbial contamination. Therefore, it is critical that the personnel working in aseptic filling operation are suitably trained and certified to work in the cleanrooms. Advanced aseptic system such as restricted access barrier system (RABS) and isolators offers separation of aseptic processing line from the external environment including operators. Automation is another way to help minimize personnel intervention, thereby reducing the risk of contamination by operators. Blow-fill-seal (BFS) technology, such as used in manufacturing of single-dose topical ophthalmic product, is considered a superior aseptic manufacturing process where advanced aseptic system and automation are combined to reduce operator interface. BFS is an automated system where container is formed in continuous process with filling and sealing. The critical area where the formed container and product is exposed to the environment before sealing is contained in a partial isolating environment supplied with high-efficiency particulate air (HEPA)-filtered air that meets ISO 5 criteria. Although BFS in general mitigates the major contamination risk by reducing human intervention, a further risk assessment should be performed to identify appropriate operation controls that is in alignment with overall contamination control strategy. Such control may include integrity testing of BFS product pathways, microbial control of plastic polymer starting material, and environmental control and monitoring. In summary, as provided in these examples, sterility assurance should not solely rely on sterility testing, but instead be achieved through sound manufacturing control strategy.

Sterile Endophthalmitis Linked to Endotoxin Contamination

Ophthalmic pharmaceuticals are sterile products that are intended to be applied onto ocular structures or used in conjunction with an ophthalmic device. While all ophthalmic products have a pharmacopoeial requirement for sterility (Code of Federal Regulations 2018a), pyrogens in ophthalmic products can be a safety concern depending on the intended use of the ophthalmic product. A pyrogen is any fever inducing substance derived from various origins. Pyrogens can be classified into two groups: microbial (e.g., bacteria, viruses, and fungi) and non-microbial (e.g., drugs and antigens). Contamination by pyrogens is considered a serious health hazard and can result in symptoms ranging from vascular alterations to shock or even death. Among these fever-inducing pyrogens, bacterial endotoxins are the most common cause of pyrogenicity in parenteral pharmaceutical products. Endotoxin is also known as lipopolysaccharide (LPS) which is part of the outer membrane of Gram-negative bacterial cell wall, including non-pathogenic Gram-negative organisms. Bacterial endotoxin is heat stable and may not be completely removed by most conventional sterilization processes such as autoclaving and irradiation. Due to its resistance to heat, it is hard to be removed from contaminated products and should therefore be tightly controlled in all products intended for injection or implantation.

The concern of endotoxin contamination in ophthalmic pharmaceuticals is raised because endotoxin is a potent ocular inflammatory agent. In recent years, an increased number of national outbreaks of **t**oxic **a**nterior **s**egment **s**yndrome (TASS) and post-**i**ntra**v**i**t**real **i**ntra**o**cular **i**nflammation (post-IVT IOI) have been linked to endotoxin contamination (Agrawal et al. 2013; Rajendran et al. 2017). Both TASS and post-IVT IOI are considered as sterile endophthalmitis, which is described as any acute intraocular inflammation that is caused by a noninfectious agent. The condition may result in significant complications and vision loss.

TASS is characterized by sterile postoperative inflammation caused by the introduction of a noninfectious agent into the anterior segment (Fig. 1) of the eye after intraocular surgery (Park et al. 2018). It is most commonly associated with cataract surgery in the absence of infectious agents and typically occurs within 48 h of surgery. Most cases of TASS are mild and readily resolved. However, some cases of TASS are severe enough to require secondary surgical interventions including glaucoma surgery and corneal transplantation. Potential causative agents of TASS are not always known and a variety of substances are implicated. In many cases, bacterial endotoxin contamination of balanced salt solution used in cataract surgery and ophthalmic viscosurgical device as well as endotoxin accumulated in ultrasonic bath used for cleaning ophthalmologic instrument were responsible for the TASS outbreaks. Endotoxin was found to be an intrinsic contaminant of the balance salt solution that caused a TASS outbreak in 2005 involving 112 patients. The endotoxin level detected in the recalled balanced salt solution (0.908 EU/mL) was higher than the allowable limit endotoxin level for ophthalmic irrigation products (0.5 EU/mL) (Kutty et al. 2008). In 2008 another TASS outbreak was observed with Ophthalmic

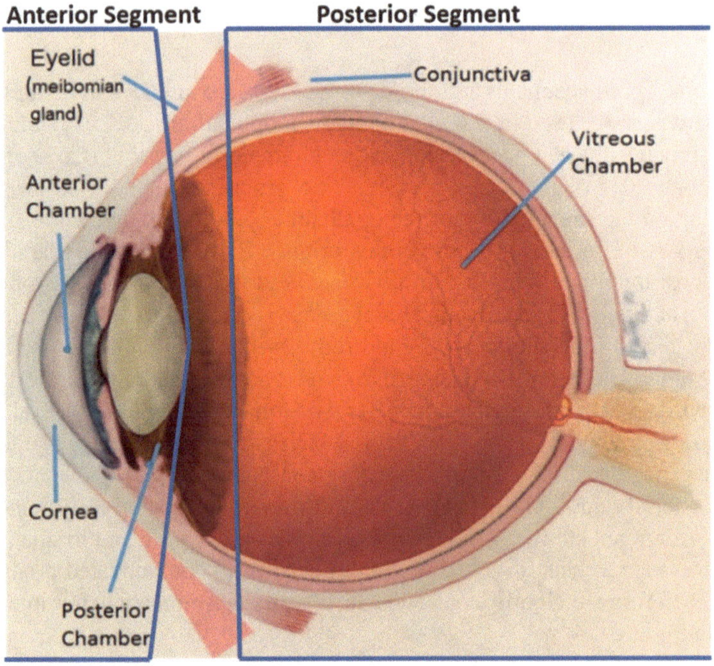

Fig. 1 Anterior and posterior segments of the eye. This illustration is a modification of an image on the website of the National Eye Institute of the National Institutes of Health

Viscosurgical Device (OVD) products. Tests of the affected lot revealed elevated levels of endotoxin. Endotoxin levels above the maximum USP level may be a potential cause of an inflammatory response and/or TASS in patients following surgery (US FDA Medical Device Recalls 2008). As mentioned above, many cases of TASS have been linked to endotoxin contamination of ophthalmic devices. Several studies have shown that the eye is sensitive to the effects of endotoxin contamination in the anterior segment through intracameral route of administration or OVDs (Buchen et al. 2012a; Nussenblatt et al. 2012).

Like TASS, post-IVT IOI is also described as sterile endophthalmitis that resolves without antibiotic treatment typically required for infectious endophthalmitis. Post-IVT IOI is defined as the appearance of significant anterior chamber and/or vitreous inflammation following intravitreal injection. Unlike TASS condition with an inflammation localized to the anterior segment, post-IVT IOI occurs in both anterior segment and posterior segment (Karl and Csaky 2018) (Fig. 1). While most IVTs are routine and uncomplicated, post-IVT IOI due to the contamination of endotoxin or other biologic impurities can be a noted risk of an intravitreal injection procedure.

The etiology of sterile inflammation secondary to intravitreal injections can be due to a variety of agents. Many cases are likely due to an immunogenic mechanism in which foreign epitopes are presented by antigen-presenting cells to B and T cells

(Johnson and Sharma 2013). However, bacterial endotoxin contamination has been implicated in several post-IVT IOI outbreaks. In 2009, a large-scale investigation in Canada following several outbreaks across the country found that most cases were linked to a specific lot with higher silicone oil residues and endotoxins compared to controls (Fielden et al. 2011). In a retrospective investigation of a 2010 outbreak of inflammation after the IVT injection, for a total of 116 patients, 69% developed sterile endophthalmitis (Wang et al. 2013). Both outbreaks are related to off-label IVT injection of the same medication that is approved for intravenous use. The presence of endotoxin in vitreous specimens was confirmed by laboratory testing. Today, the intravitreal injection of anti-VEGF agents is the most common intravitreal procedure performed in ophthalmology. Endotoxin contamination following intravitreal injection of therapeutic agents also is a growing concern due to the recent increase in the intravitreal route of administration. In addition, certain biologic therapies administered through IVT may be at increased risk of contamination by endotoxin, particularly if they are produced in *E. coli*, Gram-negative bacteria. Taken together, ophthalmic pharmaceuticals manufacturers should adhere to stringent endotoxin controls.

Animal Studies to Assess the Inflammatory Potential of Endotoxin in Ophthalmic Products

As discussed above, in many TASS cases, bacterial endotoxin from medical devices has been linked to the cause of inflammation (Richburg et al. 1986; Kreisler et al. 1992). Those medical devices are vulnerable to bacterial proliferation and subsequent endotoxin contamination as they are aqueous in nature, contain an aqueous component, or are otherwise exposed to water during the manufacturing process. Hence, OVDs that are commonly used in ophthalmic surgery are prone to endotoxin contamination (U.S. Food and Drug Administration 2015). In light of rising concerns of TASS, to facilitate more effective outbreak investigation and to identify the cause of outbreaks, the FDA developed the Proactive TASS Program (PTP) in 2012 (Eydelman et al. 2012). As a part of this effort, the FDA conducted animal studies on the inflammatory potential of endotoxin contained in OVDs and aqueous medium. Historically, intravitreal injection of endotoxin in rabbit is a common animal model to induce experimental inflammation. Given that OVDs are intended for use in anterior segment surgery, the anterior segment of the rabbit eye seems to be a more appropriate test site for the assessment of inflammatory potential of these products than the vitreous cavity. In the first study conducted by the FDA Center for Devices and Radiological Health (CDRH), the inflammatory response was evaluated when OVDs spiked with a known amount of endotoxin are injected into the anterior chamber or the vitreous cavity of the rabbit. The study concluded that the rabbit intracameral assay (anterior segment) is more sensitive in detecting inflammation secondary to endotoxin contamination of OVDs than the intravitreal assay (posterior segment) (Buchen et al. 2012b). The rabbit intracameral assay should be used to assess the inflammatory potential of endotoxin contained in OVDs (U.S. Food and Drug Administration 2015).

Previously, it has been reported that the minimum amount of endotoxin in water causing inflammation when introduced intracamerally in the rabbit eye is between 0.23 and 0.60 EU (Sakimoto et al. 2009). However, the minimum inflammatory dose of endotoxin in OVDs, which are viscous materials, that induces intraocular inflammation has been unknown. It is believed that endotoxin in OVDs could elicit greater intraocular inflammation than endotoxin in aqueous solutions. The more severe inflammation associated with endotoxin in OVDs could be due to its longer contact duration with sensitive ocular tissue when compared with endotoxin in an aqueous medium. Therefore, a lower minimum inflammatory dose of endotoxin is expected in OVDs than in aqueous medium. To better understand the inflammatory potential of endotoxin in OVDs and in aqueous medium, another study was performed to assess the inflammatory response when a balanced salt solution (Dulbecco's Phosphate-Buffered Saline, DPBS) and OVDs spiked with the same levels of endotoxin were injected intracamerally into the eyes of New Zealand white rabbits (Buchen et al. 2012a). The inflammation was evaluated by assessing the response parameters including corneal clouding, anterior chamber (AC) flare, cells and fibrin, vitreous haze and cells, cells and fibrin on lens surface, lens opacities, and onset time. The study showed that AC inflammation was observed after intracameral injection of as little as 0.02 and 0.08 EU/eye in OVDs and DPBS, respectively. As expected, observed responses to intracamerally injected endotoxin in OVDs were more severe and of longer duration than those in aqueous medium. The study raises concerns regarding the inflammation potential associated with the original acceptable ISO endotoxin levels for OVDs (International Organization for Standardization 2013a). Based on the findings from this study, the FDA provided recommendations for endotoxin limits in its 2015 guidance document for single-use intraocular ophthalmic devices (Table 1) (U.S. Food and Drug Administration 2015).

As discussed above, adverse ocular effects of endotoxin introduced intracamerally have been well studied. In contrast, despite the concerns regarding endotoxin contamination of ophthalmology products and the emerging importance of IVT injections as a route of drug administration, little is known about the sensitivity and the time course of the response to endotoxin following IVT administration. Currently there is limited information available regarding the minimum endotoxin level causing inflammation in human following intravitreal administrations. Several animal studies have shown that even low doses of endotoxins in saline injected intravitreally elicited ocular inflammation (Bantseev et al. 2017, 2019; Fleisher and McGahan 1985; McGahan et al. 1996). However, there is no consensus on the minimum effective levels of endotoxin following intravitreal injection. Moreover, as the authors noted, the translatability of these findings to humans is unknown due to anatomical and physiological differences. It is also important to note that reference standard endotoxin preparations used in these studies may not be the best representative of the real-world endotoxin contamination that may be present during biotechnology/pharmaceutical manufacturing process (Bantseev et al. 2017).

Table 1 Endotoxin limits of ophthalmic products

Route of administration	Endotoxin limit	Recommended by
Intraocular fluids—OVDs	≤0.2 EU/mL	FDA ISO 15798:2013Amendment 12017
Anterior segment solid devices – Intraocular lenses – Capsular tension ring devices – Glaucoma devices – Phacofragmentation systems—The accessories of irrigation/aspiration sleeves tubing	≤0.2 EU/mL	FDA and ISO 11979-8:2017
Ophthalmic irrigation products	NMT 0.5 EU/mL ≤0.2 EU/mL	USP <771> ISO 16771: 2015 Amendment 12017
Injected or implanted drug products	NMT 2 EU/dose/eye	USP <771>

Furthermore, in all these animal studies, the ocular inflammatory response is only assessed on the endotoxin injected intravitreally in aqueous solution. No data is available on ocular inflammatory response to endotoxin introduced intravitreally in non-aqueous matrix. Additional research is required to better understand the minimum endotoxin level causing inflammation when it is introduced intravitreally in non-aqueous matrix.

Compendial and International Guidance for Endotoxin Limits

Although the exact endotoxin levels that are responsible for ocular inflammation are generally unknown, the animal results highlight the importance of endotoxin testing for ophthalmic injection and devices. When performing endotoxin testing to determine the acceptability of a product, it is apparent that there must be an endotoxin limit above which the product should be rejected. The endotoxin limits for parenteral drugs are established on the basis of dose/kg/h using calculations recommended in USP <85> (United States Pharmacopeia and National Formulary (USP 41-NF 36) 2018c). However, the calculation is based on the intravenous, intramuscular, or intrathecal routes of administration. Therefore, the calculation method for the limit may not be appropriate for ophthalmic injections, given that eyes typically require a much lower injection volume. Currently, neither the FDA nor the European health authorities provide recommendations on the endotoxin limits for injections in the eye. In the absence of any recommended limit, all the ophthalmic manufacturers should adhere to a more stringent endotoxin control in injectable ophthalmic products due to the concerns with endotoxin-mediated sterile inflammation. In regard to this, USP <771> states that all injected ophthalmic drug products shall be prepared in a manner designed to minimize bacterial endotoxins. It recommends NMT 0.5 EU/mL for ophthalmic irrigation product, aqueous solution used during ophthalmic surgery. However, in ISO 16671 2017 amendment 1, the endotoxin limit for irrigating solutions was changed to 0.2 EU/mL (International Organization for Standardization 2015). For injected or implanted drug products, a NMT 2.0 EU/dose/eye is prescribed in USP <771>(United States Pharmacopeia and National Formulary (USP 41-NF 36) 2018a) which is deemed appropriate given that there is a larger volume in the vitreous chamber.

Apart from the ophthalmic devices and intraocular injections, there is no endotoxin testing requirement for topical eye drops. Historically, endotoxin testing was not considered for topical products because the surface of the eye itself is a natural barrier to endotoxin. An "increasing frequency" for endotoxin-mediated TASS outbreaks had prompted the FDA to require endotoxin tests for topical eye drops. In 2014, however, the FDA changed the policy on endotoxin testing of topical products based on a compelling safety data of topical ophthalmic products and given that the conjunctival and corneal epithelia serve as physical barriers against endotoxin entry (Metcalfe 2014; Cooper 2009). In line with the FDA's current thinking, USP <771> states that the endotoxin test is typically not required for topically applied ophthalmic products. As a result, the endotoxin testing is no longer required for topical ophthalmic solutions and ointments. However, some topical ophthalmic products could be intended for application on the eye with a compromised corneal surface. In this instance, the sponsor should consider application of endotoxin controls as the risk to an endotoxin-dependent immune response is high when the ocular surface barrier is breached (Pearlman et al. 2008; Blais et al. 2005). USP <771> chapter does not address the endotoxin limits for solid medical devices that are injected or implanted (United States Pharmacopeia and National Formulary (USP 41-NF 36) 2018a).

Endotoxin limits for ophthalmic solid medical devices are mentioned in the 2012 FDA guidance (U.S. Food and Drug Administration 2012). It states that "For devices that are in direct or indirect contact with the intraocular environment, a lower endotoxin limit may apply." In 2015, the FDA provided recommendations for endotoxin limits in its guidance document for single-use intraocular ophthalmic devices (U.S. Food and Drug Administration 2015). Based on the findings from animal studies, the FDA 2015 guidance states that the agency does not recognize the endotoxin limits specified in the International Organization for Standardization (ISO) standards ISO 15798:2013 (Ophthalmic implants-Ophthalmic viscosurgical devices (OVDs) (International Organization for Standardization 2013a) and ISO 11979-8:2011 (Ophthalmic implants-Intraocular Lenses) (International Organization for Standardization 2011), which are ≤0.5 EU/mL and ≤0.5 EU per device, respectively. Recently, the endotoxin limit for OVDs in ISO 15798 was changed to ≤0.2 EU/mL (ISO 15798:2013 Amendment 1) (International Organization for Standardization 2013b). In ISO 11979-8:2017 version (International Organization for Standardization 2017), the endotoxin limit for intraocular lens was updated to ≤0.2 EU/lens. Similarly, the FDA does not recognize the endotoxin limits set in the American National Standards Institute (ANSI) series of standards for intraocular lenses (American National Standard 2011). The agency recommends a limit of $\leq$0.2 EU/mL for OVDs. For anterior segment solid devices, an endotoxin limit of $\leq$0.2 EU/device is given. In Table 1, endotoxin limits of ophthalmic products recommended by the FDA and other regulatory or compendial bodies are listed.

Compendial Endotoxin Testing Methods

According to USP <1> (United States Pharmacopeia and National Formulary (USP 41-NF 36) 2018d), all products intended for parenteral administration should be prepared in a manner designed to limit bacterial endotoxins. Per 21 CFR 211.167(a), "For each batch of drug product purporting to be sterile and/or pyrogen-free, there shall be appropriate laboratory testing to determine conformance to such requirements (Code of Federal Regulations 2018e)." In the European Pharmacopoeia, according to the general monograph 0520 "Parenteral preparation,"(European Pharmacopeia 2016a) pharmaceutical preparations to be used parenterally must comply with the test for bacterial endotoxins. Furthermore, per 21 CFR 610.13 (Code of Federal Regulations 2018h), the test for pyrogenic substances by Rabbit Pyrogen Test (RPT, USP <151>) (United States Pharmacopeia and National Formulary (USP 41-NF 36) 2018e) is required for all biologic products intended for use by injection. However, the pyrogenic testing requirement for biologics may be fulfilled by a method if its equivalency to the rabbit test is demonstrated in accordance with 21 CFR 610.9 (Code of Federal Regulations 2018i). The limulus amebocyte lysate (LAL) bacterial endotoxins test (BET) is a compendial method that is equivalent to the RPT with regard to endotoxin pyrogen detection. In 1983, the FDA provided the draft guidance for determining endotoxins using a LAL test, and the USP LAL test method was also revised (U.S. Food and Drug Administration 2012). Although there are a small number of non-endotoxin pyrogens, the BET is considered appropriate based on the rationale that the absence of bacterial endotoxins in a substance or product implies the absence of pyrogenic components, provided that the presence of non-endotoxin pyrogens can be ruled out. In order to rule out the presence of non-endotoxin pyrogens, the manufacturer should perform a risk assessment. If the risk assessment indicates that non-endotoxin pyrogens may be present, it may be more appropriate to use the RPT. For biological products, RPT can be used to rule out the presence of non-endotoxin pyrogens. In this regard, the European Pharmacopoeia encourages using monocyte activation test (MAT) as an alternative for the RPT, as stated in the EP 2.6.30 (European Pharmacopeia 2016b). In most cases, RPT has been replaced by the BET, mainly relying on the fact that bacterial endotoxin is the most common pyrogen found in the pharmaceutical products. For more than 40 years, the FDA has accepted the use of a LAL test for endotoxins as an alternative for RPT. More recently, due to the growing concerns of animal welfare, the benefit of the use of recombinant factor C (rFC) assay as an alternative to the LAL test has been increasingly recognized by the pharmaceutical industry. Currently, the rFC is regarded as a non-compendial method and should be validated as described in the USP General Chapter <1225>, Validation of Compendial Procedures (United States Pharmacopeia and National Formulary (USP 41-NF 36) 2018f).

For ophthalmic medical devices, the general principle of endotoxin testing and sample preparation, as described in USP <85> (United States Pharmacopeia and National Formulary (USP 41-NF 36) 2018c), USP<161> (United States Pharmacopeia and National Formulary (USP 41-NF 36) 2018g), and ANSI/AAMI ST72 (American National Standard 2011), can be followed. Due to the high risk of endotoxin contamination associated with intraocular ophthalmic devices, the FDA's CDRH issued guidance for single-use intraocular ophthalmic devices in 2015. This guidance document spells out the recommended endotoxin test method validation regime for OVDs and provides the recommendations on test sample preparation for solid intraocular devices (U.S. Food and Drug Administration 2015).

Generally, ophthalmic products have the same or similar requirement as injectables and implants as described in the USP <1> for the parenterals (United States Pharmacopeia and National Formulary (USP 41-NF 36) 2018d). Specifically, according to USP<771>,(United States Pharmacopeia and National Formulary (USP 41-NF 36) 2018a) all injected ophthalmic drug products shall be tested for bacterial endotoxins by following the procedure described in the USP <85> (United States Pharmacopeia and National Formulary (USP 41-NF 36) 2018c). A large portion of this general chapter is harmonized with EP 2.6.14(European Pharmacopeia 2016c) and JP 4.01 (Japanese Pharmacopeia 2016b). The analytical procedures are harmonized and can be used interchangeably. In 2012, the FDA published Pyrogen and Endotoxin Q&A Guidance that provides testing recommendations for biological product drug and devices firms (U.S. Food and Drug Administration 2012). The guidance mainly focuses on the issues that may be subject to misinterpretation and are not covered in compendial procedures.

Low Endotoxin Recovery

While the LAL-based BET assays have been widely accepted and used as a release test for drug product, in some cases it shows a lower than expected endotoxin recovery of spiked endotoxins for certain drug formulations. Typically, this low endotoxin recovery (LER) is observed when products are formulated with chelating agents (e.g., sodium citrate and phosphate buffers) in combination with non-ionic surfactants (e.g., PS20 and PS80) (Reich et al. 2016). LER was reported by Chen et al. in 2013 and is defined as the inability to recover $\geq$50% activity over time when known amount of endotoxin is added to an undiluted product (Chen and Vinther 2013; Parenteral Drug Association 2019). Starting from 2013, the concern with potential underreporting of endotoxin in finished drug product due to LER has prompted the FDA to request to perform hold-time studies for biologic formulations. The hold-time studies are to assess the effect of hold time on endotoxin recovery by spiking a known amount of endotoxin into undiluted drug product and then testing for recoverable endotoxin over time. The most recently published PDA technical report further defined that LER hold-time studies should be conducted under conditions which are relevant for the manufacturing process (Parenteral Drug Association 2019). If LER is observed, the FDA recommends using RPT as an interim test for the release of final drug product until an appropriate endotoxin method is developed, provided that the endotoxin-spiked product resulted in a pyrogenic effect in rabbits. Otherwise, a risk-based approach can be used as an interim measure until a valid de-masking method becomes available (Candau-Chacon 2017).

Endotoxin Control Strategy

In addition to directly monitoring the endotoxin level in the final drug product, based on the quality by design concepts, a comprehensive microbiological control strategy should be built into the design of manufacturing process to reduce the risk for endotoxin contamination. In recent years, the importance of microbiological control has become evident due to the wide acceptance of risk-based approaches within regulatory and industrial communities. It is also known that the final product microbial testing has its inherent weaknesses as only an extremely small amount of product is sampled and tested, as well as the test methods are limited in their detection and quantitation capabilities. Therefore, product quality assurance cannot solely rely on the end-product testing. Several regulations and guidance documents provide a basic framework for the development of appropriate microbial contamination strategy. A microbial control strategy should be developed based on a comprehensive risk assessment on all possible entry points of microbial and endotoxin contamination. The microbiological control strategy should be implemented to minimize the risk of microbial contamination of the product, facility, and equipment. These controls include validated equipment cleaning, sanitization, and depyrogenation procedures; defined equipment and in-process hold times; validated filter sterilization and aseptic manufacturing process; in-process bioburden and endotoxin testing and monitoring; personnel gowning and standard operating procedures; environmental monitoring programs; and endotoxin controls on raw materials, container closure, and product contact surfaces of the manufacturing process. In addition, given that water for injection (WFI) plays important roles in ophthalmic manufacturing process, it is advisable that the endotoxin limit of WFI for ophthalmic injectable formulation should be more stringent than NMT 0.25 EU/mL prescribed by USP (United States Pharmacopeia and National Formulary (USP 41-NF 36) 2018h). Furthermore, microbial quality of water systems at manufacturing site should be closely monitored and be trended regularly. These controls either directly reduce endotoxin levels or, as part of the sterility assurance program, help to minimize the introduction of Gram-negative bacteria and thereby further reduce the risk for endotoxin contamination. In conclusion, a risk of endotoxin contamination can only be minimized when the proper control strategy for endotoxin throughout the manufacturing process in combination with end-product endotoxin testing is in place.

Preservatives' Role and Function

Topical ophthalmic products are prescribed for the treatment and management of ocular diseases including conjunctivitis, dry eye, inflammation, uveitis diseases, ocular hypertension, and glaucoma, among other diseases. Treatment can range from short-term to chronic use and presented in unpreserved unit-dose or multi-dose preserved products. Antimicrobial preservatives are added to aqueous ophthalmic products to (i) inhibit microbial growth (microbiostatic), (ii) kill microbial contaminants (microbicidal), and (iii) suppress microbial degradation of the product. All useful preservatives are toxic substances. Patients using preserved products may experience adverse effects of irritation, dryness, and surface epithelial cell loss due to the toxicity of the preservative. The level of preservative in a formulation must provide adequate protection that may arise from microbial contamination or proliferation during the use of the product but be below a level that will be toxic to patients. A balance between preservative efficacy and safety is required.

A variety of different preservatives are utilized in ophthalmic preparations. Historically, preservatives are classified into two categories: detergent preservatives and oxidizing preservatives (Table 2) (Freeman and Kahook 2009). Benzalkonium chloride (BAK), a detergent preservative, was one of the first and most commonly used preservative in topical ophthalmic products (McPherson Jr. and Wood 1949). Due to the chemical and physical properties of the preservative, it may also function as a solubilizer of active pharmaceutical ingredients. Microbicidal activity of detergent preservatives can be attributed to lipid-membrane instability and leakage of cellular contents. Oxidative preservatives act by penetrating the membrane and altering the DNA, protein, and lipid components of the bacterial cell. Examples of oxidizing preservatives include stabilized oxychloro complex (SOC or Purite®), sodium perborate (GenAqua®), and the ionic-buffered preservative (SofZia®).

Formulation Development with Preservative

The selection of a preservative system for a formulated drug product should be performed by a team of formulators, chemists, and microbiologists. The selection criterion must take into consideration the type of preservative, concentration that will be utilized in the formulation, and compatibility with the drug product components. For comfort to the patient, a preservative that maintains activity in the ocular physiological pH range of 6–8 pH units should be selected. Literature searches pertaining to the preservatives' chemical compatibility and minimum inhibitory concentrations (MIC) may be ascertained prior to formulating the drug product. MIC data can

Table 2 Common preservatives in ophthalmic products

Class	Preservative	Examples	Preservative concentration
Detergent	Benzalkonium chloride (BAK)	Azarga® Ganfort® Simbrinza® Lumigan® Xalatan® Azopt® Timolol® Cosopt®	0.1% 0.05% 0.03% 0.02% 0.02% 0.01% 0.01% 0.075%
	Polyquad® (polyquaternium-1)	Travatan® Systane® Ultra	0.01% 0.001%
	Polyquad®/Aldox® (myristamidopropyl dimethylamine)	Opti-Free Express®	Polyquad® 0.001% and Aldox® 0.0005%
	Chlorobutanol	TobraDex®	0.5%
Oxidative	Purite® (stabilized oxychloro complex, SOC)	Alphagan-P® Refresh Tears®	0.005% 0.005%
	GenAqua® (sodium perborate)	Genteal®	0.028%
	SofZia® (borate, sorbitol, propylene glycol, and zinc chloride)	Travatan Z®	Ionic-buffered system

indicate whether the preservative agent is broad spectrum or targets specific microorganisms. While literature searches are important tools in the preliminary assessment of preservatives, the preservative stability and chemical compatibility with the drug substance and other excipients within the formulation need to be determined. For example, BAK is a surface-active agent that can react with other ingredients in the formulation, especially surfactants such as polysorbates (Liu et al. 2009). Surfactants are used to increase drug solubility. However, above critical micelle concentrations (CMC), the surfactants form aggregates or micelles that can trap BAK, thereby reducing the preservative activity (Liu et al. 2009). This phenomenon may lead to BAK ineffectiveness against microorganisms (especially against pseudomonads) (Adair et al. 1971; Kampf 2018).

For the purpose of screening multiple preservative systems, an alternative initial decimal reduction time (*D* values) may be taken to the traditional, time-consuming preservative effectiveness test. *D* value refers to the time required at a given condition to kill 90% (or 1 log) of the exposed organism. *D* values of not more than 2 h for bacteria can predict that the preservative system will pass acceptance criteria stipulated in the European Pharmacopoeia, which is a more rigorous criteria than the USP criteria for preservative effectiveness (Akers et al. 1984). Once a preservative system is selected, Antimicrobial Preservative Efficacy Test (APET), a 28-day challenge test, should be performed to meet the regulatory authority expectation in which the product will be submitted for approval.

The preservative concentration must be justified in terms of efficacy and safety; the minimum concentration of preservative should be determined to provide efficacy in the formulation (European Medicines Agency 1997; International Council for Harmonisation (ICH) 2009) and the maximum concentration should be toxicologically qualified. The minimum preservative concentration must be determined experimentally using the compendial preservative effectiveness test to robustly meet the low-end shelf specification. Concentrations below the low-end shelf specifications should be tested to determine the preservative dose response as certain preservatives may not demonstrate effectiveness against all microorganisms at increasing concentrations. In order to provide safety margin to account for inherent microbial variability, a reasonably higher concentration of the preservative than the minimum effective concentration should be considered for a robust preservative system. If the product is to be commercialized globally, then the acceptance criteria should meet the global requirements for ophthalmic products. Therefore, the effectiveness of a preservative must be demonstrated in the formulated drug product by APET during product development and stability testing.

Preservative Effectiveness Chapters and International Guidelines

Preservative testing of multi-use aqueous drug products is required by the three regulatory authorities and testing is performed according to their respective pharmacopoeia: USP <51>, EP 5.1.3, and JP Preservatives-Effectiveness Tests. For the test, aqueous is defined as a water activity of more than 0.6. Currently, the respective chapters are not harmonized with respect to the acceptance criteria for the individual product categories (Table 3); the testing procedure is largely harmonized.

If sample volumes permit, testing should be performed in original product containers to show that the functional properties of the preservative are not compromised in the final packaging and meet the acceptance criteria. Alternative containers may be used for testing if the product was stored in the immediate container for market over shelf life and the contents dispensed into the alternative container just prior to initiation of testing for effectiveness. Care should be taken to avoid using materials that can interact with the preservative in the containers that are used for antimicrobial effectiveness testing.

In general, a specified quantity of the drug product is inoculated separately with a prescribed titer of microorganism from a panel of organisms representing Gram-positive and Gram-negative bacteria, yeast, and mold. As stated in the USP<51>, supplemental challenge organisms may be added if it is deemed useful to measure the biological activity of the preservative system for a specific product. The sponsor has the sole discretion to select supplemental challenge organisms based on risk of the preservative system. The sponsor may consider challenging the preservative system with clinically relevant organisms for the route of admiration (i.e., *Staphylococcus epidermis* for topical ophthalmic products, *Burkholderia cepacia* for nasal products). Even if the selected supplemental organism does not pass the compendial preservative acceptance criteria, the preservative may provide stasis. The information gathered may serve as a risk assessment tool for the preservative system.

Table 3 Acceptance criteria for ophthalmic products by the three major pharmacopoeias

	Log reduction for bacteria				
Compendia—category	6 h	24 h	7 days	14 days	28 days
EP—A	2	3	–	–	NR
EP—B	–	1	3	–	NI
USP—1	–	–	1.0	3.0	NI
JP—1A	–	–	1.0	3.0	NI

	Log reduction for fungi		
Compendia—category	7 days	14 days	28 days
EP—A	2	–	NI
EP—B	–	1	NI
USP—1	NI	NI	NI
JP—1A	NI	NI	NI

NR no recovery, *NI* no increase

At intervals specified by the compendial chapter, an aliquot of the test article is removed, the preservative is neutralized according to a validated procedure to allow recovery of any surviving microorganisms, and the sample is incubated on the appropriate media to enumerate microorganisms. The surviving population of each organism is calculated at each time interval and compared to the initial population, and the log unit reduction is reported and compared to the compendial acceptance criteria. Table 3 shows acceptance criteria for ophthalmic products by the three major pharmacopoeias. For EMA, the preservative system is expected to comply with level A criteria unless otherwise justified (European Medicines Agency 1998). The agencies may accept B criteria if justified based on risk/benefit of safety versus efficacy. If criteria A is met for all but one organism (e.g., mold), then the sponsor can seek regulatory approval for an exception of only this specified organism for the expected criteria. For example, if an increase in preservative concentration in the formulation does not provide proportional kill of a microorganism, then the sponsor may justify selecting lower levels of the preservative considering the safety profile of the preservative.

Shelf Stability Testing

Preservative effectiveness should be assessed throughout product development, scale-up, and primary stability batches. Based on ICH Q1A(R2), Stability Testing of New Drug Substances and Products, a single primary stability batch of the drug product should be tested for antimicrobial preservative effectiveness (in addition to preservative content) at the proposed shelf life for verification purposes, regardless of whether there is a difference between the release and shelf life acceptance criteria for preservative content at the storage condition (International Council for Harmonisation (ICH) 2003). The preservative's efficacy may be influenced by other chemical and physical changes in the final product formulation. EMA also requires additional preservative effectiveness testing at the lower end of the preservative shelf-life specification irrespective of chemical stability of the preservative over shelf life (European Medicines Agency 1997). Once a correlation between preservative effectiveness test and preservative content is established, the critical release and shelf life attribute of preservative system should be assessed using chemical assay of the preservative content. Testing of APET at accelerated stability may not provide useful information with respect to preservative effectiveness of the formulation. At elevated temperatures, the preservative or even other excipients may degrade but may provide transiently even greater preservative effectiveness due to unstable, short-lived intermediates that have greater antimicrobial activity than the preservative, i.e., preservative content is inversely correlated with preservative effectiveness. This artifact, however, may not be reflective of the properties of the preservative at long-term storage condition where direct correlation between preservative content and preservative effectiveness can be expected, thus justifying APET testing of formulations at long-term storage condition.

ICH guidance provides the preservative effectiveness requirements for the three regulatory bodies; however, in order to obtain approval in different global markets, extensive regulatory research must be performed for each country's requirement for preservative effectiveness testing. For example, for product approvals in Australian government Department of Health Therapeutic Goods Administration (TGA), in addition to the one primary stability batch tested at the proposed shelf, the TGA requires preservative effectiveness test data from at least two (preferably three) separately manufactured batches tested at the beginning of stability (Australian Government, Department of Health, Therapeutic Good Administration 2013).

In-Use and Discard Date Studies

For product registration, the EMA and TGA also require in-use stability testing, a period during which multi-dose product can be used, once the container is opened (typically 28 days for topical ophthalmic products). EMA suggests selecting a minimum of two pilot scale batches, one batch being toward the end of the shelf life for in-use testing. If a batch is not available at the time of filing, the testing should be performed on the last submitted stability studies (European Medicines Agency 2001b). For EMA, only the chemical content of the preservative is necessary to be demonstrated after the in-use period.

TGA requires one of the following information to support an open shelf-life period (Australian Government, Department of Health, Therapeutic Good Administration 2013):

- Results of preservative effectiveness tests that involve repeated microbial challenges over 28 days according to International Standard ISO 14730—Ophthalmic

lens care products—Antimicrobial preservative efficacy testing and guidance on determining discard date (International Organization for Standardization 2014).
- Results of preservative efficacy tests on the contents of containers of the medicine after stimulated in-use.
- Results of sterility tests on the contents of containers of the medicine after stimulated in-use.
- Results of the microbial content tests on the contents from patients used for the full duration of the open shelf life.

Currently, there is no guidance from the FDA on in-use testing for preserved ophthalmic products.

In the US, for ophthalmic (contact lens care) products that fall under the 510(k) submissions, preservative effectiveness needs to be demonstrated initially and at shelf life with microbial re-challenge on day 14 per ISO 14730 for a 28-day discard date (International Organization for Standardization 2014; U.S. Food and Drug Administration 1997). The ISO guidance also provides additional procedures for prolonged discard date determination in the subsequent Annexes B to E.

For ophthalmic products (over-the-counter, drug-device) that fall under the requirement for CE (Conformité Européenne) marking per the European Medical Device Directive, discard date studies should be performed per ISO 14730.

Forward Thinking and Design Space for New Preservatives

A balance between antimicrobial efficacy and ocular safety is essential when seeking new preservatives for ophthalmic preparations. While preservatives protect multi-dose bottles from contaminations once opened, they do have various effects on the ocular surface. As the inclusion of preservatives in multi-dose products is a requirement by regulatory authorities, there is a need for safe and efficacious preservatives that meet global requirements. There is a limited number of approved preservatives for ophthalmic use. New preservatives are needed for formulation compatibility issues. The most common ophthalmic preservative is benzalkonium chloride, which is known to have ocular effects and cannot be formulated with certain excipients. When developing formulations, consider excipients that have multiple functions (like SofZia). Select gentle preservatives that do not cause extensive ocular tear film damage (such as Purite). However, the conundrum of gentle preservatives is that most often they do not meet the stringent EP-A criteria and if EP-A is required for a global product, then gentle preservatives cannot be used.

Alternatives to preservatives are sought for products that cannot be formulated with preservatives due to formulation incompatibility issues or for a patient population that is sensitive to preservatives. Unpreserved formulations are ideal for this target group that requires ophthalmic products for chronic use for conditions such as glaucoma or dry eye. Different container closure presentations of ophthalmic products are an alternative for excluding preservatives in formulations such as blow-fill-seal (BFS) single-dose vials, multi-dose preservative free bottles (i.e., ABAK® by Thea, Ophthalmic Squeeze Dispenser (OSD) by Aptar Pharma, Novelia® by Nemera), and multi-dose pump systems (i.e. Easygrip® by Thea, Comod® by AeroPump, Replenish, Inc.). Established in the 1970s, BFS single-dose vials are the most widely used technology for unpreserved eye drops. Although readily used, this technology is considerably more expensive compared to multi-dose dispensers due to substantial overfill during processing (fill volume approximately 0.3–1 mL) and is considered less environmentally friendly with the amount of plastic generated. The innovative multi-dose preservative-free bottles have different physical barrier features (filtration, uni-directional valve, pump system) that prevent microorganisms from entering the bottle during patient use. One example of a multi-dose preservative-free container is ABAK® by Thea. This system utilizes a 0.2-μm membrane that filters the solution before dispensing a drop. When the pressure is released after actuation, the residual solution is reabsorbed and filtered from bacteria and air, ensuring the protection of the solution throughout its use. Another system is Aptar's OSD which prevents microorganisms from entering the container with a spring-loaded uni-directional valve. The tip seal keeps the system closed until a defined pressure is reached by actuation. Once the drop is actuated, the tip seal will immediately close preventing any backflow of potential microbial contaminants. In 2011, the first over-the-counter product using Aptar's OSD system was introduced into the market with VISMED® MULTI eye drops (TRB Chemedica, Int.). In 2016, the US FDA approved Allergan's Restasis MultiDose™ as the first prescription medication using Aptar's OSD. Approvals require extensive design, function, and microbial quality data to demonstrate suitability of multi-dose containers for preservative-free eye drops. Other technologies include antimicrobial additives (silver ions, PyClear®) in packaging material for ophthalmic formulations. To exert its biocidal effects, silver ions must be released from the packaging into the formulation. One must consider the risk of addition of such additives to containers as it may impact product compatibility (efficacy, stability, and quality) and safety to the patient.

The in-use microbiological quality of the preservative-free multi-dose containers must be demonstrated to show equivalency to the preserved product. EMA provides guidance on the Q&A section for studies necessary for approval in the EU (European Medicines Agency 2018b). The preservative-free container system and the drug product must be safe given the importance of vision to one's quality of life.

Conclusions

Given the sensitivity of eyes to microbial contamination and severity of consequence, a thorough evaluation of microbial control strategy is a key to assure product quality at release, over the shelf life, and during in-use period. During development, a risk-based manufacturing control and QC release testing approach for sterility and endotoxins are established to provide a holistic microbial control strategy for commercial manufacturing. Preservative and container closure systems are also evaluated as part of formulation development to control for potential microbial contamination during in-use period. While regulatory requirements lay out the framework for quality assurance, a product-specific control strategy that is developed and tested throughout the course of development integrates microbiological quality in the ophthalmic product.

References

Adair FW, Geftic SG, Gelzer J. Resistance of Pseudomonas to quaternary ammonium compounds. II. Cross-resistance characteristics of a mutant of Pseudomonas aeruginosa. Appl Microbiol. 1971;21(6):1058–63.

Agrawal S, Joshi M, Christoforidis JB. Vitreous inflammation associated with intravitreal anti-VEGF pharmacotherapy. Mediat Inflamm. 2013;2013:943409.

Akers MJ, Boand AV, Binkley DA. Preformulation method for parenteral preservative efficacy evaluation. J Pharm Sci. 1984;73(7):903–5.

American National Standard. ANSI/AAMI ST72:2011(R2016): bacterial endotoxins - test methods, routine monitoring, and alternatives to batch testing. Arlington: Association for the Advancement of Medical Instrumentation (AAMI); 2011.

Parenteral Drug Association. Technical report no. 82 low endotoxin recovery; 2019.

Australian Government, Department of Health, Therapeutic Good Administration. Guidance 17: microbial quality of prescription and over-the-counter medicines; 2013.

Bantseev V, Miller PE, Bentley E, Schuetz C, Streit TM, Christian BJ, Farman C, Booler H, Thackaberry EA. Determination of a no-observable effect level for endotoxin following a single intravitreal administration to Dutch belted rabbits. Invest Ophthalmol Vis Sci. 2017;58(3):1545–52.

Bantseev V, Miller PE, Nork TM, Rasmussen CA, McKenzie A, Christian BJ, Booler H, Thackaberry EA. Determination of a no observable effect level for endotoxin following a single intravitreal Administration to Cynomolgus monkeys. J Ocul Pharmacol Ther. 2019;35(4):245–53.

Blais DR, Vascotto SG, Griffith M, Altosaar I. LBP and CD14 secreted in tears by the lacrimal glands modulate the LPS response of corneal epithelial cells. Invest Ophthalmol Vis Sci. 2005;46(11):4235–44.

Buchen SY, Calogero D, Hilmantel G, Eydelman MB. Rabbit ocular reactivity to bacterial endotoxin contained in aqueous solution and ophthalmic viscosurgical devices. Ophthalmology. 2012a;119(7):e4–e10.

Buchen SY, Calogero D, Hilmantel G, Eydelman MB. Detecting endotoxin contamination of ophthalmic viscosurgical devices: intracameral versus intravitreal assays in rabbits. Ophthalmology. 2012b;119(7):e11–8.

Candau-Chacon R. PDA endotoxins workshop; 2017.

Carter J. Evaluation of recovery filters for use in bacterial retention testing of sterilizing-grade filters. PDA J Pharm Sci Technol. 1996;50(3):147–53.

Chen J, Vinther A. Low endotoxin recovery in common biologics products. Presented at the PDA Annual Meeting, Orlando, FL. April 15-17, 2013.

Code of Federal Regulations. 21 CFR 200.50 - Ophthalmic preparations and dispensers; 2018a.

Code of Federal Regulations. 21 CFR 610.12 – Sterility; 2018b.

Code of Federal Regulations. 21 CFR 349.3; 2018c.

Code of Federal Regulations. 21 CFR 800.10 - Contact lens solutions; sterility; 2018d.
Code of Federal Regulations. 21 CFR 211.167 - Special testing requirements; 2018e.
Code of Federal Regulations. 21 CFR 211.113 - Control of microbiological contamination; 2018f.
Code of Federal Regulations. 21 CFR 211.84 - Testing and approval or rejection of components, drug product containers, and closures; 2018g.
Code of Federal Regulations. 21 CFR 610.13 – Purity; 2018h.
Code of Federal Regulations. 21 CFR 610.9 - Equivalent methods and processes; 2018i.
Cooper JF. A case against endotoxin limits for topical ophthalmic solutions. PDA Global Conference on Microbiology; 2009.
European Commission, EudraLex. The rules governing medicinal products in the European Union, vol. 4. EU guidelines to good manufacturing practice medicinal products for human and veterinary use. Annex 1 manufacture of sterile medicinal products; 2008.
European Commission, EudraLex. The rules governing medicinal products in the European Union, vol. 4. EU guidelines to good manufacturing practice medicinal products for human and veterinary use. Annex 17: Real Time Release Testing and Parametric Release. 4; 2018.
European Medicines Agency. Note for guidance on inclusion of antioxidants and antimicrobial preservatives in medicinal products. CPMP/CVMP/QWP/115/95; 1997.
European Medicines Agency. Note for guidance on development pharmaceutics. CPMP/QWP/155/96; 1998.
European Medicines Agency. Decision trees for the selection of sterilization methods (annex to note for guidance on development pharmaceutics) CPMP/QWP/054/98 Corr; 2000.
European Medicines Agency. Note for Guidance on Parametric Release. CPMP/QWP/3015/99; 2001a.
European Medicines Agency. Note for guidance on in-use stability testing of human medicinal products. CPMP/QWP/2934/99; 2001b.
European Medicines Agency. Guideline on the sterilization of the medicinal product, active substance, excipient and primary container – draft. EMA/CHMP/CVMP/QWP/BWP/850374/2015; 2016.
European Medicines Agency. Quality of medicines questions and answers: Part 2, Stability – Endotoxin testing and sterility testing at the end of shelf-life; 2018a.
European Medicines Agency. Quality of medicines questions and answers: part 2. Quality data requirements to demonstrate suitability of multidose containers for preservative free eye drops; 2018b.
European Pharmacopeia. 0520 parenteral preparations. 9th ed. Strasbourg: Council of Europe; 2016a.
European Pharmacopeia. 2.6.30 Monocyte-activation test. European pharmacopoeia. 9th ed. Strasbourg, Council of Europe; 2016b.
European Pharmacopeia. 2.6.14 Bacterial endotoxins. European pharmacopoeia. 9th ed. Strasbourg, Council of Europe; 2016c.
European Pharmacopeia. 2.6.1 sterility. 9th ed. Strasbourg: Council of Europe; 2019a.
European Pharmacopeia. 5.1.1 Methods of preparation of sterile products. European pharmacopoeia. 9th ed. Strasbourg: Council of Europe; 2019b.
Eydelman MB, Tarver ME, Calogero D, Buchen SY, Alexander KY. The Food and Drug Administration's proactive toxic anterior segment syndrome program. Ophthalmology. 2012;119(7):1297–302.
Federal Register. Notice to Manufactures and repackers of ophthalmic solutions, vol. 18, no. 11; 1953 p. 11.
Fielden M, Nelson B, Kherani A. Acute intraocular inflammation following intravitreal injection of bevacizumab--a large cluster of cases. Acta Ophthalmol. 2011;89(8):e664–5.
Fleisher LN, McGahan MC. Endotoxin-induced ocular inflammation increases prostaglandin E2 synthesis by rabbit lens. Exp Eye Res. 1985;40(5):711–9.
Freeman PD, Kahook MY. Preservatives in topical ophthalmic medications: historical and clinical perspectives. Exp Rev Ophthalmol. 2009;4(1):59–64.

Goldberg RA, Flynn HW Jr, Isom RF, Miller D, Gonzalez S. An outbreak of streptococcus endophthalmitis after intravitreal injection of bevacizumab. Am J Ophthalmol. 2012;153(2):204–208.e201.

International Council for Harmonisation (ICH). Q5C Quality of biotechnological products: Stability testing of biotechnological/biological products; 1995.

International Council for Harmonisation (ICH). Q6A specifications: test procedures and acceptance criteria for new drug substances and new drug products: chemical substances; 1999.

International Council for Harmonisation (ICH). Q1A(R2) stability testing of new drug substance and products; 2003.

International Council for Harmonisation (ICH). Q8 (R2) Pharmaceutical development; 2009.

International Organization for Standardization. ISO 11979-8:2011, ophthalmic implants -- intraocular lenses -- part 8: fundamental requirements. Geneva: International Standards Organization; 2011.

International Organization for Standardization. ISO 15798:2013 ophthalmic implants – ophthalmic viscosurgical device. Geneva: International Standards Organization; 2013a.

International Organization for Standardization. ISO 15798:2013 amendment 1 2017 ophthalmic implants – ophthalmic viscosurgical device. Geneva: International Standards Organization; 2013b.

International Organization for Standardization. ISO 14730 – ophthalmic lens care products – antimicrobial preservative efficacy testing and guidance on determining discard date; 2014.

International Organization for Standardization. ISO 16671:2015/A1:2017 ophthalmic implants — irrigating solutions for ophthalmic. Geneva: International Standards Organization; 2015.

International Organization for Standardization. ISO 11979-8:2017, Ophthalmic implants -- Intraocular lenses -- Part 8: Fundamental requirements. Geneva: International Standards Organization; 2017.

Japanese Pharmacopeia. 4.06 sterility test. 17th ed. Tokyo: Pharmaceuticals and Medical Devices Agency; 2016a.

Japanese Pharmacopeia. 4.01 Bacterial endotoxins test. 17th ed. Tokyo: Pharmaceuticals and Medical Devices Agency; 2016b.

Johnson D, Sharma S. Ocular and systemic safety of bevacizumab and ranibizumab in patients with neovascular age-related macular degeneration. Curr Opin Ophthalmol. 2013;24(3):205–12.

Jokl DH, Wormser GP, Nichols NS, Montecalvo MA, Karmen CL. Bacterial contamination of ophthalmic solutions used in an extended care facility. Br J Ophthalmol. 2007;91(10):1308–10.

Kampf G. Adaptive microbial response to low-level benzalkonium chloride exposure. J Hosp Infect. 2018;100(3):e1–e22.

Karl G, Csaky KG. Ocular inflammation after intravitreal injections: a review of data on incidence and implications. Retin Physitian. 2018;15(1):26–32.

Kreisler KR, Martin SS, Young CW, Anderson CW, Mamalis N. Postoperative inflammation following cataract extraction caused by bacterial contamination of the cleaning bath detergent. J Cataract Refract Surg. 1992;18(1):106–10.

Kutty PK, Forster TS, Wood-Koob C, Thayer N, Nelson RB, Berke SJ, Pontacolone L, Beardsley TL, Edelhauser HF, Arduino MJ, Mamalis N, Srinivasan A. Multistate outbreak of toxic anterior segment syndrome, 2005. J Cataract Refract Surg. 2008;34(4):585–90.

Liu J, Lu GW, Sandoval M, Ciringh Y, Xue G, Jaeger D, Kompanik K, Jiao J, Gelotte KM. Determination of benzalkonium chloride partition in micelle solutions using ultrafiltration method. AAPS PharmSciTech. 2009;10(4):1216–23.

Mayo MS, Schlitzer RL, Ward MA, Wilson LA, Ahearn DG. Association of Pseudomonas and Serratia corneal ulcers with use of contaminated solutions. J Clin Microbiol. 1987;25(8):1398–400.

McPherson S Jr, Wood RM. Self-sterilizing ophthalmic solutions. Am J Ophthalmol. 1949;32(5):675–8.

McGahan MC, Grimes AM, Fleisher LN. Hemoglobin exacerbates the ocular inflammatory response to endotoxin. Graefes Arch Clin Exp Ophthal. 1996;234(10):643–7.

Metcalfe JW. Urban myths and pharmaceutical products: a CDER microbiology regulatory perspective. PDA Global Conference on Microbiology; 2014.

Nussenblatt RB, Calogero D, Buchen SY, Leder HA, Goodkin M, Eydelman MB. Rabbit intraocular reactivity to endotoxin measured by slit-lamp biomicroscopy and laser flare photometry. Ophthalmology. 2012;119(7):e19–23.

Park CY, Lee JK, Chuck RS. Toxic anterior segment syndrome-an updated review. BMC Ophthalmol. 2018;18(1):276.

Pearlman E, Johnson A, Adhikary G, Sun Y, Chinnery HR, Fox T, Kester M, McMenamin PG. Toll-like receptors at the ocular surface. Ocul Surf. 2008;6(3):108–16.

Rajendran V, Al Aboody M, Alturaiki W, Sandle T. Review on endotoxin mediated toxic anterior segment syndrome (TASS)-Outbreaks, product recall and testing limits. Eur J Parent Pharm Sci. 2017;22:20–5.

Reich J, Lang P, Grallert H, Motschmann H. Masking of endotoxin in surfactant samples: effects on Limulus-based detection systems. Biologicals. 2016;44(5):417–22.

Richburg FA, Reidy JJ, Apple DJ, Olson RJ. Sterile hypopyon secondary to ultrasonic cleaning solution. J Cataract Refract Surg. 1986;12(3):248–51.

Sakimoto A, Sawa M, Oshida T, Sugaya S, Hirono T, Ishimori A. Minimum endotoxin concentration causing inflammation in the anterior segment of rabbit eyes. Jpn J Ophthalmol. 2009;53(4):425–32.

Snyder RW, Glasser DB. Antibiotic therapy for ocular infection. West J Med. 1994;161(6):579–84.

Theodore FH, Feinstein RR. Preparation and maintenance of sterile ophthalmic solutions. J Am Med Assoc. 1953;152(17):1631–3.

Toguchi H. Sterility assurance of microspheres. Journal of Controlled Release. 1999;62(1–2):51–5.

U.S. Food and Drug Administration. Guidance for Industry Premarket Notification [510 (k)] guidance document for contact lens care products. Issued May 1; 1997.

U.S. Food and Drug Administration. Guidance for industry—sterile drug products produced by aseptic processing. Current good manufacturing practice. Silver Spring: FDA; 2004.

U.S. Food and Drug Administration. Guidance for industry: container and closure system integrity testing in lieu of sterility testing as a component of the stability protocol for sterile products. Rockville: Food and Drug Administration; 2008.

U.S. Food and Drug Administration. Guidance for industry: pyrogen and endotoxins testing: questions and answers. Silver Spring:11; 2012.

U.S. Food and Drug Administration. Avastin unit dose syringes by clinical specialties: recall—potential for serious eye infection. FDA Medwatch Alerts; 2013.

U.S. Food and Drug Administration. Endotoxin testing recommendations for single-use intraocular ophthalmic devices-guidance for industry and Food and Drug Administration Staff; 2015.

United States Pharmacopeia and National Formulary (USP 41-NF 36). <771> Ophthalmic product quality tests. Rockville: United States Pharmacopeial Convention; 2018a.

United States Pharmacopeia and National Formulary (USP 41-NF 36). <71> Sterility tests. Rockville, United States Pharmacopeial Convention; 2018b.

United States Pharmacopeia and National Formulary (USP 41-NF 36). <85> Bacterial endotoxin test. Rockville: United States Pharmacopeia and National Formulary: United States Pharmacopeial Convention; 2018c.

United States Pharmacopeia and National Formulary (USP 41-NF 36). <1> Injections and implanted drug products (parenterals)-product quality tests. Eur J Parent Pharm Sci. Rockville: United States Pharmacopeial Convention; 2018d.

United States Pharmacopeia and National Formulary (USP 41-NF 36). <151> Pyrogen Test. Rockville: United States Pharmacopeial Convention; 2018e.

United States Pharmacopeia and National Formulary (USP 41-NF 36). <1225> Validation of compendial procedures. Rockville: United States Pharmacopeial Convention; 2018f.

United States Pharmacopeia and National Formulary (USP 41-NF 36). <161> Medical devices-Bacterial endotoxin and pyrogen tests. Rockville: United States Pharmacopeial Convention; 2018g.

United States Pharmacopeia and National Formulary (USP 41-NF 36). Monograph water for injection. Rockville: United States Pharmacopeial Convention; 2018h.

US FDA Medical Device Recalls. Advanced medical optics, Inc Class 1 Device Recall Healon D; 2008.

Wang F, Yu S, Liu K, Chen FE, Song Z, Zhang X, Xu X, Sun X. Acute intraocular inflammation caused by endotoxin after intravitreal injection of counterfeit bevacizumab in Shanghai, China. Ophthalmology. 2013;120(2):355–61.

Packaging Development: Multi-Dose Container Closure for Preservative Free Products, Extractable/Leachables from Packaging, New Technologies

Seungyil Yoon and Michael Regn

Abstract Most ophthalmic drugs are currently delivered to highly sensitive eyes using a dropper. There is high concern around packaging for ophthalmic solutions and suspensions in terms of interaction between the product and packaging material. The dropper should be developed with no adverse effects caused by packaging materials. In addition, the dropper should instill the drug into the eyes accurately with no technical difficulties. The dispensing function should be evaluated by users. The dropper with a dispensing system is developed to meet quality system regulations and design control if the system is classified as a medical device.

Unit-dose vial and multidose bottle have been improved to multidose preservative-free (MDPF) bottles that do not need preservatives like benzalkonium chloride and can save the drug from multiple uses. There are still patients that have technical difficulties attempting to instill drops in their eyes. Elderly patients especially do not feel comfortable with self-dispensing and require assistance. Some novel dispensing systems are being developed to deliver the drug precisely and conveniently.

Keywords CCS · Container closure system · cGMP · Design verification · Dispensing system · Drug delivery system · MDPF · Multidose preservative-free · Ophthalmic drug · QbD · Quality by design · Quality system · Unit-dose vial

S. Yoon
Device Development, Samsung Bioepis, Incheon, South Korea

M. Regn (✉)
Packaging Engineering, AbbVie Inc., Chicago, IL, USA

S. Neervannan, U. B. Kompella (eds.), *Ophthalmic Product Development*, AAPS Advances in the Pharmaceutical Sciences Series 37,
https://doi.org/10.1007/978-3-030-76367-1_10

Introduction

Aging population with diseases such as diabetic retinopathy, dry eye, glaucoma, and age-related macular degeneration (AMD) has resulted in increased growth in the eye care market worldwide. Ophthalmic drug or device is administered to the front of eye directly by dropping the solution or injecting the drug to the back of the eye. Most ophthalmic products are applied to the eye directly. Ninety-five percent of all ophthalmic products are delivered using an eye drop bottle, and the rest of ophthalmic products are delivered by injection, ointment, and implant by a device. Therefore, eye drop bottle development will be focused in this packaging development chapter. Other systems like prefilled syringe and tube are not specialized for ophthalmic drugs but they are the same with other drug products.

The eyes are one of the most important and highly sensitive organs in the human body. The primary CCS (container closure system) containing ophthalmic products should be developed with no adverse effects caused by the CCS. The FDA guidance "Container Closure Systems for Packaging Human Drugs and Biologics" (FDA 1999) indicates there is high concern around packaging for ophthalmic solutions and suspensions in terms of interaction between the product and packaging material. Therefore, various studies should be performed to ensure no interaction between product and packaging materials. Impurity profiles should be carefully examined to identify leachable compounds from packaging materials and degradation products from the interaction between product and chemicals from packaging materials.

There are several types of eye droppers—unit-dose vial, multidose bottle, and multidose preservative-free (MDPF) bottle. The choice of a unit dose or multidose depends on the efficacy of the drug, product stability, and the course of treatment. Safe and user-friendly multidose bottle is expected to grow and leading the market in future. The global ophthalmic packaging market size is estimated at USD 7.9 billion in 2019 and is expected to reach USD 8.7 billion in 2020 (Grand View Research 2019).

Most of eye droppers are made of plastics such as LDPE (low-density polyethylene) or HDPE (high-density polyethylene). Ophthalmic products in eye drop bottle are applied to the eye directly, so the products must be sterile. Currently, antimicrobial preservative is used for multidose bottles but the preservative like benzalkonium chloride may cause ocular toxicity, and it results in discomfort, blurred vision, and irritation. Therefore, a new drug dispensing technology, MDPF bottle has been developed. MDPF bottle is an improved system but there are still unmet needs for dispensing device and CCS for ophthalmic products. This packaging development chapter provides a development strategy for ophthalmic product's CCS and device and also reviews current eye drop dispensing systems.

CCS Development Strategy

The quote "The container closure system should be suitable for its intended use" is being used all the time to develop and qualify the CCS for new pharmaceutical products. It is the most important requirement to develop safe and efficient drug products for patients. Current drug cGMP and quality system regulations and related guidelines are developed to ensure the drug safety and efficacy. In this section, a CCS development strategy is introduced to comply with regulations and guidelines.

When a CCS is developed, the type of products—drug in a container, lubricant (device) in a container, or drug in a device—must be understood. The US and EU regulations for drug product and medical device require different approaches to ensure the product safety and efficacy. Table 1 shows the types of ophthalmic products with US and EU regulations. When the product is a combination product, the primary mode of action (PMOA) should be identified to decide the center of government agency for jurisdiction of products.

Eye drop bottle is currently recognized as a container. An eye drop bottle containing a drug is developed according to the drug cGMP regulation. While the eye drop bottle containing an eye lubricant is considered as a medical device, it is developed according to quality system (QS) regulation (FDA 1997). A MDPF bottle is considered as a medical device. MDPF containing a drug is a combination product as defined in the US FDA (2015). Therefore, this combination product should be developed according to both drug cGMP and quality system regulations, while EU considers it as a drug product.

Drug cGMP and quality system regulations require different provisions to ensure the product safety and efficacy. For example, QS regulation requires design controls. The design process is controlled to assure that the device meets user needs, intended uses, and specified requirements. This design control process can improve and prevent future issues. For another example for drug cGMP, in 2005, the US FDA initiated a pilot program of the quality by design (QbD) concept to comply with the drug cGMP regulation more scientifically. The QbD is a systematic approach to develop a high-quality drug product and robust processes continuously.

Table 1 Types of ophthalmic products

Product	Container or device	Product type	US regulation	EU regulation
Drug	Eye drop bottle or vial (container)	Drug	21 CFR 210/211 (Drug cGMPs)	Directive 2001/83/EC (Medical products for human use)
Lubricant	Eye drop bottle or vial (container)	Device	21 CFR 820 (Quality System Regulation)	MDR 2017/745 (Medical device regulation)
Drug	Multidose preservative-free bottle (device)	Combination (PMOA is drug)	21 CFR 210/211 + 21 CFR 820	Directive 2001/83/EC

These new paradigms emphasize more to understand product and process with predefined objectives or user needs, but the old paradigm, trial-and-error approaches, is relying on experimental results more, which should be minimized.

Drug in Eye Drop Bottle Development According to QbD

Quality by design (QbD) paradigm is a systematic approach to ensure the quality of pharmaceutical products. Objectives and design processes should be defined to produce robust products. First, the quality target product profile (QTPP) needs to be defined based on the properties of the ophthalmic drug product and intended user population. Table 2 shows an example of QTPP for ophthalmic drug products.

Next, critical quality attributes (CQAs) are identified based on the severity of harm to patient resulting from failure to meet that quality attributes of the drug product. Table 3 summarizes the quality attributes of the ophthalmic drug product and indicates which attributes are classified as CQAs.

The identified CQAs have the potential to be impacted by the material and process of CCS. Therefore, they should be considered in developing the design of CCS and the sealing process parameters. Risk assessment is used to identify those potentially high risk from CCS material, design, and process. Various studies need to be

Table 2 QTPP for an ophthalmic drug product (simplified example)

QTPP elements		Target
Dosage form		Solution
Route of administration		Eye drop
Stability		24 months at room temperature
Product quality attribute	Assay	Meet the criteria
	Sterility	
	Leachable	
	Drop size	
Container closure system		Multi-use plastic bottle

Table 3 Critical quality attributes (CQAs) of ophthalmic drug product (simplified example)

Product quality attribute	Target	CQA?	Justification
Assay	90–110% w/w of label claim	Yes	Assay variability will affect safety and efficacy
Sterility	Pass sterility test and CCIT	Yes	Sterility breach will affect safety
Leachable	<5 ppm	Yes	Leachable will affect safety
Drop size	30 μL ± 3.0 μL	Yes	Drop size will affect efficacy

Table 4 Critical material and process parameter of CCS (simplified example)

Product critical quality attribute	Critical material parameter	Critical process parameter
Assay	The water loss through the vial can increase the concentration of the drug, so it affects the assay value. To minimize the water loss from the container, the water vapor transmission rate (WVTR) should be low to maintain the assay value within a specification	Sealing process parameters
Sterility	The container should be sealed hermetically, or the physical interference fit should provide acceptable integrity	Aseptic filling process parameters Sealing process parameters
Leachable	To minimize toxicological risk, individual leaching chemical should be low; also the ingredients should meet 21CFR175.300 indirect food additives: Adhesives and components of coatings	N/A
Drop size	The inner diameter of applicator tip should be defined to obtain the desired weight of the drop	Molding process parameters

performed to understand more about the identified risks and to develop a control strategy. Table 4 shows a simplified example of the critical material and process parameters. Design analysis and experimental studies need to perform to understand what range of parameters is acceptable. The acceptable range of parameters are monitored to produce robust products. High knowledge of materials (or design) and processes will prevent critical failures and will resolve unexpected failures quickly.

Qualification of Eye Drop Bottle

The ranges of critical material and process parameters are used to develop and qualify CCS. If the CCS is qualified to meet them, it satisfies the CCS is safe to patient, compatible to product, and protective for product and functioning as intended for users.

Safety

A CCS should not leach harmful or undesirable amounts of chemicals. An extractable study should be performed for the CCS to identify compounds that can be extracted from the materials of CCS. Extracted compounds must be compliant with regulatory guidelines.

Compatibility

A CCS should be evaluated experimentally to ensure that the CCS is compatible with the product. All ingredients in materials will leach to some degree under certain conditions. Compatibility issues are often indicated by a pH shift, product degradation, and aggregation. As a part of the product stability, leachable studies (e.g., three batches) must be performed to monitor any changes. The sources of leachable compounds are the primary container and closure (i.e., plasticizer, lubricant, pigment, stabilizer, antioxidant, binding agent), label (i.e., ink, adhesive, varnish), and secondary tray, carton, or processes (i.e., sterilization agent, preservatives, sealant, ink).

Protection

A CCS should be designed and evaluated experimentally to ensure adequate protection from environmental hazards (e.g., temperature, light, moisture, oxygen, microbial contamination) that cause a degradation and sterility breach in the quality of products. The design of CCS should consider a design space of a tolerance of part dimensions, storage conditions, and process control parameters.

Performance

A CCS should be evaluated to ensure it functions for its intended use using performance tests. This evaluation should consider critical design factors (e.g., tolerance of part dimensions, storage and performance conditions, human factors). For ophthalmic products, drop size is the main factor, which needs to be evaluated considering the tolerance of applicator tip size, surface tension of product solution on the applicator material, finger force from users, and position of bottle (e.g., vertical or 45° angle).

Quality Control

Quality control measures should be developed to ensure the consistency of container closure components. As a part of quality control, component suppliers must be qualified to ensure they produce robust components. In addition, incoming quality check is performed for every batch and critical manufacturing process parameters are monitored. CAPA (corrective and preventive action) must be performed for incoming components that do not meet QC acceptance criteria and for manufacturing process parameters that do not meet the process acceptance criteria.

Device in Eye Drop Bottle Development According to Design Control

Some ophthalmic products are developed to keep the eye moist. This product is not a drug but is a lubricant to help relieve dryness. The lubricant is classified as a class II medical device by the US FDA. The medical devices obey the quality system regulations, and they require design controls in device quality (FDA 1997). Design controls are an interrelated set of practices and procedures that are incorporated into the design and development process. This systematic assessment can identify deficiencies in design input requirements and discrepancies between the proposed designs and requirements. Therefore, the design can be corrected earlier in the development process. In addition, any potential failure modes are assessed in terms of application, design, and process of the device. The application failure mode effect analysis (aFMEA), design FMEA (dFMEA), and process FMEA (pFMEA) can be developed to minimize potential failures. This design control process helps to understand the device design systematically. Therefore, the high level of knowledge can improve the device design, prevent issues, and find root causes appropriately for unexpected failures in future. Figure 1 illustrates the influence of design controls on a design process. Initially, user needs are obtained, and design requirements are developed to meet the user requirements. Designs of device including primary CCS and secondary packaging are developed and verified to meet the design requirements. In practice, multi-functional groups work together and provide feedbacks. The important thing from this process is an iterative process between input and

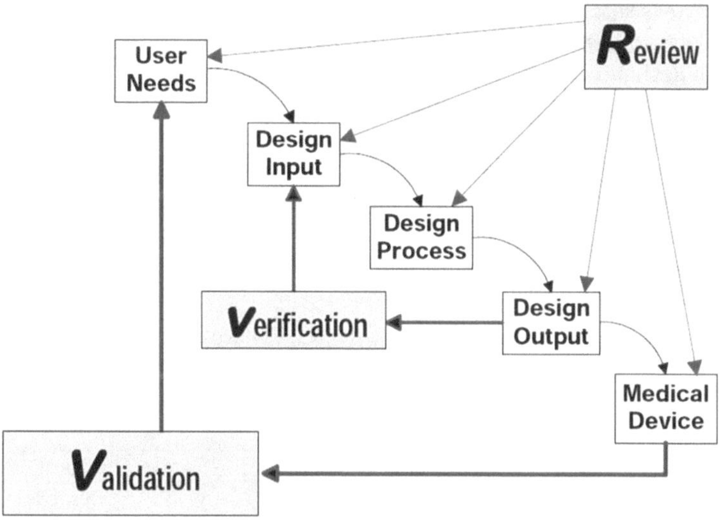

Fig. 1 Application of design controls to waterfall design process. (Medical Devices Bureau, Health Canada)

output. For example, user needs input to develop a new design output. This output is verified as confirming to the input. This output then becomes the design input for another step in the design process.

Drug in Multidose Preservative-Free (MDPF) Eye Drop Bottle

In 2017, the first MDPF prescription drug was approved by the US FDA. The MDPF bottle was developed as a medical device. As explained in the previous section, the medical device like MDPF bottle must obey the quality system regulations.

Device Verification and Validation

The systematic design control approach needs to put significant efforts to obtain user needs, which include regulatory and quality requirement. The user needs must be clearly understood, and their information is used to develop design requirements. A new device is developed to meet all design requirements. Proposed device components and systems are continuously evaluated theoretically and experimentally. Tolerance stack-up analysis, material characteristics, and functionality tests are performed to ensure the design satisfies the design requirements. Device part and system drawings and specifications are finalized as design outputs. At this stage, the probability of success is high, and the design verification/validation should be used to confirm the proposed design experimentally with well-organized study protocols. Table 5 shows an example of user needs for ophthalmic prescription drug.

The defined user needs are used to develop design requirements. Table 6 shows the list of design requirements for ophthalmic prescription drug.

Based on these design requirements, a new design may need to be developed or off-the-shelf designs may satisfy. Various studies and analyses are performed, and the drawings and specifications of components and system are finalized. Lastly, the proposed device will be confirmed that it is designed to meet user needs by performing design verification and validation studies.

Table 5 User needs for ophthalmic prescription drug (simplified example)

ID	User need	Priority
UN1	The drug should be self-administered directly to the eye	Critical
UN2	The drug should be delivered accurately	Critical
UN3	The system should be multiple uses	Desired
UN4	The drug must be kept sterile during in-use period	Critical

Table 6 Design requirements (DR) for ophthalmic prescription drug (simplified example)

ID	DR	UN	Rationale	Test (sample size)	EDO
DR1	The actuation force must be <15 N	UN1	The user can smoothly squeeze the bottle with the force <15 N	Actuation force (10)	Yes
DR2	The weight of the drop must be 30 μL ± 3.0 μL	UN2	This is the required dose for the indication	Dose accuracy (10)	Yes
DR3	The total volume of product must be >3 mL	UN3	The total volume of product is calculated including 10% overage for indicated multiple uses	Extractable volume (10)	No
UN4	Sterility test must be passed with used samples	UN4	If the product is contaminated, the sterility test shows microorganism growth	Sterility (60)	Yes

EDO essential design output

CCS for Ophthalmic Products

Ninety-five percent of all ophthalmic drugs are delivered using an eye drop bottle, and the rest of drug products are delivered by injection, ointment, and implant by a device. Injectable ophthalmic drug is packaged in a vial or prefilled syringe. Implant in device is packaged in a bag. There are no unique features in these CCS used for ophthalmic drug products. In this section, only eye drop bottles are introduced. Most of eye droppers are made of plastics such as LDPE (low-density polyethylene) or HDPE (high-density polyethylene). They are flexible, are easily squeezable, and returned to the original shape of container.

In addition to the material and design of containers, the uniform use of a color system for the container caps and package label is strongly recommended by the American Academy of Ophthalmology (AAO). The distinguished cap and label colors help patients to minimize the risk of selecting incorrect medication. The cap color system is established by AAO and endorsed by the FDA, for each class of therapeutic. This policy was adopted due to reports to the Academy and the National Registry of Drug-Induced Ocular Side Effects of serious adverse events resulting from patient difficulty in distinguishing among various ocular medications. Table 7 shows the recommended color codes for topical ocular medications (https://www.aao.org/about/policies/color-codes-topical-ocular-medications).

Unit-Dose Vial

The unit-dose vial is for one-time use and it is manufactured by a blow-fill-seal technology. The blow-fill-seal technology is an automated packaging process whereby plastic containers are blow-molded, filling needles draw the volume of

Table 7 Recommended color codes for topical ocular medications

Class	Color	Pantone number
Adrenergic agonist combinations	Light green	373C
Adrenergic agonists	Purple	2583
Anti-infectives	Tan	467
Anti-inflammatory, nonsteroidal	Gray	4
Anti-inflammatory, steroids	Pink	197
Anti-inflammatory, immunomodulators	Olive green	5763C
Beta-blocker combinations	Dark blue	281
Beta-blockers	Yellow	Yellow C
Carbonic anhydrase inhibitors	Orange	1585
Cytotoxic	Black	6
Miotics	Dark green	348
Mydriatics and cycloplegics	Red	1797
Prostaglandin analogues	Turquoise	326

sterile product into the container, and the upper part of the mold closes to form and seal the container in one continuous protected operation. The product in this unit-dose vial does not need antimicrobial preservative, so it minimizes the risk of allergic reactions for patients.

Figure 2 shows a typical design of unit-dose vial. This vial is manufactured using a blow-fill-seal process. It is made of LDPE or PP (polypropylene) and consists of a tubular, compressible body and a pointed cone with a twist-off closure. Usually multiple containers are packaged in aluminum pouch to prevent water loss or oxygen gain from permeation through the plastic vial.

The size of drops dispensed from unit-dose containers will depend on the dimensions of the opening created from the twist-off closure.

Table 8 shows the critical material and process parameters which should be considered to develop the unit-dose vial and packaging processes. The water vapor transmission rate (WVTR) of the vial should be less than 0.07 mg/day/container/saturate vapor pressure at 25 °C to maintain the assay value within the specification. The unit-dose vial should be sealed hermetically to maintain the product sterility. The plastic material should meet 21CFR175.300 indirect food additives, and individual leaching compound should be less than 5 ppm. The inner diameter of the applicator tip after twist-off should be 2.2 mm ± 0.25 to provide a desired weight of the drop.

Multidose Bottle

Multidose bottle is for multi-uses and the product solution is dispensed through the dropper tip. The ophthalmic multidose bottle is a sterile product and needs to contain an antimicrobial preservative such as benzalkonium chloride to maintain the

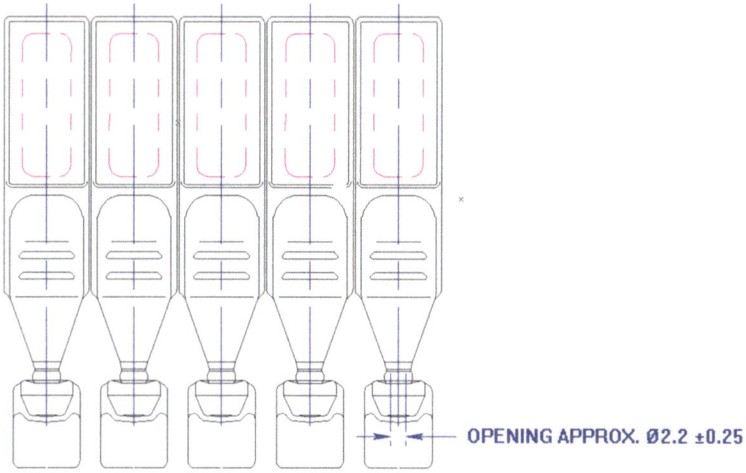

Fig. 2 Unit-dose vial (example)

Table 8 Critical material and process parameter for unit-dose vial in aluminum pouch (simplified example)

Product critical quality attribute	Critical material control parameter	Critical process control parameter
Assay	WVTR <0.07 mg/day/container/saturate vapor pressure at 25 °C	Form-fill-seal process parameters
Sterility	Hermetic seal	Form-fill-seal process parameters
Leachable	Ingredients: Meet 21CFR175.300 and individual compound <5 ppm	N/A
Drop size	The applicator tip inner diameter is 2.2 mm ± 0.25 mm before twist-off	Form-fill-seal process parameters

sterility until the product is consumed. It has advantages to use the product multiple times, more convenience, and less bulky than unit-dose vials. However, preservatives may cause ocular toxicity and it results in discomfort, blurred vision, and irritation.

The dropper tip should be designed to control the amount of drops accurately. Figure 3 shows an advanced design of dropper tip (Santvliet and Ludwig 2004). In this tip, there is an elongated narrow central duct which can prevent a jet from bottle squeezing. This ensures a drop dispensing the product. Also, the duct is tapered from narrower inner to wider outer orifice, which can control the flow of liquid entering the duct. The outer orifice is shaped with hemispherical end. It helps the appearance that the tip has more smooth edge.

The size of drops dispensed from dropper tips is influenced by several factors: the design of the dropper tip, the physicochemical properties of the product, surface tension of the product on the surface of the tip, and the user's manner of handling the dropper bottle.

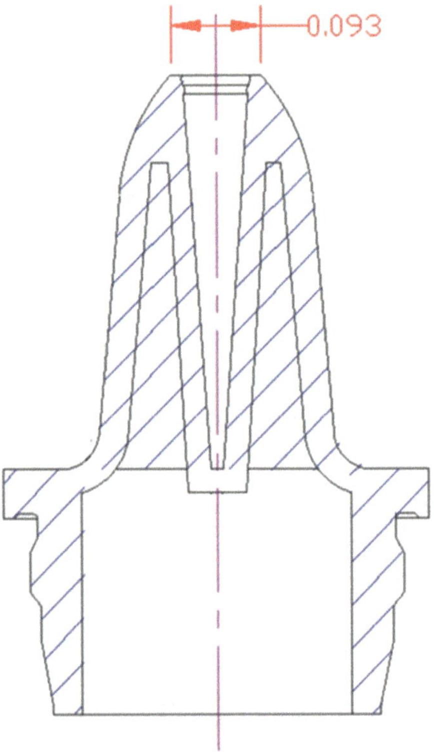

Fig. 3 Dropper tip used for multidose dropper bottle (example)

Table 9 shows the critical material and process parameters which should be considered to develop the multidose bottle and packaging processes. The water vapor transmission rate (WVTR) of the bottle should be less than 0.7 mg/day/container/saturate vapor pressure at 25 °C to maintain the assay value within the specification. The closure torque should be monitored to ensure the physical seal of the bottle. In general, the multidose bottle is sealed by an interference fit between container and closure. The interference fit should be analyzed using the worst-case scenario and should be greater than 0.01 mm. The plastic material should meet 21CFR175.300 indirect food additives, and individual leaching compound should be less than 5 ppm. The inner diameter of the orifice diameter should be 2.4 mm ± 0.1 mm to provide a desired weight of the drop.

Multidose Preservative-Free (MDPF) System

MDPF bottle has functions of device and CCS. There are advantages over unit-dose vials and multidose bottles. The MDPF bottle does not require antimicrobial preservative to maintain the product sterility, and the sterility can be maintained from multiple uses until the product is consumed. In 2017, the first MDPF prescription drug was approved by the US FDA.

Table 9 Critical material and process parameter for multidose bottle (simplified example)

Product critical quality attribute	Critical material parameter	Critical process parameter
Assay	WVTR: <0.7 mg/day/container/saturate vapor pressure at 25 °C for 10 mL container	Capping process—application torque
Sterility	Physical interference fit: >0.01 mm considering the worst case	Capping process—application torque
Leachable	Ingredients: meet 21CFR175.300 and individual compound <5 ppm	N/A
Drop size	Tip orifice diameter: 2.4 mm ± 0.1 mm	N/A

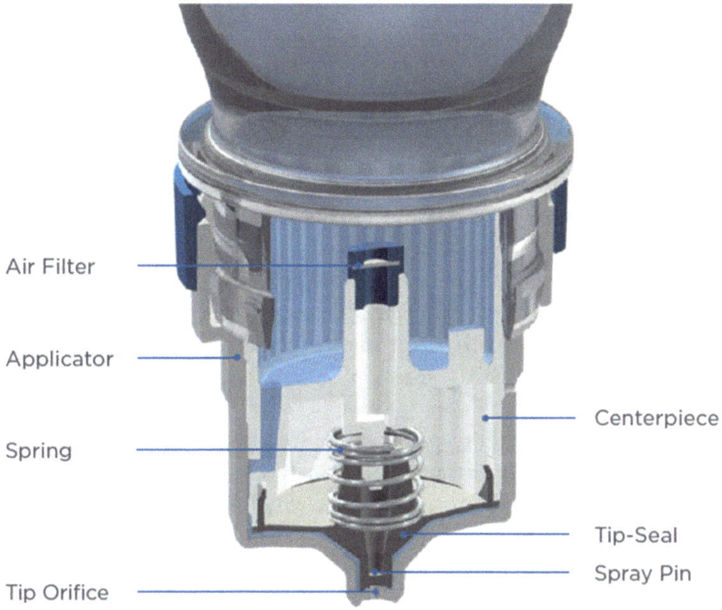

Fig. 4 Cross section of a MDPF applicator, Ophthalmic Squeeze Dispenser. (https://pharma.aptar.com/sites/default/files/products/marketing_sheet/files/pds_osd_pds_digital_aa.pdf)

Figure 4 shows an example of the cross section of MDPF applicator. When the bottle is squeezed, the inside pressure increases, the solution flows, and the tip-seal rises until the sealing membrane opens. The opening is released and the solution flows into a drop. The squeezed bottle is inflated from the air sucked back to the inside bottle via the air filter. The air filter prevents the ingress of microorganisms but the air can flow through the air filter.

As shown, the MDPF system is relatively complicated and has some technical challenges to prove that the air filter and membrane opening do not provide any

Table 10 Critical material and process parameter for multidose preservative-free (MDPF) bottle (simplified example)

Product critical quality attribute	Critical material parameter	Critical process parameter
Assay	WVTR: <0.7 mg/day/container/saturate vapor pressure at 25 °C for 10 mL container	N/A
Sterility	Physical interference fit: >0.01 mm considering the worst case In-use period Air filter <0.2 μm	Aseptic filling process parameters Capping process parameters
Leachable	Ingredients: meet 21CFR175.300 and individual compound <5 ppm	N/A
Drop size	Tip outer orifice diameter: 2 mm ± 0.1 mm	N/A

microorganism contamination. In addition, the unit cost is high to produce components and assembly.

Table 10 shows the critical material and process parameters which should be considered to develop the MDPF bottle and packaging processes. The required process parameters are similar to the multidose bottle except for in-use sterility. The MDPF does not contain an antimicrobial preservative, so it is required to show the product sterility for in-use period. The MDPF bottle is incubated, drops are collected on agar plates periodically as intended use, and the drops are incubated to see the microorganism's growth. The air filter should be able to sieve particle greater than 0.2 μm.

Other examples of multidose preservative-free systems include Novelia® from Nemera, which features similar uni-directional valve with air filter as the Aptar system; 3K® pump system from Aero Pump, which has convenient actuation features; and Iridya™ from Silgan Dispensing, which uses the elongated tip for greater accuracy (Fig. 5).

Novel Dispensing Systems to Aid Instillation to the Eye

There are many studies showing patients have technical difficulty to instill drops in the eyes and patient's lack of compliance (Tatham et al. 2013; Abelson et al. 2006). The patients, especially for elderly patients, do not feel comfortable with self-dispensing and require assistance. Simple operations such as opening the cap and squeezing the bottle to accurately instill a drop in the eye can be a source of major concern to the lack of compliance.

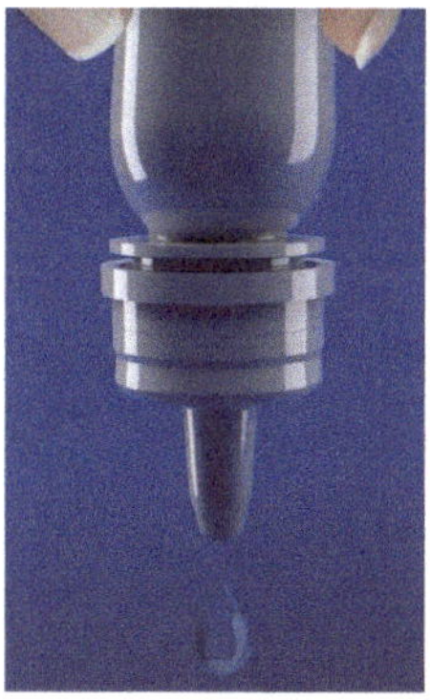

(A) Iridya™ ophthalmic multidose system [9]

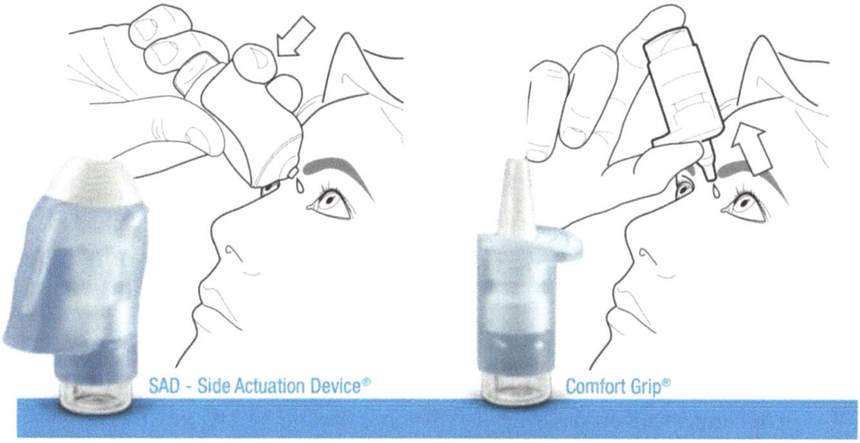

(B) Aero Pump ophthalmic multidose systems

Fig. 5 Various types of multidose preservative-free systems (https://www.aeropump.de/fileadmin/downloads/AP_ophthalmic_200131.pdf). (**a**) Iridya™ ophthalmic multidose system (https://silgandispensing.com/products/iridya). (**b**) Aero pump ophthalmic multidose systems

To improve the ease of use, enhance compliance, and deliver drug accurately, innovative drug dispensing systems are being developed. Figure 6 shows Acustream™ that can deliver drug to the eye accurately with 80% reduction in dose compared to standard drops, OptiMyst™ that can deliver drug accurately using an ultrasonic nebulizer, and VersiDoser® that uses cartridge with unit-dose type packaging for accurate delivery and features a dose counter to assist compliance.

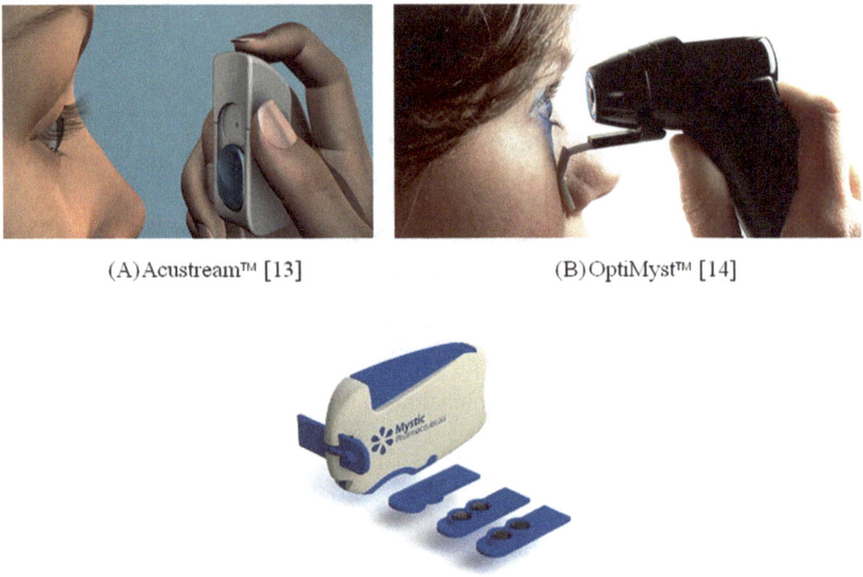

(A) Acustream™ [13] (B) OptiMyst™ [14]

(C) VersiDoser® [15]

Fig. 6 Novel dispensing systems to enable precise dosing. (**a**) Acustream™ (http://www.kedalion-thera.com). (**b**) OptiMyst™ (http://www.altitudeinc.com). (**c**) VersiDoser® (http://www.mys-ticpharmaceuticals.com)

Conclusion

Most ophthalmic drug products are self-administered to highly sensitive eyes using a dropper. The primary CCS must be qualified to meet cGMP requirements and the dispensing device must be verified according to the quality system regulations. The qualified primary CCS should be safe to use for ophthalmic drugs and the verified dispensing system is acceptable to users.

There have been a lot of efforts to improve the primary CCS of ophthalmic products. Unit-dose vial and multidose bottle have been improved to multidose preservative-free (MDPF) bottle. In addition, novel dispensing systems are being developed to deliver the drug precisely without technical difficulties. The novel dispensing systems could improve the medication adherence.

Users continuously require enhanced eye dropper systems since world's older population continues to grow. New user needs are applied to enhance dispensing systems, and they should be designed more senior friendly to meet their requirements.

References

Abelson M, Torkildsen G, Fink K. Taking steps toward better compliance: a look at the impediments in the way of good compliance and how to overcome them. Rev Ophthalmol. 2006.
FDA guidance, design control guidance for medical device manufacturers; 1997.
FDA guidance, container closure systems for packaging human drugs and biologics; 1999.
FDA guidance, current good manufacturing practice requirements for combination products; 2015.
Market analysis report, ophthalmic packaging market size, share & trends analysis report. Grand View Research; 2019.
Santvliet LV, Ludwig A. Determinants of eye drop size. Surv Ophthalmol. 2004;49:197–213.
Tatham AJ, Sarodia U, Gatrad F, Awan A. Eye drop instillation technique in patients with glaucoma. Eye. 2013;27(11):1293–8.

Part III
Pharmaceutical Product Development

Liquid Ophthalmic Drug Products: Physicochemical Properties, Formulations, and Manufacturing Considerations

Hovhannes J. Gukasyan and Richard Graham

Abstract Liquid ophthalmic drug products are the most common presentation for pharmacotherapy used to treat a variety of anterior and posterior segment diseases of the eye. Their attributes largely mirror those of parenteral formulations, but specifically consider certain qualities for drug substance and product from a perspective of compatibility and delivery to a biologically and physiologically distinct environment in and around the eye. Features such as formulation pH and osmolarity, or properties of all inactive ingredients, play a critical role when considering the route of ocular administration. This chapter provides an overview of physical chemistry, formulation, and manufacturing considerations as they relate to the anatomical characteristics and physiology of the eye from a pragmatic, historical, case-study-driven, and biosystem-based perspective.

Keywords Sterile liquid · pH · Osmolarity · Ocular tolerability · Sterile manufacturing · Anterior segment · Posterior segment

Abbreviations

AMD	Age-related macular degeneration
DCE-GS	*S*-(1,2-dicarboxyethyl)glutathione
FDA	Food and Drug Administration
GSH	Glutathione
IOP	Intraocular pressure
K_{sp}	Equilibrium constant for a solubility product
p*I*	Isoelectric point

H. J. Gukasyan (✉) · R. Graham
Pharmaceutical Development, Allergan plc, Irvine, CA, USA
e-mail: gukasyan@usc.edu

S. Neervannan, U. B. Kompella (eds.), *Ophthalmic Product Development*, AAPS Advances in the Pharmaceutical Sciences Series 37,
https://doi.org/10.1007/978-3-030-76367-1_11

pK_a	Acid dissociation constant
pK_b	Base dissociation constant
$P_{o/w}$	Oil/water partition coefficient
ROS	Reactive oxygen species
TRIS	Triethanolamine
USP	United States Pharmacopeia
UVA	Ultraviolet long-wavelength light radiation
UVB	Ultraviolet short-wavelength light radiation
β	Buffering capacity

Preface

Liquid ophthalmic products are categorized as parenteral formulations; however, they are within a highly specialized subclass of their own. To provide a sensible and comprehensive analysis of physical chemistry, compounding pharmacy or formulation, and manufacturing science that's entailed within the larger scope of all liquid ophthalmic drug products, the objectives of this chapter are twofold: first, to briefly visit key topics and critical attributes, which in turn (second) provide examples with references to benefit newcomers into the field for subsequent development of deeper expertise. Liquid ophthalmic drug products can be defined and classified by several differentiating attributes from other dosage forms that are administered into the body. From a global perspective, these characteristics stem out of three ocular biopharmaceutics blueprint attributes. Features include the qualities of drug substance or active pharmaceutical ingredient; the drug product or formulation from a perspective of aqueous solution pH, total concentration of osmolytes, and properties related to the actual vehicle composition taking into account all inactive ingredients (e.g., excipients); and finally, precise route of administration into the eye (e.g., topical eye drops vs. intraocular or periorbital injections) as it relates to the anatomical characteristics and physiology of this organ.

Considerations for Drug Substance

Relating to the active ingredient or drug substance, an inclusive examination of precedence in liquid ophthalmic products (see Table 1) suggests existence of two general categories. There are some liquid ophthalmic products that stem from pure leads. In other words, they contain an active ingredient that was discovered and developed solely for an ophthalmic indication. Moreover, most liquid ophthalmic products are carefully designed reformulations of existing active ingredients repurposed from other therapeutic indications. Molecular-drug profiling for ophthalmic repositioning, in this case, involves development of a preexisting compound into a

Table 1 Commercial liquid ophthalmic products and some off-label used parenterals[||] in an ocular setting and their critical formulation attributes (PDR Network LLC. 2016; Physicians' desk reference for ophthalmic medicines 2000; Lexi-Comp Inc. and American Pharmacists Association)

Drug name	Liquid, ophthalmic formulation attributes	Original indications	Ophthalmic indications
Acular LS® (ketorolac tromethamine)	Acular LS® ophthalmic solution is supplied as a sterile isotonic aqueous 0.4% solution, with a pH of approximately 7.4. Acular LS® ophthalmic solution contains a racemic mixture of R-(+) and S-(−)- ketorolac tromethamine. Ketorolac tromethamine may exist in three crystal forms. All forms are equally soluble in water. The pK_a of ketorolac is 3.5. This white to off-white crystalline substance discolors on prolonged exposure to light. The osmolality of Acular LS® ophthalmic solution is 290 mOsmol/kg. Each mL of Acular LS® ophthalmic solution contains active, ketorolac tromethamine 0.4%; preservative, benzalkonium chloride 0.006%; and inactives, edetate disodium 0.015%, octoxynol 40, purified water, sodium chloride, and hydrochloric acid and/or sodium hydroxide to adjust the pH		Ophthalmic solution is indicated for the reduction of ocular pain and burning/stinging following corneal refractive surgery
Acular® (ketorolac tromethamine)	Acular® ophthalmic solution is supplied as a sterile isotonic aqueous 0.5% solution, with a pH of 7.4. Acular® ophthalmic solution contains a racemic mixture of R-(+) and S-(−)- ketorolac tromethamine. Ketorolac tromethamine may exist in three crystal forms. All forms are equally soluble in water. The pK_a of ketorolac is 3.5. This white to off-white crystalline substance discolors on prolonged exposure to light. The molecular weight of ketorolac tromethamine is 376.41. The osmolality of Acular® ophthalmic solution is 290 mOsmol/kg. Each mL of Acular® ophthalmic solution contains active, ketorolac tromethamine 0.5%; preservative, benzalkonium chloride 0.01%; and inactives, edetate disodium 0.1%, octoxynol 40, purified water, sodium chloride, hydrochloric acid, and/or sodium hydroxide to adjust the pH		Ophthalmic solution is indicated for the temporary relief of ocular itching due to seasonal allergic conjunctivitis and also indicated for the treatment of postoperative inflammation in patients who have undergone cataract extraction

(continued)

Table 1 (continued)

Drug name	Liquid, ophthalmic formulation attributes	Original indications	Ophthalmic indications
Acuvail® (ketorolac tromethamine)	Acuvail solution is supplied as a sterile isotonic aqueous 0.45% preservative-free solution, with a pH of approximately 6.8. Acuvail solution contains a racemic mixture of R-(+) and S-(−)- ketorolac tromethamine. Ketorolac tromethamine may exist in three crystal forms. All forms are equally soluble in water. The pK_a of ketorolac is 3.5. This white to off-white crystalline substance discolors on prolonged exposure to light. The osmolality of Acuvail solution is approximately 285 mOsmol/kg. Each mL of Acuvail ophthalmic solution contains active, ketorolac tromethamine 0.45%, and inactives, carboxymethylcellulose sodium, sodium chloride, sodium citrate dehydrate, and purified water with sodium hydroxide and/or hydrochloric acid to adjust the pH		Ophthalmic solution is indicated for the treatment of pain and inflammation following cataract surgery
AK-con-A®	Naphazoline hydrochloride, an ocular vasoconstrictor, is an imidazoline derivative sympathomimetic amine. It occurs as a white, odorless crystalline powder having a bitter taste and is freely soluble in water and in alcohol. Active: Naphazoline HCl 1 mg (0.1%). Preservative: Benzalkonium chloride 0.1 mg (0.01%) Inactives: Boric acid, edetate disodium, purified water, sodium chloride, sodium carbonate, and hydrochloric acid may be added to adjust the pH (5.5–7.0)		Naphazoline constricts the vascular system of the conjunctiva. It is presumed that this effect is due to direct stimulation of the drug upon the alpha-adrenergic receptors in the arterioles of the conjunctiva, resulting in decreased conjunctival congestion. Naphazoline belongs to the imidazoline class of sympathomimetics
Akten® (lidocaine hydrochloride)	Akten® contains 35 mg of lidocaine hydrochloride per mL as the active ingredient. It also contains hypromellose, sodium chloride, and water for injection as inactive ingredients in the 1 mL tube configuration. Akten® contains hypromellose, sodium chloride, and water for injection as inactive ingredients in the 5 mL in 10 mL bottle configuration. The pH may be adjusted to 5.5–7.5 with hydrochloric acid and/or sodium hydroxide		Indicated for ocular surface anesthesia during ophthalmologic procedures

(continued)

Table 1 (continued)

Drug name	Liquid, ophthalmic formulation attributes	Original indications	Ophthalmic indications
Alaway® (ketotifen fumarate)	Ketotifen 0.025% (equivalent to ketotifen fumarate 0.035%), benzalkonium chloride 0.01%, glycerin, hydrochloric acid and/or sodium hydroxide, water for injection		Temporary relief of itchy eyes due to ragweed, pollen, grass, animal hair and dander
Alocril® (nedocromil sodium)	Each mL contains active, nedocromil sodium 20 mg/mL (2%); preservative, benzalkonium chloride 0.01%; and inactives, edetate disodium 0.05%, purified water, and sodium chloride 0.5%. It has a pH range of 4.0–5.5 and an osmolality range of 270–330 mOsm/kg		Indicated for the treatment of itching associated with allergic conjunctivitis
Alomide® (lodoxamide tromethamine)	Each mL of Alomide® (lodoxamide tromethamine ophthalmic solution) 0.1% contains active, 1.78 mg lodoxamide tromethamine equivalent to 1 mg lodoxamide; preservative, benzalkonium chloride 0.007%; and inactive, mannitol, hypromellose 2910, sodium citrate, citric acid, edetate disodium, tyloxapol, hydrochloric acid and/or sodium hydroxide (adjust pH), and purified water		Indicated in the treatment of the ocular disorders referred to by the terms vernal keratoconjunctivitis, vernal conjunctivitis, and vernal keratitis
Alphagan P® (brimonidine tartrate)	In solution, Alphagan® P (brimonidine tartrate ophthalmic solution) has a clear, greenish-yellow color. It has an osmolality of 250–350 mOsmol/kg and a pH of 7.4–8.0 (0.1%) or 6.6–7.4 (0.15%). Brimonidine tartrate appears as an off-white to pale-yellow powder and is soluble in both water (0.6 mg/mL) and in the product vehicle (1.4 mg/mL) at pH 7.7. Each mL of Alphagan® P contains the active ingredient brimonidine tartrate 0.1% (1.0 mg/mL) or 0.15% (1.5 mg/mL) with the inactive ingredients sodium carboxymethylcellulose, sodium borate, boric acid, sodium chloride, potassium chloride, calcium chloride, magnesium chloride, PURITE® 0.005% (0.05 mg/mL) as a preservative, purified water, and hydrochloric acid and/or sodium hydroxide to adjust the pH		An alpha-adrenergic agonist indicated for lowering intraocular pressure (IOP) in patients with open-angle glaucoma or ocular hypertension

(continued)

Table 1 (continued)

Drug name	Liquid, ophthalmic formulation attributes	Original indications	Ophthalmic indications
Alrex® (loteprednol etabonate)	Each mL contains active, loteprednol etabonate 2 mg (0.2%), and inactives, edetate disodium, glycerin, povidone, purified water, and tyloxapol. Hydrochloric acid and/or sodium hydroxide may be added to adjust the pH. The suspension is essentially isotonic with a tonicity of 250–310 mOsmol/kg. Preservative added: Benzalkonium chloride 0.01%		Ophthalmic suspension indicated for the temporary relief of the signs and symptoms of seasonal allergic conjunctivitis
Altacaine® (tetracaine hydrochloride)	Tetracaine hydrochloride 0.5% is a sterile topical ophthalmic solution useful in producing surface anesthesia of the eye. Active: Tetracaine hydrochloride 0.5%. Preservative: Chlorobutanol. Inactive: Boric acid, edetate disodium, potassium chloride, water for injection, USP. Hydrochloric acid and/or sodium hydroxide may be added to adjust the pH		For procedures in which a rapid and short-acting topical ophthalmic anesthetic is indicated such as in tonometry, gonioscopy, removal of corneal foreign bodies, conjunctival scraping for diagnostic purposes, suture removal from the cornea or conjunctiva, and other short corneal and conjunctival procedures
Amikin® (amikacin sulfate)‖	Vials contain 250 mg of active and 50 mg sodium citrate and 4.8 mg sodium metabisulfite, according to pharmaceutical details provided in package inserts from certain countries *(*no pH or osmolarity spec.)*	Treatment of infections due to gram-negative bacteria, treatment of *Mycobacterium avium complex* (oral inhalation)	Bacterial endophthalmitis by intravitreal injection (Jackson and Williamson 1999)
Ancef®, Kefzol® (cefazolin)	Intraocular dosage, adults 100 mg by subconjunctival injection or 1–2.5 mg by intracameral injection, is optional at the end of the procedure. Perioperative antisepsis with povidone-iodine is recommended. Preservative-free, pH 4.0–6.0, 290 mOsm/kg, as a sodium salt		For ophthalmic surgical infection prophylaxis

(continued)

Table 1 (continued)

Drug name	Liquid, ophthalmic formulation attributes	Original indications	Ophthalmic indications
Artificial tears®	Polyvinyl alcohol 1.4%; carboxymethylcellulose sodium 1%; glycerin 0.2%, hypromellose 0.2%, polyethylene glycol 400 1%; benzalkonium chloride, edetate disodium, NaCl, sodium phosphate, dibasic anhydrous sodium phosphate, monobasic, anhydrous, water, NaOH/HCl		Eye lubricants
Avastin® (bevacizumab)‖	Bevacizumab has an approximate molecular weight of 149 kDa and is produced in a mammalian cell (Chinese hamster ovary) expression system. Avastin (bevacizumab) injection for intravenous use is a sterile, clear to slightly opalescent, colorless to pale brown solution. Avastin is supplied in 100 and 400 mg preservative-free, single-dose vials to deliver 4 or 16 mL of Avastin (25 mg/mL) The 100 mg product is formulated in 240 mg α,α-trehalose dihydrate, 23.2 mg sodium phosphate (monobasic, monohydrate), 4.8 mg sodium phosphate (dibasic, anhydrous), 1.6 mg polysorbate 20, and water for injection, USP. The 400 mg product is formulated in 960 mg α,α-trehalose dihydrate, 92.8 mg sodium phosphate (monobasic, monohydrate), 19.2 mg sodium phosphate (dibasic, anhydrous), 6.4 mg polysorbate 20, and water for injection, USP *(*no pH or osmolarity spec.)*	Metastatic colorectal cancer	Neurovascular age-related macular degeneration (Bevacizumab (Avastin). Lower cost does not justify taking risks 2015; Lalwani et al. 2008)

(continued)

Table 1 (continued)

Drug name	Liquid, ophthalmic formulation attributes	Original indications	Ophthalmic indications
Azopt® (brinzolamide)	Azopt (brinzolamide ophthalmic suspension) 1% is supplied as a sterile, aqueous suspension of brinzolamide which has been formulated to be readily suspended and slow settling, following shaking. It has a pH of approximately 7.5 and an osmolality of 300 mOsm/kg. Each mL of Azopt (brinzolamide ophthalmic suspension) 1% contains active ingredient,brinzolamide 10 mg; preservative, benzalkonium chloride 0.1 mg; and inactives, mannitol, carbomer 974P, tyloxapol, edetate disodium, sodium chloride, and purified water, with hydrochloric acid and/or sodium hydroxide to adjust the pH		A carbonic anhydrase inhibitor indicated for the treatment of elevated intraocular pressure in patients with ocular hypertension or open-angle glaucoma
Bepreve™ (bepotastine besilate)	Bepreve™ ophthalmic solution is supplied as a sterile, aqueous 1.5% solution, with a pH of 6.8 and osmolality of approximately 290 mOsm/kg Each mL of Bepreve™ (bepotastine besilate ophthalmic solution) 1.5% contains active, bepotastine besilate15 mg (equivalent to 10.7 mg bepotastine); preservative, benzalkonium chloride 0.005%; and inactives monobasic sodium phosphate dihydrate, sodium chloride, sodium hydroxide to adjust the pH, and water for injection, USP		Treatment of itching associated with allergic conjunctivitis

(continued)

Table 1 (continued)

Drug name	Liquid, ophthalmic formulation attributes	Original indications	Ophthalmic indications
Besivance™ (besifloxacin)	Besivance™ (besifloxacin ophthalmic suspension) 0.6% is a sterile ophthalmic suspension of besifloxacin formulated with DuraSite® (polycarbophil, edetate disodium dihydrate, and sodium chloride). Each mL of Besivance™ contains 6.63 mg besifloxacin hydrochloride equivalent to 6 mg besifloxacin base. Active: Besifloxacin 0.6% (6 mg/mL). Preservative: Benzalkonium chloride 0.01%. Inactives: Polycarbophil, mannitol, poloxamer 407, sodium chloride, edetate disodium dihydrate, sodium hydroxide, and water for injection. Besivance™ is an isotonic suspension with an osmolality of approximately 290 mOsm/kg		Quinolone antimicrobial indicated for the treatment of bacterial conjunctivitis caused by susceptible isolates of the following bacteria: CDC coryneform group G, *Corynebacterium pseudodiphtheriticum*, *Corynebacterium striatum*, *Haemophilus influenzae*, *Moraxella lacunata*, *Staphylococcus aureus*, *Staphylococcus epidermidis*, *Staphylococcus hominis*, *Staphylococcus lugdunensis*, *Streptococcus mitis* group, *Streptococcus oralis*, *Streptococcus pneumoniae*, and *Streptococcus salivarius*
Betagan® (levobunolol hydrochloride)	Betagan® (levobunolol hydrochloride ophthalmic solution, USP) sterile is a noncardioselective beta-adrenoceptor blocking agent for ophthalmic use. The solution is colorless to slightly light yellow in appearance with an osmolality range of 250–360 mOsm/kg. The shelf life pH range is 5.5–7.5. Contains active levobunolol HCl 0.5%. Preservative: Benzalkonium chloride 0.004%. Inactives: Edetate disodium; polyvinyl alcohol 1.4%; potassium phosphate, monobasic; purified water; sodium chloride; sodium metabisulfite; sodium phosphate, dibasic; and hydrochloric acid or sodium hydroxide to adjust the pH		Effective in lowering intraocular pressure and may be used in patients with chronic open-angle glaucoma or ocular hypertension

(continued)

Table 1 (continued)

Drug name	Liquid, ophthalmic formulation attributes	Original indications	Ophthalmic indications
Betimol® (timolol)	Betimol® (timolol ophthalmic solution), 0.25% and 0.5%, is a non-selective beta-adrenergic antagonist for ophthalmic use. Each mL of Betimol® 0.25% contains 2.56 mg of timolol hemihydrate equivalent to 2.5 mg timolol Each mL of Betimol® 0.5% contains 5.12 mg of timolol hemihydrate equivalent to 5.0 mg timolol. Inactive ingredients: Monosodium and disodium phosphate dihydrate to adjust the pH (6.5–7.5) and water for injection, benzalkonium chloride 0.01% added as preservative. The osmolality of Betimol® is 260–320 mOsmol/kg		Treatment of elevated intraocular pressure in patients with ocular hypertension or open-angle glaucoma
Betoptic S® (betaxolol hydrochloride)	Ophthalmic suspension contains 0.25% betaxolol hydrochloride in a sterile resin suspension formulation. Each mL of Betoptic S® ophthalmic suspension contains active, betaxolol HCl 2.8 mg equivalent to 2.5 mg of betaxolol base; preservative, benzalkonium chloride 0.01%; and inactive, mannitol, poly(styrene-divinylbenzene) sulfonic acid, carbomer 934P, edetate disodium, hydrochloric acid or sodium hydroxide (to adjust the pH), and purified water		Treatment of elevated intraocular pressure in patients with chronic open-angle glaucoma or ocular hypertension
Bleph-10® (sulfacetamide sodium)	Bleph®-10 (sulfacetamide sodium ophthalmic solution, USP) 10% is a sterile, topical antibacterial agent for ophthalmic use. Contains active, sulfacetamide sodium 10% (100 mg/mL); preservative, benzalkonium chloride 0.005%; and inactives edetate disodium, polysorbate 80, polyvinyl alcohol 1.4%, purified water, sodium phosphate dibasic, sodium phosphate monobasic, sodium thiosulfate, hydrochloric acid, and/or sodium hydroxide to adjust the pH (6.8–7.5)		Indicated for the treatment of conjunctivitis and other superficial ocular infections due to susceptible microorganisms and as an adjunctive in systemic sulfonamide therapy of trachoma: *Escherichia coli*, *Staphylococcus aureus*, *Streptococcus pneumoniae*, *Streptococcus* (viridans group), *Haemophilus influenzae*, *Klebsiella* species, and *Enterobacter* species

(continued)

Table 1 (continued)

Drug name	Liquid, ophthalmic formulation attributes	Original indications	Ophthalmic indications
Blephamide® (prednisolone acetate, sulfacetamide sodium)	Blephamide® ophthalmic suspension is a sterile, topical anti-inflammatory/ anti-infective combination product for ophthalmic use. Each mL of Blephamide® ophthalmic suspension contains actives sulfacetamide sodium 10% and prednisolone acetate (microfine suspension) 0.2%. Inactives: Benzalkonium chloride (0.004%); edetate disodium; polysorbate 80; polyvinyl alcohol 1.4%; potassium phosphate, monobasic; purified water; sodium phosphate, dibasic; sodium thiosulfate; hydrochloric acid and/or sodium hydroxide to adjust the pH (6.6–7.2)		Blephamide® ophthalmic suspension is a steroid/ anti-infective combination drug indicated for steroid-responsive inflammatory ocular conditions for which a corticosteroid is indicated and where superficial bacterial ocular infection or a risk of bacterial ocular infection exists. Ocular corticosteroids are indicated in inflammatory conditions of the palpebral and bulbar conjunctiva, cornea, and anterior segment of the globe where the inherent risk of corticosteroid use in certain infective conjunctivitis is accepted to obtain diminution in edema and inflammation. They are also indicated in chronic anterior uveitis and corneal injury from chemical, radiation, or thermal burns or penetration of foreign bodies
Blink tears®	Polyethylene glycol 400 0.25%; boric acid; calcium chloride; magnesium chloride; potassium chloride; water; sodium borate; sodium chloride; sodium chlorite; hyaluronate sodium		Lubricating eye drops
Boiron, Optique 1®	Eye drops, single-use doses; purified water and 0.9% sodium chloride; HPUS ingredients officially included in the homeopathic Pharmacopœia of the United States—Calcarea fluorica, calendula officinalis, cineraria maritima, euphrasia officinalis, kali muriaticum, magnesia carbonica, silicea		Temporary relief of minor eye irritation due to fatigue or airborne irritants such as ragweed, other pollens, and dust; soothes red, dry, itchy, gritty, burning or tired eyes

(continued)

Table 1 (continued)

Drug name	Liquid, ophthalmic formulation attributes	Original indications	Ophthalmic indications
CEQUA® (cyclosporine A)	Cequa (cyclosporine ophthalmic solution) 0.09% contains a topical calcineurin inhibitor immunosuppressant. Cequa is supplied as a sterile, clear, colorless ophthalmic solution for topical ophthalmic use. It has an osmolality of 160–190 mOsmol/kg and a pH of 6.5–7.2. Each mL of Cequa contains active, cyclosporine 0.09%, and inactives: Polyoxyl hydrogenated castor oil, Octoxynol-40, polyvinylpyrrolidone, sodium phosphate monobasic dihydrate, sodium phosphate dibasic anhydrous, water for injection, and sodium hydroxide or hydrochloric acid to adjust the pH		Cequa ophthalmic solution is a calcineurin inhibitor immunosuppressant indicated to increase tear production in patients with keratoconjunctivitis sicca (dry eye)
Ciloxan® (ciprofloxacin hydrochloride)	Ciloxan® (ciprofloxacin HCL ophthalmic solution) is a synthetic, sterile, multiple dose, antimicrobial for topical use. Each mL of Ciloxan ophthalmic solution contains active, ciprofloxacin HCl 3.5 mg equivalent to 3 mg base; preservative, benzalkonium chloride 0.006%; and inactives, sodium acetate, acetic acid, mannitol 4.6%, edetate disodium 0.05%, hydrochloric acid and/or sodium hydroxide (to adjust the pH), and purified water. The pH is approximately 4.5 and the osmolality is approximately 300 mOsm		Ciprofloxacin is a fluoroquinolone antibacterial active against a broad spectrum of gram-positive and gram-negative ocular pathogens

(continued)

Table 1 (continued)

Drug name	Liquid, ophthalmic formulation attributes	Original indications	Ophthalmic indications
Combigan® (brimonidine tartrate, timolol maleate)	In solution, Combigan® (brimonidine tartrate/timolol ophthalmic solution) 0.2%/0.5% has a clear, greenish-yellow color. It has an osmolality of 260–330 mOsmol/kg and a pH during its shelf life of 6.5–7.3. Brimonidine tartrate appears as an off-white or white to pale-yellow powder and is soluble in both water (1.5 mg/mL) and in the product vehicle (3 mg/mL) at pH 7.2. Timolol maleate appears as a white, odorless, crystalline powder and is soluble in water, methanol, and alcohol. Each mL of Combigan® contains the active ingredients brimonidine tartrate 0.2% and timolol 0.5% with the inactive ingredients benzalkonium chloride 0.005%; sodium phosphate, monobasic; sodium phosphate, dibasic; purified water; and hydrochloric acid and/or sodium hydroxide to adjust the pH		Combigan® is an alpha-adrenergic receptor agonist with a beta-adrenergic receptor inhibitor indicated for the reduction of elevated intraocular pressure (IOP) in patients with glaucoma or ocular hypertension who require adjunctive or replacement therapy due to inadequately controlled IOP; the IOP lowering of Combigan® dosed twice a day was slightly less than that seen with the concomitant administration of timolol ophthalmic solution, 0.5% dosed twice a day, and brimonidine tartrate ophthalmic solution 0.2% dosed three times per day
Cosopt® Dorzolamide hydrochloride with timolol maleate	Cosopt is supplied as a sterile, clear, colorless to nearly colorless, isotonic, buffered, slightly viscous, aqueous solution. The pH of the solution is approximately 5.65, and the osmolarity is 242–323 mOsM. Each mL of Cosopt contains 20 mg dorzolamide (22.26 mg of dorzolamide hydrochloride) and 5 mg timolol (6.83 mg timolol maleate). Inactive ingredients are sodium citrate, hydroxyethyl cellulose, sodium hydroxide, mannitol, and water for injection. Benzalkonium chloride 0.0075% is added as a preservative		Cosopt® is indicated for the reduction of elevated intraocular pressure (IOP) in patients with open-angle glaucoma or ocular hypertension who are insufficiently responsive to beta-blockers (failed to achieve target IOP determined after multiple measurements over time). The IOP lowering of Cosopt administered twice a day was slightly less than that seen with the concomitant administration of 0.5% timolol administered twice a day and 2% dorzolamide administered three times a day

(continued)

Table 1 (continued)

Drug name	Liquid, ophthalmic formulation attributes	Original indications	Ophthalmic indications
Cromolyn® (cromolyn sodium)	Cromolyn sodium ophthalmic solution USP, 4%, is a clear, colorless, sterile solution intended for topical ophthalmic use. Each mL contains active, cromolyn sodium 40 mg (4%); preservative, benzalkonium chloride 0.01%; and inactives, edetate disodium 0.1% and purified water. Hydrochloric acid and/or sodium hydroxide may be added to adjust the pH (4.0–7.0)		Mast cell stabilizer indicated in the treatment of vernal keratoconjunctivitis, vernal conjunctivitis, and vernal keratitis
Cyclogyl® (cyclopentolate hydrochloride)	Each mL of Cyclogyl® (cyclopentolate hydrochloride ophthalmic solution, USP) contains active, cyclopentolate hydrochloride 0.5%, 1%, or 2%; preservative, benzalkonium chloride 0.01%; and inactives boric acid, edetate disodium, potassium chloride (except 2% strength), sodium carbonate and/or hydrochloric acid (to adjust the pH), and purified water. The pH range is between 3.0 and 5.5		Used to produce mydriasis and cycloplegia
Cystaran® (cysteamine)	Cystaran is a sterile ophthalmic solution containing 6.5 mg/mL of cysteamine hydrochloride, equivalent to 4.4 mg/mL of cysteamine (0.44%) as the active ingredient. Cysteamine is a cystine-depleting agent which lowers the cystine content of cells in patients with cystinosis. Each milliliter of Cystaran contains active, cysteamine 4.4 mg (equivalent to cysteamine hydrochloride 6.5 mg); preservative, benzalkonium chloride 0.1 mg; and inactive ingredients sodium chloride, hydrochloric acid and/or sodium hydroxide (to adjust the pH to 4.1–4.5), and purified water		A cystine-depleting agent indicated for the treatment of corneal cystine crystal accumulation in patients with cystinosis
Durezol® (difluprednate)	Durezol (difluprednate ophthalmic emulsion) 0.05% is a sterile, topical, anti-inflammatory corticosteroid for ophthalmic use. Each mL of Durezol contains active, difluprednate 0.5 mg (0.05%); inactive, boric acid, castor oil, glycerin, polysorbate 80, water for injection, sodium acetate, edetate disodium, and sodium hydroxide (to adjust the pH to 5.2–5.8) (the emulsion is essentially isotonic with a tonicity of 304 to 411 mOsm/kg); and preservative, sorbic acid 0.1%		For the treatment of inflammation and pain associated with ocular surgery and endogenous anterior uveitis

(continued)

Table 1 (continued)

Drug name	Liquid, ophthalmic formulation attributes	Original indications	Ophthalmic indications
Elestat® (epinastine hydrochloride)	Each mL contains active, epinastine HCl 0.05% (0.5 mg/mL) equivalent to epinastine 0.044% (0.44 mg/mL); preservative, benzalkonium chloride 0.01%; and inactives, edetate disodium, purified water, sodium chloride, sodium phosphate (monobasic), and sodium hydroxide and/or hydrochloric acid (to adjust the pH). Elestat® has a pH of approximately 7 and an osmolality range of 250–310 mOsm/kg		Indicated for the prevention of itching associated with allergic conjunctivitis
Emadine® (emedastine difumarate)	Each mL of Emadine® (emedastine difumarate ophthalmic solution) 0.05% contains active, 0.884 mg emedastine difumarate equivalent to 0.5 mg emedastine; preservative, benzalkonium chloride 0.01%; and inactives, tromethamine, sodium chloride, hypromellose, hydrochloric acid/sodium hydroxide (adjust pH), and purified water. It has a pH of approximately 7.4 and an osmolality of approximately 300 mOsm/kg		Indicated for the temporary relief of the signs and symptoms of allergic conjunctivitis
EYLEA® (aflibercept)	EYLEA (aflibercept) injection is a sterile, clear, and colorless to pale-yellow solution. EYLEA is supplied as a preservative-free, sterile, aqueous solution for intravitreal injection in a single-dose, glass vial designed to deliver 0.05 mL (50 μl) of solution containing 2 mg of EYLEA (40 mg/mL in 10 mM sodium phosphate, 40 mM sodium chloride, 0.03% polysorbate 20, and 5% sucrose, pH 6.2)		Age-related macular degeneration, diabetic macular edema, diabetic retinopathy, macular edema following retinal vein occlusion
FML Forte® (solution) Fluorometholone (0.25%)	Active: Fluorometholone 0.25%. Preservative: Benzalkonium chloride 0.005%. Inactives: Edetate disodium; polysorbate 80; polyvinyl alcohol 1.4%; purified water; sodium chloride; sodium phosphate, dibasic; sodium phosphate, monobasic; and sodium hydroxide to adjust the pH. FML Forte® suspension is formulated with a pH from 6.2 to 7.5		Indicated for the treatment of corticosteroid-responsive inflammation of the palpebral and bulbar conjunctiva, cornea, and anterior segment of the globe

(continued)

Table 1 (continued)

Drug name	Liquid, ophthalmic formulation attributes	Original indications	Ophthalmic indications
FML® (solution) Fluorometholone (0.1%)	Active: Fluorometholone 0.1%. Preservative: Benzalkonium chloride 0.004%. Inactives: Edetate disodium; polysorbate 80; polyvinyl alcohol 1.4%; purified water; sodium chloride; sodium phosphate, dibasic; sodium phosphate, monobasic; and sodium hydroxide to adjust the pH. FML® suspension is formulated with a pH from 6.2 to 7.5. It has an osmolality range of 290–350 mOsm/kg		Indicated for the treatment of corticosteroid-responsive inflammation of the palpebral and bulbar conjunctiva, cornea, and anterior segment of the globe
Fortaz® (ceftazidime) ‖	Fortaz in sterile crystalline form is supplied in vials equivalent to 500 mg, 1 g, 2 g, or 6 g of anhydrous ceftazidime and in ADD-vantage® vials equivalent to 1 or 2 g of anhydrous ceftazidime. Solutions of Fortaz range in color from light yellow to amber, depending on the diluent and volume used. The pH of freshly constituted solutions usually ranges from 5 to 8. Fortaz is available as a frozen, isosmotic, sterile, nonpyrogenic solution with 1 or 2 g of ceftazidime as ceftazidime sodium premixed with approximately 2.2 or 1.6 g, respectively, of hydrous dextrose, USP. Dextrose has been added to adjust the osmolality. Sodium hydroxide is used to adjust the pH and neutralize ceftazidime pentahydrate free acid to the sodium salt. The pH may have been adjusted with hydrochloric acid. Solutions of premixed Fortaz range in color from light yellow to amber. The solution is intended for intravenous (IV) use after thawing to room temperature. The osmolality of the solution is approximately 300 mOsmol/kg, and the pH of thawed solutions ranges from 5 to 7.5	Bacterial septicemia, bone and joint infections, CNS infections, empiric therapy in immuno compromised patient, gynecologic infections, intra-abdominal infections, lower respiratory tract infections, skin and skin-structure infections, urinary tract infections	Bacterial endophthalmitis by intravitreal injection (Jackson and Williamson 1999)

(continued)

Table 1 (continued)

Drug name	Liquid, ophthalmic formulation attributes	Original indications	Ophthalmic indications
Genteal® (hydroxypropyl methylcellulose)	Dextran 70 0.1%, glycerin 0.2%, and hypromellose 0.3% all acting as lubricants		Temporary relief of burning and irritation due to dryness of the eye, as a protectant against further irritation, and temporary relief of discomfort due to minor irritations of the eye or to exposure to wind or sun
HUMIRA® (adalimumab)	Adalimumab is a tumor necrosis factor blocker. It consists of 1330 amino acids and has a molecular weight of approximately 148 kDa Each 80 mg/0.8 mL prefilled syringe or prefilled pen delivers 0.8 mL (80 mg) of drug product. Each 0.8 mL of HUMIRA contains adalimumab (80 mg), mannitol (33.6 mg), polysorbate 80 (0.8 mg), and water for injection, USP Each 40 mg/0.4 mL prefilled syringe or prefilled pen delivers 0.4 mL (40 mg) of drug product. Each 0.4 mL of HUMIRA contains adalimumab (40 mg), mannitol (16.8 mg), polysorbate 80 (0.4 mg), and water for injection, USP. HUMIRA® (adalimumab) citrate-free is specifically indicated for ophthalmic use, although not clear what is the final pH or osmolarity. General description says, "the solution of HUMIRA is clear and colorless, with a pH of about 5.2"		HUMIRA is indicated or the treatment of noninfectious intermediate, posterior, and panuveitis in adults and pediatric patients 2 years of age and older
Iopidine® (apraclonidine hydrochloride)	Each mL of Iopidine 0.5% ophthalmic solution contains active, apraclonidine hydrochloride 5.75 mg equivalent to apraclonidine base 5 mg, and inactives: Sodium chloride, sodium acetate, sodium hydroxide and/or hydrochloric acid (pH 4.4–7.8), purified water, and benzalkonium chloride 0.01% (preservative)		Relatively selective alpha2-adrenergic agonist that reduces elevated, as well as normal, intraocular pressure, whether or not accompanied by glaucoma

(continued)

Table 1 (continued)

Drug name	Liquid, ophthalmic formulation attributes	Original indications	Ophthalmic indications
Isopto Carpine® (pilocarpine hydrochloride)	Each mL of Isopto Carpine® (pilocarpine hydrochloride ophthalmic solution) contains active: Pilocarpine hydrochloride 1% (10 mg/mL), 2% (20 mg/mL), or 4% (40 mg/mL)		Reduction of elevated intraocular pressure (IOP) in patients with open-angle glaucoma or ocular hypertension management of acute angle-closure glaucoma prevention of postoperative elevated IOP associated with laser surgery induction of miosis
ISOPTO® atropine (atropine sulfate)	Each mL of ISOPTO® atropine 1% contains 10 mg of atropine sulfate monohydrate equivalent to 9.7 mg/mL of atropine sulfate or 8.3 mg of atropine. pH of 3.5–6.0. Preservative: Benzalkonium chloride 0.01%. Inactive ingredients: Hypromellose, boric acid, sodium hydroxide and/or hydrochloric acid (to adjust the pH), purified water		A muscarinic antagonist indicated for mydriasis, cycloplegia, penalization of the healthy eye in the treatment of amblyopia
Jetrea® (ocriplasmin)	Ocriplasmin is a recombinant truncated form of human plasmin with a molecular weight of 27.2 kDa produced by recombinant DNA technology in a *Pichia pastoris* expression system. Jetrea is a sterile, clear, and colorless solution with no preservatives in a single-use glass vial containing 0.5 mg ocriplasmin in 0.2 mL solution for intravitreal injection after dilution Each vial contains 0.5 mg ocriplasmin (active) and 0.21 mg citric acid, 0.75 mg mannitol, sodium hydroxide (for pH adjustment), and water for injection. The pH of the solution is 3.1		Proteolytic enzyme indicated for the treatment of symptomatic vitreomacular adhesion
Kenalog® (triamcinolone acetonide)ǁ	Each mL of the sterile aqueous suspension provides 40 mg triamcinolone acetonide, with 0.65% sodium chloride for isotonicity, 0.99% (w/v) benzyl alcohol as a preservative, 0.75% carboxymethylcellulose sodium, and 0.04% polysorbate 80. Sodium hydroxide or hydrochloric acid may be present to adjust the pH to 5.0–7.5. At the time of manufacture, the air in the container is replaced by nitrogen	Rheumatoid arthritis	Treatment of sympathetic ophthalmia, temporal arteritis, and uveitis, diabetic macular edema (Fazelat and Lashkari 2011; Kovacs et al. 2012; Young et al. 2001)

(continued)

Table 1 (continued)

Drug name	Liquid, ophthalmic formulation attributes	Original indications	Ophthalmic indications
Lastacaft® (alcaftadine)	Active: Alcaftadine 0.25% (2.5 mg/mL) Inactives: Benzalkonium chloride 0.005% as a preservative; edetate disodium; sodium phosphate, monobasic; Purified water; sodium chloride; sodium hydroxide and/or hydrochloric acid (to adjust the pH). The drug product has a pH of approximately 7 and an osmolality of approximately 290 mOsm/kg		Lastacaft® is an H1 histamine receptor antagonist indicated for the prevention of itching associated with allergic conjunctivitis
Latisse® (bimatoprost)	Bimatoprost is a powder, which is very soluble in ethyl alcohol and methyl alcohol and slightly soluble in water. Latisse® is a clear, isotonic, colorless, sterile ophthalmic solution with an osmolality of approximately 290 mOsmol/kg. Contains active bimatoprost 0.3 mg/mL, preservative benzalkonium chloride 0.05 mg/mL, and inactives sodium chloride; sodium phosphate, dibasic; citric acid; and purified water. Sodium hydroxide and/or hydrochloric acid may be added to adjust the pH. The pH during its shelf life ranges from 6.8 to 7.8		A prostaglandin analog indicated to treat hypotrichosis of the eyelashes by increasing their growth including length, thickness, and darkness
Lotemax® (solution, loteprednol etabonate)	Each mL contains active loteprednol etabonate 5 mg (0.5%); inactives edetate disodium, glycerin, povidone, purified water, and tyloxapol (hydrochloric acid and/or sodium hydroxide may be added to adjust the pH. The suspension is essentially isotonic with a tonicity of 250–310 mOsmol/kg); and preservative added benzalkonium chloride 0.01%		Indicated for the treatment of steroid-responsive inflammatory conditions of the palpebral and bulbar conjunctiva, cornea, and anterior segment of the globe such as allergic conjunctivitis, acne rosacea, superficial punctate keratitis, herpes zoster keratitis, iritis, cyclitis, and selected infective conjunctivitis, when the inherent hazard of steroid use is accepted to obtain an advisable diminution in edema and inflammation, and for the treatment of postoperative inflammation following ocular surgery

(continued)

Table 1 (continued)

Drug name	Liquid, ophthalmic formulation attributes	Original indications	Ophthalmic indications
Lucentis® (ranibizumab)	Sterile, colorless, to pale-yellow solution in a single-use glass vial. Lucentis is supplied as a preservative-free, sterile solution in a single-use glass vial designed to deliver 0.05 mL of 10 mg/mL Lucentis (0.5 mg dose vial) or 6 mg/mL Lucentis (0.3 mg dose vial) aqueous solution with 10 mM histidine HCl, 10% α,α-trehalose dihydrate, 0.01% polysorbate 20, pH 5.5		Neovascular (wet) age-related macular degeneration, macular edema following retinal vein occlusion, diabetic macular edema
Lumify® (brimonidine tartrate)	Active ingredient: Brimonidine tartrate (0.025%). Inactive ingredients: Benzalkonium chloride, boric acid, calcium chloride dihydrate, glycerin, potassium chloride, sodium borate decahydrate, sodium chloride, water for injection. Hydrochloric acid and/or sodium hydroxide may be used to adjust the pH		Redness reliever, over the counter
Lumigan® (bimatoprost)	Bimatoprost is a powder, which is very soluble in ethyl alcohol and methyl alcohol and slightly soluble in water. Lumigan® 0.01% and 0.03% is a clear, isotonic, colorless, sterile ophthalmic solution with an osmolality of approximately 290 mOsmol/kg. Lumigan® 0.01% contains active bimatoprost 0.1 mg/mL, preservative benzalkonium chloride 0.2 mg/mL, and inactives sodium chloride; sodium phosphate, dibasic; citric acid; and purified water. Sodium hydroxide and/or hydrochloric acid may be added to adjust the pH. The pH during its shelf life ranges from 6.8 to 7.8		A prostaglandin analog indicated for the reduction of elevated intraocular pressure in patients with open-angle glaucoma or ocular hypertension

(continued)

Table 1 (continued)

Drug name	Liquid, ophthalmic formulation attributes	Original indications	Ophthalmic indications
Luxturna® (voretigene neparvovec-rzyl)	Each single-dose (preservative-free) vial of LUXTURNA contains 5E12 vector genomes (vg) per mL, and the excipients 180 mM sodium chloride, 10 mM sodium phosphate, and 0.001% Poloxamer 188 (pH 7.3), in a 0.5 mL extractable volume. Luxturna requires a 1:10 dilution prior to administration. After dilution, each dose of Luxturna consists of 1.5E11 vg in a deliverable volume of 0.3 mL. Luxturna may also contain residual components of HEK293 cells including DNA and protein and trace quantities of fetal bovine serum		Adeno-associated virus vector-based gene therapy indicated for the treatment of patients with confirmed biallelic RPE65 mutation-associated retinal dystrophy. Patients must have viable retinal cells as determined by the treating physician(s)
Macugen® (pegaptanib sodium)	Sterile, aqueous solution containing pegaptanib sodium for intravitreal injection is formulated to have an osmolality of 280–360 mOsm/kg and a pH of 6–7, supplied in a single-dose, prefilled syringe, as a 3.47 mg/mL solution measured as the free acid form of the oligonucleotide. The active ingredient is 0.3 mg of the free acid form of the oligonucleotide without polyethylene glycol, in a nominal volume of 90 μL. This dose is equivalent to 1.6 mg of pegaptanib sodium (pegylated oligonucleotide) or 0.32 mg when expressed as the sodium salt form of the oligonucleotide moiety. The product is a sterile, clear, preservative-free solution containing sodium chloride, monobasic sodium phosphate monohydrate, dibasic sodium phosphate heptahydrate, hydrochloric acid, and/or sodium hydroxide to adjust the pH and water for injection		Treatment of neovascular (wet) age-related macular degeneration

(continued)

Table 1 (continued)

Drug name	Liquid, ophthalmic formulation attributes	Original indications	Ophthalmic indications
Maxidex® (dexamethasone)	Each mL contains active dexamethasone 0.1%, preservative benzalkonium chloride 0.01%, vehicle hypromellose 0.5%, and inactives: Sodium chloride, dibasic sodium phosphate, polysorbate 80, edetate disodium, citric acid and/or sodium hydroxide (to adjust the pH), and purified water		Inflammatory conditions of the palpebral and bulbar conjunctiva, cornea, and anterior segment of the globe such as allergic conjunctivitis, acne rosacea, superficial punctate keratitis, herpes zoster keratitis, iritis, cyclitis, selected infective conjunctivitis when the inherent hazard of steroid use is accepted to obtain an advisable diminution in edema and inflammation; corneal injury from chemical, radiation, or thermal burns, or penetration of foreign bodies
Maxitrol® (neomycin sulfate, polymyxin B sulfate, dexamethasone)	Each mL of Maxitrol® (neomycin and polymyxin B sulfates and dexamethasone ophthalmic suspension) contains actives neomycin sulfate equivalent to neomycin 3.5 mg, polymyxin B sulfate 10,000 units, and dexamethasone 0.1% and inactives: Hypromellose 2910 0.5%, sodium chloride, polysorbate 20, hydrochloric acid and/or sodium hydroxide (to adjust the pH), purified water, and benzalkonium chloride 0.004% (preservative)		For steroid-responsive inflammatory ocular conditions for which a corticosteroid is indicated and where bacterial infection or a risk of bacterial infection exists. Ocular corticosteroids are indicated in inflammatory conditions of the palpebral and bulbar conjunctiva, cornea, and anterior segment of the globe where the inherent risk of corticosteroids use in certain infective conjunctivitis is accepted to obtain a diminution in edema and inflammation. They are also indicated in chronic anterior uveitis and corneal injury from chemical, radiation, or thermal burns, or penetration of foreign bodies

(continued)

Table 1 (continued)

Drug name	Liquid, ophthalmic formulation attributes	Original indications	Ophthalmic indications
Miochol®-E (acetylcholine chloride intraocular solution)	Packaged in a blister pack containing one vial and one ampoule. The vial contains 20 mg acetylcholine chloride and 56 mg mannitol. The accompanying ampoule contains 2 mL of a modified diluent of sodium acetate trihydrate, potassium chloride, magnesium chloride hexahydrate, calcium chloride dihydrate, and sterile water for injection. The reconstituted liquid will be a sterile isotonic solution (275–330 milliosmoles/kg) containing 20 mg acetylcholine chloride (1:100 solution) and 2.8% mannitol. The pH range is 5.0–8.2		Obtain miosis of the iris in seconds after delivery of the lens in cataract surgery, in penetrating keratoplasty, iridectomy, and anterior segment surgery where rapid miosis may be required
MOXEZA™ (moxifloxacin hydrochloride ophthalmic solution)	Each mL of MOXEZA™ solution contains 5.45 mg moxifloxacin hydrochloride, equivalent to 5 mg moxifloxacin base. Inactives: Sodium chloride, xanthan gum, boric acid, sorbitol, tyloxapol, purified water, and hydrochloric acid and/or sodium hydroxide to adjust the pH. MOXEZA™ is a greenish-yellow, isotonic solution with an osmolality of 300–370 mOsm/kg and a pH of approximately 7.4. Moxifloxacin hydrochloride is a slightly yellow to yellow crystalline powder		Indicated for the treatment of bacterial conjunctivitis caused by susceptible strains
Muro-128® (solution)	Sodium chloride 2%		Temporary relief of corneal edema

(continued)

Table 1 (continued)

Drug name	Liquid, ophthalmic formulation attributes	Original indications	Ophthalmic indications
Mydfrin® (phenylephrine HCl)	Medicinal ingredient: Phenylephrine HCl 2.5% w/v. non-medicinal ingredients: Benzalkonium chloride 0.01% w/v (preservative), boric acid, sodium bisulfite, edetate disodium, sodium hydroxide and/or hydrochloric acid (to adjust the pH), purified water		A vasoconstrictor, decongestant, and mydriatic in a variety of ophthalmic conditions and procedures; for pupillary dilatation in uveitis (to prevent posterior synechia formation), for multiple ophthalmologic surgical procedures (including phacoemulsification, intracapsular and extracapsular cataract extraction, vitrectomy, etc.), and for refraction without cycloplegia (as an adjunct to increase pupillary dilatation); fundoscopy, multiple ophthalmic diagnostic procedures and examination
Mydriacyl® (tropicamide)	Mydriacyl® (tropicamide ophthalmic solution, USP) is an anticholinergic prepared as a sterile topical ophthalmic solution in two strengths. Each mL of Mydriacyl® (tropicamide ophthalmic solution, USP) contains active tropicamide 0.5 or 1%, preservative benzalkonium chloride 0.01%, and inactives: Sodium chloride, edetate disodium, hydrochloric acid and/or sodium hydroxide (to adjust the pH), and purified water; pH range 4.0–5.8		For mydriasis and cycloplegia for diagnostic procedures
Naphcon-A®	Naphazoline hydrochloride 0.025%, redness reliever; pheniramine maleate 0.3%, antihistamine; benzalkonium chloride, boric acid, edetate disodium, purified water, sodium borate, sodium chloride, sodium hydroxide and/or hydrochloric acid		Benzalkonium chloride, boric acid, edetate disodium, purified water, sodium borate, sodium chloride, sodium hydroxide and/or hydrochloric acid

(continued)

Table 1 (continued)

Drug name	Liquid, ophthalmic formulation attributes	Original indications	Ophthalmic indications
Neosporin® (solution; neomycin sulfate, polymyxin B sulfate, gramicidin)	Neosporin ophthalmic solution (neomycin and polymyxin B sulfates and gramicidin ophthalmic solution) is a sterile antimicrobial solution for ophthalmic use. Each mL contains neomycin sulfate equivalent to 1.75 mg neomycin base, polymyxin B sulfate equivalent to 10,000 polymyxin B units, and gramicidin 0.025 mg. The vehicle contains alcohol 0.5%, thimerosal 0.001% (added as a preservative), and the inactive ingredients propylene glycol, polyoxyethylene polyoxypropylene compound, sodium chloride, and water for injection		Neosporin ophthalmic solution is indicated for the topical treatment of superficial infections of the external eye and its adnexa caused by susceptible bacteria. Such infections encompass conjunctivitis, keratitis and keratoconjunctivitis, blepharitis, and blepharoconjunctivitis
Neo-Synephrine® (phenylephrine)	(phenylephrine) 2.5% eye drops		This medication is used to dilate the pupils for eye examinations or procedures and to treat certain eye conditions. It belongs to a class of drugs known as decongestants. Phenylephrine works by narrowing the blood vessels
Nevanac® (nepafenac)	Nevanac 0.1% is supplied as a sterile, aqueous suspension with a pH approximately of 7.4. The osmolality of Nevanac 0.1% is approximately 305 mOsm/kg. Each mL of Nevanac 0.1% contains active nepafenac 0.1% and inactives boric acid, propylene glycol, carbomer 974P, sodium chloride, tyloxapol, edetate disodium, benzalkonium chloride 0.005% (preservative), sodium hydroxide and/or hydrochloric acid to adjust the pH, and purified water, USP		Indicated for the treatment of pain and inflammation associated with cataract surgery

(continued)

Table 1 (continued)

Drug name	Liquid, ophthalmic formulation attributes	Original indications	Ophthalmic indications
Ocufen® (flurbiprofen sodium)	Contains active flurbiprofen sodium 0.03% (0.3 mg/mL), preservative thimerosal 0.005%, and inactives: Citric acid, edetate disodium, polyvinyl alcohol 1.4%, potassium chloride, purified water, sodium chloride, and sodium citrate. May also contain hydrochloric acid and/or sodium hydroxide to adjust the pH. The pH of Ocufen® ophthalmic solution is 6.0–7.0. It has an osmolality of 260–330 mOsm/kg		A sterile topical nonsteroidal anti-inflammatory product for ophthalmic use indicated for the inhibition of intraoperative miosis
Ocuflox® (ofloxacin)	Contains active ofloxacin 0.3% (3 mg/mL), preservative benzalkonium chloride (0.005%), and inactives sodium chloride and purified water. May also contain hydrochloric acid and/or sodium hydroxide to adjust the pH. Ocuflox® solution is unbuffered and formulated with a pH of 6.4 (range 6.0–6.8). It has an osmolality of 300 mOsm/kg		Ocuflox® ophthalmic solution is indicated for the treatment of infections caused by susceptible strains of certain bacteria in the conditions of conjunctivitis and corneal ulcers
Ocupress® (carteolol hydrochloride)	Each mL contains 10 mg carteolol HCl and the inactive ingredients—Benzalkonium chloride 0.05 mg (0.005%) as a preservative; sodium chloride; sodium phosphate, dibasic; sodium phosphate, monobasic; and water for injection, USP. The product has a pH of 6.2–7.2		Effective in lowering intraocular pressure and may be used in patients with chronic open-angle glaucoma and intraocular hypertension
Omidria® (ketorolac phenylephrine)	Omidria is a sterile aqueous solution concentrate containing phenylephrine hydrochloride 12.4 mg/mL equivalent to 10.16 mg/mL of phenylephrine and ketorolac tromethamine 4.24 mg/mL equivalent to 2.88 mg/mL of ketorolac, as a clear, colorless, sterile solution concentrate with a pH of approximately 6.3. Inactives: Citric acid monohydrate; sodium citrate dihydrate; water for injection; may include sodium hydroxide and/or hydrochloric acid for pH adjustment		Maintain pupil size by preventing intraoperative miosis, and reducing postoperative pain, added to an irrigation solution used during cataract surgery or intraocular lens replacement

(continued)

Table 1 (continued)

Drug name	Liquid, ophthalmic formulation attributes	Original indications	Ophthalmic indications
Omnipred® (prednisolone acetate)	Each mL contains active prednisolone acetate 1.0%, preservative benzalkonium chloride 0.01% (prednisolone acetate ophthalmic suspension is an adrenocortical steroid product prepared as sterile ophthalmic suspension), vehicle hypromellose, and inactives: Dibasic sodium phosphate, polysorbate 80, edetate disodium, glycerin, citric acid and/or sodium hydroxide (to adjust the pH), and purified water		Steroid-responsive inflammatory conditions of the palpebral and bulbar conjunctiva, cornea, and anterior segment of the globe such as allergic conjunctivitis, acne rosacea, superficial punctate keratitis, herpes zoster keratitis, iritis, cyclitis, selected infective conjunctivitis, when the inherent hazard of steroid use is accepted to obtain an advisable diminution in edema and inflammation; corneal injury from chemical, radiation, or thermal burns, or penetration of foreign bodies
Opcon-A®	Naphazoline HCl (0.02675%); pheniramine maleate (0.315%); benzalkonium chloride, boric acid, edetate disodium, hypromellose, purified water, sodium borate, sodium chloride. Hydrochloric acid may be used to adjust the pH		Temporarily relieves itching and redness caused by pollen, ragweed, grass, animal hair, and dander
OptiPranolol® (metipranolol hydrochloride)	Metipranolol ophthalmic solution 0.3% is a sterile solution that contains metipranolol, a non-selective beta-adrenergic receptor blocking agent. Each mL of metipranolol ophthalmic solution, for ophthalmic administration, contains 3 mg metipranolol. Inactives: Povidone, glycerol, hydrochloric acid, sodium chloride, edetate disodium, and purified water. Sodium hydroxide may be added to adjust the pH. Preservative added: Benzalkonium chloride 0.004%		Indicated to treat increased intraocular pressure in patients with ocular hypertension or open-angle glaucoma

(continued)

Table 1 (continued)

Drug name	Liquid, ophthalmic formulation attributes	Original indications	Ophthalmic indications
Optivar® (azelastine hydrochloride)	Each mL of Optivar® contains active 0.5 mg azelastine hydrochloride, equivalent to 0.457 mg of azelastine base; preservative 0.125 mg benzalkonium chloride; and inactives: Disodium edetate dihydrate, hypromellose, sorbitol solution, sodium hydroxide, and water for injection. It has a pH of approximately 5.0–6.5 and an osmolarity of approximately 271–312 mOsmol/L		A relatively selective histamine H_1 antagonist and an inhibitor of the release of histamine and other mediators from cells (e.g., mast cells) involved in the allergic response
Pataday® (olopatadine hydrochloride)	Each mL of Pataday™ solution contains active 2.22 mg olopatadine hydrochloride equivalent to 2 mg olopatadine and inactives: Povidone, dibasic sodium phosphate, sodium chloride, edetate disodium, benzalkonium chloride 0.01% (preservative), hydrochloric acid/ sodium hydroxide (adjust pH), and purified water. It has a pH of approximately 7 and an osmolality of approximately 300 mOsm/kg		Indicated for the treatment of ocular itching associated with allergic conjunctivitis
Patanol® (olopatadine hydrochloride)	Each mL of Patanol (olopatadine hydrochloride ophthalmic solution) 0.1% contains active 1.11 mg olopatadine hydrochloride equivalent to 1 mg olopatadine, preservative benzalkonium chloride 0.01%, and inactives: Dibasic sodium phosphate, sodium chloride, hydrochloric acid/ sodium hydroxide (adjust pH), and purified water. It has a pH of approximately 7 and an osmolality of approximately 300 mOsm/kg		Indicated for the treatment of the signs and symptoms of allergic conjunctivitis

(continued)

Table 1 (continued)

Drug name	Liquid, ophthalmic formulation attributes	Original indications	Ophthalmic indications
Pazeo® (olopatadine hydrochloride)	Each mL of Pazeo solution contains an active ingredient [7.76 mg of olopatadine hydrochloride (7 mg olopatadine)] and the following inactive ingredients: Povidone, hydroxypropyl-gamma-cyclodextrin, polyethylene glycol 400, hypromellose, boric acid, mannitol, benzalkonium chloride 0.015% (preservative), hydrochloric acid/ sodium hydroxide (to adjust the pH), and purified water. Pazeo solution has a pH of approximately 7.2 and an osmolality of approximately 300 mOsm/kg		Indicated for the treatment of ocular itching associated with allergic conjunctivitis
Polytrim® (polymyxin B sulfate, trimethoprim)	Polytrim® (polymyxin B sulfate and trimethoprim ophthalmic solution, USP) is a sterile antimicrobial solution for topical ophthalmic use. It has a pH of 4.0–6.2 and osmolality of 270–310 mOsm/kg. Contains actives polymyxin B sulfate 10,000 units/mL and trimethoprim sulfate equivalent to 1 mg/mL, preservative benzalkonium chloride 0.04 mg/mL, and inactives: Purified water, sodium chloride, and sulfuric acid. May also contain sodium hydroxide to adjust the pH		Indicated in the treatment of surface ocular bacterial infections, including acute bacterial conjunctivitis, and blepharoconjunctivitis, caused by several susceptible strains of microorganisms

(continued)

Table 1 (continued)

Drug name	Liquid, ophthalmic formulation attributes	Original indications	Ophthalmic indications
Pred-G® (solution, gentamicin sulfate and prednisolone acetate)	Chemical names: Prednisolone acetate: 11β,17,21-trihydroxypregna-1,4-diene-3,20-dione 21-acetate. Gentamicin sulfate is the sulfate salt of gentamicin C1, gentamicin C2, and gentamicin C1A which are produced by the growth of *Micromonospora purpurea*. Contains actives gentamicin sulfate equivalent to 0.3% gentamicin base and prednisolone acetate (microfine suspension) 1%; preservative benzalkonium chloride 0.005%; and inactives: Edetate disodium; hypromellose; polyvinyl alcohol 1.4%; polysorbate 80; purified water; sodium chloride; and sodium citrate, dihydrate. May contain sodium hydroxide and/or hydrochloric acid to adjust the pH (5.4–6.6). Pred-G® suspension is formulated with a pH from 5.4 to 6.6 and its osmolality ranges from 260 to 340 mOsm/kg		Pred-G® suspension is indicated for steroid-responsive inflammatory ocular conditions for which a corticosteroid is indicated and where superficial bacterial ocular infection or a risk of bacterial ocular infection exists
Prefrin Liquifilm®	The active substance is phenylephrine hydrochloride 1.2 mg/ml. The preservative is benzalkonium chloride 0.005% w/v. the other ingredients are polyvinyl alcohol (Liquifilm), sodium phosphate dibasic anhydrous, sodium phosphate monobasic, disodium edetate, sodium acetate anhydrous, sodium thiosulfate anhydrous, and purified water. Sodium hydroxide or hydrochloric acid may be added to adjust the pH		Lubricating decongestant that whitens the eyes and is used for the relief of minor eye irritations caused by colds, hay fever, dust, smog, hard contact lenses, sun, swimming, and wind, when no infection is present

(continued)

Table 1 (continued)

Drug name	Liquid, ophthalmic formulation attributes	Original indications	Ophthalmic indications
Prolensa® (bromfenac)	Each mL of Prolensa contains 0.805 mg bromfenac sodium sesquihydrate (equivalent to 0.7 mg bromfenac free acid). Bromfenac sodium is a yellow to orange crystalline powder. The molecular weight of bromfenac sodium is 383.17. Prolensa ophthalmic solution is supplied as a sterile aqueous 0.07% solution, with a pH of 7.8. The osmolality of Prolensa ophthalmic solution is approximately 300 mOsmol/kg. Each mL contains bromfenac sodium sesquihydrate 0.0805%, which is equivalent to bromfenac-free acid 0.07%. Preservative: Benzalkonium chloride 0.005% Inactives: Boric acid, edetate disodium, povidone, sodium borate, sodium sulfite, tyloxapol, sodium hydroxide to adjust the pH and water for injection, USP		A nonsteroidal anti-inflammatory drug (NSAID) indicated for the treatment of postoperative inflammation and reduction of ocular pain in patients wWho have undergone cataract surgery
Quixin® (levofloxacin)	Each mL of Quixin® contains 5.12 mg of levofloxacin hemihydrate equivalent to 5 mg levofloxacin. Contains active levofloxacin 0.5% (5 mg/mL), preservative benzalkonium chloride 0.005%, and inactives sodium chloride and water. May also contain hydrochloric acid and/or sodium hydroxide to adjust the pH to approximately 6.5. Quixin® solution is isotonic with an osmolality of approximately 300 mOsm/kg		Indicated for the treatment of corneal ulcer caused by susceptible strains of the following bacteria: Gram-positive bacteria—*Corynebacterium* species *Staphylococcus aureus*, *Staphylococcus epidermidis*, *Streptococcus pneumonia*, and viridans group streptococci—and gram-negative bacteria *Pseudomonas aeruginosa* and *Serratia marcescens*

(continued)

Table 1 (continued)

Drug name	Liquid, ophthalmic formulation attributes	Original indications	Ophthalmic indications
Refresh Liquigel®	Carboxymethylcellulose sodium (1%)		Artificial tear substitute
Refresh Optive gel drops®	Carboxymethylcellulose sodium (1%) and glycerin (0.9%)		Artificial tear substitute
Refresh Optive Mega-3®	Carboxymethylcellulose sodium (0.5%), glycerin (1%), and polysorbate 80 (0.5%)		Artificial tear substitute
Refresh redness relief®	Formula: Redness reliever (phenylephrine, 0.12%) and lubricant		Removes redness and instantly moisturizes to soothe and protect dry, irritated eyes
Refresh repair/ refresh Optive®	Carboxymethylcellulose sodium (0.5%) and glycerin (0.9%)		Artificial tear substitute
Refresh® tears	Active ingredients: Carboxymethylcellulose sodium (0.5%). Purpose: Eye lubricant. Inactive ingredients: Boric acid, calcium chloride, magnesium chloride, potassium chloride, purified water, Purite™ (stabilized oxychloro complex), sodium borate, and sodium chloride. May also contain hydrochloric acid and/or sodium hydroxide to adjust the pH. 260–330 mOsm/kg		Artificial tear substitute
RESTASIS® (cyclosporine A)	RESTASIS® (cyclosporine ophthalmic emulsion) 0.05% contains a topical calcineurin inhibitor immunosuppressant with anti-inflammatory effects. Cyclosporine is a fine white powder. RESTASIS® appears as a white opaque to slightly translucent homogeneous emulsion. It has an osmolality of 230–320 mOsmol/kg and a pH of 6.5–8.0. Each mL of RESTASIS® ophthalmic emulsion contains active, cyclosporine 0.05%; and inactives: Glycerin, castor oil, polysorbate 80, carbomer copolymer type A, purified water, and sodium hydroxide to adjust the pH		Indicated to increase tear production in patients whose tear production is presumed to be suppressed due to ocular inflammation associated with keratoconjunctivitis sicca

(continued)

Table 1 (continued)

Drug name	Liquid, ophthalmic formulation attributes	Original indications	Ophthalmic indications
Rhopressa® (netarsudil dimesylate)	Rhopressa (netarsudil ophthalmic solution) 0.02% is supplied as a sterile, isotonic, buffered aqueous solution of netarsudil dimesylate with a pH of approximately 5 and an osmolality of approximately 295 mOsmol/kg. It is intended for topical application in the eye. Each mL of Rhopressa contains 0.2 mg of netarsudil (equivalent to 0.28 mg of netarsudil dimesylate). Benzalkonium chloride, 0.015%, is added as a preservative. The inactive ingredients are boric acid, mannitol, sodium hydroxide to adjust the pH, and water for injection		Indicated for the reduction of elevated intraocular pressure (IOP) in patients with open-angle glaucoma or ocular hypertension
Rohto cooling eye drops®	Naphazoline hydrochloride 0.012%; polysorbate 80 0.2%; alcohol (0.1%), benzalkonium chloride, boric acid, chlorobutanol, edetate disodium, menthol, purified water, sodium borate		Relieves redness of the eye due to minor eye irritations; temporarily relieves burning and irritation due to dryness of the eye
Tetcaine® (tetracaine hydrochloride)	Tetracaine hydrochloride ophthalmic solution 0.5% has a pH of 3.7–5.5. Active ingredient: Tetracaine hydrochloride 0.5% w/v (equivalent to 0.44% w/v tetracaine). Inactive ingredients: Sodium chloride, sodium acetate trihydrate, acetic acid (to adjust the pH approximately 4.5), water for injection, USP		Tetracaine hydrochloride ophthalmic solution 0.5%, an ester local anesthetic, is indicated for procedures requiring a rapid and short-acting topical ophthalmic anesthetic

(continued)

Table 1 (continued)

Drug name	Liquid, ophthalmic formulation attributes	Original indications	Ophthalmic indications
Tetravisc Forte® (tetracaine hydrochloride)	Tetracaine hydrochloride 0.5% is a sterile topical ophthalmic solution useful in producing surface anesthesia of the eye. Boric acid; edetate disodium; hypromellose; potassium chloride; sodium borate; sodium chloride; water for injection USP, hydrochloric acid and/or sodium hydroxide to adjust the pH		For procedures in which a rapid and short-acting topical ophthalmic anesthetic is indicated such as in tonometry, gonioscopy, removal of corneal foreign bodies, conjunctival scraping for diagnostic purposes, suture removal from the cornea or conjunctiva, other short corneal and conjunctival procedures
Tetravisc® (tetracaine hydrochloride)	Tetracaine hydrochloride 0.5% is a sterile topical ophthalmic solution useful in producing surface anesthesia of the eye. Active: Tetracaine HCI 0.5%. Preservative: Benzalkonium chloride (0.01%). Inactive: Boric acid, edetate disodium, hypromellose, potassium chloride, sodium borate, sodium chloride, water for injection USP, hydrochloric acid and/or sodium hydroxide to adjust the pH		For procedures in which a rapid and short-acting topical ophthalmic anesthetic is indicated such as in tonometry, gonioscopy, removal of corneal foreign bodies, conjunctival scraping for diagnostic purposes, suture removal from the cornea or conjunctiva, other short corneal and conjunctival procedures
TheraTears® Lubricant eye drops	Carboxymethylcellulose sodium (0.25%), 170 mOsm/kg; published pH 9.01 and 145 mmol/kg osmolarity (Chen et al. 2009)		Artificial tear substitute

(continued)

Table 1 (continued)

Drug name	Liquid, ophthalmic formulation attributes	Original indications	Ophthalmic indications
Timoptic® (timolol maleate)	Timolol maleate ophthalmic solution is supplied in two formulations: Ophthalmic solution Timoptic (timolol maleate ophthalmic solution), which contains the preservative benzalkonium chloride, and ophthalmic solution Timoptic (timolol maleate ophthalmic solution), the preservative-free formulation. Preservative-free ophthalmic solution Timoptic is supplied in OCUDOSE, a unit dose container, as a sterile, isotonic, buffered, aqueous solution of timolol maleate in two dosage strengths: Each mL of preservative-free Timoptic in OCUDOSE 0.25% contains 2.5 mg of timolol (3.4 mg of timolol maleate). The pH of the solution is approximately 7.0, and the osmolarity is 252–328 mOsm. Each mL of preservative-free Timoptic in OCUDOSE 0.5% contains 5 mg of timolol (6.8 mg of timolol maleate). Inactive ingredients: Monobasic and dibasic sodium phosphate, sodium hydroxide to adjust the pH, and water for injection		Treatment of elevated intraocular pressure in patients with ocular hypertension or open-angle glaucoma

(continued)

Table 1 (continued)

Drug name	Liquid, ophthalmic formulation attributes	Original indications	Ophthalmic indications
Timoptic-XE® (gel-forming timolol maleate)	Timoptic-XE sterile ophthalmic gel forming solution is supplied as a sterile, isotonic, buffered, aqueous solution of timolol maleate in two dosage strengths. The pH of the solution is approximately 7.0, and the osmolarity is 260–330 mOsm. Each mL of Timoptic-XE 0.25% contains 2.5 mg of timolol (3.4 mg of timolol maleate). Each mL of Timoptic-XE 0.5% contains 5 mg of timolol (6.8 mg of timolol maleate). Inactive ingredients: Gellan gum, tromethamine, mannitol, and water for injection. Preservative: Benzododecinium bromide 0.012%. The gel-forming solution contains a purified anionic heteropolysaccharide derived from gellan gum. An aqueous solution of gellan gum, in the presence of a cation, has the ability to gel. Upon contact with the precorneal tear film, Timoptic-XE forms a gel that is subsequently removed by the flow of tears		Treatment of elevated intraocular pressure in patients with ocular hypertension or open-angle glaucoma
Tobradex® (dexamethasone, tobramycin)	Tobradex® (tobramycin and dexamethasone ophthalmic suspension) is a sterile, multiple dose antibiotic and steroid combination for topical ophthalmic use. Each mL of Tobradex® (tobramycin and dexamethasone ophthalmic suspension) contains actives, tobramycin 0.3% (3 mg) and dexamethasone 0.1% (1 mg); preservative, benzalkonium chloride 0.01%; and inactives: Tyloxapol, edetate disodium, sodium chloride, hydroxyethyl cellulose, sodium sulfate, sulfuric acid and/or sodium hydroxide (to adjust the pH), and purified water		For steroid-responsive inflammatory ocular conditions for which a corticosteroid is indicated and where superficial bacterial ocular infection or a risk of bacterial ocular infection exists

(continued)

Table 1 (continued)

Drug name	Liquid, ophthalmic formulation attributes	Original indications	Ophthalmic indications
Travatan Z® (travoprost)	Travatan Z® (travoprost ophthalmic solution) 0.004% is supplied as sterile, buffered aqueous solution of travoprost with a pH of approximately 5.7 and an osmolality of approximately 290 mOsmol/kg Travatan Z® contains active, travoprost 0.04 mg/mL, and inactives: Polyoxyl 40 hydrogenated castor oil, SofZia® (boric acid, propylene glycol, sorbitol, zinc chloride), sodium hydroxide and/or hydrochloric acid (to adjust the pH), and purified water, USP. Preserved in the bottle with an ionic buffered system, SofZia®		Treatment of elevated intraocular pressure in patients with ocular hypertension or open-angle glaucoma
Triesence® (triamcinolone acetonide)	Each mL of the sterile, aqueous suspension provides 40 mg of triamcinolone acetonide, with sodium chloride for isotonicity, 0.5% (w/v) carboxymethylcellulose sodium, and 0.015% polysorbate 80. It also contains potassium chloride, calcium chloride (dihydrate), magnesium chloride (hexahydrate), sodium acetate (trihydrate), sodium citrate (dihydrate), and water for injection. Sodium hydroxide and hydrochloric acid may be present to adjust the pH to a target value 6–7.5		Sympathetic ophthalmia, temporal arteritis, uveitis, and ocular inflammatory conditions unresponsive to topical corticosteroids and visualization during vitrectomy
TRUSOPT® (dorzolamide hydrochloride)	TRUSOPT sterile ophthalmic solution is supplied as a sterile, isotonic, buffered, slightly viscous, aqueous solution of dorzolamide hydrochloride the pH of the solution is approximately 5.6, and the osmolarity is 260–330 mOsM. Each mL of TRUSOPT 2% contains 20 mg dorzolamide (22.3 mg of dorzolamide hydrochloride). Inactive ingredients are hydroxyethyl cellulose, mannitol, sodium citrate dihydrate, sodium hydroxide (to adjust the pH), and water for injection. Benzalkonium chloride 0.0075% is added as a preservative		Treatment of elevated intraocular pressure in patients with ocular hypertension or open-angle glaucoma

(continued)

Table 1 (continued)

Drug name	Liquid, ophthalmic formulation attributes	Original indications	Ophthalmic indications
Vancocin® (vancomycin)[‖]	Vancomycin hydrochloride for injection, USP, intravenous, is a chromatographically purified tricyclic glycopeptide antibiotic derived from *Amycolatopsis orientalis* (formerly Nocardia orientalis). The molecular weight is 1485.74; 500 mg of the base is equivalent to 0.34 mmol, 750 mg of the base is equivalent to 0.51 mmol, and 1 g of the base is equivalent to 0.67 mmol. When reconstituted with sterile water for injection, USP, vancomycin hydrochloride forms a clear, light to dark tan solution with a pH of 4.0 (2.5–4.5). This product is oxygen sensitive	Endocarditis, enterocolitis, staphylococcal infections	Endophthalmitis (Gan et al. 2001)
Vexol® (rimexolone)	Vexol® 1% ophthalmic suspension is a sterile, multidose topical ophthalmic suspension containing the corticosteroid, rimexolone. Each mL contains active ingredient rimexolone 10 mg (1%); preservative, benzalkonium chloride 0.01%; and inactive ingredients: Carbomer 974P, polysorbate 80, sodium chloride, edetate disodium, sodium hydroxide and/or hydrochloric acid (to adjust the pH), and purified water. The pH of the suspension is 6.0–8.0 and the tonicity is 260–320 mOsmol/kg		Indicated for the treatment of postoperative inflammation following ocular surgery and in the treatment of anterior uveitis
Viroptic® (trifluridine)	Viroptic sterile ophthalmic solution contains 1% trifluridine in an aqueous solution with acetic acid and sodium acetate (buffers), sodium chloride, and thimerosal 0.001% (added as a preservative). The pH range is 5.5–6.0 and osmolality is approximately 283 mOsm		Activity against herpes simplex virus, types 1 and 2 and vaccinia virus, and some strains of adenoviruses
Visine®	Inactive ingredients: Glycerin, hypromellose, polyethylene glycol 400; in Visine A® 3 mg/mL pheniramine maleate and 0.25 mg/mL naphazoline hydrochloride		Artificial tear substitute or allergy and redness relief

(continued)

Table 1 (continued)

Drug name	Liquid, ophthalmic formulation attributes	Original indications	Ophthalmic indications
Visudyne® (verteporfin)	Visudyne® (verteporfin for injection) is a light-activated drug used in photodynamic therapy. The finished drug product is a lyophilized dark green cake. Each mL of reconstituted Visudyne contains active verteporfin, 2 mg, and inactives ascorbyl palmitate, butylated hydroxytoluene, dimyristoyl phosphatidylcholine, egg phosphatidylglycerol, and lactose		Indicated for the treatment of patients with predominantly classic subfoveal choroidal neovascularization due to age-related macular degeneration, pathologic myopia, or presumed ocular histoplasmosis
Voltaren® (diclofenac sodium)	Voltaren ophthalmic (diclofenac sodium ophthalmic solution) 0.1% solution is a sterile, topical, nonsteroidal, anti-inflammatory product for ophthalmic use. Voltaren ophthalmic is available as a sterile solution which contains diclofenac sodium 0.1% (1 mg/mL). Inactive ingredients: Polyoxyl 35 castor oil, boric acid, tromethamine, sorbic acid (2 mg/mL), edetate disodium (1 mg/mL), and purified water. Diclofenac sodium is a faintly yellow-white to light beige, slightly hygroscopic crystalline powder. It is freely soluble in methanol, sparingly soluble in water, very slightly soluble in acetonitrile, and insoluble in chloroform and in 0.1 N hydrochloric acid. Its molecular weight is 318.14. Voltaren ophthalmic 0.1% is an iso-osmotic solution with an osmolality of about 300 mOsmol/1000 g, buffered at approximately pH 7.2. Voltaren ophthalmic solution has a faint characteristic odor of castor oil		Treatment of postoperative inflammation in patients who have undergone cataract extraction and for the temporary relief of pain and photophobia in patients undergoing corneal refractive surgery

(continued)

Table 1 (continued)

Drug name	Liquid, ophthalmic formulation attributes	Original indications	Ophthalmic indications
Vyzulta® (latanoprostene bunod)	Vyzulta™ (latanoprostene bunod ophthalmic solution) 0.024% is a prostaglandin analog formulated as a sterile topical ophthalmic solution. Vyzulta contains the active ingredient latanoprostene bunod 0.24 mg/mL, the preservative benzalkonium chloride 0.2 mg/mL, and the following inactive ingredients: Polysorbate 80, glycerin, EDTA, and water. The formulation is buffered to pH 5.5 with citric acid/sodium citrate		Indicated for the reduction of intraocular pressure (IOP) in patients with open-angle glaucoma or ocular hypertension
Xiidra® (lifitegrast)	Xiidra (lifitegrast ophthalmic solution) 5% is a lymphocyte function-associated antigen-1 (LFA-1) antagonist supplied as a sterile, clear, colorless to slightly brownish-yellow colored, isotonic solution of lifitegrast with a pH of 7.0–8.0 and an osmolality range of 200–330 mOsmol/kg. Active: Lifitegrast 50 mg/mL. Inactives: Sodium chloride, sodium phosphate dibasic anhydrous, sodium thiosulfate pentahydrate, sodium hydroxide and/or hydrochloric acid (to adjust the pH), and water for injection		Indicated for the treatment of the signs and symptoms of dry eye disease
Xolair® (omalizumab)[‖]	Xolair is a sterile, white, preservative-free, lyophilized powder contained in a single-use vial that is reconstituted with sterile water for injection (SWFI), USP, and administered as a subcutaneous (SC) injection. A Xolair vial contains 202.5 mg of omalizumab, 145.5 mg sucrose, 2.8 mg L-histidine hydrochloride monohydrate, 1.8 mg L-histidine, and 0.5 mg polysorbate 20 and is designed to deliver 150 mg of omalizumab, in 1.2 mL after reconstitution with 1.4 mL SWFI, USP *(*no pH or osmolarity spec.)*	Asthma, chronic idiopathic urticaria	Vernal keratoconjunctivitis (El-Qutob 2016)

(continued)

Table 1 (continued)

Drug name	Liquid, ophthalmic formulation attributes	Original indications	Ophthalmic indications
Zaditor® (ketotifen fumarate)	Ketotifen (0.025%) (equivalent to ketotifen fumarate 0.035%); Systane® polyethylene glycol 400 4 mg/mL, propylene glycol 3 mg/mL, benzalkonium chloride 0.01%, glycerol, purified water; boric acid, calcium chloride, hydroxypropyl guar, magnesium chloride, potassium chloride, sodium chloride, zinc chloride. May contain hydrochloric acid and/or sodium hydroxide to adjust the pH		Temporarily relieves itchy eyes due to pollen, ragweed, grass, animal hair, and dander
Zerviate® (cetirizine hydrochloride)	Each mL of Zerviate contains an active ingredient [cetirizine 2.40 mg (equivalent to 2.85 mg of cetirizine hydrochloride)] and the following inactive ingredients: Benzalkonium chloride 0.010% (preservative); glycerin; sodium phosphate, dibasic; edetate disodium; polyethylene glycol 400; polysorbate 80; hypromellose; hydrochloric acid/sodium hydroxide (to adjust the pH); and water for injection. Zerviate solution has a pH of approximately 7.0 and osmolality of approximately 300 mOsm/kg		A sterile ophthalmic solution containing cetirizine, which is a histamine-1 receptor antagonist, for topical administration to the eyes for the treatment of ocular itching associated with allergic conjunctivitis
Zioptan® (tafluprost)	Zioptan (tafluprost ophthalmic solution) 0.0015% is supplied as a sterile, preservative-free, solution of tafluprost with a pH range of 5.5–6.7 and an osmolality range of 260–0 mOsmol/kg. Zioptan contains active, tafluprost 0.015 mg/ml, and inactives, glycerol, sodium dihydrogen phosphate dihydrate, disodium edetate, polysorbate 80, hydrochloric acid and/ or sodium hydroxide (to adjust the pH), and water for injection		Prostaglandin analog indicated for reducing elevated intraocular pressure in patients with open-angle glaucoma or ocular hypertension

(continued)

Table 1 (continued)

Drug name	Liquid, ophthalmic formulation attributes	Original indications	Ophthalmic indications
Zirgan® (ganciclovir)	Each gram of gel contains active, ganciclovir 1.5 mg (0.15%); inactives, carbomer homopolymer, water for injection, sodium hydroxide (to adjust the pH to 7.2–7.6), and mannitol; and preservative benzalkonium chloride 0.075 mg (0.0075)		Indicated for the treatment of acute herpetic keratitis (dendritic ulcers)
Zithromax® (azithromycin)‖ AzaSite®	AzaSite (azithromycin ophthalmic solution) is a 1% sterile aqueous topical ophthalmic solution of azithromycin formulated in DuraSite® (polycarbophil, edetate disodium, sodium chloride). AzaSite is an off-white, viscous liquid with an osmolality of approximately 290 mOsm/kg. Preservative: 0.003% benzalkonium chloride. Inactives: Mannitol, citric acid, sodium citrate, poloxamer 407, polycarbophil, edetate disodium (EDTA), sodium chloride, water for injection, and sodium hydroxide to adjust the pH to 6.3	Chancroid, chronic obstructive pulmonary disease, *Mycobacte-rium avium* complex, acute otitis media, community-acquired pneumonia, skin and skin structure infections obtained from *Staphylococcus aureus*, *Streptococcus pyogenes* or *Streptococcus agalactiae*, streptococcal pharyngitis, urethritis, cervicitis	Bacterial conjunctivitis, treatment of meibomian gland dysfunction (Liu et al. 2014)

(continued)

Drug name	Liquid, ophthalmic formulation attributes	Original indications	Ophthalmic indications
Zylet® (loteprednol etabonate, tobramycin)	Each mL contains actives loteprednol etabonate 5 mg (0.5%) and tobramycin 3 mg (0.3%) and inactives edetate disodium, glycerin, povidone, purified water, tyloxapol, and benzalkonium chloride 0.01% (preservative). Sulfuric acid and/or sodium hydroxide may be added to adjust the pH to 5.7–5.9. The suspension is essentially isotonic with a tonicity of 260–320 mOsm/kg		A topical anti-infective and corticosteroid combination for steroid-responsive inflammatory ocular conditions for which a corticosteroid is indicated and where superficial bacterial ocular infection or a risk of bacterial ocular infection exists; inflammatory conditions of the palpebral and bulbar conjunctiva, cornea, and anterior segment of the globe such as allergic conjunctivitis, acne rosacea, superficial punctate keratitis, herpes zoster keratitis, iritis, cyclitis, and where the inherent risk of steroid use in certain infective conjunctivitis is accepted to obtain a diminution in edema and inflammation, chronic anterior uveitis, and corneal injury from chemical, radiation, or thermal burns, or penetration of foreign bodies
Zymaxid® (gatifloxacin)	Zymaxid® is a clear, pale-yellow, sterile, preserved aqueous solution with an osmolality of 260–330 mOsm/kg and a pH of 5.1–5.7. Zymaxid® contains the active ingredient gatifloxacin 0.5% (5 mg/mL) and the inactive ingredients benzalkonium chloride 0.005%, edetate disodium, sodium chloride, and purified water. Zymaxid® may contain hydrochloric acid and/or sodium hydroxide to adjust the pH		Indicated for the treatment of bacterial conjunctivitis caused by susceptible strains of the following organisms: Aerobic gram-positive bacteria (*Staphylococcus aureus*, *Staphylococcus epidermidis*, *Streptococcus* mitis group, *Streptococcus oralis*, *Streptococcus pneumoniae*) and aerobic gram-negative bacteria (*Haemophilus influenzae*)

new liquid ophthalmic drug product for use in an ocular disease. Such compound repurposing capitalizes on the fact that approved drugs and many compounds in the pipeline (note that clinical development candidates that are not yet approved could come from active or even abandoned programs in the pipeline) have achieved human testing and are accompanied with an understanding of pharmacology, defined systemic pharmacokinetics and safety data, and possibly a proof (or in vivo validation) of a mechanism of action. While there are close to 600 ophthalmic drug products captured in the current edition of the FDA's Orange Book (https://www.hhs.gov/ 2019), about 80% of these are drug repositioning examples underpinned by the fact that common molecular pathways contribute to different disease phenotypes. Furthermore, approximately the same proportion of *Orange Book* listed ocular products (~80%) are variations on ophthalmic formulations of the same drug or active ingredient, with more than half of those (approximately 200 reference listed drugs) qualifying as liquid ophthalmic drug products (https://www.hhs.gov/ 2019).

Physical and Chemical Considerations

Conventional physicochemical characterization approaches also apply to all active pharmaceutical ingredients used in liquid ophthalmic products; however, other distinctive requirements exist. Physical and chemical properties include those of small organic molecules as well as large macromolecules derived from biotechnology (e.g., biophysical considerations). Understanding of crystal structure and disposition thereof, single crystal data (molecular orientation and long-range packing, or that of salts and hydrates/solvates from the same perspective as isolated from the final step in process chemistry), solid state polymorphisms and solid form as it impacts thermodynamic stability and solubility in aqueous liquids, drug substance morphology including particle size distributions, and other properties which are related to manufacturability of a downstream product—e.g., melting point or glass transition temperature, hygroscopic tendencies, absolute density of substance, and any latent process chemistry or recombinant/fermentation-related impurities (Hilfiker et al. 2006). Intuitively, the aforementioned properties relate to the quality of a downstream product, e.g., controls around stability and purity; however, in some cases they can also directly impact performance and hence potentially affect safety and efficacy. Furthermore, drug substance chemical and biophysical properties in a selected ophthalmic candidate must also be fully characterized as they can relate to and influence the nature of previously listed physical considerations. Chemistry and (bio)physics can also impact the biopharmaceutical aspects which typically address liquid formulations and absorption mechanisms for a given dose and route of ocular delivery: for example, the balance between equilibrium solubility values in an aqueous environment vs. in oil, e.g., the oil/water partition coefficient—$P_{o/w}$ (Schoenwald and Huang 1983b; Wang et al. 1991); the ionization constant if the molecule has one within the relevant ocular physiological pH range (discussed later)—pK_a, pK_b, or pI values for acids, bases, and zwitterions, resp.

(Pawar et al. 2013; Schoenwald and Huang 1983b); and finally, the molecules' absolute or thermodynamic aqueous solubility with a defined pH-dependent solubility profile, or an equilibrium solubility product rate constant (K_{sp}) if ionic drug substance is being considered (Breda et al. 2009; Diehl and Markuszewski 1985; Maren et al. 1990; Pawar et al. 2013; Scozzafava et al. 1999; Shirasaki 2008; Shoghi et al. 2013; Sieg and Robinson 1977; Zhang et al. 2013). For classes of liquid ophthalmic suspension products, e.g., PRED FORTE® (https://www.accessdata.fda.gov/ 1973) 10 mg/mL prednisolone acetate topical microfine suspension indicated for treatment of steroid-responsive inflammation in anterior ocular segment tissues or TRIESSENCE® (https://www.accessdata.fda.gov/ 2007) 40 mg/mL injectable triamcinolone acetonide suspension indicated for posterior ocular inflammatory conditions unresponsive to topical corticosteroids and visualization during vitrectomy, the final particle size distribution plays a key role in precorneal residence time (a combination of turnover due to tear fluid secretion and nasolacrimal drainage) and intensity plus durability of intravitreal exposure, respectively (Missel et al. 2010; Sieg and Robinson 1975). Particle size characterization studies in topical liquid ophthalmic suspensions support the belief that moderate dilution of a suspension of a poorly soluble drug (such as the steroidal anti-inflammatory examples given earlier) does not diminish aqueous humor drug levels or, conversely, that the use of a higher drug particle count within a suspension increases aqueous humor (typical ocular pharmacokinetic sampling compartment) drug concentration-time profiles (Sieg and Robinson 1975). An order-of-magnitude lower dose (vs. PRED FORTE® (https://www.accessdata.fda.gov/ 1973)), 0.1% fluorometholone suspension, compared to a saturated solution of the same drug did not produce sustaining pharmacokinetic effects, suggesting that the conjunctival cul-de-sac retains suspended particles within a topical liquid ophthalmic eye drop and contributes significantly to the overall extent of steroid penetrating across the cornea (Sieg and Robinson 1975). Furthermore, investigations of various particle sizes and concentrations (e.g., 77–428 μm and 40–160 mg/mL) and their effect on intraocular residence time suggested that performance of liquid intravitreal-injectable suspension depots is insensitive to these physical and pharmaceutical parameters (Missel et al. 2010).

Chemical Characteristics

For small molecules, information on the lipophilicity, ionization state, and aqueous solubility forms a trifecta of physicochemical properties relating to the oil/water partition coefficient ($P_{o/w}$ or more commonly reported as log P). A known relationship exists with permeability across various ocular epithelial tissue barriers (note here one must consider the actual physiological route of administration for rationale in the final selection of drug substance for a liquid ophthalmic product design) or in other words absorption into the eye and intraocular target tissues (Chien et al. 1990; Edward and Prausnitz 2001; Friedrich et al. 1997; Hamalainen et al. 1997; Kidron et al. 2010; Pitkanen et al. 2005; Prausnitz and Noonan 1998; Ramsay et al. 2018,

2017; Schoenwald and Huang 1983a, b; Tai et al. 2003; Wang et al. 1991; Yoshida and Topliss 1996; Ahmed et al. 1987; Shirasaki 2008; Gukasyan et al. 2019a, b). The hydrophobic or hydrophilic nature of the active pharmaceutical ingredient can also be carefully used in delivery vehicle design, choice, and respective amounts of inactive ingredients used, and (bio)chemical specifications such as final pH, buffer capacity, and ionic strength or osmolyte content (Breda et al. 2009; He et al. 2003; Leibowitz et al. 1978; Mitra 1993; Palkama et al. 1985; Pawar et al. 2013; Sieg and Robinson 1975, 1977, 1979; Zhang et al. 2013; Singh et al. 2009). Formulation design, at least partly related to choices of inactive ingredient selection, will be discussed in detail in the following sections; however, it is noteworthy to mention that physicochemical properties like log P and pK_a (or log D which combines log P value with an acid or base dissociation constant at a particular pH) are important toward the selection of appropriate solubilizing excipients. Ionization constants (e.g., pK_a or pK_b) are similarly related to multiple biopharmaceutical dimensions as they influence molecules' final dose and overall absorption efficiency into the eye (Gukasyan et al. 2019b; Shirasaki 2008). Chiefly, these include the required dose and its inherent (pH)-solubility ratio, and also dissolved active pharmaceutical ingredient fraction within a total dose that's molecularly and thermodynamically eligible and available to present a chemical driving force (gradient) for flux across ocular tissue barriers (Mortimer and Eyring 1980). It is generally accepted that the neutral form of any drug substance is favored in terms of transcellular flux across biological membrane barriers; hence, within this context the physiological properties of tear fluid and intraocular compartments must be considered in conjunction with formulation attributes and how they would influence the degree of ionization of a molecule (if any) temporally from the time point of introduction into ocular space (Hogben et al. 1959; Kansy et al. 1998; Mortimer and Eyring 1980). This is an important theoretical concept with several practical examples in liquid ophthalmic dosage forms (e.g., those of brimonidine (https://www.accessdata.fda.gov/ 2001, 2006, 1996)) which will be discussed in the drug product pH considerations section.

Since the eye is exposed to direct light, as it relates to the circadian rhythm, diurnal and nocturnal changes in several physiological factors, esp. in topical ocular drug delivery, it is important to understand the chemical photosensitivity of liquid ophthalmic candidates. On a molecular level in solution, the absorbance of sunlight energy in the visible, UVA, and partially UVB radiation range is a common characteristic which can potentially lead to photoirritation and photoallergy. Hence, it is essential to characterize light absorbance profiles of liquid ophthalmic candidates and identify wavelengths within the relevant spectrum which achieve the maximum absorption (e.g., at λ_{max} value the molar extinction coefficient > 1000 L × mol^{-1} × cm^{-1} (https://www.ich.org/products/guidelines/ 2019)), and if needed evaluate the prevalence and phototoxic activity of light-excitable drug substances. Several mechanisms for light-induced ocular drug toxicity have been proposed and are equally helpful in in vitro or in vivo ocular models designed toward simple, inexpensive testing of developmental stage compounds as a screen for their potential ocular phototoxicity (Fishman 1986; Roberts 2002). For example, fluoroquinolone class of antibiotics commonly used in topical ophthalmic formulations, and via intravitreal or intracameral injections, are known to cause various degrees of phototoxicity (with an established structure activity relationship for their potential to cause photo-instability and photocarcinogenic effect, as well as chemical mechanisms of action) when exposed to ultraviolet (UV) light (Pawar et al. 2013; Thompson 2007). The UV-fluoroquinolone phototoxicity is associated with the formation of reactive oxygen species (ROS), where excitation by light energy produces both singlet oxygen and superoxide, followed by ocular-cellular damage (Thompson 2007). A related (e.g., via ROS mechanism) notable mechanism of action is the effect of such drugs *(or even some inactive ingredients found in liquid ophthalmic compositions, to be discussed in the subsequent section)* in liquid ophthalmic drug products on equilibrium concentrations of reduced glutathione (GSH) within physiological ocular fluids (e.g., tear fluid, aqueous humor, vitreous humor) or cells that comprise tissues which come into immediate contact with the product (Aguirre et al. 2012; Gurbay and Hincal 2004). For example, reduction of tear fluid or aqueous humor GSH concentration is known to trigger undesirable changes in corneal endothelial cell permeability (Green et al. 2001). Similarly, *S*-(1,2-dicarboxyethyl)glutathione (DCE-GS), which is biosynthesized in an enzyme-mediated reaction utilizing reduced glutathione and L-malate, is found at highest known concentrations in mammalian lens tissue and thought to play several key ocular physiological roles (Green et al. 2001; Tsuboi et al. 1990a, b). Within the context of liquid ophthalmic products, the extent of phototoxic damage would be a function of both the drug concentration (which is a known factor for the fluoroquinolone class) and total UV-light dose. Moreover, despite the availability of relatively more photostable fluoroquinolones such as 8-methoxy analogs of gatifloxacin and moxifloxacin vs. the photo-unstable ciprofloxacin, plus a paucity of data supporting human fluoroquinolone-induced photocarcinogenicity, in clinical use an advisory to avoid sunlight exposure for the duration of therapy with these agents is persistent (Thompson 2007; Gurbay and Hincal 2004).

Physical Characteristics

Drug substance solid form is an important consideration for liquid ophthalmic formulation development, and it warrants a brief discussion using a case study to exemplify challenges in drug repurposing for ophthalmic use as well as bridging and bioequivalence understanding form a pharmaco- and toxico-kinetic point of view. Studies with gatifloxacin (Table 1, fluoroquinolone broad-spectrum antibiotic) have provided the pharmaceutical industry with ample reasons and rationale to devote enough attention to identification and understanding of inter-relationships between all possible crystalline solid forms and how the polymorph landscape would impact the desired dosage form and development plans. Gatifloxacin was initially discovered as a hemihydrate crystallized from methanol (Masuzawa et al. 1991). Since this particular crystal form displayed poor characteristics for tableting, e.g., extremely hygroscopic with slow disintegration and dissolution for original therapeutic indication using enteral delivery route, this directed several subsequent polymorph screens and identification of 14 additional solid forms for gatifloxacin (Matsumoto et al. 1999; Raghaven et al. 2002). Briefly, all these studies added considerable challenges to the overall development pathway of the molecule to an oral product, called Tequin®, which was ironically withdrawn from major markets in 2006 for systemic safety reasons. As an appropriate segue to the next section, a highly soluble sesquihydrate (Raghaven et al. 2002) of gatifloxacin was ultimately chosen/repurposed and utilized for production of ophthalmic topical solutions called Zymar® followed by Zymaxid® (which differs at least based on label claim in active ingredient concentration, 0.3% (3 mg/mL) and 0.5% (5 mg/mL) gatifloxacin, resp., with benzalkonium chloride at 0.005%, EDTA, purified water, and sodium chloride in both), and as the compound went off-patent sometime in 2010, the generic maker Apotex Inc. started using the hemihydrate in their version of the topical drug product (Newman and Wenslow 2016). While several reports exist, the aqueous solubility relationship among known forms of gatifloxacin is understood to parallel its thermodynamic stability, with the pentahydrate having the lowest solubility at 25 °C (Raghaven et al. 2002). As a general best practice, an approach which evaluates (or identifies, if unknown) the risks and benefits associated with all solid forms of a given drug substance being considered for liquid ophthalmic product development should be adopted within the context of the proposed ocular dose and route of delivery. While it would be prudent to identify the form with lowest free energy and propose a process of isolating it from the last step in drug substance synthesis, for liquid ophthalmic products it is also important to address any risks of potentially forming less soluble hydrates or salts from common physiological or buffer ions. A full polymorphic landscape analysis will dictate also the complete interconversion mechanisms between known solid forms, ideally allowing for establishment of tight process controls and analytical methodology to produce crystalline material with high homogeneity (i.e., no detectable presence of other known polymorphs). If lower solubility forms exist than the one used in liquid ophthalmic product development, a potential supersaturated state is rendered and conversion during storage (or after introduction into intraocular compartments) toward lower-soluble forms can occur. While this is a temporally kinetic phenomenon, it is a risk which could impact the quality (e.g., formation of a precipitate) and performance (e.g., dissolution and absorption) of a liquid ophthalmic product. Unless there is a clear reason related to a medical benefit which suggests that a metastable or amorphous form for a drug substance is desired for product development, only the most stable solid form should be selected/developed. If the former exception is not applicable, and a less thermodynamically stable form is used for manufacturing ease (or other nonscience-related or regulatory strategic reasons), then it is incumbent upon the pharmaceutical developer to minimize and mitigate risk to patients from a performance and quality point of view (Singhal and Curatolo 2004).

Drug Product Considerations

The next layer of classification in liquid ophthalmic products relates to the design of delivery vehicle itself. While several strata of complexity exist in liquid ophthalmic formulation design from a physiologically based route of administration perspective, here the focus will be agnostic of site of ocular drug deposition. Progressive understanding of barriers presented by ocular anatomical features on drug delivery impart parallel protective mechanisms that help this organ to perform its primary function of ensuring proper vision. These protective mechanisms include clearance of exogenous chemicals (such as drug molecules) into the systemic circulation via fluid drainage and lacrimation. Liquid ophthalmic formulation design must consider these physiological attributes and find a logical balance between those and physicochemical ones that govern boundaries in product design. While a finite collection of different configurations exists, a deep understanding of all overlapping physiological and physicochemical characteristics is required to nominate possible formulation contenders for initial in vivo prototyping and testing.

All liquid ophthalmic dosage forms face a primary challenge that's related to the limited amount of space available for drug delivery to the eye. A typical eye drop volume is thought to be approximately 30 μL, although reports indicate a range between 25 and 56 μL with a key importance on dropper tip inner/outer diameter (as opposed to liquid formulation properties like viscosity or surface tension) (Brown and Lynch 1986; Lederer and Harold 1986). There is a restricted limit in the size of a dose that can be applied to, injected, and tolerated by ocular sites of drug deposition, and in the duration over which an applied dose stays in contact with absorptive surfaces of the eye (whether they are topical or intraocular). From this perspective, it is important to guarantee through proper liquid formulation design that the complete dose is either solubilized in a liquid product or fully available for accurate delivery in the case of solid, semi-solid, or colloidal suspended particulates within a liquid delivery vehicle. The formulation vehicle composition, e.g., pH, ionic content, and strength, as well as the presence of any inactive ingredients, plays a critical role since the allowed practical volumes for ocular delivery of liquid dosage forms lie within 30–100 μL range (depending on the route of administration) (Ghate and Edelhauser 2006; Lee and Robinson 1986). The three main ocular physiological fluids with which liquid ophthalmic formulations come into contact and mix with are tear fluid, aqueous humor, and vitreous humor, while estimations of the ionic content, nature of electrolytes, and pH of these fluids have been of interest from a basic science perspective for nearly a century according to early published records (Meyer and Palmer 1936). In contrary to initial hypothesis that these biological fluids had origins of dialysates (e.g., from blood circulation), their ionic content, presence of hyaluronic acid, and pH which is generally 0.1–0.3 units lower than that of blood suggested more complex biological regulation mechanisms in these ocular compartments and highlighted the importance of understanding their characteristics for drug delivery purposes (Meyer and Palmer 1936).

pH, Buffers, and Buffering Capacity

Furthermore, the pH range of aqueous preparations for ocular administration requires tight control and optimized buffering capacity (β). The latter, e.g., β, has been investigated in several eye-related fluids and displays considerable intersubject variability in ocular biosystems, depending on the methods used, e.g., acid or base titration. For example, local zones of enhanced buffering by human tear fluid across the entire pH spectrum were identified, reflecting multiple endogenous buffering components, primarily bicarbonate and a heterogeneous tear film protein population, among others (Carney et al. 1989). Baseline tear fluid pH values from several reports indicate a range from 7 to 7.5, which is highly dependent on several factors: diurnal fluctuations, e.g., tears are more acidic as sampled from eyes during waking hours of the day (average pH 7.25) than later in the day (pH 7.45) (Carney and Hill 1976); the dynamics attributed to these fluctuations could be related to metabolic byproducts associated with anaerobic conditions during sleep as well as differences in carbon dioxide activity in the eyelids-open vs. eyelids-closed configurations (1 h eyelid patching resulted in a significant acidic shift from pH 7.20 to 7.06 (Coles and Jaros 1984)), and also gender and age, especially in females where tear film pH increases significantly, e.g., 7.06 vs. 7.28, for <40 years of age vs. >40 years of age, respectively (Coles and Jaros 1984). Vitreous humor pH has been estimated in several instances and species, as it is thought to play a role during intraocular hypoxia, acidosis, and optic nerve cell health. Baseline vitreous pH in normotensive eyes is reported to be approximately 7.3, while it can decrease by as much as 0.4 pH units in cases of acute intraocular pressure (IOP) elevation (however, it is reversible if IOP is returned to normal levels within 2 h) (Lu et al. 2001). While the mechanisms of vitreous humor pH regulation are not well known, the influence of liquid intravitreal-injectable ophthalmic formulations for retinal disease treatment on posterior tissue circulation and vitreous pH is of great importance. Within an exploratory context, liquid intravitreal injections of pH 3–8 range have been evaluated and characterized as acceptable or tolerable from a post-hoc histopathological examination perspective (Aguirre et al. 2012). These studies employed specific buffers (at pH 3–4 range with a relatively low β) and counterions to prepare intravitreal liquid vehicles targeted for delivery of new chemical entities (e.g., small-molecule inhibitors of angiogenesis being repurposed from an oral route of delivery in oncology indications for the treatment of wet neovascular age-related macular degeneration (AMD)) (Aguirre et al. 2012; Marra et al. 2011). Specific counterions entertained within this wide pH range included sulfate, maleate, malate, fumarate, citrate, and phosphate; their molar concentrations were maintained in the 10–30 mM range with the intention to allow for rapid pH adjustment in the vitreous chamber microenvironment as the exact buffering mechanism and capacity of the compartment was not well defined (Aguirre et al. 2012, 2018; Marra et al. 2011). The selection of counterions from ionic chemical drug substances, which could subsequently behave as buffers in liquid ophthalmic formulations, or additional buffering agents for setting and controlling final drug product pH, is another important consideration from an ocular safety point of view. While traditional selection and use criteria for pharmaceutical salts can be considered as a starting point (Stahl et al. 2011), there are several physiologically unique principles which may be limitations in an ophthalmological setting. For example, in research formulation development work for a potent, selective vascular endothelial growth factor receptor tyrosine kinase inhibitor, PF-00337210, under consideration for the treatment of age-related macular degeneration, twofold changes were made to maximize safety and ocular delivery properties. Switching from an oral immediate release tablet in an oncology indication, PF-00337210 bismaleate (a rapidly dissolving salt form of the original drug substance) was recrystallized as a stable free-base polymorph to avoid use of maleate counterion intravitreally, thought to elicit retinal tissue toxicity partially through GSH depletion (Aguirre et al. 2012). Furthermore, to optimize the unique physicochemical properties of the drug which would allow for a sterile liquid parenteral injectable product to be developed for early testing (i.e., deliver up to 6 mg of PF-00337210 in a 0.1 mL intravitreal injection), the aqueous solubility was increased to >800 mg/mL using crystalline free base in a safer citrate buffer system at pH 3 with low β (10 mM citrate, β 0.001–0.003) to allow for rapid in situ neutralization of pharmaceutical pH (Marra et al. 2011). Buffering zone offered instantaneous intravitreal neutralization (i.e., from pH 3 to 7) of PF-00337210 doses by the endogenous ampholytes present in vitreous humor allowing for a spontaneous in situ formation of a drug substance precipitate which acted as a dose depot to reduce the frequency of intravitreal injections, expected by virtue of known rapid elimination of small molecules from this intraocular compartment (Aguirre et al. 2012; Raghava et al. 2004).

Liquid ophthalmic formulation preparations whose pH or tonicity is non-physiological are known to stimulate tear turnover, changes in aqueous humor dynamics, and transient ion solute exchange, thereby accelerating drug loss or potential compromise of ocular tissue integrity (Ghate and Edelhauser 2006; Mitra 1993; Shen et al. 2018). Early investigations, however largely based on subjective comfort indices, of appropriate formulation pH for ophthalmic use already suggested that deviating away from eyes' physiological pH caused non-productive drug losses as opposed to desirable absorption, accompanied by damage to ocular tissues in extreme cases. Furthermore, various buffering agent effects were studied as a function of lacrimation presumably based on human tolerance (Hind and Goyan 1947; Martin and Mims 1950). Plausibly the earliest quantitative approach which utilized dacryoscintigraphy as a method of detecting lacrimation, in direct proportionality to tear drainage rate constant, showed that alkaline and acid pH in liquid formulations decreased ocular bioavailability—for both nonionizable and ionizable drugs (Conrad et al. 1978). Furthermore, changing aspects (diurnal and nocturnal fluctuations) of tear film and ocular surface pH have been explored, and the mechanisms of tear fluid pH regulation have been carefully studied. pH challenges can affect formulation vehicle toleration, drug effectiveness, and clinical signs in disease-related endpoints. Specifically, the buffering capacity of tears shows considerable differences from those seen in the blood, large intrasubject variability, especially toward acidic-range titration. Local ocular zones of enhanced micro-buffering across the pH spectrum have been identified, presumably suggesting the existence of multiple buffering components (bicarbonate, protein, and others) present in ocular fluids (Carney et al. 1989; Coles and Jaros 1984). Perfusion of intraocular aqueous humor containing compartments with solutions of varying pH range revealed that outside of the pH range of 6.5–8.5, morphological and cell-physiology-related alterations occur, including direct cellular damage, as well as disruption of tight-junctional complexes, leading to loss in barrier function integrity within ocular and blood-systemic compartments. Furthermore, analysis of the extent of this breakdown has been shown to be dependent upon the magnitude and the exposure time to altered pH (Gonnering et al. 1979).

Estimations of pH have been performed in tears and aqueous and vitreous humor, reported at 7.25–7.45, 7.5, and 7.32, respectively, and the endogenous buffering capacity of each compartment is estimated to be significantly lower than that of blood in terms of the presence of species which act as buffers and recovery turnover time to baseline pH value following an exogenous stressor (Carney and Hill 1976; Carney et al. 1989; Lu et al. 2001; Paterson et al. 1975). Classical pH-partition hypothesis partially explains the influence of physiological pH (specifically the hydrogen ion concentration normally found in tear fluid or other ocular fluids where liquid dosage forms are deposited) for drugs with an acid dissociation constant (e.g., pK_a) on the extent of drug transfer, partitioning, or absorption across the phospholipid bilayer of cells. The concept reasons that when a drug is ionized, it will not be able to get through a lipid membrane, while keeping in mind that the ionized form of a drug is also in a pK_a-governed simultaneous equilibrium with its neutral form (Shore et al. 1957). For liquid ophthalmic drug products, the final pH of the formulation has exclusive control over the ratio of drugs' non-ionized vs. ionized states and therefore has a transient influence on proportion of species with higher lipid solubility. Pioneering reports indicated that the extent of ocular absorption of ionizable drugs must consider pH-dependent lacrimation in addition to the classical pH-partition explanation. Within this context, detailed pharmacokinetic ocular absorption studies of early glaucoma drug, pilocarpine, were able to fully corroborate quantitative estimations illustrating a plateau within the pH-dependent absorption into aqueous humor plot, only by taking into account both lacrimation and pH-partition hypothesis as two opposing effects above physiological pH and pK_a of the drug (Conrad et al. 1978). The enhanced delivery of brimonidine is apparent from a comparison of ALPHAGAN® (brimonidine tartrate ophthalmic solution) 0.2% at a pH of 5.6–6.6 (https://www.accessdata.fda.gov/ 1996) vs. ALPHAGAN® P (brimonidine tartrate ophthalmic solution) 0.1% at a pH of 7.4–8.0 (https://www.accessdata.fda.gov/ 2001, 2006), where the 50% lower concentration of brimonidine equivalents in ALPHAGAN® P at a more alkaline pH provides bioequivalence (comparable to aqueous humor, iris ciliary body exposures, and intraocular pressure lowering). By buffering the pH in ALPHAGAN® P to slightly basic and near 7.4–7.8, e.g., at approximately the pK_a of brimonidine (Bhagav et al. 2010), the ocular penetration is further enhanced partially due to the tendency of the drug to efficiently diffuse through lipid membranes under such circumstances where dissolved brimonidine species are predominantly unionized in neutral to alkaline formulation environments (Olejnik July 14, 2000). Increasing the pH of vehicles can promote increased corneal penetration for pilocarpine as well in accordance with the pH-partition hypothesis (Shore et al. 1957), while analogous series of experiments with nonionizable drugs and glycerin have been reported to give similar results (Sieg and Robinson 1977). Here, there is additional consideration around an extent of pH-induced lacrimation by the liquid topical ophthalmic vehicle, and the effect on precorneal drug concentration was determined to partially increase pilocarpine absorption at neutral to slightly alkaline pH. Comparisons against neutral, nonionizable controls suggested a primary relationship to pilocarpine's unique solubility characteristics coupled with less irritation and lacrimation, rather than a direct pH effect on the molecule (Sieg and Robinson 1977). Analogously, previous studies provided support for further development of L-carnosine as a functionally synergistic buffer for topical ophthalmic use, with pharmaceutical compatibility in the context of dosage forms displaying in situ gel-formation properties following eye drop mixing with resident tear fluid. L-Carnosine was shown to have higher buffering capacity (its buffer capacity, b, ranged from 0.002 to 0.01 at 7.5–44 mM of the dipeptide) when compared to tromethamine (e.g., TRIS) at pH values of 6.5–7.6, and superior stability (L-carnosine appeared to be 3–4 times more resistant to thermal acid/base-driven decomposition under most limiting conditions) when assessed against L-histidine (e.g., a common biologic buffer). For ophthalmic pharmacology and therapeutics, where a broad spectrum of topical (or injectable) ophthalmic agents require chronic dosing because of disease etiology or pharmacological mechanism of action, use of L-carnosine as a buffer was proposed to enable applications of emerging sustained delivery technologies which utilize osmotic or ionic in situ gel formation to slow down the clearance of small molecules or biologics from ocular compartments (Singh et al. 2009). Overall, based on the comparatively lower physiological buffering capacity of ocular fluids than that of blood *(*i.e. *blood, plasma, and red blood cells combined*—e.g., *the typical central compartment for drug distribution—in contrast have virtually unlimited buffering capacity* (Salenius 1957)*)*, the final pH and chemical buffer content of liquid ophthalmic products have to be carefully controlled. A global examination of known liquid ophthalmic products (Table 1) indicates that pH is targeted close to neutrality and the concentrations of exogenous buffers used in the product are maintained to a level sufficient to guarantee product quality and not interfere with endogenous ocular physiological pH (which can cause irritation and inter-ocular compartment boundary compromise) (Aguirre et al. 2018; Marra et al. 2011; Younis et al. 2008). Limited examples of drug products displaying a final pH (or range) significantly away from 7 exist, and despite the fact that these come with a strong case from a drug product quality point of view, the adequacy of such digressions from guidance criteria set forth by ocular physiological constraints is contextual, i.e., related to the nature of disease conditions and almost exclusively acute duration of treatment (as opposed to chronic conditions).

Osmolarity and Osmolality

In addition to the pH specification in liquid ophthalmic drug products, the final osmolarity of formulations (typically estimated using freezing point depression approach (Tomlinson et al. 2010)) is another essential biophysical and physiological compatibility attribute. Total solute content has been demonstrated to play a key role in injectable and topical ophthalmic liquid products. Formerly called osmolarity, by definition an osmotic concentration is the product of the osmolality and the mass density of water, in which osmolality is the quotient of the negative natural logarithm of the rational activity of water and the molar mass of water (McNaught and Wilkinson 1997, 2006). Conrad et al. published one of the earliest plausible investigations on the influence of tonicity (in addition to previously discussed pH and local ocular or systemic anesthesia) on lacrimation and topical ocular drug bioavailability. Employing the state-of-the-art microscintigraphy monitoring systems at the time, radiotracer signal dilution was detected in the tear film with hypertonic liquid formulations, suggesting considerable increase in lacrimation. The same was not evident with hypotonic formulations. Furthermore, this relationship of osmolarity and lacrimation had a proxy to ocular pharmacokinetic exposures, in an inverse relationship, where greatest ocular bioavailability was observed with deionized formulations containing a probe/drug, and hypertonicity (up to four times isotonic) giving the lowest (Conrad et al. 1978). Additional influential factors over extents and peak exposures elucidated from these studies were found to depend on precorneal contact time and mixing efficiency with the resident tear film (Conrad et al. 1978; Patton and Robinson 1975; Sieg and Robinson 1975, 1977; Singh et al. 2009). Limiting mechanisms which are apparently exerted by total solute concentration in liquid ophthalmic dosage forms are relative to the tonicity of the blood. While several different explanations exist, in the scenario where formulation osmolarity exceeds physiological tonicity, another phenomenon of rapid fluid extraction from ocular compartments into the vicinity of instilled dose can occur, effectively diluting the total dose in situ and decreasing the driving force for passive diffusive mass transfer to surrounding ocular compartments (Maurice 1971, 1980). From a liquid formulation design perspective, this can have implications on maximal amounts of inactive (esp. solubilizers, co-solvents, buffers, cyclodextrins, surfactants) and active ingredients that can act as solutes or osmolytes, which should be considered during ophthalmic safety and efficacy evaluations. Since excipients often make up a majority of the weight to volume ratio in liquid ophthalmic products, their contribution to osmolarity and final pH is also of paramount importance (Aguirre et al. 2018). Prolonged ocular dosing compartment exposure to hypertonic solutions, e.g., topical or intravitreal ophthalmic delivery, has been shown to be benign on epithelial barrier permeability. However, the opposite is true for hypoosmotic compositions introduced to ocular tissue compartments, which are reported to elicit transient increases in epithelial permeability from a topical delivery perspective, or microscopic findings manifesting themselves as mild retinal degeneration with emergence of eosinophilic bodies from an intravitreal delivery perspective (Aguirre et al. 2018; Maurice 1980).

The ionic content of ocular fluids is known to be modulated on a molecular and cellular level by several endogenous and pharmacological factors of relevance in the eye. Liquid ophthalmic dosage forms which are administered into various compartments of the eye require fine-tuning of their pharmaceutical and pharmacological properties that directly or indirectly influence osmolyte balance to further ensure compatibility, safety, and efficacy. In the anterior segment of the eye, epithelial tissues which line the entire ocular surface and come into full contact with topical liquid ophthalmic dosage forms have been characterized in terms of active and passive net fluid transfer rates across corneal and conjunctival epithelial cells. Chloride is the most abundant physiological anion, and its movement across cell membranes and mucosa/serosa of epithelial tissue layers is known to be tightly coupled to the osmotically driven flux of sodium (an abundant, physiological extracellular cation) (Mobasheri et al. 2005; Pusch and Jentsch 1994). Characterization of active ion transport in the presence and absence of molecules known to affect chloride secretion and sodium absorption in corneal and conjunctival epithelial tissues indicated that the cornea is primarily a sodium absorptive tissue, while the conjunctiva plays a largely chloride secretory role (Chang-Lin et al. 2005; Kompella et al. 1993; Shiue et al. 1998, 2000). This asymmetrical transfer of physiological ions to and from tear fluid by ocular epithelial tissues is thought to modulate composition and concentration of drugs and other solutes within the context of topical ophthalmic liquid dosage forms. While transient perturbation of this osmotic balancing mechanism by extremes in liquid formulation solute content has been shown to result in changes in drug permeability across ocular epithelia (Scholz et al. 2002), the absolute osmolarity of endogenous tear film present on ocular surface is also known to behave as a biomarker for prognosis at various degrees (e.g., mild to moderate) of dry eye disease (Tomlinson et al. 2006; Rocha et al. 2017). Toward addressing the latter, several liquid formulations of secretagogues have been tested in the treatment of ocular surface inflammation relief and tear film dysfunction, most prominent of which maybe diquafosol (Nichols et al. 2004), a purinergic receptor agonist which stimulates chloride coupled net fluid flow into the tear film (Hosoya et al. 2005; Dartt 2002; Shiue et al. 1998; Kompella et al. 1993). Osmotically driven fluid flux also plays a key role in the production of aqueous humor by ciliary epithelial cells. Here, the presence of bicarbonate exchange mechanisms found in the non-pigmented ciliary epithelium has been capitalized pharmaceutically, evidenced by well-documented slowing in the rate of aqueous humor production elicited by carbonic anhydrase inhibitors (e.g., compounds found in liquid ophthalmic drug products like AZOPT® (https://www.accessdata.fda.gov/ 1998) and TRUSOPT® (https://www.accessdata.fda.gov/ 1994)) which reduce the supply of ciliary epithelial cell cytoplasmic bicarbonate (Delamere 2005). Lastly, in the anterior chamber of the eye, fluid (possibly also by virtue of aquaporin water channels (Thiagarajah and Verkman 2002)) coupled anion secretion requires transcorneal endothelial cell net flux of chloride, bicarbonate, and/or lactate, the modulation of which through endogenous factors—such as aging—or exogenous factors which can be introduced through intracameral introduction of various ophthalmic drug products can play a role in cause or therapy for corneal stromal swelling or edema (Bonanno 2012). In the posterior segment of the eye, hypertonicity in liquid injectable ophthalmic preparations has been shown to exert macroscopic changes on a cellular level in retinal tissues in pathology reports (Aguirre et al. 2018). Furthermore, pharmacological findings suggested that INS37217 (a structural analog diquafosol, a secretagogue discussed earlier in the anterior segment setting) was able to stimulate fluid secretion from vitreous-to-choroid direction by activating similar chloride coupled osmotic movement mechanisms in retinal pigmented epithelial cells enhancing the rates of subretinal fluid reabsorption in certain experimentally induced retinal detachments (Maminishkis et al. 2002). Overall, therapeutic usefulness for selective solute control in liquid ophthalmic drug products within the context of treating a variety of retinal diseases that result in fluid accumulation in various posterior segment tissue compartments requires further study to determine if the described osmolarity linked mechanisms could be additive or synergistic in nature.

Inactive Ingredients Found in Liquid Ophthalmic Products

A high-level, global survey of known liquid ophthalmic drug products (Table 1) suggests that the arsenal of excipients available for use in product development is remarkably sparse (e.g., in comparison to other routes of parenteral drug administration). Selection of optimal route for ocular delivery depends on multiple factors, intuitively including the disease condition being treated, ocular tissue physiology (e.g., retina, choroid, and iris-ciliary body) that is targeted for pharmacological intervention, desired treatment modality or duration, as well as patient-disease demographics. Selection of key excipients in liquid ophthalmic drug products involves stratified rationale considerations. Initially choices may be limited from a pragmatic perspective, for example, precedence of use and prior utilization in a reference listed ophthalmic drug product as found in the Inactive Ingredient Search for Approved Drug Products or the Orange Book (https://www.accessdata.fda.gov/ 2019; https://www.hhs.gov/ 2019), or availability of parenteral and pharmaceutical grade excipient bulk from manufacturers which perform compendia testing on the material. However, ultimate restrictions most often come from a lack of basic scientific understanding about the full tolerability and disposition of the preferred inactive ingredients within an ocular context. Secondly, selections of excipients should be driven by a conventional functional role and appropriate requirement within the context of drug product quality, safety, and consistent performance (Rowe et al. 2012). Several existing reports have done a systematic evaluation of various functional excipients from an in vivo veterinary medicine (observational tests, e.g., the Draize eye test) and post-hoc tissue histopathology perspective, although there is limitation to translation from preclinical species to humans (Abraham et al. 2003; Wilhelmus 2001). Emerging research in this specific area of excipient qualification to enable ophthalmic drug delivery and product development could be highly helpful and influential in understanding the safety limits around selection of inactive ingredients in liquid ophthalmic products for development of topical eye drops, intravitreal and sub-tenon injections, or other novel routes of administration into this organ (Aguirre et al. 2012, 2018; Blandford et al. 1992; Younis et al. 2008).

Within this context, a unique and specific consideration among preservatives in liquid ophthalmic products is worthwhile to mention. Although preservatives are technically not inactive ingredients in liquid ophthalmic products, particular basic physiological research reports about additional roles (over those of known bactericidal and bacteriostatic activity) preservatives play in liquid eye products are noteworthy. Benzalkonium chloride has probably one of the most lengthy track records of use in topical eye drop products; however, it is not devoid of limitations in safety and tolerability which have over time resulted in the advent of alternatives like Polyquad, Purite®, and SofZia® (Ammar et al. 2010; Kahook and Noecker 2008; Dong et al. 2004). Furthermore, investigations on the influence benzalkonium chloride and commonly co-employed ethylene diamine tetra-acetic acid on the permeability of several ophthalmic drugs used for management of glaucoma showed a general trend in facilitating drug transport across the cornea and conjunctiva. This was partially attributed to some level of toxic effect that benzalkonium chloride has on ocular epithelial cells, permeabilizing them possibly transiently, however not insignificantly (Ashton et al. 1991; Scholz et al. 2002).

Historical accounts of off-label use of triamcinolone acetonide (a steroidal anti-inflammatory drug substance) within liquid ophthalmic drug product space presented as Kenalog-40® (https://www.accessdata.fda.gov/ 1965) provides a compelling retrospective argument supporting the importance of careful excipient selection within this pharmaceutical development space. Before the advent of TRIESENCE™ (https://www.accessdata.fda.gov/ 2007), Kenalog-40® was widely used via intravitreal and sub-tenon injection routes to treat ocular diseases, such as varieties of noninfectious uveitis and diabetic macular edema (Jonas 2006; Kovacs et al. 2012). As Kenalog-40® evolved into the most widely injected liquid parenteral drug product for triamcinolone acetonide application in various intraocular neovascular and edematous diseases, purification of triamcinolone suspension from this product (designed for intramuscular or intra-articular use only (https://www.accessdata.fda.gov/ 1965)) became important. Once it was clear that the solvent agent was better removed, in order to avoid the potential toxic effects of the vehicle, evaluations of different techniques used to reduce benzyl alcohol (~0.9–1%w/v) from commercially prepared triamcinolone acetonide suspensions were researched and published (Garcia-Arumi et al. 2005; Jonas 2006). Subsequent, more thorough histopathological evaluations of benzyl alcohol showed that the lack of toleration following the excipient's use in liquid ophthalmic preparations was manifested as conjunctival swelling, corneal and intraglobal opacities, and corneal lesions arising from multiple concentrations and compendia/purity grades available for testing (Younis et al. 2008). Overall, it is important to take a systematic and deliberate approach in the selection and qualification of all inactive ingredients present in liquid ophthalmic drug products, keeping in mind the physiological considerations around the actual, final physiological route of administration into the eye.

Manufacturing Considerations

As introduced earlier, all liquid ophthalmic products—occurring as solutions, suspensions, or more complex dosage forms of small molecules and compounds derived from biological sources—are specialized parenteral dosage forms, e.g., sterile products, that are intended for application to ocular compartments including locations adjacent to the eye and its immediate surrounding periorbital tissues. Ophthalmic routes of administration for liquid products include, but are not limited to: topical drops, subconjunctival, sub-tenon capsule, subretinal, sub- or suprachoroidal, intracorneal, intrascleral, intravitreal, intracameral, juxtascleral, and retrobulbar injection routes (Ghate and Edelhauser 2006). While Table 1 shows a comprehensive list of liquid ophthalmic products, with several off-label used parenterals in an ocular setting, this section succinctly enumerates consolidated, common liquid ophthalmic product preparation and quality test considerations which would apply for manufacturing. The current, electronic, United States Pharmacopeia chapter 771, with encompassed references, is recommended as a helpful resource for obtaining details on new manufacturing guidelines toward de novo development of liquid ophthalmic drug product monographs (United States Pharmacopeial Convention. Committee of Revision, 1979; United States Pharmacopeial Convention).

Sterilization process considerations add one of several important product development boundaries to selected physical, chemical, and formulation attributes for liquid ophthalmic products. Depending on the drug substance, packaging selection for route of administration and final liquid delivery vehicle composition, degradation, and/or morphological changes can occur to liquid suspensions and colloidal systems during sterilization. A particle size cutoff of <0.2 μm is required to consider filtration as a method of terminal sterilization for a liquid ophthalmic drug product. While aseptic processing remains a feasible option, the manufacture of sterile liquid ophthalmic products within class 10 or 100 clean rooms could be limiting to scale and flexibility. Design considerations for the development of steam-in-place sterilization processes, by introduction of pressurized steam into the internal cavities of a vessel used for liquid ophthalmic product manufacturing, have proven to be an effective means of making sure large, stationery processing equipment is compliant with sterility guidelines. While steam-in-place sterilization has several engineering control nuances, it does offer an advantage by potentially eliminating the need for aseptic processing or individual assembly of component parts within a manufacturing line. The latter can still introduce a risk of equipment contamination due to several possible root causes. Many liquid ophthalmic products, which are unit-dose and unpreserved, are manufactured under steam-in-place system procedures which allow the flexibility of non-aseptic fabrication followed by complete sterilization of the closed system carrying the product (Myers and Chrai 1980, 1981, 1982).

Limited aqueous solubility of drug substances is typically the most common consideration leading toward the development of suspension or colloidal-emulsion ophthalmic products (as opposed to aqueous solutions). Emulsion formulation manufacture is within a unique complex drug product category, as establishment of pharmaceutical and bioequivalence between two colloidal liquid ophthalmic products carrying the same drug substance is complicated and challenging (if not, in many cases pragmatically impossible). For such liquid ophthalmic complex drug products (e.g., cyclosporin A containing dosage forms of Restasis® (0.5 mg/mL), Ikervis® (1.0 mg/mL), Papilock mini® (1.0 mg/mL), Modusik-A Ofteno® (1.0 mg/mL), Lacrinmune® (0.5 mg/mL), TJ Cyporin® (0.5 mg/mL), Cyporin® (0.5 mg/mL), and Cyclorin® (0.5 mg/mL) (Lallemand et al. 2017)), it has been documented that "the manufacturing process is the product," i.e., a well-controlled and well-understood production and scale-up procedure should be engineered to guarantee reproducible product quality, safety, and performance (de Vlieger et al. 2019; Hussaarts et al. 2017). Topical ophthalmic emulsions are generally prepared by dissolving a drug substance into an oil phase, including a suitable emulsifying agent, considering additional suspending excipients, and mixing with the liquid aqueous phase vigorously to homogenize an oil-in-water emulsion. Essentially two macroscopic phases exist, where each phase—the oil and aqueous—is normally sterilized in advance or concurrently with charging into mixing vessel. High-shear homogenization is one approach which can be used to reduce emulsion droplet sizes to (sub) micron distributions, desirable toward improving physical stability of unit micelles by slowing their coalescing rate.

Once prototypical liquid ophthalmic drug products are manufactured, procedures for testing and accepting them need to be developed. Assessment of general quality attributes, e.g., identification, potency, purity (and impurities), sterility, and particulate matter, and in vitro product performance, i.e., dissolution or drug release of the drug substance from a suspension or colloidal drug product, can be found in USP (United States Pharmacopeial Convention. Committee of Revision. 1979; United States Pharmacopeial Convention.). Quality tests assess the integrity of the dosage form, whereas the performance tests assess drug release and other attributes that relate to in vivo drug performance. For example, the aforementioned physicochemical and biophysical considerations around the final pH and solute content, specific to liquid ophthalmic dosage forms, are described in USP ⟨pH 791⟩ and⟨osmolality and osmolarity 785⟩. Additionally, liquid ophthalmic drug products are required to be essentially free of visible foreign (extrinsic or intrinsic) particulates and subvisible particles in intra- or extra-ocular injectables. Besides terminal sterilization considerations discussed earlier, further analyses of effectiveness in antimicrobial preservatives (in the case of multidose liquids ophthalmics) and minimization of bacterial endotoxins (e.g., pyrogen-free) are essential (United States Pharmacopeial Convention. Committee of Revision 1979; United States Pharmacopeial Convention).

Design and validation of specific tests is necessary to build a good understanding and proper.

control over the manufacturing process critical for a reproducible, high-quality liquid ophthalmic drug product. For colloidal systems and some suspensions, development of such tests may pose challenges. Active ingredient release testing conducted on complex liquid colloidal ophthalmic drug products or suspensions manufactured under boundary conditions and compared to drug products that are intentionally prepared with meaningful variations in formulation and manufacturing sensitive parameters (i.e., particle size distribution, dose or drug loading, types and/or amounts of inactive ingredients) maybe far from predictive in terms of ophthalmic bioequivalence. The extents and degrees of sensitivity analysis require further discussion and research; although it is pragmatically unachievable to ascertain robust in vitro-in vivo correlations with these assays in an ophthalmic setting, some in vitro release tests and in silico simulations and modeling tools still represent promising avenues for evaluating their ability to distinguish performance (de Vlieger et al. 2019; Gukasyan et al. 2019b; Hussaarts et al. 2017). Several additional specific tests which maybe discriminating from a performance of a manufactured liquid ophthalmic product perspective include those around viscosity, particle size distribution, and inactive ingredients. Inclusion of viscosity evaluations in the specification of liquid ophthalmic products should be based on the types of dosage forms and whether changes in product viscosity will affect the overall performance. For example, in liquid suspensions, depending on the vehicles' viscosity, if drug particles settle and cake, they must re-disperse promptly in users' hands to achieve proper dose uniformity and accurate delivery. As mentioned earlier, the opposite is the case for viscosity influence on reliable eye-drop volume dispensing (vs. nozzle engineering) (Brown and Lynch 1986; Lederer and Harold 1986). While particle size and distributions can impact the intensity and duration of ophthalmic pharmacokinetics, the potential for any changes in particle size of ophthalmic suspensions and emulsions also needs to be evaluated. Lastly suitable substances may be added to ophthalmic products to increase stability, provided they are benign in the amounts administered and do not interfere with therapeutic efficacy or with responses to the specified manufacturing-related assays and quality tests (United States Pharmacopeial Convention.; United States Pharmacopeial Convention. Committee of Revision 1979).

In recent years the field of ophthalmic drug discovery and development has witnessed what some experts in the field call a renaissance (Yerxa 2018). With the advent of gene therapies which promise to be thus far the most curative solutions to several genetically inherited retinal diseases, and several new chemical entities being introduced as novel pharmacological mechanisms for management of glaucoma and dry eye disease, the importance of pharmaceutical development of liquid ophthalmic dosage forms remains essential (Gukasyan et al. 2019a). Discovery efforts continue toward treatment of rare genetic ocular diseases, neuroprotection from damage caused by glaucoma at the optic nerve head, and prevention of neovascular wet age-related macular degeneration (AMD) through inhibition and reversal of dry AMD, demand for additional pharmaceutical technology research, and development to support novel drugs in the pipeline. Considerations discussed here for drug substance (any modality), drug product blueprint attributes, and sterile manufacturing guidelines will remain vital and fundamental in clinical testing and commercialization for future progressive liquid ophthalmic drug products.

References

Abraham MH, Hassanisadi M, Jalali-Heravi M, Ghafourian T, Cain WS, Cometto-Muniz JE. Draize rabbit eye test compatibility with eye irritation thresholds in humans: a quantitative structure-activity relationship analysis. Toxicol Sci. 2003;76(2):384–91. https://doi.org/10.1093/toxsci/kfg242.

Aguirre SA, Collette W III, Gukasyan HJ, Huang W. An assessment of the ocular safety of excipient maleic acid following intravitreal injection in rabbits. Toxicol Pathol. 2012;40(5):797–806. https://doi.org/10.1177/0192623312441400.

Aguirre SA, Gukasyan HJ, Younis HS, Huang W. Safety assessment of formulation vehicles following intravitreal administration in rabbits. Pharm Res. 2018;35(9):173. https://doi.org/10.1007/s11095-018-2450-1.

Ahmed I, Gokhale RD, Shah MV, Patton TF. Physicochemical determinants of drug diffusion across the conjunctiva, sclera, and cornea. J Pharm Sci. 1987;76(8):583–6.

Ammar DA, Noecker RJ, Kahook MY. Effects of benzalkonium chloride-preserved, polyquad-preserved, and sofZia-preserved topical glaucoma medications on human ocular epithelial cells. Adv Ther. 2010;27(11):837–45. https://doi.org/10.1007/s12325-010-0070-1.

Ashton P, Podder SK, Lee VH. Formulation influence on conjunctival penetration of four beta blockers in the pigmented rabbit: a comparison with corneal penetration. Pharm Res. 1991;8(9):1166–74.

Bevacizumab (AVASTIN) and age-related macular degeneration. Lower cost does not justify taking risks. Prescrire Int. 2015;24(163):201–4.

Bhagav P, Deshpande P, Pandey S, Chandran S. Development and validation of stability indicating UV spectrophotometric method for the estimation of brimonidine tartrate in pure form, formulations and preformulation studies. Pharm Lett. 2010;2(3):106–22.

Blandford DL, Smith TJ, Brown JD, Pearson PA, Ashton P. Subconjunctival sustained release 5-fluorouracil. Invest Ophthalmol Vis Sci. 1992;33(12):3430–5.

Bonanno JA. Molecular mechanisms underlying the corneal endothelial pump. Exp Eye Res. 2012;95(1):2–7. https://doi.org/10.1016/j.exer.2011.06.004.

Breda SA, Jimenez-Kairuz AF, Manzo RH, Olivera ME. Solubility behavior and biopharmaceutical classification of novel high-solubility ciprofloxacin and norfloxacin pharmaceutical derivatives. Int J Pharm. 2009;371(1–2):106–13. https://doi.org/10.1016/j.ijpharm.2008.12.026.

Brown RH, Lynch MG. Drop size of commercial glaucoma medications. Am J Ophthalmol. 1986;102(5):673–4.

Carney LG, Hill RM. Human tear pH. Diurnal variations. Arch Ophthalmol. 1976;94(5):821–4.

Carney LG, Mauger TF, Hill RM. Buffering in human tears: pH responses to acid and base challenge. Invest Ophthalmol Vis Sci. 1989;30(4):747–54.

Chang-Lin JE, Kim KJ, Lee VH. Characterization of active ion transport across primary rabbit corneal epithelial cell layers (RCrECL) cultured at an air-interface. Exp Eye Res. 2005;80(6):827–36. https://doi.org/10.1016/j.exer.2004.12.012.

Chen SL, Png E, Tong L. Comparison of two artificial tear formulations using aberrometry. Clin Exp Optom. 2009;92(6):519; author reply 519. https://doi.org/10.1111/j.1444-0938.2009.00421.x.

Chien DS, Homsy JJ, Gluchowski C, Tang-Liu DD. Corneal and conjunctival/scleral penetration of p-aminoclonidine, AGN 190342, and clonidine in rabbit eyes. Curr Eye Res. 1990;9(11):1051–9. https://doi.org/10.3109/02713689008997579.

Coles WH, Jaros PA. Dynamics of ocular surface pH. Br J Ophthalmol. 1984;68(8):549–52.

Conrad JM, Reay WA, Polcyn RE, Robinson JR. Influence of tonicity and pH on lacrimation and ocular drug bioavailability. J Parenter Drug Assoc. 1978;32(4):149–61.

Dartt DA. Regulation of mucin and fluid secretion by conjunctival epithelial cells. Prog Retin Eye Res. 2002;21(6):555–76.

Delamere NA. Ciliary body and ciliary epithelium. Adv Organ Biol. 2005;10:127–48. https://doi.org/10.1016/S1569-2590(05)10005-6.

Diehl H, Markuszewski R. Studies on fluorescein-II the solubility and acid dissociation constants of fluorescein in water solution. Talanta. 1985;32(2):159–65.
Dong JQ, Babusis DM, Welty DF, Acheampong AA, Tang-Liu D, Whitcup SM. Effects of the preservative purite on the bioavailability of brimonidine in the aqueous humor of rabbits. J Ocul Pharmacol Ther. 2004;20(4):285–92. https://doi.org/10.1089/1080768041725326.
Edward A, Prausnitz MR. Predicted permeability of the cornea to topical drugs. Pharm Res. 2001;18(11):1497–508.
El-Qutob D. Off-label uses of Omalizumab. Clin Rev Allergy Immunol. 2016;50(1):84–96. https://doi.org/10.1007/s12016-015-8490-y.
Fazelat A, Lashkari K. Off-label use of intravitreal triamcinolone acetonide for diabetic macular edema in a pregnant patient. Clin Ophthalmol. 2011;5:439–41. https://doi.org/10.2147/OPTH.S14584.
FDA. Kenalog®-40 injection. 1965. https://www.accessdata.fda.gov/, https://www.accessdata.fda.gov/drugsatfda_docs/label/2014/014901s042lbledt.pdf.
FDA. PRED FORTE®. 1973. https://www.accessdata.fda.gov/, https://www.accessdata.fda.gov/drugsatfda_docs/label/2017/017011s047lbl.pdf.
FDA. TRUSOPT®. 1994. https://www.accessdata.fda.gov/, https://www.accessdata.fda.gov/drugsatfda_docs/label/2010/020408s047lbl.pdf.
FDA. ALPHAGAN® (brimonidine tartrate ophthalmic solution) 0.2%. 1996. https://www.accessdata.fda.gov/, https://www.accessdata.fda.gov/drugsatfda_docs/label/2016/020613s031lbl.pdf.
FDA. Azopt®. 1998. https://www.accessdata.fda.gov/, https://www.accessdata.fda.gov/drugsatfda_docs/label/2006/020816s009lbl.pdf.
FDA. ALPHAGAN® P (brimonidine tartrate ophthalmic solution) 0.1% or 0.15%. 2001, 2006. https://www.accessdata.fda.gov/, https://www.accessdata.fda.gov/drugsatfda_docs/label/2010/021262s020,021770s004lbl.pdf.
FDA. TRIESENCE™. 2007. https://www.accessdata.fda.gov/, https://www.accessdata.fda.gov/drugsatfda_docs/label/2007/022223,022048lbl.pdf.
FDA. (2019) Inactive ingredient search for approved drug products. https://www.accessdata.fda.gov/, https://www.accessdata.fda.gov/scripts/cder/iig/index.cfm.
Fishman GA. Ocular phototoxicity: guidelines for selecting sunglasses. Surv Ophthalmol. 1986;31(2):119–24.
Food and Drug Administration, Center for Drug Evaluation and Research. Orange book: approved drug products with therapeutic equivalence evaluations. 2019. https://www.accessdata.fda.gov/scripts/cder/ob/index.cfm.
Friedrich S, Saville B, Cheng YL. Drug distribution in the vitreous humor of the human eye: the effects of aphakia and changes in retinal permeability and vitreous diffusivity. J Ocul Pharmacol Ther. 1997;13(5):445–59. https://doi.org/10.1089/jop.1997.13.445.
Gan IM, van Dissel JT, Beekhuis WH, Swart W, van Meurs JC. Intravitreal vancomycin and gentamicin concentrations in patients with postoperative endophthalmitis. Br J Ophthalmol. 2001;85(11):1289–93. https://doi.org/10.1136/bjo.85.11.1289.
Garcia-Arumi J, Boixadera A, Giralt J, Martinez-Castillo V, Gomez-Ulla F, Corcostegui B, Garcia-Arumi E. Comparison of different techniques for purification of triamcinolone acetonide suspension for intravitreal use. Br J Ophthalmol. 2005;89(9):1112–4. https://doi.org/10.1136/bjo.2005.067744.
Ghate D, Edelhauser HF. Ocular drug delivery. Expert Opin Drug Deliv. 2006;3(2):275–87. https://doi.org/10.1517/17425247.3.2.275.
Gonnering R, Edelhauser HF, Van Horn DL, Durant W. The pH tolerance of rabbit and human corneal endothelium. Invest Ophthalmol Vis Sci. 1979;18(4):373–90.
Green K, Kearse EC, Yokogaki S, Awata T. Human and rabbit corneal endothelial permeability after different chemical forms of glutathione. Ophthalmic Res. 2001;33(3):151–5. https://doi.org/10.1159/000055662.
Gukasyan HJ, Hailu S, Karami TK. Ophthalmic drug discovery and development. Pharm Res. 2019a;36(5):69. https://doi.org/10.1007/s11095-019-2606-7.

Gukasyan HJ, Hailu S, Karami TK, Graham R. Ocular biopharmaceutics: impact of modeling and simulation on topical ophthalmic formulation development. Drug Discov Today. 2019b; https://doi.org/10.1016/j.drudis.2019.04.002.

Gurbay A, Hincal F. Ciprofloxacin-induced glutathione redox status alterations in rat tissues. Drug Chem Toxicol. 2004;27(3):233–42. https://doi.org/10.1081/DCT-120037504.

Hamalainen KM, Kananen K, Auriola S, Kontturi K, Urtti A. Characterization of paracellular and aqueous penetration routes in cornea, conjunctiva, and sclera. Invest Ophthalmol Vis Sci. 1997;38(3):627–34.

He Y, Li P, Yalkowsky SH. Solubilization of fluasterone in cosolvent/cyclodextrin combinations. Int J Pharm. 2003;264(1–2):25–34.

Hilfiker R, Blatter F, von Raumer M. Relevance of solid-state properties for pharmaceutical products, in polymorphism: in the pharmaceutical industry. Weinheim: Wiley; 2006.

Hind HW, Goyan FM. A new concept of the role of hydrogen ion concentration and buffer system in the preparation of ophthalmic solutions. J Am Pharm Assoc Am Pharm Assoc. 1947;36(2):33–41.

Hogben CA, Tocco DJ, Brodie BB, Schanker LS. On the mechanism of intestinal absorption of drugs. J Pharmacol Exp Ther. 1959;125(4):275–82.

Hosoya K, Lee VH, Kim KJ. Roles of the conjunctiva in ocular drug delivery: a review of conjunctival transport mechanisms and their regulation. Eur J Pharm Biopharm. 2005;60(2):227–40. https://doi.org/10.1016/j.ejpb.2004.12.007.

Hussaarts L, Muhlebach S, Shah VP, McNeil S, Borchard G, Fluhmann B, Weinstein V, Neervannan S, Griffiths E, Jiang W, Wolff-Holz E, Crommelin DJA, de Vlieger JSB. Equivalence of complex drug products: advances in and challenges for current regulatory frameworks. Ann N Y Acad Sci. 2017;1407(1):39–49. https://doi.org/10.1111/nyas.13347.

Jackson TL, Williamson TH. Amikacin retinal toxicity. Br J Ophthalmol. 1999;83(10):1199–200. https://doi.org/10.1136/bjo.83.10.1194f.

Jonas JB. Intravitreal triamcinolone acetonide: a change in a paradigm. Ophthalmic Res. 2006;38(4):218–45. https://doi.org/10.1159/000093796.

Kahook MY, Noecker RJ. Comparison of corneal and conjunctival changes after dosing of travoprost preserved with sofZia, latanoprost with 0.02% benzalkonium chloride, and preservative-free artificial tears. Cornea. 2008;27(3):339–43. https://doi.org/10.1097/ICO.0b013e31815cf651.

Kansy M, Senner F, Gubernator K. Physicochemical high throughput screening: parallel artificial membrane permeation assay in the description of passive absorption processes. J Med Chem. 1998;41(7):1007–10. https://doi.org/10.1021/jm970530e.

Kidron H, Vellonen KS, del Amo EM, Tissari A, Urtti A. Prediction of the corneal permeability of drug-like compounds. Pharm Res. 2010;27(7):1398–407. https://doi.org/10.1007/s11095-010-0132-8.

Kompella UB, Kim KJ, Lee VH. Active chloride transport in the pigmented rabbit conjunctiva. Curr Eye Res. 1993;12(12):1041–8.

Kovacs K, Wagley S, Quirk MT, Ceron OM, Silva PA, Singh RJ, Gukasyan HJ, Arroyo JG. Pharmacokinetic study of vitreous and serum concentrations of triamcinolone acetonide after posterior sub-tenon's injection. Am J Ophthalmol. 2012;153(5):939–48. https://doi.org/10.1016/j.ajo.2011.10.021.

Lallemand F, Schmitt M, Bourges JL, Gurny R, Benita S, Garrigue JS. Cyclosporine A delivery to the eye: a comprehensive review of academic and industrial efforts. Eur J Pharm Biopharm. 2017;117:14–28. https://doi.org/10.1016/j.ejpb.2017.03.006.

Lalwani GA, Berrocal AM, Murray TG, Buch M, Cardone S, Hess D, Johnson RA, Puliafito CA. Off-label use of intravitreal bevacizumab (Avastin) for salvage treatment in progressive threshold retinopathy of prematurity. Retina. 2008;28(3 Suppl):S13–8. https://doi.org/10.1097/IAE.0b013e3181644ad2.

Lederer CM Jr, Harold RE. Drop size of commercial glaucoma medications. Am J Ophthalmol. 1986;101(6):691–4.

Lee VH, Robinson JR. Topical ocular drug delivery: recent developments and future challenges. J Ocul Pharmacol. 1986;2(1):67–108.
Leibowitz HM, Kupferman A, Stewart RH, Kimbrough RL. Evaluation of dexamethasone acetate as a topical ophthalmic formulation. Am J Ophthalmol. 1978;86(3):418–23.
Lexi-Comp Inc., American Pharmacists Association. Pediatric & neonatal dosage handbook. Lexi Comp's drug reference handbooks. LexiComp; American Pharmacists Association, Hudson, OH, Washington, DC.
Liu Y, Kam WR, Ding J, Sullivan DA. One man's poison is another man's meat: using azithromycin-induced phospholipidosis to promote ocular surface health. Toxicology. 2014;320:1–5. https://doi.org/10.1016/j.tox.2014.02.014.
Lu DW, Chang CJ, Wu JN. The changes of vitreous pH values in an acute glaucoma rabbit model. J Ocul Pharmacol Ther. 2001;17(4):343–50. https://doi.org/10.1089/108076801753162753.
Maminishkis A, Jalickee S, Blaug SA, Rymer J, Yerxa BR, Peterson WM, Miller SS. The P2Y(2) receptor agonist INS37217 stimulates RPE fluid transport in vitro and retinal reattachment in rat. Invest Ophthalmol Vis Sci. 2002;43(11):3555–66.
Maren TH, Bar-Ilan A, Conroy CW, Brechue WF. Chemical and pharmacological properties of MK-927, a sulfonamide carbonic anhydrase inhibitor that lowers intraocular pressure by the topical route. Exp Eye Res. 1990;50(1):27–36.
Marra MT, Khamphavong P, Wisniecki P, Gukasyan HJ, Sueda K. Solution formulation development of a VEGF inhibitor for intravitreal injection. AAPS PharmSciTech. 2011;12(1):362–71. https://doi.org/10.1208/s12249-011-9591-4.
Martin FN Jr, Mims JL. Preparation of ophthalmic solutions with special reference to hydrogen ion concentration and tonicity. AMA Arch Ophthalmol. 1950;44(4):561–72.
Masuzawa K, Suzue S, Hirai K, Ishizaki T. 8-Alkoxyquinolonecarboxylic acid and salts thereof. US 5,043,450; 1991.
Matsumoto T, Hara M, Miyashita K, Kato Y. 8-Alkoxyquinolonecarboxylic acid hydrate with excellent stability and process for producing the same. US5880283; 1999.
Maurice DM. The tonicity of an eye drop and its dilution by tears. Exp Eye Res. 1971;11(1):30–3.
Maurice DM. Factors influencing the penetration of topically applied drugs. Int Ophthalmol Clin. 1980;20(3):21–32.
McNaught AD, Wilkinson A. IUPAC. Compendium of chemical terminology. 2nd edn (the "gold book"). Blackwell Scientific Publications, Oxford; 1997, 2006. https://doi.org/10.1351/goldbook.
Meyer K, Palmer JW. On the nature of the ocular fluids. Am J Ophthalmol. 1936;19(10):859–65. https://doi.org/10.1016/S0002-9394(36)92723-X.
Missel PJ, Horner M, Muralikrishnan R. Simulating dissolution of intravitreal triamcinolone acetonide suspensions in an anatomically accurate rabbit eye model. Pharm Res. 2010;27(8):1530–46. https://doi.org/10.1007/s11095-010-0163-1.
Mitra AK. Ophthalmic drug delivery systems. Drugs and the pharmaceutical sciences, vol. 58. New York: Dekker; 1993.
Mobasheri A, Barrett-Jolley R, Shakibaei M, Canessa CM, Martin-Vasallo P. Enigmatic roles of the epithelial sodium channel (ENaC) in articular chondrocytes and osteoblasts: mechanotransduction, sodium transport or extracellular sodium sensing? In: Kamkin A, Kiseleva I, editors. Mechanosensitivity in cells and tissues. Moscow: Academia; 2005.
Mortimer RG, Eyring H. Elementary transition state theory of the Soret and Dufour effects. Proc Natl Acad Sci U S A. 1980;77(4):1728–31.
Myers T, Chrai S. Parenteral fundamentals. Basic principles and applications of bioindicators. J Parenter Drug Assoc. 1980;34(3):234–43.
Myers T, Chrai S. Design considerations for development of steam in-place sterilization processes. J Parenter Sci Technol. 1981;35(1):8–12.
Myers T, Chrai S. Steam-in-place sterilization of cartridge filters in-line with a receiving tank. J Parenter Sci Technol. 1982;36(3):108–12.

Newman A, Wenslow R. Solid form changes during drug development: good, bad, and ugly case studies. AAPS Open. 2016;2(1):2. https://doi.org/10.1186/s41120-016-0003-4.
Nichols KK, Yerxa B, Kellerman DJ. Diquafosol tetrasodium: a novel dry eye therapy. Expert Opin Investig Drugs. 2004;13(1):47–54. https://doi.org/10.1517/13543784.13.1.47.
Olejnik OK, E.D.S. Compositions containing α-2-adrenergic agonist components. USA Patent; 14 July 2000.
Palkama A, Uusitalo H, Raij K, Uusitalo R. Comparison of the effects of adrenergic agonists and alpha-, beta 1-, beta 2-antagonists on the intraocular pressure and adenylate cyclase activity in the ciliary processes of the rabbit. Acta Ophthalmol. 1985;63(1):9–15.
Paterson CA, Pfister RR, Levinson RA. Aqueous humor pH changes after experimental alkali burns. Am J Ophthalmol. 1975;79(3):414–9.
Patton TF, Robinson JR. Influence of topical anesthesia on tear dynamics and ocular drug bioavailability in albino rabbits. J Pharm Sci. 1975;64(2):267–71.
Pawar P, Katara R, Mishra S, Majumdar DK. Topical ocular delivery of fluoroquinolones. Expert Opin Drug Deliv. 2013;10(5):691–711. https://doi.org/10.1517/17425247.2013.772977.
PDR Network LLC. PDR drug information handbook. Montvale: PDR, LLC; 2016.
Photosafety evaluation of pharmaceuticals. International conference on harmonisation of technical requirements for registration of pharmaceuticals for human use. 2019. https://www.ich.org/products/guidelines/.
Physicians' desk reference for ophthalmic medicines. Medical Economics, Montvale; 2000.
Pitkanen L, Ranta VP, Moilanen H, Urtti A. Permeability of retinal pigment epithelium: effects of permeant molecular weight and lipophilicity. Invest Ophthalmol Vis Sci. 2005;46(2):641–6. https://doi.org/10.1167/iovs.04-1051.
Prausnitz MR, Noonan JS. Permeability of cornea, sclera, and conjunctiva: a literature analysis for drug delivery to the eye. J Pharm Sci. 1998;87(12):1479–88.
Pusch M, Jentsch TJ. Molecular physiology of voltage-gated chloride channels. Physiol Rev. 1994;74(4):813–27. https://doi.org/10.1152/physrev.1994.74.4.813.
Raghava S, Hammond M, Kompella UB. Periocular routes for retinal drug delivery. Expert Opin Drug Deliv. 2004;1(1):99–114. https://doi.org/10.1517/17425247.1.1.99.
Raghaven KS, Ranadive SA, Gougoutas JZ, DiMarco JD, Parker WL, Davidovich M, Newman A. Gatifloxacin pentahydrate US6413969; 2002.
Ramsay E, Ruponen M, Picardat T, Tengvall U, Tuomainen M, Auriola S, Toropainen E, Urtti A, Del Amo EM. Impact of chemical structure on conjunctival drug permeability: adopting porcine conjunctiva and cassette dosing for construction of in silico model. J Pharm Sci. 2017;106(9):2463–71. https://doi.org/10.1016/j.xphs.2017.04.061.
Ramsay E, Del Amo EM, Toropainen E, Tengvall-Unadike U, Ranta VP, Urtti A, Ruponen M. Corneal and conjunctival drug permeability: systematic comparison and pharmacokinetic impact in the eye. Eur J Pharm Sci. 2018;119:83–9. https://doi.org/10.1016/j.ejps.2018.03.034.
Roberts JE. Screening for ocular phototoxicity. Int J Toxicol. 2002;21(6):491–500. https://doi.org/10.1080/10915810290169918.
Rocha G, Gulliver E, Borovik A, Chan CC. Randomized, masked, in vitro comparison of three commercially available tear film osmometers. Clin Ophthalmol. 2017;11:243–8. https://doi.org/10.2147/OPTH.S127035.
Rowe RC, Sheskey PJ, Cook WG, Fenton ME, American Pharmacists Association. In: Rowe RC, Sheskey PJ, Cook WG, Fenton ME, editors. Handbook of pharmaceutical excipients. 7th ed. London: APhA/Pharmaceutical Press; 2012.
Salenius P. A study of the pH and buffer capacity of blood, plasma and red blood cells. Scand J Clin Lab Invest. 1957;9(2):160–7.
Schoenwald RD, Huang H-S. Corneal penetration behavior of β-blocking agents I: physicochemical factors. J Pharm Sci. 1983a;72(11):1266–72. https://doi.org/10.1002/jps.2600721108.
Schoenwald RD, Huang HS. Corneal penetration behavior of beta-blocking agents I: physiochemical factors. J Pharm Sci. 1983b;72(11):1266–72.

Scholz M, Lin JE, Lee VH, Keipert S. Pilocarpine permeability across ocular tissues and cell cultures: influence of formulation parameters. J Ocul Pharmacol Ther. 2002;18(5):455–68. https://doi.org/10.1089/10807680260362731.

Scozzafava A, Menabuoni L, Mincione F, Briganti F, Mincione G, Supuran CT. Carbonic anhydrase inhibitors. Synthesis of water-soluble, topically effective, intraocular pressure-lowering aromatic/heterocyclic sulfonamides containing cationic or anionic moieties: is the tail more important than the ring? J Med Chem. 1999;42(14):2641–50. https://doi.org/10.1021/jm9900523.

Shen J, Lu GW, Hughes P. Targeted ocular drug delivery with pharmacokinetic/pharmacodynamic considerations. Pharm Res. 2018;35(11):217. https://doi.org/10.1007/s11095-018-2498-y.

Shirasaki Y. Molecular design for enhancement of ocular penetration. J Pharm Sci. 2008;97(7):2462–96. https://doi.org/10.1002/jps.21200.

Shiue MH, Kim KJ, Lee VH. Modulation of chloride secretion across the pigmented rabbit conjunctiva. Exp Eye Res. 1998;66(3):275–82. https://doi.org/10.1006/exer.1997.0459.

Shiue MH, Kulkarni AA, Gukasyan HJ, Swisher JB, Kim KJ, Lee VH. Pharmacological modulation of fluid secretion in the pigmented rabbit conjunctiva. Life Sci. 2000;66(7):PL105–11.

Shoghi E, Fuguet E, Bosch E, Rafols C. Solubility-pH profiles of some acidic, basic and amphoteric drugs. Eur J Pharm Sci. 2013;48(1–2):291–300. https://doi.org/10.1016/j.ejps.2012.10.028.

Shore PA, Brodie BB, Hogben CA. The gastric secretion of drugs: a pH partition hypothesis. J Pharmacol Exp Ther. 1957;119(3):361–9.

Sieg JW, Robinson JR. Vehicle effects on ocular drug bioavailability i: evaluation of fluorometholone. J Pharm Sci. 1975;64(6):931–6.

Sieg JW, Robinson JR. Vehicle effects on ocular drug bioavailability II: evaluation of pilocarpine. J Pharm Sci. 1977;66(9):1222–8.

Sieg JW, Robinson JR. Vehicle effects on ocular drug bioavailability III: shear-facilitated pilocarpine release from ointments. J Pharm Sci. 1979;68(6):724–8.

Singh SR, Carreiro ST, Chu J, Prasanna G, Niesman MR, Collette Iii WW, Younis HS, Sartnurak S, Gukasyan HJ. L-carnosine: multifunctional dipeptide buffer for sustained-duration topical ophthalmic formulations. J Pharm Pharmacol. 2009;61(6):733–42. https://doi.org/10.1211/jpp/61.06.0005.

Singhal D, Curatolo W. Drug polymorphism and dosage form design: a practical perspective. Adv Drug Deliv Rev. 2004;56(3):335–47. https://doi.org/10.1016/j.addr.2003.10.008.

Stahl PH, Wermuth CG, International Union of Pure and Applied Chemistry. Handbook of pharmaceutical salts: properties, selection, and use. 2nd rev ed. Weinheim: Wiley-VCH; 2011.

Tai MC, Lu DW, Chiang CH. Corneal and scleral permeability of quinolones--a pharmacokinetics study. J Ocul Pharmacol Ther. 2003;19(6):547–54. https://doi.org/10.1089/108076803322660468.

Thiagarajah JR, Verkman AS. Aquaporin deletion in mice reduces corneal water permeability and delays restoration of transparency after swelling. J Biol Chem. 2002;277(21):19139–44. https://doi.org/10.1074/jbc.M202071200.

Thompson AM. Ocular toxicity of fluoroquinolones. Clin Exp Ophthalmol. 2007;35(6):566–77. https://doi.org/10.1111/j.1442-9071.2007.01552.x.

Tomlinson A, Khanal S, Ramaesh K, Diaper C, McFadyen A. Tear film osmolarity: determination of a referent for dry eye diagnosis. Invest Ophthalmol Vis Sci. 2006;47(10):4309–15. https://doi.org/10.1167/iovs.05-1504.

Tomlinson A, McCann LC, Pearce EI. Comparison of human tear film osmolarity measured by electrical impedance and freezing point depression techniques. Cornea. 2010;29(9):1036–41. https://doi.org/10.1097/ICO.0b013e3181cd9a1d.

Tsuboi S, Kobayashi M, Nanba M, Imaoka S, Ohmori S. S-(1,2-dicarboxyethyl)glutathione and activity for its synthesis in rat tissues. J Biochem. 1990a;107(4):539–45. https://doi.org/10.1093/oxfordjournals.jbchem.a123082.

Tsuboi S, Ohnaka M, Ohmori S, Sakaue T, Ogata K, Itano T, Hatase O. Inhibition of platelet aggregation by S-(1,2-dicarboxyethyl)glutathione, intrinsic tripeptide in liver, heart, and lens. Arch Biochem Biophys. 1990b;279(1):146–50.

United States Pharmacopeial Convention. Committee of Revision. The United States pharmacopeia. Rockville: United States Pharmacopeial Convention, Inc.; 1979.

United States Pharmacopeial Convention. USP. NF: the official compendia of drug standards information on disk. Rockville: USPC.

de Vlieger JSB, Crommelin DJA, Tyner K, Drummond DC, Jiang W, McNeil SE, Neervannan S, Crist RM, Shah VP. Report of the AAPS guidance forum on the FDA draft guidance for industry: "drug products, including biological products, that contain nanomaterials". AAPS J. 2019;21(4):56. https://doi.org/10.1208/s12248-019-0329-7.

Wang W, Sasaki H, Chien DS, Lee VH. Lipophilicity influence on conjunctival drug penetration in the pigmented rabbit: a comparison with corneal penetration. Curr Eye Res. 1991;10(6):571–9.

Wilhelmus KR. The Draize eye test. Surv Ophthalmol. 2001;45(6):493–515.

Yerxa B. Progress in inherited retinal disease drug discovery and development: a foundation's perspective. Pharm Res. 2018;35(11):239. https://doi.org/10.1007/s11095-018-2514-2.

Yoshida F, Topliss JG. Unified model for the corneal permeability of related and diverse compounds with respect to their physicochemical properties. J Pharm Sci. 1996;85(8):819–23. https://doi.org/10.1021/js960076m.

Young S, Larkin G, Branley M, Lightman S. Safety and efficacy of intravitreal triamcinolone for cystoid macular oedema in uveitis. Clin Exp Ophthalmol. 2001;29(1):2–6.

Younis HS, Shawer M, Palacio K, Gukasyan HJ, Stevens GJ, Evering W. An assessment of the ocular safety of inactive excipients following sub-tenon injection in rabbits. J Ocul Pharmacol Ther. 2008;24(2):206–16. https://doi.org/10.1089/jop.2007.0099.

Zhang Y, Ren K, He Z, Li H, Chen T, Lei Y, Xia S, He G, Xie Y, Zheng Y, Song X. Development of inclusion complex of brinzolamide with hydroxypropyl-beta-cyclodextrin. Carbohydr Polym. 2013;98(1):638–43. https://doi.org/10.1016/j.carbpol.2013.06.052.

Ocular Suspension and Nanosuspension Products: Formulation Development Considerations

Vivek Agrahari and Onkar N. Singh

Abstract Drug delivery to the ocular diseases requires strategic approaches due to the existence of anatomical/static and physiological/dynamic barriers. Several ophthalmic conventional topical formulations are designed as solutions, suspensions, ointments, or emulsions to achieve an effective drug dose to the ocular tissues. In addition, various novel nanoformulation-based delivery systems have been explored through various routes of administration and showed promising results. In this book chapter, we briefly reviewed the routes of administration, ocular barriers with an emphasis on the topical route of drug administration, and the development of ocular suspension and nanosuspension formulations. The considerations in formulation development of suspension dosage forms are discussed to facilitate the development of safe, stable, and efficacious drug products. Moreover, the factors affecting the stability of suspensions/nanosuspensions and approaches to develop a stable product are summarized.

Keywords Ophthalmic formulations · Suspension · Nanosuspension · Ocular barrier · Topical delivery · Stability · Formulation development · Excipients

Abbreviations

API	Active pharmaceutical ingredient
EDTA	Ethylenediaminetetraacetic acid
LDPE	Low-density polyethylene
PSD	Particle size distribution
RPE	Retinal pigment epithelium

V. Agrahari · O. N. Singh (✉)
CMC/Mfg/Tech Ops, Alentia Therapeutics, Fort Worth, TX, USA
e-mail: onsingh2@yahoo.com

S. Neervannan, U. B. Kompella (eds.), *Ophthalmic Product Development*,
AAPS Advances in the Pharmaceutical Sciences Series 37,
https://doi.org/10.1007/978-3-030-76367-1_12

Introduction

It is of paramount importance to understand the anatomy and physiology of human eye from both the front of the eye (anterior) and back of the eye (posterior) perspectives in order to develop effective ophthalmic products. The eye is a complicated organ and protected from external materials by impermeable epithelium, tear secretion, and ocular drainage pathways to clear any foreign object. The eye is comprised of connective, vascular, and neural tissues. The connective tissue consists of the transparent cornea connected to the sclera through the limbus. The vascular tissue is composed of the choroid and ciliary body connected by the iris in front of the globe. The retina constitutes the neural tissue, which transmits the electrical impulse to the brain through the optic nerve. Anatomically, the eye is subdivided into anterior and posterior segments. The anterior segment includes the cornea, pupil, iris, ciliary body, conjunctiva, lens, and aqueous humor, whereas the posterior segment consists of the sclera, choroid, and retina, surrounding the vitreous cavity filled with the vitreous humor. The aqueous humor provides nutrients for the lens and cornea and maintains the intraocular pressure (Janagam et al. 2017). The structure of the eye is schematically represented in Fig. 1 (reproduced with permission from (Barar et al. 2016)).

Clinically, the anterior segment diseases are often treated by using solutions, suspensions, or ointments, however, the existence of anatomical/static (conjunctiva, cornea, sclera, blood aqueous, and retinal) and physiological/dynamic (choroid blood flow, efflux transporters, tear washing, nasolacrimal drainage) barriers limits the efficacy/bioavailability of these conventional dosage forms (Agrahari et al. 2017; Barar et al. 2016; Gaudana et al. 2010; Janagam et al. 2017). In addition to the eye's intrinsic ability to exclude external molecules, the undesirable physicochemical properties, such as the low aqueous solubility of drugs, impose a significant challenge to ensure a high therapeutic efficacy. Furthermore, an ideal ocular formulation should be self-administered (for topically applied dosage forms) and nonirritating to ensure high patient compliance.

Routes of Ocular Drug Administration and the Associated Barriers to Consider for Developing Ophthalmic Products

There are several routes of drug administration to the anterior segment of the eye: topical, intracameral, subconjunctival, and systemic (Janagam et al. 2017). The typical routes assessed for the posterior segment drug delivery are intraocular (suprachoroidal, intravitreal), topical, systemic, and periocular (subconjunctival, sub-Tenon, retrobulbar) (Agrahari et al. 2017; Peptu et al. 2015). These administration routes are briefly discussed below and schematically represented in Fig. 2. Depending on the route of administration, one or more ocular barriers need to be bypassed for drugs to reach the anterior or posterior segments.

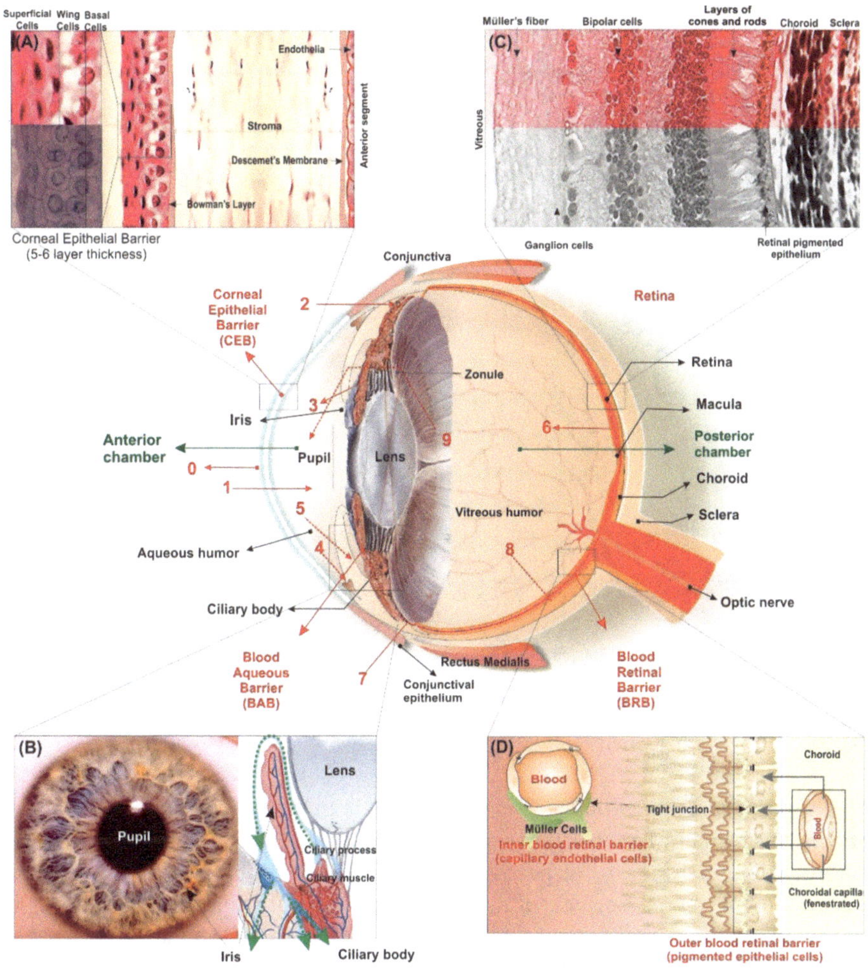

Fig. 1 Schematic demonstration of the anatomy and the biological membranes and barriers of the eye. Panels (**A**, **B**, **C**, and **D**) represent the corneal epithelial barrier (CEB), the blood-aqueous barrier (BAB), the biostructures of the retina, and the blood-retinal barriers (BRB), both the inner endothelial and outer pigmented epithelial barriers. (Reproduced with permission from Barar et al. (2016))

Intracameral administration is a local drug delivery method for direct injection into the anterior segment, avoiding the first-pass metabolism, cornea, conjunctiva, and blood-aqueous barriers (Janagam et al. 2017).

Subconjunctival administration administers drugs into the subconjunctival space around the outside of the sclera (Janagam et al. 2017). The drug then penetrates through the sclera and reaches to the anterior segment. It is a minimally invasive and local route avoiding the cornea and blood-aqueous barriers and the first-pass metabolism (Janagam et al. 2017).

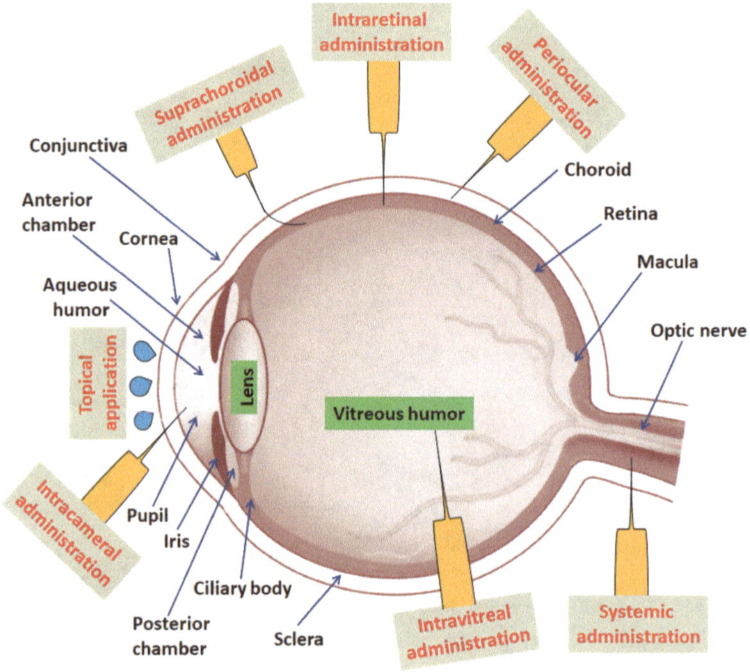

Fig. 2 Administration routes for delivering therapeutics to the anterior and posterior segments of the eye. (Adapted from Agrahari et al. (2017))

Systemic administration can deliver drugs to both the anterior and posterior segments, but with low bioavailability due to the presence of the blood-aqueous barrier and blood-retinal barrier, respectively (Agrahari et al. 2017; Barar et al. 2016; Janagam et al. 2017). Because of the presence of the tight junctional complexes in the two layers comprising the blood-aqueous barrier, it restricts the penetration of drugs from the blood into the aqueous humor. Thus, high doses are required to achieve therapeutic drug levels in the aqueous humor, which can cause adverse effects. In addition, efflux transporters expressed on the apical and basolateral cell membranes of the human retinal pigment epithelium (RPE) limit drug permeation from the choroid to the retina after systemic administration (Barar et al. 2016; Janagam et al. 2017).

Intravitreal delivery has the potential to provide the highest intraocular bioavailability by circumventing several barriers of the posterior eye segment due to its proximity to the retina, choroid, and RPE (Rowe-Rendleman et al. 2014). However, intravitreal administration is invasive and painful, and repeated injections are associated with risks of hemorrhage, retinal detachment, increased intraocular pressure, cataract formation, bacterial endophthalmitis, and degeneration of photoreceptors (Falavarjani and Nguyen 2013).

Periocular delivery is an emerging, less-invasive route and utilizes the transscleral pathway to deliver drugs next to the choroid. However, drug losses via conjunctival, episcleral blood, and lymphatic flow are the limiting factors to periocular administration (Peptu et al. 2015; Raghava et al. 2004).

Suprachoroidal injection is one of the most suitable routes to reach the choroid and vitreous humor (Rai Udo et al. 2015; Hartman and Kompella 2018) since the suprachoroidal space lies internal to the sclera and provides a natural route for drugs injected across the sclera along the inner surface of the eye into the posterior segment.

Topical route of drug administration to the eye is the most convenient and self-administrable route for the anterior segment and provides high patient compliance and minimal side effects. Depending on the formulation and drug physiochemical characteristics, drugs can reach to the cornea, conjunctiva, sclera, aqueous humor, iris, ciliary body, vitreous humor, and retina sites after topical instillation (Janagam et al. 2017). However, precorneal factors and anatomical barriers adversely affect the bioavailability of topical formulations (Fig. 3) (reproduced with permission from (Janagam et al. 2017)). Precorneal factors include solution drainage, blinking, the tear film, tear turnover, and corneal/conjunctival barriers. Due to these factors, only ~1–7% of the topically administered drugs can reach to the aqueous humor (Janagam et al. 2017; Ghate and Edelhauser 2006). The tear film, composed of water, electrolytes, and various proteins, is the first obstacle for topically administered drugs and consists of three layers: an outermost lipid layer, a thicker aqueous middle layer, and an innermost mucin layer. Human tear volume is about 3.4–10.7 μl per eye (Scherz et al. 1974) with a turnover rate of 0.5–2.2 μl/min (Janagam et al. 2017; Worakul and Robinson 1997; Mishima et al. 1966). The tear film has a rapid

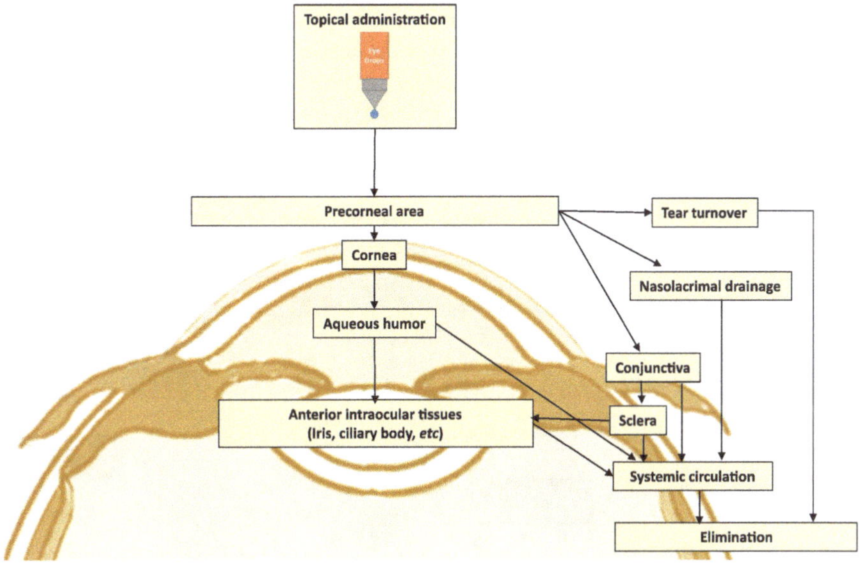

Fig. 3 Schematic representation of disposition of drug in the eye following topical administration. (Reproduced with permission from Janagam et al. (2017))

restoration time of 2–3 min (Janagam et al. 2017; Worakul and Robinson 1997). Due to the fast turnover rate of the tear film, the topically administered doses are quickly washed away and drained into the nasolacrimal duct after instillation. Due to these factors, the contact time of topically administered formulations with the ocular membranes is low, and less than 5% of the applied dose permeates the eye and reaches the intraocular tissues (Gaudana et al. 2010; Ali et al. 2016).

The cornea also limits the penetration of exogenous substances into the eye. It is composed of five layers: the epithelium, Bowman's membrane, stroma, Descemet's membrane, and endothelium (Janagam et al. 2017; Sridhar 2018). Each layer has a different polarity, and the epithelium, stroma, and endothelium layers form substantial barriers to drug penetration (Gaudana et al. 2010; Janagam et al. 2017). The corneal epithelium limits the permeation of hydrophilic molecules due to the hydrophobicity of the epithelium and the presence of tight junctional proteins between the corneal epithelial cells. The highly hydrated stroma poses a significant barrier for the penetration of lipophilic drugs. Most of the topical drugs permeate across the cornea to the aqueous humor, and from there, drugs distribute to the trabecular meshwork, iris, and ciliary body. However, the physical lenticular barrier, blood flow of the iris-ciliary body, and aqueous humor turnover limit drug distribution further to the vitreous and retina. Topically administered drugs can also be absorbed into the anterior segment through a non-corneal conjunctiva/sclera pathway (Ahmed et al. 1989; Ahmed and Patton 1985). The sclera has a large surface area and comparatively high permeability than the cornea. The trans-scleral permeation primarily depends on the size of the molecules rather than the lipophilicity. A schematic representation of the disposition of drug in the eye following various routes after ocular administration is provided in Fig. 4.

To improve the ocular bioavailability, various conventional (suspension, emulsion, ointments, aqueous gel) and novel drug delivery systems (nanosuspension, nanomicelle, nanoparticle, liposome, dendrimer, implant, microneedle, and in situ

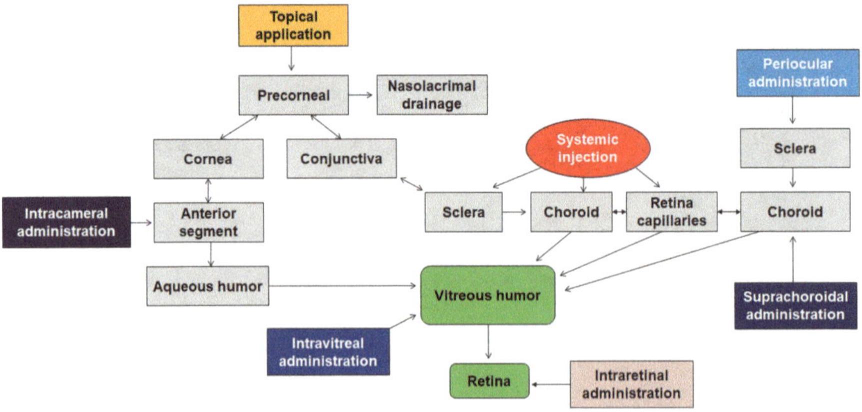

Fig. 4 Pathways for distribution of drug to the eye following different delivery routes. (Adapted from Agrahari et al. (2017))

thermosensitive gel) are explored through various routes of administration and showed promising clinical/nonclinical results. An ophthalmic topical formulation could be designed as solution, suspension, ointment, or emulsion. Since the duration of drug action from eye drop solutions is relatively short, frequent drug administration is needed. Therefore, patient compliance is low, and thus, patient-friendly and long-acting topical delivery systems are needed. Several novel ocular drug delivery systems, as discussed by Barar et al. (2016), represent the recently developed products/devices for the treatment of anterior and posterior segment diseases. However, considering the scope of this book chapter, only the development of suspension and nanosuspension formulations for the anterior/posterior segment eye diseases is discussed.

Ophthalmic Suspension Formulation

Suspension dosage forms are dispersions of finely divided undissolved drug particles in an aqueous vehicle containing suitable suspending and dispersing agents. Suspension dosage forms offer distinct advantages in increasing the corneal contact time of drugs and thus provide a more sustained therapeutic action compared to solutions (Patel et al. 2013; Kaur and Kanwar 2002). These may also improve the stability, bioavailability, and efficacy of hydrophobic molecules. However, the formulation of an ophthalmic suspension is complex, challenging, and requires understanding of the properties of the dispersed phase and the dispersion medium.

Target Product Profile (TPP) and Desirable Attributes

Ophthalmic suspension formulations must fulfill the crucial requirements of safety, efficacy, stability, manufacturability, and bioavailability. In addition to these, special attention should be given to other formulation factors (components) that may affect ocular tolerability and safety. A typical ophthalmic product is sterile, nearly isotonic, contains preservatives, and is packaged into a suitable tamper-evident multidose dispensing system or form-fill-seal (FFS) package for unit dose. The ophthalmic suspension product development criteria below describe critical steps and necessary studies to develop the formulation with desired attributes, meeting pharmaceutical and regulatory requirements. In general, the desirable attributes of an ophthalmic suspension product are:

- Safe, effective, and stable during the shelf life of the product
- A particle size $\leq$10 μm in order to minimize the ocular irritation. Ideally, the particles (based on their shape, size, etc.) should not cause irritation to the eye
- Physical attributes, such as particle size, size distribution, and formulation viscosity, should remain uniform throughout the shelf life of the product

- The drug should not have a quick sedimentation rate and must suspend easily upon shaking without forming a cake
- Resuspension should produce a homogeneous mix of drug particles to provide a reproducible content uniformity with each dose administered
- The formulation viscosity must promote uniform flow from the container
- Multidose product must meet regulatory criteria for preservative effectiveness
- Must be sterile and endotoxin free for both anterior and posterior eye segment usage

The first step in rational product developmentis to construct the quality TPP that identifies quality attributes critical for product performance. The elements of a good TPP for an ophthalmic suspension formulation should consider:

- Route of administration
- Safety and efficacy
- Target pH
- Drug/formulation stability
- Preservation for multidose products
- Package type (bottle size, fill volume, types of plug)
- Dosing frequency and dosing protocol (administration with or without shaking)
- Ease of manufacturing process
- Scalable and cGMP manufacturing capability
- Shelf life and storage conditions
- Sterility and endotoxin levels (<0.5 EU/mL)
- Target population and distribution market

Key Considerations in the Development of Ophthalmic Suspension and Nanosuspension Formulations

In order to design an ophthalmic suspension product that addresses the above TPP and desirable attributes, a systematic approach is needed in identifying a prototype formulation during the product development phase. The important aspects when considering the development of dosage forms for ocular therapeutics are duration of therapy, intended targeted tissue, safety, and patient compliance. The first step in suspension product development, once the TPP and desired formulation profiles/attributes are identified, is establishing its physical and chemical attributes such as appearance, viscosity, osmolarity, resuspendability, and pH. Understanding of the interfacial, wetting, particle interaction, surface charge, aggregation, sedimentation, and rheological properties is required to formulate an effective and aesthetically good suspension. The choice of excipients and pH in formulation development should be based on the physiological comfort, product stability, and efficacy requirements. Accordingly, the formulation factors and processing parameters affecting physical and chemical stability should be considered. The critical factors that need to be considered during the formulation of ocular suspensions are discussed below.

Physical Properties of the Active Pharmaceutical Ingredient (API)

The critical issues in the development of a suspension formulation related to the physicochemical properties are non-homogeneity of the dosage form, settling, cake formation, aggregation of the suspended particles, and resuspendability issues. A continuous mixing is often required during the manufacturing and filling process to assure homogeneity of the dosage form. Considering these issues, one of the early preformulation activities in suspension formulation development is to evaluate the drug physicochemical properties, such as pKa, LogP, solubility in various solvents, dissolution rate, chemical stability of the solid and solution (pH-dependent) forms of the drug, polymorphism, melting point, density, particle size, hygroscopicity, surface area, and flow characteristics. The ionization constant is an important parameter in ocular absorption of drugs since it is predominantly the unionized form determines the extent of bioavailability, though both the ionizable and the unionized forms may diffuse across ocular membranes. Therefore, selecting the functional groups that maximize the unionized fraction at physiological pH without compromising solubility, stability, and potency, is important. The interfacial properties of the suspended drugs are also important, and the low interfacial tension particles are easily wetted by water and can be easily suspended. However, high interfacial tension particles are not easily wetted and need surfactants to increase the wettability of the particles by reducing the surface tension. Ideally, the drug should be insoluble in the continuous phase to develop a suspension dosage form; however, since many drugs are suitably soluble in the continuous phase, the problem is a consequence of storage temperature variations, which can lead to supersaturation and crystal growth (Ostwald ripening). This can be neutralized by the use of crystallization inhibitors such as povidone. Drug storage temperature, humidity, and packaging materials require evaluation as part of the formulation development process. Preformulation studies are important to carry out in this regard to characterize the drug substance. A list of such preformulation studies is summarized in Table 1.

Particle Size of the API

The particle size used in ocular suspensions is of primary importance due to its relationship with the ocular irritation and in formulating physically stable suspension. Drug particle size influences product appearance, settling rates, drug solubility, rate and extent of dissolution, in vivo absorption, resuspendability, and overall

Table 1 Preformulation studies in ophthalmic suspension formulation

Parameters	Rationale
Ionization constant (p*K*a)	To determine aqueous solubility, assess drug classification in terms of its solubility and ocular tissue permeability, and identify the best formulation feasible candidate
Log *P*	To determine aqueous solubility, assess drug classification in terms of its solubility and tissue permeability, and identify the best formulation feasible candidate
Interfacial properties of the drugs	To determine wettability by water and therefore suspendibility
pH-dependent solubility	To determine intrinsic solubility and pH solubility at ocular pH conditions
Solvent compatibility of drug	To determine solubility in various solvents/buffers
Excipient compatibility	To determine the best excipient for a particular formulation and drug
Effect of common ion on drug solubility	To determine intrinsic solubility and pH solubility at ocular pH conditions
Physicochemical stability at various pH and temperature conditions	To determine the best buffer conditions and storage/packaging criteria and excipient selection criteria
Particle size of the drug	To determine potential ocular irritation and in formulating physically stable suspension
Polymorphism	To determine the crystal structure and effect of the manufacturing and processing parameters on drug particle size, safety, bioavailability, and drug/formulation stability
Photostability	To determine the light effect on storage, efficacy, and stability of drugs including the excipients
Sterilization effect	To determine the best method of sterilization and any detrimental effect on the drug and excipient properties
Preservative compatibility and efficacy testing	To determine concentration and storage condition dependent preservative effectivity against a wide spectrum of microorganisms and compatibility with formulation and packaging components and formulation characteristics
Packaging compatibility	To assess packaging compatibility drug, preservative, excipients, and final formulation

stability. In general, the drug particle size of <10 μm is recommended for ophthalmic suspension formulations to facilitate patient comfort, to minimize the damage to the cornea (Kaur and Kanwar 2002; Missel 2012), and to ensure that the suspension does not lead to irritation (foreign body sensation) of the sensitive ocular tissues. This is also important to ensure that uniform dosage is delivered to the eye since the drug solubility is favored by smaller particle size. However, other factors such as particle concentration, density, and shape may also contribute to the comfort threshold of the patients.

The processes to achieve the desired particle size distribution (e.g., grinding, air-jet micronization, wet milling with ceramic beads, spray drying, precipitation from supercritical fluid, and controlled crystallization) can affect the properties of the drug product. For example, comminution (grinding or milling) methods may generate heat that can cause polymorphic changes and the size of the drug particles, which can affect the dissolution and drug delivery features. The comminution of the particles results in the increase of the surface area and, hence, free surface energy, which makes the suspension thermodynamically unstable. In addition, the effect of proposed sterilization methods on the drug properties should be assessed. These preliminary evaluations indicate the optimal particle size of the API, size reduction and solubilization methods, and buffer pH range to provide a stable suspension formulation. The observations during the preformulation development are important and need to be considered in designing the scale-up manufacturing activities.

Drug particles may also exist in different crystalline forms (polymorphism) in a suspension dosage form. A change in crystal structure and particle size may occur during storage or manufacturing process, causing alterations in the solubility and bioavailability. Hence, the size distribution of particles and aggregates of drugs in a suspension formulation should be controlled in order to provide uniformity in the dosing and reproducible drug delivery characteristics. Thus, the potential for any changes in particle size due to Ostwald ripening or particle agglomeration needs to be evaluated. However, it is desirable to keep the particles below the recommended size of <10 μm (d_{90}) for topical ocular administration (Kaur and Kanwar 2002; Missel 2012) if a prolonged drug delivery duration is desired to minimize potential irritation. For injectable formulation to the back of the eye, the optimum particle size is larger than the optimal size for topical administration of 10 μm (d_{90}). The preferred size for injectable formulation is generally between 30 and 100 μm, and the preferred shape is rod-shaped particles (Thackaberry et al. 2017). The duration of drug action for suspension is particle size dependent. In addition to controlling the particle size, the drug crystal form selected should be thermodynamically the most stable form. Hence, performing the polymorphism form conversion studies under various processing, storage, and stability conditions should be a part of the preformulation activities. The most commonly used analytical techniques to characterize polymorphic conversion are X-ray powder diffraction (XRPD), differential scanning calorimetry (DSC), and microscopy-based methods.

While the ocular retention increases with an increase in the particle size, the rate of dissolution of the suspended drug increases with decreasing particle size. Thus, an optimum particle size has to be selected based on the therapeutic agent used. The compendial requirements of particle size specification in EP, JP, and USP are:

EP: Particles with diameter 20–50 μm should be 20 or less per 10 μg active ingredient. Particles with diameter 50–90 μm should be 2 or less per 10 μg active ingredient. Particles with diameter 90 μm or more should not be observed per 10 μg active ingredient.

JP: No particles >75 μm.

USP: Solid particles must be smaller than 5–10 μm to avoid ocular discomfort or irritation.

Role of Excipients

Ophthalmic suspension contains several inactive excipients, such as dispersing and wetting agents, suspending agents, buffering agents, tonicity agents, and preservatives. The selection of these materials is generally based on the route of administration, drug dose, drug physicochemical characteristics, excipient safety, and any possible adverse effects. Depending on their physicochemical properties, excipients such as surface-active agents can play several roles, ranging from wetting agents, stabilizers, solubilizers, preservatives (antimicrobial agents), to, potentially, corneal permeation enhancers (Ibrahim 2019). However, sometimes, the excipient use is limited by their potential toxicity, irritancy to the ocular tissues, and unwanted interactions with other excipients or drug. Hence, understanding the mechanisms of their different roles and the interactions with other formulation components can help determine their safe and effective concentration intended for ocular application. The amount of surfactant should be carefully evaluated, as excessive amounts can lead to eye irritation, foam formation during manufacturing and upon shaking the product, or affect the interactions with other excipients. A summary of various excipients and their recommended levels in ophthalmic suspension formulation is provided in Table 2.

Viscosity-Modifying (Enhancing) Agents

Viscosity of ocular topical (suspension) formulation is one of the most important factors. Increasing the formulation viscosity may reduce the drainage rate, prolong the precorneal residence time, enhance ocular absorption, and control the rate at which the drop flows out of the container, thus enhancing the ease of application. The viscosity of ocular formulations must be maintained to a certain level to avoid any blockage of the lacrimal ducts. The reported critical formulation viscosity threshold is 55 mPa/s, and no further increase in contact time between the dosage form and the eye occurs above this level (Jones 2016). The viscosities of commercially available products are frequently <30 mPa/s; otherwise, discomfort due to blurred vision and foreign body sensation occurs, resulting in a faster elimination due to reflex tears and blinks (Salzillo et al. 2016; Jones 2016). Polymers such as methylcellulose, polyvinyl alcohol, polyvinylpyrrolidone, polyethylene glycol, sodium carboxymethyl cellulose, and hydroxypropyl methylcellulose are common viscosity-enhancing agents of ocular formulations.

Wetting and Solubilizing Agents

Surface-active agents are predominantly employed in suspension to effectively disperse the drug during manufacture and product use and to enhance the physical stability of the dispersed particles. The wetting and solubilizing agents (to lower the

Table 2 Excipients and their recommended levels in ophthalmic suspension formulation

Category	Excipients	Recommended levels
Wetting/ solubilizing agents	Tweens (polysorbates 20/40/60/80)	1% w/w
	Spans (sorbitan monolaurate/ monooleate/monopalmitate)	
	Sodium lauryl sulfate	0.1–2% w/v
Viscosity modifiers/ suspending agents	Hydroxypropyl methylcellulose (hypromellose)	0.45–1.0% w/w
	Methylcellulose	2% w/w
	Poly(vinyl alcohol)	0.25–3.00% w/w (concentration dependent on the molecular weight and typically used at 1.4% w/w)
	Poly(acrylic acid)	
	Polyvinylpyrrolidone or povidone	1.7% w/w
	Hydroxyethyl cellulose	0.8% w/w
pH-modifying buffers	Citrate, phosphate, borate, or acetate buffers	Variable
Preservatives	Benzalkonium chloride	0.002–0.02% w/v (typically 0.01% w/v)
	Benzethonium chloride	0.01–0.02% w/v
	Cetrimonium bromide	0.005% w/v
	Esters of parahydroxybenzoates (parabens); mixtures of methyl and propyl esters of parahydroxybenzoic acid	Typically at a combined concentration of 0.2% w/w
	Organic mercurial compounds	0.001–0.002% w/v for phenylmercuric acetate, 0.002% w/v for phenylmercuric nitrate, and 0.001–0.004% w/v for thimerosal
	Organic alcohols (phenoxyethanol, chlorobutanol, phenylethyl alcohol)	0.25–0.5% w/v
Antioxidants	Sodium metabisulfite	0.1%
	Ethylenediaminetetraacetic acid (EDTA)	0.1%
	Sodium bisulfite	0.1%
	Thiourea	0.1%
Tonicity agents	Dextrose	5.51% (isoosmotic conc.)
	Glycerin	2.6% (isoosmotic conc.)
	Sodium chloride	0.9% (isoosmotic conc.)
Clarifying agents	Polysorbate 20/80	Max 1% w/w
	HPMC	Max 1% w/w

contact angle between the solid surface and the wetting liquid and improve the solubility of poorly water-soluble drugs) that are generally used include Tweens (polysorbates), Spans (sorbitan monolaurate/monooleate/monopalmitate), and sodium lauryl sulfate. Nonionic surfactants are generally preferred because of their less toxicity compared to ionic surfactants.

Suspending Agents

Suspending agents prevent sedimentation by affecting the rheology of suspensions. These polymers adsorbed on the surface of the particle, creates a steric effect by preventing the individual particles from getting sufficiently close to each other so that they are prevented from getting to the primary minimum (DLVO theory, explained later in this chapter), and thus coagulate/aggregate and settle out as a deflocculated sediment that is difficult to redisperse. Since the driving force for the adsorption of these polymers would be a reduction in interfacial energy, the polymers that do adsorb onto the surface of the particles must be able to bridge the energy gap. Thus, polymers that are amphiphilic in nature (have both hydrophilic and lipophilic groups) (e.g., poloxamers) are required. In ophthalmic suspensions, methylcellulose, sodium carboxymethylcellulose, hydroxypropyl methylcellulose, and synthetic polymers such as carbomers, poloxamers, and polyvinyl alcohol are generally used as suspending agents.

pH Buffering Agents

The ocular formulation pH is an important determinants of the stability of the drugs and the drug absorption across the cornea. Ideally, the pH of the ocular suspension should be controlled at or around 7.4 (physiological pH of tear fluid) (Missel 2012) using the appropriate buffer system or vehicle while not causing any physical or chemical instability to the drugs. However, the outer surfaces of the eye can tolerate a wide pH range of 3.5–8.5 (Ammar et al. 2009), but the normal range to prevent corneal damage is 6.5–8.5. The drug p*K*a determines the ionization of the therapeutic agent at defined pH values. To be effectively absorbed, the drug must exhibit in the ionized and non-ionized forms. The non-ionized form is required to partition into and diffuse across the lipid-rich outer layer of the cornea (the epithelium), whereas the inner layer of the cornea (the stroma) is predominantly aqueous, and therefore, the ionized form of the drug is needed. The non-ionized drug then diffuses to the endothelium/aqueous humor interface where ionization and dissolution into the aqueous humor occur.

Tonicity Agents

An isotonic ophthalmic formulation is with the tonicity equal to that of a 0.9% NaCl solution (290 mOsm). However, the osmotic pressure of the aqueous intraocular fluid is slightly higher than that of normal tears (~305 mOsm) (Missel 2012). The external eye is much more tolerant of tonicity variations and usually can tolerate solutions equivalent to 0.5–1.8% NaCl (Missel 2012). However, tear fluid in some cases of dry eye keratoconjunctivitis sicca is reported to be hypertonic, and a hypotonic artificial-tear product is used to counteract this condition.

Clarifying Agents

Ophthalmic formulations must be free from foreign particles, which are generally accomplished by filtration (helps to achieve clarity of the product). Particles in ophthalmic formulations can cause damage to the eye by causing abrasions to the cornea or the eyelid membranes. Suitable clarifying agents such as polysorbate 20/80 and HPMC may be used in ocular formulations.

Preservatives

The addition of preservatives is required to prevent the growth of the microorganisms since the products can be contaminated with microorganisms during the therapeutic uses (for multiuse products) and manufacturing/filling processes. In general, an ideal preservative should be effective at low concentration against a wide spectrum of microorganisms, soluble in the formulation at the desired concentration, nontoxic, compatible with formulation and packaging components, not have any effect on the viscosity or formulation characteristics, and stable over a wide pH range and temperature conditions. The commonly used preservatives are cationic, surface active and ionizable, and as a result their performance can be affected by the pH, ionic strength, presence of ionized components, and the adsorption of the preservative to the surface of the suspended solid particles. Therefore, the compatibility of the preservatives with suspension vehicle, excipient, and drug needs to be assessed in advance with a suitable pH range, ionic strength, surfactant, and polymer, to significantly reduce the formulation development time. In addition to the compatibility, it is necessary to study the effect of other formulation excipients on preservative effectiveness as well as the physical stability. The efficacy of the preservatives must be assessed using the appropriate pharmacopoeial method. The concentration of the preservative should be optimized to provide adequate efficacy with minimal concentration-dependent toxicity. Other factors such as the loss of preservative to sorption in processing, adequate control of pH and temperature during processing, and the order of component addition can affect the preservative efficacy. The key preservatives of ophthalmic suspension products are briefly discussed below (Missel 2012; Kulshreshtha et al. 2010; Ibrahim 2019; Jones 2016).

Cationic preservatives: The common cationic preservatives are benzalkonium and benzethonium chlorides. Benzalkonium chloride is typically used at a concentration of 0.01% w/v in ocular suspensions (range between 0.002% and 0.02% w/v). However, the resistance of certain microorganisms to benzalkonium chloride (e.g., *Pseudomonas aeruginosa*) has been observed. Therefore, 0.1% w/v disodium edetate (disodium EDTA) is used to enhance the antimicrobial activity of benzalkonium chloride by chelating divalent cations in the outer membrane of the bacterial cell, thereby rendering the bacteria more permeable to the diffusion of the antimicrobial agent. Also, the antimicrobial properties of benzalkonium chloride decrease at $pH < 5.0$ (Jones 2016). Benzethonium chloride exhibits lower antimicrobial activity than benzalkonium chloride and commonly used within the concentration range of 0.01–0.02% w/v.

Esters of parahydroxybenzoates (parabens): Mixtures of methyl and propyl parabens, typically at a combined concentration of 0.2% w/w, are used in ophthalmic formulations. The concern regarding their ocular usage is the irritancy of the parabens, which limits their use in ophthalmic preparations.

Organic alcohols: Chlorobutanol and phenylethyl alcohol are the two primary agents in this category. Under alkaline conditions, hydrolysis of chlorobutanol occurs, releasing HCl, thus preferred to be used only for acidic ophthalmic preparations. Also, the formulations must be stored in glass containers since chlorobutanol is volatile and lost from solution if stored in polyolefin containers. Another issue with the use of chlorobutanol is its limited solubility (typically used at the level of 0.5% v/v and saturation solubility is 0.7% w/v at room temperature). Therefore, below room temperature, precipitation of the preservatives may occur. Phenylethyl alcohol has similar properties and issues as of chlorobutanol, such as poor solubility, volatility, and partitioning into plastic containers. The typical concentration of phenylethyl alcohol used in ophthalmic preparations is 0.25–0.50% v/v.

Organic mercurials: Phenylmercuric nitrate, phenylmercuric acetate, and thimerosal are compounds in this category. The phenylmercurics are reported to have deposited in the lens of the eye, and thimerosal has been associated with ocular sensitization, thus not the first options as preservative (Jones 2016).

Antioxidants: Antioxidants, such as sodium metabisulfite, sodium sulfite, and EDTA, are added to ocular suspensions to enhance the stability of drugs that are susceptible to oxidation or degradation by free radicals. However, the acceptance criteria for antioxidant content should be established based on the levels necessary to maintain the product's stability throughout its proposed usage and shelf life.

Chelating agents: Most commonly used chelating agent in ophthalmic suspension is ethylenediaminetetraacetic acid (EDTA). With this mechanism, EDTA can also enhance stability of the active drug by sequestering the heavy metal ions and thus serves as an antioxidant for drugs that have their oxidation catalyzed by heavy metals. EDTA has multiple functions, as a buffer for free divalents and preventing their buildup in the cornea while also enhancing the antimicrobial action of other preservatives.

Preservative's Safety and Efficacy Assessment in Ocular Formulation

Appropriate care should be taken on selecting preservatives at lowest possible but effective concentration because of the high sensitivity of ocular tissues. Quaternary ammonium compounds such as benzalkonium chloride are capable of destroying bacteria and mycoplasma by binding to their negatively charged cell membrane followed by dissociation and leakage of cellular contents. Unfortunately, this effect is also capable of causing injuries even to ocular epithelial cells, especially at high concentrations. However, not only the preservative but also several other factors interplay to determine formulation's toxicity including the types and concentrations of the excipients, dosing frequency, and the residence time on the ocular surface. Additional formulation factors that can be adjusted to affect the preservative efficacy and toxicity at low concentrations are the storage temperature, processing parameters, and packaging components; thus, there is also a need of an appropriate optimization of these parameters. The ophthalmic suspension formulations must meet regulatory jurisdictional requirement of preservative effective test (PET) for multidose products at initial and throughout the product shelf life (Tables 3, 4, 5, and 6). Out of the regulatory criteria, the EPA PET criteria is the most astringent in the order of EPA > EPB > USP > JP.

Table 3 Preservative effectiveness test (PET) as per regulatory jurisdictions. USP Criteria for Tested Microorganisms (United States Pharmacopeia. USP <51>. Antimicrobial effectiveness testing. Rockville, MD)

For category 1 (sterile parenteral) products	
Bacteria	Not less than 1.0 log reduction from the initial calculated count at 7 days, not less than 3.0 log reduction from the initial count at 14 days, and no increase from the 14 days' count at 28 days
Yeast and molds	No increase from the initial calculated count at 7, 14, and 28 days

Table 4 Preservative effectiveness test (PET) as per regulatory jurisdictions. EP criteria for tested microorganisms (European Pharmacopeia. EP <5.1.3>Efficacy of antimicrobial preservatives)

Log reduction						
		6 h	24 h	7 day	14 day	28 day
Bacteria	A	2	3	–	–	NR
Bacteria	B	–	1	3	–	NI
Fungi	A	–	–	2	–	NI
Fungi	B	–	–	–	1	NI

NR no recovery, *NI* no increase, *A* recommended, *B* mandatory

Table 5 Preservative effectiveness test (PET) as per regulatory jurisdictions. JP criteria for tested microorganisms (Japanese Pharmacopeia. JP <19> Preservative effectiveness tests)

For category 1A (sterile parenteral) products	
Bacteria	14 days: reduction of 0.1% of inoculum count or less
Bacteria	28 days: Same or less than level after 14 days
Yeast and molds	14 days: Same or less than level after 14 days
Yeast and molds	28 days: Same or less than level after 14 days

Table 6 Preservative effectiveness test (PET) as per regulatory jurisdictions. The temperature of incubation and duration of microbial growth on the petri plates for each organism, defined in each compendia

Cultures	Temperature	Duration USP	Duration EP	Duration JP
Bacteria	30–35°	3–5 days	3–5 days (refers to the microbial limit test)	≤3 days (refers to the microbial limit test)
S. aureus				
P. aeruginosa				
E. coli				
Yeast	20–25°	3–5 days	5–7 days (refers to the microbial limit test)	≤5 days (refers to the microbial limit test)
C. albicans				
Mold	20–25°	3–7 days	5–7 days (refers to the microbial limit test)	≤5 days (refers to the microbial limit test)

Sterility

Ophthalmic suspensions must possess appropriate sterility with consideration given to preservation, osmolality, buffering capacity, viscosity, and packaging. Suspension products may pose challenges during manufacturing to achieve a sterile product since the possibilities of either degradation or morphological changes may occur during the sterilization process. Hence, the effect of sterilization methods (e.g., dry heat, autoclaving, ethylene oxide treatment, and gamma irradiation) on the drug and formulation properties should be assessed. The selection of sterilization procedure depends upon the nature of the dosage form, and a combination of methods can be used for ophthalmic products. Although it is preferable to sterilize ophthalmic formulations in their final container by autoclaving, this method may not be a suitable approach for thermally unstable drugs or formulations. As alternative aseptic manufacturing methods such as aseptic filtration, irradiation, or formulation of dosage form under aseptic conditions may also be applied.

The commonly used techniques in the formulation of a sterile suspension product are autoclaving (wet steam), dry heat, aseptic filtration, ethylene oxide, and irradiation. These all have their specific advantages and limitations; for example, the autoclaving and dry heat methods can only be used for thermostable products because of the high heat involved in these approaches. The aseptic filtration (generally through a 0.22 μm size filter) cannot be efficiently utilized for suspension products due to non-filterability of suspended particles especially of larger sized ones and of the higher viscosity products. The ethylene oxide method is advantageous for thermolabile molecules/products; however, the elimination of residual ethylene oxide from the product is time-consuming and challenging. The gamma radiation method can have degradative impact on the drug or excipients including the safety concern for human uses and therefore is not used much.

Container/Closure System

The container/closure characteristics of an ocular suspension product should be evaluated with a prototype formulation in order to demonstrate suitability of the final container/closure system. The tests to evaluate the protection for the formulation provided by the container/closure, the safety and compatibility of the container/closure, and the performance of the container/closure system are: light transmission, water vapor permeation, seal integrity/leakage, monitoring of extractable/leachable, evaluation of loss of excipients, and dosing uniformity. The most common container (bottle) is made of low-density polyethylene (LDPE), either natural/clear or opaque color, so it can be easily squeezed to deliver the required dose. In general, the round plug made of polyethylene and polypropylene closure is used in ophthalmic suspension products.

Nanosuspensions

As discussed earlier, ocular drug delivery is challenging, and the most conventional formulations are unable to efficiently deliver the drugs into the targeted areas due to the presence of several complex barriers and elimination mechanisms, which resulted in a significantly low bioavailability of the drugs. Nanotechnology became a common approach for pharmaceutical product development, including for suspension dosage form. Nanosuspensions are colloidal dispersion of submicron particles stabilized by polymers or surfactants. These systems are emerged as promising strategy for delivery of hydrophobic drugs in enhancing the retention time in precorneal tissues and improving the drug bioavailability due to the high solubility of nanosuspension formulations. Nanosuspensions can be solid or crystalline drug nanosuspensions (consist of crystalline nano-sized drug particles, stabilized with the help of surfactants or polymers) or polymer-coated drug nanosuspensions (drug is coated or encapsulated within a polymer matrix). Crystalline or solid nanosuspensions are preferred in terms of the minimal requirement for excipients, high drug loading, and ease of scale-up manufacturing. Nanosuspensions can be manufactured using the top-down techniques (using high-pressure homogenization, media milling, etc.) and bottom-up approaches (molecules are assembled to form nanoparticles using solvent-antisolvent method, emulsification solvent evaporation technique, lipid emulsion/microemulsion template, super critical fluid process, etc.) (Lai et al. 2015; Patravale et al. 2004; Rabinow 2004).

One of the primary reasons for a wide drug delivery application of nanosuspensions is their ability to provide formulations of poorly soluble drugs with higher dissolution rates because of their small particle size and thus high surface area. In general, the nanosuspension approach offers the following advantages in ocular drug delivery (Maharjan et al. 2019; Patravale et al. 2004; Rabinow 2004; Sutradhar et al. 2013; Yadollahi et al. 2015): (1) ease of application, (2) lesser eye irritation as smaller nano-sized particles are better tolerated than larger particles, (3) enhancement in the bioavailability of the drugs and thus reduction in the amount of dose, (4) increased precorneal residence time, and (5) enhancement of the physicochemical drug stability. Thus, nanosuspension is an effective and convenient approach in ocular drug delivery, offering maintained therapeutic drug concentration, reduced administration frequency, and increased patient compliance.

Manufacturing Process of Nanosuspension Formulations

Nanosuspensions can be manufactured by several processes broadly categorized as top-down, bottom-up, and combination of these two processes. The top-down approach consists of breaking the bigger particles into smaller ones using different milling techniques, such as media milling, high-pressure homogenization, and microfluidization. Though there is no use of toxic/harsh solvents and high drug loading can be achieved, these methods are high-energy processes with the generation of a lot of heat, and therefore, the processing of thermolabile materials is challenging. The bottom-up (precipitation) processes refer to the generation of small nano-sized particles from their molecular solutions using various approach such as solvent-antisolvent, supercritical fluid, emulsification-solvent evaporation, and spray drying. These can be carried out at ambient temperatures, and therefore, thermolabile molecules can be processed. A combination of precipitation and high-pressure homogenization methods (e.g., Nanoedge® technology) can also be applied. Several reviews described the methods of pharmaceutical nanosuspension production including their advantages and disadvantages.

Application of Nanosuspension Formulations in Ocular Drug Delivery

A number of studies have proved the efficacy of nanosuspension in improving ocular bioavailability of corticosteroids. Corticosteroids such as prednisone, dexamethasone, and hydrocortisone are the first choice for treatment of anterior segment inflammation, however, using these drugs in a large dose frequently may lead to cataracts, glaucoma, and optic nerve injury. Therefore, delivery of corticosteroid by nanosuspensions to improve its bioavailability is an attractive option. Kassem et al. found that the corticosteroids, such as hydrocortisone, prednisolone, and dexamethasone, coated by nanosuspensions resulted in an enhanced rate and extent of ophthalmic drug absorption, as well as a considerably higher intensity of drug action with extended duration of drug effect compared to solutions and microcrystalline suspensions (Kassem et al. 2007). In another study, Ali et al. used hydrocortisone nanosuspension for the treatment of inflammation disorders of the eye, and the results showed a better bioavailability of hydrocortisone in the form of nanosuspension (Ali et al. 2011). Nanosuspension can also deliver other drugs successfully. For instance, Abrego et al. prepared nanosuspensions and nanoparticles as ophthalmic delivery of pranoprofen (Abrego et al. 2014). The result showed that the release profiles of pranoprofen from the primary nanosuspensions and rehydrated nanoparticles (the primary nanosuspension was freeze-dried and rehydrated in water) were similar and exhibited a sustained drug delivery pattern. Another work showed the potential of nanosuspension in ocular drug delivery of ketotifen fumarate (Soltani et al. 2016). Nanosuspension has been able to localize the drug into the cornea ex vivo with an enhanced in vitro ocular drug delivery. The results from these studies concluded that nanosuspension could be an efficient ophthalmic drug delivery system. A list of approved nanosuspension or suspension products for the ocular diseases is provided in Table 7.

Manufacturing Consideration in Scale-Up Development of Ocular Suspension Dosage Form

Scale-up manufacturing of sterile ocular suspensions or nanosuspensions requires thorough understanding of the factors that influence their physicochemical stability and other critical attributes. For example, the drug particle morphology is a key factor in suspension product dissolution rate, resuspendability, and syringeability. The type and concentration of surfactants can impact resuspendability and chemical stability of the product. Additionally, the drug particle size reduction methods may impact the drug quality and should be evaluated in advance.

The scale-up manufacturing process development of suspension products should determine the operating conditions applicable to large-scale batches with no compromise of the quality in assuring the therapeutic effectiveness and stability of the product. The physical properties of the drug, such as particle size, polymorphism, and ionization characteristics, are key factors influencing the scale-up production of suspension dosage form. Specifications to ocular multidose suspension products should include particle size/size distribution of the drug, assay, degradation products (impurities), resuspendability, pH, viscosity, sterility, and preservative effectiveness test (PET, not required for unit dose suspension products). The particle size distribution is a very critical attribute and should be examined during each manufacturing step. Resuspendability of the product over the shelf life must also be assessed to assure in obtaining a precise dose after shaking of the suspension product bottle before use.

In suspensions, the insoluble drug is uniformly dispersed throughout the liquid phase with excipients using homogenization. However, suspensions are susceptible to changes in equipment speed, time, and processing temperature to produce desired dispersion of the drug. Depending on the types of homogenizing equipment as well

Table 7 List of approved suspension or nanosuspension products for the ocular diseases

Approved ophthalmic suspension products	Company name
Lotemax™ (loteprednol etabonate ophthalmic suspension) 0.5%	Bausch & Lomb
Simbrinza® (brinzolamide/brimonidine tartrate ophthalmic suspension) 1%/0.2%	Alcon (now Novartis)
Neomycin/polymyxin B sulfates and dexamethasone ophthalmic suspension	Falcon (now Sandoz, a Novartis company)
Alrex® (loteprednol etabonate ophthalmic suspension) 0.2%	Bausch & Lomb
Brinzagan™ (brinzolamide ophthalmic suspension 1% w/v)	Allergan
INVELTYS™ (loteprednol etabonate ophthalmic suspension 1%)	Kala pharmaceuticals Inc.
Prednisolone acetate ophthalmic suspension, USP 1%	Sandoz, a Novartis company
ILEVRO® (nepafenac ophthalmic suspension, 0.3%)	Alcon (now Novartis)
BETOPTIC S® (betaxolol hydrochloride 0.25% as a base) ophthalmic suspension	Alcon (now Novartis)
Zylet™ (loteprednol etabonate 0.5% and tobramycin 0.3% ophthalmic suspension)	Bausch & Lomb
Flarex® (fluorometholone ophthalmic suspension) USP	Allergan
Cortisporin ophthalmic suspension (neomycin and polymyxin b sulfates and hydrocortisone ophthalmic suspension)	Pfizer
Azopt® (brinzolamide ophthalmic suspension) 1%	Alcon (now Novartis)
DEXYCU™ (dexamethasone intraocular suspension) 9%	Icon BioScience, Inc.
Maxidex® (dexamethasone ophthalmic suspension) 0.1%	Alcon (now Novartis)
Maxitrol® (neomycin and polymyxin B sulfates and dexamethasone ophthalmic suspension)	Alcon (now Novartis)
Natacyn® (natamycin ophthalmic suspension) 5%	Novartis
Nevanac® (nepafenac ophthalmic suspension) 0.1%	Alcon (now Novartis)
Omnipred® (prednisolone acetate ophthalmic suspension)	Novartis
Tobradex® (tobramycin and dexamethasone ophthalmic suspension)	Alcon (now Novartis)
Obradex® ST (tobramycin/dexamethasone ophthalmic suspension) 0.3%/0.05%	Alcon (now Novartis)
Triesence® (triamcinolone acetonide injectable suspension) 40 mg/mL	Alcon (now Novartis)
Besivance® (besifloxacin ophthalmic suspension) 0.6%	Bausch & Lomb
Vexol®1% (rimexolone ophthalmic suspension)	Alcon

as the processes, the results may vary in generating uniformly dispersed particles. Hence, the transition from lab-scale R&D batches produced using small-capacity equipment to large-scale homogenizer demands precise control of settings between various equipment models to generate desired results. In this regard, multiple small-scale batches are required to assure the success of large-scale manufacturing. The process validation may also require real-time sampling and in-process testing of the products relative to targeted specifications. Ophthalmic suspension products may also pose challenges during sterile manufacturing since the possibilities of either degradation or morphological changes may occur during the sterilization process. Hence, the effect of various sterilization methods on drug and formulation attributes should be assessed. If the suspension products cannot be manufactured by terminal sterilization methods due to stability issues; an alternative approach is aseptic manufacturing.

Stability Consideration of the Suspension and Nanosuspension Dosage Forms

In order to understand the role of excipients in ocular suspension and nanosuspension formulations, it is important to understand the stability and process by which these formulations are stabilized. Suspension dosage forms are kinetically stable but inherently thermodynamically unstable systems. When left undisturbed for a long time, the suspension particles aggregate, sediment, and finally cake, hence must redisperse readily to achieve dosage uniformity. A higher viscosity of dispersion medium offers the advantage of slower sedimentation of the particles; however, it may compromise spreadability for topical ophthalmic suspensions. Thus, the shear thinning is necessary so that the suspension is highly viscous with slow sedimentation during storage, i.e., when minimal shear is present but has low viscosity after agitation (high shear) to facilitate ease of pourability from the storage containers. In general, the properties and stability of the suspension are influenced by the physicochemical characteristics of the dispersed phase, the dispersion medium/vehicle, and their interactions when mixed. There are three important attributes for the stability of the suspension drug product: chemical stability, physical stability, and microbiological stability (preservative efficacy) (Missel 2012; Kulshreshtha et al. 2010).

Physical Stability

The common physical stability issues of suspension formulation include agglomeration, sedimentation/creaming, crystal growth, and change of crystallinity (polymorphism) (Wu et al. 2011). Ideally, the particles in physically stable suspension remain uniformly distributed throughout the dispersion. However, the large surface area of small particles creates high total surface energy, which is thermodynamically unfavorable. Thus, the system tends to decrease the surface area in order to minimize the free energy by formation of agglomerates. This may lead to flocculation or aggregation, dependent on the attractive and repulsive forces within the system. Agglomeration can cause rapid settling/creaming, crystal growth, and inconsistent dosing of the dosage form. The most common strategy to solve this is the use of stabilizers to reduce interfacial tension and prevent agglomeration to generate a stable nanosuspension formulation. The common stabilizers are phospholipids, polymers, surfactants (ionic and nonionic), or a combination of these materials.

Flocculated and deflocculated suspension: When the particles are held together in a loose open structure, the system is in the state of flocculation (Kulshreshtha et al. 2010). The loose aggregates have a larger size compared to the single particle and, thus, higher sedimentation rate. The loose structure of the rapidly settling flocs contains a significant amount of entrapped medium; thus, the final flocculation volume is relatively large, and the flocs can be easily redispersed by simple agitation, which is highly desirable to ensure uniform dosing. In deflocculated suspension, the particles settle as small individual particles, resulting in a slow sedimentation rate. This leads to a high-density sediment that may be difficult to redisperse as the energy barrier is much higher compared with a flocculated suspension (Kulshreshtha et al. 2010). A deflocculated suspension remains dispersed for a longer time; however, it leads to formation of a close-packed arrangement, resulting in cake formation in case of sedimentation. A comparison of deflocculated and flocculated suspension is provided in Table 8.

Role of particle size distribution: Particle size distribution (PSD) plays a key role in the physical stability of the suspension products. The rate of sedimentation, agglomeration, suspendability, and thus the bioavailability of APIs and rheological behavior of formulation are directly affected by the particle size. The particles, through random motion over time, aggregate because of the natural tendency to decrease the large specific surface area and excess surface energy. The frequency of particle-particle collision depends on PSD, particle concentration, dispersion medium viscosity, and temperature. Stokes' law (Eq. 1). This indicates the important role of particle size, medium viscosity, and density differences between medium and dispersed phase on the particle sedimentation rate (Kulshreshtha et al. 2010).

Table 8 Comparative property of deflocculated and flocculated suspension

Properties	Deflocculated suspension	Flocculated suspension
Particle existence	Separate entities	Loose aggregates
Particle size	Small compared to flocculated suspension	Large
Rate of sedimentation	Slow	Fast
Sediment structure	Compact	Scaffold-like loose
Redispersion properties	Difficult	Easy
Final flocculation volume	Small	Relatively large
Appearance	Because the suspended material remains suspended for a relatively long time, product looks good in appearance. The supernatant also remains cloudy, even when settling is apparent	Because of rapid sedimentation and the presence of an obvious, clear supernatant region, the product is somewhat unsightly

$$v = {}^{2}/_{9}\left(d_1 - d_2\right) g r^2 / \eta \tag{1}$$

Here, r is the radius of the particle/sphere, η is the viscosity of the liquid, v is the flow velocity, d_1 is the density of the particle/sphere, d_2 is the density of the liquid, and g is the gravitational constant.

The Stokes' equation applies to dilute suspensions and assumes spherical and monodisperse particles, which may not be encountered in real systems. Equation 2 gives the changed sedimentation velocity (Alexander et al. 1990).

$$v' = v\varepsilon^{\eta} \tag{2}$$

Here, v' is the hindered sedimentation velocity, v is the sedimentation velocity from Eq. (1), ε is the porosity of the system, and n is the measure of hindering.

According to the Stokes' law, reduction of particle size leads to a decrease in the rate of sedimentation of the suspended particles. However, reducing particle size beyond a certain limit may lead to formation of a compact cake upon sedimentation. Hence, there should be a balance between particle size distribution, viscosity of the continuous phase, and the difference in density between the dispersed and the continuous phases. The other approaches to alleviate sedimentation issues are by matching the drug particle density with the medium or by increasing the medium viscosity. Figure 5 (reproduced with permission from (Wu et al. 2011)) shows different sedimentation types (deflocculated, flocculated, and open floc based) that occur in suspension formulations.

The attraction and repulsion between the particles depend on the potential energy barrier between them and arise from the difference in the extent of repulsive forces in comparison with attractive electrostatic forces. Colloidal suspensions can be stabilized in both aqueous and nonaqueous medium through electrostatic repulsion and steric stabilization. Stability due to electrostatic repulsion can be explained by the classical Derjaguin-Landau-Verwey-Overbeek (DLVO) theory; according to which, there are two major forces acting on colloidal particles in a dispersion medium: electrostatic repulsive forces due to overlap of electrical double layers (EDL) and van der Waals attractive forces. The EDL arises because of the charge at the solid-liquid interface. To maintain electrical neutrality of the system, counter ions present in the media are attracted toward the surface to form a double layer of ions: a tightly bound first layer of ions, also known as the Stern layer; and a diffuse layer of ions, also called the Gouy or Gouy-Chapman layer (Fig. 6, reproduced with permission from (Wu et al. 2011)). The possible lowest electrolyte concentration should be used since as the ionic strength of the medium increases, the thickness of EDL decreases, and the force of repulsion becomes smaller due to screening of the surface charge.

In steric stabilization mechanism, the high concentrations of polymers added to the suspension or nanosuspensions get adsorbed onto the surfaces of newly formed particles of the hydrophobic drug with the hydrophobic parts of the polymer attached to the particle surface and the hydrophilic chains extending into the aqueous environment (Fig. 7, reproduced with permission from (Wu et al. 2011)). Due to steric

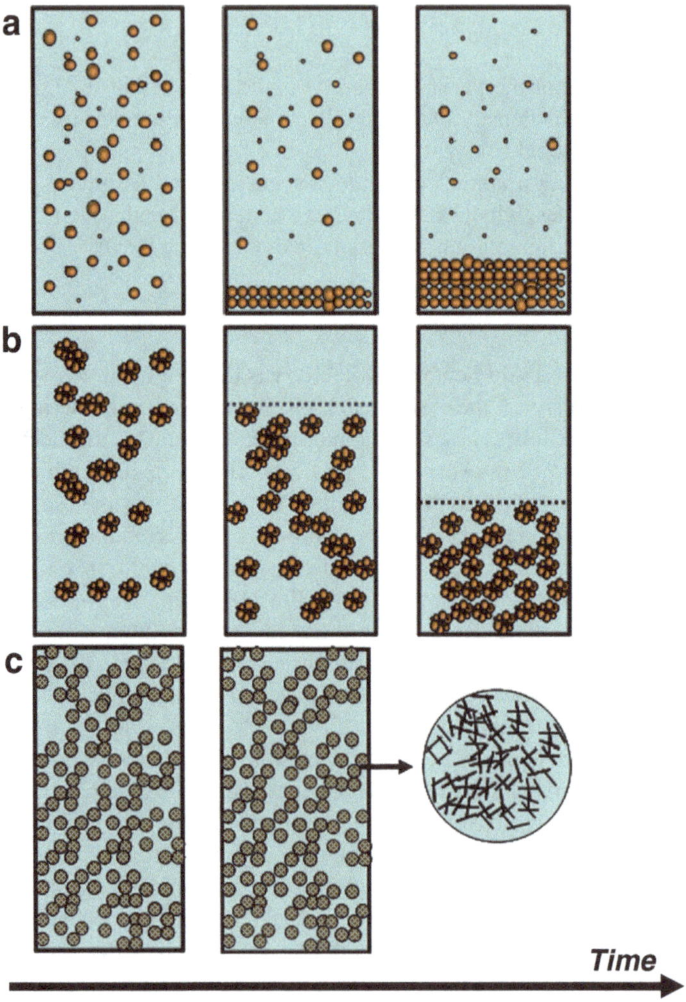

Fig. 5 Sedimentation in (**a**) deflocculated suspension, (**b**) flocculated suspension, and (**c**) open floc-based suspension. (Reproduced with permission from Wu et al. (2011))

effects, the long polymeric chains extended into the water prevent the two particles from coming very close to each other. Thus, the dispersion medium must be a good solvent for the adsorbed macromolecule to allow the polymer chains to extend into bulk solution. In practice, the most common steric stabilizers are block and graft copolymers, composed of two parts: one is insoluble in the dispersion medium and firmly anchors the stabilizing moiety, and the second is soluble in the dispersion medium, providing the steric repulsion. In comparison, electrostatic stabilization is more susceptible to the ionic strength of the dispersion medium, and the high concentrations of ions in the dispersion medium lead to the screening of the surface

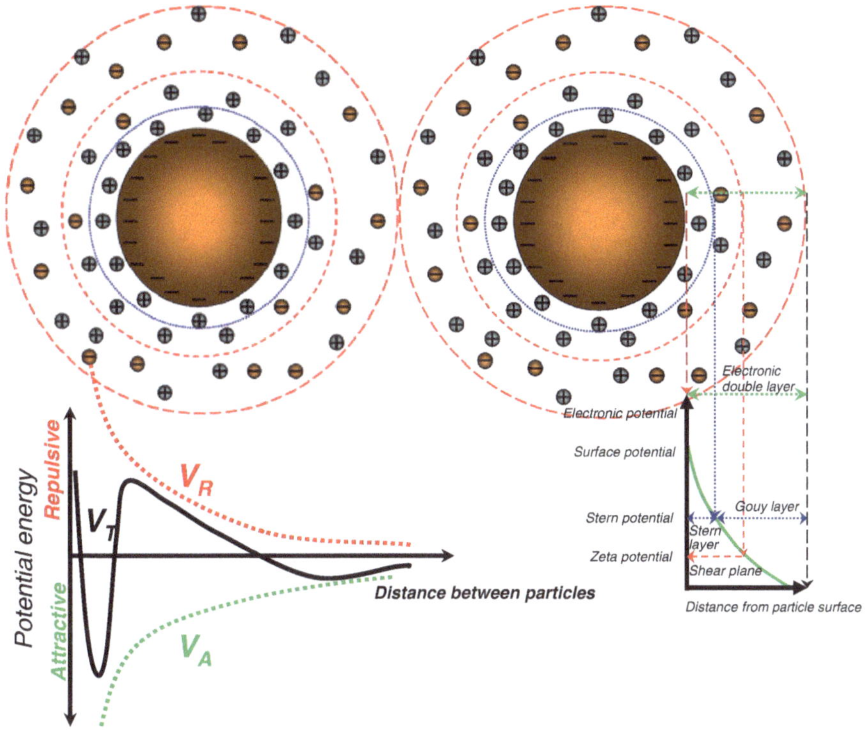

Fig. 6 Illustration of classical DLVO theory. Attractive forces are dominant at very small and large distances, leading to primary and secondary minimum, while repulsive forces are prevailing at intermediate distances and create net repulsion between the dispersed particles, thus preventing particle agglomeration. (Reproduced with permission from Wu et al. (2011))

charge, which decreases the thickness of the diffuse double layer. The depleted double layer makes the dispersed particles susceptible to aggregation. On the other hand, the hydration of the polymers is more susceptible to temperature changes. Hence, sterically stabilized suspensions are more prone to destabilization by temperature fluctuations. Therefore, a combination of both ionic surfactants and a polymeric stabilizer reduces the self-repulsion between the ionic surfactants, facilitating close packing of the stabilizer molecule layer around the particle, a more efficient approach in preventing particle agglomeration.

Crystal growth: Crystal growth in colloidal suspensions is generally known as Ostwald ripening (Wu et al. 2011; Kulshreshtha et al. 2010), which is a process where large particles grow at the expense of smaller particles and subsequently leads to a shift in the particle size and size distribution to a higher range. According to Ostwald-Freundlich equation, small particles have higher saturation solubility than larger particles (Wu et al. 2011), creating a drug concentration gradient between them. Consequently, molecules diffuse from the higher concentration surrounding small particles to around larger particles with lower drug concentration, generating

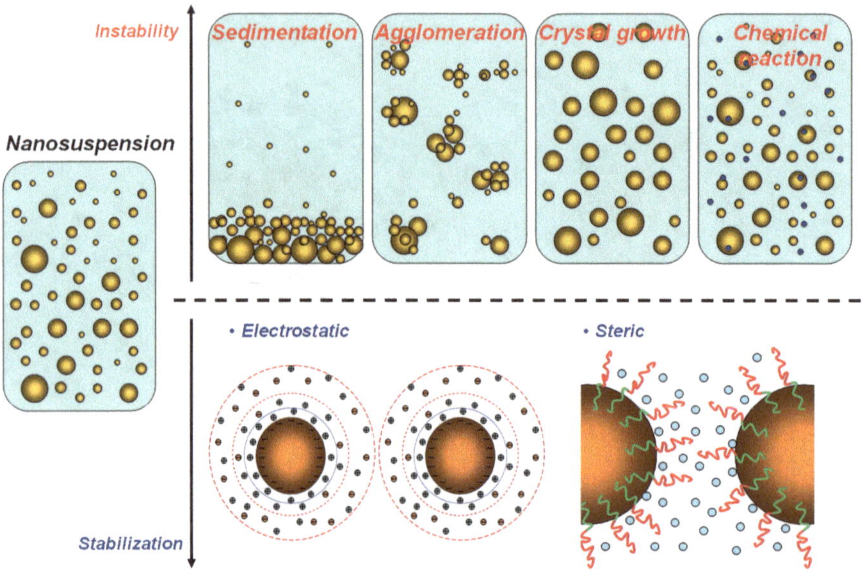

Fig. 7 Schematic summary of instability issues and general stabilization mechanisms of nanosuspension/suspension products. (Reproduced with permission from Wu et al. (2011))

supersaturated solution around the large particles, leading to drug crystallization. This process leaves an unsaturated solution surrounding the small particles, causing dissolution of the drug molecules from the small particles into the bulk medium. A narrow particle size distribution can minimize the saturation solubility difference and drug concentration gradients within the medium and, thus, help to inhibit the occurrence of Ostwald ripening. Stabilizers being absorbed on the particles surface can reduce the interfacial tension between the solid particles and liquid medium, thus preventing the Ostwald ripening. Solubility, temperature, and mechanical agitation may also affect the Ostwald ripening process. A summary of instability issues and general stabilization mechanisms of nanosuspension and suspension products is schematically represented in Fig. 7 (reproduced with permission from (Wu et al. 2011)).

In summary, the formulation factors that can be adjusted to affect the physical stability of the formulation include (Kulshreshtha et al. 2010):

- ***Flocculation/deflocculation:*** (a) Add charged surface-active polymer or surfactant, (b) add oppositely charged flocculation agent (to shield the surface charges of the particles and to reduce the zeta potential to zero, at which point flocculation is observed), (c) add nonionic surface-active polymer or surfactant, (d) adjust ionic strength of vehicle (high concentrations of ions in the dispersion medium lead to the screening of the surface charge, which decreases the thickness of the diffuse double layer and makes the dispersed particles susceptible to aggregation), and (e), depending on the drug pKa, adjust pH to modify the surface charge.

- ***Sedimentation rate:*** (a) Increase the viscosity of the vehicle (polymer stabilizers adsorbed on the surface of the particle, create a steric effect by preventing the individual particles from getting sufficiently close to each other, and help in settling out as a deflocculated sediment that is difficult to redisperse) and (b) decrease the particle size of the drug (leads to a decrease in the rate of sedimentation of the suspended particles; however, reducing the particle size beyond a certain limit may lead to formation of a compact cake upon sedimentation).
- ***Ostwald ripening and crystal growth:*** (a) Generation of narrow particle size distribution (a narrow particle size distribution can minimize the saturation solubility difference around large and small particles and drug concentration gradients within the medium and, thus, help to inhibit the occurrence of Ostwald ripening), (b) addition of stabilizers (stabilizers being absorbed on the particles surface can reduce the interfacial tension between the solid particles and liquid medium, thus preventing the Ostwald ripening), and (c) optimize the solubility, temperature, and mechanical agitation (all these can lead to supersaturation in solubility and crystal growth (Ostwald ripening)).

Chemical Stability

The chemical stability for suspension products is mostly drug specific since each drug molecule has specific functional groups that affect the stability. Several factors such as storage temperature and pH, chemical stability of entrapped drugs, as well as the type and molecular weight of the polymer used can lead to the chemical instability of suspension. There are primarily three frequently encountered chemical stability issues: hydrolysis, oxidation, and photodegradation (Kulshreshtha et al. 2010). The formulation parameters that can be adjusted to address these chemical stability issues are (Kulshreshtha et al. 2010):

- ***Hydrolysis:*** (a) Reduce solubility of the drug in the vehicle, (b) adjust the pH to avoid acid or base catalysis, or (c) reduce the storage temperature.
- ***Oxidation:*** (a) Add an antioxidant to the formulation, (b) remove oxygen from the manufacturing process and package, (c) use a more protective package, or (d) reduce the storage temperature.
- ***Photodegradation:*** (a) Reduce the solubility of the drug in the vehicle (if photodegradation occurs to drug in solution), or (b) use a more protective package and/or storage condition.

In summary, the topical drug delivery to the ocular diseases requires strategic approaches due to the presence of several anatomical/static and physiological/dynamic barriers. Considering this, the development of conventional or nanoformulation-based delivery systems requires appropriate selection of the excipients and formulation development strategy to achieve an effective drug dose to the ocular tissues. In this book chapter, we emphasized on the topical route of drug administration and the development of ocular suspension and nanosuspension formulations. The considerations in the formulation development approaches summarized here may help in facilitating the development of safe, stable, and efficacious ocular drug products.

References

Abrego G, Alvarado HL, Egea MA, Gonzalez-Mira E, Calpena AC, Garcia ML. Design of nanosuspensions and freeze-dried PLGA nanoparticles as a novel approach for ophthalmic delivery of pranoprofen. J Pharm Sci. 2014;103(10):3153–64. https://doi.org/10.1002/jps.24101.

Agrahari V, Agrahari V, Mandal A, Pal D, Mitra AK. How are we improving the delivery to back of the eye? Advances and challenges of novel therapeutic approaches. Expert Opin Drug Deliv. 2017;14(10):1145–62. https://doi.org/10.1080/17425247.2017.1272569.

Ahmed I, Patton TF. Importance of the noncorneal absorption route in topical ophthalmic drug delivery. Invest Ophthalmol Vis Sci. 1985;26(4):584–7.

Ahmed I, Francoeur ML, Thombre AG, Patton TF. The kinetics of timolol in the rabbit lens: implications for ocular drug delivery. Pharm Res. 1989;6(9):772–8.

Alexander KS, Azizi J, Dollimore D, Uppala V. Interpretation of the hindered settling of calcium carbonate suspensions in terms of permeability. J Pharm Sci. 1990;79(5):401–6. https://doi.org/10.1002/jps.2600790508.

Ali HSM, York P, Ali AMA, Blagden N. Hydrocortisone nanosuspensions for ophthalmic delivery: a comparative study between microfluidic nanoprecipitation and wet milling. J Control Release. 2011;149(2):175–81. https://doi.org/10.1016/j.jconrel.2010.10.007.

Ali J, Fazil M, Qumbar M, Khan N, Ali A. Colloidal drug delivery system: amplify the ocular delivery. Drug Deliv. 2016;23(3):710–26. https://doi.org/10.3109/10717544.2014.923065.

Ammar HO, Salama HA, Ghorab M, Mahmoud AA. Nanoemulsion as a potential ophthalmic delivery system for dorzolamide hydrochloride. AAPS PharmSciTech. 2009;10(3):808–19. https://doi.org/10.1208/s12249-009-9268-4.

Barar J, Aghanejad A, Fathi M, Omidi Y. Advanced drug delivery and targeting technologies for the ocular diseases. Bioimpacts. 2016;6(1):49–67. https://doi.org/10.15171/bi.2016.07.

Falavarjani KG, Nguyen QD. Adverse events and complications associated with intravitreal injection of anti-VEGF agents: a review of literature. Eye (Lond). 2013;27(7):787–94. https://doi.org/10.1038/eye.2013.107.

Gaudana R, Ananthula HK, Parenky A, Mitra AK. Ocular drug delivery. AAPS J. 2010;12(3):348–60. https://doi.org/10.1208/s12248-010-9183-3.

Ghate D, Edelhauser HF. Ocular drug delivery. Expert Opin Drug Deliv. 2006;3(2):275–87. https://doi.org/10.1517/17425247.3.2.275.

Hartman RR, Kompella UB. Intravitreal, subretinal, and Suprachoroidal injections: evolution of microneedles for drug delivery. J Ocul Pharmacol Ther. 2018;34(1–2):141–53. https://doi.org/10.1089/jop.2017.0121.

Ibrahim SS. The role of surface active agents in ophthalmic drug delivery: a comprehensive review. J Pharm Sci. 2019;108(6):1923–33. https://doi.org/10.1016/j.xphs.2019.01.016.

Janagam DR, Wu L, Lowe TL. Nanoparticles for drug delivery to the anterior segment of the eye. Adv Drug Deliv Rev. 2017;122:31–64. https://doi.org/10.1016/j.addr.2017.04.001.

Jones DS. FASTtrack: pharmaceutics - dosage form and design. Chicago: Pharmaceutical Press; 2016.

Kassem MA, Abdel Rahman AA, Ghorab MM, Ahmed MB, Khalil RM. Nanosuspension as an ophthalmic delivery system for certain glucocorticoid drugs. Int J Pharm. 2007;340(1–2):126–33. https://doi.org/10.1016/j.ijpharm.2007.03.011.

Kaur IP, Kanwar M. Ocular preparations: the formulation approach. Drug Dev Ind Pharm. 2002;28(5):473–93. https://doi.org/10.1081/DDC-120003445.

Kulshreshtha AK, Singh ON, Wall GM. Pharmaceutical suspensions, from formulation development to manufacturing. New York: Springer; 2010.

Lai F, Schlich M, Pireddu R, Corrias F, Maria Fadda A, Sinico C. Production of nanosuspensions as a tool to improve drug bioavailability: focus on topical delivery. Curr Pharm Des. 2015;21(42):6089–103.

Maharjan P, Cho KH, Maharjan A, Shin MC, Moon C, Min KA. Pharmaceutical challenges and perspectives in developing ophthalmic drug formulations. J Pharm Investig. 2019;49(2):215–28. https://doi.org/10.1007/s40005-018-0404-6.

Mishima S, Gasset A, Klyce SD Jr, Baum JL. Determination of tear volume and tear flow. Investig Ophthalmol. 1966;5(3):264–76.

Missel MC. Remington: essentials of pharmaceutics. Chicago: Pharmaceutical Press; 2012.

Patel A, Cholkar K, Agrahari V, Mitra AK. Ocular drug delivery systems: an overview. World J Pharmacol. 2013;2(2):47–64. https://doi.org/10.5497/wjp.v2.i2.47.

Patravale VB, Date AA, Kulkarni RM. Nanosuspensions: a promising drug delivery strategy. J Pharm Pharmacol. 2004;56(7):827–40. https://doi.org/10.1211/0022357023691.

Peptu CA, Popa M, Savin C, Popa RF, Ochiuz L. Modern drug delivery systems for targeting the posterior segment of the eye. Curr Pharm Des. 2015;21(42):6055–69.

Rabinow BE. Nanosuspensions in drug delivery. Nat Rev Drug Discov. 2004;3(9):785–96. https://doi.org/10.1038/nrd1494.

Raghava S, Hammond M, Kompella UB. Periocular routes for retinal drug delivery. Expert Opin Drug Deliv. 2004;1(1):99–114. https://doi.org/10.1517/17425247.1.1.99.

Rai Udo J, Young SA, Thrimawithana TR, Abdelkader H, Alani AW, Pierscionek B, Alany RG. The suprachoroidal pathway: a new drug delivery route to the back of the eye. Drug Discov Today. 2015;20(4):491–5. https://doi.org/10.1016/j.drudis.2014.10.010.

Rowe-Rendleman CL, Durazo SA, Kompella UB, Rittenhouse KD, Di Polo A, Weiner AL, Grossniklaus HE, Naash MI, Lewin AS, Horsager A, Edelhauser HF. Drug and gene delivery to the back of the eye: from bench to bedside. Invest Ophthalmol Vis Sci. 2014;55(4):2714–30. https://doi.org/10.1167/iovs.13-13707.

Salzillo R, Schiraldi C, Corsuto L, D'Agostino A, Filosa R, De Rosa M, La Gatta A. Optimization of hyaluronan-based eye drop formulations. Carbohydr Polym. 2016;153:275–83. https://doi.org/10.1016/j.carbpol.2016.07.106.

Scherz W, Doane MG, Dohlman CH. Tear volume in normal eyes and keratoconjunctivitis sicca. Albrecht Von Graefes Arch Klin Exp Ophthalmol. 1974;192(2):141–50.

Soltani S, Zakeri-Milani P, Barzegar-Jalali M, Jelvehgari M. Comparison of different nanosuspensions as potential ophthalmic delivery Systems for ketotifen fumarate. Adv Pharm Bull. 2016;6(3):345–52. https://doi.org/10.15171/apb.2016.046.

Sridhar MS. Anatomy of cornea and ocular surface. Indian J Ophthalmol. 2018;66(2):190–4. https://doi.org/10.4103/ijo.IJO_646_17.

Sutradhar KB, Khatun S, Luna IP. Increasing possibilities of Nanosuspension. J Nanotechnol. 2013;2013:12. https://doi.org/10.1155/2013/346581.

Thackaberry EA, Farman C, Zhong F, Lorget F, Staflin K, Cercillieux A, Miller PE, Schuetz C, Chang D, Famili A, Daugherty AL, Rajagopal K, Bantseev V. Evaluation of the toxicity of intravitreally injected PLGA microspheres and rods in monkeys and rabbits: effects of depot size on inflammatory response. Invest Ophthalmol Vis Sci. 2017;58(10):4274–85. https://doi.org/10.1167/iovs.16-21334.

Worakul N, Robinson JR. Ocular pharmacokinetics/pharmacodynamics. Eur J Pharm Biopharm. 1997;44(1):71–83. https://doi.org/10.1016/S0939-6411(97)00064-7.

Wu L, Zhang J, Watanabe W. Physical and chemical stability of drug nanoparticles. Adv Drug Deliv Rev. 2011;63(6):456–69. https://doi.org/10.1016/j.addr.2011.02.001.

Yadollahi R, Vasilev K, Simovic S. Nanosuspension technologies for delivery of poorly soluble drugs. J Nanomater. 2015;2015:13. https://doi.org/10.1155/2015/216375.

Emulsions for Topical Eye Delivery: State of the Art and Future Perspectives

Frederic Lallemand and Jean-Sebastien Garrigue

Abstract Ophthalmic emulsions are formulation described in the literature since more than 30 years and officially listed in the US and European pharmacopeias as appropriated ocular dosage forms. However, while once was expecting that emulsion would be used as drug vehicle, surprisingly a new category of products emerged from the market, the artificial tears based on emulsions. Those products are the new generation of artificial tears being now as widely used in place of hydrogels. These eye drops provide several advantages over hydrogels or saline solutions as they supplement the tears with lipids acting as lubricant and more importantly as a barrier against evaporation and tear film stabilizer. On the other hand, emulsions as drug carriers were very rare to reach the market. About 35 active ingredients were tested in emulsions and described in more than 55 scientific articles, leading to only four prescription products in the USA and Europe. Restasis reached the US market in 2003, followed by Durezol in 2006, Ikervis in 2015 in Europe, and Xelpros in the USA in 2018. Some other products under clinical stage may reach the market within the next 5 years. This chapter is giving an outlook on the field of ophthalmic emulsions, providing the last scientific updates for the two categories of products. The different products marketed and under development are described and discussed, and the technical gaps to fill are pointed out to facilitate other developments to emphasize the importance of that complex but very useful dosage form.

Keywords Emulsion · Ophthalmology · Formulation · Manufacturing · Restasis · Ikervis · Durezol · Artificial tears · Drug delivery · Ocular surface

---

F. Lallemand (✉)
SAS eXinov, Rambouillet, France

J.-S. Garrigue
Santen Pharmaceuticals, Paris, France
e-mail: Jean-Sebastien.Garrigue@santen.com

S. Neervannan, U. B. Kompella (eds.), *Ophthalmic Product Development*, AAPS Advances in the Pharmaceutical Sciences Series 37,
https://doi.org/10.1007/978-3-030-76367-1_13

Abbreviations

AT	Artificial tears
AUC	Area under curve
BAK	Benzalkonium chloride
BUT	Tear breakup time
CK	Cetalkonium chloride
CsA	Cyclosporin A
DED	Dry eye disease
EMA	European Medicines Agency
FDA	Food and Drug Administration
HMW	High molecular weight
HPMC	Hydroxypropyl methylcellulose
MCT	Medium-chain triglycerides
MGD	Meibomian gland dysfunction
NSAIDs	Nonsteroidal anti-inflammatory drugs
o/w	Oil in water
PC	Phosphatidylcholine
SEDDSs	Self-emulsifying drug delivery systems
SH	Sodium hyaluronate
TFLL	Tear film lipid layer
v/v	Volume/volume
VKC	Vernal keratoconjunctivitis
w/v	Weight/volume
w/w	Weight/weight

Introduction

In their ophthalmic preparation monographs, the European Pharmacopeia (monograph 1163) and the United States Pharmacopeia (monograph 771) present emulsions alongside solutions, suspensions, and ointments, as possible dosage forms for the topical ocular delivery of drugs. Although solutions and ointments are well-known, emulsions are used in a limited number of ophthalmic products, which is surprising considering the numerous biopharmaceutical advantages of this dosage form.

With four main authorized pharmaceutical products (Restasis, Durezol, Ikervis and Xelpros) and several artificial tears (Europe and US), oil-in-water (o/w) emulsions are now included in the panorama of ophthalmic treatments. Emulsions are ideal vehicles to transport lipophilic and poorly soluble drugs, can protect drugs from hydrolysis by preventing direct contact of the molecule with water (Tamilvanan and Benita 2004), may attenuate the irritant potential of drugs (Liang et al. 2008), and can increase penetration and retention time on the ocular surface.

Even though emulsions have been used for a very long time in dermatology, their use in ophthalmology is more recent. As stated by Robin and Ellis (1978) in the 1970s, the gold standard for administering topically lipophilic compounds was the use of ointments. At that time, Robin noted that the corneal passage of lipophilic drugs was low due to a higher affinity for vehicle (lanolin, waxes, petrolatum, mineral oil, etc.) than for the cornea. In addition, this dosage form is not convenient to administer because of its high viscosity (finger application), and it is not comfortable (blurry vision, sticky, and oily) so that application at night is recommended.

In the 1980s, a major review of ophthalmic drug delivery systems (Shell 1984) cited topical emulsions as a vehicle for pilocarpine but gave no tangible example.

In the 1990s, this dosage form gained ground in ophthalmology. The first articles were published by Gallarate et al. (1988) and Gasco et al. (1989) in 1988 and 1989, respectively. In these articles, the authors showed that the corneal penetration of timolol in the rabbit eye was 4.2 times greater when formulated in emulsion than in solution. In 1990, Professor Maichuk treated patients suffering from ocular mycoses with an emulsion of ketoconazole (Maichuk Iu et al. 1990).

Following those precursors, numerous molecules have been tested in ophthalmic emulsions by different teams and companies (Table 1), with today four pharmaceutical products on the European and US markets and others under development plus several lipid-containing artificial tears based on this technology.

Major reviews have addressed technical considerations in a very detailed manner. Differences between the microemulsion and emulsion, formulation issues (Peng et al. 2011), choice of excipients (oil and surfactants) (Vandamme 2002), mode of production (Tiwari et al. 2018), characterization (Tamilvanan and Benita 2004), and the effect of the zeta potential of emulsion droplets on ocular penetration (Lallemand et al. 2012) have been examined. Consequently, this chapter will not review all these aspects but will instead focus on current marketed products and provide informative examples of development.

Currently, the use of emulsions in eye care can be divided into two main categories: emulsions as artificial tears and emulsions as drug carriers. The leading and most common is the use of emulsion itself as artificial tears in slight to moderate dry eye and seasonal allergy and as a lubricant to alleviate corneal lesions. A dedicated section in this chapter will describe the different products used as artificial tears, their status, their mode of action, and their composition. In the second part, we will discuss the use of emulsion as a drug carrier to administer active drugs to the ocular surface, making use of their capacity to solubilize poorly soluble or lipophilic molecules by presenting the main products on the market today and those still being developed. We will end this chapter by examining the scientific challenges remaining to be solved or investigated to reach a full understanding of these complex colloidal systems to support current and future developments.

Table 1 Use of emulsions as lipid-based artificial tears

Emulsion type	Brand name	Main components	Refs.
Anionic emulsion	Emustil®	Soybean oil 7%, egg yolk phospholipids 3%, glycerol 1.8%, in Tris-HCl buffer	Scifo et al. (2010)
	Soothe XP®	Mineral oils 5.5%, octoxynol-40, polysorbate 80, boric acid, edetate disodium, sodium borate, purified water	Fogt et al. (2016)
	Lipimix®	Medium-chain triglycerides 1.25%, soybean oil 1.25%, phospholipids 0.3%, glycerol 2.25%, EDTA 0.05%, alpha-tocopherol 0.02%, water	Aydemir et al. (2008)
	Systane Balance®	Mineral oil 1%, dimyristoyl phosphatidylglycerol 0.005%, PEG 40 stearate 0.38%, sorbitan tristearate 0.29%, hydroxypropyl guar 0.05%, propylene glycol, 0.6% boric acid 1%, EDTA 0.001%, sorbitol 0.7%, Polyquad 0.001%, qs HCl, NaOH, purified water	Aguilar et al. (2014)
	Under development	Castor oil 1% or 5%, saline solution; 0.1%, 0.5% or 1% SH solution	Hasegawa et al. (2014)
	Under development	Castor oil 10%, polysorbate 80 8%; loaded with indomethacin (0.1% w/v) and: (1) Uncoated (2) Coated with 1.5% chitosan (3) Coated with 0.5% HPMC	Yamaguchi et al. (2009)
Cationic emulsion	Cationorm® (Retaine® MGD)	Mineral oil 1%, CKC 0.002%, tyloxapol 0.2%, poloxamer 1880.01%, glycerol 1.5%, purified water	Kinnunen et al. (2014), Liang et al. (2008)
	Cationorm Pro®/Plus	MCT 2%, CKC 0.005%, tylopaxol 0.2%, glycerol 1.5–2.5%, purified water	Lallemand (et al. 2012)

AT artificial tears, *Ch* cholesterol, *CKC* cetalkonium chloride, *HA* hyaluronic acid, *HMW* high molecular weight, *HPMC* hydroxypropyl methylcellulose, *MCT* medium-chain triglyceride, *o/w* oil in water, *PC* phosphatidylcholine, *PLGA* poly(lactic-*co*-glycolic acid), *PS* phosphatidylserine, *SH* sodium hyaluronate, *v/v* volume/volume, *w/v* weight/volume, *w/w* weight/weight

Products on the Market and Being Developed for Topical Delivery

Emulsions Used as Artificial Tears

Introduction

The tear film was first described by Wolff et al. as being composed of three layers—the outer lipid layer, the middle aqueous layer, and the inner mucin layer—which interacts with the corneal epithelium (Bron et al. 2004). This film plays an important role in the physiological hydration of the ocular surface and ocular surface homeostasis. At the air–water interface, the outer layer, also called the tear film lipid layer

(TFLL), is primarily composed of lipids that are assumed to help prevent water evaporation and ocular surface dewetting as well as provide a smooth optical surface (Bron et al. 2004; Foulks 2007; Mishima and Maurice 1961; Pucker and Haworth 2015). These lipids are released by the meibomian glands located in the eyelids. The TFLL was found to be composed primarily of nonpolar lipids (wax and cholesterol esters), small amounts of polar lipids (Butovich et al. 2008; Rantamaki et al. 2011; Cwiklik 2016), as well as fatty acids and fatty alcohols that act as surfactants to stabilize the lipid–water mixture (Pucker and Haworth 2015; Butovich et al. 2008). More recently, amphiphilic lipids (i.e., surfactants), such as cholesteryl sulfate, *O*-acyl-ω-hydroxy fatty acids, various sphingolipids, and phospholipids, were also found in the human TFLL (Lam et al. 2014). All components of the TFLL play a key role in tear surface tension. Alterations in tear fluid rheology, differences in lipid composition, or downregulation of specific tear proteins occur in various types of dry eye disease (DED).

Artificial tears (AT) are used as first-line treatment for DED and to replace or supplement tears (Moshirfar et al. 2014), but lipid-containing eye drops have been recently proposed as a step closer to natural tears, because they more closely mimic the aqueous and lipid layers (Rieger 1990; Garrigue et al. 2017).

The US Food and Drug Administration (FDA) Code of Federal Regulations Title 21 part 349 lists some lipids as active ingredients that can be used as "emollients" or "lubricants" in ophthalmic over-the-counter (OTC) products (e.g., lanolin, light mineral oil, mineral oil, paraffin, petrolatum, white ointment, white wax, and yellow wax) indicated for "the temporary relief of burning and irritation due to dryness of the eye."

Among lipid-based products (e.g., ointments, liposomal sprays), oil-in-water emulsions were specifically designed to mimic the tear film and provide long-lasting lubrication to the eye surface. They consist of oily droplets stabilized by surfactants or emulsifiers dispersed in an aqueous medium. For topical ophthalmic use, most emulsions contain submicron-sized particles prepared with well-tolerated oils (e.g., sesame oil, castor oil, soya oil, paraffin oil, paraffin light, lanolin, corn oil, glycerin monostearate, medium-chain monoglycerides, and medium-chain triglycerides) and emulsifiers (e.g., phospholipids [Lipoid], polysorbate 80 (Tween® 80), Cremophor® RH, poloxamer 407, poloxamer 188, and tyloxapol).

These emulsions can be anionic (negatively charged oil nanodroplets) or cationic (positively charged oil nanodroplets), depending on the components added to the formulation during the emulsion process. This surface charge and its intensity may affect the physicochemical and biological behavior of the emulsions (Tamilvanan and Benita 2004).

Anionic Emulsions

The anionic artificial tear emulsion eye drop was first tested in human by Prof. Tsubota and his team (Goto et al. 2002). The product tested comprised 2% castor oil, 5% polyoxyethylene castor oil, 0.3% sodium chloride, 0.15% potassium chloride, 0.5% boric acid, and distilled water. The result was probably a coarse emulsion but the trial gave interesting results. Forty eyes of 20 patients suffering from Meibomian gland dysfunction (MGD) (a defect in lipid tear secretion) were instilled six times daily for 2 weeks. The emulsion was positive on all the outcomes measured: symptoms, tear interference grade, tear evaporation, fluorescein and rose Bengal scores, tear breakup time (BUT), and Meibomian gland orifice obstruction. The mechanism of action was hypothesized to be an improvement of tear stability as a result of lipid spreading, ease of meibum expression, prevention of tear evaporation, and the lubricating effect of the oil eye drops. Since then, many other artificial tears based on emulsion were developed and marketed.

A number of anionic emulsions are commercially available as artificial tear products in some countries: Optive Plus® (Allergan), Systane® Balance (Alcon), Soothe® XP (Bausch and Lomb), Emustil® (SIFI), Refresh Endura® (Allergan), Aquarest®/ Liposic® (Bausch and Lomb), and Lipimix® (Pharma Stulln) (see Table 1). They have demonstrated a favorable tolerability profile in DED patients, helping decrease eye irritation and vision disturbances, confirming preclinical safety and efficacy:

- Emustil® (an anionic emulsion containing natural phospholipids) significantly improved tear volume and reduced corneal damage when applied four times a day for 7 days either as a monotherapy or in combination with sodium hyaluronate in a mouse model of dry eye (Scifo et al. 2010).
- An anionic emulsion containing sodium hyaluronate and castor oil tested in a porcine short-term dry eye model also showed a protective effect against corneal desiccation (Hasegawa et al. 2014). Interestingly, coating emulsions with chitosan was shown to prolong the residence time of the emulsion in the tear fluid of rabbits (Yamaguchi et al. 2009).

Clinical studies evaluating the efficacy of anionic emulsions on DED symptoms and TFLL stability have shown positive results.

- In patients with mild to moderate DED treated with an anionic emulsion of 1.25% castor oil or 0.32% hypromellose three times daily for 1 month, both treatments resulted in decreased tear evaporation rates, but as expected only the emulsion improved lipid layer structure at day 30 (Khanal et al. 2007).
- A large multicenter observational study including over 1000 DED patients followed for 4 weeks after starting or switching to Optive Plus® showed improvements in DED level of severity, tear BUT, and Schirmer test scores relative to baseline (Kaercher et al. 2014).
- Emustil® administered four times daily for 90 days in evaporative DED patients improved tear stability and decreased both tear osmolarity and corneal staining to a greater extent than single-dose 0.15% sodium hyaluronate and 0.3% hydroxypropyl methylcellulose (McCann et al. 2012).
- Systane® Balance moderately improved corneal staining and tear film BUT in patients with dry eye associated with MGD (Sindt and Foulks 2013). In a randomized, controlled, investigator-masked comparison study in patients with lipid-deficient DED, Systane® Balance (four times daily for 4 weeks) was found

to restore tear film stability and improve ocular surface staining and Meibomian gland functionality (Aguilar et al. 2014).
- DED patients treated with Soothe® (anionic emulsion) demonstrated increased lipid layer thickness (Scaffidi and Korb 2007).
- A castor oil-based anionic emulsion was well tolerated and showed improved symptom scores, tear interference grade, tear evaporation, corneal staining, tear film BUT, and orifice obstruction scores in MGD patients when administered six times daily for 2 weeks (Goto et al. 2002).

In the published clinical trials listed herein, the most common adverse event related to the instillation of anionic emulsions was blurred vision. Few adverse events were reported, underlying the overall good tolerance and comfort of these oil-in-water emulsions.

Cationic Emulsions

Cationic emulsions are biphasic formulations of positively charged oil nanodroplets (oily phase) dispersed in water (the continuous phase) (Lallemand et al. 2012; Daull et al. 2014). The positively charged nanodroplets interact with the negatively charged ocular surface mucins and cell membranes (Daull et al. 2014; Rabinovich-Guilatt et al. 2004; Royle et al. 2008). This bioadhesive property prolongs eye drop residence time and improves the residence time of lipophilic drug molecules delivered via these emulsions. The physicochemical parameters of a cationic emulsion influence its stability and physiological biocompatibility; for instance, the optimal properties of nanodroplets include a size less than 200 nm, a pH between 5 and 7, and an osmolarity of approximately 270–300 mOsm/kg (i.e., isotonic).

Choosing the appropriate cationic agent is also essential. Several agents, such as stearylamine, oleylamine, poly(ethylenimine), poly(L-lysine), 1,2-di-(9Z-octadecenoyl)-sn-glycero-3-phosphoethanolamine (DOPE), 1,2-di-(9Z-octadecenoyl)-3-trimethylammonium propane (DOTAP), and benzalkonium chloride (BAK) derivatives are available, but their use as cationic agents is limited by toxicity, stability, or regulatory issues. These limitations encouraged the development of safe and well-tolerated ophthalmic cationic emulsions. Cetalkonium chloride (CKC), a C16 BAK derivative, has been selected as the preferred cationic agent in cationic emulsions, because it has superior lipophilicity that limits it to the oily phase of the oil-in-water nanoemulsion, thereby minimizing toxicity.

In emulsions, quaternary ammoniums such as CKC, which are not soluble in the aqueous phase, are used solely as cationic agents and do not exert any detergent activity or preservative action (Lallemand et al. 2012; Daull et al. 2014; Kurup et al. 1992; Sznitowska et al. 2002). Additionally, cationic emulsions provide better spreading coefficients on the cornea and conjunctiva than conventional eye drops and anionic emulsions, thereby improving ocular surface wettability. This has been demonstrated with Cationorm® (Santen Pharmaceuticals, also known as Retaine® MGD in the USA) in contact angle and surface tension studies (see Table 1).

Additionally, a cationic emulsion containing a combination of phospholipids (Lipoid E 80), poloxamer 188 (Pluronic F68), and stearylamine as emulsifiers had a favorable tolerability profile when administered (40 single-drop instillations per day for 5 days) to rabbit eyes, despite the potential of stearylamine to create corneal lesions (Tamilvanan and Benita 2004). In other preclinical studies conducted with human corneal epithelial cells (HCE-2 cells), incubation with Cationorm® or Systane® for up to 30 min was not associated with any significant change in cell morphology or viability, and a reduced inflammatory response was elicited when compared to BAK (Kinnunen et al. 2014). Moreover, cationic emulsions utilizing BAK (0.02%) and CKC (0.002%) as cationic agents are less toxic than BAK and CKC aqueous solutions at the same concentrations in a severe eye irritation rabbit model (15 instillations at 5-min intervals) (Liang et al. 2008). In this study, a CKC-containing cationic emulsion caused a similar degree of redness, chemosis, and conjunctiva secretions as the saline control. The tolerability of cationic emulsions was also confirmed with a latanoprost-loaded emulsion in HCE cells, in a rat corneal wound model and a rabbit ocular tolerance model (Liang et al. 2012; Daull et al. 2012). Cationic lipids have also demonstrated the potential for exhibiting both antimicrobial and anti-inflammatory properties (Myint et al. 2015; Daull et al. 2018). Cationorm® was shown to reduce stromal inflammatory cell infiltration as determined by in vivo confocal microscopy (Daull et al. 2016). More recently, CKC, when combined with MCT or mineral oils, was shown to stabilize the TFLL (Nencheva et al. 2018; Georgiev et al. 2017, 2016; Cwiklik et al. 2017).

In patients with mild to moderate DED, instillation of Cationorm® (1 drop per eye four times daily for 28 days) significantly improved tear film BUT as early as day 7 and was well tolerated by patients (Amrane et al. 2014). In this study, a significant improvement in lissamine green staining was observed after 28 days of treatment, suggesting that this cationic emulsion reinforces the lipid layer, minimizes evaporation, and stabilizes the tear film to protect the ocular surface. In another clinical trial comparing Cationorm® versus Vismed® in patients with moderate to severe DED, although both eye drops showed similar safety profiles and efficacy in terms of improvements in objective signs of dry eye, Cationorm® was superior in improving the global symptoms score of ocular discomfort (Robert et al. 2016). The data from these studies suggest the use of cation emulsions because ophthalmic drug delivery systems are generally well tolerated. These cationic emulsions are lipid-based drug nanocarriers and promising new vehicles for ocular drug delivery, e.g., delivery of cyclosporine to treat various ocular surface diseases, including dry eye (Lallemand et al. 2012; Leonardi et al. 2016).

The use of chitosan as a cationic agent (Kesavan et al. 2013) may be a good alternative due to its numerous biological properties and its good safety (Irimia et al. 2018). However, to date no cationic emulsion based on chitosan has been tested in human, probably due to difficulties in stabilizing and manufacturing these prototypes.

Conclusion

Emulsion artificial tears are about to supplant hydrogels in the management of slight to moderate DED. This success is linked to the combination of multiple actions on the different features of dry eye and its main cause, tear evaporation (DEWS 2007; Lemp et al. 2012). Ophthalmologists have a wide range of products with different types and amounts of oils, surface charge, osmolality, and additional beneficial features. The new generation of emulsions will combine emulsified oil, hydrogel, and osmoprotectant to address most symptoms of dry eye, and future products will evolve with a better understanding of dry eye physiopathology.

Emulsions Used as Active Ingredient Vehicle

In addition to the beneficial use of ophthalmic emulsions as artificial tears, emulsions possess a huge potential as drug carriers for topical delivery. Emulsions can solubilize lipophilic compounds in their dispersed oily phase, protect the drug from contact with water, improve penetration by improving residence time on the ocular surface, and also provide a better penetration in ocular tissues by a huge surface of contact between oil droplets and the eye (Tamilvanan and Benita 2004; Lallemand et al. 2012).

Topical ocular delivery of cyclosporine A (CsA) is the preferred method for CsA delivery as a treatment for ocular inflammatory diseases, such as uveitis, vernal keratoconjunctivitis (VKC), and DED. However, due to the high molecular weight and very hydrophobic nature of CsA combined with the natural protective mechanisms of the eye against xenobiotics, achieving therapeutic levels of CsA in ocular tissues can be difficult. Given that CsA is highly soluble in various oils, e.g., medium-chain triglycerides (MCTs), castor oil, and olive oil, lipid-based formulations are particularly well suited for CsA delivery including the use of lipid-based emulsions (Lallemand et al. 2003, 2017).

Conventional oil-in-water emulsions have been developed to deliver CsA to ocular tissues located in the front of the eye (cornea, conjunctiva). These formulations have a number of advantages and features: they spread readily over the ocular surface and maximize the specific contact surface for drug absorption given the colloidal nature.

In addition, because the drug is already dispersed at the molecular level (i.e., solubilized) and encapsulated within the oil droplet, the risk of precipitation of CsA on the ocular surface is very limited. As described above, the TFLL itself is composed of lipids in its outer layer. Therefore, it is likely that a fraction of the oil droplet encapsulating CsA may merge with the TFLL, entrapping a portion of active ingredients. Since the TFLL turnover is much longer than that of the aqueous phase (Mochizuki et al. 2009), TFLL may act as a drug reservoir and allow for sustained release of CsA, although this has not been experimentally confirmed.

In 2003, Allergan was the first company to bring a licensed formulation of CsA eye drops to market. Restasis®, a preservative-free anionic oil-in-water nanoemulsion, contains CsA dissolved in castor oil with polysorbate 80 as the emulsifying agent. The resulting emulsion is further stabilized by carbomer copolymer. Marketing authorization for Restasis® in the USA was granted by the FDA in 2002 (US Food and Drug Administration 2002). However, Restasis® is presently not approved in the European Union (Lallemand et al. 2017; Labbe et al. 2017).

Bausch and Lomb launched Lacrinmune® in Argentina, which has a composition similar to that of Restasis®, except for the addition of sodium hyaluronate to the emulsion (Bausch and Lomb 2009). Hyaluronic acid derivatives are well-known and are widely used as artificial tears (AT) in the management of mild to moderate DED. The addition of sodium hyaluronate (SH) increases the viscosity of the emulsion, resulting in a longer residence time on the ocular surface than that of conventional eye drops.

In 2015, Ikervis® (Santen) was granted marketing authorization by the European Medicines Agency (EMA) for use in Europe (European Medicines Agency 2015).

We will first review these major emulsion products and then the emulsions under development.

Marketed Products

Restasis

In 1992, when Allergan started the development of Restasis, severe DED was largely an unmet medical need. CsA, an immunosuppressive drug, was already described as a potential candidate in the management of this condition (Faulds et al. 1993). However, due to its lipophilicity, cyclosporine was only available as an oily solution (Lallemand et al. 2003). Emulsion was therefore the most appropriate vehicle for this molecule. Allergan paved the way for ocular emulsions by providing full development from animal studies (Acheampong et al. 1999; Angelov et al. 1998) to clinical trials. In the first trial, the authors showed that CsA ophthalmic emulsions (0.05%, 0.1%, 0.2%, and 0.4%) were safe and well tolerated and significantly improved the ocular signs and symptoms of moderate to severe DED and that 0.05% and 0.1% should be selected for the phase III clinical trials (Stevenson et al. 2000). Two phase III clinical trials enrolled a total of 877 patients, and the medication proved to be safe and effective in the treatment of moderate to severe DED, yielding improvements in both objective and subjective measures with a significant improvement in subjective measures for the 0.05% emulsion (Sall et al. 2000). Allergan also showed that after long-term administration of the emulsion twice a day for 6 months at 0.05 and 0.1%, the systemic exposure was negligible, proving the systemic safety of this emulsion (Small et al. 2002). It received approval (December 2002) from the US FDA for Restasis (cyclosporine ophthalmic emulsion, 0.05%) as the first and only therapy for patients with keratoconjunctivitis sicca (i.e., DED) whose lack of tear production is presumed to be due to ocular inflammation (US Food and Drug Administration 2002). The product is administered twice a day.

The emulsion is composed of cyclosporine 0.05%, castor oil 1.25%, polysorbate 80 1%, Pemulen 0.05%, glycerine 2.2%, sodium hydroxide qs pH 7.2–7.6, and purified water qs 100% (US patent 8,629,111). Its physicochemical description and manufacturing process could not be found in the literature; however, US patent 8,633,162 describes a simple stirring and high shear mixing after mixing both phases. This process is possible due to the high amount of polysorbate (1%) compared to the oil amount (1.25%). With this ratio, it is not necessary to bring high energy to the system. The sterilization is described in the patent as being either heat sterilization, sterile filtration or aseptic manufacture. Details on the physicochemical properties of Restasis would have been very helpful to better understand its efficacy and why the 0.05% solution was more efficient than higher concentrations in the treatment of DED. This absence of a dose response may stem from several different parameters: either physicochemical properties such as different droplet sizes and surface charges or the distribution of cyclosporine within the emulsion, in the oily phase or in the aqueous phase or more simply by a side effect induced by higher amounts of CsA in the formulation (reported side effects are stinging and burning sensation on the ocular surface).

Restasis is now very widely used in the world for off-label inflammatory eye conditions, making this product one of the rare blockbusters in ophthalmology.

The US FDA has edited a "June 2013 Draft CsA Bioequivalence guidance," which stated that bioequivalence to Restasis could be based on in vitro and physicochemical assessment only (Walenga et al. 2019) while Allergan may request a clinical efficacy study to obtain approval (Gore et al. 2017). Allergan's position is based on the complexity of the emulsion system, which is not totally elucidated, and also on the fact that Restasis acts on DED through its vehicle in addition to CsA, as discussed above in the artificial tears section of this chapter. This complexity makes it difficult to compare two products solely based on the physicochemical parameters. At the time this chapter is written, no generics are yet authorized in the USA.

Durezol®

Difluprednate ophthalmic emulsion 0.05% (0.5 mg/mL) (Durezol®, Alcon Laboratories, Inc., Fort Worth, TX, USA) is another ophthalmic lipid emulsion commercially available in the USA (Donnenfeld 2011). This corticosteroid was approved by the US FDA in 2008 for the treatment of pain and inflammation associated with ocular surgery (cataract) (US Food and Drug Administration 2008).

Difluprednate is not soluble in water (9.7 μg/mL at 25 °C) (Wang et al. 2018), and its octanol–water partition coefficient (LogP) is expected to be between 2 and 4 (Dong et al. 2019).

The excipients, as listed on the package insert and label information, are boric acid, castor oil, glycerin, polysorbate 80, water for injection, sodium acetate, edetate disodium, and sodium hydroxide (to adjust the pH to 5.2–5.8). The emulsion is essentially isotonic with a tonicity of 304–411 mOsm/kg and is preserved by sorbic acid 0.1%. According to Yamaguchi et al. (2005), castor oil was chosen over other lipids because of better solubility in this oil, and the emulsion has oil phase at 5% w/w and polysorbate 80 at 4% w/w (Kimura et al. 2000). With 5% castor oil and 4% surfactant, this emulsion is clearly not optimized and could have been developed with less oil and less surfactant. It is manufactured using a classical two-step emulsification process, including high-shear mixing followed by high-pressure homogenization and is filter-sterilized. The droplet is about 100 nm in diameter (Yamaguchi et al. 2005; Patil et al. 2019).

Durezol is routinely used and greatly appreciated by ophthalmologists for the management of a wide range of ocular inflammation conditions, because it has shown enhanced penetration, better bioavailability in ocular tissues, rapid local metabolism, and strong efficacy, with a low incidence of adverse effects (Mulki and Foster 2011). However, like all corticoids, difluprednate is associated with elevated intraocular pressure (IOP) (Meehan et al. 2010). The emulsion did not prevent this side effect.

Generics are already competing with Durezol (Mercado-Sesma et al. 2017; Popovic et al. 2018).

Ikervis®

Ikervis® is a cationic nanoemulsion indicated for the treatment of severe keratitis in adult patients with DED that has not improved despite treatment with tear substitutes (Kinnunen et al. 2014). It has to be administered only once a day (while Restasis is administered twice a day). The same product was also registered in 2018 under a centralized procedure at EMA as Verkazia® for the treatment of severe vernal keratoconjunctivitis (European Medicines Agency 2015, 2018). The cationic nanoemulsion is a patented technology based on the Novasorb™ platform developed by Novagali Pharma, France (now Santen SAS). Due to the net positive charge of the oil nanodroplets, the residence time and the ocular penetration of CsA are higher with the cationic emulsion than with other formulations. In a pharmacokinetic study on rabbit eyes, corneal exposure to CsA after a single dose was 1.84 times greater for a 0.05% CsA cationic emulsion (Novasorb™ formulation) than for a 0.05% CsA anionic emulsion (Restasis®), with AUC_{0-72} h of 26,703.0 ng h/g and 14,333.2 ng h/g, respectively (Daull et al. 2013a). Correspondingly, the corneal clearance of CsA was 57% less for the 0.05% CsA cationic emulsion than for the 0.05% CsA anionic emulsion (0.8 g/h and 1.4 g/h, respectively). It is assumed that the residence time of CsA in Ikervis® (Novasorb™ CsA cationic nanoemulsion) is greater than that in Restasis® (CsA anionic nanoemulsion) because of electrostatic interactions between the positively charged droplets and negatively charged mucus protein of the corneal epithelium (Lallemand et al. 2012). This mechanism of action is thought to work in conjunction with the hypothesized reservoir effect of the TFLL. The combination of these effects, as well as higher dosage strength, could very likely explain the difference in dosing regimen between once-a-day Ikervis®

and twice-a-day Restasis®. Kuwano et al. compared the ocular pharmacokinetics of three CsA formulations in rabbit eyes: a castor oil solution (Oil-CsA), a micellar solution (Aq-CsA) with a composition similar to Papilock Mini®, and an oil-in-water emulsion (Em-CsA) with a composition similar to Restasis®. This study reported that the AUC_{0-12} of Em-CsA and Aq-CsA were, respectively, 9.2- and 28.5-fold higher than the AUC_{0-12} of Oil-CsA in corneal stroma endothelium. The same pattern was observed in the bulbar conjunctiva, wherein the AUC_{0-12} of Em-CsA and Aq-CsA were 2.4- and 5.1-fold higher, respectively, than the AUC_{0-12} of Oil-CsA (Kuwano et al. 2002).

Ikervis is composed of cyclosporine at 0.1%, 2% medium-chain triglycerides, 0.2% tyloxapol and 0.01% poloxamer 188, 2.2% glycerin, and a cationic agent cetalkonium chloride at 0.005%. The emulsion is produced by a two-step conventional process using high-shear mixing and a high-pressure homogenizer. The system needs energy because the surfactant only amounts to 0.215% of the formulation for 2% oil. The emulsion is sterilized by autoclave 20 min at 120 °C, which is proof of the system's excellent stability (Lallemand et al. 2012). The emulsion is very well characterized in terms of its physicochemical properties.

Improving the penetration of CsA in ocular tissues is obviously required to maximize the clinical efficacy of CsA products. Nevertheless, DED is a very complex disease with various etiologies and manifestations (signs and symptoms) so that an optimal treatment may require additional interventions alongside pharmacological therapy to achieve optimal outcomes. In this respect, emulsions, such as lipid-containing vehicles, favorably interact with the TFLL and the ocular surface to help relieve DED signs and symptoms and provide treatment benefits beyond the pharmacokinetic and formulation features.

Lacrinmune®

Bausch and Lomb launched Lacrinmune in Argentina, which has a composition similar to that of Restasis, except for the addition of sodium hyaluronate to the emulsion (Bausch and Lomb 2009). The addition of sodium hyaluronate increases the viscosity of the emulsion, resulting in a longer residence time on the ocular surface than that of aqueous eye drops. Unfortunately, no clinical results have been published on this product.

Xelpros®

Xelpros is the most recent emulsion product approved in the USA (September 2018). It is an emulsion of latanoprost at 0.005%, preservative free, and was developed with SPARC's Swollen Micelle microemulsion technology. In a head-to-head comparative study in glaucoma patients, Xelpros™ was found to be as efficacious and safe as Xalatan. For the moment, very little information is available on this product (https://www.sparc.life/research-programs/xelpros). The composition is described in Table 2.

Table 2 Use of emulsions as drug delivery vehicles for drug products

Emulsion type	Brand name	*API*/main components	Refs.
Anionic emulsion	Restasis® (Allergan)	*Cyclosporine 0.05%*, castor oil 1.25%, polysorbate 80 1%, Pemulen 0.05% Glycerine 2.2%, sodium hydroxide qs pH 7.2–7.6, and purified water qs 100%	US 8,629,111 (Acheampong et al. 2014)
	Xelpros® (Sun Pharma)	*Latanoprost (0.005%)*, castor oil, sodium borate, boric acid, propylene glycol, edetate disodium, polyoxyl 15 hydroxystearate, sodium hydroxide, hydrochloric acid, and water for injection	NDA 206185 (2018)
	Durezol® (Alcon)	*Difluprednate 0.05%*, castor oil 5%, polysorbate 80 4%, sorbic acid 0.1%, glycerin, sodium acetate, edetate disodium, sodium hydroxide, water for injection	Yamaguchi et al. (2005)
	Lacrinmune® (Bausch and Lomb)	*Cyclosporine 0.05%*, glycerin 0.5% castor oil 0.5%, polysorbate 80 0.2%, hyaluronic acid 0.3%, potassium sorbate 0.180%, NaOH qs pH, water for injection	Bausch and Lomb (2009)
	Under development	*Indomethacin* 0.1%, castor oil 10%, and polysorbate 80 8.0%; and: (1) uncoated (2) coated with 1.5% chitosan (3) coated with 0.5% HPMC	Yamaguchi et al. (2009)
Cationic emulsion	Ikervis®/Verkazia® (Santen)	*Cyclosporine* 0.1%, MCT 2%, CKC 0.005%, tylopaxol 0.2%, glycerol 1.5–2.5%, purified water	Leonardi et al. (2016)
	DE130—Catioprost® (Santen)	Latanoprost 0.005% MCT 2%, CKC 0.005%, tylopaxol 0.2%, glycerol 1.5–2.5%, purified water	Liang et al. (2012), Daull et al. (2012, 2017)
Polyaphron	PADciclo™ (MC2 Therapeutics)	*Cyclosporine* 0.03–0.06%, MCT, laureth-4, poloxamer 188, polyacrylic acid (carbomer 980), glycerol, NaOH demineralized water	Steele (2012)

CKC cetalkonium chloride, *CsA* cyclosporin A, *HMW* high molecular weight, *HPMC* hydroxypropyl methylcellulose, *MCT* medium-chain triglyceride, *o/w* oil in water, *PC* phosphatidylcholine, *SH* sodium hyaluronate, *v/v* volume/volume, *w/v* weight/volume, *w/w* weight/weight

Products Under Development

PADciclo™

PADciclo™, being developed by the Danish biotech company MC2 Therapeutics, comprises a dispersion of polyaphrons encapsulating CsA within oily micrometer-sized aphrons dispersed in a hydrogel of carbomer (Steele 2012). Polyaphrons are lipid-based formulations first described in the late 1970s (Sebba 1979). The authors of this chapter have considered polyaphrons as emulsions due to the presence of dispersed oil, surfactant, and water. The main difference of polyaphrons over the usual emulsions is the high concentration of the dispersed oil phase, up to nearly 90% in proprietary formulations, and the amount of surfactant necessary to maintain stability of the oily phase is very low, usually less than 0.5%. These attributes make polyaphrons a good vehicle for ocular delivery. However, their cream-like viscosity can make them unsuitable for topical ocular administration (Lallemand et al. 2017). The CsA concentrations in PADciclo are slightly lower than those of existing licensed products (Restasis 0.05% CsA bid and Ikervis 0.1% CsA qd), suggesting that the polyaphron technology may improve ocular delivery of CsA. Recent preclinical pharmacokinetic data demonstrated negligible systemic CsA exposure, but after multiple topical PADcicloTM administrations, the conjunctival and corneal penetration of CsA was up to fivefold greater than that achieved with Restasis (Praestegaard et al. 2016). A phase II clinical trial (EudraCT 2015–000937-54) evaluating the safety and efficacy of two dosages of PADciclo administered once daily (0.03% and 0.06% w/ w CsA) was launched in the summer of 2015 and successfully completed in Q3 2018 (n = 265). In this randomized, double-masked, parallel-arm, controlled 6-month trial enrolling 263 adult patients, two active treatment arms (MC2-03 0.06% CsA and MC2-03 0.03% CsA) were compared to two control arms (MC2-03 vehicle and lubricant) as an add-on therapy to a standard-of-care lubricant. Active MC2-03 eye drop treatment resulted in a strong clinical improvement in severe dry eye patients over control arms both on the corneal fluorescence staining response (p = 0.0090) and the mean change from baseline to month 6 (p = 0.0496).

A comprehensive biomarker study confirms anti-inflammatory efficacy, including a reduction of HLA-DR-positive cells from baseline to month 6 (p = 0.028). The data demonstrate a favorable safety and tolerability profile, including low incidence of eye irritation (data presented at Ophthalmology Innovation Summit OIS October25th, 2018, Chicago, IL, USA). Discussions with authorities on the next development steps are ongoing.

Catioprost®

Glaucoma is one of the leading causes of blindness for people over the age of 60. It can occur at any age but is more common in older adults. This disease is caused by abnormally high pressure in the eye, damaging the optic nerve. One of the main antiglaucoma/ocular antihypertensive drugs is latanoprost, a prostaglandin F2alpha, which acts as a selective agonist at the prostaglandin F receptor, increasing outflow of aqueous fluid, thus lowering IOP (Digiuni et al. 2012). Sold under the brand Xalatan by Pfizer, although an excellent product, it possesses two drawbacks. Firstly, this product has to be kept at 5 °C because of a hydrolysis process in water; secondly, it contains benzalkonium chloride, a potent preservative that is deleterious when used on a daily basis, leading to dry eye symptoms. Glaucoma and DED commonly occur together. It is estimated that 60% of glaucoma patients have symptoms of ocular surface disease. Ophthalmic emulsion may solve the problem of stability of latanoprost by encapsulating the compound in oily droplets and, like artificial tears, may treat dry eye symptoms by lubricating and healing the corneal surface. Since latanoprost is soluble in oil (Log *D*(pH 7) = 4.28 (Rodriguez-Aller et al. 2015), it was formulated in a cationic emulsion. This emulsion Catioprost (at 0.005%) is preservative-free and based on the Ikervis formulation (Lallemand et al. 2012). The prototype was applied to cultured corneal cells and in a corneal scrapping rat model. In both models, Catioprost showed improvement in corneal healing versus comparators (Liang et al. 2012). In another primate model, Catioprost was as effective as Xalatan in lowering IOP, with an improved ocular tolerance profile after one instillation per day for 5 days (Daull et al. 2012). Following these promising results, a phase II clinical trial was conducted on glaucoma patients suffering from dry eye symptoms, comparing IOP lowering and healing of ocular surface lesions between patients treated with Travatan Z (a soft preserved prostaglandin) and Catioprost. The clinical data demonstrated that Catioprost reduced IOP to the same level as Travatan Z, induced less conjunctival hyperemia, and showed a reduction of corneal fluorescence staining compared to the Travatan Z group. The results suggest that beyond the simple removal of preservative, Catioprost acts by restoring and stabilizing the tear film along with its anti-inflammatory and wound-healing properties (Daull et al. 2017). A multinational phase III clinical trial has received authorization to start in 2019 (EudraCT Number: 2017-004262-95) (Santen 2017). This trial has the objective of demonstrating that the IOP-reducing effect of DE-130A (Catioprost) is non-inferior to that of Xalatan®. Results should be available by 2021.

It was not possible to review all the attempts and prototypes that are described in the literature. We will only look at a few molecules that were formulated by different teams. Table 3 updated from Peng 2011 lists all the studies conducted with ophthalmic emulsions with the type of study and the reference indicating where more information can be found.

Tacrolimus

Tacrolimus is a potent immunosuppressive agent with limited corneal penetration and low aqueous solubility. It can be used in the prevention of corneal graft rejection and in some ocular inflammatory diseases (Zhai et al. 2011). A first attempt at formulation was made in 2010 but no follow-up could be found with this team (Wang et al. 2010). A more recent prototype was described by Silva-Cunha et al. (2014).

Table 3 Ophthalmic drugs explored in emulsion/microemulsion systems

Drug	Type of study	References
Adaprolol maleate	In vitro	Anselem et al. (1993)
Amphotericin B	In vivo	Cohen et al. (1996)
Azithromycin	In vitro, in vivo	Liu et al. (2009)
Ciprofloxacin	In vitro	Pandey et al. (2019)
Chloramphenicol	In vitro	Ashara and Shah (2017)
Cyclosporine	In man	Lallemand et al. (2003, 2017)
Delta-8-tetrahydrocannabinol	In vivo	Muchtar et al. (1992)
Dexamethasone	In vitro, in vivo, in man	Kesavan et al. (2013), Suresh and Dewangan (2011), Fialho and da Silva-Cunha (2004), Li et al. (2016), Daull et al. (2013b)
Diclofenac	In vitro	Siebenbrodt and Keipert (1993)
Difluprednate	In vitro, in vivo, in man	Yamaguchi et al. (2005), Kakimoto et al. (2018), Garg et al. (2016)
Everolimus	Ex vivo	Baspinar et al. (2008)
Fisetin	In vivo	Joussen et al. (2000)
Genistein	In vivo	Joussen et al. (2000)
Fluconazole	Ex vivo, in vivo	Pathak et al. (2013)
Flurbiprofen axetil	In vivo	Shen et al. (2011, 2010)
HU-211 synthetic cannabinoid	In vitro, in vivo	Naveh et al. (2000)
Indomethacin	In vitro, in vivo	Yamaguchi et al. (2009), Calvo et al. (1996a), Klang et al. (2000), Czajkowska-Kosnik et al. (2012), Muchtar et al. (1997)
Ketoconazole	In man	Maichuk Iu et al. (1990)
Latanoprost	In vitro, in vivo, in man	Liang et al. (2012), Daull and Garrigue (2013)
Levobunolol	In vitro	Gallarate et al. (1993)
Lidocaine	In vitro	Gulsen and Chauhan (2005, 2004)
Luteolin	In vivo	Joussen et al. (2000)
Miconazole	In vitro	Wehrle et al. (1996)
Palmatine	In vitro, in vivo	Yin et al. (2016)
Pilocarpine	In vitro, in vivo, in man	Garty et al. (1994), Naveh et al. (1994), Beilin et al. (1995)
Piroxicam	In vitro	Klang et al. (1996)
Poly-anionic oligonucleotide For VEGFR-2-(17 MER)	In vitro, in vivo	Hagigit et al. (2008, 2010)

(continued)

Table 3 (continued)

Drug	Type of study	References
Prednisolone	In vitro	Ibrahim et al. (2009)
Retinoic acid	In man	Selek et al. (2000)
Sirolimus	Ex vivo	Buech et al. (2007)
Tacrolimus	In vitro, in vivo	Wang et al. (2010), Silva-Cunha et al. (2014)
Timolol	In vitro, ex vivo	Gallarate et al. (1993, 2013), Li et al. (2007)

Updated table from Peng J. Drug Del. Sci. Tech., 21 (1) 111–121, 2011

This team used isopropyl myristate as the oil phase (6.0% w/v) and (polysorbate 80) and propylene glycol as stabilizers. A transparent formulation containing 1.0% w/v of tacrolimus was obtained after two steps of high-shear mixing. This emulsion showed good ocular tolerance in the rabbit eye but no further development could be found in the literature.

Dexamethasone

Dexamethasone is one of the most powerful corticoids currently used and several ophthalmic preparations were tested (Suresh and Dewangan 2011). It is therefore logical that some teams have tried to formulate dexamethasone in ocular emulsions.

The first study describes the development and characterization of an oil-in-water microemulsion containing dexamethasone and the evaluation of its pharmacokinetics in rabbits after topical ocular application (Fialho and da Silva-Cunha 2004). The oily phase (isopropyl myristate) and then the surfactant (Cremophor EL) and the water phase were added and rotated, using a high-shear mixer; then, a cosurfactant (propylene glycol) was added to the macroemulsion and high-shear mixed. The emulsion contained 15.0% w/w of Cremophor EL, 5.0% w/w of isopropyl myristate as the oil phase, ultrafiltrated water with benzalkonium chloride (0.01% w/w) as the aqueous phase, and 15.0% w/w of propylene glycol as the cosurfactant. Dexamethasone was used at a concentration of 0.1% w/v. The microemulsion-based dexamethasone showed good tolerance in the rabbit eye (even though a very high amount of surfactant is present) and seemed to provide a higher degree of ocular tissue penetration compared to conventional aqueous solution with $AUC_{0-540\ min}$ of 325.60 ± 36.51 and 121.67 ± 10.16, respectively, after one instillation. Unfortunately, development was not pursued.

The next study describes the use of mucoadhesive chitosan-coated cationic microemulsions (CH-MEs) for ophthalmic delivery of dexamethasone at 0.1% to treat uveitis (Kesavan et al. 2013). Tween 80 was used as the surfactant, isopropyl myristate as the oily phase, and chitosan as the interfacial cationic agent to obtain emulsion by simple magnetic stirring, with a mean droplet size from 50 to 17 nm and zeta potential approximately +25 mV. The prototypes showed acceptable physicochemical behavior and good stability for 3 months and exhibited sustained drug release. Furthermore, these emulsions possess good mucoadhesive properties. In vivo study results indicated that mucoadhesive emulsions performed better in retaining dexamethasone than did the marketed drug solution. Again, a promising formulation but with no other work was published.

Li et al. also used a cationic nanoemulsion to deliver dexamethasone combined with polymyxin B (Li et al. 2016). Nanoemulsions with the lipid phase octyldodecanol-phosphatidylcholine (70%:30%) containing 0.05% (w/w) dexamethasone were produced by high-pressure homogenization, followed by dissolving the hydrophilic molecules in the water phase, e.g., polymyxin B (0.1%, w/w), cetylpyridinium chloride (0.01%, w/w), and glycerol (2.6%, w/w). The particles were below 200 nm with a narrow size distribution. The zeta potential of the optimized formulation was shifted from approximately +9 mV. This prototype did not go further into development. The zeta potential is not very high, and the octyldodecanol is not commonly used in pharmaceutics but more commonly in cosmetics. Also, the use of a preservative as the cationic agent at a concentration where it can be deleterious is not the best choice.

Finally, an emulsion was used to administer dexamethasone in the form of a palmitate prodrug in the vitreous by intravitreal injection (Daull et al. 2013b). This emulsion contained medium-chain triglycerides such as oil and lecithin as surfactant for a droplet size of about 200 nm. No information on the amount of oil and surfactant or on the mode of production could be found. The emulsion loaded with dexamethasone palmitate showed an excellent safety profile (in cat, rat and rabbit) and provided a therapeutic concentration of corticoid in the retina and choroid for a period of 9 months in the eye of a rabbit. This emulsion underwent a phase I clinical trial for the evaluation of safety and tolerability of a single injection administered to patients with diabetic macular edema secondary to diabetic retinopathy in 2010, but the results were not published (Hamdi et al. 2015).

Indomethacin

Ocular inflammation is a common eye disorder. Nonsteroidal anti-inflammatory drugs (NSAIDs) such as indomethacin are increasingly being used to treat such inflammations (Bucolo et al. 2014).

Back in the 1990s, Calvo used indomethacin as a model molecule to evaluate the effect of nanocarriers on corneal penetration, one of the nanocarriers being a microemulsion about 200 nm in size and a zeta potential of −40 mV (Calvo et al. 1996a, b). The interesting conclusions of these studies were that the nanocarriers were well tolerated following topical ocular administration, and whatever indomethacin-loaded colloidal systems were used, they significantly improved the corneal penetration of indomethacin in comparison to a commercial formulation, meaning that the small size is the driving factor for penetration.

A few years later, Klang et al. also evaluated the penetration into the cornea of a cationic emulsion composed of medium-chain triglycerides, lecithin, poloxamer

188, and stearylamine as a cationic agent (Klang et al. 2000). This study showed that the cationic charge of the emulsion could improve the penetration of indomethacin in ocular tissue. This prototype has led to the cationic platform used in Novasorb, as discussed above for Ikervis and Cationorm.

Yamaguchi from Senju Pharmaceuticals also described a cationic emulsion that uses chitosan as a cationic agent, castor oil (5%) and polysorbate 80. The production is a two-step emulsifying process by high-shear mixing and high-pressure homogenization to obtain a particle size of about 120 nm and a zeta potential of +23 mV. This prototype is very close to Durezol and was developed by the same team. The pharmacokinetic study in rabbits indicated higher distribution of the cationic emulsion on the ocular surface than the negative emulsion. The drug concentrations for the cationic emulsion in cornea, conjunctiva, and aqueous humor were clearly higher than for the negative control 1 h after instillation, giving a promising prototype (Yamaguchi et al. 2009).

Self-Emulsifying Drug Delivery Systems (SEDDSs)

Self-emulsifying drug delivery systems (SEDDSs) are preparations used to improve the bioavailability of poorly soluble drugs. SEDDSs are mixtures of oil and surfactant that emulsify upon contact with water and under gentle mixing. Examples of a commercially available oral SEDDS are Sandimmun Neoral (CsA), Norvir (ritonavir), and Fortovase (saquinavir). This concept was applied to ocular delivery of indomethacin and hydrocortisone by Czajkowska-Koanik et al. (Czajkowska-Kosnik et al. 2012). After topical application, SEDDSs form, with lacrimal fluid and upon gentle agitation (blinking), an oil-in-water emulsion. Despite the increased drug solubility (up to 0.6% indomethacin and up to 1% hydrocortisone), SEDDSs do not allow for faster in vitro drug release and penetration through a dialysis membrane, because the thermodynamic activity of the drug was not improved. The team used MCT as oil and Span 80 (sorbitan monooleate), Tween 20 (polyoxyethylene sorbitan monolaurate), and Cremophor EL (polyethoxylated castor oil) as surfactants. Even though the approach is interesting because it avoids emulsification steps and problems of emulsion stability, SEDDSs will need a large amount of work to be tested in human.

Discussion and Conclusion

As we have seen in this chapter, emulsions may provide several advantages to ocular drug delivery and use as artificial tears. Then, why are so few products on the market and being developed?

The first reason is that there are few lipophilic, oil-soluble molecules. Most ocular drugs were chemically developed to be soluble in water for the conventional aqueous eye drop dosage form. Another reason may be a sort of fear that may arise toward a complex formulation, complex manufacturing processes such as sterility, and the possible issue of stability or fear of tolerance issues. However, a more detailed understanding of the emulsion's physicochemical features would benefit its use. The US FDA (Rahman Z et al.) listed the physicochemical parameters than could influence directly the safety and efficacy of an ophthalmic emulsions: particle size, turbidity, zeta potential, viscosity, osmolality, surface tension, contact angle, pH, and drug diffusion (Rahman et al. 2014). Those parameters are appropriate but may not be sufficient to fully characterize an emulsion, and a deeper understanding on the influence of excipient and manufacturing process on them would greatly impact the development of this dosage form.

Size

The mean globule diameter and the polydispersity index (PI) of emulsions are two important parameters for predicting the physical stability as well as the possibility of using the globules or droplets as drug carriers with a biological effect (Charman et al. 1992). The size is a key driver of drug penetration in ocular tissue. From a biological point of view, as oral emulsions, fine emulsification enhances absorption in tissue due to a huge specific area, increasing contact with target tissue. However, a study comparing the effect of different droplet sizes has never been published. This type of study would help formulators target the most appropriate droplet size and droplet size distribution to either stay on the surface or penetrate into the anterior chamber.

Drug Distribution into the Complex Emulsion System

As discussed by Gore et al., the location of the drug dissolved in oil is of major importance for stability but most particularly for drug absorption (Gore et al. 2017). In fact, the drug can partition in the oily phase, in the aqueous phase, either free or under micelles when excess surfactant is present, or at the oil–water interface. The distribution of the drug to these different locations can influence the penetration in ocular tissue. Unfortunately, the literature reviewed produced no studies investigating the partition of a drug and its influence on the efficacy of the emulsion. This distribution is influenced by the log p of the drug as well as the emulsion's mode of production (Sila-on et al. 2008).

Mechanism of Penetration

Several studies have shown an increased penetration of drugs in ocular tissues when formulated into emulsion versus aqueous solution. However, the exact mechanism of penetration is not known: Is it a passive diffusion of the molecule, an active mechanism through endocytosis, or perhaps both? A pharmacokinetic study after topical administration with labeled excipients and labeled drug in the system would make it possible to determine whether the drug is penetrating in presence of excipient or alone in the cornea and conjunctiva, thus disclosing the mechanism of penetration. Knowing that, the formulation would be different with, for example, a higher concentration of drug in the oil droplet or smaller droplets, etc.

Fate of the Emulsion on the Ocular Surface

Once instilled, we do not know if the emulsion breaks, leaving the oil in the lipid layer of the tear film, or if the oil droplets remain stable. Also, for emulsions with bioadhesive properties (cationic emulsion, viscosified emulsions), residence time and tear film distribution would contribute valuable information for the development of novel emulsions.

Possible Improvements Would Contribute to Extending the Use of Emulsions

Multiple water-in-oil-water emulsions would make it possible to use emulsions for hydrophilic drug substances such as proteins included in aqueous droplets dispersed in oily droplets dispersed in water (Glasser et al. 2016). This type of formulation would provide new physical and biological properties to topical emulsions.

Surfactant-free emulsions will probably be the new paradigm in emulsion formulation. Emulsions can now be stabilized by proteins (which could be therapeutic proteins), polysaccharides (Bouyer et al. 2012), or other natural polymers such as cellulose (Schulz and Daniels 2000) and guar (Mafi et al. 2014). The wide range of natural polymers opens a new field of investigation for totally natural and well-tolerated ophthalmic emulsions.

Other Routes of Administration

In this chapter, we have cited an example of an emulsion injected intravenously (Daull et al. 2013b). Other than a few examples, emulsions are not used for other ocular routes such as subconjunctival, intravitreal, and palpebral administration. These formulations would provide new biological properties such as sustained released at the site of administration.

Process of Manufacture

The final physicochemical characteristics of an emulsion are obviously based not only on the choice of excipients and their concentration but also on the manufacturing process. This process is of great importance and can directly affect efficacy of the emulsion by modifying some of the driving safety and efficacy parameters, such as viscosity, pH, zeta potential, mean droplet size and droplets size distribution, active ingredient distribution within the emulsion (aqueous, oily, or interfacial phases), and presence of micelles in the aqueous phase. The manufacturing process is a question in the development of ophthalmic emulsions, but unfortunately the literature is very poor on this topic. Sila-on et al. have demonstrated the importance of order of incorporation of excipients on the partitioning of active drug within the emulsion (Sila-on et al. 2008). One can either use a high amount of surfactant to self-emulsify or an emulsion with simple stirring. If the formulator is willing to optimize the amount of surfactant and decrease this amount as low as possible, then energy must be brought to the formulation. This is currently the most commonly used approach. It usually includes several steps of emulsification by high-shear mixing followed sometimes by high-pressure homogenization. These processes may be long for larger commercial batches and create a stress for the product. Formulator must therefore anticipate the choice of excipients and their ratio to obtain appropriate physicochemical properties. Self-emulsions have not yet shown good efficacy after topical application (Czajkowska-Kosnik et al. 2012). Therefore, new processes are being investigated (ultrasound, Pickering, spontaneous emulsification (Lefebvre et al. 2017), membrane emulsification (Gehrmann and Bunjes 2017)), endowing new features to emulsion, such as a new partition of the drug with the system, new droplet size profile, etc.

Eye drops are to be sterile, therefore submitted to a sterilization process which can greatly impact the physicochemical parameters of emulsions. A liquid can be sterilized by three manners: filtration at 0.22 μm, autoclaving, and aseptic production. The sterilizing filtration will retain the droplets which have a diameter above 0.22 μm and modify the droplets distribution and therefore penetration profile and behavior on the ocular surface and potentially the amount of active ingredient. The autoclave is the easiest and most suitable way to sterilize emulsion contrary to what most people think. A submicron emulsion with appropriate excipients can easily support the thermic conditions of an autoclave, usually 110–125 °C during 10–30 min. Such amount of energy will modify the thermodynamics of the system in positive and negative ways. The droplets size and size distribution will change significantly not only either by increasing or reducing the diameter of the droplets but also sometimes by tightening the size distribution. The viscosity can be decreased due to the degradation of certain viscosifying polymers, leading the formulator to anticipate this degradation when choosing the amount of polymer and its chain length. Zeta potential, interfacial tension, and pH can also be modified due to the apparition of several degradation products and loss of active ingredient. But if well controlled, this sterilizing process can easily be used for emulsion. Aseptic process is costly but useful when emulsion cannot be submitted to heat stress. The different excipients have to be sterilized separately either by filtration or by autoclave.

For a same qualitative and quantitative excipient and active ingredient composition but different manufacturing processes, the safety and efficacy profile can be significantly modified.

To Conclude

Emulsions are now a well-established tool for the treatment of anterior-segment diseases. Topical administration has been routine either as artificial tears or as an active substance carrier for over two decades. They are now about to supplant hydrogels in the management of moderate dry eye and have replaced ointments to administer lipophilic compounds. Health authorities, ophthalmologists, and patients are satisfied with emulsion-based eye drops. As discussed above, several improvements and a deeper understanding of the biology related to physicochemistry would contribute to the emergence of several new products in the coming years.

References

Acheampong AA, Shackleton M, Tang-Liu DD, Ding S, Stern ME, Decker R. Distribution of cyclosporin a in ocular tissues after topical administration to albino rabbits and beagle dogs. Curr Eye Res. 1999;18(2):91–103.

Acheampong A, Tang-Liu DD, Chang JN, Power DF. Methods of providing therapeutic effects using cyclosporin components. United States. Patent US8629111B2; 2014.

Aguilar AJ, Marquez MI, Albera PA, Tredicce JL, Berra A. Effects of Systane((R)) balance on noninvasive tear film break-up time in patients with lipid-deficient dry eye. Clin Ophthalmol. 2014;8:2365–72.

Amrane M, Creuzot-Garcher C, Robert PY, Ismail D, Garrigue JS, Pisella PJ, Baudouin C. Ocular tolerability and efficacy of a cationic emulsion in patients with mild to moderate dry eye disease - a randomised comparative study. J Fr Ophtalmol. 2014;37(8):589–98.

Angelov O, Wiese A, Yuan Y, Andersen J, Acheampong A, Brar B. Preclinical safety studies of cyclosporine ophthalmic emulsion. Adv Exp Med Biol. 1998;438:991–5.

Anselem S, Beilin M, Garty N. Submicron emulsion as ocular delivery system for adaprolol maleate, a soft b-blocker. Pharm Res. 1993;10(Suppl):S205.
Ashara KC, Shah KV. Emulsion of chloramphenicol: an overwhelming approach for ocular delivery. Folia Med (Plovdiv). 2017;59(1):23–30.
Aydemir O, Demir T, Türkçüoğlu P, Turgut B, Çeliker Ü (2008) Efficiency of additional topical ophthalmic phospholipid-based microemulsion (Lipimix) in the treatment of chronic blepharitis with dry eye. Paper presented at the 9th congress of the International Society of Dacryology and dry eye, Istanbul, Turkey,
Baspinar Y, Bertelmann E, Pleyer U, Buech G, Siebenbrodt I, Borchert HH. Corneal permeation studies of everolimus microemulsion. J Ocul Pharmacol Ther. 2008;24(4):399–402.
Bausch & Lomb. Lacrinmune (product information; in Spanish); 2009.
Beilin M, Bar-Ilan A, Amselem S. Ocular retention time of submicron emulsion (SME) and the miotic response to pilocarpine delivered in SME. Invest Ophthalmol Vis Sci. 1995;36:S166.
Bouyer E, Mekhloufi G, Rosilio V, Grossiord J-L, Agnely F. Proteins, polysaccharides, and their complexes used as stabilizers for emulsions: alternatives to synthetic surfactants in the pharmaceutical field? Int J Pharm. 2012;436(1–2):359–78.
Bron AJ, Tiffany JM, Gouveia SM, Yokoi N, Voon LW. Functional aspects of the tear film lipid layer. Exp Eye Res. 2004;78(3):347–60.
Bucolo C, Marrazzo G, Platania CB, Romano GL, Drago F, Salomone S. Effects of topical indomethacin, bromfenac and nepafenac on lipopolysaccharide-induced ocular inflammation. J Pharm Pharmacol. 2014;66(7):954–60.
Buech G, Bertelmann E, Pleyer U, Siebenbrodt I, Borchert HH. Formulation of sirolimus eye drops and corneal permeation studies. J Ocul Pharmacol Ther. 2007;23(3):292–303.
Butovich IA, Millar TJ, Ham BM. Understanding and analyzing meibomian lipids--a review. Curr Eye Res. 2008;33(5):405–20.
Calvo P, Alonso MJ, Vila-Jato JL, Robinson JR. Improved ocular bioavailability of indomethacin by novel ocular drug carriers. J Pharm Pharmacol. 1996a;48(11):1147–52.
Calvo P, Vila-Jato JL, Alonso MJ. Comparative in vitro evaluation of several colloidal systems, nanoparticles, nanocapsules, and nanoemulsions, as ocular drug carriers. J Pharm Sci. 1996b;85(5):530–6.
Charman SA, Charman WN, Rogge MC, Wilson TD, Dutko FJ, Pouton CW. Self-emulsifying drug delivery systems: formulation and biopharmaceutic evaluation of an investigational lipophilic compound. Pharm Res. 1992;9(1):87–93.
Cohen T, Sauvageon-Martre H, Brossard D, D'Hermies F, Bardin C, Chast F, Chaumeil JC. Amphotericin B eye drops as a lipidic emulsion. Int J Pharm. 1996;137:249–54.
Cwiklik L. Tear film lipid layer: a molecular level view. Biochim Biophys Acta. 2016;1858:2421–30.
Cwiklik L, Melcrova A, Daull P, Garrigue J. Tear film break-up: a molecular level view by employing in silico approach. Paper presented at the ARVO, Baltimore; 2017.
Czajkowska-Kosnik A, Sznitowska M, Mirkowska K. Self-emulsifying oils for ocular drug delivery. II. In vitro release of indomethacin and hydrocortisone. Acta Pol Pharm. 2012;69(2):309–17.
Daull P, Garrigue JS. reservative-free cationic nanoemulsion of latanoprost. Ophthalmology times Europe. 2013 [Epub ahead of print].
Daull P, Buggage R, Lambert G, Faure MO, Serle J, Wang RF, Garrigue JS. A comparative study of a preservative-free Latanoprost cationic emulsion (Catioprost) and a BAK-preserved Latanoprost solution in animal models. J Ocul Pharmacol Ther. 2012;28(5):515–23.
Daull P, Lallemand F, Philips B, Lambert G, Buggage R, Garrigue JS. Distribution of cyclosporine A in ocular tissues after topical administration of cyclosporine A cationic emulsions to pigmented rabbits. Cornea. 2013a;32(3):345–54.
Daull P, Paterson CA, Kuppermann BD, Garrigue JS. A preliminary evaluation of dexamethasone palmitate emulsion: a novel intravitreal sustained delivery of corticosteroid for treatment of macular edema. J Ocul Pharmacol Ther. 2013b;29:258–69.

Daull P, Lallemand F, Garrigue JS. Benefits of cetalkonium chloride cationic oil-in-water nanoemulsions for topical ophthalmic drug delivery. JPP. 2014;66(4):531–41.
Daull P, Feraille L, Elena PP, Garrigue JS. Comparison of the anti-inflammatory effects of artificial tears in a rat model of corneal scraping. J Ocul Pharmacol Ther. 2016;32:109–18.
Daull P, Amrane M, Garrigue JS. Novasorb® cationic nanoemulsion and latanoprost: the ideal combination for glaucoma management? Glauc Open Access. 2017;2(1):107.
Daull P, Guenin S, Hamon de Almeida V, Garrigue JS. Anti-inflammatory activity of CKC-containing cationic emulsion eye drop vehicles. Mol Vis. 2018;24:459–70.
DEWS. The epidemiology of dry eye disease: report of the epidemiology Subcommittee of the International dry eye WorkShop (2007). Ocul Surf. 2007;5:93–107.
Digiuni M, Fogagnolo P, Rossetti L. A review of the use of latanoprost for glaucoma since its launch. Expert Opin Pharmacother. 2012;
Dong Y, Hengst L, Patel D, Hunt R, Qu H, Choi S, Ashraf M, Cruz CN, Xu X. A kinetic approach to determining drug distribution in complex biphasic systems. J Pharm Sci. 2019;108:2002–11.
Donnenfeld ED. Difluprednate for the prevention of ocular inflammation postsurgery: an update. Clin Ophthalmol. 2011;5:811–6.
European Medicines Agency. Ikervis authorisation. 2015. https://www.ema.europa.eu/documents/assessment-report/ikervis-epar-public-assessment-report_en.pdf. Accessed 11 Feb 2019.
European Medicines Agency. Verkazia authorisation. 2018. https://www.ema.europa.eu/en/medicines/human/EPAR/verkazia. Accessed 11 Feb 2019.
Faulds D, Goa KL, Benfield P. Cyclosporin. A review of its pharmacodynamic and pharmacokinetic properties, and therapeutic use in immunoregulatory disorders. Drugs. 1993;45(6):953–1040.
Fialho SL, da Silva-Cunha A. New vehicle based on a microemulsion for topical ocular administration of dexamethasone. Clin Exp Ophthalmol. 2004;32(6):626–32.
Fogt JS, Kowalski MJ, King-Smith PE, Epitropolous AT, Hendershot AJ, Lembach C, Maszczak JP, Jones-Jordan LA, Barr JT. Tear lipid layer thickness with eye drops in meibomian gland dysfunction. Clin Ophthalmol. 2016;10:2237–43.
Foulks GN. The correlation between the tear film lipid layer and dry eye disease. Surv Ophthalmol. 2007;52(4):369–74.
Gallarate M, Gasco MR, Trotta M. Influence of octanoic acid on membrane permeability of timolol from solutions and from microemulsions. Acta pharmaceutica technologica. 1988;34(2):102–5.
Gallarate M, Gasco MR, Trotta M, Chetoni P, Saettone M. Preparation and evaluation in vitro of solutions and o/w microemulsions containing levobunolol as ion-pair. Int J Pharm. 1993;100(1–3):219–25.
Gallarate M, Chirio D, Bussano R, Peira E, Battaglia L, Baratta F, Trotta M. Development of O/W nanoemulsions for ophthalmic administration of timolol. Int J Pharm. 2013;440(2):126–34.
Garg P, Tuteja N, Qayum S. To study the efficacy of difluprednate ophthalmic emulsion and prednisolone acetate ophthalmic suspension on post-operative inflammation in cataract surgery. J Clin Diagn Res. 2016;10(12):NC05–8.
Garrigue JS, Amrane M, Faure MO, Holopainen JM, Tong L. Relevance of lipid-based products in the management of dry eye disease. J Ocul Pharmacol Ther. 2017;33(9):647–61.
Garty N, Lusky M, Zalish M, Rachmiel R, Greenbaum A, Desatnik H, Neumann R, Howes J, Melamed S. Pilocarpine in submicron emulsion formulation for treatment of ocular hypertension-a phase-ii clinical-trial. In: Investigative ophthalmology & visual science, vol. 4. Philadelphia: Lippincott-Raven; 1994. p. 2175.
Gasco MR, Gallarate M, Trotta M, Bauchiero L, Gremmo E, Chiappero O. Microemulsions as topical delivery vehicles: ocular administration of timolol. J Pharm Biomed Anal. 1989;7(4):433–9.
Gehrmann S, Bunjes H. Preparation of nanoemulsions by premix membrane emulsification: which parameters have a significant influence on the resulting particle size? J Pharm Sci. 2017;106(8):2068–76.
Georgiev GA, Yokoi N, Ivanova S, Tonchev V, Daull P. Surface chemistry of the interactions of cationic nanoemulsions with human meibum films. Paper presented at the ARVO, Seattle, 1–5 May 2016.

Georgiev GA, Yokoi N, Nencheva Y, Peev N, Daull P. Surface chemistry interactions of Cationorm with films by human Meibum and tear film compounds. Int J Mol Sci. 2017;18(7):1558.
Glasser CA, Vila MM, Pereira JC, Chaud MV, Oliveira Junior JM, Tubino M, Balcao VM. Development of a water-in-oil-in-water multiple emulsion system integrating biomimetic aqueous-core lipid nanodroplets for protein entity stabilization. Part II: process and product characterization. Drug Dev Ind Pharm. 2016;42(12):1990–2000.
Gore A, Pujara C, Attar M, Neervannan S. Ocular emulsions and dry eye: a case study of a non-biological complex drug product delivered to a complex organ to treat a complex disease. GABI J. 2017;6(1):13–23.
Goto E, Shimazaki J, Monden Y, Takano Y, Yagi Y, Shimmura S, Tsubota K. Low-concentration homogenized castor oil eye drops for noninflamed obstructive meibomian gland dysfunction. Ophthalmology. 2002;109(11):2030–5.
Gulsen D, Chauhan A. Ophthalmic drug delivery through contact lenses. Invest Ophthalmol Vis Sci. 2004;45(7):2342–7.
Gulsen D, Chauhan A. Dispersion of microemulsion drops in HEMA hydrogel: a potential ophthalmic drug delivery vehicle. Int J Pharm. 2005;292(1–2):95–117.
Hagigit T, Nassar T, Behar-Cohen F, Lambert G, Benita S. The influence of cationic lipid type on in-vitro release kinetic profiles of antisense oligonucleotide from cationic nanoemulsions. Eur J Pharm Biopharm. 2008;70(1):248–59.
Hagigit T, Abdulrazik M, Orucov F, Valamanesh F, Lambert M, Lambert G, Behar-Cohen F, Benita S. Topical and intravitreous administration of cationic nanoemulsions to deliver antisense oligonucleotides directed towards VEGF KDR receptors to the eye. J Control Release. 2010;145(3):297–305.
Hamdi Y, Lallemand F, Benita S. Drug-loaded nanocarriers for back-of-the-eye diseases-formulation limitations. J Drug Deliv Sci Technol. 2015;30:331–41.
Hasegawa T, Amako H, Yamamoto T, Tazawa M, Sakamoto Y. Corneal-protective effects of an artificial tear containing sodium hyaluronate and castor oil on a porcine short-term dry eye model. J Vet Med Sci. 2014;76(9):1219–24.
Ibrahim SS, Awad GA, Geneidi A, Mortada ND. Comparative effects of different cosurfactants on sterile prednisolone acetate ocular submicron emulsions stability and release. Colloids Surf B Biointerfaces. 2009;69(2):225–31.
Irimia T, Dinu-Pirvu CE, Ghica MV, Lupuleasa D, Muntean DL, Udeanu DI, Popa L. Chitosan-based in situ gels for ocular delivery of therapeutics: a state-of-the-art review. Mar Drugs. 2018;16(10):373.
Joussen AM, Rohrschneider K, Reichling J, Kirchhof B, Kruse FE. Treatment of corneal neovascularization with dietary isoflavonoids and flavonoids. Exp Eye Res. 2000;71(5):483–7.
Kaercher T, Thelen U, Brief G, Morgan-Warren RJ, Leaback R. A prospective, multicenter, non-interventional study of Optive Plus((R)) in the treatment of patients with dry eye: the prolipid study. Clin Ophthalmol. 2014;8:1147–55.
Kakimoto H, Takamura Y, Arimura S, Miyake S, Matsumura T, Gozawa M, Iwasaki K, Morioka M, Yamada Y, Inatani M. Effect of 0.05% difluprednate ophthalmic emulsion on proinflammatory cytokine levels after retinal laser photocoagulation in rabbits. J Ocul Pharmacol Ther. 2018;34(5):410–5.
Kesavan K, Kant S, Singh PN, Pandit JK. Mucoadhesive chitosan-coated cationic microemulsion of dexamethasone for ocular delivery: in vitro and in vivo evaluation. Curr Eye Res. 2013;38(3):342–52.
Khanal S, Tomlinson A, Pearce EI, Simmons PA. Effect of an oil-in-water emulsion on the tear physiology of patients with mild to moderate dry eye. Cornea. 2007;26(2):175–81.
Kimura M, Yasueda S-I, Yamaguchi M, Inada K. Compositions containing difluprednate. Patent EP0878197 of 1998; 2000.
Kinnunen K, Kauppinen A, Piippo N, Koistinen A, Toropainen E, Kaarniranta K. Cationorm shows good tolerability on human HCE-2 corneal epithelial cell cultures. Exp Eye Res. 2014;120:82–9.

Klang S, Baszkin A, Benita S. The stability of piroxicam incorporated in a positively-charged submicron emulsion for ocular administration. Int J Pharm. 1996;132(1–2):33–44.

Klang S, Abdulrazik M, Benita S. Influence of emulsion droplet surface charge on indomethacin ocular tissue distribution. Pharm Dev Technol. 2000;5(4):521–32.

Kurup TRR, Wan LSC, Chan LW. Preservative requirements in emulsions. Pharm Acta Helv. 1992;67(7):204–8.

Kuwano M, Ibuki H, Morikawa N, Ota A, Kawashima Y. Cyclosporine a formulation affects its ocular distribution in rabbits. Pharm Res. 2002;19(1):108–11.

Labbe A, Baudouin C, Ismail D, Amrane M, Garrigue JS, Leonardi A, Figueiredo FC, Van Setten G, Labetoulle M. Pan-European survey of the topical ocular use of ciclosporin a. J Fr Ophtalmol. 2017;40:187–95.

Lallemand F, Felt-Baeyens O, Besseghir K, Behar-Cohen F, Gurny R. Cyclosporine a delivery to the eye: a pharmaceutical challenge. Eur J Pharm Biopharm. 2003;56(3):307–18.

Lallemand F, Daull P, Benita S, Buggage R, Garrigue JS. Successfully improving ocular drug delivery using the cationic nanoemulsion Novasorb. J Drug Deliv. 2012;2012:604204. https://doi.org/10.1155/2012/604204.

Lallemand F, Schmitt M, Bourges JL, Gurny R, Benita S, Garrigue JS. Cyclosporine a delivery to the eye: a comprehensive review of academic and industrial efforts. Eur J Pharm Biopharm. 2017;117:14–28.

Lam SM, Tong L, Duan X, Petznick A, Wenk MR, Shui G. Extensive characterization of human tear fluid collected using different techniques unravels the presence of novel lipid amphiphiles. J Lipid Res. 2014;55(2):289–98.

Lefebvre G, Riou J, Bastiat G, Roger E, Frombach K, Gimel JC, Saulnier P, Calvignac B. Spontaneous nano-emulsification: process optimization and modeling for the prediction of the nanoemulsion's size and polydispersity. Int J Pharm. 2017;534(1–2):220–8.

Lemp MA, Crews LA, Bron AJ, Foulks GN, Sullivan BD. Distribution of aqueous-deficient and evaporative dry eye in a clinic-based patient cohort: a retrospective study. Cornea. 2012;31(5):472–8.

Leonardi A, Van Setten G, Amrane M, Ismail D, Garrigue J-S, Figueiredo FC, Baudouin C. Efficacy and safety of 0.1% cyclosporine a cationic emulsion in the treatment of severe dry eye disease: a multicenter randomized trial. Eur J Ophthalmol. 2016;26(4):287–96.

Li CC, Abrahamson M, Kapoor Y, Chauhan A. Timolol transport from microemulsions trapped in HEMA gels. J Colloid Interface Sci. 2007;315(1):297–306.

Li X, Muller RH, Keck CM, Bou-Chacra NA. Mucoadhesive dexamethasone acetate-polymyxin B sulfate cationic ocular nanoemulsion--novel combinatorial formulation concept. Pharmazie. 2016;71(6):327–33.

Liang H, Brignole-Baudouin F, Rabinovich-Guilatt L, Mao Z, Riancho L, Faure MO, Warnet JM, Lambert G, Baudouin C. Reduction of quaternary ammonium-induced ocular surface toxicity by emulsions: an in vivo study in rabbits. Mol Vis. 2008;14:204–16.

Liang H, Baudouin C, Daull P, Garrigue JS, Brignole-Baudouin F. In vitro and in vivo evaluation of a preservative-free cationic emulsion of latanoprost in corneal wound healing models. Cornea. 2012;31(11):1319–29.

Liu Y, Lin X, Tang X. Lipid emulsions as a potential delivery system for ocular use of azithromycin. Drug Dev Ind Pharm. 2009;35(7):887–96.

Mafi R, Gray C, Pelton R, Ketelson H, Davis J. On formulating ophthalmic emulsions. Colloids Surf B: Biointerfaces. 2014;122:7–11.

Maichuk Iu F, Karimov MK, Lapshina NA. [Ketoconazole in the treatment of ocular mycoses]. Vestn oftalmol. 1990;106(1):44–6.

McCann LC, Tomlinson A, Pearce EI, Papa V. Effectiveness of artificial tears in the management of evaporative dry eye. Cornea. 2012;31(1):1–5.

Meehan K, Vollmer L, Sowka J. Intraocular pressure elevation from topical difluprednate use. Optometry. 2010;81(12):658–62.

Mercado-Sesma A, Contreras-Rubio A, Baiza-Duran L, Olvera-Montano O, Miranda-Robles M, Bonilla-Garcia J. Bioavailability of generic 0.05% difluprednate emulsion in the aqueous humor, cornea, and conjunctiva of New Zealand rabbits after a single dose compared with commercial difluprednate. J ophthalmic Inflamm Infect. 2017;7(1):10.

Mishima S, Maurice DM. The oily layer of the tear film and evaporation from the corneal surface. Exp Eye Res. 1961;1:39–45.

Mochizuki H, Yamada M, Hatou S, Tsubota K. Turnover rate of tear-film lipid layer determined by fluorophotometry. Br J Ophthalmol. 2009;93(11):1535–8.

Moshirfar M, Pierson K, Hanamaikai K, Santiago-Caban L, Muthappan V, Passi SF. Artificial tears potpourri: a literature review. Clin Ophthalmol. 2014;8:1419–33.

Muchtar S, Almog S, Torracca MT, Saettone MF, Benita S. A submicron emulsion as ocular vehicle for delta-8-tetrahydrocannabinol: effect on intraocular pressure in rabbits. Ophthalmic Res. 1992;24(3):142–9.

Muchtar S, Abdulrazik M, Frucht-Pery J, Benita S. Ex-vivo permeation study of indomethacin from a submicron emulsion through albino rabbit cornea. J Control Release. 1997;44(1):55–64.

Mulki L, Foster CS. Difluprednate for inflammatory eye disorders. Drugs Today (Barc). 2011;47(5):327–33.

Myint M, Bucki R, Janmey PA, Diamond SL. Synthesis and structure-activity relationships of novel cationic lipids with anti-inflammatory and antimicrobial activities. Bioorg Med Chem Lett. 2015;25(14):2837–43.

Naveh N, Muchtar S, Benita S. Pilocarpine incorporated into a submicron emulsion vehicle causes an unexpectedly prolonged ocular hypotensive effect in rabbits. J Ocul Pharmacol. 1994;10(3):509–20.

Naveh N, Weissman C, Muchtar S, Benita S, Mechoulam R. A submicron emulsion of HU-211, a synthetic cannabinoid, reduces intraocular pressure in rabbits. Graefes Arch Clin Exp Ophthalmol. 2000;238(4):334–8.

NDA 206185 (Revised: 09/2018) XelPros(r): full prescribing information. https://www.accessdata.fda.gov/drugsatfda_docs/label/2018/206185s000lbl.pdf. Accessed 22 Feb 2019.

Nencheva Y, Olzynska A, Melcrova A, As Georgiev G, Daull P, Garrigue J-S, Cwiklik L. Improving stability of tear film lipid layer via concerted action of two drug molecules: a biophysical view. Biophys J. 2018;114(3):104a.

Pandey D, Kesharwani P, Jain D. Entrapment of drug-sorbate complex in submicron emulsion: a potential approach to improve antimicrobial activity in bacterial corneal infection. J Drug Deliv Sci Technol. 2019;49:455–62.

Pathak MK, Chhabra G, Pathak K. Design and development of a novel pH triggered nanoemulsified in-situ ophthalmic gel of fluconazole: ex-vivo transcorneal permeation, corneal toxicity and irritation testing. Drug Dev Ind Pharm. 2013;39:780–90.

Patil SM, Li V, Peng J, Kozak D, Xu J, Cai B, Keire DA, Chen K. A simple and noninvasive DOSY NMR method for droplet size measurement of intact oil-in-water emulsion drug products. J Pharm Sci. 2019;108(2):815–20.

Peng CC, Bengani LC, Jung HJ, Leclerc J, Gupta C, Chauhan A. Emulsions and microemulsions for ocular drug delivery. J Drug Deliv Sci Technol. 2011;21(1):111–21.

Popovic M, Chan C, Lattanzio N, El-Defrawy S, Schlenker MB. Comparative cost evaluation of brand name and generic ophthalmology medications in Ontario. Can J Ophthalmol. 2018;53(2):173–87.

Praestegaard M, Steele F, Gomez F. Pharmacokinetic characterization of a novel ocular formulation of ciclosporin. Invest Ophthalmol Vis Sci. 2016;57(12):5388.

Pucker AD, Haworth KM. The presence and significance of polar meibum and tear lipids. Ocul Surf. 2015;13(1):26–42.

Rabinovich-Guilatt L, Couvreur P, Lambert G, Dubernet C. Cationic vectors in ocular drug delivery. J Drug Target. 2004;12(9–10):623–33.

Rahman Z, Xu X, Katragadda U, Krishnaiah YS, Yu L, Khan MA. Quality by design approach for understanding the critical quality attributes of cyclosporine ophthalmic microemulsion. Mol Pharm. 2014;11(3):787–99.
Rantamaki AH, Telenius J, Koivuniemi A, Vattulainen I, Holopainen JM. Lessons from the biophysics of interfaces: lung surfactant and tear fluid. Prog Retin Eye Res. 2011;30(3):204–15.
Rieger G. Lipid-containing eye drops: a step closer to natural tears. Ophthalmologica. 1990;201(4):206–12.
Robert P-Y, Cochener B, Amrane M, Ismail D, Garrigue J-S, Pisella P-J, Baudouin C. Efficacy and safety of a cationic emulsion in the treatment of signs and symptoms of moderate-to-severe dry eye disease: a prospective randomised controlled study. Eur J Opthalmol. 2016;26:546–55.
Robin JS, Ellis PP. Ophthalmic ointments. Surv Ophthalmol. 1978;22(5):335–40.
Rodriguez-Aller M, Guinchard S, Guillarme D, Pupier M, Jeannerat D, Rivara-Minten E, Veuthey JL, Gurny R. New prostaglandin analog formulation for glaucoma treatment containing cyclodextrins for improved stability, solubility and ocular tolerance. Eur J Pharm Biopharm. 2015;95(Pt B):203–14.
Royle L, Matthews E, Corfield A, Berry M, Rudd PM, Dwek RA, Carrington SD. Glycan structures of ocular surface mucins in man, rabbit and dog display species differences. Glycoconj J. 2008;25(8):763–73.
Sall K, Stevenson OD, Mundorf TK, Reis BL. Two multicenter, randomized studies of the efficacy and safety of cyclosporine ophthalmic emulsion in moderate to severe dry eye disease. CsA phase 3 study group. Ophthalmology. 2000;107(4):631–9.
Santen SAS. Phase III, Multinational, Multicenter, Investigator-Masked, Randomised, Active-Controlled Trial, comparing the efficacy and safety of DE-130A with Xalatan® in Patients with Open-Angle Glaucoma or Ocular Hypertension over a 3-Month period, followed by a 12-Month Follow-Up with Open-Label DE-130A Treatment. 2017. https://www.clinicaltrialsregister.eu/ctr-search/search?query=2017-004262-95. Accessed 15 Feb 2019.
Scaffidi RC, Korb DR. Comparison of the efficacy of two lipid emulsion eyedrops in increasing tear film lipid layer thickness. Eye Contact Lens. 2007;33(1):38–44.
Schulz MB, Daniels R. Hydroxypropylmethylcellulose (HPMC) as emulsifier for submicron emulsions: influence of molecular weight and substitution type on the droplet size after high-pressure homogenization. Eur J Pharm Biopharm. 2000;49(3):231–6.
Scifo C, Barabino S, De Pasquale G, Blanco AR, Mazzone MG, Rolando M. Effects of a new lipid tear substitute in a mouse model of dry eye. Cornea. 2010;29(7):802–6.
Sebba F. The behaviour of minute oil droplets encapsulated in a water film. Colloid Polym Sci. 1979;257(4):392–6.
Selek H, Unlu N, Orhan M, Irkec M. Evaluation of retinoic acid ophthalmic emulsion in dry eye. Eur J Ophthalmol. 2000;10(2):121–7.
Shell JW. Ophthalmic drug delivery systems. Surv Ophthalmol. 1984;29(2):117–28.
Shen JQ, Gan Y, Gan L, Zhu CL, Zhu JB. [Ion-sensitive nanoemulsion-in situ gel system for ophthalmic delivery of flurbiprofen axetil]. Yao Xue Xu Bao. 2010;45(1):120–5.
Shen J, Gan L, Zhu C, Zhang X, Dong Y, Jiang M, Zhu J, Gan Y. Novel NSAIDs ophthalmic formulation: flurbiprofen axetil emulsion with low irritancy and improved anti-inflammation effect. Int J Pharm. 2011;412(1–2):115–22.
Siebenbrodt I, Keipert S. Poloxamer-systems as potential Ophthalmics II. Microemulsïons. Eur J Pharm Biopharm. 1993;39(1):25–30.
Sila-on W, Vardhanabhuti N, Ongpipattanakul B, Kulvanich P. Influence of incorporation methods on partitioning behavior of lipophilic drugs into various phases of a parenteral lipid emulsion. AAPS PharmSciTech. 2008;9(2):684–92.
Silva-Cunha A, da Silva GR, de Castro WV, Fialho SL. Evaluation of the pharmacokinetics and ocular tolerance of a microemulsion containing tacrolimus. J Ocul Pharmacol Ther. 2014;30(1):59–65.
Sindt CW, Foulks GN. Efficacy of an artificial tear emulsion in patients with dry eye associated with meibomian gland dysfunction. Clin Ophthalmol. 2013;7:1713–22.

Small DS, Acheampong A, Reis B, Stern K, Stewart W, Berdy G, Epstein R, Foerster R, Forstot L, Tang-Liu DD. Blood concentrations of cyclosporin a during long-term treatment with cyclosporin a ophthalmic emulsions in patients with moderate to severe dry eye disease. J Ocul Pharmacol Ther. 2002;18(5):411–8.

Steele F. An ophthalmic composition GB. Patent; 2012.

Stevenson D, Tauber J, Reis BL. Efficacy and safety of cyclosporin a ophthalmic emulsion in the treatment of moderate-to-severe dry eye disease: a dose-ranging, randomized trial. The Cyclosporin a phase 2 study group. Ophthalmology. 2000;107(5):967–74.

Suresh PK, Dewangan D. Ophthalmic delivery system for dexamethasone: an overview. Int J Innov Pharm Res. 2011;2(4):161–5.

Sznitowska M, Janicki S, Dabrowska EA, Gajewska M. Physicochemical screening of antimicrobial agents as potential preservatives for submicron emulsions. Eur J Pharm Sci. 2002;15(5):489–95.

Tamilvanan S, Benita S. The potential of lipid emulsion for ocular delivery of lipophilic drugs. Eur J Pharm Biopharm. 2004;58(2):357–68.

Tiwari R, Pandey V, Asati S, Soni V, Jain D. Therapeutic challenges in ocular delivery of lipid based emulsion. Egypt J Basic Appl Sci. 2018;5(2):121–9.

US Food and Drug Administration. Restasis drug approval package. 2002. https://www.accessdata.fda.gov/drugsatfda_docs/nda/2003/21-023_Restasis.cfm. Accessed 11 Feb 2019.

US Food and Drug Administration. Durezol drug approval package. 2008. https://www.accessdata.fda.gov/drugsatfda_docs/nda/2008/022212s000TOC.cfm. Accessed 11 Feb 2019.

Vandamme TF. Microemulsions as ocular drug delivery systems: recent developments and future challenges. Prog Retin Eye Res. 2002;21(1):15–34.

Walenga RL, Babiskin AH, Zhang X, Absar M, Zhao L, Lionberger RA. Impact of vehicle physicochemical properties on modeling-based predictions of cyclosporine ophthalmic emulsion bioavailability and tear film breakup time. J Pharm Sci. 2019;108(1):620–9.

Wang Y-D, Tang X-L, Ye C-T, Wu W. Determination of tacrolimus ophthalmic emulsion by HPLC. Chin Pharm J. 2010;45(3):221–3.

Wang X, Patil SM, Keire DA, Xu X, Chen K. Application of ultra-centrifugation and bench-top (19)F NMR for measuring drug phase partitioning for the ophthalmic oil-in-water emulsion products. AAPS PharmSciTech. 2018;19(4):1647–51.

Wehrle P, Korner D, Benita S. Sequential statistical optimization of a positively-charged submicron emulsion of miconazole. Pharm Dev Technol. 1996;1(1):97–111.

Yamaguchi M, Yasueda S, Isowaki A, Yamamoto M, Kimura M, Inada K, Ohtori A. Formulation of an ophthalmic lipid emulsion containing an anti-inflammatory steroidal drug, difluprednate. Int J Pharm. 2005;301(1–2):121–8.

Yamaguchi M, Ueda K, Isowaki A, Ohtori A, Takeuchi H, Ohguro N, Tojo K. Mucoadhesive properties of chitosan-coated ophthalmic lipid emulsion containing indomethacin in tear fluid. Biol Pharm Bull. 2009;32(7):1266–71.

Yin J, Xiang C, Lu G. Cationic lipid emulsions as potential bioadhesive carriers for ophthalmic delivery of palmatine. J Microencapsul. 2016;33(8):718–24.

Zhai J, Gu J, Yuan J, Chen J. Tacrolimus in the treatment of ocular diseases. BioDrugs. 2011;25(2):89–103.

Microparticle Products for the Eye: Preformulation, Formulation, and Manufacturing Considerations

Vanessa Andrés-Guerrero, Irene T. Molina-Martínez, Irene Bravo-Osuna, and Rocío Herrero-Vanrell

Abstract Posterior segment diseases are one of the major causes of blindness. Their treatments require successive intraocular injections being associated to adverse effects. One of the main challenges in the ophthalmological therapy is to decrease the number of interventions. Intraocular drug delivery systems (IDDS) emerge as a therapeutic tool for long-term delivery of therapeutic molecules including as main components biopolymers and active substances (single or in combination). Also, other components (mainly additives) can be added to improve the technological properties of the formulation. Devices with a low surface area, such as implants (>1 mm) or microparticles (1–1000 μm), are preferred as they are able to deliver the therapeutic cargo for long periods of time. Biodegradable microparticles (mainly microspheres) prepared with poly(lactic-*co*-glycolic) (PLGA) polymers are being extensively evaluated as IDDS. They can encapsulate different kinds of active substances (biotechnological products and low molecular weight molecules) alone or combined in the same device, which then can provide drug delivery for days or even months.

Keywords Microparticles · PLGA · Preformulation · Formulation · Microencapsulation techniques · Intraocular · Microspheres

Abbreviations

AMD	Age-related macular degeneration
anti-VEGF	Anti-vascular endothelial growth factor
CNV	Choroidal neovascularization
CLSM	Confocal laser scanner microscopy
DX	Dexametasone

V. Andrés-Guerrero · I. T. Molina-Martínez · I. Bravo-Osuna · R. Herrero-Vanrell (✉)
Faculty of Pharmacy, Department of Pharmaceutics and Food Technology, Complutense University of Madrid, Madrid, Spain
e-mail: rociohv@ucm.es

S. Neervannan, U. B. Kompella (eds.), *Ophthalmic Product Development*, AAPS Advances in the Pharmaceutical Sciences Series 37,
https://doi.org/10.1007/978-3-030-76367-1_14

DR	Diabetic retinopathy
DSC	Differential scanning calorimetry
EE	Encapsulation efficiency
GPC	Gel permeation chromatography
Tg	Glass transition temperatures
GDNF	Glial cell line-derived neurotropic factor
HA	Hyaluronic acid
HPMC	Hydroxypropyl methylcellulose
IDDS	Intraocular drug delivery systems
LC	Loading capacity
Tm	Melting points
MSs	Microspheres
O/W	Oil-in-water emulsion
O_1/O_2	Oil-in-oil emulsion
PGSS	Particles from gas-saturated solutions
PEG	Polyethylene glycol
PGA	Poly(glycolic) acid
PLA	Poly(lactic) acid
PLGA	Poly(lactic-*co*-glycolic) acid
PDR	Proliferative diabetic retinopathy
RESS	Rapid expansion of supercritical solutions
RP	Retinitis pigmentosa
SEM	Scanning electron microscopy
SLMs	Solid lipid microparticles
S/O/W	Solid-in-oil-in-water emulsion
SAS	Supercritical antisolvent
$scCO_2$	Supercritical carbon dioxide
SCF	Supercritical fluid
TEM	Transmission electron microscopy
TA	Triamcinolone acetonide
$W_1/O/W_2$	Water-in-oil-in-water emulsion
W/O	Water-in-oil emulsion
XRD	X-ray diffraction

Introduction

The treatment of ophthalmic diseases can benefit from the use of novel therapeutic systems based on nano- and microtechnologies. In the eye, pharmaceutical dosage forms can be administered by different routes: topical, periocular, and intraocular (Herrero-Vanrell et al. 2013a). If the drug has to reach the anterior segment of the eye, topical administration is usually preferred, commonly in the form of eye drops. However, despite the advantages of using this noninvasive route, topical administration is limited by poor ocular bioavailability (only about 5% of the instilled drug is able to reach the intraocular tissues) (Cape et al. 2008).

Pathologies affecting the back of the eye are one of the major causes of blindness. In most of these disorders, the optic nerve and the retina are affected. Neurodegenerative chronic diseases of the posterior segment of the eye include devastating diseases as age-related macular degeneration (AMD), diabetic retinopathy (DR), and glaucoma, among others (Liew et al. 2017; McGuinness et al. 2017; Tian et al. 2015). Due to the chronicity of these pathologies, effective drug concentrations in the target tissue have to be maintained for extended periods of time. The access of the drug to the back of the eye when administered topically or by oral administration is limited by the blood-aqueous and blood-retinal barriers, respectively. Intraocular or periocular injections are used to avoid these physiological effective barriers (Herrero-Vanrell et al. 2014). Despite the advantages of the administration of the active compound near the target tissue, repeated injections are needed to achieve and maintain effective drug levels for prolonged periods of time. For this reason, one of the main challenges facing ophthalmic therapies is to decrease the number of interventions. This can be done with the use of intraocular drug delivery systems (IDDS). A lot of efforts are being made in the development of IDDS (Bravo-Osuna et al. 2016). These systems include as main components a combination of one or more biopolymers and active substances (single or combined). Also, other excipients (mainly additives) can be included in the formulations to improve their properties. Depending on the biomaterial employed, these devices can disappear from the site of administration (biodegradable) or remain during the patient's life or until their removal (non biodegradable). According to their size, drug delivery systems are commonly named implants (>1 mm), microparticles (1–1000 μm), or nanoparticles (1–1000 nm). For long-term delivery, implants and microparticles are preferred as they are able to release the drug for long periods of time. Nonbiodegradable implants (Vitrasert™, Retisert™, and Iluvien™) and one biodegradable implant (Ozurdex®) have been already developed for intraocular administration (Bravo-Osuna et al. 2016). Microparticles can encapsulate different kinds of active substances (biological macromolecules and low molecular weight molecules), alone or in combination, which can be delivered for days or months (Arranz-Romera et al. 2019a; Bravo-Osuna et al. 2018; Checa-Casalengua et al. 2011). Depending on the method of preparation, the structure of microparticles results in a reservoir or in a matrix type of delivery system. If the polymer or a mixture of polymers is covering the drug pellet (reservoir structure), the microdevices are known as microcapsules. In the case of microspheres (matrix structure), the drug is dispersed in the polymer matrix (Fig. 1).

Biodegradable microparticles are preferred as they gradually disappear from the site of administration. Microparticles can be injected using small-gauge needles (25–34G) by intraocular or periocular administration (Fig. 2). In ophthalmic drug delivery, the amount of microparticles can be adjusted to the patient's needs, allowing a personalized therapy (Bravo-Osuna et al. 2016).

The main objective of microparticle's design is to achieve effective drug concentrations at the site of action with the desired therapeutic response in a personalized therapy. To ensure the quality of the product, several parameters have to be taken

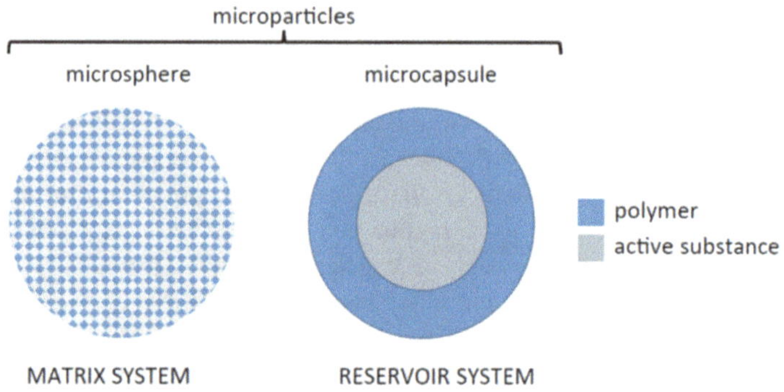

Fig. 1 Structure of microparticles (microspheres and microcapsules)

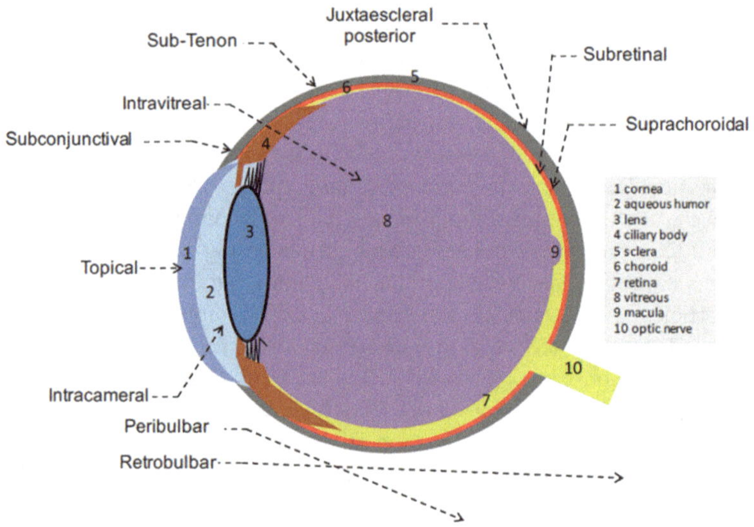

Fig. 2 Eye anatomy and ocular administration routes

into account. Among them, chemical and physical stability, uniformity of dosage, dissolution or release profile, and suitable preservation are required (USP 2019).

Components of Microparticles

Active Substances

For intraocular administration, a large variety of active substances have been included in microspheres (antiproliferative drugs, anti-inflammatory agents, immunosuppressants, antibiotics, and even biologicals). These include Doxorubicin,

5-fluorouracil, and retinoic acid for proliferative retinopathy, dexamethasone (DX) and cyclosporine for uveitis, anti-vascular endothelial growth factor (anti-VEGF) for age-related macular degeneration (AMD), budesonide and celecoxib for diabetic retinopathy, triamcinolone acetonide (TA) for macular edema, acyclovir for herpes infection, ganciclovir for cytomegalovirus retinitis, neurotrophic factors for neuroprotection, and the inhibitor of protein kinase C (PKC412) for choroidal neovascularization (CNV). Neuroprotective agents (coenzyme Q10, dexamethasone, melatonin, tauroursodeoxycholic acid) for glaucoma and retinitis pigmentosa and a combination of the anti-inflammatory TA and the antibiotic ciprofloxacin to prevent ocular inflammation and infection after cataract surgery have been also employed (6–7, 10). Other interesting strategy is the use of microspheres to improve retinal repair, as co-transplantation of PLGA microspheres loaded with matrix metalloproteinase-2 and retinal progenitor cells (Yao et al. 2011). Most recently, the simultaneous co-delivery of multiple active substances from multiloaded microspheres is emerging as a therapeutic tool for the treatment of chronic and multifactorial diseases affecting the posterior segment of the eye (Arranz-Romera et al. 2019a, b; Bravo-Osuna et al. 2016, 2018; Checa-Casalengua et al. 2011; Herrero-Vanrell et al. 2013b).

Prior to the development of a microparticle device, it is essential to determine the critical physicochemical properties of the selected active substances. A recommended list of the requirements, known as preformulation, is collected in a bibliography dedicated to pharmaceutical technology (Aulton-Pharmaceutics 2002; Ramón 2016). Solubility, crystallinity, melting point, stability, morphology, identity, and purity are the main parameters to be analyzed. Pharmaceutical raw material suppliers must include information about the physicochemical properties of the active compound to allow proper identification. In the case of biological products, they can be provided in solution or freeze-dried. Providers must include information relative to the carrier and solvent used to dissolve the biotechnological substance.

Biomaterials Employed in Microencapsulation

Proteins (i.e., albumin and gelatin) and biodegradable polymers, such as polyesters (lactide and glycolide polymers and copolymers) and polyorthoesters (polycaprolactone), have been widely used to prepare microparticles for intraocular administration (Herrero-Vanrell et al. 2013a). Among them, lactic (L) and glycolic (G) acid derivatives have been the most popular for the preparation of biodegradable microspheres. Poly(lactic) acid (PLA), poly(glycolic) acid (PGA), and poly-lactic-*co*-glycolic acid (PLGA) are approved for clinical purposes. The degradation rate of these biopolymers depends on the molecular weight (low molecular weight results in a faster degradation time) and L:G ratio (degradation time PLA > PGA > PLGA). Polymers with free-COOH groups are more hydrophilic, and PEGylation increases their hydrophilicity (Swider et al. 2018).

Other Components of Microparticulate Formulations

The properties of microparticles can be tuned, thanks to the use of different excipients (or additives). For example, plasticizers are included to increase the mechanical strength of microcapsule walls. Other substances, such as propylene glycol, phthalic derivatives, glycerin, fatty acid esters, and citric acid, can also be incorporated for this purpose (Wan et al. 1992; Timilsena et al. 2019).

Cross-linking agents are useful to promote internal linkage between the components of the coating material in microparticles. For this reason, formaldehyde, glutaraldehyde, and other more tolerable substances such as genipin have been used to increase the hardness of the gelatin microcapsule walls.

The use of additives is further extended to increase drug loading and also to modulate the release rate of the drug (Martínez-Sancho et al. 2004a). As it will be described further on if an O/W emulsion is used for microsphere's manufacturing, aqueous soluble and oily substances can be added as a strategy to modulate the release rate of the drug. Gelatin has been used as an additive and, when added to the external phase of the emulsion, has demonstrated an improved release rate for the antiviral drug acyclovir (Martínez-Sancho et al. 2003). Oily additives remain inside the microspheres and can produce an increase of the encapsulation efficiency and also modulate the release rate of the therapeutic molecule. This is the case of vitamin E included in PLGA microspheres loaded with the biological agent glial cell line-derived neurotrophic factor (GDNF) for the treatment of glaucoma (Checa-Casalengua et al. 2011).

Also, amphiphilic additives have been used to optimize the encapsulation of poorly water-soluble drugs. One example is Pluronic F68 in cyclosporine-loaded PLGA microspheres, optimized and assayed for the experimental treatment of uveitis in rabbits.(He et al. 2006).

Solvents

Organic solvents are widely used in the manufacture of microparticles to dissolve the biopolymer or the drug, playing a vital role in the finished product. One of the main drawbacks of employing organic solvents is that they can be toxic remain when remaining in the final product. For this reason, their concentration must be regulated in the final pharmaceutical formulation. Some specific assumptions about residual solvents must be taken into account in the synthesis and formulation of pharmaceutical products, especially if they are used to prepare IDDS.

Depending on the possible risk to human health, solvents are classified in three groups: solvents to be avoided (class 1 solvents), solvents to be limited (class 2 solvents), and solvents with a low toxic potential (class 3 solvents) (Table 1). Most of the solvents used to prepare microparticles belong to class 2, requiring chromatographic techniques to demonstrate that the residual amount does not exceed the

Table 1 Classification of residual solvents and their assessments

Residual solvent class	Assessment
Class 1 (solvents to be avoided)	Known human carcinogens
	Strongly suspected human carcinogens Solvents particularly known to have ozone-depleting properties
Class 2 (solvents to be limited)	Nongenotoxic animal carcinogens or possible causative agents of other irreversible toxicity, such as neurotoxicity or teratogenicity
	Solvents suspected of other significant but reversible toxicities
Class 3 (solvents with low toxic potential)	Solvents with low toxic potential to humans; no health-based exposure limit is needed

limits of the maximum tolerable daily intake (TDI). Depending on the solvent used, the concentration limit (ppm) is in the range between 50 (i.e., 2-methoxyethanol) and 3880 (i.e., cyclohexane).

Class 3 solvents such as acetone (56 °C boiling point) and class 2 solvents including methanol (64.7 °C), dichloromethane (39.6 °C), and acetonitrile (82 °C) are used to fabricate PLGA microparticles. For intraocular purposes it is very important to demonstrate that the residual solvent (even in class 2 solvents with residual amounts below the maximum TDI) is not potentially toxic to retinal tissues.

Microencapsulation Techniques

Scale up production, residual solvent content, size of microparticles, and sterilization are the main critical parameters to be taken into consideration when selecting a microencapsulation method. Microencapsulation techniques can be classified into four groups: (1) physical (spray-drying, supercritical fluid precipitation, solvent evaporation, and fluid-bed microencapsulation), (2) physicochemical (coacervation and ionic gelation), (3) chemical (interfacial polymerization), and (4) others (microfluidics).

Physical Microencapsulation Techniques

Spray-Drying

Spray-drying is a rapid microencapsulation method, in which the liquid feed is transformed into a dried powder by spraying the feed into a hot fluid (air or inert gas such as nitrogen). This technique is useful to encapsulate both hydrophilic and lipophilic low molecular weight active substances as well as peptides and proteins (Ozkan et al. 2019; Swider et al. 2018; Wan and Yang 2016). One of the main advantages of this process is that droplet drying and particle formation can be accomplished within a time range from milliseconds to a few seconds, yielding solid particles. The process can be divided into three steps: (1) atomization of the (polymer/drug/solvent) as a liquid solution or suspension into small droplets using an injection nozzle, (2) drying of the droplets upon contact with the wet gas (i.e., nitrogen) at high temperature that renders the particles dry, and (3) separation of the porous dry microparticles from the drying medium by cyclone separation or bag house filtration (membrane filter) (Fig. 3). Finally, the powdered product is dispensed into a collection of vessels and/or bag filter (Davis and Walker 2018). If spray-freeze-drying is employed, the gas is introduced at low temperature. Atomizing devices include rotary atomizers, two-fluid nozzles, and pressure nozzles.

The factors that influence the procedure are drying temperature, drying airflow rate, feed flow rate, rheology of the bulk fluid, vapor pressure of the solvents and their partial pressure in the gas phase, speed of atomizer, sort of carrier agent, and concentration of the carrier agent.

The main drawback of the conventional spray-drying method is that the process can also produce agglomeration of particles or their adhesion to the walls of the apparatus, generating a significant loss of product. Furthermore, this technique provides highly porous microspheres. In ophthalmic drug delivery, spray-drying has been used to prepare microspheres loaded with triamcinolone acetonide (TA) and ciprofloxacin, which were administered by a periocular injection (Paganelli et al. 2009).

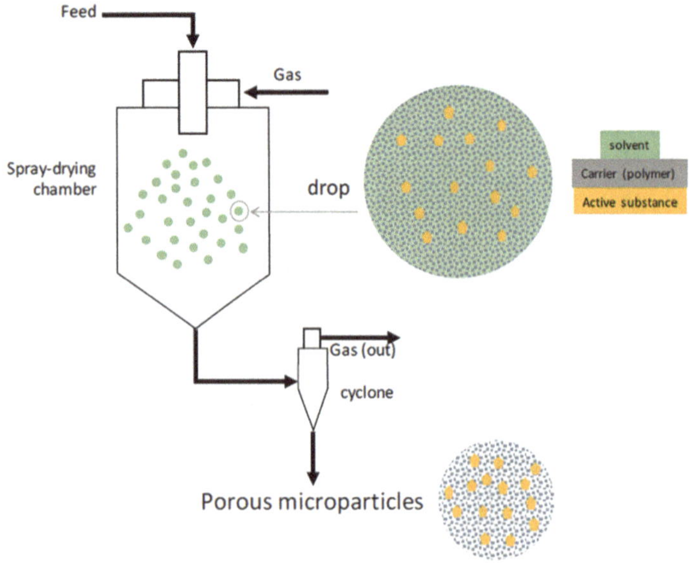

Fig. 3 Schematic diagram of spray-drying microencapsulation process

Supercritical Fluid Precipitation

Supercritical or near-critical fluid processes for generating microparticles have gained considerable attention in the past decade (Cape et al. 2008; Ozkan et al. 2019). A supercritical fluid (SCF) is when a solvent is held above its critical temperature and pressure. Under these special physical conditions, the solvent possesses properties between those of liquids (density and high solvating power) and gases (low viscosity, high diffusivities, as well as mass transfer rates). Gases such as carbon dioxide ($scCO_2$), pentane, propane, and nitrogen are commonly used as supercritical fluids. Among them, $scCO_2$ is the most widely used, particularly in the rapid expansion of supercritical solutions or RESS method. It is relatively cheap, and the critical point is easily accessible (critical temperature 31.1 °C and critical pressure 7.38 MPa), which allows the fluid to be used under mild conditions without leaving behind any harmful residues at the end of the process. The SCF microencapsulation procedure can be summarized in three steps: (1) solvent (rapid expansion of supercritical solutions or RESS), (2) antisolvent (supercritical antisolvent or SAS), and (3) solute (particles from gas-saturated solutions or PGSS).

1. For RESS, as the first step, all the components (active substance and biopolymer) are dissolved in SCF with or without a cosolvent. Then, the solution is expanded by using a small nozzle (with optional coaxial channels) into a lower pressure area, generating a precipitate of the solutes due to a decrease in solvent power of the SCF upon expansion (Fig. 4). Successful application of this technique is limited to solutes soluble in SCF.
2. In the antisolvent process, the SCF is brought into contact with the previously formed drug/polymer solution in a high-pressure chamber by injecting both components. Under these conditions, the drug solubility decreases upon contact with SCF, resulting in the formation of nano- or microparticles. After that, the organic solvent is eliminated from the particles with a continuous flow of supercritical fluid (Fig. 5).

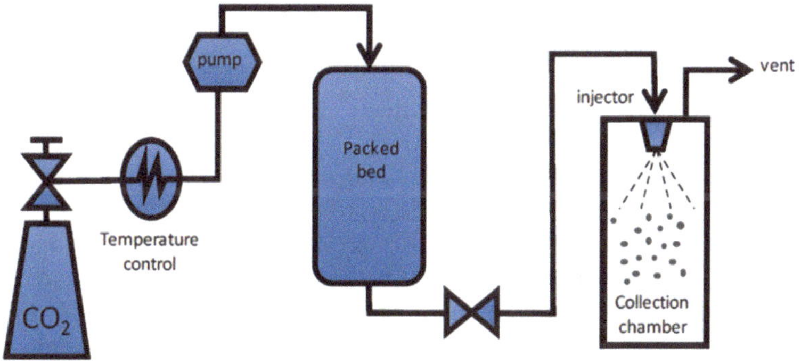

Fig. 4 Schematic diagram of rapid expansion of supercritical solutions (RESS) microencapsulation method

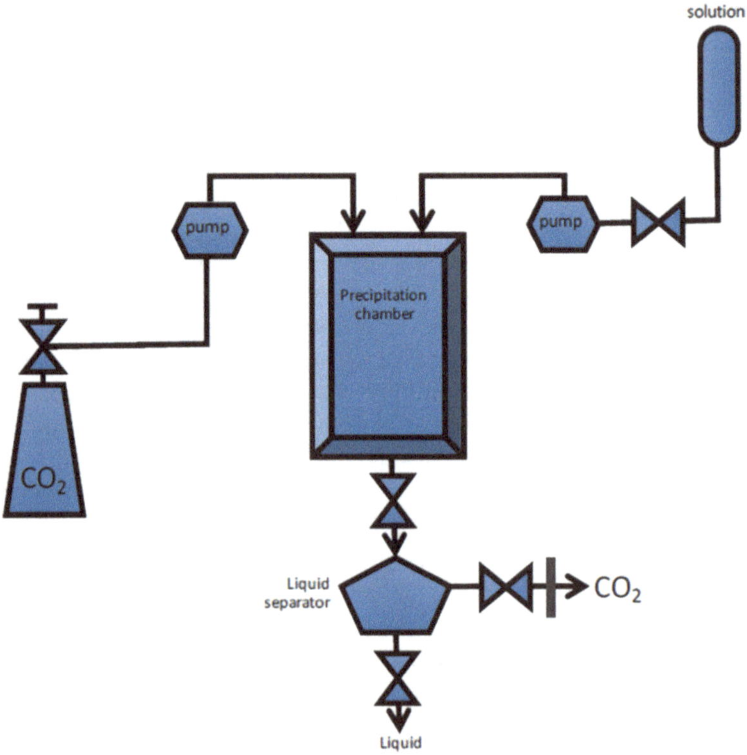

Fig. 5 Schematic diagram of supercritical antisolvent (SAS) precipitation microencapsulation method

3. In the PGSS process, SCF is used as a solute to first saturate drug/polymer solution in a suitable solvent. Then, the mixture is atomized rendering solid particles as a consequence of the cooling effect, generated by the release of the SCF and/or expansion of the solvent and the resulting reduction in drug/polymer solubility (Fig. 6).

Emulsion Solvent Evaporation/Extraction

Microencapsulation of active substances using emulsification methods is based on the preparation of single or double emulsions (Swider et al. 2018). If a single emulsion is formed, oil-in-water (O/W) or oil-in-oil (O_1/O_2) systems are used. For microencapsulation purposes, the inner phase of the emulsion is usually an organic solvent that is not soluble in the continuous phase. In this way, the polymer is first dissolved in the organic solvent. Then, the drug is incorporated in the polymeric solution. The obtained solution or dispersion is then added to the aqueous phase that contains a surfactant. Mixing of the two nonmiscible phases is performed by sonication or

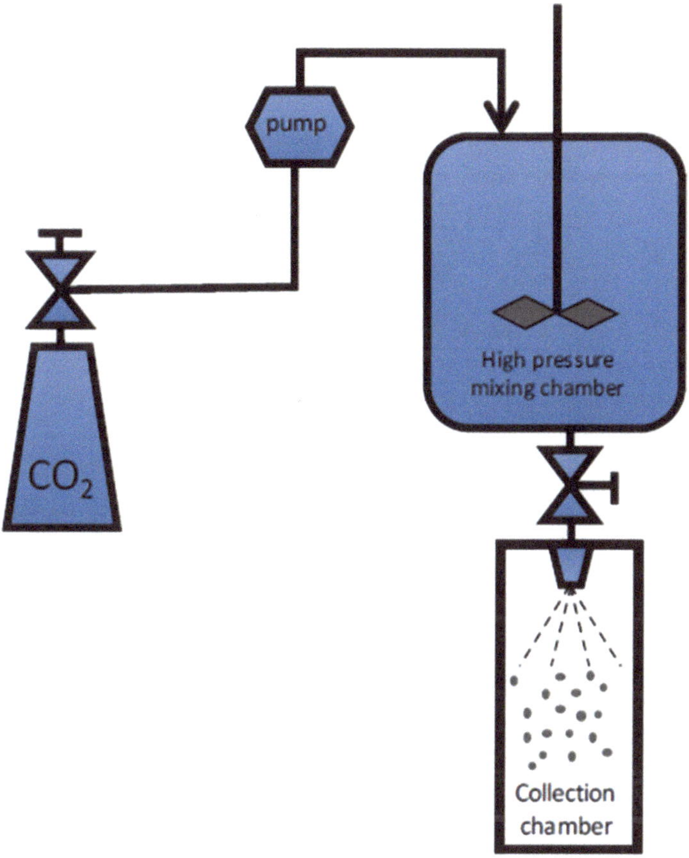

Fig. 6 Schematic diagram of particles from gas-saturated solutions (PGSS) microencapsulation method

homogenization. Once formed, the emulsion is maintained and stirred until the solvent is evaporated, leaving only solid microspheres. The obtained microparticulate product is filtered, washed, freeze-dried, and preserved until use (Fig. 7).

Microspheres prepared by the double emulsion ($W_1/O/W_2$) require a two-step emulsification procedure. In this evaporation/extraction technique, water droplets containing the active substance are first dispersed in an organic polymeric solution, forming the first emulsion (W/O). Then, this emulsion is dispersed in an aqueous phase to form a double $W_1/O/W_2$ emulsion. This method is useful to encapsulate more hydrophilic cargo (i.e., proteins) compared to W/O single emulsion and also different and even incompatible molecules (Fig. 8).

If the particles are obtained from an oil-in-oil (O_1/O_2) emulsion, the polymer is first dissolved using polar organic solvents, such as acetone, acetonitrile, or methanol. This manufacturing step is the same as the ones employed for single (W/O) emulsion and double ($W_1/O/W_2$) emulsion. But the mature microspheres from this

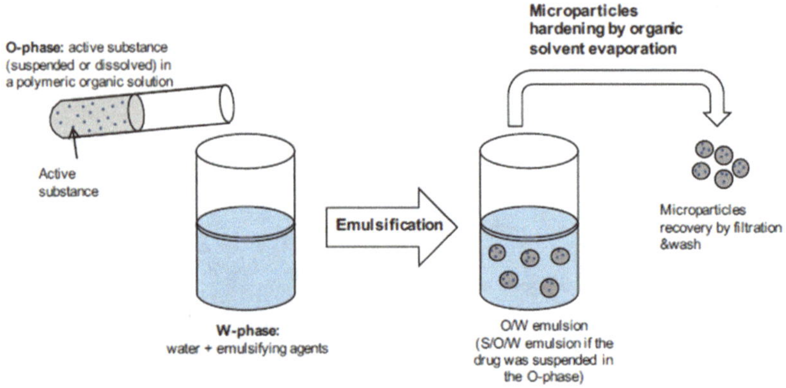

Fig. 7 Microsphere formation by O/W simple emulsion

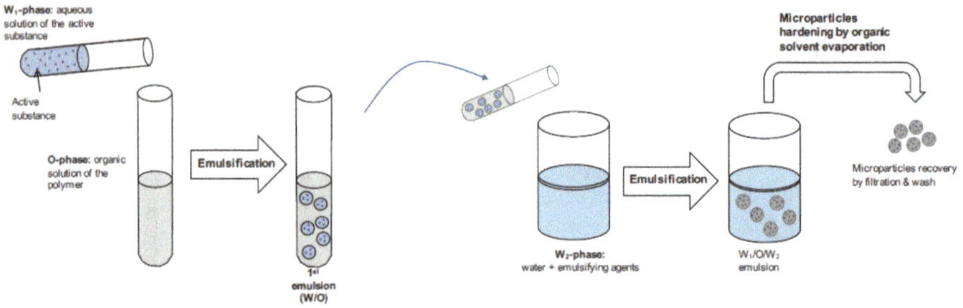

Fig. 8 Microsphere formation by W/O/W double emulsion

method are washed with nonpolar organic solvents, such as petroleum ether or cyclohexane. Also, a solid-oil-in-water (S/O/W) emulsion technique has been described for the encapsulation of biological products. For this method, the protein is first suspended in an oil (Wan et al. 1992).

Fluid-Bed Microencapsulation

Fluid-bed microencapsulation hinges on the coverage of the active substance (core) with an atomized coating material, creating microcapsules. The coating can be deposited in single or multiple layers. Particles are maintained in suspension by cyclic movements because of the airflow through the fluidized bed (Fig. 9). Spray nozzles can be located in the middle or at the bottom of the fluidized bed. The process is continued until a desired film thickness is attained.

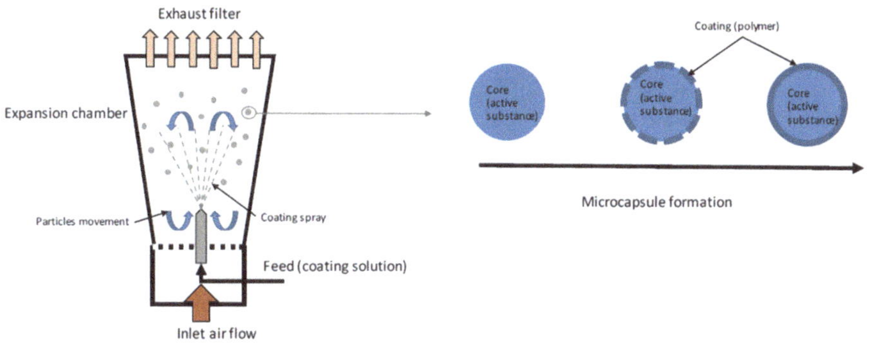

Fig. 9 Schematic diagram of fluid-bed microencapsulation process, in which the coating agent is injected in the drying chamber from the bottom

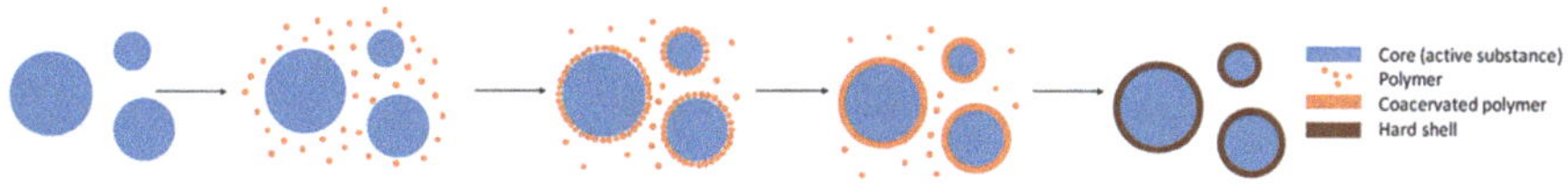

Fig. 10 Microparticle formation by simple coacervation

Physicochemical Microencapsulation Techniques

Coacervation

Coacervation is known as a procedure of phase separation in which two liquid phases are separated.

The process occurs by the formation of two liquid phases, one of them rich in colloid (coacervate) and the other one poor in colloid (supernatant). The separation process is induced by changing the media environment (change of pH, ionic strength, temperature). In the case of microencapsulation, the coacervated polymer is deposited around the active substance yielding microcapsules (Fig. 10).

Complex coacervation includes oppositely charged polymers, which can form a shell surrounding the core (active substance) (Timilsena et al. 2019). In this procedure, aqueous solution of two or more polymers that are usually above their gelling temperatures is employed. One of the most popular polymers is gelatin in combination with other polysaccharides (alginate, agar gum, carboxymethylcellulose, chitosan, pectin). Each polymer is dissolved in a corresponding solvent. After that, an emulsion is formed by mixing the two polymeric solutions. Coacervation is induced by changing the environment conditions (i.e., pH in the case of gelatin). Finally, cooling and posterior hardening of the polymeric matrices are performed by increasing the temperature, desolvation, or cross-linking.

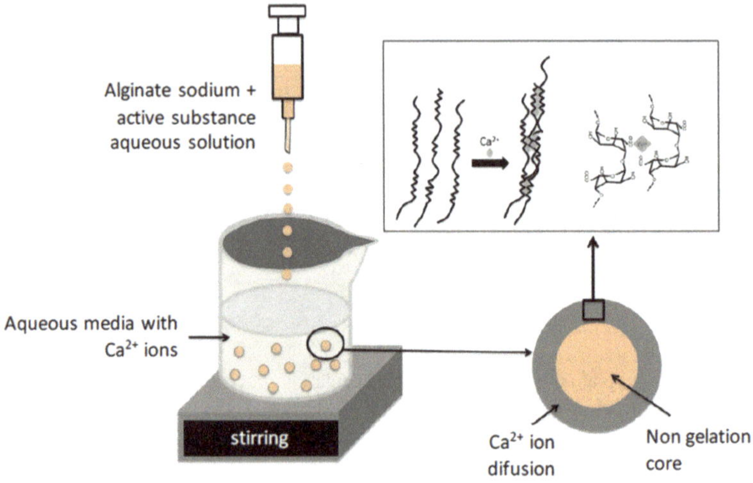

Fig. 11 Microparticle formation by external ionic gelation

Ionic Gelation

In this technique, cross-linking of polyelectrolytes is produced in the presence of multivalence ions, such as Ca^{2+}, Ba^{2+}, and Al^{3+} (Swider et al. 2018). Polyelectrolytes used in the ionic gelation include alginate, chitosan, pectin, and gellan gum, among others. Spherical gel particles are obtained by dropping an aqueous polymer solution from a syringe needle or a nozzle into an ionic solution.

Gelation can be performed externally or internally. In external gelation, the multivalent ion dissolution is maintained under agitation while the polymer solution is added dropwise (Fig. 11). The size of the microcapsules can be controlled by using different gauge needles. On the other hand, if ionic salt solution is added dropwise, the process is known as internal gelation.

Chemical Microencapsulation Techniques

Interfacial Polymerization

The "in situ" or "interfacial" polymerization phenomena forms microcapsules. In this microencapsulation method, a functional monomer is dissolved in a liquid, which is then dispersed in an aqueous phase that contains a dispersant. Microcapsules are formed by the addition of a co-reactive multifunctional amine. In the case of "interfacial" polymerization, an emulsion is formed, and reactive monomers or oligomers are included in the continuous and discontinuous phases. When both phases are mixed, the polymerization occurs with the use of an initiator of some kind (usually heating or UV irradiation) (Fig. 12). Special care should be taken to

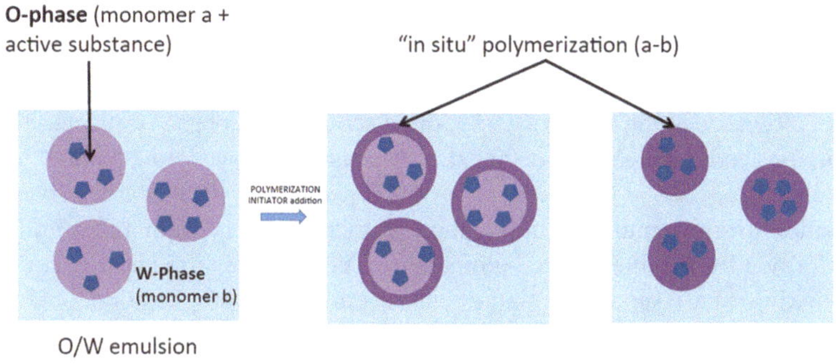

Fig. 12 Microparticle formation by interfacial polymerization

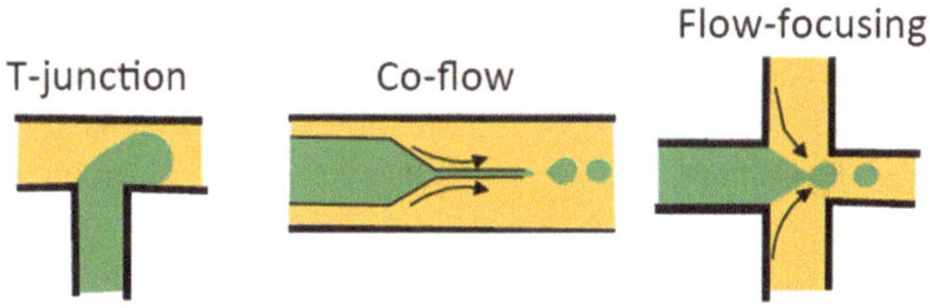

Fig. 13 Schematic illustration of various channel geometries including cross-flow (T-junction), co-flow, and flow focusing

maintain the desired temperature as it can modify the structure and activity of proteins. One of the major drawbacks of this method is the generation of free radicals during the polymerization that can cause toxicity.

Others

Microfluidics

Most conventional microencapsulation methods usually result in microparticles with large polydispersity (Li et al. 2018). Droplet-based microfluidics is emerging as one interesting tool that allows composition and particle size tuning. Microfluidics has been widely employed in the solvent-emulsion evaporation technique. Unlike the conventional methods of forming emulsions in which the droplet breakup is generated by agitation, microfluidic devices have the advantage in that they can produce one drop at a time. Microfluidic geometries include cross-flow, co-flow, and flow focusing (Fig. 13). In the cross-flow, also called T-junction, the inner phase is broken off in the T-shaped junction (angle between 0 and 180°). This device is very useful for single emulsions rendering monodisperse microparticles. In a co-flow geometry (coaxial junction), the system is formed via two channels (channel 1 including the external phase is inserted and aligned with channel 2 that includes the inner phase).

Monodisperse droplets are formed when both phases flow in parallel through the channels. Finally, in the flow focusing, the discontinuous phase is injected through a capillary feed tube. Multiple nozzles are also employed to increase production yield.

Besides forming emulsions, fabrication of microparticles using microfluidics can be performed by polymerization, temperature-induced gelation and freezing, ionic cross-linking, phase separation, interfacial reaction, and complexation.

Sterilization

Selecting a sterilization method is one of the critical steps in the manufacture of drug delivery systems for ophthalmology. The choice of the most appropriate sterilization technique depends on the components of the formulation. Steam sterilization and dry heat sterilization are adequate for heat-resistant materials. This is not the case for microparticle drug delivery systems prepared with heat-sensible materials as the temperature of sterilization usually is higher than the melting point or glass transition temperature of the polymers. Also, the drug must be temperature stable to use these sterilization methods. Chemical sterilization involves gas exposure of products, such as ethylene oxide, and it is frequently employed for the sterilization of healthcare devices and instruments. However, this gas can be adsorbed in biomaterials, causing potential toxicity after their administration if it is not totally removed. For this reason this sterilization is avoided in polymeric drug delivery systems. Gamma radiation possesses high penetration power, and it is usually employed to sterilize polymeric devices. The dose required to assure sterilization of a product is 25 kGy (European Guideline 3AQ4a 1992). Despite the advantages of using final gamma radiation sterilization of bioresorbable polyesters, irradiation exposure can induce some non-desired effects, mainly the risk of alteration of the properties of the final formulation (Nijsen et al. 2002). In fact, gamma radiation of polyester PLGA induces dose-dependent chain scission with a consequential reduction of molecular weight, which decreases the transition temperature of the polymer. Furthermore, the degradation of PLGA has been linked to the formation of free radicals that increase the polymer reactivity. Also, changes in particle size of the final microparticle formulation due to aggregation have been associated with gamma sterilization. These problems can be partially resolved by using low temperatures during the irradiation exposure (Martínez-Sancho et al. 2004b).

Characterization of Microparticles

Morphological Studies

Once prepared, external morphology of microparticles can be observed using light microscopy and scanning electron microscopy (SEM) (Fig. 14). For SEM, nonconductive materials, such as polymeric dry particles, require a carbon and/or metal coating. A sputter coating for SEM analysis with metals can include noble metals, such as gold, gold/palladium, platinum, and silver.

Additionally, ultrathin sections of the particles can be observed by transmission electron microscopy (TEM) (Fig. 15). This technique is very useful to characterize microparticles as the nature of the active substance or additives plays an important role in their inner structure. For example, microspheres containing only the polymer (PLGA) show a homogeneous structure inside the particles. However, heterogeneous inner morphologies appear when the microspheres include crystalline drugs or oils.

Also, fluorescence image techniques, such as confocal laser scanner microscopy (CLSM), allow the acquisition of images with a controllable shallow depth of the field (Fig. 16).

Particle Size Analysis and Distribution

Microparticle sizes can be measured using different methods. One of the most widely used is the electrical stream sensing zone technique (Coulter counter). In this method, for size determination, a small amount of particles are dispersed in a

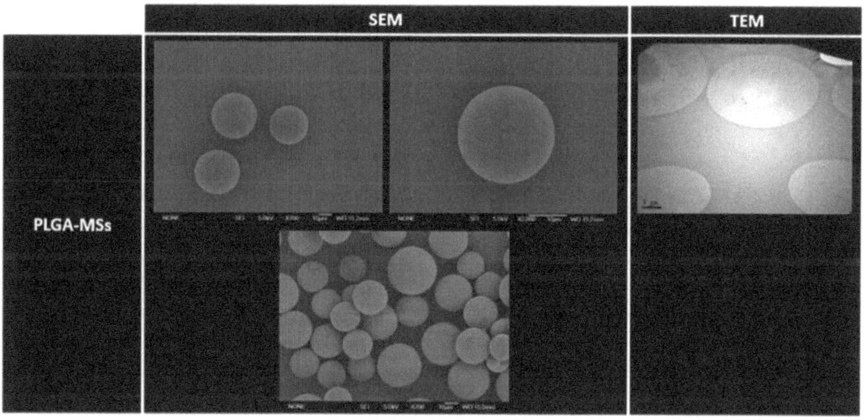

Fig. 14 Morphological evaluation by SEM (×700 and ×2000), TEM, and particle size distribution of PLGA microspheres (Resomer RG503® PLGA 50:50, mean particle size: 25 ± 0.8 μm)

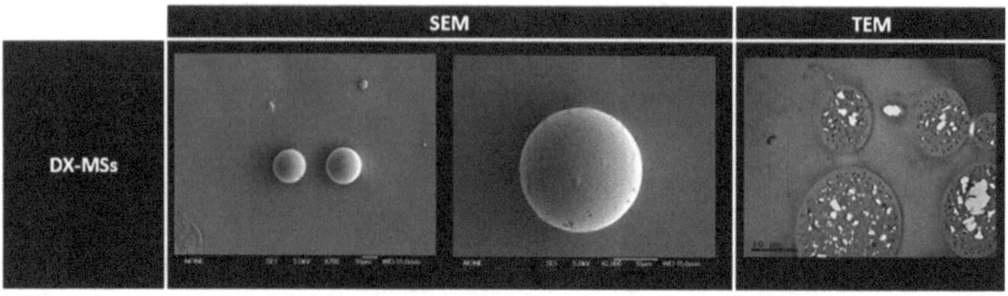

Fig. 15 Morphological evaluation by SEM (×700 and ×2000) and TEM of dexamethasone (DX)-loaded microspheres [DX: PLGA Resomer RG503® (2:10); mean particle size (27.5 ± 0.6 μm)]

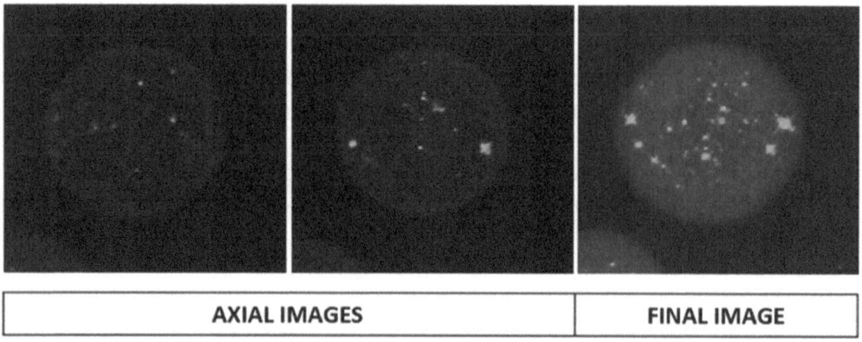

Fig. 16 Confocal microscopy images of microspheres elaborated with PLGA 50:50, Resomer RG503® vitamin E, and BSA-FITC. Granulometric fraction (20–38 μm)

solution with an electrolyte (Fig. 17). The diluted suspension is then forced to pass through a measuring orifice of an electrical stream sensing zone apparatus. The electrodes, situated on either side of the aperture and surrounded by the electrolyte solution, monitor the change in electrical signal that occurs when a microparticle crosses through the orifice and displaces its own volume of electrolyte solution. The change in electrical resistance between the electrodes is proportional to the volume of the particle, yielding the volume-equivalent sphere diameter.

The laser light scattering technique is based on the interaction of laser light with particles. Microparticles, which have a wavelength much higher than light, produce only a small change in the forward ray direction. On the other hand, small particles produce a higher dispersion as their sizes are lower than the wavelength of light (Amrite et al. 2006).

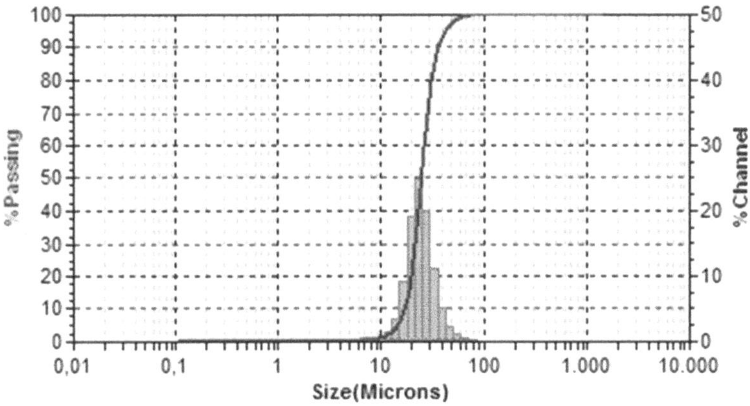

Fig. 17 Particle size distribution of microspheres measured by light scattering using a Microtrac® S3500 laser diffraction particle size analyzer. Particles were prepared with PLGA 50:50, vitamin E, and glial cell-derived neurotropic factor (GDNF)

Infrared Absorption Spectrophotometry

An infrared spectrum allows characterizing of the active substance and polymer as single raw materials and after forming the microstructures (microspheres or microcapsules). In this way, chemical interactions between the components of microparticles are studied by means of the infrared light interaction with the molecules.

Differential Scanning Calorimetry (DSC)

Differential scanning calorimetry (DSC) is useful to characterize the raw materials (active substance, polymer, and other components) and to find potential physicochemical interactions between the components in the microparticle formulation. The glass transition temperatures (Tg) for amorphous substances and melting points (Tm) of crystalline products can be identified. Comparison between the scans obtained for raw materials, physical mixtures, and the final formulation allows interpretation of possible interactions between the components of the microparticles.

X-Ray Diffraction (XRD)

Crystallinity grades of substances can be identified by X-ray diffraction (XRD). As cited previously for infrared absorption spectroscopy and DSC, this technique allows identification of changes in the structure of the microparticle's components.

The analysis is performed by comparing X-ray diffraction of raw materials and their physical mixtures with the spectra obtained in the analysis of the final formulation.

Gel Permeation Chromatography (GPC)

Gel permeation chromatography (GPC) is used for the characterization of the polymer in terms of molecular weight and polydispersity. Possible changes due to the microencapsulation procedure and sterilization of the polymer can also be studied. This technique can also be very useful to study the changes in the polymers forming the particles during a release study (Martínez-Sancho et al. 2004b).

Determination of Drug Loading Efficiency

Encapsulation efficiency is calculated as the ratio of the actual active substance content in the microparticles over the theoretical drug loading according to the following equation:

$$EE(\%) = \left(\frac{\text{Amount of actual drug}}{\text{Amount of theoretical drug} + \text{amount of polymer}} \right) \cdot 100$$

The loading capacity (LC) of the drug in the microparticles is expressed as a percentage or as the amount of the active substance (usually expressed as ng or µg) per mg of microparticles as follows:

$$LC(\%) = \left(\frac{\text{Amount of entrapped drug}}{\text{Amount of microparticles}} \right) \cdot 100$$

$$LC = \frac{\text{Amount of drug}}{\text{Amount of particles}}$$

In Vitro Release Studies

Drug release studies are performed in "sink conditions" for realistic drug dissolution. For the in vitro release experiments, the particles are suspended in an aqueous media (usually phosphate buffer at 37 °C) while being maintained with agitation during the time of the study. At fixed times, and after centrifugation, all the supernatant is removed, and the concentration of the active substance is quantified by an appropriate technique. In most cases, the volume that was removed is then replaced with fresh medium to continue the study (Herrero-Vanrell and Refojo 2001). In other experiments, only part of the media is removed and then replaced with the same volume of fresh liquid. Several authors have described the use of surfactants, such as sodium azide, included in the release media to avoid contamination (García-Caballero et al. 2017).

Factors such as the polymer composition, its molecular weight, as well as the solubility, and encapsulated drug loading influence the release rate profiles. In the case of PLGA microspheres, of special relevance is the PLA:PGA ratio, since the 50:50 ratio is the most biodegradable. Low molecular weight polymers deliver the encapsulated drug faster than high molecular weight ones. As the size of the particles is increased (low surface area), the release rate of the drug decreases.

Administration of Microparticles: Efficacy Studies

Microspheres for ocular administration can be administered by topical, intraocular, and periocular routes. Injection of particles is a critical issue. Extemporaneous preparation of microsphere's suspension in an aqueous vehicle (e.g., saline solution, balanced salt solution, or phosphate buffer) has to be performed without blocking the suspension flow. As described by several authors, injection of the particles is usually made using needles ranging from 25 to 34 gauge or even from a glass micropipette (Herrero-Vanrell and Refojo 2001).

One of the major problems associated with the use of saline solutions as vehicles for microspheres is their adherence to the syringe walls. This problem can be solved with the use of aqueous solutions of polymers, such as hyaluronic acid (HA) or hydroxypropyl methylcellulose (HPMC), as vehicles (Herrero-Vanrell and Refojo 2001).

Topical Administration

Topical administration of microspheres is not very common because they are of a potential risk to cause discomfort to the patient after instillation. Solid lipid microparticles (SLMs) with a size up to 15 μm loaded with cyclosporine have been proposed for topical administration in rabbits' eyes (Wolska et al. 2018). Microparticles prepared using polymeric materials, such as chitosan, polycarbophil, and PLGA, have been considered for topical administration. One example is the use of tetracaine-loaded bovine-serum albumin-chitosan microparticles (size 4 μm), which were able to increase the anesthetic effect of the drug when compared with the commercial formulation (Addo et al. 2010). Another example is the mucoadhesive microspheres prepared with polycarbophil, which were loaded with the antibiotic sulfacetamide (Sensoy et al. 2009) for the treatment of keratitis. Both of these formulations were prepared using the spray-drying technique. The last example involves the use of PLGA microspheres with the addition of polyethylene glycol

(PEG) in a topical formulation. This formulation was able to increase the bioavailability of the hypotensive drug brimonidine compared to the commercial formulation (Park et al. 2015).

Intraocular Administration

Intraocular administration includes intracameral, intravitreal, or subretinal injections. Intravitreal administration of microparticles (mainly microspheres) has been the most explored area with the aim being to treat posterior segment of the eye diseases (proliferative diabetic retinopathy (PDR), age-related macular degeneration (AMD), diabetic retinopathy (DR), uveitis, macular edema, cytomegalovirus retinitis, neuroprotection in glaucoma, and retinitis pigmentosa (RP), among others) (Herrero-Vanrell et al. 2014). To this, a great number of active substances with low molecular weights (i.e., antiproliferative active substances, anti-inflammatory drugs, immunosuppressants, and antibiotics) or biotechnological products have been encapsulated into microparticles by using the solvent emulsion evaporation technique. Among the biodegradable polymers intended for intravitreal administration, PLGA is the most employed.

One of the main concerns regarding the intravitreal injection of microspheres is that they can interfere with the visual pathway. However, an intravitreal injection of PLGA microparticles in humans has been described to leave a free visual axis because the microspheres have the tendency to aggregate at the site of injection (Cardillo et al. 2006).

Regarding tolerance, some intraocular injections of PLGA microspheres have been associated with a mild localized foreign body reaction that decreases with time (Khoobehi et al. 1991; Visscher et al. 1985). The signs observed were described to be similar to the ones reported for sutures that disappeared 2–4 weeks after surgery (Giordano et al. 1995). Histopathologic studies performed in rabbits after an intravitreal injection of PLGA microspheres revealed mononuclear cells and multinucleated giant cells. No other signs were described as the retina was not affected (Veloso et al. 1997).

It is worthy of noting that the foreign body reactions observed after injection of microparticles can be linked to other components of a formulation. Some of these reactions have been known to be caused by the solvent employed during the particle manufacturing process. An evaluation of the toxicity of the drug delivery systems must include the tolerance of any potential toxic substance employed in the manufacturing process that can be associated with unexpected side effects. Dichloromethane (DCM or methylene chloride) (b.p. @40 °C) has been extensively used in the manufacture of microparticles. This halogenated solvent has been reported to cause inflammation of the conjunctiva and eyelids (Ballantyne et al. 1976). Other authors described the use of nonhalogenated solvents that are potentially less toxic, such as isopropyl formate (b.p. 68.2 °C) or ethyl acetate (b.p. 77.1 °C), to dissolve the PLGA using the solvent evaporation method to prepare microparticles (Jang and Sah 2011). Despite the advantages of using these nonhalogenated solvents, they exhibited a slower rate of evaporation compared to dichloromethane, with the potential risk of residual solvent entrapment in the microparticle formulation. In fact, some authors have evaluated the toxicity of non-loaded PLGA microspheres in cynomolgus monkeys, which revealed some unexpected immune reactions. In this work, particles were prepared by the solvent evaporation technique from water-in-oil-in-water emulsion, with ethyl acetate as the polymer solvent (Thackaberry et al. 2017). The authors attributed the toxicity to the particles although they described that low levels of residual solvent were present in the microparticulate drug delivery systems. Although residual solvents in drug products are acceptable under established lower limits, the tissues of the eye are highly sensitive, and more extensive tolerance studies need to be performed. It is clear that further studies of the intraocular toxicity of the solvents employed to prepare microparticles will be necessary to clarify the potential adverse effects attributed to the particles.

Periocular Route

Periocular administration arises as an alternative route to avoid the secondary effects related to intravitreal injections. Microparticles for the treatment of ocular diseases have been injected by sub-Tenon and subconjunctival routes. The microparticulate drug delivery systems should be available in the periocular space for a long time in order to provide sustained levels of the active substance and enhance drug penetration into the intraocular tissues. The retention of particles in the sclera has been demonstrated to be linked to particle size. While 2 μm microspheres and 200 nm nanoparticles were retained at the site of administration for at least 60 days, smaller sizes (20 nm) were rapidly cleared by systemic and lymphatic circulation (Amrite et al. 2008).

PLGA microspheres prepared by the solvent evaporation method from an O/W emulsion loaded with celecoxib (3.9 ± 0.6 μm) have been injected by subconjunctival injection for the treatment of diabetes. The effects of Adriamycin-loaded PLA microspheres (100 μg dose) using an O/O emulsion technique after subconjunctival injection in rabbits were demonstrated to prevent fibrosis after glaucoma filtering surgery (Kimura et al. 1992). A sub-Tenon injection of a combination of microspheres (2 mg) loaded with ciprofloxacin hydrochloride (1.07 ± 0.35 μm) and a triamcinolone acetonide solution (25 mg) was able to prevent infection and inflammation after cataract surgery in humans. The microspheres were prepared by the spray-drying technique (Paganelli et al. 2009).

Conclusions

Microparticulate drug delivery systems are emerging as therapeutic tools for the treatment of posterior segment eye diseases. Microspheres represent an alternative to repeated intraocular administrations as they are able to release the drug in a controlled fashion for long periods of time. The injection of microparticles can be performed with small-gauge needles (30–34 G), circumventing the need for more invasive surgical procedures. Furthermore, if they are prepared from biodegradable polymers, such as PLA and PLGA, microparticles disappear from the site of injection after delivering the encapsulated drug. These polymers have been widely used to encapsulate a variety of active substances (low, medium, and high molecular weight molecules as well as biologicals) for the treatment of intraocular diseases. PLA and PLGA microspheres have demonstrated the tendency to aggregate in the vitreous and then behave more like an implant instead of microparticle. This proves to be desirable tendency for intraocular drug delivery systems because there is no interference with the visual pathway. Microspheres can be loaded with a single drug or a combination of several active substances encapsulated in the same microdevice. Multiloaded microspheres are able to release the therapeutic substances simultaneously, resulting in the treatment of multifactorial neurodegenerative posterior segment eye diseases, such as age-related macular degeneration, diabetic retinopathy, and glaucoma. Furthermore, as multiloaded microdevices contain several drugs, the amount of injected polymer is significantly reduced. The administration of the optimal dose of the active molecule for an individual patient (personalized therapy) can be achieved by adjusting the amount of microparticles to be administered. In summary, microspheres offer therapeutic innovative solutions and new indications in the treatment of chronic and multifactorial diseases affecting the back of the eye.

Acknowledgments The authors acknowledge Spanish Ministry of Economy, Industry, and Competitiveness (MAT-2017-83858-C2-1-R), ISCIII-FEDER "Una manera de hacer Europa" RETICS: RD16/0008, and UCM Research Group 920415 "Innovation, Therapy, and Pharmaceutical Development in Ophthalmology" (INNOFTAL).

References

Addo RT, Siddig A, Siwale R, Patel NJ, Akande J, Uddin AN, D'Souza MJ. Formulation, characterization and testing of tetracaine hydrochloride-loaded albumin-chitosan microparticles for ocular drug delivery. J Microencapsul. 2010;27:95–104. https://doi.org/10.3109/02652040903010638.

Amrite AC, Ayalasomayajula SP, Cheruvu NP, Kompella UB. Single periocular injection of celecoxib-PLGA microparticles inhibits diabetes-induced elevations in retinal PGE2, VEGF, and vascular leakage. Invest Ophthalmol Vis Sci. 2006;47:1149–60. https://doi.org/10.1167/iovs.05-0531.

Amrite AC, Edelhauser HF, Singh SR, Kompella UB. Effect of circulation on the disposition and ocular tissue distribution of 20 nm nanoparticles after periocular administration. Mol Vis. 2008;14:150–60. pii:v14/a20.

Arranz-Romera A, et al. Simultaneous co-delivery of neuroprotective drugs from multi-loaded PLGA microspheres for the treatment of glaucoma. J Control Release. 2019a;297:26–38. https://doi.org/10.1016/j.jconrel.2019.01.012.

Arranz-Romera A, Esteban-Pérez S, Garcia-Herranz D, Aragón-Navas A, Bravo-Osuna I, Herrero-Vanrell R. Combination therapy and co-delivery strategies to optimize treatment of posterior segment neurodegenerative diseases. Drug Discov Today. 2019b; https://doi.org/10.1016/j.drudis.2019.03.022.

Aulton-Pharmaceutics. The science of dosage form design. London: Churchill Livingstone, Harcourt Publisher Limited; 2002.

Ballantyne B, Gazzard MF, Swanston DW. The ophthalmic toxicology of dichlorømethane. Toxicology. 1976;6:173–87.

Bravo-Osuna I, Andres-Guerrero V, Pastoriza Abal P, Molina-Martinez IT, Herrero-Vanrell R. Pharmaceutical microscale and nanoscale approaches for efficient treatment of ocular diseases. Drug Deliv Transl Res. 2016;6:686–707. https://doi.org/10.1007/s13346-016-0336-5.

Bravo-Osuna I, Andrés-Guerrero V, Arranz-Romera A, Esteban-Pérez S, Molina-Martínez IT, Herrero-Vanrell R. Microspheres as intraocular therapeutic tools in chronic diseases of the optic nerve and retina. Adv Drug Deliv Rev. 2018;126:127–44. https://doi.org/10.1016/j.addr.2018.01.007.

Cape SP, Villa JA, Huang ET, Yang TH, Carpenter JF, Sievers RE. Preparation of active proteins, vaccines and pharmaceuticals as fine powders using supercritical or near-critical fluids. Pharm Res. 2008;25:1967–90. https://doi.org/10.1007/s11095-008-9575-6.

Cardillo JA, Souza-Filho AA, Oliveira AG. Intravitreal Bioerudivel sustained-release triamcinolone microspheres system (RETAAC). Preliminary report of its potential usefulnes for the treatment of diabetic macular edema. Arch Soc Esp Oftalmol. 2006;81(675–677):679–81.

Checa-Casalengua P, Jiang C, Bravo-Osuna I, Tucker BA, Molina-Martínez IT, Young MJ, Herrero-Vanrell R. Retinal ganglion cells survival in a glaucoma model by GDNF/Vit E PLGA microspheres prepared according to a novel microencapsulation procedure. J Control Release. 2011;156:92–100. https://doi.org/10.1016/j.jconrel.2011.06.023.

Davis M, Walker G. Recent strategies in spray drying for the enhanced bioavailability of poorly water-soluble drugs. J Control Release. 2018;269:110–27. https://doi.org/10.1016/j.jconrel.2017.11.005.

European Guideline 3AQ4a. The use of ionising radiation in the manufacture of medicinal products. Official Publications in the European Communities, London; 1992.

García-Caballero C, et al. Six month delivery of GDNF from PLGA/vitamin E biodegradable microspheres after intravitreal injection in rabbits. Eur J Pharm Sci. 2017;103:19–26. https://doi.org/10.1016/j.ejps.2017.02.037.

Giordano GG, Chevez-Barrios P, Refojo MF, Garcia CA. Biodegradation and tissue reaction to intravitreous biodegradable poly(D,L-lactic-co-glycolic) acid microspheres. Curr Eye Res. 1995;14:761–8. https://doi.org/10.3109/02713689508995797.

He Y, Wang JC, Liu YL, Ma ZZ, Zhu XA, Zhang Q. Therapeutic and toxicological evaluations of cyclosporine a microspheres as a treatment vehicle for uveitis in rabbits. J Ocul Pharmacol Ther. 2006;22:121–31. https://doi.org/10.1089/jop.2006.22.121.

Herrero-Vanrell R, Refojo MF. Biodegradable microspheres for vitreoretinal drug delivery. Adv Drug Deliv Rev. 2001;52:5–16.

Herrero-Vanrell R, Vicario de la Torre M, Andres-Guerrero V, Barbosa-Alfaro D, Molina-Martinez IT, Bravo-Osuna I. Nano and microtechnologies for ophthalmic administration, an overview. J Drug Deliv Sci Technol. 2013a;23:75–102.

Herrero-Vanrell R, Vicario de la Torre M, Andrés-Guerrero V, Barbosa-Alfaro D, Molina-Martínez IT, Bravo-Osuna I. Nano and microtechnologies for ophthalmic administration, an overview. J Drug Deliv Sci Technol. 2013b; https://doi.org/10.1016/S1773-2247(13)50016-5.

Herrero-Vanrell R, Bravo-Osuna I, Andrés-Guerrero V, Vicario-de-la-Torre M, Molina-Martínez IT. The potential of using biodegradable microspheres in retinal diseases and other intraocular pathologies. Prog Retin Eye Res. 2014;42:27–43. https://doi.org/10.1016/j.preteyeres.2014.04.002.

Jang J, Sah H. Nonhalogenated solvent-based solvent evaporation process useful in preparation of PLGA microspheres. J Microencapsul. 2011;28:490–8. https://doi.org/10.3109/02652048.2011.586066.

Khoobehi B, Stradtmann MO, Peyman GA, Aly OM. Clearance of sodium fluorescein incorporated into microspheres from the vitreous after intravitreal injection. Ophthalmic Surg. 1991;22:175–80.

Kimura H, Ogura Y, Moritera T, Honda Y, Wada R, Hyon SH, Ikada Y. Injectable microspheres with controlled drug release for glaucoma filtering surgery. Invest Ophthalmol Vis Sci. 1992;33:3436–41.

Li W, et al. Microfluidic fabrication of microparticles for biomedical applications. Chem Soc Rev. 2018;47:5646–83. https://doi.org/10.1039/c7cs00263g.

Liew G, Wong VW, Ho IV. Mini review: changes in the incidence of and progression to proliferative and sight-threatening diabetic retinopathy over the last 30 years. Ophthalmic Epidemiol. 2017;24:73–80. https://doi.org/10.1080/09286586.2016.1259638.

Martínez-Sancho C, Herrero-Vanrell R, Negro S. Poly (D,L-lactide-co-glycolide) microspheres for long-term intravitreal delivery of aciclovir: influence of fatty and non-fatty additives. J Microencapsul. 2003;20:799–810. https://doi.org/10.1080/02652040310001600532.

Martínez-Sancho C, Herrero-Vanrell R, Negro S. Optimisation of aciclovir poly(D,L-lactide-co-glycolide) microspheres for intravitreal administration using a factorial design study. Int J Pharm. 2004a;273:45–56. https://doi.org/10.1016/j.ijpharm.2003.12.006.

Martínez-Sancho C, Herrero-Vanrell R, Negro S. Study of gamma-irradiation effects on aciclovir poly(D,L-lactic-co-glycolic) acid microspheres for intravitreal administration. J Control Release. 2004b;99:41–52. https://doi.org/10.1016/j.jconrel.2004.06.004.

McGuinness MB, Karahalios A, Finger RP, Guymer RH, Simpson JA. Age-related macular degeneration and mortality: a systematic review and meta-analysis. Ophthalmic Epidemiol. 2017;24:141–52. https://doi.org/10.1080/09286586.2016.1259422.

Nijsen JF, et al. Influence of neutron irradiation on holmium acetylacetonate loaded poly(L-lactic acid) microspheres. Biomaterials. 2002;23:1831–9.

Ozkan G, Franco P, De Marco I, Xiao J, Capanoglu E. A review of microencapsulation methods for food antioxidants: Principles, advantages, drawbacks and applications. Food Chem. 2019;272:494–506. https://doi.org/10.1016/j.foodchem.2018.07.205.

Paganelli F, et al. A single intraoperative sub-tenon's capsule injection of triamcinolone and ciprofloxacin in a controlled-release system for cataract surgery. Invest Ophthalmol Vis Sci. 2009;50:3041–7. https://doi.org/10.1167/iovs.08-2920.

Park CG, et al. Mucoadhesive microparticles with a nanostructured surface for enhanced bioavailability of glaucoma drug. J Control Release. 2015;220:180–8. https://doi.org/10.1016/j.jconrel.2015.10.027.

Ramón MP. Tratado de Tecnología Farmacéutica. Volumen II: Operaciones básicas. Madrid: Editorial Síntesis; 2016.

Sensoy D, Cevher E, Sarici A, Yilmaz M, Ozdamar A, Bergişadi N. Bioadhesive sulfacetamide sodium microspheres: evaluation of their effectiveness in the treatment of bacterial keratitis caused by Staphylococcus aureus and Pseudomonas aeruginosa in a rabbit model. Eur J Pharm Biopharm. 2009;72:487–95. https://doi.org/10.1016/j.ejpb.2009.02.006.

Swider E, Koshkina O, Tel J, Cruz LJ, de Vries IJM, Srinivas M. Customizing poly(lactic-co-glycolic acid) particles for biomedical applications. Acta Biomater. 2018;73:38–51. https://doi.org/10.1016/j.actbio.2018.04.006.

Thackaberry EA, et al. Evaluation of the toxicity of intravitreally injected PLGA microspheres and rods in monkeys and rabbits: effects of depot size on inflammatory response. Invest Ophthalmol Vis Sci. 2017;58:4274–85. https://doi.org/10.1167/iovs.16-21334.

Tian K, Shibata-Germanos S, Pahlitzsch M, Cordeiro MF. Current perspective of neuroprotection and glaucoma. Clin Ophthalmol. 2015;9:2109–18. https://doi.org/10.2147/OPTH.S80445.

Timilsena YP, Akanbi TO, Khalid N, Adhikari B, Barrow CJ. Complex coacervation: Principles, mechanisms and applications in microencapsulation. Int J Biol Macromol. 2019;121:1276–86. https://doi.org/10.1016/j.ijbiomac.2018.10.144.

USP. United States Pharmacopeia and National Formulary. Rockville: United States Pharmacopeial Convention; 2019.

Veloso AA, Zhu Q, Herrero-Vanrell R, Refojo MF. Ganciclovir-loaded polymer microspheres in rabbit eyes inoculated with human cytomegalovirus. Invest Ophthalmol Vis Sci. 1997;38:665–75.

Visscher GE, Robison RL, Maulding HV, Fong JW, Pearson JE, Argentieri GJ. Biodegradation of and tissue reaction to 50:50 poly(DL-lactide-co-glycolide) microcapsules. J Biomed Mater Res. 1985;19:349–65. https://doi.org/10.1002/jbm.820190315.

Wan F, Yang M. Design of PLGA-based depot delivery systems for biopharmaceuticals prepared by spray drying. Int J Pharm. 2016;498:82–95. https://doi.org/10.1016/j.ijpharm.2015.12.025.

Wan LS, Heng PW, Chia CG. Plasticizers and their effects on microencapsulation process by spray-drying in an aqueous system. J Microencapsul. 1992;9:53–62. https://doi.org/10.3109/02652049209021222.

Wolska E, et al. Ocular irritation and cyclosporine A distribution in the eye tissues after administration of solid lipid microparticles in the rabbit model. Eur J Pharm Sci. 2018;121:95–105. https://doi.org/10.1016/j.ejps.2018.05.015.

Yao J, Tucker BA, Zhang X, Checa-Casalengua P, Herrero-Vanrell R, Young MJ. Robust cell integration from co-transplantation of biodegradable MMP2-PLGA microspheres with retinal progenitor cells. Biomaterials. 2011;32:1041–50. https://doi.org/10.1016/j.biomaterials.2010.09.063.

Nanoparticle Products for the Eye: Preformulation, Formulation, and Manufacturing Considerations

E. Sánchez-López, Eliana B. Souto, M. Espina, A. Cano, M. Ettcheto, Antoni Camins, and M. L. García

Abstract Recent developments in nanotechnology have provided an unprecedented opportunity to improve the prevention and treatment of ocular diseases. Because of their inherent valuable properties such as high surface-to-volume ratio, ease of surface functionalization with desired ligands, reduction of drug side effects, and potential ability to cross biological barriers, nanoparticles are increasingly recognized as promising candidates for ocular therapy of anterior and posterior eye disorders. Nanoparticle products have become one of the most promising strategies in order to treat the eye globe. There are several administration routes for ocular drug delivery with eye drops being one of the most widely used. Moreover, intravit-

E. Sánchez-López (✉) · A. Cano · M. L. García (✉)
Faculty of Pharmacy and Food Sciences, Department of Pharmacy and Pharmaceutical Technology and Physical Chemistry, University of Barcelona, Barcelona, Spain

Institute of Nanoscience and Nanotechnology (IN2UB), University of Barcelona, Barcelona, Spain

Networking Research Centre of Neurodegenerative Disease (CIBERNED), Instituto de Salud Juan Carlos III, Madrid, Spain
e-mail: marisagarcia@ub.edu

E. B. Souto
Faculty of Pharmacy, Department of Pharmaceutical Technology, University of Coimbra, Coimbra, Portugal

CEB—Centre of Biological Engineering, University of Minho, Braga, Portugal

M. Espina
Faculty of Pharmacy and Food Sciences, Department of Pharmacy and Pharmaceutical Technology and Physical Chemistry, University of Barcelona, Barcelona, Spain

Institute of Nanoscience and Nanotechnology (IN2UB), University of Barcelona, Barcelona, Spain

M. Ettcheto · A. Camins
Networking Research Centre of Neurodegenerative Disease (CIBERNED), Instituto de Salud Juan Carlos III, Madrid, Spain

Faculty of Pharmacy and Food Sciences, Department of Pharmacology and Therapeutic Chemistry, University of Barcelona, Barcelona, Spain

S. Neervannan, U. B. Kompella (eds.), *Ophthalmic Product Development*, AAPS Advances in the Pharmaceutical Sciences Series 37,
https://doi.org/10.1007/978-3-030-76367-1_15

real injection and implants will also be analyzed throughout this chapter. Among the different types of nanocarriers, biodegradable nanoparticles such as lipid or polymeric nanoparticles have been extensively studied for ocular drug delivery. These drug delivery systems belong to the nanometer range and are able to encapsulate drugs in order to carry and deliver them on the target tissue using different strategies such as targeting with peptides or monoclonal antibodies.

The selective release of the drug in the target organ to exert its effective therapeutic action will remain a distant reality unless the design of the nanocarrier takes into account the biological barriers. Biodegradable multifunctional carriers can be prepared by several different methods that facilitate the development of a new generation of chemically functionalized polymeric or lipid nanoparticles with physicochemical characteristics suitable to cross the different eye barriers and achieve a biopharmaceutical profile appropriate for the ocular route of administration.

This chapter focuses on the application of innovative technological strategies for the development and manufacturing, under quality by design criteria, of nanoparticle carriers for the management of ocular disorders, which undoubtedly constitute one of the most important health challenges of today's society.

Keywords Nanoparticles · PLGA nanoparticles · Lipid nanoparticles · Polymeric nanoparticles · Ocular drug delivery

Abbreviations

AMD	Age-related macular degeneration
BRB	Blood-retinal barrier
CS	Chitosan
CLSM	Confocal laser scanner microscopy
CD	Cyclodextrins
DoE	Design of experiments
DR	Diabetic retinopathy
DSC	Differential scanning calorimetry
EE	Encapsulation efficiency
FDA	Food and Drug Administration
Tg	Glass transition temperatures
HA	Hyaluronic acid
HPMC	Hydroxypropyl methylcellulose
iBRB	Inner BRB
IDDS	Intraocular drug delivery systems
LNPs	Lipid nanoparticles
LDCs	Lipid-drug conjugates
LNCs	Lipid nanocapsules
LC	Loading capacity
Tm	Melting points

NPs	Nanoparticles
NEs	Nanospheres
NLC	Nanostructured lipid carriers
oBRB	Outer BRB
PI	Polydispersity index
PEG	Polyethylene glycol
PGA	Poly(glycolic)
PLA	Poly(lactic) acid
PLGA	Poly(lactic-*co*-glycolic)
PNP	Polymeric nanoparticles
PCL	Polycaprolactone
PDR	Proliferative diabetic retinopathy
RP	Retinitis pigmentosa
RPE	Retinal pigment epithelium
SEM	Scanning electron microscopy
SLN	Solid lipid nanoparticles
TEM	Transmission electron microscopy
TA	Triamcinolone acetonide
XRD	X-ray diffraction
ZP	Zeta potential

Introduction

Topical administration of drugs to the eye is the most common approach for treating eye diseases, performed by dropping the formulation into the conjunctival cul-de-sac (Fangueiro et al. 2016). In this approach, the dosage forms applied are liquid (solutions, suspensions, in situ forming gels), intended for single or for multiple applications (preserved solutions), or semisolid (ointments). Physicochemical properties and stability of the drug and dosage form, in conjunction with the physiological parameters, are the main factors that determine ocular drug bioavailability, efficacy, and tolerability of eye drops. Although eye drops are a common vehicle for administering drugs to the eye, their bioavailability in the internal ocular tissues is limited due to the presence of ocular barriers (cornea, blood-aqueous, and blood-retinal barriers), tear turnover, and nonproductive conjunctival adsorption (Sánchez-López et al. 2017a, b). Therefore, only a slow amount (lesser than 5%) of drug instilled into the conjunctival fornix is able to reach the internal eye tissues. As a consequence, the frequency of instillation must be increased, thereby increasing the risk of systemic side effects and hindering patient adherence to treatment (Sánchez-López et al. 2017a, b).

Recently, nanotechnology-based delivery systems (such as mucoadhesive polymers, liposomes, microemulsions, lipid, and polymeric nanoparticles) are receiving considerable attention as suitable approaches to increase the residence time of the drugs on the ocular surface and to improve corneal penetration of the drug, overcoming the chemical and mechanical barriers. Among the biodegradable drug delivery systems, nanoparticles (NPs) constitute a field of intense biomedical research, due to their potential as diagnostic or therapeutic agents (Sánchez-López et al. 2016b; Ramos et al. 2016; Rincón et al. 2018; Gonzalez-Mira et al. 2012). Several eye disorders of anterior or posterior eye segment could potentially be treated using different types of NPs.

The main advantages of using NPs as an ocular drug delivery system include (Nayak and Misra 2018):

- These systems can be manufactured to control morphometric and surface characteristics, avoiding ocular size-dependent irritation and improving ocular penetration. This could be used for both passive and active drug targeting (Nayak and Misra 2018).
- NPs can be prepared in order to obtained a sustained drug release. In addition, they have the potential to target ocular tissues at minimum cost and high therapeutic value (Nagarwal et al. 2009). NPs can modify the distribution and drug clearance from the eye obtaining an increase in drug therapeutic efficacy as well as a reduction in side effects (Nayak and Misra 2018).
- Choosing an appropriate matrix, these systems can increase the therapeutic efficacy while reducing side effects.
- Targeted drug products suitable for extraocular or intraocular administration may be developed, selecting the administration route corresponding to the segment of the eye (anterior or posterior) affected.

Moreover, an increasing number of pharmaceutical companies are involved in the development and commercialization of nanomaterials in biological and medical applications (Salata 2004). The majority of these companies (such as Merck, Wyeth or Abbott) are developing pharmaceutical applications, mainly for drug delivery (Junghanns and Müller 2008).

This chapter summarizes the state-of-the-art of nanoparticles for ocular drug delivery, describing their preformulation, formulation, and manufacturing requirements, and discusses their commercialization prospects.

Overcoming Ocular Barriers

The ocular globe possesses several barriers that drugs and drug delivery systems should overcome in order to achieve a suitable therapeutic efficacy. These barriers are localized in the anterior and posterior eye segments and include the cornea and anterior segment barriers, the sclera and Bruch's-choroid complex, as well as the blood-retinal barrier (BRB) (Huang et al. 2018). At the same time, they restrict fluids uptake of and in turn prevent penetration of foreign bodies. As a defensive function, ocular barriers also prevent the ocular tissues against the penetration of active pharmaceutical ingredients (Sánchez-López et al. 2017a, b).

In eye drop formulations, the type of drug and the patient's condition determine the frequency of instillation. In some serious infectious or inflammatory conditions, the drops may need to be used frequently (several times per day). In contrast, the most commonly used treatments for glaucoma only need to be instilled once a day.

Anatomical, physiological, and physicochemical ocular barriers limit drug delivery, increasing the difficulty to achieve therapeutic amounts at internal tissues of the eye. To overcome these, two different approaches can be followed. The first approach involves the application of alternative delivery routes (non-conventional routes), allowing for more direct access to the intended target sites. The second consists on the development of novel drug delivery systems able to provide higher permeability values and controlled drug release at the target tissue. Combinations of both strategies are being currently utilized and optimized in order to achieve optimal therapeutic efficacy with minimal side effects (Souto et al. 2019).

In this sense, ocular barriers selectively control the inward/outward transport of compounds. Consequently, a better understanding of these biological obstacles would increase the knowledge in order to focus ophthalmic drug therapy towards specified delivery/targeting with minimal adverse consequences (Barar et al. 2008). Nowadays, the design of targeted ocular drug delivery systems to overcome eye barriers still remains one of the greatest challenges in pharmaceutical science (Sánchez-López et al. 2017a, b). Examples for the most relevant ocular disorders to arrive to the target tissue overcoming ocular barriers can be observed in Table 1.

Overcoming the Tear Film

The precorneal tear film (Fig. 1) forming the interface between the air and corneal surface constitutes the first barrier of the eye (Dursun et al. 2000; Barar et al. 2008). It is highly relevant for topical administrations such as eye drops or ointments, where the majority of the solutions are eliminated within a few seconds, resulting in poor bioavailability (<5%). It consists on a high percentage of buffered fluid, and in addition, it contains mucins expressed by human ocular surface epithelia display antiadhesive properties, providing an effective barrier (Rolando and Zierhut 2001; Barar et al. 2008; Huang et al. 2018).

In order to overcome this layer, NPs should possess the ability to adhere to the mucus layer, a phenomenon known as mucoadhesion (Lai et al. 2009). Several strategies to increase mucoadhesion can be employed. The presence of tertiary amines, which may interact with anionic components of mucus, is one strategy that can be employed, but their possible toxicity and ocular irritation constitute a major drawback (Lai et al. 2009). One of the most widely used strategies is poly(ethylene glycol) (PEG) addition. PEG is an hydrophilic polymer routinely used in pharmaceutics to improve systemic circulation and minimize opsonization (Lai et al. 2009). Some authors have reported that PEG molecules may establish adhesive interactions due to their ability to interact with the mucus network by diffusion processes and undergo hydrogen bonding (Peppas et al. 1999; Huang et al. 2000; Serra et al. 2006). However, it has been recently observed that PEG molecular weight and particle surface charge have to be taken into account (Lai et al. 2009).

Table 1 Overview of nanoparticles' manufacturing by different methods for ocular administration

Ocular disorder	Nanoparticles (NPs)	Preparation method	References
Ocular surface-associated diseases	Cyclosporine A-CS NPs	Ionotropic gelation	De Salamanca et al. (2006)
	Cyclosporine A-SLN	High-shear homogenization and ultrasound method	Gökçe et al. (2009)
	Indomethacin-CS	Ionic gelation	Badawi et al. (2008)
	Indomethacin-SLN	High-pressure homogenization	Hippalgaonkar et al. (2013)
	Diclofenac-TMCS	Ionic gelation	Asasutjarit et al. (2015)
	Diclofenac-NLC	High-pressure homogenizer	Attama et al. (2008)
	Flurbiprofen-PCL	Solvent displacement	Ramos et al. (2017)
	Flurbiprofen-PLGA Flurbiprofen-PLGA-PEG Flurbiprofen-PLG-POD	Solvent displacement	Araújo et al. (2009), Vega et al. (2012), Vasconcelos et al. (2015)
	Flurbiprofen NLC	Ultrasound homogenizer High-pressure homogenization	Gonzalez-Mira et al. (2010, 2011)
	Dexibuprofen-PLGA-PEG	Solvent displacement	Sánchez-López et al. (2016b)
	Pranoprofen-PLGA	Solvent displacement	Abrego et al. (2014)
	Carprofen-PLGA	Solvent displacement	Parra et al. (2015)
	Ursolic-oleanolic acid-PLGA	Solvent displacement	Alvarado et al. (2015)
	Dexamethasone-PLGA	Solvent diffusion	Rafie et al. (2010)
	Betamethasone PLA-PEG	Emulsion-solvent diffusion	Ishihara et al. (2009)
	Progesterone-PBCA	Emulsion polymerization	Li et al. (1986)
	Progesterone-SLN	Ultrasound homogenization High-pressure homogenization	Esposito et al. (2017)
	Triamcinolone-PLGA	Emulsion-solvent diffusion	Sabzevari et al. (2013)
	Triamcinolone-NLC	High-pressure homogenization	Araujo et al. (2010, 2012)
	Fluocinolone-PLGA Fluocinolone-PLGA-PEG-peptide	Solvent displacement	Gonzalez-Pizarro et al. (2019)
	Epigallocatechin gallate-SLN	Multiple emulsion	Fangueiro et al. (2014b)
	Baicalin-SLN	Emulsification/ ultrasonication	Liu et al. (2011)
	Tobramycin-SLN	Warn microemulsion method	Cavalli et al. (2002)
	Chloramphenicol-SLN	Melt-emulsion ultrasonication (low-temperature solidification)	Hao et al. (2011)
	Levofloxacin-SLN	Solvent evaporation	Baig et al. (2016)
	Acyclovir-PLGA	Nanoprecipitation	Alkholief et al. (2019)
	Hyaluronic acid-CS	Ionotropic gelation	Contreras-Ruiz et al. (2010)

(continued)

Table 1 (continued)

Ocular disorder	Nanoparticles (NPs)	Preparation method	References
Glaucoma	Metipranolol-PCL Metipranolol-IBCA	Deposition interfacial Interfacial polymerization	Losa et al. (1993)
	Carteolol-PCL NPs (reducing IOP)		Marchal-Heussler et al. (1993)
	Melatonin-PLGA-PEG (reducing IOP)	Solvent displacement	Musumeci et al. (2013)
	Memantine-PLGA-PEG (retinal neuroprotection)	Double emulsion	Sánchez-López et al. (2018a, b)
	Dorzolamide-CS	Ionic gelation	Shinde et al. (2013)
	HA-CS NPs	Ionotropic gelation	Wadhwa et al. (2010)
	Latanoprost-PLGA NPs	Emulsification-solvent evaporation	Giarmoukakis et al. (2013)

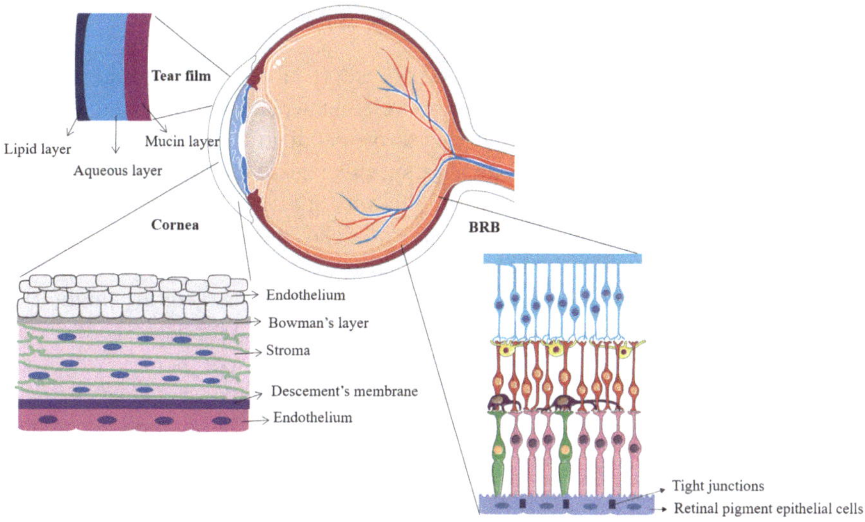

Fig. 1 Main ocular barriers involved in ocular drug delivery

In addition, PEGylation of nanoparticles may also enhance their stability in mucus (Lai et al. 2009). Moreover, due to mucoadhesive and also hydration properties, hyaluronic acid is being widely employed in order to improve the transport of nanoparticles through the tear film (Contreras-Ruiz et al. 2010). Moreover, other strategies are available, such as developing a collagen shield able to preserve formulation transparency and enhance the mucoadhesion of the formulations (Agban et al. 2016).

Overcoming the Corneal Barrier

The corneal route represents the main absorption path for the majority of the ophthalmic therapies. However, corneal absorption is also considered to be a rate-limited process due to the presence of the corneal epithelium (Barar et al. 2008). It is well-known that the cornea is a significant mechanical and chemical barrier to drug delivery. In this sense, small lipophilic molecules are able to be transported, whereas hydrophilic compounds are less likely to cross through the cornea (Huang et al. 2018). Corneal epithelium constitutes the most important obstacle for drug delivery, restricting the absorption into the eye. Stroma and endothelium, on the other hand, provide very little resistance to transcorneal permeation, because they inhibit the transport of only highly lipophilic compounds. Moreover, there are some well-known penetration enhancers that can be added to the formulations in order to ensure nanoparticle's penetration through the corneal tissue. Some examples of penetration enhancers are monoacyl phosphoglycerides, Azone (laurocapram), hexamethylenelauramide, hexamethyleneoctanamide, and decylmethylsulfoxide, which have demonstrated to be able to increase drug transport (Tai-Lee et al. 1990; Tang-Liu et al. 1994).

The mechanisms of penetration enhancers on corneal drug transport are usually related with structural modifications; e.g. brefeldin A and sodium azide enhance their effect by blocking the active transport across cellular membranes (Ottiger et al. 2009). It might be possible that the penetration enhancer at high concentrations acts by loosing the epithelial cell junctions. In this way, it would facilitate the influx of water and hydrophilic compounds but, in turn, delay the movement of lipophilic compounds by developing a highly hydrated barrier (Tang-Liu et al. 1994). However, penetration enhancers while promoting corneal permeation of drugs might also damage the cornea (Rathore and Gupta 2007). In this sense, Agarwal and colleagues were able to dissolve a hydrophobic drug in semifluorinated alkanes, which are amphiphilic liquids that can dissolve hydrophobic drugs (Agarwal et al. 2018). This new strategy can be used as an adjuvant in nanoparticulate formulations as corneal permeation enhancer. A recent strategy consists on targeting the drug delivery systems with substances such as peptides, that are able to cross the corneal tissue using the intrinsic cellular transport. As an example, PepT1 and PepT2 substrates have shown to increase drug transport across the tissue (Rathore and Gupta 2007). Some scientists have conjugated peptides to PEGylated NPs such as POD, a peptide for ocular delivery and human immunodeficiency virus transactivator. POD was conjugated with biodegradable PLGA-PEG NPs, improving ocular drug bioavailability (Vasconcelos et al. 2015).

In addition, Li and colleagues developed triblock copolymer poly(ethylene glycol)-poly(ε-caprolactone)-g-polyethyleneimine nanoparticles able to enhance the transport of lipophilic drugs (Li et al. 2015).

Overcoming the Blood-Retinal Barrier

The retinal ganglion cells (RGCs) localized in the retinal tissue are protected by blood-retinal barrier (BRB). This barrier controls the movement of compounds between the ocular vascular beds and the retinal tissues and also prevents leakage into the retina of potentially harmful agents (Kaur et al. 2008). Endothelial cells are sealed by tight junctions and surrounded by astrocytes, muller cells, and pericytes (Kaur et al. 2008). In order to arrive to this area using drug delivery systems, intravitreal, periocular, and systemic routes are the most widely used (Jiang et al. 2018). However, these repetitive injections may result in several side effects. For this reason, reduction on the number of injections is being intended using drug delivery systems. Moreover, hydrogels administered intravitreally have also been shown to successfully deliver proteins intro the retina (Delplace et al. 2019).

Recently, targeting of NPs to overcome the BRB has demonstrated to be a suitable strategy. In this sense, Bhattacharya and colleagues conjugated NPs to enzyme assisted cleavable linkers aimed to release conjugated cargo within the retinal pigment epithelial (RPE) cells (Bhattacharya et al. 2017). Also, transferrin, an 80 kDa protein able to bind to iron and transport it throughout the body, is uptaken via transferrin receptor-mediated endocytosis by retinal cells. Therefore, some authors have functionalized NPs with transferrin to deliver drugs effectively into the retinal tissues (Bisht et al. 2018). Regarding SLN, either SLN prepared using dextran or hyaluronic acid shown an improvement in the retinal drug bioavailability after intravitreal administration (Apaolaza et al. 2016; Bisht et al. 2018).

Development of Nanoparticles as Ocular Drug Delivery Systems

Preformulation

An ocular drug delivery system should accomplish several requisites, such as its sterility, isotonicity, absence of particles, suitable pH, and viscosity.

In this sense, sterility is one of the main features of ocular drug delivery systems. It can be achieved using different techniques, being the most widely used the autoclave. However, this technique causes a temperature increase that has been shown to affect some drug delivery systems, such as PNPs. PNPs can be sterilized using other methods such as gamma irradiation. According to the European Pharmacopoeia, 25 KGy is the recommended dose in order to sterilize PNPs. Moreover, in order to maintain the sterility, some antimicrobial excipients can be added. In some cases, to avoid possible irritations induced by antimicrobial excipients, preservative-free eye drops could be applied by using blow-fill-seal single-dose droppers or preservative-free multidose containers based on pump systems (Marx and Birkhoff 2017).

Osmolarity of topical formulations should also be maintained. Osmolality of formulation designed for ocular administration should be similar to lacrimal fluid, around 302–318 mOsm/kg, in order to avoid ocular irritation (Ramos et al. 2017). Osmolality can be increased either by adding salts to the formulation or some excipients that can also provide other properties, such as sucrose or mannitol. This addition of excipients in some cases can be useful also against freeze-dry stress as is the case of sucrose and mannitol. On the other hand, lacrimal fluid, although isotonic, can tolerate until 1.4% NaCl. Hypotonic solutions are able to increase the permeability of the epithelium, increasing the concentration of drug, whereas hypertonic solutions cause dehydration, leading to irritation and discomfort.

pH is other relevant feature to take into account in topical ocular formulations. Tear fluid pH is around 7.4–7.7, but it should be noted that the eye possesses buffer capacity, and therefore it is accepted that topical formulations should have a pH value from 6.5 to 8.5.

The absence of particles can be achieved using membrane filters, which will also remove big nanoparticles dispersed in the system.

Viscosity in topical formulations is an important value to take into account. Generally, the more contact between the eye and the formulation, the better the therapeutic response. However, patient comfort should be taken into account, since highly viscous substances can cause vision blurriness and ocular discomfort.

The surface tension value of the precorneal film is comprised between 43.6 and 46.6 mN/m for normal patients and 49.6 mN/m for dry eye patients. Surfactants decrease tension surface of the formula, favoring its miscibility with the precorneal film. In addition, they alter the permeability of the corneal epithelium, increasing the penetration of the active substance. In this sense, in chronic pathologies where surfactants are continously used, they could present certain toxicity on the cornea.

Active Compounds

The active substances contained in an ocular formulation should be able to accomplish the features inherent to ocular drug delivery. Therefore, one of the aims of the use of nanocarriers such as PNP or LNP is to increase the solubility of drugs that are not soluble in an aqueous media. Moreover, other characteristics such as stability are also critical. In order to increase the stability of drugs, proteins, and gene material, NPs have shown to be promising candidates. However, not only insoluble or instable drugs can be encapsulated. Some hydrophilic drugs can also improve their pharmaceutical profile using these nanosystems. In this sense, active compounds should be able to be encapsulated either in the polymeric or lipid matrix/core, should be compatible with the surfactants used, and do not develop covalent bonds with the polymer or lipid. Both the melting point and hydrophilicity of the drug play an essential role and will determine the preparation method of the drug delivery system.

In addition to nanocarriers, prodrug strategies can also be employed either themselves or in addition to nanoparticulate systems. Aiming to improve drugs' bioavailability and also to reduce adverse drug reactions, a prodrug strategy can often be used to enhance drug lipophilicity. At the same time, prodrugs would reduce the effect of the permeability barrier. Currently, 5–7% of the drugs approved worldwide can be classified as prodrugs (Rautio et al. 2008).

Prodrug strategies can be employed for different purposes, such as increase drug solubility, improve the shelf life, or stabilize a drug both chemically and metabolically (Achouri et al. 2013).

Biomaterials

The selection of NP matrix (e.g., polymer or lipid) is essential to obtain a final product with specific characteristic able for the requirements of administration route and to maintain the effect for an established duration.

Biodegradable polymeric NPs (PNPs) can be prepared from a variety of natural and synthetic polymers. The selection of the polymer depends on many factors such as the following: morphology of the NPs, properties of the drug (aqueous solubility, stability, etc.), surface modification, functionalization, degree of biodegradability and biocompatibility, and drug release profile desired for the final product.

Proteins (i.e., albumin and gelatin) and biodegradable polymers, such as polyesters (lactide and glycolide polymers and copolymers) and polyorthoesters (polycaprolactone), have been widely used to prepare PNP for intraocular administration. Among them, lactic and glycolic acid derivatives have been the most popular for the preparation of biodegradable nanospheres. Poly(lactic) acid (PLA), poly(glycolic) (PGA), and poly-lactic-*co*-glycolic acid (PLGA) are already approved for clinical purposes. The degradation rate of these biopolymers depends on the molecular weight (low molecular weight results in a faster degradation time) and L:G ratio (degradation time PLA > PGA > PLGA). Polymers with free COOH groups are more hydrophilic and PEGylation increases their hydrophilicity.

On the other hand, lipid nanoparticles represent one of the most recent advanced in drug delivery systems (Wang et al. 2015). LNPs are suitable for the incorporation of both lipophilic and hydrophilic drugs within the lipid matrix (Sánchez-López et al. 2017a, b). However, lipid structures tend to form an ordered crystalline matrix and therefore they tend to expel the drug from the lipid matrix. For this reason, lipid nanoparticles prepared using lipids containing fatty acid chains can improve, at the same time, long-term stability and drug loading capacity. The drawback of long fatty acid side chains is that it increases particles size, but this could be overcome using the combination of long- and short-chain fatty acids (Sánchez-López et al. 2017a, b). Several lipid materials are already approved by European and US regulatory authorities in order to prepare lipid nanoparticles for ocular drug delivery. The most common are (1) dicaprylocaprate (Labrafac); (2) esters of behenic acid with glycerol (Compritol®888ATO); (3) Diglycerides (dipalmitin, distearin); (4) Monoglycerides (glyceryl monostearate, glyceryl palmitostearate e.g., Precirol ATO®; (5) Aliphatic alcohols (cetylic alcohol, stearylic alcohol); (6) Fatty acids of C10–C12 chains (decanoic acid, linoleic acid); (7) Polyalcohol esters, cholesterol and esters (cholesteryl hemisuccinate, cholesteryl butyrate, and cholesteryl palmitate).

Other lipids, such as cationic lipids, can be employed developing cationic LNPs. They have been recently investigated for targeting ocular mucosa in order to achieve the posterior segment of the eye (e.g., retina) (Leonardi et al. 2015). Cationic lipids are able to interact using electrostatic interactions with the negative surface charge of ocular mucosa. Therefore, this lipid improves the ocular retention time and also increases the nanoparticle's bioadhesion (Fangueiro et al. 2014a). Moreover, cationic LNPs are being used also for the adsorption of gene material (Wang et al. 2015). Using this approach, the most common compounds added to the formulation are dioleoyl trimethylammonium propane (DOTAP) and cetyltrimethylammonium (CTAB) (Sánchez-López et al. 2017a, b). However, these cationic lipids have shown certain toxicity, and only a small amount can be used in the formulations, such as 0.5%, which was the concentration of CTAB used by Fangueiro and colleagues (Fangueiro et al. 2014a). Moreover, lipids such as polyethyleneimine of polylysine have also been used (Wang et al. 2015).

Excipients

During preformulation development, the choice of excipients and buffers must be based upon physiological comfort and product stability. These excipients should possess a proven track record with the Food and Drug Administration (FDA).

In this sense, the first step in product development is establishing the target points regarding the physical and chemical attributes of the formulation. In this sense, appearance, viscosity, surface tension, osmolarity, and pH should be taken into account. In order to achieve these parameters, different excipients can be added to the formulation. Ideally, excipients should be as follows: (1) safe without local or systemic side effects, (2) increase the concentration of drug on the target tissue, (3) compatible with ocular tissues and drug delivery systems, and (4) biodegradable and biocompatible (Kim et al. 2016). Therefore, several excipients may be added in order to accomplish a wide variety of purposes.

The suitable pH for an ophthalmic formulation is 7.4, the same as the tear fluid. However, the majority of the drugs are chemically unstable at this pH. In order to solve this issue, a buffer can be included to ensure constant pH. The buffer should provide a pH close to the physiological pH (although not necessary 7.4) while not causing chemical instability. Citric buffer, phosphate buffer, and borate buffer are the most widely used in ocular formulations, the latter specially in antibiotic formulations.

One of the most widely used excipients used in ophthalmic drug delivery systems is cyclodextrin (CD), which belongs to a group of cyclic oligosaccharides able to improve the solubility of lipophilic drugs by forming complexes. In ocular drug delivery formulations, coadministration of CDs with drugs have reported higher corneal penetration values and enhanced ocular absorption. CD also able to increase the therapeutic efficacy of poorly water-soluble drugs, such as dexamethasone, cyclosporin, acetazolamide, and so on (Table 2). It has been observed that CDs can

Table 2 Cyclodextrin (CD) drug delivery formulations used for topical ocular drug delivery

Cyclodextrin	Preparation method	Type of nanocarrier	Active compound	Target segment	Physicochemical parameters	In vitro/in vivo studies	References
Sulfobutylether-β-CD	Ionotropic gelation	PNPs: Chitosan/sulfobutylether-β-cyclodextrin NPs	Econazole nitrate	Anterior segment (antifungal effect)	Size: 90–673 nm ZP: + 22–33 mV EE: 13–45%	50% release after 8 h Zero-order release kinetics Antifungal effects superior to free drug	Musumeci et al. (2013)
Sulfobutylether-β-CD	Ionotropic gelation	PNPs: Chitosan/sulfobutylether-β-cyclodextrin NPs	Naringenin	Posterior segment	Size: 446.4 ± 112.8 nm ZP: +22.5 ± 4.91 mV	Sustained release effect (in vivo pharmacokinetics in aqueous humor)	Zhang et al. (2016)
α-CD β-CD γ-CD	–	Complexes of β-cyclodextrin	Dexamethasone	Anterior and posterior segment	10 nm–1 mm	Increased drug concentration in ocular tissues and solubility	Loftson and Steffánsson (2011)
Hydroxypropyl-β-CD	–	Complexes 1:1 of CD and drug	Hydrocortisone	Anterior segment	K_{app} 0.636 mM^{-1}	Increased ocular bioavailability	Davies et al. (1997)
Hydroxypropyl-β-CD	Polymer complexation using heating	Complexation of CD and HPMC and other excipients (sodium edetate, sodium chloride, and benzalkonium chloride)	Dexamethasone	Anterior segment		Enhanced solubility, permeability, and corneal bioavailability against Maxidex®	Kristinsson et al. (1996)
Hydroxypropyl-β-CD	Solvent displacement	PNP: PLGA-PEG	Flurbiprofen	Anterior segment	Freeze-dried Stable freeze-dried and at 4 °C during 3 months (liquid form)	Penetrate corneal epithelium Nonirritant	Vega et al. (2013)

(continued)

Table 2 (continued)

Cyclodextrin	Preparation method	Type of nanocarrier	Active compound	Target segment	Physicochemical parameters	In vitro/in vivo studies	References
α-CD and hydroxypropyl-β-CD	Hydration on Carbopol, addition of CD, and dispersion of the mixture into a gel	Gel LNPs: Carbopol 940, sodium hydroxide, and mannitol	Pilocarpine	Anterior segment	Increase in the corneal permeation of pilocarpine nitrate	Increased miotic enhanced corneal penetration effect and by fourfold	Aktaş et al. (2003)
Hydroxypropyl-β-CD 2-hydroxy-3-trimethyl-ammoniopropyl CD	Desolvation method	Gelatin NPs	Hydrocortisone	Anterior segment	Size: 110 to 220 nm ZP 24 and 212 mV EE: 35–45%	After 5 h, 60% of the drug is released	Vandervoort and Ludwig (2004)
ɣ-CD	Self-assembly	NPs containing HPMC and CD and other excipients	Dorzolamide	Anterior segment	Size: 21 µm Drug content: 29.7 mg/ml	NPs were no more toxic or irritating to the eye than Trusopt in rabbits after 1-month administration	Jóhannesson et al. (2014), Gudmundsdottir et al. (2014)

act carriers by surrounding the hydrophobic drug molecules and delivering them to the corneal epithelium surface where they are released (Sahoo et al. 2008).

Some excipients are able to increase contact time with the eye and maintain surface moisture. Chondroitin sulfate sodium salt (Mesoglican®) has been highly used for artificial tears. Carbopol, which also possess several applications such as a bioadhesive material, controlled release agent, emulsifying agent, emulsion stabilizer, rheology modifier, and also a stabilizing agent. Poloxamers, gellan gum, alginate, and carrageenans are used for gel-based formulations and hyaluronic acid (HA). The latter is highly relevant in formulations used for dry eye since offers suitable properties such as: high water binding capacity; rheological properties similar to those of mucus; and demonstrated safety (Kim et al. 2016). Thickening agents, such as methyl cellulose or hydroxypropyl methylcellulose (HPMC), may be added to increase the time of contact of the formulation with the ocular surface.

In order to accomplish the isotonicity, substances such as NaCl, KCl, sodium sulphate, or potassium nitrate can be added. Also, the buffer and the salts containing it should be taken into account in order to ensure isotonicity.

Surfactants constitute an essential compound of nanoparticulate formulations in order to ensure the stability. In addition, they facilitate the contact with the ocular surface. The most widely used are Tween 20, Tween®80, polyvinyl(alcohol) (PVA), and Pluronic® F68. On the preparation methods, surfactants form the aqueous phase either themselves or mixed with hydrophilic active compounds and also can exert a dual function by protecting NPs against freeze-drying stress, such as PVA, or increasing the transport through some barriers such as the BRB as is the case of Tween.

Other agents such as antioxidants can be employed either inside the nanocarrier, outside, or both. EDTA and ascorbic acid can be used for compounds that are easily oxidized and added to the formulation to ensure the stability of the active compound (Cano et al. 2019).

Optimization Studies

Using adequate substance for ocular drug delivery, the formulation can be prepared using several methods (specified in the following section). Moreover, optimization of the formulation parameters constitutes a critical step on the preparation of drug delivery systems. Therefore, the most critical parameters involved in the process should be studied in order to obtain the optimal formulation as well a safe space in which small deviations would not cause formulation modifications. Previous to the optimization, some screening models can be employed in order to ensure a suitable optimization design. Optimization of the formulation composition is essential to ensure a successful and consistent drug delivery without compromising other criteria, such as dosage form size, or wasting valuable drug. This step is crucial since an inadequate model can lead to false conclusions.

Optimization models using mathematical programs are able to represent problem choices as decision variables. These models seek values to measure objective functions of the decision variables subjected to constraints on variable values expressing the limits on possible decision. In order to optimize the formulation, the design of experiments (DoE) is an essential element in drug product development. DoE is an effective method of evaluating product parameters and identifying critical parameters that need consideration during product optimization. Unlike the conventional method of evaluating one variable/parameter at a time, which can sometimes provide incomplete information when other parameters exist, DoE allows for a simultaneous evaluation of various parameters, as well as their interactions and influence on target product parameters or response variables. In this sense, DoE approximations, using either Box-Behnken or factorial design, have become one of the most useful tools in pharmaceutical drug delivery either at laboratory or industrial scale (Singare et al. 2010; Yang et al. 2010; Sánchez-López et al. 2018a, b). Using this approximation, a matrix containing different levels of the independent factors or variables to study (e.g., pH of the formulation, amount of polymer, or amount of drug) is designed, and the interaction between these variables is studied by analyzing their effect in the formulation parameters (e.g., size, polydispersity index, zeta potential, or entrapment efficacy of the drug) (Ramos et al. 2017; Vega et al. 2008),

Nanoparticles for Ocular Drug Delivery

The main advantages derived from nanoparticles used for ocular drug delivery are (1) sustained release of the drug from the nanocarrier, (2) the possibility to overcome ocular barriers, (3) increase of the patient therapeutic adherence by reducing the frequency of administration, (4) protection of the active compounds against inactivation by lacrimal enzymes or tear proteins, and (5) enhancing corneal permeation (Sánchez-López et al. 2017a, b).

In addition to the anatomical and physiological characteristics of the eye, corneal penetration depends on several factors, such as the properties of the drug and the dosage form, the partitioning properties of the drug molecule, as well as its molecular weight and the average size and surface charge of the carrier (Sánchez-López et al. 2017a, b).

Biodegradable NPs are one of the most widely used drug delivery systems for ocular administration due to their ability to prolong corneal residence time, increasing bioavailability through biological membranes (Sánchez-López et al. 2017a, b).

Polymeric Nanoparticles

Nanoparticulate systems are chemically characterized as colloidal carriers, ranging from 10 to 1000 nm. In order to be applied in the ophthalmic tissues, required specific compositions should be done to assure safety and efficacy. In this sense, polymers can be employed in order to develop PNP, such as polylactic acid (PLA), poly(lactide-*co*-glycolide) (PLGA), chitosan, or polycaprolactone (PCL).

PLGA is a copolymer composed of PLA and glycolic acid, which has been widely used for ophthalmic drug delivery (Choi et al. 2012). PLGA is approved by the US FDA for medical purposes due to biodegradability and nontoxicity (Choi et al. 2012; Di Toro et al. 2004). PLGA is nontoxic due to the fact that in the body it is degraded by hydrolysis into lactic acid and glycolic acid, which in turn degrades to water and carbon dioxide (Choi et al. 2012; Xue 2004). By modifying molecular weight, the degradation time of PLGA can be modulated lasting from 1 week to several months (Choi et al. 2012; Sánchez-López et al. 2016b, 2018a, b; Vega et al. 2013).

Chitosan is a copolymer made of β-(1,4)-2-acetamido-D-glucose and β-(1,4)-2-aminoD glucose unit and it is produced by eliminating an acetate moiety from chitin through hydration in concentrated alkali media (Wang et al. 2011). When chitosan contacts with the mucus, the amino and carboxyl groups are combined with mucus glycoproteins, forming a hydrogen bond. This causes an adhesive effect (Wang et al. 2011). Due to these adhesive properties, chitosan can be used either to form the PNP matrix core or to cover the surface, improving the residence time and diminishing the clearance after topical administration. In a study conducted with PNP loading natamycin, PNPs were structurally made of a mixture of chitosan and lecithin. Those PNPs demonstrated high bioavailability in rabbit ocular tissues, administering lower doses in comparison to the marketed suspensions reported with the same effect (Bhatta et al. 2012). The interesting features of chitosan PNP for ocular administration of drugs especially rely in:

1. The relatively easy preparation of nanoparticles.
2. The possibility to obtain a homogeneous PNP population and being able to modulate their size and surface charge.
3. A suitable ability for the association of different kinds of active compounds (hydrophobic and hydrophilic).
4. Chitosan itself possesses wound healing and antimicrobial activity.
5. Versatility for the incorporation of other molecules (within the nanomatrix structure, being able to produce hybrid nanoparticles (Paolicelli et al. 2009).

PCL is a biodegradable polyester approved by the FDA with a low melting point and a glass transition temperature of about −60 °C. This polymer is suitable for ophthalmic drug delivery, and it is highly hydrophobic (Barbault-Foucher et al. 2002). Moreover, it can also be used in order to undertake the surface modification of PNP (Shenoy and Amiji 2005).

Also, PNP may be divided in two principal structures: nanocapsules and nanospheres (Fig. 2). Nanocapsules possess the drug encapsulated inside the polymeric matrix, whereas in nanospheres, the drug is homogeneously dispersed along the polymeric lattice (Patel et al. 2013).

The small size of nanoparticles is a key feature in order to decrease irritation phenomena in the corneal tissue and capacity to sustain drug delivery, avoiding multiple administrations. However, these formulations may be rapidly removed from the precorneal area. Therefore, PNP possessing mucoadhesive properties may be prepared, increasing extending the time that the drug remains in the precorneal

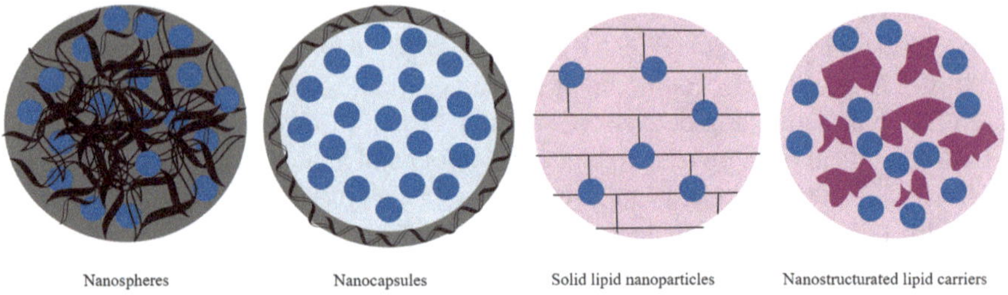

Fig. 2 Types of polymeric and lipid nanoparticles

tissue. In order to achieve this objective, compounds, such as previously mentioned chitosan, hyaluronic acid, or polyethylene glycol (PEG), can be employed. In this sense, PEG surface modifications cause the NP avoid to immune responses and reside in the blood circulation for longer time (Choi et al. 2012; Gref et al. 1994). Musumeci and colleagues compared PLGA NPs with PEG-PLGA encapsulating melatonin, showing that PEG-modified PLGA NPs were suitable to lower the intraocular pressure in rabbits (Musumeci et al. 2013).

Lipid Nanoparticles

Lipid nanoparticles (LNPs) are particles comprised in the nanometer range, from 50 until 1000 nm, whose matrix is made of biocompatible solid lipids or a mixture of solid and liquid lipids (Battaglia et al. 2016). The goal of ophthalmic drug delivery systems has traditionally been to maximize the ocular drug absorption, minimizing systemic absorption, thus avoiding side effects (Sánchez-López et al. 2017a, b). An ideal ocular dosage form should release the drug and overcome the protective ocular barriers to assure therapeutic levels of drug in the intraocular tissues (Sánchez-López et al. 2017a, b). Biocompatibility and mucoadhesive properties of lipid nanoparticles (LNPs) improve their interaction with the ocular mucosa. This may increase the drug corneal residence time obtaining high bioavailability (Sánchez-López et al. 2017a, b).

Lipid components of lipid nanoparticles can interact with the outside lipid layer of the tear film due to the fact that they show similar properties to those of the tear film. Therefore, LNPs are able to increase the residence time of carrier in the conjunctival sac, where it acts as a drug depot (Sánchez-López et al. 2017a, b).

Depending on the lipid matrix, LNPs can be divided into solid lipid nanoparticles (SLN), nanostructured lipid carriers (NLC), lipid-drug conjugates (LDCs), and lipid nanocapsules (LNCs) (Battaglia et al. 2016; Battaglia and Gallarate 2012). The key concept is that by altering the nanostructure of the lipid matrix, the encapsulation of active compounds can be increased and the drug expulsion during storage avoided (Battaglia et al. 2016).

SLNs are the most studied type of lipid nanoparticles and are constituted by a solid lipid matrix surrounded by a layer of surfactants in an aqueous dispersion (Battaglia et al. 2016; Battaglia and Gallarate 2012).

NLCs are composed by a lipid core consisting of a mixture of solid and liquid lipids, and the resulting matrix shows a melting point depression compared with the original solid lipid. However, the matrix is still solid at body temperature. According with the production method and their lipid composition, three types of NLCs can be distinguished. The imperfect type is obtained mixing spatially different lipids (Tamjidi et al. 2013). A second type is the amorphous type, where the lipid matrix is solid but not crystalline. The last kind of NLC is multiple NLC (multiple oil-in-fat-in-water (O/F/W)), where the main solid lipid matrix contains small oil compartments (Tamjidi et al. 2013). Therefore, they are obtained by mixing a solid lipid with a higher amount of oil (Battaglia et al. 2016). However, it has been reported that the oil compartments are mainly localized on the NLC surface, and this could lead to drug damage from environmental agents (Tamjidi et al. 2013).

Lipid-drug conjugate (LDC) nanoparticles were developed to overcome the limitation of low drug loading capacity of SLNs and NLCs for hydrophilic drugs (Battaglia et al. 2016; Battaglia and Gallarate 2012).

Lipid nanocapsules (LNCs) have nearly the same particle size as SLNs, but they possess a core-shell particulate structure, which constitutes a hybrid between polymeric nanocapsules and liposomes (Hoarau et al. 2004). LNCs are based on an internal liquid or semiliquid oil core combined with an external lipid layer solid at room temperature (Battaglia and Gallarate 2012).

Nanoparticle Preparation Methods

Polymeric Nanoparticles

The methods for the preparation of NP include several steps. Usually, an emulsified system is prepared on the first step, and the nanoparticles are formed in the second step that is achieved by the precipitation or the gelation of a polymer or by polymerization of monomers (Battaglia and Gallarate 2012). However, there are other methods that do not require the preparation of an emulsion prior to the obtaining of the PNP, because they are based in other mechanisms that will be explained along this section (Battaglia and Gallarate 2012).

Emulsification-Solvent evaporation

Emulsification-solvent evaporation method was the first preparation method obtained from a preformed polymer (Battaglia and Gallarate 2012). In this method, the preformed polymer is solved in an organic solvent, i.e., ethyl acetate or methylene chloride, and this phase is mixed with a water phase containing the surfactant.

The emulsion decreases their size droplets applying sonication or homogenization (Hans and Lowman 2002). Afterward, the emulsion is transformed to a nanoparticle suspension by evaporation of the solvent, which diffuses through the continuous phase of the emulsion (Battaglia and Gallarate 2012; Sánchez-López et al. 2016a). Usually, this process is carried out under vacuum using a rotaevaporator, and it comprises two evaporation phases. Firstly, a fast evaporation process decreases the droplet size to a minimal value. After this, a slow evaporation removes the last percentage of solvent, and this can cause an increase in droplet size due to coalescence phenomena. This size increase will vary according to the polymer used and its ability to adsorb on the interface between the oil and water phase of the emulsion (Battaglia and Gallarate 2012).

For the encapsulation of hydrophilic drugs, double emulsion is preferred (Andrieux et al. 2013). In this method, the drug is solved into an aqueous phase (internal water phase). A solution of the polymer in an organic solvent (usually dichloromethane or ethyl acetate) is added to the internal water phase containing the drug. This is emulsified using either Ultra-Turrax or an ultrasonic probe forming an emulsion o/w. Into this emulsion, a second water phase containing the surfactant is added, and the mixture is sonicated again forming a double emulsion w/o/w. The organic solvent is removed using either magnetic stirring overnight or rotaevaporation procedure (Ariza-Sáenz et al. 2017; Cano et al. 2018).

Emulsification-Solvent Displacement Method or Nanoprecipitation

Nanoprecipitation, or also known as solvent displacement method, involves two steps. Firstly, the polymer is solved in an organic solution, and the surfactant is solved in an aqueous phase. Secondly, the addition of the organic phase to the water phase under magnetic stirring causes polymer precipitation followed by the diffusion of the organic solvent in the aqueous medium (Nagavarma et al. 2012). The polymer used should be solved in a water-miscible solvent of intermediate polarity. This organic phase is added dropwise into a stirring aqueous solution containing a stabilizer as a surfactant. Polymer deposition on the interface between the water and the organic solvent, which is caused by fast diffusion of the solvent, leads to the immediate formation of a colloidal suspension (Nagavarma et al. 2012). Afterward, the organic solvent is usually evaporated under reduced pressure (Sánchez-López et al. 2016b; Vega et al. 2013).

The solvent displacement technique allows the preparation of both nanospheres and nanocapsules, when a small volume of nontoxic oil is incorporated in the organic phase. However, these techniques are limited to the use of water-miscible solvents with all the drawbacks that the use of these solvents such as acetone or dichloromethane have such as its toxic and irritant effects. In addition, this method is widely used for lipophilic drugs obtaining high entrapment effacement percentages. Usually more than 90% of the drugs are encapsulated into the polymeric core (Cañadas et al. 2016; Silva-Abreu et al. 2018).

Emulsification-Reverse Salting-Out

Salting-out is an emulsification preparation process, in which two phases are required. The drug and the polymer are solved in an organic solvent (e.g. acetone) emulsified into an aqueous gel containing the salting-out agent (e.g., calcium chloride, magnesium chloride, magnesium acetate, nonelectrolytes like sucrose) and a colloidal stabilizer, such as hydroxyethyl cellulose, polyvinylpyrrolidone (Priya et al. 2018). In this method, there is a separation of the water-miscible solvent from the aqueous solution due to salting-out phenomena. This method does not require high temperatures being especially useful in case of heat-sensitive or thermolabile active compounds (Priya et al. 2018).

Dialysis Technique

Dialysis technique constitutes a basic and effective way to prepare small PNPs. Firstly, the polymer is dissolved in the organic solvent. Afterward, this mixture is placed inside a dialysis bag (taking into account the suitable molecular weight for each polymer). During dialysis, the solvent loses its solubility as a result of displacement. In this way, progressive aggregation of the polymer occurs, and a homogeneous suspension of nanoparticles is obtained (Derman et al. 2018).

Supercritical Fluid Technology

Supercritical fluids are fluids that do not change in phase despite the change of pressure. Supercritical CO_2 is the most widely used supercritical fluid, because it is compatible with the critical state ($Tc = 31.1\ ^{\circ}C$, $Pc = 73.8$ bar), is nontoxic, and is nonflammable (Derman et al. 2018).

The most widely used methods are rapid expansion of supercritical solution (RESS) and supercritical anti-solvent method (SAS). In the RESS method, the drug substance is dissolved in the organic solvent and then released to the supercritical fluid. The organic phase rapidly dissolves in the supercritical solvent and remains nanoparticles that can be filtered back. In the SAS method, the active substance and polymer are dissolved in the supercritical solvent at high pressure (Derman et al. 2018).

Supercritical fluid technology offers an interesting and effective particle production technique, avoiding many of the disadvantages of traditional methods (Derman et al. 2018). Among all, the most interesting advantage of this method is the avoiding of organic solvents.

In Situ Polymerization

In situ polymerization can be carried out by using emulsion polymerization reactions. This method requires the formation of the polymer, and for this reason, a monomer is used. The monomer is added in the emulsion instead of a polymer solution, and the polymer forms by polymerization obtaining PNP (Vauthier and Bouchemal 2009). Although this technique can be applied for a wide range of monomer, only a few monomers are suitable to produce PNP for in vivo applications as drug delivery carriers (Vauthier and Bouchemal 2009). For example, some years ago, alkylcyanoacrylates were widely used and were developed using this method (Salgueiro et al. 2004).

The polymerization can also be carried out using polycondensation methods. In this method, two monomers are dissolved in two different phases (organic and aqueous), and the polymer film is formed at the interface between them. The organic phase contains a monomer, an oil, a surfactant, and a water-miscible solvent. The aqueous phase is composed of a hydrophilic monomer and a surfactant. Therefore, polycondensation phenomena occurs by mixing both emulsions, causing the formation of the oil-in-water emulsion (Vauthier and Bouchemal 2009).

Lipid Nanoparticles

Lipid nanoparticle formation methods consist on mixing the lipid phase with the water phase containing the surfactant. In this sense, different techniques in order to reduce the droplet size and obtain either SLN or NLCs have been developed. The methods can be classified in low energy or high energy methods.

Low Energy Methods

Solvent Emulsification-Evaporation Method

The lipid matrix is dissolved in an organic solvent (that should be immiscible in water), and it is emulsified in an aqueous phase. Upon evaporation of the solvent under reduced pressure, a nanoparticle dispersion is formed by precipitation of the lipid in the aqueous medium (Mutyam Pallerla and Prabhakar 2013). The main advantage of this technique is the avoidance of heat (Yadav et al. 2013).

Microemulsion-Based Method

In this method, an optically transparent mixture containing the lipids, drugs, and surfactants is heated above the lipid melting temperature (Ravanfar et al. 2016). This phase is dispersed in cold water (2–3 °C) under stirring. High-temperature gradients facilitate rapid lipid crystallization and prevent aggregation (Mutyam Pallerla and Prabhakar 2013).

Lipid Film Dispersion Method

In this method, the lipid and the drug are solved into suitable organic solvents. The mixture is rotaevaporated and a lipid film is formed. Afterward, to the lipid film, the aqueous solution containing the surfactant is added. The ultrasonic probe or Ultra-Turrax can be used to decrease and uniform particle size (Surender and Deepika 2016).

High Energy Methods

High-Shear Homogenization

This technique consists on melting the lipid phase above its melting point and mixing it with the surfactant dispersed in water at the same temperature. The mixture is homogenized using Ultra-Turrax homogenizer at predetermined time and speed. Lipid nanoparticles are formed (Gardouh 2013; Triplett and Rathman 2009). The main disadvantage of this technique is the low energy, and therefore, nanoparticles possess usually bigger nanoparticle size and elevated polydispersity index.

High-Pressure Homogenization (HPH)

Since some years ago, HPH has been proved to be a simple technique, well-established on a large scale and already available in pharmaceutical industry (Battaglia and Gallarate 2012). This approach uses high-pressure valve homogenizers, which reduce the droplet size in an already coarse emulsion but hardly can create a nanoemulsion from two separate phases. A high-shear mixer normally produces the coarse emulsion, and it later goes into a high-pressure homogenizer.

This technique consists on melting the solid lipid above its melting point and a mixture of surfactant, and water is added at the same temperature. An emulsion is obtained using the Ultra-Turrax, and then, in order to decrease the lipid droplets, the high-pressure homogenizer is used. An o/w formulation is formed and the formulation is cooled down to room temperature in order to produce either SLN or NLCs in the case of an additional liquid lipid is used (Silva et al. 2011). As the passage number and homogenization pressure increase, the size absorption increases, and the interfacial tension decreases.

The cold homogenization technique is a variation of this method aimed for thermosensitive compounds. In the cold homogenization, after the lipid is melted, it is instantaneously cooled down, creating small nanoparticles, which are homogenized using the HPH (Tapeinos et al. 2017).

Ultrasonication Technique

This preparation method consists on melting the lipid phase above its melting point and mixing it with the surfactant dispersed in water at the same temperature. The mixture is homogenized using Ultra-Turrax, and the obtained emulsion is ultrasonicated using a probe sonicator at high temperatures. After this process, which can be repeated at predetermined times and amplitude waves in order to increase the energy that is given to the system, the emulsions are cooled down obtaining lipid nanoparticles (Das et al. 2011; Thatipamula et al. 2011). Ultrasonication parameters can be regulated and optimized using the factorial design approach.

Characterization Parameters

Characterization of the nanoparticles is one of the critical steps involved in the development of ocular drug delivery systems. In this sense, the most relevant parameters are summarized below.

The average size of the particles is one of the most relevant parameters during the preparation of nanoparticles (Gaumet et al. 2008). It is usually measured using a light-scattering diffusion methodology with a parallel particle size analyzer. Photonic correlation spectroscopy is also employed to measure the effects of Brownian motion. In addition, microscopical techniques such as transmission electron microscopy (TEM) can be used. In the case of SLN or NLCs, cryoTEM offers the possibility to achieve better observations of these types of nanoparticles (Jores et al. 2004). In general, particles smaller than 10 μm are considered suitable for ophthalmic formulations in order to improve patient comfort and minimize damage to cornea.

Regarding the particle size distribution characterization, polydispersity index (PI) or heterogeneity index is a parameter used to define the size range of the nanocarrier, describing the degree of nonuniformity of a size distribution. In this sense, PI is a relevant parameter, since it can measure the size variations and dispersity that the sample includes. The values of this parameter higher than 0.2 indicate a heterogenous field. Mathematically, it is the consequence of the ratio between the standard deviation and mean size of particles. The surface charge of the NPs constitutes also a relevant parameter. It is usually measured using a ZetaPALS instrument as the zeta potential. In order to cross the ocular barriers, positive charges are more favorable due to the negative charge of the barriers (Leonardi et al. 2015; Paolicelli et al. 2009).

In addition, the entrapment efficacy of the nanoparticles is the most critical parameter. Nanocarriers should be able to load the formulation into the polymeric or lipid matrix, achieving high entrapment efficacy values. The measurement of this parameter can be carried out either directly, by dissolving the polymeric or lipid matrix and measuring the drug entrapped, or indirectly, by measuring the amount of drug not loaded into the formulation. The latter could be carried out by a previous filtration-centrifugation procedure to separate free and encapsulated drug (Sánchez-López et al. 2018a, b). However, the technique used will depend on the drug and will be the most adequate to measure the drug.

Freeze-drying of Nanoparticles

Formulation of drugs into nanoparticulate systems, made from biodegradable polymeric or lipid materials, constitutes one of the most challenging fields of application of nanostructured materials in pharmaceutics. However, colloidal systems tend to be thermodynamically unstable, and this instability phenomenon is the main obstacle limiting the applicability of nanoparticles when these aqueous suspensions are stored for extended periods (Abdelwahed et al. 2006; Schaffazick et al. 2003).

Stabilizing layer formation at the NP surface (such as those formed by surfactants, polyvinylpyrrolidone, and polyethylene glycol) has been used to improve the system stability. Due to a steric stabilization mechanism, PEGylated nanoparticles are reported to possess an increased in vitro stability against non-PEGylated counterparts (Hinrichs et al. 2006). Despite of that, the shelf life of PEGylated nanoparticles is still limited. Therefore, it would be convenient to store the native and PEGylated nanoparticles in dry conditions.

In order to improve the physical and chemical stability of these nanoparticulate systems, water needs to be removed from the formulation. The most common industrially used process, which allows converting solutions or suspensions into solids of sufficient stability for distribution and storage, in the pharmaceutical and food processing field, is the lyophilization (Holzer et al. 2009). Nevertheless, this process, which consists of removing water from a frozen sample by sublimation and desorption under vacuum, generates various stresses during freezing and drying steps, which could induce undesirable modifications on physicochemical, structural, and biopharmaceutical properties of the nanostructured systems (Schaffazick et al. 2003).

The steps involved in freeze-drying of pharmaceutical formulations, including freeze-drying, are freezing, primary drying, and secondary drying. The freezing step forms a solid sample in order to start the lyophilization process (Abdelwahed et al. 2006). This temperature decrease can be an aggressive step and can be controlled, depending on the parameters that the manufacturer is interested. Regarding drug delivery systems, usually a rapid cooling process is beneficial. At this point also, an additional step, named annealing, can be included in order to avoid sample disturbances. It has been observed that a decrease and posterior increase in temperature followed by a complete freezing of the sample can be suitable. The primary drying stage involves sublimation of ice from the frozen product. Here, the temperature is low, but the pressure is decreased, so the ice sublimes and the water vapor formed passes through the dried portion of the product to the surface of the sample. After primary drying, the secondary drying consists of removing the absorbed water from the product by maintaining the pressure but increasing the temperature until the values are around 20 °C. This is the water which did not separate out as ice during the freezing and did not sublimate.

After lyophilization, some target points should be maintained. Physical and chemical characteristics of the nanoparticles should be maintained as before freeze-dry (size, polydispersity index, entrapment efficacy and zeta potential). Also, a suitable cake appearance (homogeneous and non-cracked) with short reconstitution time would be preferable. However, this is difficult for NPs, since this process may generate many stresses that could destabilize colloidal suspension of NP (Abdelwahed et al. 2006). Therefore, in order to protect the nanosystems against freeze-dry aggression, substances such as cryoprotectants (against freezing stress) or drying stress (lyoprotectant) can be added (Table 3). The most popular cryoprotectants for freeze-drying of NPs are sugars: trehalose, sucrose, glucose, and mannitol. However, some formulations such as those containing poly(vinyl alcohol) may not need any additional protection.

Administration Routes and Applications of Nanoparticles for Ocular Drug Delivery

Anterior Eye Segment

The anterior segment constitutes the first layers that nanoparticles would encounter and comprises the most common ocular diseases, such as inflammation, infections, or dry eye, among others (Table 3). Several routes are aimed to arrive at different sites of the anterior segment.

Topical Administration: Eye Drops

Topical administration of nanoparticles has been extensively studied using different kinds of drugs. Among all, ocular inflammation is one of the most common diseases in ophthalmology, especially for disorders in which the anterior eye segment is involved. Usually, it is treated with eye drops containing nonsteroidal anti-inflammatory drugs that can reduce inflammation and pain. With the aim to avoid side effects and to improve ocular penetration, different kinds of nanoparticles containing anti-inflammatory agents have been developed in the last years (Vega et al. 2006, 2008; Araujo et al. 2009; Araújo et al. 2009; Alvarado et al. 2015).

Chitosan nanoparticles have demonstrated also suitable properties for ocular drug delivery. Chitosan with a positive charge and its interaction with negatively charged epithelial cells of the cornea have been suggested as the main mechanisms of the increased residence time of chitosan nanoparticles in the corneal tissue. In addition, it has been reported that the nanometer size of colloidal carriers plays a major role in their internalization and transport across the corneal tissues (Calvo

Table 3 Freeze-dried nanoparticles for ocular drug delivery and protectants used

Cryo-/lyoprotectant	NP matrix	Active compound	Surfactant	Ocular segment	Physicochemical parameters	References
Mannitol 5% w/v	PNP: Chitosan (hyaluronic acid addition after freeze-drying)	Dexamethasone	Cross-linker sodium tripolyphosphate (TPP) and sodium dihydrogen phosphate	Anterior segment	Size: 400.57 nm PI: 0.427 EE: 72%	Chetoni et al. (2016)
Trehalose, P188 (surfactant), and polyethylene glycol	PNPs: Poly-epsilon-caprolactone	Flurbiprofen	P188	Anterior segment	Size:188.4 nm PI: 0.087 ZP: −16.4 mV	Ramos et al. (2016, 2017)
Mannitol	PNP: Eudragit RL 100	Ketoprofen	Polyvinyl alcohol	Anterior segment	Size: 182–314 nm EE: Up to 95%	Soltani et al. (2016)
Hydroxypropyl-β-CD 5% (w/v)	PNP: PLGA	Carprofen (CP)	P188	Anterior segment	Size: 189.50 nm PI: 0.01 ZP: −22.8 mV EE: 74.70%	Soltani et al. (2016)
Surfactants protect the LNP	SLN: Stearic acid	Itraconazole	Epikuron 200 and sodium taurocholate	Anterior segment	Size: 80 nm PI: 0.15 ZP: −25.7 mV	Chetoni et al. (2016)

PI polydispersity index, *ZP* zeta potential, *EE* entrapment efficiency et al. 1996). In this sense, Chooi and colleagues prepared prednisolone-loaded ammonium palmitoyl glycol chitosan-based nanoparticles. They were able to increase the absorption of this drug on the corneal tissue, enhancing drug bioavailability (Chooi et al. 2014).

Topical Administration: Gels and In Situ Gelling Systems

Drug delivery systems for topical administration can also be dispersed in a viscous matrix that will be aimed to increase the retention time of the nanoparticles with the corneal tissue (Abrego et al. 2015). In this sense, hydrogels have shown to achieve increased bioavailability of the drug, improving the biopharmaceutical profile on the anterior segment (Abrego et al. 2015).

More recently, in situ gelling systems have emerged. In situ gelling systems contain nanoparticles dispersed in a substance, usually a polymer. An external stimulus, which can be pH modification, temperature, ions, or UV radiation, causes a phase transition, increasing the viscosity of the liquid solution forming a gel matrix. For this purpose, polymers, such as poly(lactide) and poly(glycolide), and their derivatives, polycaprolactone, chitosan, or polyethylene glycol, can be employed (Yeo et al. 2007; Gonzalez-Pizarro et al. 2019). These systems show a great capacity to reduce side effects of the therapeutic molecules, decrease the administration rate, increase the bioavailability of drugs, and extend the contact time of the drugs with the cornea tissue (Souto et al. 2019).

Intracameral Administration

Intracameral administration is a local ocular method in order to inject drugs directly into the anterior chamber of the eye. It avoids the corneal tissue, the conjunctiva and blood-aqueous barrier. It does not involve first-pass metabolism, and it provides high efficacy and bioavailability (Janagam et al. 2017). This administration route has been used also in order to treat the anterior and posterior eye segments for diseases such as glaucoma (Lai and Luo 2017).

Embedment into Contact Lenses

Contact lenses are able to cover the cornea due to their curved shape and thin surface. Once applied, they adhere to the wet surface in.

the exterior of the eye, mainly due to surface tension phenomena (Souto et al. 2019). Contact lenses allow drugs to remain in contact with the ocular tissue, decreasing the amount of drug eliminated through the lacrimal duct.

Although they can be useful for some cases, drug embedment into contact lenses possess some drawbacks, such as low and irregular drug loading. In addition, the drug release is usually fast. In order to overcome these limitations, contact lens containing drug loaded nanoparticles can be used. Therefore, the nanosystems would be dispersed in the contact lens material (Souto et al. 2019). Nasr and colleagues loaded loteprednol etabonate into nanoparticles and dispersed them into the contact lenses, demonstrating a better profile of release of drug in comparison to the usual eye drop systems (Nasr et al. 2016). A similar approach has been used by other authors, loading silver nanoparticles into contact lenses for antimicrobial applications (Fazly Bazzaz et al. 2014) or to deliver timolol-loaded nanoparticles for glaucoma (Maulvi et al. 2016).

Posterior Eye Segment

Delivery for the posterior eye segment should take into account that the drug release needs to be sustained for longer periods. Moreover, the size and surface chemical composition are the main factors influencing the penetration into the posterior ocular segment (Souto et al. 2019). Therefore, nanoparticles for this ocular area possess a sustained drug release along with a decreased drug clearance (Souto et al. 2019). Nanostructured lipid carriers (NLC) and biodegradable polymeric nanoparticles (NPs) have gained attention as promising drug delivery agents that can transport across the RBB and increase the uptake of appropriate drugs in the posterior segment of the eye. Polymeric nanoparticles made with polyesters such as PLGA-PEG are promising candidates to pass the RBB and released the drug in the internal ocular structures (Table 1).

In order to treat important diseases of the posterior eye segment, such as age-related macular degeneration, retinal degeneration, and diabetic retinopathy, intravitreal drug administration is widely used (Martens et al. 2015; Bisht and Rupenthal 2016). Intravitreal injection is the most direct approach to deliver drugs to the posterior eye segment obtaining transitory drug therapeutic levels while avoiding the toxicity associated with systemic treatment (Barcia et al. 2009). However, this invasive method has serious potential side effects, such as retinal detachment, hemorrhage, or endophthalmitis, and their probability increases with the number of injections (Barcia et al. 2009). In this sense, usually not a single but a series of injections are required. Therefore, this approach is considered potentially dangerous, and other less invasive methods constitute an unmet medical need. In order to solve this problem, nanoparticle-based drug delivery systems have shown promising results in ophthalmic research over the past 10 years.

Subconjunctival Administration

Subconjunctival administration has been proposed as a suitable alternative for ocular drug delivery to the posterior eye segment (Kang et al. 2009). This injections underneath the conjunctiva allows drugs to bypass the epithelium, one of the main barriers that limits drug entry. In this sense, Aniruddha and colleagues proved that polystyrene nanoparticles were able to remain on the injection site being this kind of administration suitable for nanoparticles and microparticles (Amrite and Kompella 2005). In this sense, they found out that NPs larger than 200 nm were almost completely retained at the site of administration for at least 2 months (Amrite and Kompella 2005). Size showed to be of high relevance, since bigger nanoparticles (e.g., 20 vs. 200 nm) tend to sustain the retinal drug delivery better than smaller nanoparticles (Amrite et al. 2008). Therefore, this administration can be used to treat several retinal diseases, such as retinoblastoma or glaucoma, among others (Kang et al. 2009; Giarmoukakis et al. 2013).

Intravitreal Administration

Intravitreal administration involves the direct administration of drug solution/suspension into vitreous cavity. After the application of the intravitreal injection, the nanoparticulated systems should be transported across several retinal layers. The capacity of the nanoparticles to travel from the vitreous humor to the retinal target tissues is mainly due to the properties of the nanoparticles surface and can be modulated using polymer chemical modifications.

Subretinal Administration

During the last years, subretinal administration is being more used in clinical applications (Peng et al. 2017). Although the knowledge level of the clinicians should be higher than other administrations such as intravitreal administration, subretinal delivery has a direct effect delivering the drug into the subretinal space (Oberkirch et al. 2019). Nanoparticles administered using this route can arrive to the retina and treat retinal disorders and had been used as carriers to transport both genetic material and drugs (Apaolaza et al. 2015; Delgado et al. 2011).

Microneedles

Microneedles constituting one of the most recent administrations have been recently introduced as a minimally invasive means for localizing drug formulation within the target ocular tissues with greater precision and accuracy than the hypodermic needles (Thakur Singh et al. 2017; Patel et al. 2013; Prausnitz et al. 2019). It constitutes one of the most novel administration strategies, and it might reduce the side effects associated with intravitreal injections (Jiang et al. 2007). Only hundreds of microns into the scleral tissue being able to avoid damage to inner ocular membranes (Patel et al. 2013). In addition, microneedles can increase the amount of drug to be delivered into the retina or choroids by overcoming the blood-retinal barrier.

In this sense, swelling microneedles fabricated using polymers are especially relevant, since they allow drug diffusion through the swollen polymeric matrix combining microneedles technology with polymeric nanoparticles (Thakur Singh et al. 2017). Moreover, Jiang and colleagues fabricated microneedles to be inserted in the scleral tissue, being able to effectively diffuse between 10 and 35 μl of nanoparticles from each microneedle (Jiang et al. 2009).

Functionalization of Nanoparticles

The possibility of coating the surface with different substances, such as PEG, avoids recognition by the macrophages of the reticuloendothelial system (RES), which increases the probability that they reach the internal eye tissues (Giannavola et al. 2003). The average size (related with fabrication method selected) and the type and number of linkers on the surface of the NPs can be modulated in their ability to cross the RBB. Therefore, targeted nanoparticles can be considered as a new tool to increase drug delivery to the eye. One of the beneficial consequences of functionalizing NPs is that the desired properties can be controlled in a predictable manner to fit the specific applications (Rameshkumar and Ramaraj 2013). After functionalization of NPs with biomolecules, such as monoclonal antibodies or peptides, through a covalent bond, the bioactivity of the targeted material should be analyzed for any undesirable modifications.

Cell-penetrating peptides (CPPs) are short, water-soluble, partly hydrophobic, and/or polybasic peptides (at most 30–35 amino acid residues) with a net positive charge at physiological pH which are able to cross, or penetrate, cell membranes being able to penetrate cell membranes (Gräslund et al. 2011). Due to their cell internalization properties, CPPs are considered promising candidates for the transport of drugs or other therapeutic substances to different ocular structures, such as the cornea or retina (Vasconcelos et al. 2015).

The physicochemical properties of the functionalized NPs (the size, distribution, surface charge), the biopharmaceutical and toxicological behavior must be controlled because it could determine their therapeutic efficacy. Functionalized nanoparticles (FNPs) provide new possibilities for the development of new multifunctional tools for biomedical and nanotechnological applications (Rameshkumar and Ramaraj 2013).

Conclusions

An effective treatment of ocular diseases constitutes a challenge in ocular pharmacotherapy, because the eye is one of the most complex organs in the human body. The anatomy, physiology, and biochemistry of the eye protect this organ from foreign substances. Therefore, delivery of ophthalmic drugs to the targeted ocular tissues is limited by dynamic, static, or metabolic ocular barriers. To circumvent ocular barriers and to achieve desired drug levels, various drug delivery strategies with a wide variety of administration routes have been developed. In this sense, nanotechnology-based biodegradable nanoparticulate delivery systems offer a promising alternative for overcoming the ocular limitations. Among these systems, polymeric and lipid nanoparticles developed a key role in the ocular drug administration.

Polyesters, such as PLA, PLGA, or PCL, are the most used polymers to obtain biodegradable polymeric nanoparticles, with physicochemical characteristics and biopharmaceutical behavior suitable for ocular administration. On the other hand, SLN and NLC are recent colloidal systems that opened a window for drugs' ocular administration. The technology used to produce LNPs is feasible in the laboratory and easily reproducible at an industrial scale.

The possibility to modify the surface of nanoparticles to avoid recognition of the reticuloendothelial system cells increases the probability that they reach the internal eye tissues. In addition, the chemical functionalization of NPs, by using monoclonal antibodies or cell-penetrating peptides, improved their properties as ocular delivery system. These modifications are being intensively studied and are nowadays the key to achieve therapeutic drug levels on the target site, avoiding side effects.

Despite the progress made in this field, especially in the treatment of the disorders of the anterior segment of the eye, new efforts must be carried out to overcome the challenge of achieving an effective therapeutic system for the treatment of diseases that affect the posterior segment of the eye.

Acknowledgments This work was supported by the Spanish Ministry of Science and Innovation (SAF-2016-33307). M.L.G., M.E., A.C., and E.S.L. belong to 2014SGR-1023. The first author, E.S.L., acknowledges the support of Institute of Nanoscience and Nanotechnology (ART2018 project). The authors want to acknowledge the Portuguese Science and Technology Foundation (FCT/MCT) and European Funds (PRODER/COMPETE) under the project UID/AGR/04033/2013, M-ERA-NET-0004/2015-PAIRED, co-funded by FEDER, under the partnership Agreement PT2020.

References

Abdelwahed W, et al. Freeze-drying of nanoparticles: formulation, process and storage considerations. Adv Drug Deliv Rev. 2006;58(15):1688–713.

Abrego G, et al. Design of nanosuspensions and freeze-dried PLGA nanoparticles as a novel approach for ophthalmic delivery of pranoprofen. J Pharm Sci. 2014;103(10):3153–64.

Abrego G, et al. Biopharmaceutical profile of pranoprofen-loaded PLGA nanoparticles containing hydrogels for ocular administration. Eur J Pharm Biopharm. 2015;95:261–70.

Achouri D, et al. Recent advances in ocular drug delivery. Drug Dev Ind Pharm. 2013;39(11):1599–617.

Agarwal P, et al. Semifluorinated alkane based systems for enhanced corneal penetration of poorly soluble drugs. Int J Pharm. 2018;538(1–2):119–29.

Agban Y, et al. Nanoparticle cross-linked collagen shields for sustained delivery of pilocarpine hydrochloride. Int J Pharm. 2016;501(1–2):96–101.

Aktaş Y, et al. Influence of hydroxypropyl β-cyclodextrin on the corneal permeation of pilocarpine. Drug Dev Ind Pharm. 2003;29(2):223–30.

Alkholief M, et al. Employing a PLGA-TPGS based nanoparticle to improve the ocular delivery of acyclovir. Saudi Pharm J. 2019;27(2):293–302.
Alvarado HL, et al. Design and optimization of oleanolic/ursolic acid-loaded nanoplatforms for ocular anti-inflammatory applications. Nanomedicine. 2015;11(3):521–30.
Amrite AC, Kompella UB. Size-dependent disposition of nanoparticles and microparticles following subconjunctival administration. J Pharm Pharmacol. 2005;57(12):1555–63.
Amrite AC, et al. Effect of circulation on the disposition and ocular tissue distribution of 20 nm nanoparticles after periocular administration. Mol Vis. 2008;14:150–60.
Apaolaza PS, et al. Solid lipid nanoparticle-based vectors intended for the treatment of X-linked juvenile retinoschisis by gene therapy: in vivo approaches in Rs1h-deficient mouse model. J Controll Release. 2015;217:273–83.
Apaolaza PS, et al. Structural recovery of the retina in a retinoschisin-deficient mouse after gene replacement therapy by solid lipid nanoparticles. Biomaterials. 2016;90:40–9.
Araujo J, et al. Nanomedicines for ocular NSAIDs: safety on drug delivery. Nanomedicine. 2009;5(4):394–401.
Araújo J, et al. Effect of polymer viscosity on physicochemical properties and ocular tolerance of FB-loaded PLGA nanospheres. Colloids Surf B Biointerfaces. 2009;72(1):48–56.
Araujo J, et al. Optimization and physicochemical characterization of a triamcinolone acetonide-loaded NLC for ocular antiangiogenic applications. Int J Pharm. 2010;393(1–2):168–76.
Araujo J, et al. Release profile and transscleral permeation of triamcinolone acetonide loaded nanostructured lipid carriers (TA-NLC): in vitro and ex vivo studies. Nanomedicine. 2012;8(6):1034–41.
Ariza-Sáenz M, et al. Penetration of polymeric nanoparticles loaded with an HIV-1 inhibitor peptide derived from GB virus C in a vaginal mucosa model. Eur J Pharm Biopharm. 2017;120:98–106.
Asasutjarit R, et al. Development and evaluation of diclofenac sodium loaded-N-trimethyl chitosan nanoparticles for ophthalmic use. AAPS PharmSciTech. 2015;16(5):1013–24.
Attama AA, Reichl S, Müller-Goymann CC. Diclofenac sodium delivery to the eye: in vitro evaluation of novel solid lipid nanoparticle formulation using human cornea construct. Int J Pharm. 2008;355(1–2):307–13.
Badawi AA, et al. Chitosan based nanocarriers for indomethacin ocular delivery. Arch Pharm Res. 2008;31(8):1040–9.
Baig MS, et al. Application of box-Behnken design for preparation of levofloxacin-loaded stearic acid solid lipid nanoparticles for ocular delivery: optimization, in vitro release, ocular tolerance, and antibacterial activity. Int J Biol Macromol. 2016;85:258–70.
Barar J, Javadzadeh AR, Omidi Y. Ocular novel drug delivery: impacts of membranes and barriers. Expert Opin Drug Deliv. 2008;5(5):567–81.
Barbault-Foucher S, et al. Design of poly-ε-caprolactone nanospheres coated with bioadhesive hyaluronic acid for ocular delivery. J Control Release. 2002;83(3):365–75.
Barcia E, et al. Downregulation of endotoxin-induced uveitis by intravitreal injection of polylactic-glycolic acid (PLGA) microspheres loaded with dexamethasone. Exp Eye Res. 2009;89(2):238–45. https://doi.org/10.1016/j.exer.2009.03.012.
Battaglia L, Gallarate M. Lipid nanoparticles: state of the art, new preparation methods and challenges in drug delivery. Expert Opin Drug Deliv. 2012;9(5):497–508.
Battaglia L, et al. Application of lipid nanoparticles to ocular drug delivery. Expert Opin Drug Deliv. 2016;13(12):1743–57.
Bhatta RS, et al. Mucoadhesive nanoparticles for prolonged ocular delivery of natamycin: in vitro and pharmacokinetics studies. Int J Pharm. 2012;432(1–2):105–12.
Bhattacharya M, et al. Differentially cleaving peptides as a strategy for controlled drug release in human retinal pigment epithelial cells. J Control Release. 2017;251:37–48.
Bisht R, Rupenthal ID. PLGA nanoparticles for intravitreal peptide delivery: optimization, characterization and toxicity evaluation. In: Pharmaceutical development and technology; 2016; 23:324–333.

Bisht R, et al. Nanocarrier mediated retinal drug delivery: overcoming ocular barriers to treat posterior eye diseases. Wiley Interdiscip Rev Nanomed Nanobiotechnol. 2018;10(2):1–21.
Calvo P, et al. Improved ocular bioavailability of indomethacin by novel ocular drug carriers. J Pharm Pharmacol. 1996;48(11):1147–52.
Cañadas C, et al. In vitro, ex vivo and in vivo characterization of PLGA nanoparticles loading pranoprofen for ocular administration. Int J Pharm. 2016;511(2):719–27.
Cano A, et al. Epigallocatechin-3-gallate loaded PEGylated-PLGA nanoparticles: a new antiseizure strategy for temporal lobe epilepsy. Nanomed Nanotechnol Biol Med. 2018;14(4): 1073–85.
Cano A, et al. Dual-drug loaded nanoparticles of Epigallocatechin-3-gallate (EGCG)/ascorbic acid enhance therapeutic efficacy of EGCG in a APPswe/PS1dE9 Alzheimer's disease mice model. J Control Release. 2019;301(March):62–75.
Cavalli R, et al. Solid lipid nanoparticles (SLN) as ocular delivery system for tobramycin. Int J Pharm. 2002;238(1–2):241–5.
Chan JM, Valencia PM, Zhang L, Langer R, Farokhzad OC. Polymeric nanoparticles for drug delivery. In Cancer nanotechnology methods and protocols. Springuer, 2010;163–177.
Chetoni P, et al. Solid lipid nanoparticles as promising tool for intraocular tobramycin delivery: pharmacokinetic studies on rabbits. Eur J Pharm Biopharm. 2016;109:214–23.
Choi JS, Seo K, Yoo JW. Recent advances in PLGA particulate systems for drug delivery. J Pharm Investig. 2012;42(3):155–63.
Chooi KW, et al. Physical characterisation and long-term stability studies on quaternary ammonium palmitoyl glycol chitosan (GCPQ) - a new drug delivery polymer. J Pharm Sci. 2014;103(8):2296–306.
Contreras-Ruiz L, et al. Ocular tolerance to a topical formulation of hyaluronic acid and chitosan-based nanoparticles. Cornea. 2010;29(5):550–8.
Das S, et al. Formulation design, preparation and physicochemical characterizations of solid lipid nanoparticles containing a hydrophobic drug: effects of process variables. Colloids Surf B: Biointerfaces. 2011;88(1):483–9.
Davies NM, Wang G, Tucker IG. Evaluation of a hydrocortisone/hydroxypropyl-β-cyclodextrin solution for ocular drug delivery. Int J Pharm. 1997;156(2):201–9.
De Salamanca AE, et al. Chitosan nanoparticles as a potential drug delivery system for the ocular surface: toxicity, uptake mechanism and in vivo tolerance. Investig Ophthalmol Vis Sci. 2006;47(4):1416–25.
Delgado D, et al. Dextran and protamine-based solid lipid nanoparticles as potential vectors for the treatment of X-linked juvenile retinoschisis. Hum Gene Ther. 2011;23(4):345–55. https://doi.org/10.1089/hum.2011.115.
Delplace V, et al. Controlled release strategy designed for intravitreal protein delivery to the retina. J Controll Release. 2019;293(August 2018):10–20.
Derman S, et al. Synthesis and characterization methods of polymeric nanoparticles. Character Appl Nanomater. 2018;1:1–7.
Di Toro R, Betti V, Spampinato S. Biocompatibility and integrin-mediated adhesion of human osteoblasts to poly(DL-lactide-co-glycolide) copolymers. Eur J Pharm Sci. 2004;21(2–3):161–9.
Dursun D, et al. The effects of experimental tear film removal on corneal surface regularity and barrier function. Ophthalmology. 2000;107(9):1754–60.
Esposito E, et al. Data on scaling up and in vivo human study of progesterone lipid nanoparticles. Data Brief. 2017;14:639–42.
Fangueiro JF, Andreani T, Egea MA, et al. Design of cationic lipid nanoparticles for ocular delivery: development, characterization and cytotoxicity. Int J Pharm. 2014a;461(1–2):64–73.
Fangueiro JF, Andreani T, Fernandes L, et al. Physicochemical characterization of epigallocatechin gallate lipid nanoparticles (EGCG-LNs) for ocular instillation. Colloids Surf B: Biointerfaces. 2014b;123:452–60.

Fangueiro JF, et al. Biopharmaceutical evaluation of epigallocatechin gallate-loaded cationic lipid nanoparticles (EGCG-LNs): in vivo, in vitro and ex vivo studies. Int J Pharm. 2016;502(1):161–9.

Fazly Bazzaz BS, et al. Preparation, characterization and antimicrobial study of a hydrogel (soft contact lens) material impregnated with silver nanoparticles. Cont Lens Anterior Eye. 2014;37(3):149–52. https://doi.org/10.1016/j.clae.2013.09.008.

Gardouh A. Design and characterization of glyceryl monostearate solid lipid nanoparticles prepared by high shear homogenization. Br J Pharm Res. 2013;3(3):326–46.

Gaumet M, et al. Nanoparticles for drug delivery: the need for precision in reporting particle size parameters. Eur J Pharm Biopharm. 2008;69(1):1–9.

Giannavola C, et al. Influence of preparation conditions on acyclovir-loaded poly-d,l-lactic acid nanospheres and effect of PEG coating on ocular drug bioavailability. Pharm Res. 2003;20(4):584–90.

Giarmoukakis A, et al. Biodegradable nanoparticles for controlled subconjunctival delivery of latanoprost acid: in vitro and in vivo evaluation. Preliminary results. Exp Eye Res. 2013;112:29–36.

Gökçe EH, et al. Cyclosporine a-loaded solid lipid nanoparticles: ocular tolerance and in vivo drug release in rabbit eyes. Curr Eye Res. 2009;34(11):996–1003.

Gonzalez-Mira E, et al. Design and ocular tolerance of flurbiprofen loaded ultrasound-engineered NLC. Colloids Surf B: Biointerfaces. 2010;81(2):412–21.

Gonzalez-Mira E, et al. Optimizing flurbiprofen-loaded NLC by central composite factorial design for ocular delivery. Nanotechnology. 2011;22(4):045101.

Gonzalez-Mira E, et al. Improved and safe transcorneal delivery of flurbiprofen by NLC and NLC-based hydrogels. J Pharm Sci. 2012;101(2):707–25.

González-Pizarro R, Parrotta G, Vera R, Sánchez-Lopéz E, Galindo R, Kjeldsen F, Badia J, Baldoma L, Espina M, García ML. Ocular penetration of fluorometholone-loaded PEG-PLGA nanoparticles functionalized with cell-penetrating peptides: in vitro and in vivo studies. Nanomedicine. 2019;14:–3089.

Gonzalez-Pizarro R, et al. In-situ forming gels containing fluorometholone-loaded polymeric nanoparticles for ocular inflammatory conditions. Colloids Surf B: Biointerfaces. 2019;175(November 2018):365–74.

Gräslund A, et al. Mechanisms of cellular uptake of cell-penetrating peptides. J Biophys. 2011;2011 https://doi.org/10.1155/2011/414729.

Gref R, et al. Biodegradable long-circulating polymeric nanospheres. Science. 1994;263(5153):1600–3.

Gudmundsdottir BS, et al. γ-Cyclodextrin nanoparticle eye drops with dorzolamide: effect on intraocular pressure in man. J Ocul Pharmacol Ther. 2014;30(1):35–41. https://doi.org/10.1089/jop.2013.0060.

Hans M, Lowman A. Biodegradable nanoparticles for drug delivery and targeting. Curr Opinion Solid State Mater Sci. 2002;6(4):319–27.

Hao J, et al. Development and optimization of solid lipid nanoparticle formulation for ophthalmic delivery of chloramphenicol using a Box-Behnken design. Int J Nanomedicine. 2011;6:683–92.

Hinrichs WLJ, et al. The choice of a suitable oligosaccharide to prevent aggregation of PEGylated nanoparticles during freeze thawing and freeze drying. Int J Pharm. 2006;311(1–2):237–44.

Hippalgaonkar K, et al. Indomethacin-loaded solid lipid nanoparticles for ocular delivery: development, characterization, and in vitro evaluation. J Ocul Pharmacol Ther. 2013;29(2):216–28.

Hoarau D, et al. Novel long-circulating lipid nanocapsules. Pharm Res. 2004;21(10):1783–9. https://doi.org/10.1023/B:PHAM.0000045229.87844.21.

Holzer M, et al. Physico-chemical characterisation of PLGA nanoparticles after freeze-drying and storage. Eur J Pharm Biopharm. 2009;72(2):428–37.

Huang Y, et al. Molecular aspects of muco- and bioadhesion: tethered structures and site-specific surfaces. J Control Release. 2000;65(1–2):63–71.

Huang D, Chen YS, Rupenthal ID. Overcoming ocular drug delivery barriers through the use of physical forces. Adv Drug Deliv Rev. 2018;126:96–112. https://doi.org/10.1016/j.addr.2017.09.008.

Ishihara T, et al. Polymeric nanoparticles encapsulating betamethasone phosphate with different release profiles and stealthiness. Int J Pharm. 2009;375(1–2):148–54.

Janagam DR, Wu L, Lowe TL. Nanoparticles for drug delivery to the anterior segment of the eye. Adv Drug Deliv Rev. 2017;122:31–64.

Jiang J, et al. Coated microneedles for drug delivery to the eye. Investig Ophthalmol Vis Sci. 2007;48(9):4038–43.

Jiang J, et al. Intrascleral drug delivery to the eye using hollow microneedles. Pharm Res. 2009;26(2):395–403. https://doi.org/10.1007/s11095-008-9756-3.

Jiang S, et al. Nanotechnology in retinal drug delivery. Int J Ophthalmol. 2018;11(6):1038–44.

Jóhannesson G, et al. Dorzolamide cyclodextrin nanoparticle suspension eye drops and trusopt in rabbit. J Ocul Pharmacol Ther. 2014;30(6):464–7.

Jores K, et al. Investigations on the structure of solid lipid nanoparticles (SLN) and oil-loaded solid lipid nanoparticles by photon correlation spectroscopy, field-flow fractionation and transmission electron microscopy. J Control Release. 2004;95(2):217–27.

Junghanns JUAH, Müller RH. Nanocrystal technology, drug delivery and clinical applications. Int J Nanomedicine. 2008;3(3):295–309.

Kang SJ, et al. Subconjunctival nanoparticle carboplatin in the treatment of murine retinoblastoma. Arch Ophthalmol. 2009;127(8):1043–7.

Kaur C, Foulds WS, Ling EA. Progress in retinal and eye research blood–retinal barrier in hypoxic ischaemic conditions: basic concepts, clinical features and management. Prog Retinal Eye Res. 2008;27(6):622–47.

Kim NJ, et al. Nanobiomaterials for ophthalmic drug delivery. New York: Springer; 2016.

Kristinsson JK, et al. Dexamethasone-cyclodextrin-polymer co-complexes in aqueous eye drops: aqueous humor pharmacokinetics in humans. Investig Ophthalmol Vis Sci. 1996;37(6):1199–203.

Lai JY, Luo LJ. Chitosan-g-poly(N-isopropylacrylamide) copolymers as delivery carriers for intracameral pilocarpine administration. Eur J Pharm Biopharm. 2017;113:140–8.

Lai SK, Wang YY, Hanes J. Mucus-penetrating nanoparticles for drug and gene delivery to mucosal tissues. Adv Drug Deliv Rev. 2009;61(2):158–71.

Leonardi A, et al. Cationic solid lipid nanoparticles enhance ocular hypotensive effect of melatonin in rabbit. Int J Pharm. 2015;478(1):180–6.

Li VHK, et al. Ocular drug delivery of progesterone using nanoparticles. J Microencapsul. 1986;3(3):213–8.

Li J, et al. Positively charged micelles based on a triblock copolymer demonstrate enhanced corneal penetration. Int J Nanomedicine. 2015;10:6027–37. https://doi.org/10.2147/IJN.S90347.

Liu Z, et al. Preparation and evaluation of solid lipid nanoparticles of baicalin for ocular drug delivery system in vitro and in vivo. Drug Dev Ind Pharm. 2011;37(4):475–81.

Loftson T, Steffánsson E. Cyclodextrin nanotechnology for ophthalmic drug delivery. 2011. US7893040B2

Losa C, et al. Design of new formulations for topical ocular administration: polymeric nanocapsules containing metipranolol. Pharm Res. 1993;10(1):80–7.

Marchal-Heussler L, et al. Poly(ε-caprolactone) nanocapsules in carteolol ophthalmic delivery. Pharm Res. 1993;10:386–90.

Martens TF, et al. Coating nanocarriers with hyaluronic acid facilitates intravitreal drug delivery for retinal gene therapy. J Controll Release. 2015;202:83–92.

Marx D, Birkhoff M. Ophthalmic squeeze dispenser. Drug Dev Deliv. 2017;17(7):40–4.

Maulvi FA, et al. In vitro and in vivo evaluation of novel implantation technology in hydrogel contact lenses for controlled drug delivery. J Controll Release. 2016;226:47–56. https://doi.org/10.1016/j.jconrel.2016.02.012.

Musumeci T, et al. Polymeric nanoparticles augment the ocular hypotensive effect of melatonin in rabbits. Int J Pharm. 2013;440(2):135–40.
Mutyam Pallerla S, Prabhakar B. A review on solid lipid nanoparticles. Int J Pharm Sci Rev Res. 2013;20(2):196–206. https://doi.org/10.5958/0975-4377.2016.00030.6.
Nagarwal RC, et al. Polymeric nanoparticulate system: a potential approach for ocular drug delivery. J Control Release. 2009;136(1):2–13.
Nagavarma BVN, et al. Different techniques for preparation of polymeric nanoparticles-a review. Asian J Pharm Clin Res. 2012;5(Suppl. 3):16–23.
Nasr FH, et al. Preparation and evaluation of contact lenses embedded with polycaprolactone-based nanoparticles for ocular drug delivery. Biomacromolecules. 2016;17(2):485–95.
Nayak K, Misra M. A review on recent drug delivery systems for posterior segment of eye. Biomed Pharmacother. 2018;107:1564–82.
Oberkirch BJ, et al. US010226379B2; 2019.
Ottiger M, et al. Efficient intraocular penetration of topical anti-TNF-α single-chain antibody (ESBA105) to anterior and posterior segment without penetration enhancer. Investig Ophthalmol Vis Sci. 2009;50(2):779–86. https://doi.org/10.1167/iovs.08-2372.
Paolicelli P, et al. Chitosan nanoparticles for drug delivery to the eye. Expert Opin Drug Deliv. 2009;6(3):239–53.
Parra A, et al. Design and elaboration of freeze-dried PLGA nanoparticles for the transcorneal permeation of carprofen: ocular anti-inflammatory applications. Colloids Surf B Biointerfaces. 2015;136:935–43.
Patel A, et al. Ocular drug delivery systems: an overview. World J Pharmacol. 2013;2(2):47–64.
Peng Y, Tang L, Zhou Y. Subretinal injection: a review on the novel route of therapeutic delivery for vitreoretinal diseases. Ophthalmic Res. 2017;58(4):217–26.
Peppas NA, et al. Poly(ethylene glycol)-containing hydrogels in drug delivery. J Control Release. 1999;62(1–2):81–7.
Prausnitz MR, Jiang N, Henry FE. USOO7918814B2; 2019.
Priya J, et al. A review on polymeric nanoparticles: a promising novel drug delivery system. J Glob Pharma Technol. 2018;10(4):10–7.
Rafie F, et al. In vivo evaluation of novel nanoparticles containing dexamethasone for ocular drug delivery on rabbit eye. Curr Eye Res. 2010;35(12):1081–9.
Rameshkumar P, Ramaraj R. Gold nanoparticles deposited on amine functionalized silica sphere and its modified electrode for hydrogen peroxide sensing. J Appl Electrochem. 2013;43(10):1005–10.
Ramos GR, et al. Influence of freeze-drying and γ -irradiation in preclinical studies of flurbiprofen polymeric nanoparticles for ocular delivery using d-(+)-trehalose and polyethylene glycol. Int J Nanomedicine. 2016;11:4093–106.
Ramos GR, et al. Freeze drying optimization of polymeric nanoparticles for ocular flurbiprofen delivery: effect of protectant agents and critical process parameters on long-term stability. Drug Dev Ind Pharm. 2017;43(4):637–51.
Rathore MS, Gupta VB. Drug delivery enhancement strategies through cornea: a review. Asian J Pharm. 2007;1(1):40–9.
Rautio J, et al. Prodrugs: design and clinical applications. Nat Rev Drug Discov. 2008;7(3):255–70. https://doi.org/10.1038/nrd2468.
Ravanfar R, et al. Preservation of anthocyanins in solid lipid nanoparticles: optimization of a microemulsion dilution method using the Placket-Burman and Box-Behnken designs. Food Chem. 2016;199:573–80.
Rincón M, et al. Development of pranoprofen loaded nanostructured lipid carriers to improve its release and therapeutic efficacy in skin inflammatory disorders. Nano. 2018;8(1022):1–28.
Rolando M, Zierhut M. The ocular surface and tear film and their dysfunction in dry eye disease. Surv Ophthalmol. 2001;45(Suppl 2):203–10.
Sabzevari A, et al. Polymeric triamcinolone acetonide nanoparticles as a new alternative in the treatment of uveitis: in vitro and in vivo studies. Eur J Pharm Biopharm. 2013;84(1):63–71.

Sahoo SK, Dilnawaz F, Krishnakumar S. Nanotechnology in ocular drug delivery. Drug Discov Today. 2008;13(3–4):144–51.
Salata OV. Applications of nanoparticles in biology and medicine. J Nanobiotechnol. 2004;6(3):1–6.
Salgueiro A, et al. Stability and ocular tolerance of cyclophosphamide-loaded nanospheres. J Microencapsul. 2004;21(2):213–23.
Sánchez-López E, Ettcheto M, et al. New potential strategies for Alzheimer's disease prevention: pegylated biodegradable Dexibuprofen Nanospheres administration to APPswe/PS1dE9. Nanomed Nanotechnol Biol Med. 2016a;13:1171–82.
Sánchez-López E, Egea MA, et al. PEGylated PLGA nanospheres optimized by design of experiments for ocular administration of dexibuprofen– in vitro, ex vivo and in vivo characterization. Colloids Surf B: Biointerfaces. 2016b;145:241–50.
Sánchez-López E, Espina M, Doktorovova S, Souto EB, et al. Lipid nanoparticles (SLN, NLC): overcoming the anatomical and physiological barriers of the eye – part I – barriers and determining factors in ocular delivery. Eur J Pharm Biopharm. 2017a;110:70–5.
Sánchez-López E, Espina M, Doktorovova S, Souto EBB, et al. Lipid nanoparticles (SLN, NLC): overcoming the anatomical and physiological barriers of the eye – part II - ocular drug-loaded lipid nanoparticles. Eur J Pharm Biopharm. 2017b;110:58–69.
Sánchez-López E, et al. Memantine-loaded PEGylated biodegradable nanoparticles for the treatment of Glaucoma. Small. 2018a;14(2):1–12.
Sánchez-López E, et al. Memantine loaded PLGA PEGylated nanoparticles for Alzheimer's disease : in vitro and in vivo characterization. J Nanobiotechnol. 2018b;16(32):1–16.
Schaffazick SR, et al. Freeze-drying polymeric colloidal suspensions: Nanocapsules, nanospheres and nanodispersion. A comparative study. Eur J Pharm Biopharm. 2003;56(3):501–5.
Serra L, Doménech J, Peppas NA. Design of poly(ethylene glycol)-tethered copolymers as novel mucoadhesive drug delivery systems. Eur J Pharm Biopharm. 2006;63(1):11–8.
Shenoy DB, Amiji MM. Poly(ethylene oxide)-modified poly(ε-caprolactone) nanoparticles for targeted delivery of tamoxifen in breast cancer. Int J Pharm. 2005;293(1–2):261–70.
Shinde U, Ahmed MH, Singh K. Development of dorzolamide loaded 6-O-carboxymethyl chitosan nanoparticles for open angle glaucoma. J Drug Deliv. 2013;2013:562727.
Silva AC, et al. Preparation, characterization and biocompatibility studies on risperidone-loaded solid lipid nanoparticles (SLN): high pressure homogenization versus ultrasound. Colloids Surf B: Biointerfaces. 2011;86(1):158–65.
Silva-Abreu M, et al. Optimization, biopharmaceutical profile and therapeutic efficacy of pioglitazone-loaded PLGA-PEG nanospheres as a novel strategy for ocular inflammatory disorders. Pharm Res. 2018;35(1):11.
Singare DS, et al. Optimization of formulation and process variable of nanosuspension: an industrial perspective. Int J Pharm. 2010;402(1–2):213–20.
Soltani S, et al. Design of eudragit RL nanoparticles by nanoemulsion method as carriers for ophthalmic drug delivery of ketotifen fumarate. Iran J Basic Med Sci. 2016;19(5):850–60.
Souto EB, et al. Advanced formulation approaches for ocular drug delivery: state-of-the-art and recent patents. Pharmaceutics. 2019;11(9):460.
Surender V, Deepika M. Solid lipid nanoparticles: a comprehensive review. J Chem Pharm Res. 2016;8(8):102–14. www.jocpr.com.
Tai-Lee K, Cooper ER, Hager DF, Keister JC. Use of monoacyl phosphoglycerides to enhance the corneal penetration of ophthalmic drugs. 1990. EP0465580A4
Tamjidi F, et al. Nanostructured lipid carriers (NLC): a potential delivery system for bioactive food molecules. Innov Food Sci Emerg Technol. 2013;19:29–43.
Tang-Liu DD, et al. Effects of four penetration enhancers on corneal permeability of drugs in vitro. J Pharm Sci. 1994;83(1):85–90.
Tapeinos C, Battaglini M, Ciofani G. Advances in the design of solid lipid nanoparticles and nanostructured lipid carriers for targeting brain diseases. J Controll Release. 2017;264:306–32.
Thakur Singh RR, et al. Minimally invasive microneedles for ocular drug delivery. Expert Opin Drug Deliv. 2017;14(4):525–37.

Thatipamula RP, et al. Formulation and in vitro characterization of domperidone loaded solid lipid nanoparticles and nanostructured lipid carriers. DARU J Pharm Sci. 2011;19(1):23–32.
Triplett MD, Rathman JF. Optimization of β-carotene loaded solid lipid nanoparticles preparation using a high shear homogenization technique. J Nanopart Res. 2009;11(3):601–14.
Vandervoort J, Ludwig A. Preparation and evaluation of drug-loaded gelatin nanoparticles for topical ophthalmic use. Eur J Pharm Biopharm. 2004;57:251–61.
Vasconcelos A, et al. Conjugation of cell-penetrating peptides with poly(lactic-co-glycolic acid)-polyethylene glycol nanoparticles improves ocular drug delivery. Int J Nanomedicine. 2015;10:609–31.
Vauthier C, Bouchemal K. Methods for the preparation and manufacture of polymeric nanoparticles. Pharm Res. 2009;26(5):1025–58.
Vega E, et al. Flurbiprofen loaded biodegradable nanoparticles for ophthalmic administration. J Pharm Sci. 2006;95(11):2393–405.
Vega, E. et al. (2008) PLGA nanospheres for the ocular delivery of flurbiprofen: drug release and interactions J Pharm Sci, 97(\), pp. 5306–5317.
Vega E, et al. Role of hydroxypropyl- β -cyclodextrin on freeze-dried and gamma-irradiated PLGA and PLGA – PEG diblock copolymer nanospheres for ophthalmic flurbiprofen delivery. Int J Nanomedicine. 2012;7:1357–71.
Vega E, Egea MA, et al. Flurbiprofen PLGA-PEG nanospheres: role of hydroxy-β-cyclodextrin on ex vivo human skin permeation and in vivo topical anti-inflammatory efficacy. Colloids Surf B: Biointerfaces. 2013;110:339–46.
Wadhwa S, et al. Hyaluronic acid modified chitosan nanoparticles for effective management of glaucoma: development, characterization, and evaluation. J Drug Target. 2010;18(4):292–302.
Wang JJ, et al. Recent advances of chitosan nanoparticles as drug carriers. Int J Nanomedicine. 2011;6:765–74.
Wang Y, Rajala A, Rajala RVS. Lipid nanoparticles for ocular gene delivery. J Funct Biomater. 2015;6(2):379–94. https://doi.org/10.3390/jfb6020379.
Xue SW. Synthesis, characterization, biodegradation, and drug delivery application of biodegradable lactic/glycolic acid polymers: part II: biodegradation. Artif Cells Blood Substit Immobil Biotechnol. 2004;32(4):575–91. https://doi.org/10.1081/BIO-200039635.
Yadav N, Khatak S, Singh Sara UV. Solid lipid nanoparticles- a review. Int J Appl Pharm. 2013;5(2):8–18.
Yang CR, et al. Preparation, optimization and characteristic of huperzine a loaded nanostructured lipid carriers. Chem Pharm Bull. 2010;58(5):656–61.
Yeo Y, et al. In situ cross-linkable hyaluronan hydrogels containing polymeric nanoparticles for preventing postsurgical adhesions. Ann Surg. 2007;245(5):819–24.
Zhang P, et al. Preparation and evaluation of naringenin-loaded sulfobutylether-β-cyclodextrin/chitosan nanoparticles for ocular drug delivery. Carbohydr Polym. 2016;149:224–30.

Advanced Hydrogel Formulations for the Eye

Hu Yang

Abstract Several novel types of ocular hydrogel formulations have been developed to address safety and ease of application. This chapter discusses supramolecular hydrogels and dendrimer hydrogels. Mechanisms underlying the formation of those hydrogels and their applications for medication of ocular diseases are reviewed. Supramolecular hydrogels form on the basis of host-guest interactions. Dendrimer hydrogels are comprised of well-defined hyperbranched dendrimer building blocks. Green chemistry including the aza-Michael addition and biorthogonal chemistry have been successfully applied to make dendrimer hydrogels. These new formulations hold promise for improving ocular disease treatment but require further assessment in pre-clinical and clinical studies for bench-to-bedside translation.

Keywords Supramolecular hydrogel · Dendrimer hydrogel · Host-guest interactions · Aza-Michael addition · Biorthogonal chemistry · Ocular drug delivery · Glaucoma · Green chemistry

Introduction

The human eye is anatomically broken down into the anterior and posterior segments. The anterior segment is comprised of the structures in front of the vitreous humor, including the cornea, tear film, conjunctiva, aqueous humor, iris, lens, and ciliary body. The posterior segment includes the vitreous humor, sclera, retina, choroid, and optic nerve. To effectively treat ocular diseases, drugs need to be delivered to the anterior or posterior segments. Various physiological barriers exist to protect

H. Yang (✉)
Linda and Bipin Doshi Department of Chemical and Biochemical Engineering, Missouri University of Science and Technology, Rolla, MO, USA
e-mail: huyang@mst.edu

S. Neervannan, U. B. Kompella (eds.), *Ophthalmic Product Development*, AAPS Advances in the Pharmaceutical Sciences Series 37,
https://doi.org/10.1007/978-3-030-76367-1_16

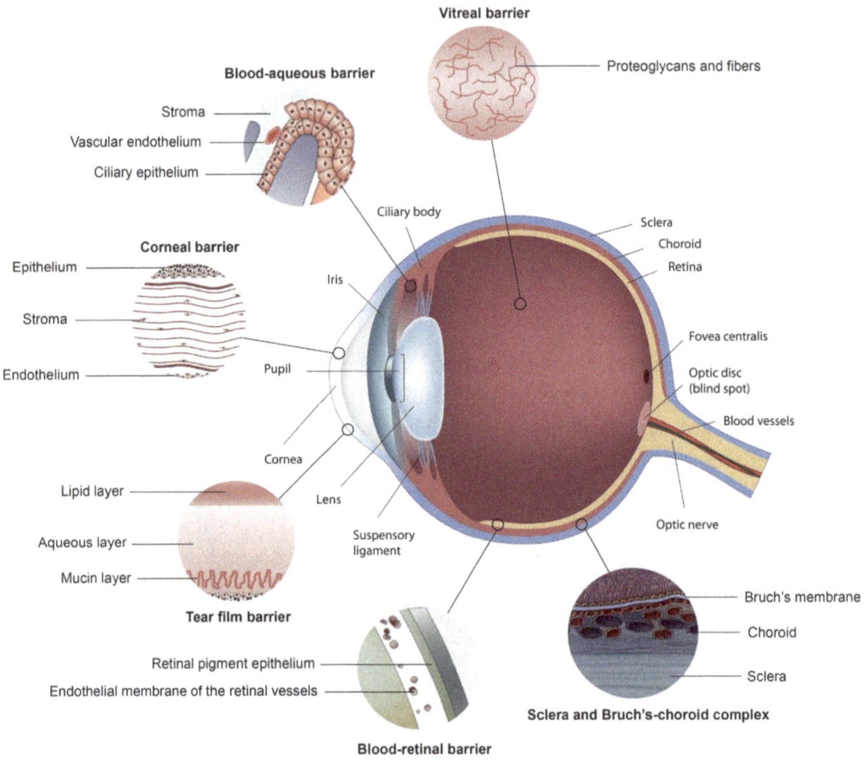

Fig. 1 Physiological barriers in ocular drug delivery. (Reprinted with permission from Huang et al. (2018). Copyright 2018 Elsevier)

the eye, posing significant challenges to the delivery of therapeutics (Fig. 1) (Cholkar et al. 2013; Huang et al. 2018).

Saline drops are by far the most convenient delivery method. However, the use of saline drops is conditional. They are not suited for treating conditions affecting the posterior segment of the eye as the therapeutics hardly reach the target tissues due to the presence of physiological barriers. Even for the medication targeting the anterior segment, saline drops are often inefficient (Urtti 2006). In situ-forming or injectable hydrogels have attracted much attention because they are highly structurally adaptable. They can deliver a broad spectrum of drugs and may overcome limitations that are commonly associated with saline solutions. A number of hydrogel-based ocular formulations have been developed. Some have entered clinical trials or have been successfully commercialized.

Traditional methods for hydrogel preparations often rely on additional chemical reactants to initiate or promote cross-linking reactions. For instance, UV curing has to use photoinitiators to produce free radicals, which are needed to initiate photopolymerization. Our studies and many others have shown that the generation of excess free radicals may cause tissue or cellular damages. In addition, this method is less controllable. The physicochemical properties of the drug formulations may change over time. Furthermore, this method may not be suitable for preparing formulations that are structurally susceptible to heat, light, or organic solvents. Ocular hydrogels have been preformulated by using a variety of chemical and physical methods. It is indisputable that ocular drug hydrogel formulations prepared on the basis of traditional methods have broad applications (Table 1) (Cooper and Yang 2019). In particular, it is compelling to prepare ocular gels on the basis of in situ gelling mechanisms, in which sol-gel transition occurs in response to environmental stimuli, such as temperature, pH, or ions (Al-Kinani et al. 2018; Sapino et al. 2019).

The latest efforts have shifted focus to the development of green methods to prepare ocular hydrogel formulations. In this chapter, we attempt to review the latest advances in ocular hydrogel development with a focus on two types of relatively green formulations—supramolecular hydrogels and dendrimer hydrogels—elucidate mechanisms underlying the preparation of those hydrogels, and discuss their applications for medication of ocular diseases.

Table 1 Representative hydrogels developed for treatment of ocular diseases in various regions of the eye. Reprinted with permission from (Cooper and Yang 2019). Copyright 2019 Elsevier

Segment of the eye	Hydrogel application		Examples
Anterior	Soft contact lenses	Vision correcting	• PHEMA • PDMS • Silicone
		Drug eluting	• PHEMA • PAMAM dendrimer/PLGA nanoparticle
	Wound healing	Corneal	• ReSure® • DuraSeal™
		Retinal	• Multi-arm modified PEG
	Intraocular lenses	Lens	• Surface-modified silicone • Drug-eluting acrylate
		Haptics	• Drug-eluting poly(L-lactide-*co*-ε-caprolactone) film
Posterior	Vitreous substitute		• Zwitterionic poly(MPDSA-co-AC) • Tetra-armed PEG; thiolated gellan and poly(methacrylamide-*co*-methacrylate)
	Intravitreal hydrogels	Drug eluting	• Modified PAMAM dendrimer (clickable) • Temperature-sensitive PNIPAAm
		Cell scaffold	• Hyaluronan and methylcellulose • RGD-alginate

Supramolecular Hydrogels

Supramolecular hydrogels can form on the basis of host-guest interactions, including van der Waals force, hydrogen-bonding interaction, $\pi - \pi$ stacking interaction, hydrophobic/hydrophilic attraction, and electrostatic interaction. Supramolecular hydrogels are an appealing class of biomaterials for making ocular drug formulations because of the ease of fabrication, avoidance of introducing additional chemicals to the product, and cost-effectiveness. Unlike covalent cross-linking strategies, host-guest interactions are dynamic and reversible in nature. Therefore, supramolecular hydrogels are structurally reversible and undergo a sol-gel phase transition in response to changes in external stimuli (Ma and Zhao 2015). Macrocyclic molecules and their derivatives have been most investigated as host molecules for making supramolecular structure-based delivery of drugs, genes, imaging agents, and many other therapeutic entities (Fig. 2).

By virtue of excellent biocompatibility, cyclodextrin (α-, β-, and γ-CDs)-based supramolecular systems have received considerable attention in biomedical and pharmaceutical applications. α-CD, β-CD, and γ-CD contain 6, 7, and 8 glucopyranose units, respectively (Loh 2014; Gigliotti et al. 2016; Hatziagapiou et al. 2017; Liu et al. 2016). An inclusion complex forms between α-cyclodextrin (α-CD) and poly(ethylene glycol) (PEG) (guest molecule) (Harada and Kamachi 1990). In this system, PEG chains penetrate into the cavities of α-CDs to form

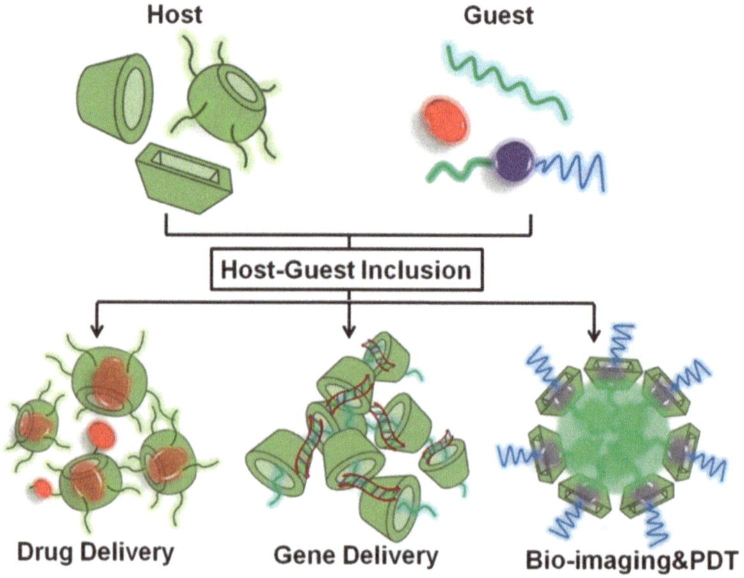

Fig. 2 Biomedical applications of supramolecular systems based on host-guest interactions. (Reprinted with permission from Ma and Zhao (2015). Copyright 2015 American Chemical Society)

pseudopolyrotaxanes. Sequentially, hydrogen bonds stabilize the resulting pseudopolyrotaxanes, leading to the formation of supramolecular hydrogels (Ceccato et al. 1997; Harada et al. 2014; Takahashi et al. 2016). One issue associated with the use of PEGs is that long PEG chains may not be cleared out efficiently due to their large hydrodynamic radii (Jeong et al. 1997). Although short PEGs ($M_n < 2$ kDa) are more pharmacokinetically appealing, they have limited ability to interact with α-CD to form a network for the same reason (Li et al. 2006). Our laboratory was the first to report success using low-molecular-weight 4-arm polyethylene glycol (4-PEG) as a guest molecule to form branched polyrotaxanes with α-CD (Fig. 3) (Wang et al. 2018b).

We used this new supramolecular hydrogel to deliver brimonidine. We found that α-CD increases the solubility of brimonidine by 60% and the release of the antiglaucoma drug brimonidine is highly dependent on shear stress. The higher the shear stress, the quicker the drug release rate is. We attribute the shear stress-dependent drug release to the rheological behavior change of the gel and the disassembly of branched polyrotaxanes (Fig. 4). The studies we conducted on this supramolecular hydrogel revealed that shear-thinning properties are thixotropic because of dynamic reversible inclusion complexation. This property has also been observed in other supramolecular hydrogels (Bairi et al. 2014; Ma and Zhang 2011; Wang et al. 2017b). Such a property allows supramolecular hydrogels to be preformulated and retain injectability (Guvendiren et al. 2012; Wu et al. 2016; Baral et al. 2016; Parisi-Amon et al. 2013; Wang et al. 2016). Glaucoma is a chronic ocular disease characterized with elevated intraocular pressure (IOP). Topical administration often raises patient compliance issues due to the demand of repeated daily dosing.

PEG-based copolymers can also engage CDs to form host-guest interactions in the form of micelles (Fig. 5). To find long-acting, safer formulations is a must. Using low-molecular-weight PEG derivatives to formulate PEG/α-CD supramolecular hydrogels is a green method to prepare ocular drug formulations and warrants further investigation for the application of ocular medication.

A recent report shows that β-CD can be used to prepare an injectable formulation, in which the antifungal agent VCZ is encapsulated into β-CD. β-CD was first polymerized and then cross-linked with polyaldehyde dextran (PAD) and

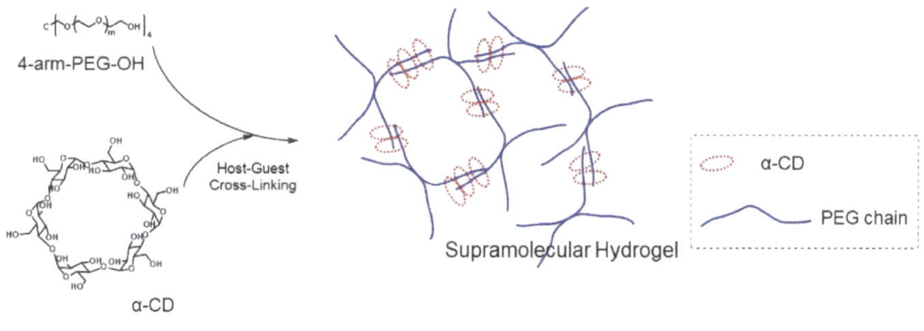

Fig. 3 Schematic illustration of the formation of supramolecular hydrogel on the basis of α-CD and 4-PEG. (Reprinted with permission from Wang et al. (2018b). Copyright 2018 Elsevier)

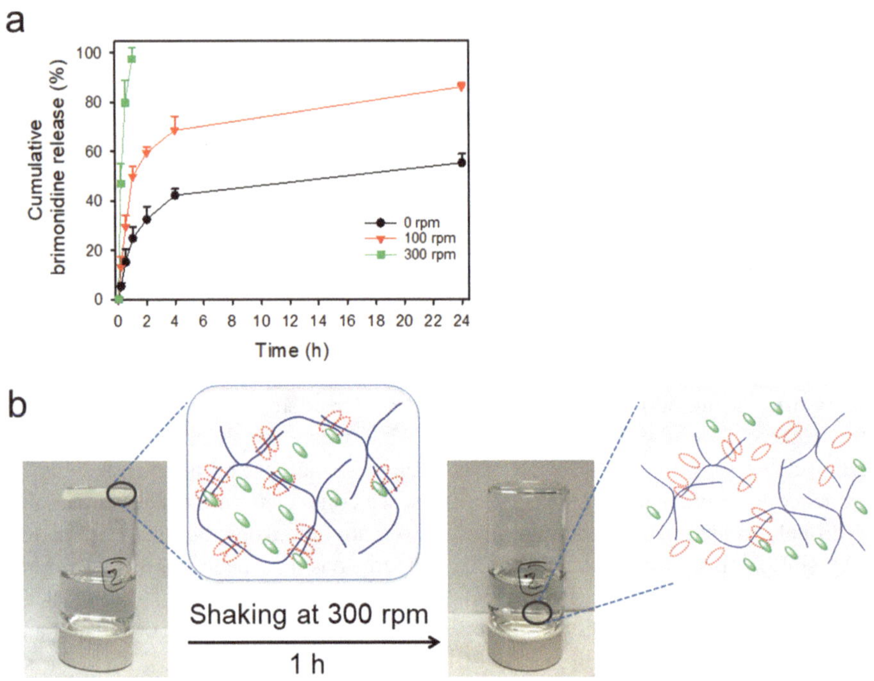

Fig. 4 (luchi vetter) In vitro release kinetics of brimonidine from the 4-arm PEG/α-CD supramolecular hydrogel at different shear rates ($n = 3$). (**b**) Schematic illustration of in vitro drug release in response to shear force change. (Reprinted with permission from Wang et al. (2018b). Copyright 2018 Elsevier)

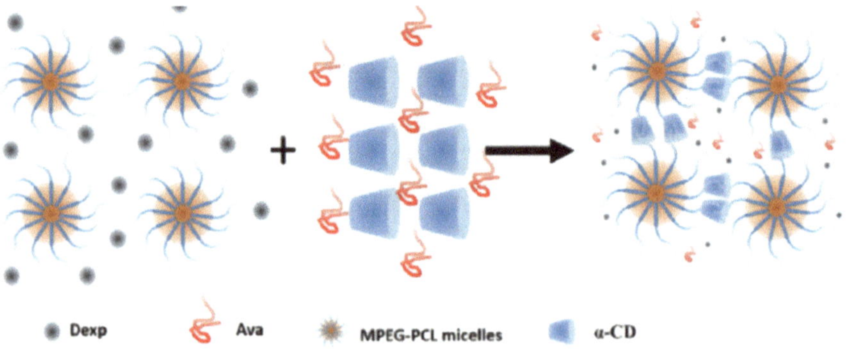

Fig. 5 Schematic formation of Dexp-Ava hydrogel. (Reprinted with permission from Huang et al. (2017). Copyright 2017 Elsevier)

carboxymethyl chitosan (CMCTS) through the Schiff base reaction (Fig. 6). The release of VCZ lasted more than a week. The sustained release was achieved presumably with the combination of degradation of poly β-CD and diffusion-driven

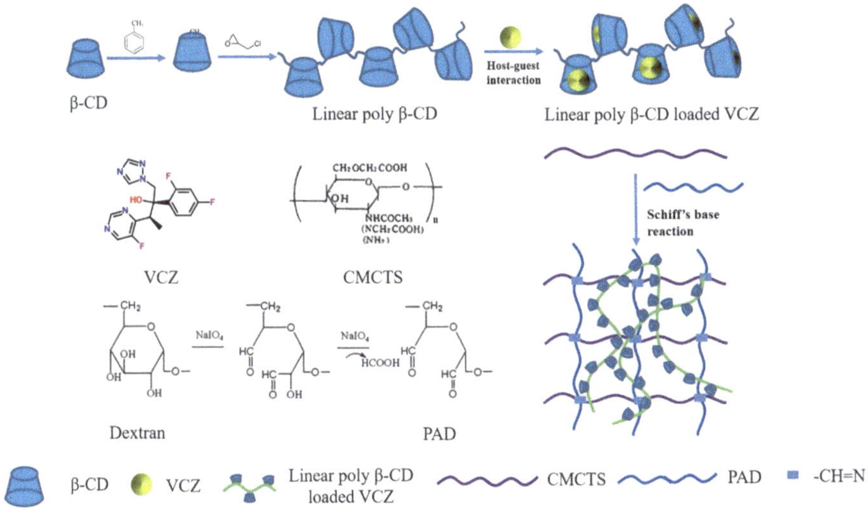

Fig. 6 Preparation of voriconazole (VCZ)-loaded injectable hydrogel. *CMCTS* carboxymethyl chitosan, *PAD* polyaldehyde dextran. (Reprinted with permission from Yang et al. (2019). Copyright 2019 Wiley)

release of VCZ from β-CD. While this application remains to be validated, the authors suggested that the hydrogel in its maximal swollen stage be injected into the vitreous cavity to treat fungal endophthalmitis.

Host-guest interactions such as hydrophobic interactions can be used to load hydrophobic compounds into the internal cavity of CDs. For instance, β-CD was functionalized with dialdehyde (β-CD-DA) and loaded with ofloxacin (OFLX), which is a highly potent antibacterial agent but has low solubility and bioavailability (Chen et al. 2018). β-CD-DA can be cross-linked with collagen to form an antibacterial film, which was shown to be effective in inhibiting both gram-positive and gram-negative bacteria. Cyclodextrin themselves can also be cross-linked to form nanosponges with expanded ability to accommodate drugs via either inclusion or noninclusion complexation (Fig. 7) (Hayiyana et al. 2016). Anti-inflammatory eye drops typically include the nonsteroidal anti-inflammatory drug (NSAID) diclofenac sodium, which has limited water solubility and poor corneal permeation. This new nanosponge was able to carry diclofenac sodium and improve the corneal permeation of diclofenac sodium by more than twofold in an ex vivo pig corneal model. The complexes have obtained significantly improved water solubility. The drug was quickly released in an hour.

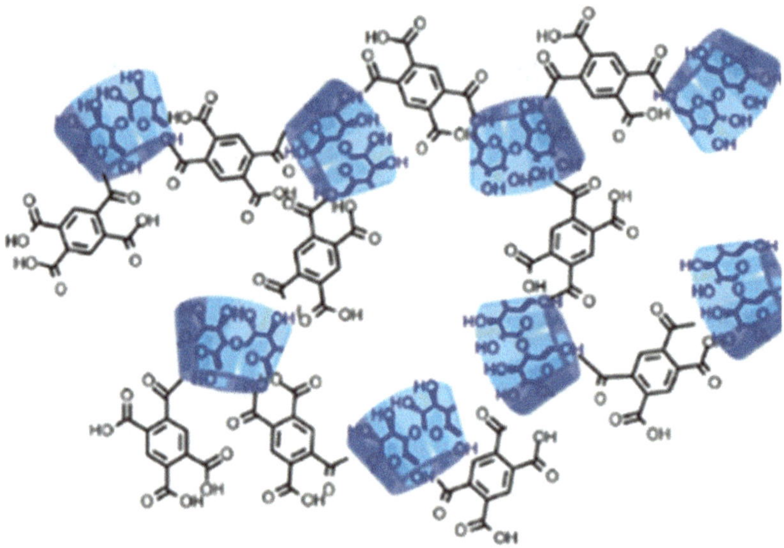

Fig. 7 Schematic representation of the synthesis reaction for cyclodextrin-based nanosponges, demonstrating the linkage of cyclodextrin molecules by pyromellitic dianhydride (PMDA) (cross-linking reaction). (Reprinted with permission from Hayiyana et al. (2016). Copyright 2016 Bentham Science Publishers Ltd)

Dendrimer Hydrogels

Dendrimers have well-defined nanoscaled structures with relatively low polydispersity. They have attracted tremendous attention for biomedical applications and drug and gene delivery because of their structural features and properties (Tomalia et al. 1985; Milhem et al. 2001; Yang and Kao 2006). A variety of moieties including therapeutic, ligands, and imaging probes can be covalently conjugated to the dendrimer surface, encapsulated into the inner core or complexed with dendrimer to form functional nanoparticles (He et al. 2018; Lancina and Yang 2017; Xu et al. 2016; Yuan et al. 2010). A novel class of hydrogels based on PAMAM dendrimers, namely, dendrimer hydrogels, has been recently developed by us. Dendrimer hydrogels are composed of inter-cross-linked hyperbranched dendrimer macromolecules (Fig. 8). We have been actively studying them for antiglaucoma drug delivery and developed several formulation strategies toward the safer use of dendrimer hydrogels.

In our first method, we chose the conventional photopolymerization to cross-link dendrimers that were coupled with PEG acrylates in the presence of photoinitiators (Desai et al. 2010). This method was very efficient in making cross-linked hydrogel networks; however, the use of photoinitiators raised safety concerns as they have shown toxic effects on intracellular signaling pathways such as AKT (Fig. 9) (Leyuan Xu et al. 2015). In particular, DMPA and I-2959 were shown to have stronger dose-dependent toxicity than EY. The inhibited AKT activity may result in a

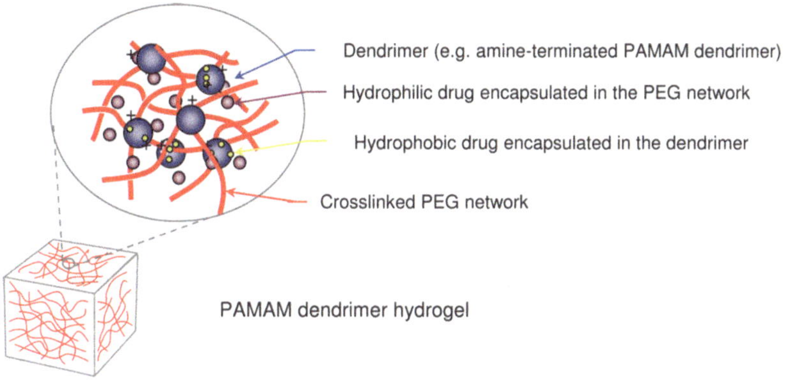

Fig. 8 Schematic of a cross-linked PAMAM dendrimer network. (Reprinted with permission from Desai et al. (2010). Copyright 2010 American Chemical Society)

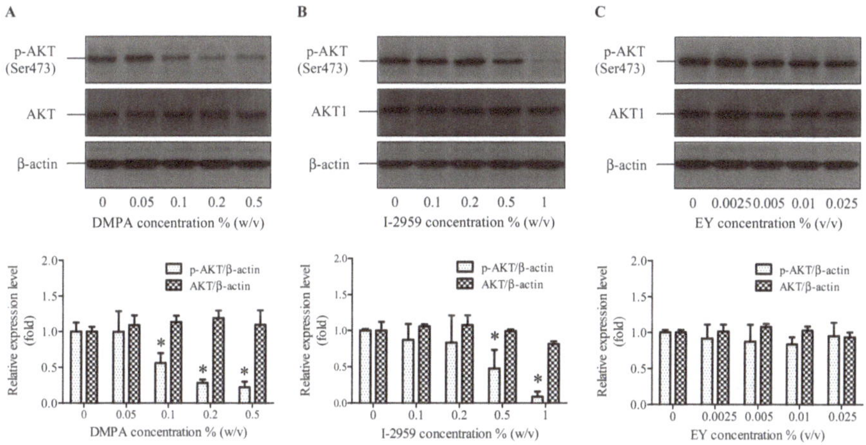

Fig. 9 Effects of DMPA (**a**), I-2959 (**b**), and EY (**c**) on intracellular AKT signaling in HN4 cells. Signaling molecule AKT and its phosphorylated form p-AKT expression levels were determined by using Western blotting. Quantitative analysis of the bands was made by densitometry using NIH ImageJ, and expression levels normalized to β-actin are presented. The data are representative of one of three independent experiments and expressed as mean ± SD (n = 3). *p < 0.05. *DMPA* 2,2-dimethoxy-2-phenylacetophenone, *I-2959* 2-hydroxy-1-[4-(2-hydroxyethoxy)phenyl]-2-methyl-1-propanone; and *EY* eosin Y photoinitiating system. (Reprinted with permission from Xu et al. (2015). Copyright 2015 The Royal Society of Chemistry)

reduction in cell viability, motility, and proliferation as well as angiogenesis. Furthermore, free radicals induced by exposing these photoinitiators to UV lights were also toxic. Cumulative toxicity of those photoinitiators may be severe, given the chronic use of hydrogels.

Recently we developed a relatively green method to prepare dendrimer hydrogels based on the highly efficient aza-Michael addition reaction. The aza-Michael

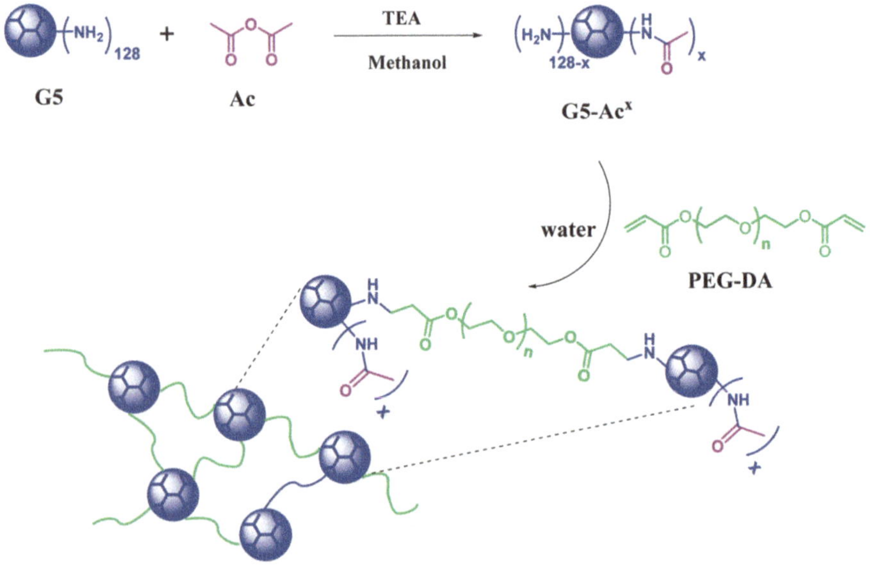

Fig. 10 Scheme of acetylated G5 (G5-AC^x) synthesis and aza-Michael addition reaction of G5 or G5-Ac with PEG-DA. *AC* acetic anhydride; the superscript x indicates the degree of functionalization, *PEG-DA* poly(ethylene glycol) diacrylate. (Reprinted with permission from Wang et al. (2017a). Copyright 2017 American Chemical Society)

addition reaction takes place between nitrogen nucleophiles of dendrimer primary amines and α,β-unsaturated esters in PEG diacrylate (Fig. 10) (Wang et al. 2017a). PAMAM dendrimer was acetylated to have various degrees of surface charge neutralization as a means to manipulate gelation kinetics and properties. This method does not require catalyst. The reaction proceeds efficiently in water at room temperature. No side products or free radicals are generated during the preparation, thus greatly easing safety concerns.

We went on to develop micrometer-sized dendrimer hydrogels (μDHs), i.e., dendrimer microgels, by combining inverse microemulsion and the aza-Michael addition (Fig. 11) (Wang et al. 2018a). Microgels are particles with a micrometer-scale three-dimensional network. μDHs possess the properties of macroscopic dendrimer hydrogels and exhibit particle features, providing greater structural flexibility for programmable drug delivery and controlled release. In this method, we prepared a water solution containing PAMAM dendrimer, and PEG-DA is dispersed in an organic phase to form microdroplets, which serve as reactors to accommodate the aza-Michael addition reaction. The microdroplets solidify to form gel particles. We observed that μDHs are pH-dependent degradable and have good cytocompatibility. Drugs can be loaded into μDHs and released following the zero-order release kinetics. Our recent study (unpublished) has produced the encouraging result to support this new type of dendrimer hydrogel as a new platform for ocular drug delivery.

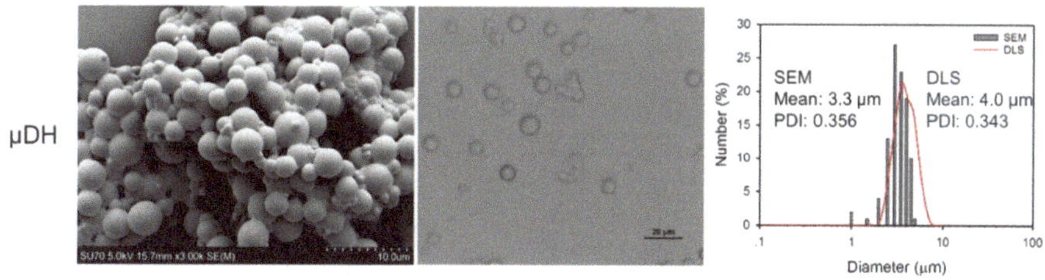

Fig. 11 SEM image (left panel), optical micrographs of gel particles in suspension (middle panel), and particle size and size distribution (right panel) of μDH. (Reprinted with permission from Wang et al. (2018a). Copyright 2018 American Chemical Society)

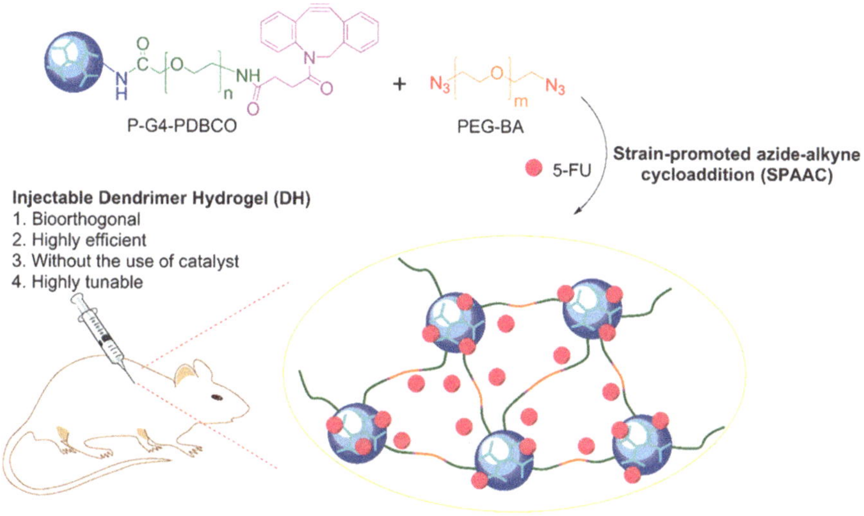

Fig. 12 Preparation of bioorthogonal dendrimer hydrogels based on the strain-promoted azide-alkyne cycloaddition (SPAAC). (Reprinted with permission from Xu et al. (2017). Copyright 2017 American Chemical Society)

In addition to the new dendrimer hydrogels prepared using the aza-Michael addition, we developed bioorthogonal DHs. We functionalized PAMAM dendrimer with strained alkyne, i.e., dibenzocyclooctyne (DBCO), and then reacted it with polyethylene glycol bisazide (PEG-BA) in water via strain-promoted azide-alkyne cycloaddition (SPAAC) to form a dendrimer-PEG cross-linked network (Fig. 12) (Xu et al. 2017). This reaction takes place rapidly at room temperature in the absence of catalysts. This formulation utilizes reactive groups that are not naturally present in the body. Therapeutics such as proteins and peptides can be loaded into the gel during the reaction and avoid cross-reacting with the dendrimer hydrogel. The safety of this new formulation for the eye has been demonstrated by us. Our ongoing studies include the test of this formulation for antiglaucoma drug delivery.

Conclusions

Recent efforts with focus on safety and ease of application have led to the development of several novel types of ocular hydrogel formulations, including supramolecular hydrogels and dendrimer hydrogels, as discussed in this book chapter. It must be pointed out that the safe application of such formulations has to be tested sufficiently in preclinical and clinical trials before they can be used clinically. To accelerate the bench-to-bedside translation of these new formulations, we also need to address issues, such as manufacturing/scaling up, packaging, long-term storage, sterility, and so on. For particular ocular disease medication, formulations need to be tested in authentic disease models to demonstrate safety and efficacy. Profiles of degradation and/or clearance of the formulation need to be established.

Acknowledgments This work was supported by the National Institutes of Health (R01EY024072).

References

Al-Kinani AA, Zidan G, Elsaid N, Seyfoddin A, Alani AWG, Alany RG. Ophthalmic gels: past, present and future. Adv Drug Deliv Rev. 2018;126:113–26. https://doi.org/10.1016/j.addr.2017.12.017.

Bairi P, Chakraborty P, Mondal S, Roy B, Nandi AK. A thixotropic supramolecular hydrogel of adenine and riboflavin-5[prime or minute]-phosphate sodium salt showing enhanced fluorescence properties. Soft Matter. 2014;10(28):5114–20. https://doi.org/10.1039/C4SM00627E.

Baral A, Roy S, Ghosh S, Hermida-Merino D, Hamley IW, Banerjee A. A peptide-based mechanosensitive, proteolytically stable hydrogel with remarkable antibacterial properties. Langmuir. 2016;32(7):1836–45. https://doi.org/10.1021/acs.langmuir.5b03789.

Ceccato M, Lo Nostro P, Baglioni P. α-Cyclodextrin/polyethylene glycol polyrotaxane: a study of the threading process. Langmuir. 1997;13(9):2436–9. https://doi.org/10.1021/la9609231.

Chen Y, Song W, Zhao X, Han Q, Ren L. An antibacterial collagen membrane crosslinked by the inclusion complex of beta-cyclodextrin dialdehyde and ofloxacin for bacterial keratitis (article). RSC Adv. 2018;8(32):18153–62. https://doi.org/10.1039/c8ra02160k.

Cholkar K, Patel SP, Vadlapudi AD, Mitra AK. Novel strategies for anterior segment ocular drug delivery. J Ocul Pharmacol Ther. 2013;29(2):106–23. https://doi.org/10.1089/jop.2012.0200.

Cooper RC, Yang H. Hydrogel-based ocular drug delivery systems: emerging fabrication strategies, applications, and bench-to-bedside manufacturing considerations. J Control Release. 2019;306:29–39. https://doi.org/10.1016/j.jconrel.2019.05.034.

Desai PN, Yuan Q, Yang H. Synthesis and characterization of photocurable polyamidoamine dendrimer hydrogels as a versatile platform for tissue engineering and drug delivery. Biomacromolecules. 2010;11(3):666–73. https://doi.org/10.1021/bm901240g.

Gigliotti CL, Minelli R, Cavalli R, Occhipinti S, Barrera G, Pizzimenti S, et al. In vitro and in vivo therapeutic evaluation of camptothecin-encapsulated beta-cyclodextrin nanosponges in prostate cancer. J Biomed Nanotechnol. 2016;12(1):114–27. https://doi.org/10.1166/jbn.2016.2144.

Guvendiren M, Lu HD, Burdick JA. Shear-thinning hydrogels for biomedical applications. Soft Matter. 2012;8(2):260–72. https://doi.org/10.1039/C1SM06513K.

Harada A, Kamachi M. Complex formation between poly(ethylene glycol) and α-cyclodextrin. Macromolecules. 1990;23(10):2821–3. https://doi.org/10.1021/ma00212a039.

Harada A, Takashima Y, Nakahata M. Supramolecular polymeric materials via cyclodextrin–guest interactions. Acc Chem Res. 2014;47(7):2128–40. https://doi.org/10.1021/ar500109h.

Hatziagapiou K, Bethanis K, Lambrou GI, Yannakopoulou K, Karpusas M, Braoudaki M, et al. Enhanced gefitinib cytotoxicity in the presence of cyclodextrins: in-vitro and biophysical studies towards potential therapeutic interventions for cancer. J Biomed Nanotechnol. 2017;13(5):522–33. https://doi.org/10.1166/jbn.2017.2374.

Hayiyana Z, Choonara Y, Makgotloe A, du Tait L, Kumar P, Pillay V. Ester-based hydrophilic cyclodextrin nanosponges for topical ocular drug delivery (article). Curr Pharm Des. 2016;22(46):6988–97. https://doi.org/10.1174/1381611811666161116113107.

He H, Yuan Q, Bie J, Wallace RL, Yannie PJ, Wang J, et al. Development of mannose functionalized dendrimeric nanoparticles for targeted delivery to macrophages: use of this platform to modulate atherosclerosis. Transl Res. 2018;193:13–30. https://doi.org/10.1016/j.trsl.2017.10.008.

Huang J, Wang W, Yu J, Yu X, Zheng Q, Peng F, et al. Combination of dexamethasone and Avastin (R) by supramolecular hydrogel attenuates the inflammatory corneal neovascularization in rat alkali burn model (article). Colloids Surf B Biointerfaces. 2017;159:241–50. https://doi.org/10.1016/j.colsurfb.2017.07.057.

Huang D, Chen YS, Rupenthal ID. Overcoming ocular drug delivery barriers through the use of physical forces. Adv Drug Deliv Rev. 2018;126:96–112. https://doi.org/10.1016/j.addr.2017.09.008.

Jeong B, Bae YH, Lee DS, Kim SW. Biodegradable block copolymers as injectable drug-delivery systems. Nature. 1997;388(6645):860–2.

Lancina M, Yang H. Dendrimers for ocular drug delivery (review). Can J Chem. 2017;95(9):897–902. https://doi.org/10.1139/cjc-2017-0193.

Li J, Li X, Ni X, Wang X, Li H, Leong KW. Self-assembled supramolecular hydrogels formed by biodegradable PEO–PHB–PEO triblock copolymers and α-cyclodextrin for controlled drug delivery. Biomaterials. 2006;27(22):4132–40. https://doi.org/10.1016/j.biomaterials.2006.03.025.

Liu Y, Pu YZ, Sun L, Yao HJ, Zhao BQ, Zhang RZ, et al. Folic acid functionalized gamma-cyclodextrin C-60, a novel vehicle for tumor-targeted drug delivery. J Biomed Nanotechnol. 2016;12(7):1393–403. https://doi.org/10.1166/jbn.2016.2275.

Loh XJ. Supramolecular host-guest polymeric materials for biomedical applications. Mater Horizons. 2014;1(2):185–95. https://doi.org/10.1039/C3MH00057E.

Ma D, Zhang L-M. Supramolecular gelation of a polymeric prodrug for its encapsulation and sustained release. Biomacromolecules. 2011;12(9):3124–30. https://doi.org/10.1021/bm101566r.

Ma X, Zhao Y. Biomedical applications of supramolecular systems based on host-guest interactions. Chem Rev. 2015;115(15):7794–839. https://doi.org/10.1021/cr500392w.

Milhem OM, Mobedi H, Day NH, McKeown NB, Attwood D, D'Emanuele A. Development and characterization of dendrimer conjugates for drug delivery. In: Proceedings - 28th international symposium on controlled release of bioactive materials and 4th consumer & diversified products conference, San Diego, 23–27 June 2001, vol 1, pp. 329–330.

Parisi-Amon A, Mulyasasmita W, Chung C, Heilshorn SC. Protein-engineered injectable hydrogel to improve retention of transplanted adipose-derived stem cells. Adv Healthc Mater. 2013;2(3):428–32. https://doi.org/10.1002/adhm.201200293.

Sapino S, Chirio D, Peira E, Abellán Rubio E, Brunella V, Jadhav SA, et al. Ocular drug delivery: a special focus on the thermosensitive approach. Nanomaterials (Basel). 2019;9(6) https://doi.org/10.3390/nano9060884.

Takahashi S, Yamada NL, Ito K, Yokoyama H. Inclusion complex of α-cyclodextrin with poly(ethylene glycol) brush. Macromolecules. 2016;49(18):6947–52. https://doi.org/10.1021/acs.macromol.6b01238.

Tomalia DA, Baker H, Dewald J, Hall M, Kallos G, Martin S, et al. A new class of polymers: starburst-dendritic macromolecules. Polym J (Tokyo). 1985;17(1):117–32.

Urtti A. Challenges and obstacles of ocular pharmacokinetics and drug delivery. Adv Drug Deliv Rev. 2006;58(11):1131–5. https://doi.org/10.1016/j.addr.2006.07.027.

Wang X, Wang C, Zhang Q, Cheng Y. Near infrared light-responsive and injectable supramolecular hydrogels for on-demand drug delivery. Chem Commun. 2016;52(5):978–81. https://doi.org/10.1039/C5CC08391E.

Wang J, He H, Cooper RC, Yang H. In situ-forming polyamidoamine dendrimer hydrogels with tunable properties prepared via Aza-Michael addition reaction. ACS Appl Mater Interfaces. 2017a;9(12):10494–503. https://doi.org/10.1021/acsami.7b00221.

Wang LL, Sloand JN, Gaffey AC, Venkataraman CM, Wang Z, Trubelja A, et al. Injectable, guest–host assembled polyethylenimine hydrogel for siRNA delivery. Biomacromolecules. 2017b;18(1):77–86. https://doi.org/10.1021/acs.biomac.6b01378.

Wang J, Cooper R, He H, Li B, Yang H. Polyamidoamine dendrimer microgels: hierarchical arrangement of dendrimers into micrometer domains with expanded structural features for programmable drug delivery and release (article). Macromolecules. 2018a;51(15):6111–8. https://doi.org/10.1021/acs.macromol.8b01006.

Wang J, Williamson GS, Yang H. Branched polyrotaxane hydrogels consisting of alpha-cyclodextrin and low-molecular-weight four-arm polyethylene glycol and the utility of their thixotropic property for controlled drug release. Colloids Surf B Biointerfaces. 2018b;165:144–9. https://doi.org/10.1016/j.colsurfb.2018.02.032.

Wu H, Liu S, Xiao L, Dong X, Lu Q, Kaplan DL. Injectable and pH-responsive silk nanofiber hydrogels for sustained anticancer drug delivery. ACS Appl Mater Interfaces. 2016;8(27):17118–26. https://doi.org/10.1021/acsami.6b04424.

Xu L, Sheybani N, Yeudall WA, Yang H. The effect of photoinitiators on intracellular AKT signaling pathway in tissue engineering application. Biomater Sci. 2015;3(2):250–5. https://doi.org/10.1039/c4bm00245h.

Xu L, Kittrell S, Yeudall WA, Yang H. Folic acid-decorated polyamidoamine dendrimer mediates selective uptake and high expression of genes in head and neck cancer cells. Nanomedicine. 2016;11(22):2959–73. https://doi.org/10.2217/nnm-2016-0244.

Xu L, Cooper RC, Wang J, Yeudall WA, Yang H. Synthesis and application of injectable bioorthogonal dendrimer hydrogels for local drug delivery. ACS Biomater Sci Eng. 2017;3(8):1641–53. https://doi.org/10.1021/acsbiomaterials.7b00166.

Yang H, Kao WJ. Dendrimers for pharmaceutical and biomedical applications. J Biomater Sci Polym Ed. 2006;17(1–2):3–19.

Yang C, Gao L, Liu X, Yang T, Yin G, Chen J, et al. Injectable Schiff base polysaccharide hydrogels for intraocular drug loading and release. J Biomed Mater Res A. 2019; https://doi.org/10.1002/jbm.a.36677.

Yuan Q, Yeudall WA, Yang H. PEGylated polyamidoamine dendrimers with bis-aryl hydrazone linkages for enhanced gene delivery. Biomacromolecules. 2010;11(8):1940–7.

Ophthalmic Product Development for Biologics

Gitanjali Sharma, Chen Zhou, Saurabh Wadhwa, Ashwin C. Parenky, Kenneth S. Graham, Amardeep S. Bhalla, Dingjiang Liu, Hunter H. Chen, and Mohammed Shameem

Abstract Due to the complex and unique anatomy and physiology of the eye, the delivery of therapeutic agents to the back of the eye remains a major challenge. The National Eye Institute estimates that the number of people affected by severe ophthalmic diseases such as glaucoma, age-related macular degeneration (AMD), and diabetic retinopathy (DR) will double by 2050, creating approximately US$139 billion economic burden associated with eye diseases alone. This presents a growing need for improved therapeutic agents for the treatment of eye disorders. The current paradigm for the treatment of diseases is shifting toward biologics that show higher specificity in comparison to the conventional small molecule therapeutics. However, the delivery of biologic molecules has specific challenges including, for example, limited stability and poor penetration across biological membranes. Adding to the difficulty of resolving these issues is the lack of preclinical models for assessing safety and pharmacokinetic profile of the formulations. An additional challenge specific to intraocular drug delivery includes the strict volume limitation for intraocular delivery, which necessitates the need for high concentration formulations and drug-device combinations in order to deliver an efficacious dose of the drug. These high concentration biologic formulations may pose additional challenges associated with high viscosity, insufficient drug solubility, product manufacturing, storage, and handling, as well as challenges to drug administration. In addition, many of the common GRAS listed excipients used to stabilize or mitigate viscosity in biologics have not been evaluated for use in ophthalmic preparations. This chapter highlights some of the formulation development and stability challenges faced by pharmaceutical scientists during the development of ophthalmic biological products and summarizes some current, relevant regulatory guidance.

Keywords Ophthalmic · Formulation · Regulatory · Biologics · Stability

G. Sharma (✉) · C. Zhou · S. Wadhwa · A. C. Parenky · K. S. Graham · A. S. Bhalla · D. Liu · H. H. Chen (✉) · M. Shameem
Formulation Development Group, Regeneron Pharmaceuticals, Inc., Tarrytown, NY, USA
e-mail: hunter.chen@regeneron.com

S. Neervannan, U. B. Kompella (eds.), *Ophthalmic Product Development*, AAPS Advances in the Pharmaceutical Sciences Series 37,
https://doi.org/10.1007/978-3-030-76367-1_17

Introduction

Need for Development of Ophthalmic Biologic Agents

More than ten million people in the United States are affected with serious retinal diseases, such as macular edema, retinal vein occlusion (RVO), diabetic macular edema (DME), and diabetic retinopathy (DR). The National Eye Institute has estimated that the number of people affected by these diseases is expected to double by 2050 with an economic burden of about US$139 billion associated with eye diseases (Cabrera et al. 2019). Prior to the use of biologics, various strategies like focal laser therapy, intravitreal steroids, photocoagulation, and photodynamic therapy (with VISUDYNE®) were used for treatment of serious ophthalmic diseases. These agents were able to delay the onset of diseases or slow down vision loss as well as reduce the probability of choroidal neovascularization (CNV) recurrences (Cabrera et al. 2019). Biologics, such as monoclonal antibodies, are designed to specifically target and bind to specific molecules, thus resulting in fewer side effects and enhanced activity. High specificity, potency, and clinically demonstrated safety associated with biologic molecules render them as the treatment of choice over small molecules (Mandal et al. 2018; Leader et al. 2008; Usmani et al. 2017). A recent report indicated that the worldwide sales of biologics for ophthalmic indications exceeded US$8 billion in the year 2016 and is expected to reach US$35.7 billion by the year 2025 (Kim et al. 2014; Mandal et al. 2018).

With the advent of anti-VEGF (vascular endothelial growth factor) agents, there has been a paradigm shift in the treatment of serious ocular diseases like diabetic macular edema (DME), diabetic retinopathy (DR), and wet age-related macular degeneration (wAMD). Most of the intravitreal anti-VEGF agents (e.g., EYLEA®, LUCENTIS®, and BEOVU®) act by inhibiting VEGF A. The efficacy of intravitreal injections of anti-VEGF agents has been reported for the treatment of various ocular diseases, like neovascular complications of wAMD, DME, diabetic retinopathy, and retinal vascular occlusions (Nicholson and Schachat 2010; Wong and Scott 2010; Campochiaro et al. 2010; Heier et al. 2012). Anti-VEGF therapies have established substantial recognition and are considered the standard of care for treating proliferative vascular diseases of the posterior segment of the eye (Xue et al. 2019).

Anti-VEGF biological molecules that are commercially approved for treating eye disorders (Table 1) (Usmani et al. 2017) are either proteins (e.g., EYLEA®, LUCENTIS®, and BEOVU®) or an aptamer (MACUGEN®, which is a PEGylated single strand nucleic acid that binds specifically to VEGF). Biological molecules, for the treatment of back of the eye disorders, that are currently in development include OPT-302 (Opthea), KSI-301 (Kodiak Sciences), and conbercept (Chengdu Kanghong Biotech Co., Ltd.).

These anti-VEGF agents work by inhibiting the binding of different isoforms of VEGF (e.g., VEGF A, VEGF C, VEGF D, and placental growth factor) to its receptors. They act as soluble decoy receptors that have higher affinity for VEGF as compared to the body's native receptors. Blockade of VEGF C/D and VEGF A by

Table 1 Summary of commercially available intravitreal products used for the treatment of ocular diseases

Product	Trade name	Molecule type	Target	Approved indications (as of March 2020)	Year of approval in the USA	Company
Aflibercept	EYLEA®	Fc fusion protein	VEGF A, placental growth factor (PlGF)	• Neovascular (wet) age-related macular degeneration (AMD) • Macular edema following retinal vein occlusion (RVO) • Diabetic macular edema (DME) • Diabetic retinopathy (DR)	2011 (vial) and 2019 (prefilled syringe)	Regeneron Pharmaceuticals, Inc. and Bayer
[a]Bevacizumab	AVASTIN®	IgG1 mAb	VEGF A	• Choroidal neovascular membrane (CNV) in AMD • Diabetic retinopathy	2004	Roche/Genentech
Brolucizumab-dbll	BEOVU®	ScFv	VEGF A ($VEGF_{110}$, $VEGF_{121}$, $VEGF_{165}$)	• Neovascular (wet) age-related macular degeneration (AMD)	2019	Novartis AG
Ocriplasmin	JETREA®	Truncated human plasmin	Proteolytic activity against vitreous proteins (e.g., laminin, fibronectin, and collagen)	• Symptomatic vitreomacular adhesion	2012	Oxurion
Pegaptanib sodium	MACUGEN®	Pegylated oligonucleotide	$VEGF_{165}$	• Neovascular (wet) age-related macular degeneration (AMD)	2004	Bausch + Lomb

(continued)

Table 1 (continued)

Product	Trade name	Molecule type	Target	Approved indications (as of March 2020)	Year of approval in the USA	Company
Ranibizumab	LUCENTIS®	Fab of IgG1 mAb	VEGF A	• Neovascular (wet) age-related macular degeneration (AMD) • Macular edema following retinal vein occlusion • Diabetic macular edema	2006 (vial) and 2018 (prefilled syringe)	Roche/Genentech
Voretigene neparvovec-rzyl	LUXTURNA®	Vector genome	AAV2 gene therapy product	• Biallelic RPE65 mutation-associated retinal dystrophy	2017	Spark Therapeutics

Note: The information listed in the table was taken from package inserts and literature
[a]Off-label use of bevacizumab (Schmucker et al. 2015)

combination therapies is being investigated. For example, OPT-302 (Opthea) is a soluble form of VEGF receptor-3 that blocks the activity of two isoforms of VEGF (VEGF C and D). OPT-302 in combination with EYLEA is currently being evaluated to treat DME. Recently, Opthea reported in Phase 2b in treatment of naïve patients with wAMD. Phase 2a trials met the primary endpoint and achieved vision gains as measured by best-corrected visual acuity (BVCA) (clinicalTrials.gov, NCT #03397264). Table 1 gives a list of commercially available anti-VEGF agents used for the treatment of serious eye disorders.

Current Landscape for Ophthalmic Biologics

Currently, there are in total six FDA-approved ophthalmic biologics in the US market and one biologic (bevacizumab) used off-label, as listed in Table 1. All of them except MACUGEN, a chemically synthesized pegylated oligonucleotide, are therapeutic protein products. Here, we will focus on the current ophthalmic protein drugs and their formulation details and briefly discuss their mechanisms of action.

EYLEA® (Aflibercept)

Aflibercept is a 115 kDa Fc fusion protein, including the Fc region of a human IgG1 mAb and an extracellular vascular endothelial growth factor (VEGF) receptor sequence (VEGF1 and VEGF2) (Eylea_Package_Insert 2011). This highly glycosylated molecule is formulated at 40 mg/mL in 10 mM sodium phosphate pH 6.2, 40 mM sodium chloride, 0.03% w/v polysorbate 20, and 5% (w/v) sucrose. The final solution is approximately isotonic. It is supplied in type I glass vials or prefilled syringes for intravitreal injection. Aflibercept competes for the binding of VEGF A and placental growth factor (PlGF) and therefore prevents the activation of cognate VEGF receptors and subsequent retinal neovascularization and vascular permeability.

AVASTIN® (Bevacizumab)

Bevacizumab is a humanized monoclonal IgG1 antibody with a molecular weight (MW) of 149 kDa (Avastin_Package_Insert 2004). It was originally approved by the FDA in 2004 for the treatment metastatic colorectal cancer. Subsequently in 2005, bevacizumab was adopted for off-label use to treat neovascular and exudative ocular diseases by intravitreal injection. The murine complementarity-determining regions (CDR) are able to bind VEGF by competing with native VEGF receptors. AVASTIN is formulated at 25 mg/mL bevacizumab in 50 mM phosphate, 6% trehalose (w/v), 0.04% (w/v) polysorbate (PS) 20, and pH 6.2. Repackaging of bevacizumab for intravitreal injection via a compounding pharmacy has been associated with serious risks which include of microbial contamination due to poor aseptic techniques or supply of spurious vials (Saoji et al. 2018; Lowes 2013) or contamination with silicone oil microdroplets (Liu et al. 2011).

BEOVU® (Brolucizumab-dbll)

Brolucizumab-dbll is a humanized monoclonal single-chain Fv (scFv) antibody fragment with a molecular weight of ~26 kDa (Beovu_Package_Insert 2019). This molecule is expressed in nonmammalian *E coli* BL21 (DE3) and consists of light chain and heavy chain fragments, which belong to human kappa and VH3 subtypes, respectively. BEOVU is the drug product with 120 mg/mL brolucizumab-dbll in 10 mM sodium citrate, 5.8% (w/v) sucrose, 0.02% (w/v) polysorbate 80, and pH 7.2 in a 2 mL single-dose glass vial for intravitreal injection. Brolucizumab-dbll is a human VEGF inhibitor and binds to the three major isoforms of VEGF A, thereby preventing the ligand interaction with the cognate VEGF receptors, VEGFR-1, and VEGFR-2, thereby suppressing endothelial cell proliferation, neovascularization, and vascular permeability.

XEOMIN® (IncobotulinumtoxinA)

IncobotulinumtoxinA is a neurotoxin belonging to botulinum toxin type A with MW of 150 kDa (Xeomin_Package_Insert 2010). The formulation of XEOMIN is a lyophilized dosage form that after reconstitution includes 50–200 unit/mL protein in 0.47% sucrose (w/v) and 0.1% human serum albumin (HSA). HSA as a protein is used in XEOMIN as a stabilization excipient. It has been used in the past both as a lyoprotectant and cryoprotectant for freeze-dried protein formulations in a number of therapeutic products, such as EPOGEN®, KOGENATE®, and RECOMBINATE™ (Wang 2000). It can also function as a surfactant to block the adsorption of protein on various interfaces, therefore enhancing protein interfacial stability. Intramuscular injection of incobotulinumtoxinA can help rebalance the actions of muscles controlling eye rotation by cleaving SNAP25 and consequently release of acetylcholine.

MACUGEN® (Pegaptanib Sodium)

Pegaptanib is a selective vascular endothelial growth factor (VEGF) antagonist. It is an aptamer, a pegylated modified oligonucleotide, which adopts a three-dimensional confirmation enabling it to bind to the extracellular VEGF (Macugen_Package_Insert 2004). MACUGEN is supplied as a single-dose prefilled syringe and contains 3.47 mg/mL of solution. It contains 0.3 mg of the active oligonucleotide free acid form, without polyethylene glycol (PEG), in a nominal volume of 90 μL. The drug product is a sterile, preservative-free formulation containing monobasic sodium phosphate monohydrate, dibasic sodium phosphate heptahydrate, hydrochloric acid, and sodium hydroxide in water for injection. The molecular weight is ~50 kDa, osmolality of 280–360 mOsm/kg, and pH of the formulation is in a range of 6–7. Pegaptanib sodium is indicated for neovascular (wet) age-related macular degeneration.

JETREA® (Ocriplasmin)

Ocriplasmin is a truncated form of recombinant human plasmin with MW of 27.2 kDa (Jetrea_Package_Insert 2012). JETREA is a liquid solution formulated with 2.5 mg/mL ocriplasmin in 5.5 mM citrate, 0.375% (w/v) mannitol, and pH 3.1 intended for intravitreal injection. Ocriplasmin acts through proteolytic activity against proteins in the vitreous body and vitreoretinal interface and therefore triggers the dissolution of the protein matrix that causes vitreomacular adhesion (a pathological condition where the vitreous gel of the eye adheres to the retina).

LUCENTIS® (Ranibizumab)

Ranibizumab is the Fab (antigen-binding fragment) of bevacizumab with MW of 48 kDa (Lucentis_Package_Insert 2006). In contrast to the common mammalian cell expression system used in mAb production, ranibizumab is produced in *E. coli* cells. The formulation of ranibizumab is supplied in a vial or prefilled syringe for intraocular injection and includes 10 mg/mL or 6 mg/mL protein with 10 mM histidine, 10% (w/v) trehalose, 0.01% (w/v) PS 20, and pH 5. Similar to bevacizumab, ranibizumab functions to bind and neutralize VEGF A receptor and thereby prevents VEGF A signaling through cellular VEGF receptor signaling.

In addition to the FDA-approved drug products listed in Table 1, most of the biological agents are challenged with biosimilars (Table 2). Biosimilars are molecules, which are similar to the existing innovator biologics with comparable pharmacokinetic, pharmacodynamic, immunogenicity, safety, and efficacy to the innovator biologic. The development of these biosimilars involves much less time (8–10 years) and lower cost (US$100–200 million) as compared to the original biologic agent (cost US$1200–2500 million and time 10–15 years) (Sharma et al. 2018). Companies developing biosimilars do not need to invest heavily on clinical trials; instead, they need robust analytical bioequivalence to prove similarity. At least one clinical trial is required to compare pharmacokinetics, and one large randomized clinical trial is required to demonstrate clinical equivalence. Six biosimilars of LUCENTIS and EYLEA are currently under clinical investigation (Sharma et al. 2018). Eight bevacizumab biosimilars are approved for oncology rather than for ophthalmic use.

Development of biologics for intraocular use comes with its unique set of challenges, including the delivery, pharmaceutical development, as well as Chemistry, Manufacturing, and Controls (CMC), and ophthalmic-specific regulatory requirements. All these aspects make for a complex set of challenges that require special

Table 2 Biosimilars for EYLEA and LUCENTIS (adapted from Sharma et al. 2018)

Commercial name of the innovator biologic	Name of the biosimilar	Name of the company	Stage of development
EYLEA	M710	Momenta Pharmaceuticals (Cambridge, MA, USA) Mylan NV (Cambridge, MA, USA)	NA (pivotal trial 2018)
	ALT-L9	Alteogen, Inc. (Daejeon, South Korea)	NA (IND filling with USA FDA 2018)
	FYB203	Formycon AG (Munich, Germany)	Expected marketing approval USA 2023 Europe 2025
	CHS-2020	Coherus BioSciences (Redwood City, CA, USA)	NA (preclinical)
	SB15	Biogen Idec (Cambridge, MA, USA) Samsung Bioepis (Incheon, South Korea)	Preclinical development
LUCENTIS	SB11	Biogen Idec (Cambridge, MA, USA) Samsung Bioepis (Incheon, South Korea)	Phase 3
	FYB 201	Formycon AG (Munich, Germany) bioeq GmbH (Holzkirchen, Germany)	Expected marketing approval USA 2020 Europe 2022 (Phase 3)
	Xlucane	Xbrane Biopharma (Solna, Sweden)	Expected marketing approval USA 2020 Europe 2022 (preclinical)
	PF582	Pfenex Inc. (San Diego, CA, USA)	NA (Phase 1/2)
	CHS3351	Coherus BioSciences (Redwood City, CA, USA)	NA (preclinical)

NA not applicable, *IND* investigational new drug application, *FDA* Food and Drug Administration, *EMA* European Medicines Agency

attention during preclinical development, manufacturing, and clinical assessment. This chapter will review such challenges that are associated with ophthalmic development of biologics (Fig. 1).

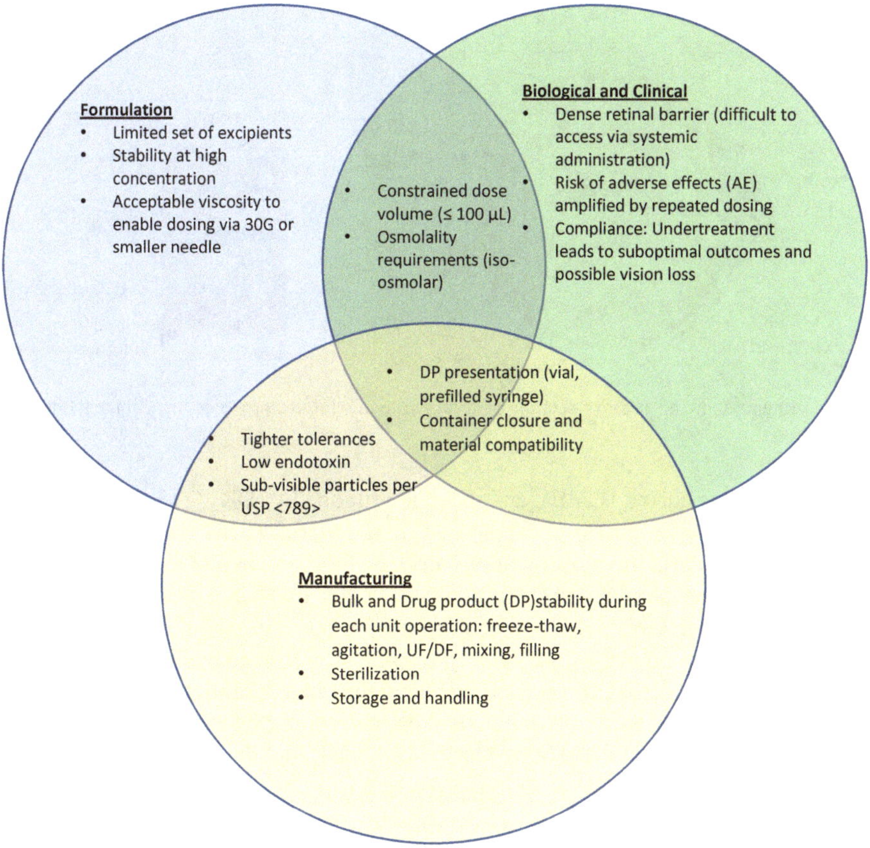

Fig. 1 Summary of major challenges associated with the development of ophthalmic biologics

Challenges Associated with the Development of Ophthalmic Biologics

Biological Challenges

The posterior segment of the eye, while highly vascularized, is not readily accessible to biologics or other molecules via the systemic circulation (Fig. 2). At the same time, topical administration does not allow for high enough drug levels to reach the posterior pole and effectively treat retinal diseases (Bruno et al. 2013). The retina and macula, which are primarily responsible for vision, are the target site for the anti-VEGF drugs. Access to these tissues presents a drug delivery challenge which complicates the treatment of serious disorders affecting vision in the aging population, such as wet age-related macular degeneration (wAMD), choroidal neovascularization (CNV), diabetic macular edema (DME), retinal vein occlusion (RVO),

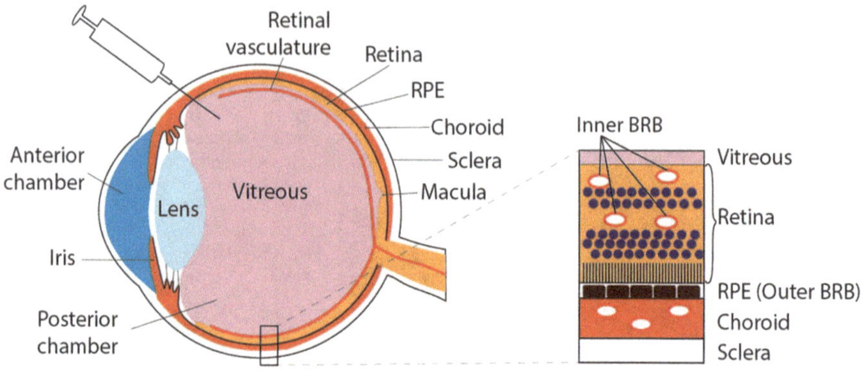

Fig. 2 Image showing a cross section of a human eye. (Figure provided by Bibiana Iglesias, Regeneron)

cystoid macular edema (CME), and diabetic retinopathy (DR) (Zajac-Pytrus et al. 2015; Wiersbitzky et al. 1985). The blood retinal barrier (BRB) is a physiological barrier that regulates the movement of nutrients, ions, etc. in and out of the eyes. It is an intricate nonvascular and clear retinal tissue forming that restricts the transport of majority of the therapeutic molecules into the posterior segment of the retina. There are two components to the BRB: the inner BRB formed by the tight junctions between retinal endothelial cell capillaries and the outer BRB that is formed by the tight junctions between retinal pigmented epithelium (RPE) cells. Biologic agents are generally hydrophilic and have a large size, low lipophilicity, and charged functional groups, which tend to inhibit penetration through dense biological membranes (Usmani et al. 2017; Bruno et al. 2013).

Intravitreal injection directly delivers anti-VEGF biologic molecules into the vitreous, typically using a syringe and 27–30-gauge needle, resulting in maximum exposure of the posterior pole to the drug. At the same time, intravitreal injections present several challenges. First, the placement and the depth of penetration of the needle are critical. Inappropriate placement can result in a traumatic cataract or damage to other internal structures of the eye. If the needle penetrates too deep into the eye, it may cause retinal damage. Second, the typical intravitreal injection volume, such as 50 μL, can be challenging to measure and deliver accurately, thus potentially putting patients at risk for under- or overdosing. The maximum volume that can be safely administered intravitreally is approximately 100 μL. Larger injection volumes (>100 μL) can cause side effects such as increased intraocular pressure (IOP) (Mandal et al. 2018). These volume restrictions require the development of high-concentration formulations, presenting additional challenges listed in the formulation development section. The posterior eye is comprised of three concentric layers: the sclera (outermost layer), the choroid (middle layer), and the retina (inner layer). The interior of the eye is divided into three chambers: anterior chamber (containing aqueous humor), posterior chamber (between the iris and the lens), and vitreous chamber. The vitreous chamber is directly accessible via intravitreal injection.

This allows drugs to reach the back of the eye in order to treat retinal diseases. The flow of molecules between the eye and the systemic circulation is tightly controlled by the blood-retina barrier (BRB). The BRB has two components: inner BRB (tight junctions between endothelial cells in retinal vessels) and outer BRB comprised of tight junctions between retinal pigmented epithelium (RPE) cells.

Challenges Associated with Formulation Development

Target concentrations for ophthalmic formulations are have been increasing from ≥10 mg/mL to ≥120 mg/mL (Table 3). The need for high-concentration formulations is driven by the inability to accommodate large dose volumes inside the vitreous of the eye without undesirable side effects (Sobolewska et al. 2017). However, exponential increases in protein concentration lead to increases in viscosity.

Table 3 Comparison of dosage regimens of intravitreal anti-VEGF agents for neovascular (Wet) age-related macular degeneration (AMD)

Product	Route of administration	Eye disease indication	Formulation	Recommended dosage and administration
EYLEA (aflibercept)	Intravitreal	Neovascular (wet) age-related macular degeneration (AMD)	40 mg/mL protein in 10 mM phosphate, 40 mM sodium chloride, 5% (w/v) sucrose, 0.03% (w/v) polysorbate 20, pH 6.2	2 mg (0.05 mL) administered by intravitreal injection every 4 weeks (approximately every 28 days, monthly) for the first 3 months, followed by 2 mg (0.05 mL) via intravitreal injection once every 8 weeks (2 months)
BEOVU (brolucizumab-dbll)	Intravitreal	Neovascular (wet) age-related macular degeneration (AMD)	120 mg/mL protein in 10 mM sodium citrate, 5.8% sucrose, 0.02% (w/v) polysorbate 80, pH 7.2	6 mg (0.05 mL of 120 mg/mL solution) monthly (approximately every 25–31 days) for the first three doses, followed by one dose of 6 mg (0.05 mL) every 8–12 weeks
LUCENTIS (ranibizumab)	Intravitreal	Neovascular (wet) age-related macular degeneration (AMD)	6 mg/mL or 10 mg/mL protein in 10 mM histidine HCl, 10% α,α-trehalose dihydrate, 0.01% polysorbate 20, pH 5.5	0.5 mg (0.05 mL) administered by intravitreal injection once a month

Note: Formulation, disease indication, and recommended dosage and administration were taken from literature and package inserts

Ophthalmic formulations need to have a viscosity (generally up to ~15 cP) to enable intravitreal injections using 30G or smaller needles. High viscosities can affect the syringe-ability, including parameters such as time required for complete injection and force required to administer the formulation with suitable needles (Mitragotri et al. 2014). There are several reasons for the increased viscosity of high-concentration protein formulations, which primarily include strong intermolecular interactions between the protein molecules, the pH and ionic strength of the solution, and the interaction of biological molecules with formulation excipients (Singh et al. 2014). Hu et al. presented a comprehensive review of FDA-approved high-concentration monoclonal drug products for subcutaneous administration and effective formulation strategies for reducing viscosity, including buffer types (acetate, citrate, histidine, and phosphate), polyols (sucrose, mannitol, trehalose, and sorbitol), and viscosity reducing amino acid excipients (glycine, proline, and arginine) (Hu et al. 2020). A number of other additives listed in the FDA inactive ingredients guide have been applied to reduce viscosity of high-concentration protein and monoclonal antibody formulations for systemic or subcutaneous delivery (Whitaker et al. 2017; U.S. Food and Drug Administration (Inactive ingredient search for approved drug products) accessed in March 2020). However due to the unique biology of the eye, the application of most of these excipients for ophthalmic intravitreal injections is not yet investigated and approved by the FDA and requires extensive safety evaluation before they can be applied for biologic ophthalmic formulations. Whitaker et al. evaluated the effect of 58 different excipients on the viscosity of high-concentration mAb formulations (Whitaker et al. 2017). Excipients tested included salts, such as sodium chloride, sodium citrate, succinate, ethylenediaminetetraacetic acid (EDTA), etc. Organic solvents (e.g., dimethyl sulfoxide (DMSO) and 5% ethanol), sugars (e.g., mannitol, sorbitol, and dextran), amino acid excipients (e.g., histidine, aspartate, glutamate, isoleucine, and methionine), and cyclodextrins were also tested. Based on their screening, they selected the lead excipients that reduced viscosity levels to <15 cP. Fourteen of the excipients evaluated were shown to have a positive concentration-dependent effect on viscosity reduction; however, the majority of these excipients have not been established as safe for intravitreal administration in clinical settings.

In addition to the challenges associated with viscosity, high-concentration biologics can present challenges with respect to protein aggregation. At high protein concentration, the propensity of intermolecular interactions increases due to crowding of protein molecules leading to product instability and aggregation (Whitaker et al. 2017). Garidel et al. described that high-concentration protein formulations (>100 mg/mL) are meant to show specific solution properties, like high viscosity, opalescence, formation of gel, or the increased tendency of protein-protein interactions, leading to aggregation (Garidel et al. 2017). Two types of instabilities are associated with protein aggregation: conformational and colloidal stability. Conformational stability is based on the difference in free energy between the native and denatured state of the protein, whereas colloidal stability is related to the ability of the molecule to stay as dispersed monomer in solution. Sugars (e.g., sucrose and trehalose), polysaccharides (e.g., dextrans), and surfactants (e.g., pluronics and polysorbates) are the excipients of choice for improving stability of biopharmaceuticals for intravenous or subcutaneous delivery (Mandal et al. 2018; Daugherty and Mrsny 2006; Sasahara et al. 2003; Kerwin 2008). Polysorbate 20 (EYLEA, LUCENTIS, and conbercept) and poloxamer 188 (LUXTURNA) are examples of surfactants used in commercially available intravitreal formulations. Due to the small vitreous volume (~4 mL compared to ~5 L of blood), there are tighter tolerance limits for the levels of surfactants and control in biologic ophthalmic formulations as compared to other injectable formulations due to exposure of excipients to much smaller vitreous volume (~4 mL) compared to total blood volume (~5 L) for humans (U.S. Food and Drug Administration (Inactive ingredient search for approved drug products) accessed in March 2020). However, due to the unique biology of the eye, not all of these are approved by the FDA for intravitreal injection and require extensive safety evaluation before they can be applied for biologic ophthalmic formulations. Intravitreal formulations also require tighter controls for osmolality (target 300 ± 30 mOsm) compared to that of that of formulations for other routes of administration (Marra et al. 2011). Ophthalmic biologics are required to be isoosmotic due to the small vitreous volume and greater sensitivity of eyes. This further restricts the use of excipients and buffer concentrations that can be evaluated for intravitreal formulations.

Although a wide variety of other excipients (e.g., amino acids like glycine and arginine) have been used for stabilizing protein formulations that are administered intravenously or subcutaneously, a much smaller subset of excipients with proven clinical experience is available for use in formulations to be administered intravitreally. In general, lower and more tightly controlled levels of excipients are required by the regulatory agencies.

Clinical Development Challenges

Risks Associated with Repeated Intravitreal Injections

Clinical intravitreal administration of biologics requires repeated dosing, such as monthly dosing, to maintain effective therapeutic levels. The most common side effects of intravitreal injection are subconjunctival hemorrhage, eye pain, and vitreous floaters. More serious adverse effects, such as retinal detachment and endophthalmitis, may also occur albeit at lower incidence rates. A standardized injection procedure can reduce the risks associated with intravitreal injections (Cabrera et al. 2019; Moshfeghi 2008). In addition to the abovementioned outcomes, there is a social burden on the patient and the patient's family who must seek a specialized healthcare professional for the administration of these drugs on a monthly to quarterly basis (Falavarjani and Nguyen 2013; Moshfeghi 2008; Borkar et al. 2018). This could lead to undertreatment and suboptimal outcomes. For example, as shown in Table 3, EYLEA at 2 mg (0.05 mL of a 40 mg/mL solution) is administered approximately every 4 weeks (~28 days) for the first 3 months followed by a maintenance dose every 8 weeks; patients may also be treated with one dose every 12 weeks after one year of effective therapy. BEOVU is dosed at 6 mg (0.05 mL of a 120 mg/mL solution) monthly (approximately every 25–31 days) for the first three doses, followed by one dose of 6 mg every 8–12 weeks.

Healthcare professionals now understand the need for personalized treatment based on individual patient's needs, reducing the frequency of dosing in patients with slow disease progression (Cabrera et al. 2019; Hussain et al. 2017). In the case of anti-VEGF therapy, several treatment regimens have been evaluated: continuous dosing, as-needed (PRN) or treat-and-extend dosing regimens (Schmucker et al. 2015). Although real-life outcomes might show that PRN dosing is preferred by both physicians and patients, the best results have been observed with regularly scheduled doses.

Determination of Pharmacokinetics (PK) After Intravitreal Injection

Information on intravitreal PK of anti-VEGF agents helps to understand the required dosing interval. There are major challenges associated with determining the pharmacokinetic profile of anti-VEGF agents in the eye. One of the major challenges is the invasiveness of sampling vitreous humor. Given these challenges, serum samples are often used to evaluate the pharmacokinetics of intravitreally administered drug molecules. However, the blood retinal barrier restricts the transport of large biologic molecules into systemic circulation. This makes the assessment of vitreous PK parameters from plasma levels challenging due to insufficient or unrepresentative levels of biologics in the plasma. Do et al. reported a clinical investigation where an aflibercept injection was performed in non-vitrectomized eyes with wet AMD in order to evaluate the intraocular and systemic concentrations of free and bound aflibercept in humans with neovascular AMD (Do et al. 2020). The aqueous and plasma samples were tested at baseline and different time points (4 h, 1, 3, 7, 14, and 28 days) after the intravitreal injection. The population size for this study was small ($N = 5$), and the median half-life of aflibercept (with high variation) was determined to be 11 days. The median plasma concentrations of free and bound aflibercept remained transient during the first week (either very low or undetectable) and became undetectable 7 days after the intravitreal injection. Since the systemic VEGF levels could not be determined, it was difficult to predict if aflibercept had any undesired systemic VEGF blockade. In another clinical investigation, Krohne et al. obtained aqueous humor from 18 non-vitrectomized eyes of 18 patients to determine the ocular pharmacokinetics and half-life after single intravitreal injection of 0.5 mg of ranibizumab (Krohne et al. 2012). The study included patients in the age group of 61–85 years that were previously diagnosed with clinically significant cataract and macular edema (secondary to wAMD, diabetic maculopathy, and retinal vein occlusion). Aqueous humor samples obtained during cataract surgery showed that the peak concentrations were reached within in 1 day (C_{max}: 56.1 μg/mL). Elimination half-life, corrected for ocular volume, was determined to be ~7.2 days (Krohne et al. 2012). This half-life was shorter than the median half-life of aflibercept (11 days) mentioned above (Do et al. 2020) and slightly shorter than the half-life of bevacizumab (9.82 days) determined previously (Krohne et al. 2012).

Translation of Results from Clinical Trials to Real-World Outcomes

Mehta et al. described the development of global registries (Fight Retinal Blindness!), which identify real-world safety and efficacy outcomes after intravitreal anti-VEGF treatment. These can be very different from the ones observed during the scheduled and enforced dosing scenario of a clinical trial (Mehta et al. 2018). Some of the aspects that were difficult to assess in a clinical trial included the following: the risk of late reactivation of the disease, different outcomes in the two eyes (e.g., second versus first treated eyes), and the higher chances of posterior capsular rupture during cataract surgery in the patients who have received intravitreal anti-VEGF therapy (Sparrow et al. 2012). Vaze et al. reported that most of the seminal clinical trials for intravitreal (IVT) anti-VEGF treatment for neovascular AMD failed to identify the reactivation of disease after discontinuation of the treatment with IVT ranibizumab and bevacizumab (Vaze et al. 2014). Ninety-one percent of the eyes developed significant decline in visual acuity after the last injection to the time of recurrence. It is important to note that sometimes, unexpected adverse events become apparent during the post-marketing phase, because it includes a larger population, and are more representative real-world clinical situations. The American Society of Retina Specialists (ASRS) Research and Safety in Therapeutics (ReST) Committee reported vasculopathy including occlusive vasculitis after intravitreal injection of BEOVU following its approval for the treatment of wAMD in October 2019 (ASRS BEOVU update, March 30, 2020). Although many examples of increased risk or negative findings that were not observed in the clinical setting have been identified occasionally, more positive observations have occurred, as in the improved visual acuity after switching anti-VEGF agents during treatment for neovascular AMD. Lee et al. showed significant improvement in visual acuity in patients who switched to aflibercept after being chronically treated with ranibizumab (Lee et al. 2018).

Challenges Associated with Masking Treatment Arms in a Clinical Study

Additional challenges associated with clinical studies testing intravitreal drugs are associated with masking of the treatment patients receive. Sham intravitreal injections are unique to ophthalmic drug products and are used for masking a study participant. The procedures mimic the real intravitreal injection, but the needle does not penetrate the eye. In comparison, placebo intravitreal injections administer an agent that has no active ingredient, e.g., saline or 0.9% sodium chloride injection.

Glassman et al. reported a study where 423 patients were randomized to different treatment procedures: prompt laser and sham injections, prompt laser in combination with intravitreal ranibizumab, deferred laser with intravitreal injection of ranibizumab, or prompt laser with intravitreal injections of triamcinolone up to every 16 weeks with intermittent administration of sham injections (Glassman et al. 2012). Participants with both study eyes had one eye randomized to sham injection and laser treatment and the other eye randomized to the treatment group. Sham injections were performed to mimic the real injections and were administered by pressing the hub of the syringe against the conjunctiva. The laser treatment was not masked. The results demonstrate that masking appears to have a lower success rate if one eye received a sham injection and the other eye received a therapeutic injection. In addition, masking was also challenging if the eye sometimes received real therapeutic injection and at other times received sham injection.

Need for Suitable Animal Models

A crucial challenge faced in the development of intraocular products is the translation of therapeutic effectiveness from preclinical animal testing to the clinical evaluation. Rodent and rabbit disease models are the most commonly used due to cost-effectiveness and ease of handling. Laser-induced choroidal neovascularization (CNV) models in mice and nonhuman primates are also used to evaluate vascular permeability blockade. However, these models are limited by the fact that a portion of the injuries self-heal. Therefore, there has been a need to identify animal models with large eyes that can mimic the chronic disposition of human eye diseases.

Cao et al. established a model of chronic retinal neovascularization in rabbits that was comparable in stability and persistence to human disease eyes with pathologic neovascularization (Cao et al. 2018). A single intravitreal injection of DL-α-aminoadipic acid (AAA) induced retinal degeneration and neovascularization in the damaged area. The disease was stable and leakage persisted for 65 weeks. In this model, IVT administration of aflibercept was able to reversibly inhibit vascular permeability. Another study using a similar animal model showed that single intravitreal injection of other anti-VEGF agents (aflibercept, bevacizumab, and ranibizumab) also resulted in complete inhibition of vascular leakage for 8–10 weeks before recurrence to pretreatment levels (Li et al. 2018).

Recent studies have reported the use of laser-induced model in nonhuman primates (NHP) for development of drugs for neovascular AMD. This model employs laser treatment to induce CNV lesions with considerable angiographic leakage, and the model is reproducible (Olvera-Montaño et al. 2019). However, predicting PK and safety based on animal models is challenging due to the small vitreous volume. Also, structural differences between rodent and human VEGF species make it hard to corelate efficacy in humans (Rodrigues et al. 2018). These physiological and anatomical differences pose a challenge in predicting clinical responses.

Additionally, the dose usually needs to be scaled allometrically to establish dose limits for first-in-human (FIH) studies. Therefore, the selection of a suitable animal model is critical for clinical success. Lichtlen et al. developed a laser-induced CNV model in monkeys to evaluate the application of suitable antibody fragments for the prevention of experimental CNV (Lichtlen et al. 2010). CNV was induced by exposing small high-energy laser spots to different areas of retina. Bevacizumab (anti-VEGF), adalimumab (anti TNF-alpha), and ESBA105 were given intravitreally 1 week prior and 1 and 3 weeks after the laser treatment. ESBA105 was also applied topically, separately. Both anti-VEGF and anti-TNF alpha agents significantly reduced CNV after intravitreal injection. The topical ESBA105 also showed reduction in CNV. Shah et al. also used a mouse laser-induced CNV model for neovascular age-related macular degeneration (Shah et al. 2015). The model is reproducible (Olvera-Montaño et al. 2019). However, predicting PK and safety based on animal models is challenging. Additionally, the dose usually needs to be scaled allometrically (which requires a minimum of two species and calculation of clearance) to establish dose limits for first-in-human (FIH) studies.

A key issue in the preclinical/toxicology studies for ranibizumab was the selection of a suitable animal model for anti-VEGF activity (Pennesi et al. 2012). The difference between rodents and human forms of VEGF led to the failure of rodent models for this humanized anti-VEGF antibody (Lu and Adelman 2009). This antibody was then tested in macaques and showed strong evidence of safety and efficacy that formed a basis for human clinical trial (Pennesi et al. 2012).

Formulation and Drug Product Development Considerations For Ophthalmic Biologics

Formulation development is critical for stability, dosage, and delivery of therapeutic molecules. Although the strategy of adapting a platform formulation (based on historical experience) for accommodating the speed to clinic is gaining acceptance in many pharmaceutical companies, the development and characterization of a suitable formulation for new therapeutic molecules to treat ocular diseases is still a recognized challenge due to the unique properties of every new biological molecule. Good Lab Practice (GLP) requirements for stability testing indicate the need to confirm stability and activity during the assessment of formulations. ICH Q8 provides guidance for pharmaceutical development by applying scientific methods and quality risk management, applicable to the development of a product and its manufacturing process. Pharmaceutical development supports the understanding of design space, specifications, and manufacturing controls and provides information that helps manage and mitigate risks to product quality. The principles of quality by design (QbD) suggest that the quality cannot be tested into a product but should be built in the product by an appropriate design of drug molecules, product, and manufacturing process with systematic experiments and studies. Critical formulation and quality attributes (e.g., pH, viscosity, opalescence, aggregates, etc.) and process parameters (like mixing, filtration, and processing time) are generally recognized by evaluating the extent to which their deviation can have an impact on the critical quality attributes, including activity of the drug product (Table 8).

Suitable analytical techniques (e.g., size exclusion chromatography, reverse phase chromatography, enzyme-linked immunosorbent assay (ELISA) binding assay, mass spectrometry, etc.) are applied to evaluate stability and bioactivity. As a part of release and stability testing, it is required to show that the activity of the biologic is maintained. Also, the toxicology evaluation of the biologic agent should be performed in a species in which the activity is confirmed.

Pragmatic Considerations for Formulation Development of Biologics

Apart from the clinical challenges, a number of technical aspects associated with IVT administration of biologics, like viscosity, osmolality, particulate matter, dosing volume, and endotoxin levels, are also critical. The requirements for IVT products are different from the topical ocular or parenteral products (intravenous, subcutaneous, etc.). Extensive drug product development studies are performed to evaluate formulation stability, light sensitivity, excipients, material compatibility, and container closures. These efforts help to define the design space, specifications, and manufacturing controls for the drug product. Additionally, physicochemical properties of the drug substance, compatibility of drug substance with excipients, the type of excipients selected and their effect on aggregation, compatibility, safety, etc. also need to be assessed. This section will cover some of the challenges associated with the pharmaceutical development of ophthalmic biologic drug products.

Proteins are macromolecules with molecular weights ranging from several thousand kDa to a million kDa. These biological molecules, composed of 20 different natural amino acids, have a three-dimensional structure including secondary structures such as alpha helix and beta pleated sheet and higher-order global structures (Manning et al. 1989; Chi et al. 2003b). Due to the dynamic nature, the native-state structure of proteins is the most thermodynamically favored ensemble of many different folded structures. Protein unfolding may only require the equivalent of a few hydrogen-bonding energies, i.e., several kcal/mol. Aggregation of therapeutic proteins from the unfolded and partially unfolded protein species can lead to undesirable consequences, such as loss of efficacy, aggregation, and immunogenicity (Carpenter et al. 2009). Unfolding and aggregation can be caused by the various stresses that proteins are exposed to during manufacturing as well as during drug administration. Some of these stresses include, but are not limited to, factors such as elevated temperatures, extreme pH, light exposure, impurities, air water interfaces, shear stress, and various contact surfaces (Mahler et al. 2009). Therefore, it is critical to develop a formulation that is not only stable during manufacturing and can also withstand all potential stress conditions over a typical 2-year shelf-life.

To achieve a stable formulation, a thorough understanding of the physicochemical properties of protein molecules during the drug candidate selection via developability assessment is a prerequisite (Jarasch et al. 2015). During this early-stage assessment, protein conformational stability, thermal stability, isoelectric point (pI), hydrophobicity, colloidal stability, and critical degradation pathways need to be evaluated. Early-stage assessment is limited by resources and timelines and therefore may not eliminate all risks associated with product development, but it allows the selection of lead candidates with fewer development risks. Techniques that can characterize a limited volume of material in a high-throughput fashion can help to provide a better risk assessment during early product development (Xu et al. 2019). All these parameters define the acceptance criteria for the quality target product profile (QTPP). Figure 3 shows a flowchart describing a typical developability evaluation prior to the selection of a lead drug molecule. With an array of developability information, formulations are built with the purpose of maximizing protein stability and other desired properties. If the candidate molecule does not meet the acceptance criteria for the desired target product profile during the screening, then the next possible candidate maybe evaluated. In the following section, the common formulation attributes such as pH and excipients such as salts, polyols, and surfactants will be reviewed regarding their effect on protein stability for intravitreal formulations. The undesirable impurities, interfaces, and photo/mechanical stresses and their impact on protein formulation will also be discussed.

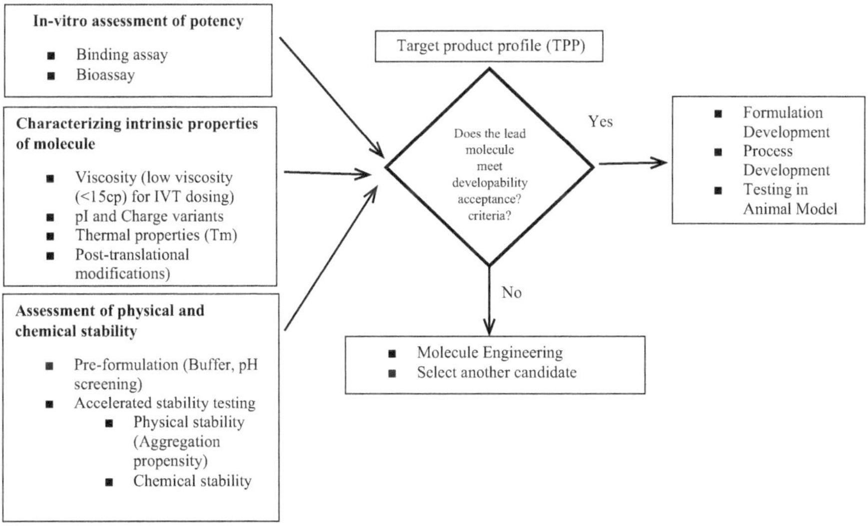

Fig. 3 Developability assessment of biologics

Effect of pH and Buffer Type on Protein Stability

Intraocular formulations require the control of buffer pH and excipients. Most approved intravitreal formulations have a pH in the range of 5–7 (Table 7). The stability of most intravitreal formulations is greatly influenced by the pH of the formulation (Manning et al. 2018), because conformation and activity of a protein have strong dependence on pH. Also, the pH can also strongly affect the formulation viscosity and protein aggregation. However, the impact of pH is different on different formulation attributes. For example, pH dependence is different for deamidation, isomerization, fragmentation, aggregation, colloidal stability, etc. Therefore, it is important to balance the pH and impact the critical stability attributes. The pH of the formulation is often selected based on the aggregation and charge variance of the molecule. Table 4 gives the formulation pH and pI values of the commercially available therapeutic protein molecules administered intravitreally. The pH of the intravitreal formulation (50–100 μL) can affect the pH of the vitreous (Sobolewska et al. 2017). Additionally, the vitreous consisting of collagenous and non-collagenous glycoproteins, glycosaminoglycans, and very low cellularity has a limited capacity to resist changes in pH due to its bicarbonate buffering system. Sobolewska et al. measured the changes in the pH of human vitreous (pH 7.3) caused by intravitreal anti-VEGF agents (aflibercept, ziv-aflibercept, ranibizumab, rituximab, and bevacizumab) (Sobolewska et al. 2017). Each of the anti-VEGF formulations and saline (control) was added at 20% v/v to the vitreous fluid in vitro and gently mixed to form a homogenous solution. The addition of saline increased the pH of the vitreous fluid by 0.46 units, while the addition of the anti-VEGF agents decreased the pH to different extents (e.g., rituximab showed the lowest decrease of 0.06 units, while

Table 4 List of commercially available intravitreal therapeutic protein drugs and the pI values of the proteins

Product	Formulation	pH	pI of the molecule	Reference
EYLEA (aflibercept)	40 mg/mL protein in 10 mM phosphate, 40 mM NaCl, 5% (w/v) sucrose, 0.03% (w/v) polysorbate 20	6.2	8.2[a]	Hirvonen et al. (2016) and Li et al. (2011)
AVASTIN (bevacizumab) (off-label use)	25 mg/mL protein in 50 mM phosphate, 6% (w/v) trehalose, 0.04% (w/v) polysorbate 20	6.2	8.8	Hirvonen et al. (2016) and Li et al. (2011)
LUCENTIS (ranibizumab)	6 mg/mL or 10 mg/mL protein in 10 mM histidine, 10% (w/v) trehalose, 0.01% (w/v) polysorbate 20	5.5	8.8	Hirvonen et al. (2016) and Li et al. (2011)

[a]Isoelectric point (pI) of EYLEA is based on the calculated value from primary sequence (Hirvonen et al. 2016). The actual pI value is variable and depends on the extent of glycosylation and sialylation

bevacizumab, ranibizumab, aflibercept, and ziv-aflibercept showed a decrease of 0.62, 0.41, 0.36, and 0.27 units, respectively).

Whitaker et al. evaluated the contribution of pH on the electroviscous effect of biologic formulations (Whitaker et al. 2017). They showed that the local electrostatic attractions between protein molecules can be modulated by increasing or decreasing the pH, and the ionic charges of the excipient molecules can shield these attractive forces, thereby reducing the intermolecular interactions between protein molecules. Many chemical degradation pathways such as deamidation and isomerization have a strong pH dependence (Ahern and Manning 1992). For example, deamidation for asparagine (Asn) can either be acid or base catalyzed. At acidic pH conditions, the deamidation of Asn is mediated via acid-catalyzed hydrolysis to aspartate. However, at neutral to basic pH, deamidation for Asn ensues via the initial formation of a tetrahedral intermediate (Peters and Trout 2006). Aspartate isomerization is a product of Asn residue to form cyclic imide intermediate and iso-Asp (Aspartate). The influence of pH on the overall physical and chemical stability of proteins suggests the importance of having suitable pH and buffers at different stages of manufacturing, including purification, formulation, and product storage.

The choice of formulation buffer is also crucial and challenging especially for the biologics that will be administered intravitreally and delivered directly into the patient's eye. The choice of buffer and its concentration should consider the following parameters: (1) the capacity of a buffer system to maintain the target pH; (2) the stability of buffer under stress conditions, e.g., histidine buffer might be preferred due to its function as a photostabilizer when exposed to light stress (Du et al. 2018); and (3) the potential effect on protein stability. Different buffers have been used to modulate the pH of biologic products, such as citrate, phosphate, acetate, histidine, succinate, glycine, and tris (Zbacnik et al. 2017). However, intravitreal formulations are limited by the choice of excipients, buffering agents, and their concentration (e.g., glycine may not be preferred due to its potential neurotransmitter effect).

Role of Excipients in Improving Protein Stability

The effect of excipients on protein stability depends on the nature of the excipient itself and its interaction with protein. The common excipients applied in protein formulations can be allocated into the following categories: (1) salt, (2) sugar, (3) surfactant, (4) amino acid, (5) protein, and (6) other polymers (Table 5). Below we will highlight the important categories of excipients and their impact on the overall stability of biologics.

Salt

The ionic strength afforded by salt ions can significantly affect the protein physical stability (Chi et al. 2003b). The three-dimensional structure of proteins is mediated by intramolecular electrostatic interactions, hydrogen bonding, hydrophobic

Table 5 Survey of excipients in commercial drug products approved by the FDA for intravitreal (IVT) dosing

Excipient category	Excipients in commercial drug product approved for IVT dosing
Tonicity (salt)	NaCl
Sugar	Sucrose, trehalose, mannitol
Buffer	Citrate, histidine, phosphate
Amino acid	Arginine
Surfactant	Polysorbate 20, polysorbate 80, poloxamer 188
Polymer	PEG, PLGA

Note: The information listed in the table was taken from MACUGEN, LUCENTIS, EYLEA, BEOVU, JETREA, and OZURDEX® package inserts and literature
PEG polyethylene glycol, *PLGA* poly D,L-lactic-co-glycolic acid

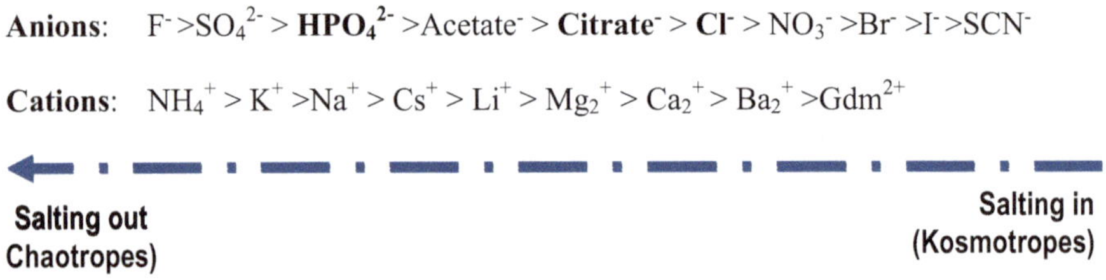

Fig. 4 Classification of ions in the Hofmeister series

interaction, and salt bridges. The electrostatic interactions can be inhibited at high ionic strengths via charge screening (Yadav et al. 2012). Protein conformational stability may be reduced due to decreased electrostatic interactions caused by ionic charges of the excipients. Additionally, the ionic strength and presence of charged ions were found to provide a reduction in protein viscosity by shielding anisotropic short-ranged electrostatic interaction (Chow et al. 2016; Singh et al. 2014). Similarly, the ability of salts to salt in (kosmotropes) or salt out (chaotropes) can influence the colloidal stability of proteins (Fig. 4).

Salts affect the protein stability via ionic strength and potentially preferential interactions (Schneider et al. 2011; Okur et al. 2017). Protein stability, solubility, and unfolding are affected by the presence of salts according to a definite trend called the Hofmeister series. These salts interact differentially with the polar and nonpolar parts of the protein molecules and have differential effects on solubility and stability of the protein molecules. The salts in the Hofmeister series are classified based on their "salting out" (or denaturing) and "salting in" effects. The denaturant effect of some salts, also called chaotropes, is ascribed to the nonspecific interactions between the ions of the salts and the polar regions of the protein. This interaction is more prominent with the protein backbone. These agents disrupt the interactions between solvent and macromolecules and decrease the overall solubility of the protein. Kosmotropes on the other hand increase the interaction between

Table 6 Osmolality of commercially available ophthalmic biologics (Sobolewska et al. 2017)

Product	Formulation	Osmolality (mOsm)
EYLEA (aflibercept)	40 mg/mL protein in 10 mM phosphate, 40 mM NaCl, 5% (w/v) sucrose, 0.03% (w/v) polysorbate 20, pH 6.2	286
BEOVU (brolucizumab-dbll)	120 mg/mL protein in 10 mM sodium citrate, 5.8% sucrose, 0.02% (w/v) polysorbate 80, pH 7.2	[a]190
AVASTIN (bevacizumab) (off-label use)	25 mg/mL protein in 50 mM phosphate, 6% (w/v) trehalose, 0.04% (w/v) polysorbate 20, pH 6.2	182
LUCENTIS (ranibizumab)	6 mg/mL or 10 mg/mL protein in 10 mM histidine, 10% (w/v) trehalose, 0.01% (w/v) polysorbate 20, pH 5.5	289

Note: The information in this table is taken from literature and package inserts
[a]Osmolality for Beovu is calculated and excludes contribution from protein

solvent and proteins, thereby increasing solubility (Tadeo et al. 2009). Figure 4 gives the classification of ions in the Hofmeister series.

Citrate, chloride, and phosphate have salting-in properties and are suitable for local or systemic dosing due to their biological compatibility and are common in commercially available ophthalmic biologic formulations. Ammonium and sulfate ions have a salting-out effect and are commonly used to precipitate protein during purification. Nevertheless, this kind of preferential effect only happens at high salt concentration, which typically exceeds the physiologically relevant osmolality range of 250–350 mOsm/kg. Intravitreal administration formulations are also limited since they have to be iso-osmotic. Table 6 gives the osmolality values for commercially available ophthalmic biologics. The excipient levels in a formulation are further restricted if the protein concentration is high and is contributing to the osmolality of the formulation.

Sugars

Nonreducing sugars, such as sucrose and trehalose, are the most common polyols used as formulation excipients for ophthalmic biologics. Reducing sugars like dextrose have the potential to undergo Maillard reaction with amino acid side-chains, thereby causing coloration of proteins (Carpenter et al. 2002) and instability which are not desirable. The selection of nonreducing sugars is driven in part because they will not react in this manner and cause discoloration or protein adducts.

According to the work by Timasheff et al., sugars are excluded from the protein surface, which consequently reduces the overall surface area of the protein and stabilizes the compacted native-state structure (Timasheff 2002). A linear dependence of protein stabilization on the degree of solution exclusion exists. Therefore, in formulation with suitable sugar concentrations, protein conformational stability is significantly enhanced. For example, many studies have found that the thermal melting temperature of various proteins was significantly improved by sucrose or trehalose during the thermal ramping experiment by biophysical techniques, such as DSC, CD, etc. (Krishnan et al. 2002; Chi et al. 2003a). Sucrose and trehalose have been able to significantly improve the long-term storage stability at refrigerated or elevated temperatures. Due to the more compact structure in the presence of sugars, the chemical degradation pathways of proteins, such as oxidation of surface-exposed methionine or metal-catalyzed oxidation, is reduced (Li et al. 1995, 1996). In some cases, sugars might promote the formation of a multimeric native-state comprising a smaller surface area and compacted structure in proteins, which is even more stable than the native state (Wang and Warne 2010). The osmolality and viscosity contributed by the sugars limit the concentration at which they can be incorporated into injectable formulations. He et al. evaluated the impact of seven different sugars: sucrose, trehalose, fructose, galactose sorbitol, glucose, fructose, and xylose on viscosity of monoclonal antibody formulation (He et al. 2011). It was shown that the increase in viscosity was dependent on the concentration of sugars as well as proteins. Increasing molar amounts of all sugars increased the viscosity of protein solution. This increase was higher with disaccharides like trehalose and sucrose as compared to monosaccharides like xylose and galactose. This is caused by the preferential hydration or exclusion of sugar that mediates interactions between sugar and protein molecules.

Surfactant

During manufacturing, proteins are exposed to a variety of stresses, like agitation, mixing, filtration, and freeze-thaw (discussed under manufacturing considerations and challenges in this book chapter, section "Manufacturing Considerations"). For example, during the mixing and filling unit operations, proteins can partially unfold and adsorb onto air-water interfaces by hydrophobic interactions (Li et al. 2019). Proteins reorient themselves and expose the hydrophobic parts of the protein molecules in order to enhance interaction with the interface, thereby increasing the intermolecular protein-protein interactions and the probability of protein aggregation (Agarkhed et al. 2013; Wang et al. 2008). Additionally, during mixing the air-water interface is constantly refreshed, and cavitation can happen which triggers the formation of hydroxyl free radical causing protein degradation (Torisu et al. 2017). In the case of filtration, the protein may adsorb on the solid membrane and potentially form a gel layer. The continuous build-up of protein will ultimately lead to membrane fouling and loss of filtration efficiency. Also, surface adsorption can result in loss of protein and can be especially critical when the protein therapeutic is formulated at low protein concentration. Depending on the nature of the interface as well as the surface properties of protein molecules, the degree of risk for protein in stability varies.

The capability of a surfactant to stabilize proteins in the presence of stress factors has warranted its popularity in biologics product formulations. However, the quality of these excipients must be considered. Various peroxide impurities have been found in lower grade polysorbate 20 and polysorbate 80. These impurities can result in chemical degradation of proteins (Tomlinson et al. 2015). These surfactants are susceptible to degradation processes such as hydrolysis. The degradation intermediates from polysorbates often trigger protein degradation, aggregation, and subvisible particle formation. Polysorbates are susceptible to autooxidation and hydrolysis at ethylene oxide subunits and hydrolysis of fatty acid bond. These mechanisms lead to the formation of peroxides and short-chain free fatty acids which may affect the stability of the drug product (Kerwin 2008). These reactions are dependent on the solution pH, heat, ultraviolet (UV) light, and the presence of oxygen, peroxides, and metal ions (e.g., copper). Kerwin reported that the maximum hydrolysis for polysorbate 80 occurred at pH less than 3 and more than 7.6 and increased with the increasing temperature (Kerwin 2008). Higher concentration of polysorbates also showed lower rates of degradation. Therefore, the level and purity of surfactants in a formulation should be strictly controlled so that the detrimental effects on protein stability can be minimized.

Non-ionic surfactants have been widely used to mitigate protein instability at interfaces. Currently, the most commonly used non-ionic surfactant for intraocular formulations includes polysorbate 20, polysorbate 80, and poloxamer 188 (Table 7). There are several mechanisms by which these surfactants can enhance protein stability: (1) compete with protein molecules for interacting with the solid surface therefore preventing protein loss at the interface, (2) co-adsorb on the interface with protein molecules as a surfactant-protein complex, (3) directly bind to hydrophobic patches on protein surface to thermodynamically stabilize protein conformation or prevent protein self-interaction, and 4) function as chemical chaperones favoring protein refolding and sterically hindering the intermolecular interactions that cause aggregation (Agarkhed et al. 2013).

Amino Acids and Polymers

Amino acids are versatile excipients that can play multiple roles such as the following: buffer, antioxidant, stabilizer, bulking agent, and viscosity reducer. Glycine, histidine, lysine, isoleucine, and arginine have been evaluated as excipients for improving the stability and reducing viscosity of biologics (Whitaker et al. 2017). However, the use of these excipients is very restricted in ophthalmic intravitreal formulations. The need for the development of stable high-concentration formulations for biologics endorses the use of viscosity reducing agents like arginine hydrochloride and stabilizing excipients like sucrose and mannitol. For example, methionine and histidine have the ability to act as scavengers of free radicals and have been used as antioxidant agents (Cai et al. 1995). Besides amino acids, polymers such as polyethylene glycol (PEG) are commonly screened for the stabilization of proteins. Schlesinger et al. reported a polycaprolactone (PCL)-based reservoir device consisting of an aflibercept formulation with PEG3350 (Schlesinger et al. 2019). The paper proposed a device and design methodology to maintain the stability and sustain the release of aflibercept. The device was well tolerated for

Table 7 Summary of ophthalmic biologics with formulations, storage conditions, container, and dosage forms

Product	Molecule type	Storage condition	Handling precautions	Formulation	Primary container	Dosage form
EYLEA (aflibercept)	Fc fusion protein	Should be refrigerated at 2 °C to 8 °C (36 °F to 46 °F)	Protect from light, do not freeze	40 mg/mL protein in 10 mM phosphate, 40 mM NaCl, 5% (w/v) sucrose, 0.03% (w/v) polysorbate 20, pH 6.2	Vial and PFS	Liquid
AVASTIN (bevacizumab) (off-label use)	IgG1 mAb	AVASTIN vials (100 mg [NDC 50242–060-01] and 400 mg [NDC 50242–061-01]) are stable at 2 °C to 8 °C (36 °F to 46 °F)	AVASTIN vials should be protected from light, do not freeze or shake	25 mg/mL protein in 50 mM phosphate, 6% (w/v) trehalose, 0.04% (w/v) polysorbate 20, pH 6.2	Vial	Liquid
BEOVU (brolucizumab-dbll)	ScFv	Store in the refrigerator between 2 °C to 8 °C (36 °F to 46 °F)	Do not freeze	120 mg/mL protein in 10 mM sodium citrate, 5.8% sucrose, 0.02% (w/v) polysorbate 80, pH 7.2	Vial	Liquid
XEOMIN (incobotulinumtoxinA)	Botulinum toxin	Unopened vials can be stored at room temperature 20 °C to 25 °C (68 °F to 77 °F), in a refrigerator at 2 °C to 8 °C (36 °F to 46 °F), or in a freezer at −20 °C to −10 °C (−4 °F to 14 °F)	None	50–200 unit/mL protein in 0.47% (w/v) sucrose, 0.1% (w/v) HSA	Vial	Lyophilized
JETREA (ocriplasmin)	Truncated human plasmin	Store frozen at or below −4 °F (−20 °C)	Protect the vials from light by storing in the original package until time of use	2.5 mg/mL protein in 5.5 mM citrate, 0.375% (w/v) mannitol, pH 3.1	Vial	Liquid

MACUGEN (pegaptanib sodium)	Pegylated oligonucleotide	Store in the refrigerator at 2 °C to 8 °C (36 °F to 46 °F)	Do not freeze or shake vigorously	3.47 mg/mL pegaptanib sodium in sodium phosphate, sodium chloride, and sodium hydroxide, pH 6–7	PFS	Liquid
LUCENTIS (ranibizumab)	Fab of IgG1 mAb	Should be refrigerated at 2 °C to 8 °C (36 °F to 46 °F)	Do not freeze	6 mg/mL or 10 mg/mL protein in 10 mM histidine, 10% (w/v) trehalose, 0.01% (w/v) polysorbate 20, pH 5.5	Vial and PFS	Liquid
LUXTURNA (voretigene neparvovec)	Vector genome	Store LUXTURNA and diluent frozen at ≤−65 °C	None	5×10^{12} vector genomes (vg) per mL, 180 mM NaCl, 10 mM phosphate, 0.001% poloxamer 188	Vial	Liquid

Note: The formulation and recommended storage condition were taken from literature and package inserts
PFS prefilled syringes, *HAS* human serum albumin

12 weeks when implanted intravitreally into the eyes of nonhuman primates. Excipients in clinical intravitreal formulations require a thorough safety evaluation. In the case of devices or sustained release formulations, the safety evaluation period must extend beyond the duration of drug release and, if biodegradable, include the tolerability of degradation products that would form in situ.

Container Closure Considerations

Type I borosilicate glass vials and single-dose glass prefilled syringes have been used for commercial intravitreal products, like EYLEA, LUCENTIS, MACUGEN, etc. Glass vials provide a convenient primary container for a biologic drug due to the industry's vast experience in fill/finish of vial-based drug products and the well-characterized performance of this type of container. However, for a drug packaged in a glass vial, dose preparation requires the use of administration components, such as disposable syringes and needles that may or may not be included with the drug. These ancillary disposable components could introduce materials into the product that negatively impact drug quality or the patient's safety.

The factors mentioned above caused a shift from vials to prefilled syringes (PFS) as the primary container for anti-VEGF agents. The use of PFS provides multiple advantages including a simpler dose preparation potentially reducing adverse effects related to the use of ancillary disposable components. Drug delivery using a PFS typically involves fewer number of steps (reduced handling) compared to a glass vial and can also reduce overfill and waste of drug product. Intravitreal products like EYLEA and LUCENTIS are available in glass prefilled syringes. Although glass is the most commonly used material for syringe manufacture, syringes manufactured using polymers such as cyclic polyolefin (COP) and copolymers of COP are also available (Waxman and Vilivalam 2017). Syringes for ophthalmic use have been mostly manufactured with a luer-cone tip (covered by a tip-cap) and a needle adapter. The staked needle syringes are generally not considered acceptable for intravitreal injection because of the "dulling" of the needle, which occurs from the needle shield used to seal the syringe. This dulling makes it difficult for the needle to penetrate the sclera of the eye during intravitreal injection. Luer-cone tip syringes are preferred for intravitreal dosing due to lower silicone oil droplet formation. The Standard Care versus Corticosteroid for Retinal Vein Occlusion (SCORE) Study compared the use of staked-on to luer-cone tip needles to reduce the silicone oil droplet formation. The rationale for the modification of syringe type was that in the staked-on needle-type syringe, the silicone oil droplets were extracted from the inner side of the syringe, while the plunger pushed through the barrel. Changing the needle to a luer-cone type created a 50 μL residual space in the needle hub, where the silicone oil droplets would get trapped. Intravitreal silicone oil was observed in 44% of the participants who were exposed to staked-on needle syringes and in 13% of the participants who were exposed to both staked-on and luer-cone-type syringes, whereas no intravitreal silicone oil was observed in the participants exposed to luer-cone-type syringes only (Scott et al. 2009).

The intravitreal injection is limited by the volume (50–100 μL) that can be administered into the eye. The limited injection volume poses an additional challenge in terms of the accuracy of the dose administered. During the administration using a PFS, the clinician first attaches a suitable administration needle and then expels any air bubbles and subsequently primes the needle by expelling the drug product (DP) from the syringe until the piston is aligned with the predetermined dose mark on the syringe barrel. Like any other medication device, it is required that the prefilled syringes deliver consistent and accurate doses upon injection by different users or health practitioners.

Due to the reasons described above, the development of an intravitreal drug product as PFS requires a comprehensive evaluation of syringe and needle. This includes selection of suitable materials for the primary container, siliconization method for the syringe barrel, tungsten content of the syringe, addition of a dose line, filling technology, and mandated terminal sterilization of the syringe exterior. Siliconization of glass syringes and/or pistons and needles is a commonly used approach to lubricate and, therefore, minimize frictional forces introduced by the movement of the piston during drug administration. However, the amount of silicone oil used and the method of siliconization cause the silicone oil to leach into the product and therefore must be considered during container selection. This could not only impact the functionality of the syringe but also induce protein instability and increase subvisible particulate counts, which may exceed the United States Pharmacopeia (USP) requirement for ophthalmology products. The presence of the silicone oil could directly induce protein adsorption onto the oil droplet surface or, in some cases, the migration of nanometer to micrometer-sized silicone oil droplets into the solution, triggering even more severe protein adsorption due to an increase in the surface area or impact on the patient by introducing too much silicone oil into the vitreous when the solution is injected. The amount of silicone oil available to leach into the product solution can be reduced by using a baked-on siliconization and covalent attachment (Gerhardt et al. 2015; Funke et al. 2016). Badkar et al. described the method of evaluating stability of a model protein after spiking it with a silicone oil/water emulsion (Badkar et al. 2011). The protein drug product was spiked with silicone oil in 2 mL type 1 glass vials with 0.5 mg, 1.0 mg, and 1.5 mg equivalent of silicone oil. These vials were then subjected to agitation stress and accelerated (25 °C) and real time (5 °C) conditions to evaluate the impact of silicone oil on aggregation levels of protein. Control arm consisting of mAb formulation vials not spiked with silicone oil was also included in this study. There was no significant difference in the levels of soluble aggregates (measured by size exclusion chromatography (SEC)) for the vials at 25 °C and 5 °C after agitation at 300 rpm for up to 3 days and 5 days, respectively. Varying silicone oil levels did not show any impact on soluble aggregates. The vials at 25 °C showed a slight increase in the levels of soluble aggregates after 1 year of storage, whereas the vials at 5 °C did not show any increase in aggregates as function of time. Aggregation levels were confirmed using SEC, HIAC, and optical imaging. Ophthalmic intravitreal formulations are presented with the challenge of maintaining low subvisible particle count to avoid interference with vision (Table 9). Silicone oil-induced aggregation is more likely to occur in formulations lacking surfactants, at high protein concentration, and formulations with pH close to pI of the protein (Saffell-Clemmer 2017).

Tungsten pins are commonly used during syringe bore formation and maintain the bore opening during luer-cone formation process. Some protein drugs may be sensitive to degradation induced by residual tungsten and tungsten oxides (Seidl et al. 2012; Jiang et al. 2009; Liu et al. 2010). Multiple ways to reduce or eliminate residual tungsten have been tested and include using pins made of other metals or alloys of platinum and nickel, ceramics, silicides, etc. (Prais et al. 2008). Compatibility of the drug must be established with all the components of a syringe delivery system, which involves the syringe barrel, silicone oil, rubber piston, and tip seal. Additionally, appropriate testing for extractables/leachables and testing compatibility of drug with tungsten and silicone oil should be performed (Jenke 2014; Jiang et al. 2009). The tungsten-induced aggregation can either be formulation-specific or related to the type of drug molecule. Tungsten spiking studies are performed to confirm the compatibility with drug product. The drug product solution is spiked with sodium tungstate at 1, 10, and 100 ppm levels and analyzed for aggregates using size exclusion chromatography, turbidity, and particle imaging (e.g., HIAC). Performing such studies are essential to understand the interaction of the intended product with the container closure system over the duration of intended storage at the appropriate temperature. The requirements of monitoring impurities in the drug product from a safety and efficacy standpoint are particularly stringent for ophthalmic applications.

In addition to conventional container and closure systems used for intravitreal injections, significant advancement has been made with devices that are designed to be minimally invasive for intravitreal delivery therapeutics to the back of the eye. The suprachoroidal space which lies between the choroid and the sclera has been used as a reservoir to deliver therapeutics to the posterior segment of the eye (Patel et al. 2011). The unique requirements for delivering therapeutics to the suprachoroidal space require innovative design solutions to engineer injector devices that can accurately deliver ophthalmic drugs to the posterior segment of the eye. Microneedles are one such devices that have the capability to deliver therapeutics to the suprachoroidal space and are often accompanied by proprietary devices (Edelhauser et al. 2014). Microneedles are often accompanied by unique devices for performing the injection, such as the syringe assembly, plunger rod, materials used in the construction of the syringe, needles, etc. These variations in container closure systems and other unique design features need to be considered when developing drug products that meet the stringent regulatory requirements for ophthalmic products. Delivery of therapeutics to the suprachoroidal space is an active area of research and a potentially promising approach to deliver therapeutic molecules to the posterior segment of the eye.

Manufacturing Considerations

Drug Product Development

Shaking, mixing, and diluting at different stages of drug product development, ranging from production and purification to shipping and administration, can cause proteins to experience a variety of stresses leading to the formation of aggregates, such as high-molecular-weight (HMW) species and proteinaceous particulates in various size ranges. These aggregates can form at any stage of manufacturing including upstream and downstream processing as well as during drug product formulation, filling, storage, and administration. The high-molecular-weight species can range from soluble aggregates to large insoluble aggregates. A well-controlled level of soluble aggregates is acceptable when supported by preclinical safety data. However, larger aggregates may lead to the formation of visible and subvisible particles ≥10 μm and ≥25 μm, respectively. For subvisible particulates in drug products, as per the United States Pharmacopeia (USP) for parenteral products, particles with size range ≥10 μm and ≥25 μm should not exceed 6000 particles and 600 particles per container, respectively (with a nominal container content of 100 mL) (Joubert et al. 2011; Cromwell et al. 2006; Paul et al. 2017). The control of subvisible and visible particles is more stringent for ophthalmic preparations due to safety concerns, as required per USP <789>, where particles with size range ≥10 μm, ≥25 μm, ≥50 μm should not exceed 50/mL, 5/mL, and 2/mL, respectively.

Table 9 gives the particle count based on light obscuration experiments as specified for the ophthalmic products in USP <789>. For intraocular application, particles may adversely affect vision and may be seen as vitreous floaters. Also, aggregation of proteins is a major concern as it may impact the safety and efficacy of the drug product. In addition, they can cause opalescence, with elevated levels of aggregates and may lead to visual inspection failures and possible rejection of the manufactured batch.

A concentrating step after ultrafiltration/dia-filtration (UF/DF) is an important step in drug substance manufacturing to achieve high protein levels or concentrations. The greater viscosity of high-concentration protein formulations leads to strong back pressures in the UF/DF systems causing a reduction in the filtration flow, thereby adding additional challenges to process development (Garidel et al. 2017; Shire 2009). Furthermore, high viscosity also affects rheological and syringeability properties of formulations. This is critical for intravitreal ophthalmic formulations, where the formulation is injected through a 30G or smaller needle, requiring greater injection force for more viscous solutions. Other process steps that are likely to be affected during the manufacturing of high-concentration protein formulations include sterile filtration, pumping, and filling.

Ophthalmic biologics are presented with tighter controls due to the increased sensitivity of the eye to impurities like aggregates and foreign contaminants. Therefore, it is more critical to monitor the process parameters and quality attributes for ophthalmic intravitreal products. Conditions that potentially present high risks to intravitreal biologics product quality and stability include freeze-thaw cycles, mixing/agitation, light exposure, and time out of refrigeration (TOR) (Table 8). These conditions can be associated with several stability challenges leading to physical degradation or covalent modification. Physical degradation is commonly

Table 8 Effect of manufacturing unit operations on protein stability

Unit operation	Potential risk factors affecting product quality	Quality attributes to be monitored
Filtration and UF/DF	• Protein or surfactant adsorption on filtration surface • Protein denaturation due to interaction with membrane • Shear stress during pumping and filtration	• Protein concentration • HMW (high molecular weight) • Critical excipients (buffer, surfactant) • Visible and subvisible particulates
Freeze-thaw of bulk DS	• Cryoconcentration • Ice-liquid interface and excipient crystallization causing pH shift • Interaction of protein and formulation excipient with container/closures, leachables, and extractables	• Protein concentration and uniformity • HMW
Formulation compounding, dilution, and mixing	• Shear during mixing • Chances of impurities being introduced from excipients • Formulation hold time • Exposure and adsorption of formulation components to different surfaces	• Protein concentration • HMW, LMW (low molecular weight) • Excipients and surfactants (e.g., polysorbate 20, buffers) • Subvisible particles
DP filling in vials and syringes	• Foaming and dripping during filling • Foreign particles • Light exposure	• Visible and subvisible particulates • HMW • Chemical stability (oxidation)
Sterilization	• Permeability of sterilant across primary container, e.g., piston, tip cap, and syringe barrel for a prefilled syringe product • Compatibility of packaging components with the sterilant	• HMW, LMW • Chemical stability (oxidation) • Visible and subvisible particulates • Functionality of the primary container (break loose/glide force of a PFS product)
Labelling and packaging	• Temperature excursions • Light exposure	• Visible and subvisible particulates • HMW • Chemical stability (oxidation)
Transport	• Shock and drop effects • Pressure changes and vibrations	• Visible and subvisible particulates • HMW

manifested by the association of monomeric units of proteins to dimers or higher-order aggregates, while covalent modification involves protein fragmentation. Unintended exposure of the protein molecule to light stress at any stage of processing, manufacturing, packaging, or storage may lead to chemical degradation and post-translational modifications (e.g., oxidation, deamidation, and covalent cross-linking of amino acid units). This presents an additional challenge for light-sensitive molecules, where the exposure needs to be controlled by covering, shrouding, and using suitable light filters at different stages of drug product manufacturing. Bulk freeze-thaw is an important step in the fill finish of many biological agents. There are several advantages to freezing a drug substance: improved stability, reduced chance of microbial growth, and the elimination of agitation-induced foaming during transportation. However, uncontrolled freezing-thawing procedures can lead to cryoconcentration of proteins and potentially precipitation of excipients or buffer components (Lashmar et al. 2007). Additionally, the thawing process can lead to ice recrystallization and potentially cause protein denaturation at ice-liquid interface (Strambini and Gabellieri 1996; Rathore and Rajan 2008). Filtration and filling also expose the drug product to various surfaces, like filtration membrane, vials, stoppers, and plungers, which can sometimes cause adsorption of proteins and surfactants leading to protein unfolding and denaturation. These quality defects can cause impurities, protein instability, and aggregates, leading to clinical adverse effects like increased intraocular pressure and inflammation after IVT injection. Tighter process controls on freeze-thawing, determining the TOR limit for miniating stability of the ophthalmic drug product, protection from light, and implementing other suitable process controls on unit operations can reduce quality defects.

Sterilization Considerations

Sterility of an ophthalmic drug product can be ensured by aseptic processing, terminal sterilization, or both. Several published guidance documents, such as ISO 13408-1:2008 and reference literature, provide an overview of aseptic processing (Lambert and Martin 2013) and will not be discussed here. Sterilization of the exterior of the container closure containing ophthalmic drugs is a unique requirement and has been discussed in documents such as 21 CFR 200.50. Terminal sterilization is usually performed on the final packaging of the product. For example, a prefilled syringe packaged in a blister pack that is permeable to the sterilant allows for sterilization of the external surface of the syringe. The blister pack should be designed to ensure package integrity post-sterilization to keep the contents sterile through the shelf-life of the product. Steam and radiation (gamma or e-beam) sterilization are usually not practical for biologic drugs due to their potential impact on product quality. Alternate methods suitable for terminal sterilization of biological ophthalmic products involve gaseous agents (sterilants) that can achieve sterility without exposure to high temperature. To protect the drug product, the primary container components used must be impermeable to the chosen sterilizing agent. Gaseous sterilizing agents include vaporized hydrogen peroxide (VHP) (McEvoy and Rowan 2019), ethylene oxide (EtO), nitrogen oxides (NO and NO_2), formaldehyde, chlorine oxides, etc. (Kohli 2019). Factors that should be considered during the development of a terminal sterilization process for an ophthalmic biologic drug product include:

1. Permeability of sterilant across the primary container, e.g., piston, tip cap, and syringe barrel for a prefilled syringe product.
2. Compatibility of packaging components with the sterilant.
3. Effect of sterilization process on the quality of the product and functionality (break loose/glide force of a PFS product) of the primary container.
4. Duration of sterilization and batch size.
5. Efficiency of sterilization (sterility assurance level [SAL]) and concentration of sterilant required to achieve SAL.
6. Ease of removal of residual sterilant after sterilization and environmental, health, and safety requirements, etc.

A typical sterilization process includes introduction of the gaseous sterilant, under suitable combination of temperature and pressure, into a sealed chamber with the packaged drug product (e.g., PFS containing drug product contained in a blister pack) for a duration that achieves SAL followed by purging/flushing of the excess sterilant out of the sterilizing chamber. SAL is defined as the probability of a single viable microorganism occurring on an item after sterilization (Bush and Gertzman 2016). A sterilization process needs to be validated to demonstrate its effectiveness. As stated in ISO 11135, the kinetics of inactivation of a pure culture of microorganisms by a sterilization treatment can be described by the exponential relationship between the number of microorganisms surviving and the extent of treatment. Invasive medical devices like PFS require SAL level of 10^{-6} (ANSI/AAMI/ISO 11135). A documented procedure for obtaining, recording, and interpreting the results is required to establish a validated sterilization method that will consistently achieve sterility without any adverse impact on the device (e.g., PFS) and the product. The validation process for sterilization usually requires testing the growth of biological indicators (such as spores test) on the product after exposure to fractional, half, and full cycle of sterilization procedure. Further, the sterilization validation is reviewed periodically to ensure product sterility. It is also important to understand and demonstrate the impact of the sterilization process on product quality which may be performed by monitoring the stability of drug product post-sterilization.

Regulatory Expectations

The growing need for improved therapeutics for the treatment of eye disorders and safety concerns using off-label products for IVT injection has led to increased scrutiny from regulatory agencies. Bevacizumab is approved by the FDA for various forms of cancer and is being used off-label for wAMD as it is substantially less expensive than the FDA-approved anti-VEGF drugs for that disease. This drug is administered intravitreally from a single vial by either withdrawing multiple injections or by dividing the vial into aliquots of multiple syringes and vials via a compounding pharmacy (Saoji et al. 2018). Considering that the vials for this drug do not contain preservatives or antimicrobial agents, there is a high risk of microbial contamination due to poor aseptic techniques or supply of spurious vials. In 2011, the FDA warned clinicians that repackaged intravitreal injections of bevacizumab caused a cluster of serious streptococci endophthalmitis infections and blindness in Miami, Florida. Similar cases also emerged at Veterans Affairs Hospital in Tennessee among patients treated for wAMD (https://www.medscape.com/viewarticle/781039).

BEOVU is another intravitreal anti-VEGF agent that came under the scrutiny of regulatory agencies. This drug product showed cases of vasculopathy, including occlusive vasculitis during Phase 3 of clinical trials. The side effects are not listed as related to the formulation in the label. The ASRS Research and Safety in Therapeutics (ReST) recommended careful evaluation of anterior and posterior segment of the eyes for any signs of inflammation prior to dosing of the drug product. The committee also suggested the use of steroids for the treatment of inflammation and complete resolution of inflammation prior to restarting treatment with BEOVU. Another incidence that shows tight regulations on ophthalmic biologics is the FDA's recent rejection of Allergan's (AbbVie's) biologics license application (BLA) for Abicipar pegol for neovascular age-related macular degeneration (nAMD). The safety concerns raised by the FDA indicated increased rate of intraocular inflammation following the administration of Abicipar pegol at a dose of 2 mg/0.06 mL (https://www.biospace.com/article/fda-reject-s-allergan-s-abicipar-pegol-for-age-related-macular-degeneration). These two examples (BEOVU and Abicipar pegol) suggest that the new modalities require thorough safety assessment.

Ophthalmic formulations are scrutinized by the regulatory agencies for the requirement of sterilization of the final product including ophthalmic preparation and dispensers (21 CFR 200.50). The guideline states that any ophthalmic preparations that fall below their professed standard of purity or quality may be unsafe. It requires that the ophthalmic preparations should be sterile at the time of preparation, filling, and sealing. The container is required to be sealed such that the contents cannot be used without breaking the seal. Section "Sterilization Considerations" shows that the ophthalmic preparations can be sterilized using various methods such as steam or gaseous agents. However, as per the regulatory guidance, the sponsor is also required to demonstrate the stability of the drug product after exposure to sterilization processes. The stability assessment will include monitoring critical quality attributes of the drug product for the desired shelf-life in the packaged form post-sterilization to confirm any change in purity, color, introduction of leachables from the container, etc.

The complexities of development and manufacturing for biologics, container closure systems, and devices present distinctive challenges for the pharmaceutical industry in the development and commercialization of ophthalmic drug products.

The devices are regulated as per 21 CFR 886 guidance, and the sponsors are required to demonstrate the safety and effectiveness of the device in clinic. In addition, for ophthalmic devices, the Medicines and Healthcare products Regulatory Authority (MHRA) has established a yellow card reporting scheme for identifying and collecting information on adverse drug reactions, defective medicines, counterfeit medicines, or suspected problems and incidents involving medical devices. Identification of problems and accumulation of adverse incident reports may require significant changes to the device or its withdrawal.

The regulatory guideline ICH Q8 defines the requirements for the development of pharmaceutical products through the application of scientific approaches and quality risk management. The critical processes are identified, and formulation components are characterized by evaluating the extent to which their variation can affect the critical quality attributes of the product. This approach provides a higher degree of understanding of the manufacturing processes and controls and facilitates drafting specifications and formulating risk-based regulatory decisions. Product specifications facilitate monitoring the general quality attributes of the product during manufacturing and stability. The physicochemical parameters of the formulation should be within the acceptable limits.

Product quality tests are performed to confirm the integrity of the dosage form, while performance tests evaluate drug release, potency, bioactivity, and other attributes that can be corelated to the in vivo performance. Together, the tests confirm the identity, safety, potency, purity, impurity content, and identification of impurities in the drug product. All ophthalmic drug products are required to be essentially free of visible particles. The subvisible particle count is regulated as per USP <789>. The particle content limit for ophthalmic preparations compared to injectable drug products is smaller (Table 9). Such stringent requirements make the development of ophthalmic preparations challenging from a manufacturing stand point, which involves exposure of biologic molecules to various stresses (described in section "Manufacturing Considerations"). Revised USP <771> Ophthalmic Products—Quality Tests lists the drug product quality universal tests and specific tests. This section highlights the drug product quality tests for biologic drug products, as specified in USP <771>:

1. Identification: This test establishes the identity of the molecule(s) present in a drug product. The analytical method or test should be specific to the molecule and should be able to distinguish between closely related molecules. Raman spectroscopy, mass spectrometry, immune assay, and chromatography are some of the tests that can identify the specific drugs.
2. Protein concentration assay: This test is desired to be specific and stability indicating. The test determines the strength (content) of the molecule in the drug

Table 9 Light obscuration particle count for ophthalmic solutions as specified in USP <789>

Diameter	≥10 μm	≥25 μm	≥50 μm
Number of particles per mL	50/mL	5/mL	2/mL

product. Biologics concentrations are usually determined by UV-based spectrophotometric methods and chromatography methods, such as reversed-phase high-performance liquid chromatography (HPLC).

3. Impurities: An effective quality control strategy is needed to ensure the level of impurities is minimized and the impact on protein stability is under control. Process-related impurities present as by-products or organic/inorganic impurities and can cause product instability and toxicity. It is essential to test for the presence of these impurities at specific manufacturing steps. A number of impurities can be introduced into therapeutic protein during various stages of manufacturing. After cell culture production, multiple purification steps such as Protein A affinity column, ionic exchange column, and size exclusion column are necessary to remove cell culture media and impurities, such as host cell DNA and host cell proteins (HCP). Host cell proteins (HCP) are common impurities that come from early stages in the manufacturing and are a major concern for protein stability and safety. The presence of HCP in the final products can also lead to severe anaphylaxis shock and, therefore, the levels of HCP are more restrictive and controlled for ophthalmic administration. Other impurities like metals might come from the cell culture media or manufacturing equipment. The presence of these impurities can lead to the oxidation of molecules and protein aggregation in drug product. Specific tests like inductively coupled plasma mass spectrometry (ICP-MS) can identify trace levels of metal impurities in biologic formulations and can help track the source of contamination. Low contamination levels are stringent requirements of intravitreal biologics.
4. Endotoxins: Another major impurity concern for intravitreal formulations. Bacterial endotoxins are lipopolysaccharides released from gram-negative bacteria and are resistant to heat and therefore difficult to eliminate from the sterile drug product and surgical instruments. Eyes are sensitive and susceptible to inflammation on exposure to endotoxin impurities; thereby, high endotoxin level can affect the safety of the drug product after intravitreal injection. Bantseev et al. evaluated the safety of single 50 μL intravitreal injections with increasing endotoxin levels ranging from 0.01 to 0.75 endotoxin units/eye (EU/eye) in Dutch belted rabbits (Bantseev et al. 2017). Intravitreal injections of endotoxin levels $\geq$0.05 EU/eye showed dose-dependent anterior segment inflammation, while intravitreal injections of 0.01 EU/eye showed no observable adverse effect.
5. Potency: This is an important assay that determines the biological activity of the drug product. In vitro cell-based assays or ELISA binding assays are preferred for rapid screening of products. However, for some products, in vivo testing of the drug product is established. For example, human growth hormone has been shown to increase the weight in rats after daily administration of hormone (Baid et al. 2011; Hoffman and Pisch-Heberle 2000).
6. Purity: Establishing the purity of the biologic drug product is of paramount importance as it defines the safety and efficacy of the drug product. The most commonly used methods for assessing stability of proteins during screening of candidate molecules and release of drug product include assessing aggregation levels and size variance using size exclusion chromatography (SEC) and SEC

coupled to multi-angle light scattering (SEC-MALS), respectively, charge variance using imaged capillary isoelectric focusing (iCIEF), and other post-translational modifications using liquid chromatography/mass spectrometry (LC/MS).

Future Outlook

Despite the success of anti-VEGF therapy in the treatment of serious back of the eye disorders, there are several practical limitations like high cost, need for frequent injections, and failure to maintain an adequate response in some patients (Al-Khersan et al. 2019). For chronic ocular diseases such as diabetic macular edema (DME) and diabetic retinopathy (DR), optimal efficacy requires that injections are performed every 4 weeks for LUCENTIS (ranibizumab) or 8–12 weeks for EYLEA (aflibercept). These treatment regimens can be burdensome for the patient, often resulting in undertreatment and loss of visual acuity. Thus, a common patient-centric goal for ophthalmic researchers in both academia and industry is to develop alternate therapies and delivery technologies that extend the therapeutic exposure time or sustain efficacy to lengthen dosing intervals and improve durability (Cao et al. 2019).

Great efforts by investigators in academia and industry are devoted to prolonging the duration of action of current anti-VEGF therapies. Innovation is being driven by increasing the retinal bioavailability by reducing the size of the protein, enhancing efficacy by binding to one or more validated targets, or increasing the vitreal half-life. At the time of this writing, a nonbiodegradable polymer-based intraocular device with a refillable reservoir (Port Delivery System [PDS], Genentech/Roche) is being investigated in late-stage clinical trials to deliver 100 mg/mL ranibizumab to treat wAMD (https://clinicaltrials.gov/ct2/show/NCT03677934). An important consideration when developing devices is demonstrating long-term safety of permanent implants and the ability to safely remove them, in case of serious adverse events. For example, vitreous hemorrhage occurred in 50% (11 of 22) of the first PDS-treated patients in the Phase 2 study, and after altering the surgical implant procedure, vitreous hemorrhage was reduced to 4.5% of PDS-treated patients (7 of 157) (Campochiaro et al. 2010). Biodegradable polymeric intraocular devices have shown promise to deliver proteins in vitro as well as in vivo in the eye for 12 weeks. These polycaprolactone-based reservoir devices were well tolerated in rabbit and nonhuman primates (Lance et al. 2016; Schlesinger et al. 2019). As an alternative to reservoir-based devices, Ocular Therapeutix is developing a biodegradable PEG-based hydrogel as a sustained release formulation for use with small molecules or proteins to treat back of the eye diseases (press release October 13, 2016). Polymer antibody conjugates are being developed by Kodiak Science, whereby a phosphorylcholine biopolymer is conjugated to a mAb via a site-specific linkage, creating a ~950 kDa molecule. This phosphorylcholine biopolymer extends the antibody half-life within the vitreous (press release October 11, 2019).

Biologic modalities to treat serious ocular diseases extend beyond proteins. LUXTURNA (voretigene neparvovec-rzyl) is a one-time gene therapy for biallelic RPE65 mutation-associated retinal dystrophy (LUXTURNA Label 2020). RPE65 gene replacement treatment happens via subretinal injection of an adeno-associated virus serotype 2 (AAV2). At the moment, gene therapy is the only pharmacologic treatment option for this disease. A subretinal injection is more invasive than intravitreal or topical delivery, but it is considered acceptable as a one-time procedure. REGENXBIO is conducting clinical trials to deliver adeno-associated virus (AAV)-mediated anti-VEGF treatment (RGX-314) by subretinal injection, and Phase 2 studies are expected to begin in 2020 (https://clinicaltrials.gov/ct2/show/NCT03066258). Suprachoroidal delivery is a less invasive route of administration being investigated for gene therapy. Delivery of adeno-associated virus serotype 8 (AAV8) induced widespread green fluorescent protein (GFP) transgene expression in the RPE and photoreceptors in animal models (Campochiaro et al. 2010). Gene therapy or nucleic acids (siRNA, antisense oligonucleotide, etc.) are examples of an avenue of research and development that can potentially deliver meaningful benefit to patients beyond what is possible with current protein-based modalities or drug delivery systems.

Combination therapy of anti-VEGF agents like antiplatelet derived growth factor (anti-PDGF) and anti-angiopoetin-2 (Ang-2) has been evaluated in the past. Anti-PDGF therapy is aimed at stripping the pericytes that cover blood vessels, thereby making them more susceptible to anti-VEGF agents. Rinucumab is an anti-PDGF receptor-β antibody that was co-formulated with aflibercept (REGN2176–3, Regeneron) and tested to treat patients with wet AMD. Similarly, nesvacumab is an anti-Ang2 mAb that was co-formulated with aflibercept (REGN910–3, Regeneron) and tested to treat patients with wet AMD and DME. Faricimab (FOVISTA®, Roche/Genentech) is a bispecific antibody that simultaneously binds to Ang-2 and VEGF A and was evaluated in clinic for wet AMD and DME (https://clinicaltrials.gov/ct2/show/study/NCT03622580). Phase 3 of clinical testing failed to meet the primary end point of mean change in visual acuity at 12 months. The combination treatment using 1.5 mg FOVISTA (pegpleranib) anti-PDGF therapy in combination with EYLEA (aflibercept) or AVASTIN (bevacizumab) anti-VEGF therapy did not show superiority over monotherapy with aflibercept for the treatment of wAMD (https://www.businesswire.com/news/home/20170814005286/en/Ophthotech-Announces-Results-Phase-3-Trial-Fovista).

For chronic diseases, topical delivery is an attractive noninvasive route of administration, but there are many biological barriers that impede delivery to the back of the eye to treat retinal diseases (Rodrigues et al. 2018). PanOptica recently completed Phase 1/2 clinical trials with a topical eye drop formulation of an anti-VEGF (PAN-90806) small molecule to treat wAMD as a monotherapy (PanOptica Press Release 10 Oct 2019). To achieve higher concentrations on the ocular surface to treat postoperative inflammation and pain, Kala Pharmaceuticals developed a nanoparticle that enhances the movement of drug particles through the tear film enabling twice daily dosing (press release August 23, 2018).

Biological barriers to the target tissue should be considered when selecting the ideal route of administration or drug delivery system. In addition to the unique challenges of intraocular delivery as described in this chapter, any new excipient, device material, or polymer must be thoroughly evaluated along with the active molecule for safety during preclinical and clinical development. Therefore, an innovative and comprehensive approach is required to address the unique challenges of developing novel drug products to treat serious ocular diseases.

Acknowledgments The authors would like to thank David Hauss, Carl Romano, Bibiana Iglesias, Candace Drumma, Michele Leone, James Sutherland, and Karen Chu for reviewing the manuscript. Special thanks to Bibiana Iglesias for providing the image showing the anatomy of the eye.

References

Agarkhed M, O'Dell C, Hsieh M-C, Zhang J, Goldstein J, Srivastava A. Effect of polysorbate 80 concentration on thermal and photostability of a monoclonal antibody. AAPS PharmSciTech. 2013;14(1):1–9. https://doi.org/10.1208/s12249-012-9878-0.

Ahern TJ, Manning MC. Stability of protein pharmaceuticals. Part A. New York: Plenum Press; 1992.

Al-Khersan H, Hariprasad SM, Salehi-Had H. Dexamethasone and anti-VEGF combination therapy for the treatment of diabetic macular edema. Ophthalmic Surg Lasers Imaging Retina. 2019;50(1):4–7. https://doi.org/10.3928/23258160-20181212-01.

Avastin_Package_Insert. 2004. https://www.accessdata.fda.gov/drugsatfda_docs/label/2009/125085s0169lbl.pdf.

Badkar A, Wolf A, Bohack L, Kolhe P. Development of biotechnology products in pre-filled syringes: technical considerations and approaches. AAPS PharmSciTech. 2011;12(2):564–72. https://doi.org/10.1208/s12249-011-9617-y.

Baid R, Tyagi P, Durazo SA, Kompella UB. Protein drug delivery and formulation development. In: Kompella UB, Edelhauser HF, editors. Drug product development for the back of the eye. Boston, MA: Springer US; 2011. p. 409–48. https://doi.org/10.1007/978-1-4419-9920-7_17.

Bantseev V, Miller PE, Bentley E, Schuetz C, Streit TM, Christian BJ, Farman C, Booler H, Thackaberry EA. Determination of a no-observable effect level for endotoxin following a single intravitreal administration to Dutch belted rabbits. Invest Ophthalmol Vis Sci. 2017;58(3):1545–52. https://doi.org/10.1167/iovs.16-21356.

Beovu_Package_Insert. 2019. https://www.accessdata.fda.gov/drugsatfda_docs/label/2019/761125s000lbl.pdf.

Borkar DS, Obeid A, Su DC, Storey PP, Gao X, Regillo CD, Kaiser RS, Garg SJ, Hsu J. Endophthalmitis rates after bilateral same-day intravitreal anti-vascular endothelial growth factor injections. Am J Ophthalmol. 2018;194:1–6. https://doi.org/10.1016/j.ajo.2018.06.022.

Bruno BJ, Miller GD, Lim CS. Basics and recent advances in peptide and protein drug delivery. Ther Deliv. 2013;4(11):1443–67. https://doi.org/10.4155/tde.13.104.

Bush K, Gertzman AA. Chapter 5. Process development and manufacturing of human and animal acellular dermal matrices. In: Albanna MZ, Holmes Iv JH, editors. Skin tissue engineering and regenerative medicine. Boston: Academic Press; 2016. p. 83–108. https://doi.org/10.1016/B978-0-12-801654-1.00005-X.

Cabrera FJ, Wang DC, Reddy K, Acharya G, Shin CS. Challenges and opportunities for drug delivery to the posterior of the eye. Drug Discov Today. 2019;24(8):1679–84. https://doi.org/10.1016/j.drudis.2019.05.035.

Cai Q, Takemura G, Ashraf M. Antioxidative properties of histidine and its effect on myocardial injury during ischemia/reperfusion in isolated rat heart. J Cardiovasc Pharmacol. 1995;25(1):147–55. https://doi.org/10.1097/00005344-199501000-00023.

Campochiaro PA, Heier JS, Feiner L, Gray S, Saroj N, Rundle AC, Murahashi WY, Rubio RG. Ranibizumab for macular edema following branch retinal vein occlusion: six-month primary end point results of a phase III study. Ophthalmology. 2010;117(6):1102–1112.e1101. https://doi.org/10.1016/j.ophtha.2010.02.021.

Cao J, MacPherson TC, Iglesias BV, Liu Y, Tirko N, Yancopoulos GD, Wiegand SJ, Romano C. Aflibercept action in a rabbit model of chronic retinal neovascularization: reversible inhibition of pathologic leakage with dose-dependent duration. Invest Ophthalmol Vis Sci. 2018;59(2):1033–44. https://doi.org/10.1167/iovs.17-22897.

Cao Y, Samy KE, Bernards DA, Desai TA. Recent advances in intraocular sustained-release drug delivery devices. Drug Discov Today. 2019;24(8):1694–700. https://doi.org/10.1016/j.drudis.2019.05.031.

Carpenter JF, Chang BS, Garzon-Rodriguez W, Randolph TW. Rational design of stable lyophilized protein formulations: theory and practice. In: Carpenter JF, Manning MC, editors. Rational design of stable protein formulations: theory and practice. Boston, MA: Springer US; 2002. p. 109–33. https://doi.org/10.1007/978-1-4615-0557-0_5.

Carpenter JF, Randolph TW, Jiskoot W, Crommelin DJA, Middaugh CR, Winter G, Fan Y-X, Kirshner S, Verthelyi D, Kozlowski S, Clouse KA, Swann PG, Rosenberg A, Cherney B. Overlooking subvisible particles in therapeutic protein products: gaps that may compromise product quality. J Pharm Sci. 2009;98(4):1201–5. https://doi.org/10.1002/jps.21530.

Chi EY, Krishnan S, Kendrick BS, Chang BS, Carpenter JF, Randolph TW. Roles of conformational stability and colloidal stability in the aggregation of recombinant human granulocyte colony-stimulating factor. Protein Sci. 2003a;12(5):903–13. https://doi.org/10.1110/ps.0235703.

Chi EY, Krishnan S, Randolph TW, Carpenter JF. Physical stability of proteins in aqueous solution: mechanism and driving forces in nonnative protein aggregation. Pharm Res. 2003b;20(9):1325–36. https://doi.org/10.1023/a:1025771421906.

Chow CK, Allan BW, Chai Q, Atwell S, Lu J. Therapeutic antibody engineering to improve viscosity and phase separation guided by crystal structure. Mol Pharm. 2016;13(3):915–23. https://doi.org/10.1021/acs.molpharmaceut.5b00817.

Cromwell ME, Hilario E, Jacobson F. Protein aggregation and bioprocessing. AAPS J. 2006;8(3):E572–9. https://doi.org/10.1208/aapsj080366.

Daugherty AL, Mrsny RJ. Formulation and delivery issues for monoclonal antibody therapeutics. Adv Drug Deliv Rev. 2006;58(5–6):686–706. https://doi.org/10.1016/j.addr.2006.03.011.

Do DV, Rhoades W, Nguyen QD. Pharmacokinetic study of intravitreal aflibercept in humans with neovascular age-related macular degeneration. Retina (Philadelphia, PA). 2020;40(4):643–7. https://doi.org/10.1097/iae.0000000000002566.

Du C, Barnett G, Borwankar A, Lewandowski A, Singh N, Ghose S, Borys M, Li ZJ. Protection of therapeutic antibodies from visible light induced degradation: use safe light in manufacturing and storage. Eur J Pharm Biopharm. 2018;127:37–43. https://doi.org/10.1016/j.ejpb.2018.02.007.

Edelhauser HF, Verhoeven RS, Burke B, Struble CB, Patel SR. Intraocular distribution and targeting of triamcinolone acetonide suspension administered into the suprachoroidal space. Invest Ophthalmol Vis Sci. 2014;55(13):5259.

Eylea_Package_Insert. 2011. https://www.accessdata.fda.gov/drugsatfda_docs/label/2017/125387s054lbl.pdf.

Falavarjani KG, Nguyen QD. Adverse events and complications associated with intravitreal injection of anti-VEGF agents: a review of literature. Eye (Lond). 2013;27(7):787–94. https://doi.org/10.1038/eye.2013.107.

Funke S, Matilainen J, Nalenz H, Bechtold-Peters K, Mahler HC, Friess W. Silicone migration from baked-on silicone layers. Particle characterization in placebo and protein solutions. J Pharm Sci. 2016;105(12):3520–31. https://doi.org/10.1016/j.xphs.2016.08.031.

Garidel P, Kuhn AB, Schäfer LV, Karow-Zwick AR, Blech M. High-concentration protein formulations: how high is high? Eur J Pharm Biopharm. 2017;119:353–60. https://doi.org/10.1016/j.ejpb.2017.06.029.

Gerhardt A, Nguyen BH, Lewus R, Carpenter JF, Randolph TW. Effect of the siliconization method on particle generation in a monoclonal antibody formulation in pre-filled syringes. J Pharm Sci. 2015;104(5):1601–9. https://doi.org/10.1002/jps.24387.

Glassman AR, Stockdale CR, Beck RW, Baker C, Bressler NM, Diabetic Retinopathy Clinical Research N. Evaluation of masking study participants to intravitreal injections in a randomized clinical trial. Arch Ophthalmol. 2012;130(2):190–4. https://doi.org/10.1001/archophthalmol.2011.387.

He F, Woods CE, Litowski JR, Roschen LA, Gadgil HS, Razinkov VI, Kerwin BA. Effect of sugar molecules on the viscosity of high concentration monoclonal antibody solutions. Pharm Res. 2011;28(7):1552–60. https://doi.org/10.1007/s11095-011-0388-7.

Heier JS, Brown DM, Chong V, Korobelnik JF, Kaiser PK, Nguyen QD, Kirchhof B, Ho A, Ogura Y, Yancopoulos GD, Stahl N, Vitti R, Berliner AJ, Soo Y, Anderesi M, Groetzbach G, Sommerauer B, Sandbrink R, Simader C, Schmidt-Erfurth U. Intravitreal aflibercept (VEGF trap-eye) in wet age-related macular degeneration. Ophthalmology. 2012;119(12):2537–48. https://doi.org/10.1016/j.ophtha.2012.09.006.

Hirvonen LM, Fruhwirth GO, Srikantha N, Barber MJ, Neffendorf JE, Suhling K, Jackson TL. Hydrodynamic radii of ranibizumab, aflibercept and bevacizumab measured by time-resolved phosphorescence anisotropy. Pharm Res. 2016;33(8):2025–32. https://doi.org/10.1007/s11095-016-1940-2.

Hoffman H, Pisch-Heberle S. Analytical methods and stability testing of biopharmaceuticals. In: McNally EJ, editor. Protein formulation and delivery. New York: Informa Healthcare; 2000. p. 71–110.

Hu Q, Jiang B, Liu D, Tang X, Daly T, Shameem M. Chapter 14: Practical considerations in high concentration formulation development for monoclonal antibody drug products. In: Jameel F, Skoug JW, Nesbitt RR, editors. Development of biopharmaceutical drug-device products. Cham: Springer International; 2020. p. 343–72. https://doi.org/10.1007/978-3-030-31415-6_14.

Hussain RM, Hariprasad SM, Ciulla TA. Treatment burden in neovascular AMD: visual acuity outcomes are associated with anti-vegf injection frequency. Ophthalmic Surg Lasers Imaging Retina. 2017;48(10):780–4. https://doi.org/10.3928/23258160-20170928-01.

Jarasch A, Koll H, Regula JT, Bader M, Papadimitriou A, Kettenberger H. Developability assessment during the selection of novel therapeutic antibodies. J Pharm Sci. 2015;104(6):1885–98. https://doi.org/10.1002/jps.24430.

Jenke DR. Extractables and leachables considerations for prefilled syringes. Expert Opin Drug Deliv. 2014;11(10):1591–600. https://doi.org/10.1517/17425247.2014.928281.

Jetrea_Package_Insert. 2012. https://www.accessdata.fda.gov/drugsatfda_docs/label/2012/125422s000lbl.pdf.

Jiang Y, Nashed-Samuel Y, Li C, Liu W, Pollastrini J, Mallard D, Wen ZQ, Fujimori K, Pallitto M, Donahue L, Chu G, Torraca G, Vance A, Mire-Sluis T, Freund E, Davis J, Narhi L. Tungsten-induced protein aggregation: solution behavior. J Pharm Sci. 2009;98(12):4695–710. https://doi.org/10.1002/jps.21778.

Joubert MK, Luo Q, Nashed-Samuel Y, Wypych J, Narhi LO. Classification and characterization of therapeutic antibody aggregates. J Biol Chem. 2011;286(28):25118–33. https://doi.org/10.1074/jbc.M110.160457.

Kerwin BA. Polysorbates 20 and 80 used in the formulation of protein biotherapeutics: structure and degradation pathways. J Pharm Sci. 2008;97(8):2924–35. https://doi.org/10.1002/jps.21190.

Kim YC, Chiang B, Wu X, Prausnitz MR. Ocular delivery of macromolecules. J Control Release. 2014;190:172–81. https://doi.org/10.1016/j.jconrel.2014.06.043.

Kohli R. Chapter 11. Applications of gas-phase cleaning for removal of surface contaminants. In: Kohli R, Mittal KL, editors. Developments in surface contamination and cleaning: applications of cleaning techniques. Amsterdam: Elsevier; 2019. p. 423–77. https://doi.org/10.1016/B978-0-12-815577-6.00011-6.

Krishnan S, Chi EY, Webb JN, Chang BS, Shan D, Goldenberg M, Manning MC, Randolph TW, Carpenter JF. Aggregation of granulocyte colony stimulating factor under physiological conditions: characterization and thermodynamic inhibition. Biochemistry. 2002;41(20):6422–31. https://doi.org/10.1021/bi012006m.

Krohne TU, Liu Z, Holz FG, Meyer CH. Intraocular pharmacokinetics of ranibizumab following a single intravitreal injection in humans. Am J Ophthalmol. 2012;154(4):682–686.e682. https://doi.org/10.1016/j.ajo.2012.03.047.

Lambert B, Martin J. Chapter III. 1.2. Sterilization of implants and devices. In: Ratner BD, Hoffman AS, Schoen FJ, Lemons JE, editors. Biomaterials science. 3rd ed. San Diego, CA: Academic Press; 2013. p. 1339–53. https://doi.org/10.1016/B978-0-08-087780-8.00125-X.

Lance KD, Bernards DA, Ciaccio NA, Good SD, Mendes TS, Kudisch M, Chan E, Ishikiriyama M, Bhisitkul RB, Desai TA. In vivo and in vitro sustained release of ranibizumab from a nanoporous thin-film device. Drug Deliv Transl Res. 2016;6(6):771–80. https://doi.org/10.1007/s13346-016-0298-7.

Lashmar UT, Vnanderburgh M, Little SJ. Bulk freeze-thaw of macromolecules—effect of cryoconcentration on their formulation and stability. BioProcess Int. 2007;5(6):44–54.

Leader B, Baca QJ, Golan DE. Protein therapeutics: a summary and pharmacological classification. Nat Rev Drug Discov. 2008;7(1):21–39. https://doi.org/10.1038/nrd2399.

Lee CS, Kim AJ, Baughman D, Egan C, Bailey C, Johnston RL, Natha S, Khan R, Brand C, Akerele T, McKibbin M, Downey L, Al-Husainy S, Lee AY, Tufail A. Visual acuity improvement when switching from ranibizumab to aflibercept is not sustained. Retina (Philadelphia, PA). 2018;38(5):951–6. https://doi.org/10.1097/IAE.0000000000001637.

Li S, Schoneich C, Borchardt RT. Chemical instability of protein pharmaceuticals: mechanisms of oxidation and strategies for stabilization. Biotechnol Bioeng. 1995;48(5):490–500. https://doi.org/10.1002/bit.260480511.

Li S, Patapoff TW, Nguyen TH, Borchardt RT. Inhibitory effect of sugars and polyols on the metal-catalyzed oxidation of human relaxin. J Pharm Sci. 1996;85(8):868–72. https://doi.org/10.1021/js9504550.

Li SK, Liddell MR, Wen H. Effective electrophoretic mobilities and charges of anti-VEGF proteins determined by capillary zone electrophoresis. J Pharm Biomed Anal. 2011;55(3):603–7. https://doi.org/10.1016/j.jpba.2010.12.027.

Li J, Krause ME, Chen X, Cheng Y, Dai W, Hill JJ, Huang M, Jordan S, LaCasse D, Narhi L, Shalaev E, Shieh IC, Thomas JC, Tu R, Zheng S, Zhu L. Interfacial stress in the development of biologics: fundamental understanding, current practice, and future perspective. AAPS J. 2019;21(3):44. https://doi.org/10.1208/s12248-019-0312-3.

Li Y, Busoy JM, Zaman BAA, Tan QSW, Tan GSW, Barathi VA, Cheung N, Wei JJ, Hunziker W, Hong W, Wong TY, Cheung CMG. A novel model of persistent retinal neovascularization for the development of sustained, anti-VEGF therapies, Exp Eye Res. 2018;174:98–106. https://doi.org/10.1016/j.exer.2018.05.027. Epub 2018 May 28.

Lichtlen P, Lam TT, Nork TM, et al. Relative contribution of VEGF and TNF-α in the cynomolgus laser-induced CNV model: comparing the efficacy of bevacizumab, adalimumab, and ESBA105. Invest Ophthalmol Vis Sc. 2010;51(9):4738–4745. https://doi.org/10.1167/iovs.09-4890.

Liu W, Swift R, Torraca G, Nashed-Samuel Y, Wen ZQ, Jiang Y, Vance A, Mire-Sluis A, Freund E, Davis J, Narhi L. Root cause analysis of tungsten-induced protein aggregation in pre-filled syringes. PDA J Pharm Sci Technol. 2010;64(1):11–9.

Liu L, Ammar DA, Ross LA, Mandava N, Kahook MY, Carpenter JF. Silicone oil microdroplets and protein aggregates in repackaged bevacizumab and ranibizumab: effects of long-term storage and product mishandling. Invest Ophthalmol Vis Sci. 2011;52(2):1023–34. https://doi.org/10.1167/iovs.10-6431.

Lowes R. Compounding pharmacy crackdown continues with avastin recall. 2013. https://www.medscape.com/viewarticle/781039.

Lu F, Adelman RA. Are intravitreal bevacizumab and ranibizumab effective in a rat model of choroidal neovascularization? Graefes Arch Clin Exp Ophthalmol. 2009;247(2):171–7. https://doi.org/10.1007/s00417-008-0936-y.

Lucentis_Package_Insert. 2006. https://www.accessdata.fda.gov/drugsatfda_docs/label/2014/125156s105lbl.pdf.

Luxturna Label. 2020. https://www.fda.gov/media/109906/download.

Macugen_Package_Insert. 2004. https://www.accessdata.fda.gov/drugsatfda_docs/label/2011/021756s018lbl.pdf.

Mahler HC, Friess W, Grauschopf U, Kiese S. Protein aggregation: pathways, induction factors and analysis. J Pharm Sci. 2009;98(9):2909–34. https://doi.org/10.1002/jps.21566.

Mandal A, Pal D, Agrahari V, Trinh HM, Joseph M, Mitra AK. Ocular delivery of proteins and peptides: challenges and novel formulation approaches. Adv Drug Deliv Rev. 2018;126:67–95. https://doi.org/10.1016/j.addr.2018.01.008.

Manning MC, Patel K, Borchardt RT. Stability of protein pharmaceuticals. Pharm Res. 1989;6(11):903–18.

Manning MC, Liu J, Li T, Holcomb RE. Rational design of liquid formulations of proteins. Adv Protein Chem Struct Biol. 2018;112:1–59. https://doi.org/10.1016/bs.apcsb.2018.01.005.

Marra MT, Khamphavong P, Wisniecki P, Gukasyan HJ, Sueda K. Solution formulation development of a VEGF inhibitor for intravitreal injection. AAPS PharmSciTech. 2011;12(1):362–71. https://doi.org/10.1208/s12249-011-9591-4.

McEvoy B, Rowan NJ. Terminal sterilization of medical devices using vaporized hydrogen peroxide: a review of current methods and emerging opportunities. J Appl Microbiol. 2019;127(5):1403–20. https://doi.org/10.1111/jam.14412.

Mehta H, Tufail A, Daien V, Lee AY, Nguyen V, Ozturk M, Barthelmes D, Gillies MC. Real-world outcomes in patients with neovascular age-related macular degeneration treated with intravitreal vascular endothelial growth factor inhibitors. Prog Retin Eye Res. 2018;65:127–46. https://doi.org/10.1016/j.preteyeres.2017.12.002.

Mitragotri S, Burke PA, Langer R. Overcoming the challenges in administering biopharmaceuticals: formulation and delivery strategies. Nat Rev Drug Discov. 2014;13(9):655–72. https://doi.org/10.1038/nrd4363.

Moshfeghi AA. Rate of endophthalmitis after anti-VEGF intravitreal injection. Retina Today. 2008:75–6.

Nicholson BP, Schachat AP. A review of clinical trials of anti-VEGF agents for diabetic retinopathy. Graefes Arch Clin Exp Ophthalmol. 2010;248(7):915–30. https://doi.org/10.1007/s00417-010-1315-z.

Okur HI, Hladílková J, Rembert KB, Cho Y, Heyda J, Dzubiella J, Cremer PS, Jungwirth P. Beyond the Hofmeister series: ion-specific effects on proteins and their biological functions. J Phys Chem B. 2017;121(9):1997–2014. https://doi.org/10.1021/acs.jpcb.6b10797.

Olvera-Montaño O, Baiza-Duran L, Quintana-Hau JD, Quiñonez-Alvarado MG, Zeng W, Gong L, Muñoz-Villegas P. Comparing the efficacy of an anti-human VEGF—a neutralizing antibody versus bevacizumab on a laser-induced choroidal neovascularization (CNV) rhesus monkey model. Drug Des Devel Ther. 2019;13:3813–21. https://doi.org/10.2147/DDDT.S219350.

PanOptica Press Release. https://www.panopticapharma.com/wp-content/uploads/2019/10/PanOptica-PAN-90806-Data-Summary-Release-Clean-Final-Oct-09-2019.pdf. Accessed 10 Oct 2019.

Patel SR, Lin AS, Edelhauser HF, Prausnitz MR. Suprachoroidal drug delivery to the back of the eye using hollow microneedles. Pharm Res. 2011;28(1):166–76. https://doi.org/10.1007/s11095-010-0271-y.

Paul AJ, Bickel F, Röhm M, Hospach L, Halder B, Rettich N, Handrick R, Herold EM, Kiefer H, Hesse F. High-throughput analysis of sub-visible mAb aggregate particles using automated fluorescence microscopy imaging. Anal Bioanal Chem. 2017;409(17):4149–56. https://doi.org/10.1007/s00216-017-0362-2.

Pennesi ME, Neuringer M, Courtney RJ. Animal models of age related macular degeneration. Mol Aspects Med. 2012;33(4):487–509. https://doi.org/10.1016/j.mam.2012.06.003.

Peters B, Trout BL. Asparagine deamidation: pH-dependent mechanism from density functional theory. Biochemistry. 2006;45(16):5384–92. https://doi.org/10.1021/bi052438n.

Prais AW, Cocheteux B, Cortes A, Delabie P, Wales E, Luedtke RD, Schaecher R, Vulliet D. Method for reducing or eliminating residue in a glass container and a glass container made in accordance therewith. Google Patents, US 2008/0103438, 2008.

Rathore N, Rajan RS. Current perspectives on stability of protein drug products during formulation, fill and finish operations. Biotechnol Prog. 2008;24(3):504–14. https://doi.org/10.1021/bp070462h.

Rodrigues GA, Lutz D, Shen J, Yuan X, Shen H, Cunningham J, Rivers HM. Topical drug delivery to the posterior segment of the eye: addressing the challenge of preclinical to clinical translation. Pharm Res. 2018;35(12):245. https://doi.org/10.1007/s11095-018-2519-x.

Saffell-Clemmer W. Impact of PFS and filling process selection on biologic product stability. BioPharm Int. 2017;30(2):28–32.

Saoji K, Trehan H, Narayanan R, Verma L. A study on the contamination of injection bevacizumab on storage of multidose vials. Indian J Ophthalmol. 2018;66(2):252–5. https://doi.org/10.4103/ijo.IJO_969_16.

Sasahara K, McPhie P, Minton AP. Effect of dextran on protein stability and conformation attributed to macromolecular crowding. J Mol Biol. 2003;326(4):1227–37. https://doi.org/10.1016/s0022-2836(02)01443-2.

Schlesinger EB, Bernards DA, Chen HH, Feindt J, Cao J, Dix D, Romano C, Bhisitkul RB, Desai TA. Device design methodology and formulation of a protein therapeutic for sustained release intraocular delivery. Bioeng Transl Med. 2019;4(1):152–63. https://doi.org/10.1002/btm2.10121.

Schmucker CM, Rücker G, Sommer H, Virgili G, Loke YK, Oeller P, Agostini H, Ehlken C. Treatment as required versus regular monthly treatment in the management of neovascular age-related macular degeneration: a systematic review and meta-analysis. PLoS One. 2015;10(9):e0137866. https://doi.org/10.1371/journal.pone.0137866.

Schneider CP, Shukla D, Trout BL. Arginine and the Hofmeister series: the role of ion–ion interactions in protein aggregation suppression. J Phys Chem B. 2011;115(22):7447–58. https://doi.org/10.1021/jp111920y.

Scott IU, Oden NL, VanVeldhuisen PC, Ip MS, Blodi BA, Antoszyk AN. SCORE Study Report 7: incidence of intravitreal silicone oil droplets associated with staked-on vs luer cone syringe design. Am J Ophthalmol. 2009;148(5):725–732.e727. https://doi.org/10.1016/j.ajo.2009.06.004.

Seidl A, Hainzl O, Richter M, Fischer R, Böhm S, Deutel B, Hartinger M, Windisch J, Casadevall N, London GM, Macdougall I. Tungsten-induced denaturation and aggregation of epoetin alfa during primary packaging as a cause of immunogenicity. Pharm Res. 2012;29(6):1454–67. https://doi.org/10.1007/s11095-011-0621-4.

Shah RS, Soetikno BT, Lajko M, Fawzi AA. A mouse model for laser-induced choroidal neovascularization. J Vis Exp. 2015;106:e53502. https://doi.org/10.3791/53502.

Sharma A, Reddy P, Kuppermann BD, Bandello F, Lowenstein A. Biosimilars in ophthalmology: "Is there a big change on the horizon?". Clin Ophthalmol. 2018;12:2137–43. https://doi.org/10.2147/OPTH.S180393.

Shire SJ. Formulation and manufacturability of biologics. Curr Opin Biotechnol. 2009;20(6):708–14. https://doi.org/10.1016/j.copbio.2009.10.006.

Singh SN, Yadav S, Shire SJ, Kalonia DS. Dipole-dipole interaction in antibody solutions: correlation with viscosity behavior at high concentration. Pharm Res. 2014;31(9):2549–58. https://doi.org/10.1007/s11095-014-1352-0.

Sobolewska B, Heiduschka P, Bartz-Schmidt K-U, Ziemssen F. pH of anti-VEGF agents in the human vitreous: low impact of very different formulations. Int J Retina Vitreous. 2017;3:22. https://doi.org/10.1186/s40942-017-0075-x.

Sparrow JM, Taylor H, Qureshi K, Smith R, Birnie K, Johnston RL. The Cataract National Dataset electronic multi-centre audit of 55,567 operations: risk indicators for monocular visual acuity outcomes. Eye (Lond). 2012;26(6):821–6. https://doi.org/10.1038/eye.2012.51.

Strambini GB, Gabellieri E. Proteins in frozen solutions: evidence of ice-induced partial unfolding. Biophys J. 1996;70(2):971–6. https://doi.org/10.1016/s0006-3495(96)79640-6.

Tadeo X, López-Méndez B, Castaño D, Trigueros T, Millet O. Protein stabilization and the Hofmeister effect: the role of hydrophobic solvation. Biophys J. 2009;97(9):2595–603. https://doi.org/10.1016/j.bpj.2009.08.029.

Timasheff SN. Protein-solvent preferential interactions, protein hydration, and the modulation of biochemical reactions by solvent components. Proc Natl Acad Sci U S A. 2002;99(15):9721–6. https://doi.org/10.1073/pnas.122225399.

Tomlinson A, Demeule B, Lin B, Yadav S. Polysorbate 20 degradation in biopharmaceutical formulations: quantification of free fatty acids, characterization of particulates, and insights into the degradation mechanism. Mol Pharm. 2015;12(11):3805–15. https://doi.org/10.1021/acs.molpharmaceut.5b00311.

Torisu T, Maruno T, Hamaji Y, Ohkubo T, Uchiyama S. Synergistic effect of cavitation and agitation on protein aggregation. J Pharm Sci. 2017;106(2):521–9. https://doi.org/10.1016/j.xphs.2016.10.015.

U.S. Food and Drug Administration. Inactive ingredient search for approved drug products. https://www.accessdata.fda.gov/scripts/cder/iig/indexcfm, https://www.accessdata.fda.gov/scripts/cder/iig/indexcfm. Accessed Mar 2020.

Usmani SS, Bedi G, Samuel JS, Singh S, Kalra S, Kumar P, Ahuja AA, Sharma M, Gautam A, Raghava GPS. THPdb: database of FDA-approved peptide and protein therapeutics. PLoS One. 2017;12(7):e0181748. https://doi.org/10.1371/journal.pone.0181748.

Vaze A, Fraser-Bell S, Gillies M. Consequences of long-term discontinuation of vascular endothelial growth factor inhibitor therapy in the patients with neovascular age-related macular degeneration. Acta Ophthalmol. 2014;92(8):e697–8. https://doi.org/10.1111/aos.12417.

Wang W. Lyophilization and development of solid protein pharmaceuticals. Int J Pharm. 2000;203(1–2):1–60.

Wang W, Warne NW. Approaches to managing protein aggregation in product development. In: Aggregation of therapeutic proteins; 2010. p. 333–65. https://doi.org/10.1002/9780470769829.ch8.

Wang W, Wang YJ, Wang DQ. Dual effects of Tween 80 on protein stability. Int J Pharm. 2008;347(1–2):31–8. https://doi.org/10.1016/j.ijpharm.2007.06.042.

Waxman L, Vilivalam VD. A comparison of protein stability in prefillable syringes made of glass and plastic. PDA J Pharm Sci Technol. 2017;71(6):462–77. https://doi.org/10.5731/pdajpst.2016.007146.

Whitaker N, Xiong J, Pace SE, Kumar V, Middaugh CR, Joshi SB, Volkin DB. A formulation development approach to identify and select stable ultra–high-concentration monoclonal antibody formulations with reduced viscosities. J Pharm Sci. 2017;106(11):3230–41. https://doi.org/10.1016/j.xphs.2017.06.017.

Wiersbitzky S, Ballke EH, Burghardt R, Spangenberg U, Joswig T, Baufeld W, Ordt HA, Paul W. Long-term study of various immunologic functions in children with chronic nonspecific lung diseases. Z Erkr Atmungsorgane. 1985;164(3):241–53.

Wong TY, Scott IU. Clinical practice. Retinal-vein occlusion. N Engl J Med. 2010;363(22):2135–44. https://doi.org/10.1056/NEJMcp1003934.

Xeomin_Package_Insert. 2010. https://www.accessdata.fda.gov/drugsatfda_docs/label/2010/125360lbl.pdf.

Xu Y, Wang D, Mason B, Rossomando T, Li N, Liu D, Cheung JK, Xu W, Raghava S, Katiyar A, Nowak C, Xiang T, Dong DD, Sun J, Beck A, Liu H. Structure, heterogeneity and developability assessment of therapeutic antibodies. MAbs. 2019;11(2):239–64. https://doi.org/10.1080/19420862.2018.1553476.

Xue K, Zhao X, Zhang Z, Qiu B, Tan QSW, Ong KH, Liu Z, Parikh BH, Barathi VA, Yu W, Wang X, Lingam G, Hunziker W, Su X, Loh XJ. Sustained delivery of anti-VEGFs from thermogel depots inhibits angiogenesis without the need for multiple injections. Biomater Sci. 2019;7(11):4603–14. https://doi.org/10.1039/C9BM01049A.

Yadav S, Shire SJ, Kalonia DS. Viscosity behavior of high-concentration monoclonal antibody solutions: correlation with interaction parameter and electroviscous effects. J Pharm Sci. 2012;101(3):998–1011. https://doi.org/10.1002/jps.22831.

Zajac-Pytrus HM, Pilecka A, Turno-Krecicka A, Adamiec-Mroczek J, Misiuk-Hojlo M. The dry form of age-related macular degeneration (AMD): the current concepts of pathogenesis and prospects for treatment. Adv Clin Exp Med. 2015;24(6):1099–104. https://doi.org/10.17219/acem/27093.

Zbacnik TJ, Holcomb RE, Katayama DS, Murphy BM, Payne RW, Coccaro RC, Evans GJ, Matsuura JE, Henry CS, Manning MC. Role of buffers in protein formulations. J Pharm Sci. 2017;106(3):713–33. https://doi.org/10.1016/j.xphs.2016.11.014.

Part IV
Specialty Products and Generic Development

Implantable Devices to Treat Ophthalmic Conditions: Drug Delivery Systems

Susan Lee, Laszlo Romoda, and Michael Robinson

Abstract Drug delivery for ophthalmic diseases is often difficult due to the anatomy and physiology of the eye. Limited absorption and the layers of ocular tissue inhibit adequate amounts of topically applied ocular drops from reaching the target tissues of the eye. Similarly, the blood–aqueous and blood–retinal barriers prevent an adequate amount of intravenously administered drugs from reaching the eye. While intravitreal injection can bypass some of these limitations, these injections are invasive and must be performed often. Implantable drug delivery devices have been designed to overcome many of the difficulties associated with other ocular treatment options. These devices can be biodegradable, soluble, or nonbiodegradable and can be placed in different parts of the eye depending on the target tissue. For example, they can be placed intracamerally, intravitreally, or in the cul-de-sac of the eye. To optimize the ocular drug delivery system, various factors such as water solubility, toxicity, efficacy, and chemical and biopharmaceutical properties are taken into consideration. Although these drug delivery devices also have limitations, they are a great alternative that can reduce treatment burden, provide more targeted delivery, and minimize systemic side effects.

Keywords Ocular · Implants · Inserts · Intravitreal · Intracameral · Sustained release · Drug delivery

S. Lee (✉) · L. Romoda · M. Robinson
Allergan, an AbbVie company, Irvine, CA, USA
e-mail: susan.s.lee@abbvie.com; laszlo.romoda@abbvie.com

S. Neervannan, U. B. Kompella (eds.), *Ophthalmic Product Development*, AAPS Advances in the Pharmaceutical Sciences Series 37,
https://doi.org/10.1007/978-3-030-76367-1_18

Abbreviations

AMD	Age-related macular degeneration
BRVO	Branch RVO
CRVO	Central RVO
DME	Diabetic macular edema
EVA	Ethylene vinyl acetate
GA	Geographic atrophy
PAH	Polyanhydrides
PCL	Polycaprolactones
PDS	Port delivery system
PGA	Polyglycolic acid
PLA	Polylactic acid
PLGA	Polylactic-co-glycolic acid
POE	Polyorthoesters
PVA	Polyvinyl alcohol
RMP	Replenish MicroPump
RVO	Retinal vein occlusion
VEGF	Vascular endothelial growth factor

Introduction

Treatment of ophthalmic diseases is inherently challenging due to the anatomy and physiology of the eye. Although topical medical therapy is the most common initial administration route, physical and dynamic barriers, such as ocular tissue layers, tear turn over, and clearance mechanisms, restrict delivery to the anterior and posterior segments of the eye and lower bioavailability (Edelhauser et al. 2010; Gaudana et al. 2010). A small amount of a topically applied dose is absorbed into the anterior segment, and only a percentage of that can move into the posterior segment (Lee and Robinson 2001). This makes treating diseases of the posterior segment such as diabetic retinopathy, age-related macular degeneration (AMD), and retinal vein occlusion (RVO), particularly challenging due to the greater diffusional distance (Lee and Robinson 2001; Yasukawa et al. 2006). In order to maintain minimum therapeutic concentrations, topical ocular drugs need to be administered frequently, often resulting in poor patient compliance (Hermann et al. 2010; Nordmann et al. 2010; Salyani and Birt 2005). Oral or IV administration can also be inadequate because the blood–aqueous and blood–retinal barriers limit the entry of the drug into the eye from the bloodstream (Cunha-Vaz 1979). Although some can be effective, they are accompanied by significant systemic side effects (Farkouh et al. 2016). In the 1980s, the first clinical studies of intravitreal (IVT) injected medications, 5-fluorouracil and ganciclovir, were conducted with the objective of providing more effective and targeted therapy. The injected substances were found to move throughout the vitreous fairly readily. The success led to the evaluation of a number of other

IVT agents and the pace of development of new applications for IVT injection has continued (Jager et al. 2004). Although intravitreal injection is the preferred method for drug delivery to the posterior segment, there are drawbacks such as its invasiveness, the requirement for frequent administration, and its association with retinal detachment, cataract, endophthalmitis, and increased intraocular pressure (IOP) (Jager et al. 2004; Macha and Mitra 2002).

Due to these challenges, biomaterial and biotechnology advances have translated into the development of unique alternative treatment approaches in the form of implantable drug devices (Ghate and Edelhauser 2006; Patel et al. 2013; Yasukawa et al. 2006). This chapter presents a comprehensive view of controlled drug delivery devices that are either approved/marketed or currently in development and highlights their distinguishing features.

Ocular Drug Delivery Systems

Alternatives to Typical Ophthalmic Drug Administration Routes

Implantable ocular drug delivery systems can provide localized, controlled drug release over an extended period of time, which lowers the number of treatments required, possibly reducing the number of physician office visits and overall treatment costs. Systemic side effects are minimized due to the blood–retina/blood–aqueous barriers, high peak drug concentrations associated with pulsed dosing that are avoided, an increase in treatment adherence, and a potential reduction in treatment-related AEs (Ghate and Edelhauser 2008; Lee et al. 2011; Patel et al. 2013; Streilein 2003; Yasukawa et al. 2006). Some of the advantages and limitations of the different ocular drug delivery methods are summarized in Table 1. Various factors such as water solubility, toxicity, efficacy, chemical and biopharmaceutical properties, as well as the anatomy of the targeted ocular tissue are taken into consideration when optimizing the ocular drug delivery system (Avitabile et al. 2001). These devices can be inserted in the conjunctival cul-de-sac or punctum of the eye, or implanted in the subconjunctival, episcleral, intravitreal, or intracameral regions, and are categorized according to their degree of invasiveness and route of administration (Fig. 1) (Jervis 2017). In this chapter, we focus on biodegradable and non-biodegradable polymer-based implants, implantable drug pumps, and ocular inserts that are currently approved/marketed or under investigation.

Biodegradable Ocular Drug Delivery Systems

Biodegradable systems have been developed for intracameral and intravitreal placement and typically consist of a homogeneous polymeric pellet containing a therapeutic agent/drug and biodegradable polymer (Lee et al. 2010). The medication is released from the pores and skeleton as the hydrophobic polymer is converted into a

Table 1 Limitations and benefits of ocular drug delivery methods (Edelhauser et al. 2010; Gaudana et al. 2010; Ghate and Edelhauser 2006; Lee et al. 2002, 2010; Lee and Robinson 2001; Yasukawa et al. 2006)

Method	Advantages	Disadvantages
Topical administration	• Convenient to use • Low cost	• Limited uptake • Short acting • Poor adherence to therapy • Systemic side effects
Intravitreal injection	• Targeted delivery	• Invasive/inconvenient/short lasting • Potential adverse events related to injection
Systemic administration	• Ease of administration	• Limited ocular penetration • Systemic toxicity
Biodegradable implants	• Sustained targeted drug delivery • Do not require removal • Improves patient compliance	• May require invasive surgery • Potential for erratic drug release • Shorter duration of action vs. nonbiodegradable • Cost
Nonbiodegradable implants	• Sustained targeted drug delivery • Controlled drug release profile • Improves patient compliance	• May require invasive surgery • Require removal • Potential adverse events related to implantation or removal surgery • Cost

water-soluble material via bulk or surface erosion. Bulk erosion occurs when water penetrates the matrix at a rate that is greater than that of polymer hydrolysis, while surface erosion occurs when water penetrates at a rate lower than that of polymer hydrolysis (Fig. 2) (Kuno and Fujii 2010; Robinson and Whitcup 2012). The drug release rate is influenced by the type of polymer and its biodegradation kinetics, the total surface area of the implant, and the percentage of loaded drug (Anderson and Shive 1997). Some advantages of the biodegradable system are that they do not elicit permanent chronic foreign body reactions or require surgical removal after the drug supply has been exhausted (Conway 2008; Jain 2000; Kimura and Ogura 2001). Some limitations of the biodegradable system that could impact effectiveness and/or safety are variable drug release profiles, a shorter duration of action when compared with nonbiodegradable implants, and destruction of the loaded medication before release is possible (Alhalafi 2017; Lee 2015; Miller et al. 1977). Biodegradable implants in clinical use and under investigation are summarized in Table 2.

Common biodegradable polymers include polylactic acid (PLA), polyglycolic acid (PGA), polylactic-co-glycolic acid (PLGA), polycaprolactones (PCL), and polyanhydrides (PAH), which break down into nontoxic byproducts and can be eliminated safely by the human body (Lee et al. 2011). They are aliphatic polyesters that belong to the "bulk-eroding" class of polymers and are the most widely studied of the synthetic biodegradable polymers, due to their mechanical properties, low immunogenicity and toxicity, excellent biocompatibility, and predictable biodegradation kinetics (Jain 2000). They are degraded predominantly by nonenzymatic

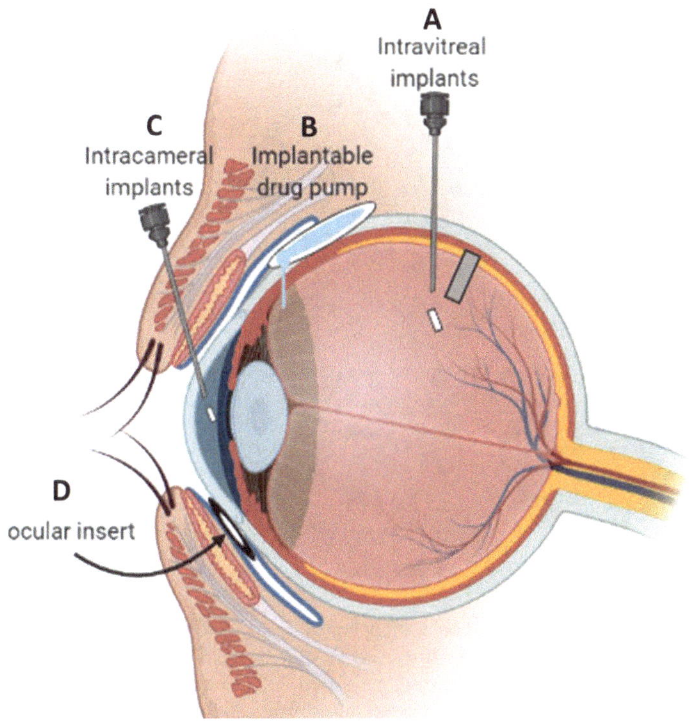

Fig. 1 Types and locations of ocular drug delivery implants. (**a**) Both biodegradable and nonbiodegradable implants, that can be free-floating or anchored, can be placed intravitreally. (**b**) Implantable drug pumps are permanent devices that can be refilled. They are typically anchored subconjunctivally or in the pars plana. (**c**) Intracameral implants are typically free floating, biodegradable, and placed in the anterior chamber of the eye. (**d**) Ocular inserts are typically placed in the conjunctival cul-de-sac or punctum of the eye. Inserts can be insoluble, soluble, or biodegradable

hydrolysis to form lactic and glycolic acid monomers that are ultimately metabolized to carbon dioxide and water via the Krebs cycle (Swati and Oshin 2018). When these polymers are used, drugs are released through pseudo first-order kinetics, which occurs in three phases: burst release, diffusion and chain scission, and biodegradation and mass loss. In the initial burst phase, a rapid release of drug molecules is associated with the surface. Next, water is able to infuse in and cause random hydrolytic scission of bonds leading to polymer degradation (Lee et al. 2010, 2011).

Other biodegradable polymers under investigation for ocular implant use are polyorthoesters (POEs), polycyclic aromatic hydrocarons (PAHs), and polycarprolactone (PCL). POEs are synthetic, hydrophobic polymers whose degradation rate can be controlled by incorporating acidic or basic excipients into the polymer matrix. PAHs are hydrophobic polymers with hydrolytically labile anhydride linkages. The tendency of these linkages to react with drugs containing free amino groups limits the use of PAHs (Park and Lakes 2007). PCL is a semicrystalline,

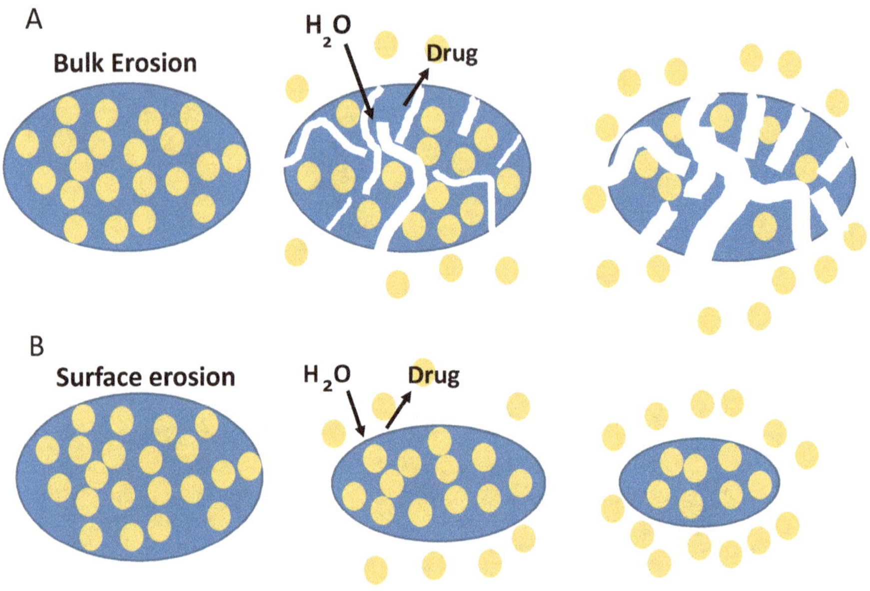

Fig. 2 Drug release mechanisms and biodegradation of matrix implants. In biodegradable implants, the drug (yellow circles) is dispersed in a biodegradable matrix (blue oval). As water penetrates the pores of the matrix, the drug molecules diffuse out. (**a**) Illustrates the bulk erosion process. Water molecules enter into the core of the implant and drug molecules exit from the core. The polymer begins to break down from internal cavitation. (**b**) Illustrates the surface erosion process. Drug and polymer are solubilized and released only on the surface of the implant. Over time, the implant reduces in volume and surface area (Lee et al. 2010, 2011)

hydrophobic polymer in which initial degradation by bulk erosion is followed by a second, slow phase characterized by mass loss due to chain cleavage and drug diffusion from the polymer matrix (Silva-Cunha et al. 2009). Overall, PCL contains fewer $(CH)_2$ units in the main chain than PAHs and is characterized by slow degradation (≥1 year) and high drug permeability, making it an ideal candidate for drug delivery systems (Park and Lakes 2007; Swati and Oshin 2018).

Biodegradable Implants in Clinical Use

Ozurdex/Posurdex

Ozurdex® (Allergan, an AbbVie company) is an intravitreal implant containing 0.7 mg of dexamethasone in a PLGA-based matrix (Novadur®, Allergan, an AbbVie company) that undergoes biphasic degradation, providing an initial loading/peak dose for 2 months, followed by a lower dose for up to 4 months. A specially designed preloaded, single-use applicator with a 22-gauge needle facilitates injection into the vitreous (Lee et al. 2010). It was approved by the United States Food and Drug

Table 2 Polymeric drug delivery systems (approved or under investigation) used for the treatment of chronic ocular diseases

Name	Characteristics	Material	Active agent	Duration	Ocular disease	Approval	Reference
Ozurdex (Allergan, an AbbVie company)	Biodegradable, intravitreal implant	Novadur® platform	Dexamethasone	6 months	Macular edema following branch RVO, non-infectious uveitis, DME	FDA approved	"Development History and FDA Approval Process for Ozurdex" (n.d.) and Lee et al. 2010)
DEXYCU (Eyepoint Pharmaceuticals, Inc)	Biodegradable, intracameral implant	Verisome®	Dexamethasone	Up to 30 days	Inflammation: postoperative cataract surgery	FDA approved	"EyePoint Pharmaceuticals Announces U.S. Commercial Launch of DEXYCU (dexamethasone intracoular suspensio) 9%" (2019)
Bimatoprost SR (Allergan, an AbbVie company)	Biodegradable, intracameral implant	Novadur® platform	Bimatoprost	6 months	OAG	Investigational	Lewis et al. (2017)
Brimonidine DDS (Allergan, an AbbVie company)	Biodegradable, intravitreal implant	Novadur® platform	Brimonidine	3 months	Retinitis pigmentosa, glaucomatous optic neuropathy, GA due to age-related macular degeneration, and rhegmatogenous retinal detachment	Investigational	Freeman (2016), Freeman et al. (2019) and Lee et al. 2010)

(continued)

Table 2 (continued)

Name	Characteristics	Material	Active agent	Duration	Ocular disease	Approval	Reference
GB-102 (Graybug Vision, Inc)	Biodegradable, intravitreal implant	PLGA and mPEG	Sunitinib malate	6 months	Wet AMD	Investigational	"Novel anti-VEGF for wet AMD meets safety, tolerability endpoint" (2019)
OTX-TKI (Ocular Therapeutix, Inc)	Biodegradable, intravitreal implant	Hydrogel	Tyrosine kinase inhibitor	9 months	Wet AMD	Investigational	"Ocular Therapeutix Announces Dosing of First Patient in Phase 1 Clinical Trial for the Treatment of Wet AMD" (2019)
OTX-TIC (Ocular Therapeutix, Inc)	Biodegradable, intracameral implant	Hydrogel	Travoprost	4–6 months	Glaucoma	Investigational	"Ocular Therapeutix™ announces treatment of first patient in phase 1 Clinical Trial of OTX-TIC (travoprost intracameral implant) for the Treatment of Glaucoma and Ocular Hypertension" (2018)
AR-13503 (Aerie Pharmaceuticals, Inc)	Biodegradable, intravitreal implant	Polyesteramide polymer	Multi-kinase (rho kinase/ protein kinase C) inhibitor	Months	nAMD and DME	Investigational	"Aerie Pharmaceuticals Announces Acceptance of Its Investigational New Drug Application for AR-13504 Sustained Release Implant" (2019)

Name	Characteristics	Material	Active agent	Duration	Ocular disease	Approval	Reference
AR-1105 (Aerie Pharmaceuticals, Inc)	Biodegradable, intravitreal implant	PRINT ®	Dexamethasone	6 months	Macular edema due to RVO	Investigational	"Drug Delivery Sustained Release Technology" (2017), "Aerie Pharmaceuticals Initiates Phase 2 Clinical Trials of AR-1105 in Patients with Macular Edema Associated with RVO" (2019) and Sandahl et al. (2018)
PA5108 (PolyActiva Pty Ltd)	Biodegradable, intracameral implant	Polytriazole hydrogel	Latanoprost	6 months	Glaucoma	Investigational	"PolyActiva Commences Its First Phase I Clinical Trial with Potential to Improve Daily Lives of Millions of Glaucoma Patients" (2018)
Retisert (Bausch & Lomb)	Nonbiodegradable, intravitreal implant	Silicone reservoir with PVA membrane coated release pore	Fluocinolone	Up to 3 years	Chronic noninfectious uveitis of the posterior segment	FDA approved	Lee et al. (2010)
Iluvien™ (Alimera Sciences)	Nonbiodegradable, intravitreal implant	PVA matrix that is encased in a polyimide tube	Fluocinolone Acetate	18–30 months	DME	FDA approved	Campochiaro et al. (2011, 2012) and "Iluvien Prescribing information" (2019)
YUTIQ™ (EyePoint Pharmaceuticals)	Nonbiodegradable, intravitreal implant	Polyamide polymer-based reservoir, capped with silicone and a PVA membrane	Fluocinolone Acetate	3 years	Chronic, non-infectious uveitis of the posterior segment	FDA approved	EyePoint (2021) and FDA (2019)

Administration (FDA) in 2009 for the treatment of macular edema following branch retinal vein occlusion (RVO) and central RVO (CRVO), in 2010 for the treatment of noninfectious uveitis, and in 2014 for the treatment of diabetic macular edema (DME) ("Ozurdex approval history, Development History, and FDA Approval Process for Ozurdex" 2010). In phase 2 and phase 3 studies, conducted in patients with macular edema, the implant demonstrated significant improvements in visual acuity, vascular leakage, and central retinal thickness, compared with untreated or sham-controlled groups. Although the implant was shown to be well tolerated overall (Boyer et al. 2014; Haller et al. 2010, 2011; Kuppermann et al. 2007; Williams et al. 2009), the risk of cataract progression was increased in patients who received more than one implant (Boyer et al. 2014; Haller et al. 2011), and previous pars plana vitrectomy and the absence of lens capsule were found to be significant risk factors for implant migration into the anterior chamber (Khurana et al. 2014).

DEXYCU

Dexamethasone intraocular suspension 9% (DEXYCU) (EyePoint Pharmaceuticals, Inc.) was FDA approved in 2018 for the postoperative cataract surgery treatment of inflammation. It is administered by intracameral injection and uses the biodegradable sustained release Verisome® technology ("EyePoint Pharmaceuticals Announces U.S. Commercial Launch of DEXYCU (dexamethasone intracoular suspensio) 9%" 2019).

Bimatoprost Implant

Bimatoprost implant (Allergan, an AbbVie company) is an intracamerally injected, biodegradable implant that relies on an ophthalmic drug delivery system (DDS) to provide sustained-release of bimatoprost (Fig. 3). Bimatoprost implant is designed to provide long-term IOP-lowering to patients with open-angle glaucoma (OAG). In a phase 1/2 study, a single implant provided similar IOP lowering to topical bimatoprost 0.03% at 4 months. Overall, 91% of study eyes did not require rescue with topical bimatoprost or retreatment with an implant at 4 months. Moreover, bimatoprost implant lasted 6 months in 71% of patients. Most AEs reported in study eyes occurred within 2 days post-injection and were transient. Conjunctival hyperemia, the most frequent AE associated with bimatoprost and prostaglandin analogues, was less frequent with bimatoprost implant (6.7%) than the topical solution (17.3%) (Lewis et al. 2017). It recently gained FDA approval as the first intracameral, biodegradable sustained-release implant for the treatment of OAG or ocular hypertension (Allergan 2020).

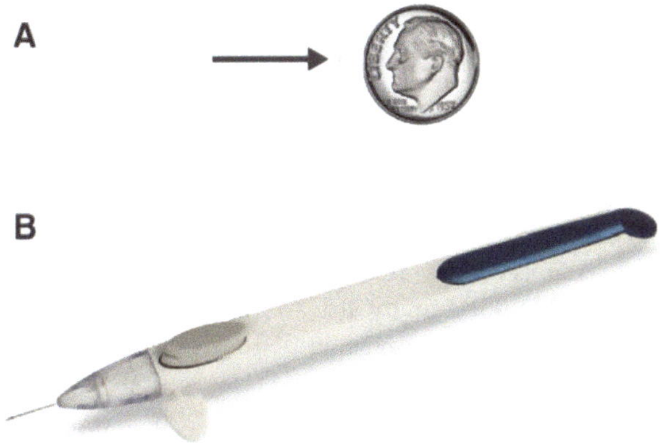

Fig. 3 Bimatoprost implant. (**a**) The Bimatoprost implant itself, next to a dime for size comparison. (**b**) The single-use implant applicator system for intracameral injection with a 28 gauge needle (Seal et al. 2019)

Investigational Biodegradable Implants

Brimonidine DDS

The injected intravitreal implant Brimonidine DDS, based off of Novadur® technology, provides sustained release of brimonidine. It has been evaluated in phase 2 studies involving patients with retinitis pigmentosa, glaucomatous optic neuropathy, geographic atrophy (GA) due to age-related macular degeneration, and rhegmatogenous (macula-off) retinal detachment (ClinicalTrials.gov identifiers: NCT00661479, NCT00693485, NCT00658619, and NCT00972374, respectively) (Lee et al. 2010). Results from the study of patients with GA indicated that treatment with the brimonidine implant, 132 μg or 264 μg, on day 1 and month 6 consistently produced smaller mean changes in GA lesion from baseline than sham treatment. At month 12, the primary endpoint, 18.8% and 27.5% reductions in GA progression rates were observed in the two treatment groups, respectively. Treatment-related ocular AEs were predominantly attributed to the injection procedure, the most common being conjunctival hemorrhage and conjunctival hyperemia (Freeman 2016). Although a study comparing the brimonidine implant (400 μg) and sham treatment every 3 months up to month 21 (ClinicalTrials.gov identifier: NCT02087085) was discontinued due to overall slow GA progression, the implant reduced the progression rate of GA by 10% at month 24 ($P = 0.047$) and 12% at month 30 ($P = 0.017$), compared with sham treatment (Freeman et al. 2019).

GB-102

GB-102 (Graybug Vision, Inc. Redwood City, CA) is a biodegradable intravitreal injection that is under investigation for the treatment of wet AMD. It is composed of microparticles made from PLGA and methoxy-polyethylene glycol (mPEG)-PLGA and sunitinib malate, a small molecule receptor tyrosine kinase inhibitor of vascular endothelial growth factor (VEGF)-1, -2, and -3. As the microparticles start to biodegrade, sunitinib malate is slowly released ("Graybug Vision Presents Top Line Results of Phase 1/2a ADAGIO Study at Hawaiian Eye & REtina 2019" 2019). In the phase 1/2a study (ClinitalTrial.gov identifier: NCT03249740), it met its primary endpoint of safety and tolerability. Although no serious adverse events were reported, there were reports of eye pain, photophobia, blurred vision, and vitreous haze. A single dose of GB-102 was able to maintain 90% of patients at 3 months and 70% of patients at 6 months ("Novel anti-VEGF for wet AMD meets safety, tolerability endpoint" 2019). Graybug is also developing drug-encapsulated microparticles for glaucoma treatments that will be injected in the subconjunctival space (GB-201, 202, 203) ('GB-201, GB-202 and GB-203 – Glaucoma Products' n.d.).

OTX-TKI

OTX-TKI (Ocular Therapeutix, Inc. Bedford, MA) is an investigational treatment for wet AMD that is delivered by intravitreal injection. It is composed of a bioresorbable hydrogel fiber containing tyrosine kinase inhibitor (TKI) particles that can potentially deliver the drug for a period of up to 9 months. A phase 1 clinical trial (ClinicalTrial.gov identifier: NCT03630315) is currently underway ("Ocular Therapeutix Announces Dosing of First Patient in Phase 1 Clinical Trial for the Treatment of Wet AMD" 2019).

OTX-TIC

Ocular Therapeutix has another investigational product that uses the bioresorbable material mentioned above. OTX-TIC is injected intracamerally (Fig. 4) and is composed of travoprost for the treatment of glaucoma. Preclinical studies have suggested that it can release drug for up to 4–6 months ("Ocular Therapeutix™ Announces Treatment of First Patient in Phase 1 Clinical Trial of OTX-TIC (travoprost intracameral implant) for the Treatment of Glaucoma and Ocular Hypertension" 2018).

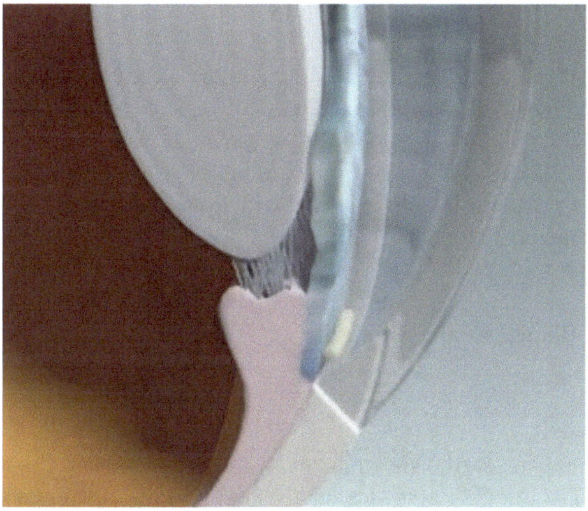

Fig. 4 OTX-TIC. The image shows the intracameral placement of OTX-TIC in the eye. (Image from Blizzard et al. (2019) and used under the terms of the Creative Commons Attribution 4.0 International License)

AR-13503

AR-13503 (Aerie Pharmaceuticals, Bedminster, NJ) was approved to start first-in-human clinical studies in 2019 (ClinicalTrial.gov identifier: NCT03835884). The novel drug is composed of a bioerodible polyesteramide polymer that releases a proprietary multi-kinase (Rho kinase/Protein kinase C) inhibitor. It is to be injected intravitreally for the treatment of nAMD and DME ("Aerie Pharmaceuticals Announces Acceptance of Its Investigational New Drug Application for AR-13504 Sustained Release Implant" 2019).

AR-1105

Also from Aerie Pharmaceuticals, AR-1105 is an implant that releases dexamethasone for the treatment of macular edema due to retinal vein occlusion (RVO). It is in phase 2 trials (ClinicalTrial.gov identifier: NCT03739593), and it uses the biodegradable PRINT® (Particle Replication In Non-Wetting Template) technology, which is a proprietary polymer that can be molded in the shape of a micropatterned template along with the drug. This implant can be injected intravitreally ("Aerie Pharmaceuticals Initiates Phase 2 Clinical Trials fo AR-1105 in Patients with Macular Edema Associated with RVO" 2019; "Drug Delivery Sustained Release Technology" 2017; Sandahl et al. 2018).

PA5108

PA5108 (PolyActiva Pty Ltd, Melbourne Australia) is composed of the drug latanoprost within a biodegradable polytriazole hydrogel system. It is placed intracamerally with a 27G needle for the treatment of glaucoma and is currently in Phase 1 clinical trials (ClinicalTrial.gov identifier: NCT03604328) ("PolyActiva Commences Its First Phase I Clinical Trial with Potential to Improve Daily Lives of Millions of Glaucoma Patients" 2018).

Discontinued Biodegradable Implants

Surodex

Surodex™ (Allergan, an AbbVie company, Dublin, Ireland) is a rod-shaped implant consisting of PLGA, 60 μg of dexamethasone an anti-inflammatory agent, and hydroxypropyl methylcellulose, which ensured drug release at a constant rate over 7–10 days (Lee et al. 2011). Following cataract surgery, the implant was inserted in the anterior chamber, without suture fixation, to control postoperative inflammation (Jain 2000; Lee 2015; Lee et al. 2011). In clinical studies, the implant was well tolerated and was shown to reduce anterior chamber cells and flare in the postoperative period. It exhibited anti-inflammatory properties that were at least as effective as those of topical steroids, while being less toxic (Kimura and Ogura 2001; Seah et al. 2005; Tan et al. 1999).

ENV515

ENV515 (Travoprost XR; Envisia Therapeutics, Durham, NC) is an intracameral ocular implant designed to release travoprost using the PRINT® technology to provide sustained IOP lowering to patients with primary OAG ("Envisia Therapeutics releases interim ENV515 (Travoprost XR) phase 2 data demonstrating 11-month duration-of-action after a single dose in patients with glaucoma" 2017). Interim analysis of a phase 2 study (ClinicalTrial.gov identifier: NCT02371746) showed the IOP-lowering effect of travoprost XR was comparable to that of topical prostaglandin analogues (latanoprost [Xalatan®, Pfizer] and bimatoprost [Lumigan®, Allergan, an AbbVie company]) used before study initiation, as well as topical timolol maleate 0.5% ophthalmic solution. The mean IOP reduction from baseline (26.1 mmHg) following a single injection was reportedly 25% (6.7 mmHg) at 11 months, and the most common AE was early onset hyperemia ("Envisia Therapeutics releases interim ENV515 (Travoprost XR) phase 2 data demonstrating 11-month duration-of-action after a single dose in patients with glaucoma" 2017).

IBI 20089

IBI 20089 (EyePoint Pharmaceuticals, formerly ICON Biosciences, Inc.) is an intravitreal drug delivery implant that relied on a proprietary, nanopolymer-based technology (Verisome™; Ramscor, Inc., Menlo Park, CA) designed for sustained release of a broad range of pharmaceutical agents, including small molecules, peptides, proteins, and monoclonal antibodies (Haghjou et al. 2011; Lee et al. 2010). The technology was reportedly highly versatile and could be formulated into a biodegradable solid, gel, or liquid substance (Haghjou et al. 2011). Moreover, degradation was reportedly independent of the reactant(s) concentration (Lee et al. 2010). A liquid formulation that delivers 6.9 mg or 13.8 mg of triamcinolone acetonide was recently evaluated for safety and efficacy in an open-label phase 1 trial that enrolled 10 patients with cystoid RVO-associated macular edema RVO (Lim et al. 2010). The formulation was well tolerated, with two reports of AEs: elevated IOP, treated with an Ahmed glaucoma valve, and panretinal coagulation (Lim et al. 2010). Despite seemingly encouraging data, it appears the implant is no longer being investigated.

Nonbiodegradable Ocular Drug Delivery Systems

One shortcoming often associated with biodegradable polymers, the initial and final medication release bursts, can be avoided by storing the drug in a reservoir. A nonbiodegradable polymer that is semipermeable or has fixed openings for smaller areas of diffusion can surround the drug in order to avoid release bursts (Fig. 5) (Bourges et al. 2006; Liechty et al. 2010; Yasin et al. 2014). Alternatively, the drug can be stored in a nonbiodegradable matrix, although an initial burst can be observed in this system (Conway 2008; Yasin et al. 2014). With nonbiodegradable polymers, the main advantage is long-term drug release due to near zero-order kinetics, which means a consistent amount of the drug is released over time (Bourges et al. 2006; Lee et al. 2010; Liechty et al. 2010; Patel et al. 2013). These implants can be designed to be a free-floating pellet injected intravitreally or intracamerally or anchored to the sclera. Some disadvantages are that their placement may require a large incision and sutures or some other form of anchoring, the need to surgically remove the depleted implant, and the extra costs and increased risk of AEs associated with these additional procedures (Lee et al. 2010; Patel et al. 2013). Nonbiodegradable implants in clinical use and under investigation are summarized in Table 2.

Polymers typically used for fabricating nonbiodegradable implants include polyvinyl alcohol (PVA), ethylene vinyl acetate (EVA), and silicon (Jervis 2017). While EVA and silicon are relatively impermeable hydrophobic polymers often used as drug-restricting membranes, PVA is a hydrophilic and more permeable polymer (Conway 2008; Kearns and Williams 2009; Lee et al. 2010). The combination of both an EVA/silicon membrane and a PVA are often used. Drug release occurs when

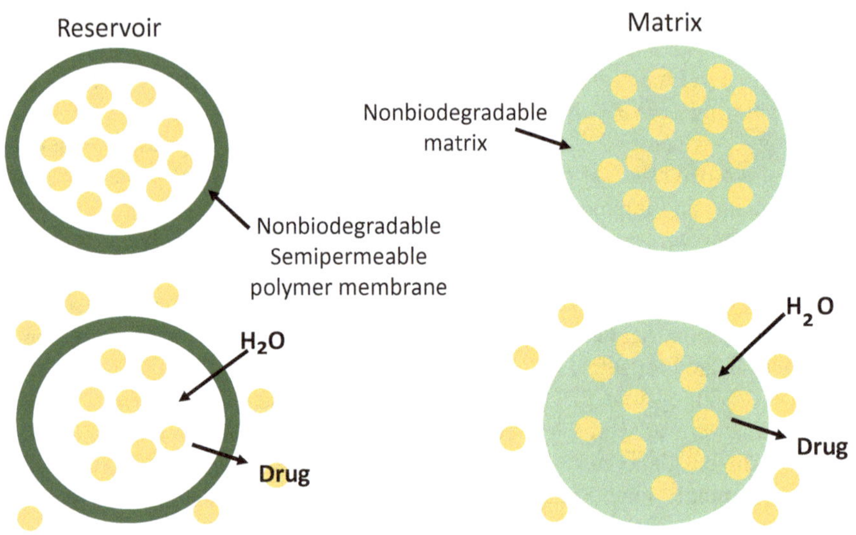

Fig. 5 Nonbiodegradable drug delivery system. In a reservoir system, the drug (yellow circles) is surrounded by a nonbiodegradable membrane (green ring) that is semipermeable. This membrane acts as the framework of the implant and regulates the rate of release, allowing it to be constant. As water diffuses into the device, the drug pellet is dissolved and a saturated solution is released by diffusion out of the device. In the matrix system, the drug (yellow circles) is dispersed in a nonbiodegradable matrix (light green circle) and released through diffusion (Bourges et al. 2006; Conway 2008; Liechty et al. 2010; Yasin et al. 2014)

water diffuses through the outer EVA/silicon coating and partially dissolves the enclosed drug, forming a saturated drug solution that is then released into the surrounding tissue via diffusion (Conway 2008; Kearns and Williams 2009). The drug release rate can be slowed by increasing the surface area or thickness of the drug-restricting membrane and can be increased by maximizing the surface area available for drug diffusion or by using a more permeable membrane (Lee et al. 2010).

Nonbiodegradable Intravitreal Implants in Clinical Use

Retisert

Retisert® (Bausch & Lomb 2016) is an intravitreal disc-shaped implant that was FDA-approved in 2005 for the treatment of chronic noninfectious uveitis of the posterior segment (Jervis 2017; Lee et al. 2011). It consists of a fluocinolone tablet encased in a silicone reservoir containing a single release pore, coated with a PVA membrane (Durasert™, EyePoint Pharmaceuticals, Watertown, MA). The implant is attached to a 5.5-mm silicone tab used to suture the implant into the vitreous at the pars plana near the ciliary processes Lee et al. 2011). In randomized studies involving patients with noninfectious uveitis, the implant improved visual acuity

(Callanan et al. 2008; Jaffe et al. 2006; Kempen et al. 2011; Sangwan et al. 2015), reduced recurrence rates of uveitis (Callanan et al. 2008; Jaffe et al. 2019; Sangwan et al. 2015), and required less adjunctive systemic immunosuppression therapy than nonimplanted eyes (Sangwan et al. 2015). Another study conducted in patients with DME showed that the implant significantly improved visual acuity and retinopathy severity scores, reduced macular edema, and prevented retinal thickening for up to 3 years when compared with standard care (Pearson et al. 2011). Some drawbacks include an elevated risk of cataract surgery and glaucoma (Callanan et al. 2008; Jaffe et al. 2006; Kempen et al. 2011; Pearson et al. 2011; Sangwan et al. 2015) and spontaneous intraocular dissociation of the implant occurring years after placement (Rofagha et al. 2013). Another limitation for the clinical use of the Retisert implant is the high cost when compared to the oral prednisone therapy (Mohammad et al. 2007).

Iluvien

Iluvien™ (Alimera Sciences, Alpharetta, GA) is an intravitreal implant consisting of 190 μg of fluocinolone acetate embedded in a PVA matrix that is encased in a polyimide tube (Durasert™, EyePoint Pharmaceuticals, Watertown, MA; formerly Medidur™, PSivida). The implant is capped at one end and open at the other end, which allows for diffusion of water into the matrix and drug release (Fig. 6) ("Iluvien Prescribing information" 2019). Although two doses of the implant have been evaluated, with an average delivery rate of 0.2 μg/day (lasting 24–36 months) and 0.5 μg/day (lasting 18–24 months) (Campochiaro et al. 2010, 2011; Lee et al. 2011; Lee and Robinson 2009), only the lower-dose version was FDA-approved in 2014 to treat DME ("Iluvien Approval History" 2019). The lower dose produced similar levels of visual improvement as the higher dose, but with a lower rate of side effects (Campochiaro et al. 2012). Patients with persistent DME, who had at least one

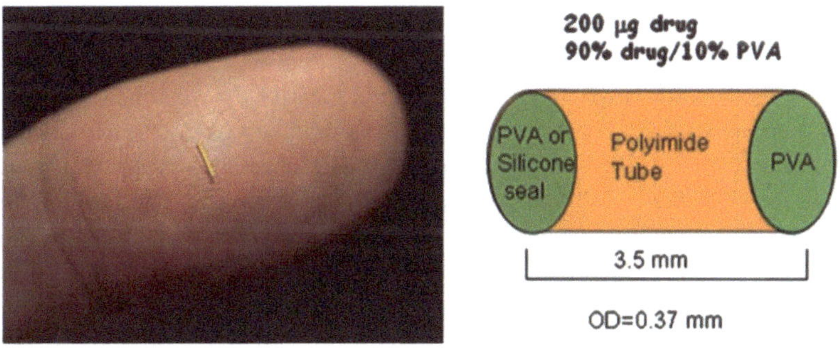

Fig. 6 Schematic view of Iluvien, a nonbiodegradable intravitreal implant. This implant consists of a polyimide tube that is capped on one end and open on the other. Within the tube, fluocinolone acetate is embedded in a PVA matrix, which allows water to diffuse in and release the drug (Haghjou et al. 2011)

previous laser photocoagulation treatment, were shown to improve visual acuity 3, 6, and 12 months after a single insertion, although the change from baseline was not statistically significant at 12 months (Campochiaro et al. 2010). An analysis of 2 randomized, controlled studies showed that, when compared with sham treatment, there was statistically significant improvement of visual acuity at 24 months. There was also improvements in retinal thickness, and patients were less likely to receive additional DME treatments (Campochiaro et al. 2011). Additionally, it was found that the treatment benefit was maintained over 3 years, with maximum benefit at 30 months (Campochiaro et al. 2012). Nonetheless, as expected with steroidal implants, the incidence of elevated IOP and cataract surgery was increased in Iluvien treated patients (Campochiaro et al. 2010, 2011).

YUTIQ

YUTIQ™ (EyePoint Pharmaceuticals) is an intravitreal implant (3.5 × 0.37 mm) that was FDA-approved for the treatment of chronic, noninfectious uveitis of the posterior segment in 2018 (EyePoint 2021; FDA 2019). The implant consists of 180 μg of fluocinolone acetonide contained in a polyamide polymer-based reservoir, capped with silicone on the one end and a permeable PVA membrane on the other end (Durasert™, EyePoint Pharmaceuticals, Watertown, MA) (Jaffe et al. 2019; FDA 2019). It is designed to release fluocinolone acetonide over 3 years, at an initial rate of 0.25 μg/day ('Iluvien Prescribing information' 2019; EyePoint 2021). Two studies comparing YUTIQ vs sham indicated that YUTIQ reduced recurrence rates of uveitis and chronic noninfectious posterior uveitis (EyePoint 2021; Jaffe et al. 2006, 2019).

Investigational Nonbiodegradable Implants

Travoprost Intraocular Implant

The Travoprost Intraocular Implant (Glaukos Corporation, San Clemente, CA) uses the iDose™ delivery system and is being developed for the treatment of OAG and ocular hypertension. It consists of an implantable reservoir housed in a titanium implant that is secured in the iridocorneal angle. The reservoir releases travoprost through a membrane over 1 year and can be replaced. Interim data from a phase 2 study indicated that the implant resulted in 30% IOP reduction from baseline at 12 months (Varma 2018; Stephenson 2018) and the safety profile appeared favorable with no cases of conjunctival hyperemia reported (Stephenson 2018). Two ongoing phase 3 studies (ClinicalTrial.gov identifiers: NCT02754596 and NCT03868124) are expected to be completed in 2020 and 2023, respectively.

Targeted Episcleral Delivery System (Episcleral Implant/Reservoir)

This device (3T Ophthalmics Targeted Therapy Technologies, LLC Irvine, CA) is placed under the conjunctiva attached to the episcleral, but does not penetrate the eye. It is impermeable, except for the side that is open on the sclera. It acts as a reservoir that maintains a concentration gradient favoring diffusion of the drug through the sclera and also allows for the sustained release of drug (Pontes de Carvalho et al. 2006). The device has been tested preclinically with a variety of drugs including brimonidine (De Carvalho et al. 2014), topotecan, melphalan (Carvalho et al. 2016), and celecoxib (Lima et al. 2018). In addition, a phase 1 clinical trial was started in 2019 using dexamethasone for the treatment of DME (ClinicalTrial.gov identifier: NCT04005430).

Discontinued Nonbiodegradable Implants

Vitrasert

Vitrasert® (Bausch & Lomb, Rochester, NY), a controlled-release (5–8 months) intravitreal implant that consisted of a ganciclovir tablet in a PVA matrix surrounded by a nonbiodegradable EVA coating (Durasert™ technology [formerly Medidur]), EyePoint Pharmaceuticals, Watertown, MA), was FDA-approved (1996) for the treatment of cytomegalovirus retinitis associated with acquired immune deficiency syndrome (Patel et al. 2013). Ganciclovir was released via passive diffusion through a small opening in the EVA membrane (at the base of the implant) (Jervis 2017). Notably, the implant was shown to be twice as effective at slowing disease progression, compared with intravenous ganciclovir (Musch et al. 1997). It was also thought to be the best treatment choice, compared with the oral or intravenous prodrug, for cytomegalovirus retinitis lesions that pose an immediate risk to vision (Kedhar and Jabs 2007). Nonetheless, the implant was discontinued in 2013 following patent expiration ("Psivida Corp Annual report 2016").

I-Vation

I-Vation™ (SurModics, Eden Prairie, MN) consisted of a titanium helical coil coated with triamcinolone acetonide and encased in a proprietary blend of nonbiodegradable polybutyl methacrylate and EVA, the composition of which controlled the delivery rate (Jervis 2017; SurModics 2013). The device had a sharpened tip, which was used to make the incision for intravitreal implantation. Its helical shape maximized the surface area for drug coating while enabling secure anchoring to the pars plana/sclera (Conway 2008). The implant had been under investigation for the treatment of DME, but results from a phase 1 clinical trial indicated relatively high incidences of conjunctival hemorrhage (90%) and lenticular opacities (35%) (Kiernan and Mieler 2009). Perhaps as a consequence of these findings, enrollment in a phase 2 trial (ClinicalTrials.gov identifier: NCT00692614) ("A Study of MK0140 in Diabetic Patients With Macular Edema (0140-001)" n.d.) was low, and the study was terminated in 2008 (Kiernan and Mieler 2009).

Lumitect

Lumitect® (Lux Biosciences Inc., Jersey City, NJ), also known as LX201, was a silicone-based, drug-eluting, episcleral implant that was originally developed at the National Eye Institute (BioSpace 2006). It was designed to deliver cyclosporine for up to 3 years. However, they failed to meet the primary endpoint of a phase 3 study (ClinicalTrials.gov Identifier: NCT00447642) conducted to evaluate its effectiveness in preventing corneal allograft rejection/failure, and the study was terminated. The company is reportedly developing an alternative oral drug (voclosporin [Luveniq]) (Weintraub 2012).

NT-503

NT-503 (NeuroTech Pharmaceuticals, Inc. Cumberland, RD) used encapsulated cell therapy (ECT) to deliver a soluble anti-VEGF receptor protein for the treatment of wet AMD. ECT was implanted into the vitreous for up to 2 years and held in place with sutures. When more than the expected number of patients needed rescue medication, the phase 2 clinical study (ClinicalTrial.gov identifier: NCT02228304) was discontinued ("NT-502 ECT" 2016).

Implantable Drug Pumps

Implantable drug pumps are also nonbiodegradable implants. In contrast to other ocular drug delivery devices that hold only a predetermined amount of drug, implantable drug pumps are refillable (Pearce et al. 2015).

Replenish MicroPump

The Replenish MicroPump (RMP; Replenish, Pasadena, CA) is a surgically implantable drug pump. It is a subconjunctival/episcleral implant with a reservoir that is refillable by a transconjunctival injection (Pearce et al. 2015). It is designed to release nanoliter doses of medication at a preprogrammed interval through an intraocular cannula implanted in the pars plana. Anterior and posterior platforms that can target both ocular segments are in development (Lo et al. 2009; Saati et al. 2010). There were no intraoperative complications, no serious AEs, and no worsening of visual acuity or central foveal thickness, compared with baseline during the first-in-man 90-day safety study of the RMP in patients with DME. The RMP delivered the programmed ranibizumab dosage in 7/11 patients, while the remaining four patients received a lower dose, and treatment was complemented with standard intravitreal injection (Humayun et al. 2014).

Port Delivery System

The Port Delivery System (PDS, ForSight VISION4, Inc., acquired by Genentech/Roche Holding AG, Basel, Switzerland) is in development as a refillable drug delivery device (Joseph et al. 2017; Pearce et al. 2015). PDS is composed of polysulfone and is placed through a scleral incision in the pars plana. The device has a semipermeable titanium membrane that allows constant passive diffusion of the drug into the vitreous (Fig. 7). In a phase 2 study, the PDS (filled with ranibizumab 10, 40, or 100 mg/mL) was evaluated in 220 patients with neovascular AMD who had received ≥2 prior anti-VEGF intravitreal injections and were responsive to treatment. Only patients with the PDS 100-mg/mL treatment had improvements in best-corrected visual acuity and central foveal thickness that were comparable with the monthly intravitreal ranibizumab 0.5 mg injections. The optimized PDS implant insertion and refill procedures were generally well tolerated, with a rate of postoperative vitreous hemorrhage of 4.5% (7/157, including 1 serious event) (Campochiaro et al. 2019). The potential reduction in treatment burden supported further development

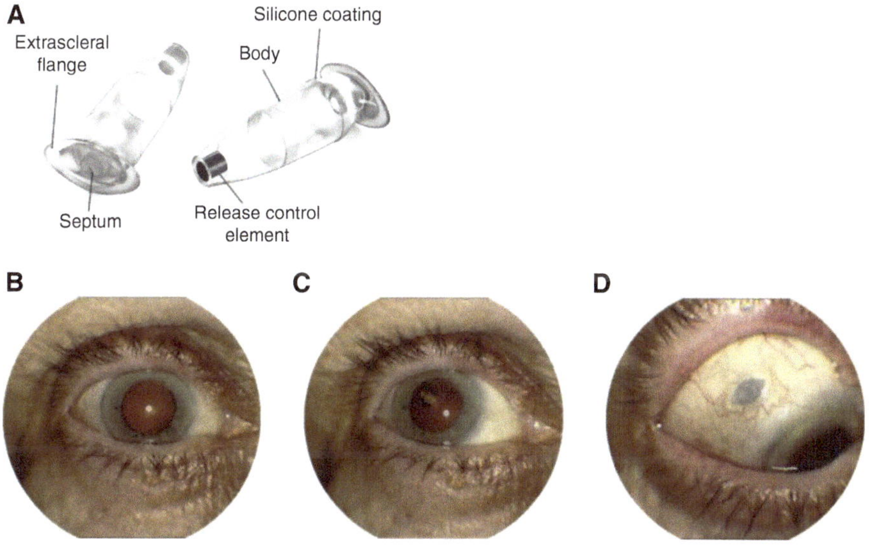

Fig. 7 Schematic of the Port Delivery System (PDS) implanted in the eye. The device is anchored in the pars plana, and the drug reservoir portion of the device is in the vitreous. The device can be refilled multiple times with a special needle. The semipermeable titanium membrane allows continuous, passive diffusion of the drug in the reservoir into the vitreous. (Image used from Campochiaro et al. (2019) with permission conveyed through Copyright Clearance Center, Inc)

in ongoing phase 3 studies that are expected to be completed in 2022 (ClinicalTrials.gov identifiers: NCT03677934 and NCT03683251) ("A Phase 3 Study to Evaluate the Port Delivery System Implant with Ranibizumab Compared with Monthly Ranibizumab Injections in Participants with Wet Age-Related Macular Degeneration (Archway)" 2021; "Extension Study for the Port Delivery System With Ranibizumab (Portal)" 2021). However, a phase 1/2 study of the PDS (filled with methotrexate 0.6 or 2.3 mg) in noninfectious uveitis (ClinicalTrials.gov identifier: NCT02125266) was terminated due to an unacceptable frequency of drug-related AEs ("Safety and Preliminary Efficacy Study of V404 PDS in Uveitis" n.d.). Assuming proper functioning of the PDS, further investigation of the device with other drugs might be warranted.

Ocular Inserts

Designed for placement in the cul-de-sac, conjunctival sac, or punctum of the eye, ocular inserts aim to increase the contact time between the medication and conjunctival tissue to ensure sustained release suited for topical or systemic treatment. Although less invasive and easier to place than intravitreal implants, ocular inserts have several potential disadvantages. For example, they can interfere with vision and cause foreign body sensation in the eye. This can cause discomfort leading to irritation and excessive tearing which dilutes the drug. The insert can also become lost during sleep or while rubbing the eye and movement around the eye can complicate the removal if an insoluble insert migrates to the upper fornix. Additionally, due to its rigidity, the insert can be difficult to place and remove. There are three types of ocular inserts: (1) insoluble inserts, (2) soluble inserts, and (3) biodegradable inserts (Kumari et al. 2010).

Insoluble Ocular Inserts in Clinical Use or In Development

Insoluble ocular inserts can be further broken down into three categories: diffusion insert, osmotic insert, and soft contact lens. Diffusion inserts typically consist of a drug reservoir enclosed in a specially designed semipermeable or microporous membrane. As tear fluid permeates through the membrane and inside the reservoir, internal pressure drives the drug out at a controlled rate. Osmotic inserts generally have two basic designs. In one, the drug is the central part of the insert and is surrounded by the polymer as discrete small deposit. While in the other, the drug and osmotic solutes are placed in two separate reservoirs surrounded by an elastic impermeable membrane and semipermeable membrane, respectively. Both types of osmotic inserts are covered with a peripheral film made of an insoluble semipermeable polymer and as tears diffuse through the semipermeable membrane of the reservoir, the osmotic pressure increases, causing the polymer matrix to rupture and form apertures through which the drug is released through zero-order kinetics (Kumari et al. 2010). Soft contact lens consists of covalently cross-linked hydrophilic or hydrophobic polymers forming a three-dimensional matrix capable of retaining water, aqueous solution, or solid components. They do not deliver drugs as uniformly as other insoluble ophthalmic systems, and generally, the drug release is very rapid at the beginning then declines exponentially with time. The initial release rate can be decreased by adding hydrophobic components (Kumari et al. 2010).

Mydriasert

Mydriasert (Thea Laboratories, Clermont-Ferrand, France) is an insoluble rod-shaped ophthalmic insert composed of tropicamide and phenylephrine hydrochloride. It is placed in the upper/lower conjunctival fornix and delivers mydriasis prior to surgery (Bertens et al. 2018). In studies comparing Mydriasert versus phenylephrine and tropicamide eye drops, the mydriatic effect was comparable after 60 min, but was superior after 90 min and maintained good pupil dilation during cataract surgery (Saenz-de-Viteri et al. 2013). Mydriasert has also been compared with eye drops in patients undergoing retinal angiography. Although they have similar efficacy, the low total drug dose administered with the insert may potentially reduce the risk of cardiovascular side effects (Cagini et al. 2014).

Punctal Plugs

Several punctal plugs of varying shapes and dimensions are currently under investigation, including some containing prostaglandin analogues for the reduction of IOP in the management of glaucoma. Mati Therapeutics (Austin, TX) is developing a product, known as latanoprost-punctal plug delivery system (L-PPDS), based on its proprietary punctal plug delivery system (Evolute®) to deliver latanoprost. Evolute's drug core reportedly allows sustained, unidirectional drug release into the tear film, thus minimizing systemic absorption. Vistakon Pharmaceuticals (a division of Johnson & Johnson Vision Care, Inc.; Jacksonville, FL) is testing a different platform to deliver bimatoprost (Whitcup and Azar 2017). Phase 2 studies evaluating those plugs associated with sustained release of latanoprost or bimatoprost in glaucoma patients are either ongoing or have been completed but have not yet been published (ClinicalTrials.gov Identifiers: NCT00855517 and NCT01229982).

Bimatoprost Ring

The Bimatoprost Ring (Allergan, an AbbVie company; formerly Helios, ForSight) is a 1-mm-thick ring with a diameter of 24–29 mm consisting of an internal polypropylene support covered with bimatoprost-loaded silicone for the management of glaucoma. It is placed in the fornix and releases bimatoprost over 6 months (Fig. 8) (Varma 2018). In a phase 2 study comparing the ring to regular unpreserved timolol

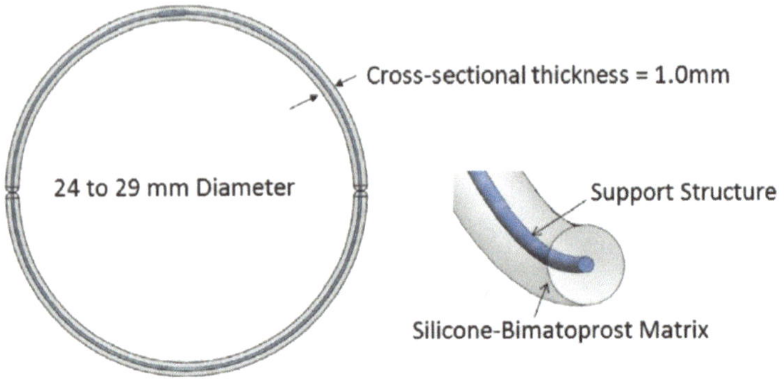

Fig. 8 Schematic of the bimatoprost ring ocular insert. The soft insert is constructed of a bimatoprost and silicone-matrix polymer with an internal polypropylene support structure. It is placed on the ocular surface of the eye (Brandt et al. 2016)

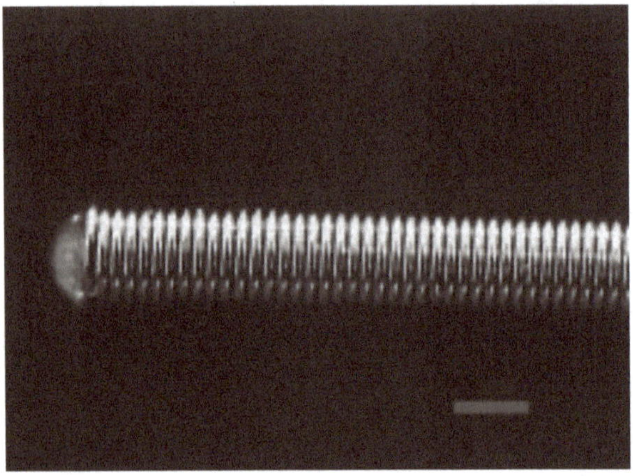

Fig. 9 Schematic of OphthaCoil. The coiled stainless-steel wire device can be loaded with drugs and placed in the lower conjunctival sac. (Image used from Pijls et al. (2005) with permission conveyed through Copyright Clearance Center, Inc)

0.5% ophthalmic solution over 6 months, IOP reduction was not significantly different and drop-out rate was high. Retention rate was improved after 13 months as patients gain experience using the ring (Macha and Mitra 2002).

OphthaCoil

OphthaCoil is a coiled stainless-steel wire device being developed in the Netherlands for placement in the lower conjunctival sac (Fig. 9). Drugs, such as antibiotics or mydriatic agents, can be loaded on microspheres or filaments that are placed in the device lumen or deposited on the outside SlipSkin® surface as coating (Bertens et al. 2018). In human pilot trials, short-term high tolerance and comfort of the device was demonstrated for a period of 2 h (Pijls et al. 2005, 2007). Preclinical and clinical trials are being conducted to further explore the potential of an ocular coil as an ocular drug delivery device for an extended period of time, up to 28 days (Bertens et al. 2018).

TODDD

The topical ophthalmic drug delivery device (TODDD™, Amorphex Therapeutics, Andover, MA) is an "eight-shaped" elastomer, viscoelastic polymer (20-mm long, about 8-mm wide, and 1-mm thick) containing timolol or prostaglandin (Bertens et al. 2018). It is intended for placement on the sclera, below the upper eyelid of patients with glaucoma. The timolol-loaded device has demonstrated an IOP reduction of 16%–22% after 6 months (Bethke 2015).

Soluble Ocular Inserts

Soluble inserts utilize either natural or synthetic/semisynthetic polymers that release drug by diffusion when tears penetrate the insert and form a layer of gel around the core of the insert (Kumari et al. 2010). The advantages of these devices are that they are completely soluble and do not need to be removed (Calles et al. 2015; Kumari et al. 2010). Drawbacks include rapid penetration of the lacrimal fluid into the device, blurred vision due to solubilization of insert components, and the glassy constitution of the insert increases the risk of expulsion (Calles et al. 2015).

Lacrisert

Lacrisert® (Aton Pharma, Inc./Bausch Health Companies, Lawrenceville, NJ), which was introduced in 1981 for the treatment of dry eye, is a translucent, rod-shaped, water-soluble insert made of hydroxypropyl cellulose (HPC), a physiologically inert substance designed for daily placement into the inferior cul-de-sac of the eye. Preclinical dissolution studies have shown that the HPC inserts become softer within 1 h of placement and completely dissolved in 14–18 h. The release of HPC stabilizes and thickens the precorneal tear film, prolonging the tear breakup time (Lee et al. 2011). In a multicenter, crossover study, once-daily treatment with the insert was generally well tolerated with more patients preferring the insert and reporting greater comfort with the insert than with artificial tears. Side effects, which were typically mild and transient, include blurred vision, ocular discomfort/irritation, matting/stickiness of eyelashes, photophobia, hypersensitivity, eyelid edema, and hyperemia (Hill 1989; Bausch and Lomb 2016).

OTX-TP

OTX-TP (Ocular Therapeutix, Bedford, MA) is a cylindrical, resorbable punctal plug that contains preservative-free travoprost and is intended for glaucoma treatment. It expands when hydrated and releases drug over 3 months, by which time 90% of the plug has dissolved and drained into the nasolacrimal duct. Phase 2 studies have demonstrated 88% retention at 75 days and slightly less IOP lowering than timolol at 90 days (Varma 2018).

Bioerodible/Biodegradable Ocular Inserts

Biodegradable ocular inserts are typically composed of a homogeneous drug dispersion coated by a hydrophobic, polymeric matrix, such as POEs and orthocarbonates that are impermeable to the drug (Lee et al. 2011). Drug release occurs as the tear fluid makes contact with the device and induces superficial bioerosion of the matrix causing the insert to dissolve in the eye in days or months (Kumari et al. 2010).

NODS

The New Ophthalmic Delivery System (NODS®; Smith and Nephew Pharmaceuticals Ltd., Essex, UK) is made from water-soluble PVA and placed in the cul-de-sac of the lower eyelid. It has been loaded with drugs such as pilocarpine, chloramphenicol, and tropicamide (Bertens et al. 2018). Although commercially available, there have been reports of intense miosis (Greaves et al. 1992), as well as problems with the detachment of the insert from its applicator (Diestelhorst and Krieglstein 1994).

Investigational Ocular Inserts

Brimonidine-Based Insert

Development of a brimonidine tartrate-releasing insert for the treatment of glaucoma has been initiated, as evidenced by in vitro and in vivo findings published between 2011 and 2014 (Bhagav et al. 2011; Ravindran et al. 2014). Although various matrix types are being evaluated, none have been reported for investigation in clinical studies (Mealy et al. 2014).

Discontinued Ocular Inserts

Dextenza

Dextenza, the resorbable PEG punctal plug (Ocular Therapeutix, Bedford, MA), was designed for the treatment of inflammatory eye conditions. It initially showed encouraging results in phase 2 and 3 trials in terms of improving signs and symptoms of allergic conjunctivitis. However, in July 2017, the FDA rejected the company's new drug application (NDA) due to deficiencies in the manufacturing process and analytical testing identified during inspection of a manufacturing facility (Bertens et al. 2018).

Gelfoam

Gelfoam® discs (Pharmacia & Upjohn Compnay LLC, Peapack, New Jersey, USA) are biodegradable inserts made of resorbable gelatin, which could be infused with mydriatic drugs or insulin and inserted in the lower conjunctival fornix to treat various ocular diseases (Bertens et al. 2018). In clinical studies involving volunteers, some developed a palpebral conjunctival infection (hyperemia), while others developed superficial punctate erosion (Lee et al. 2002; Niegvesky et al. 2000). There have been no reports that this ocular insert was ever commercialized, and it does not appear to be under further investigation (Bertens et al. 2018).

Ocusert Pilo

Ocusert Pilo (Alza Corporation, acquired by Johnson & Johnson, New Brunswick, NJ) is an oblong-shaped, sustained-release, nonbiodegradable ocular insert that was placed in the conjunctiva and was approved in 1974 for the treatment of glaucoma (Bertens et al. 2018). It consisted of pilocarpine hydrochloride and alginic acid, contained within a reservoir enclosed by two release-controlling EVA membranes and surrounded by a titanium oxide ring to aid in positioning and placement (Conway 2008; Ghate and Edelhauser 2006). Originally available in two doses, it was the first marketed device to achieve zero-order kinetics, but was discontinued in 1998 due to unexpected burst release and dislocation problems (Bertens et al. 2018; Ghate and Edelhauser 2006).

Conclusion

This chapter has provided an overview of ocular implantable drug devices, which have been engineered to overcome some of the challenges with treating ocular diseases. While topically administered drops are convenient, they have limited uptake, must be applied often, and have a risk of poor patient adherence. Systemic administration is easy but has restricted ocular penetration and systemic side effects. While intravitreal injections have targeted delivery, they required frequent injections due to short lasting treatment. Implantable drug devices are longer lasting, decreasing the treatment burden, and typically have more targeted delivery when compared to other methods of ocular delivery. Some shortcomings of nonbiodegradable implants are that they require invasive surgeries for implantation and removal, both processes have associated adverse events. While biodegradable implants do not have to be removed, they have the potential for erratic drug release and typically do not last as long as nonbiodegradable implants. Biological and technological advances continue to improve treatment options for ocular disease, and continual effort is being made to improve ocular drug delivery systems to minimize adverse events and other issues while maximizing the benefits.

References

A study of MK0140 in diabetic patients with macular edema (0140-001) a study of MK0140 in diabetic patients with macular edema (0140-001). Available from: https://clinicaltrials.gov/ct2/results?term=NCT00692614&Search=Search. Accessed 27-Oct-2021.

Aerie Pharmaceuticals Announces Acceptance of Its Investigational New Drug Application for AR-13504 Sustained Release Implant. BioSpace; 2019.

Aerie Pharmaceuticals Initiates Phase 2 Clinical Trials of AR-1105 in Patients with Macular Edema Associated with RVO. Eyewire; 2019.

Alhalafi A. Applications of polymers in intraocular drug delivery systems. Oman J Ophthalmol. 2017;10:3–8.

Allergan. Allergan receives FDA approval for DURYSTA™ (BIMATOPROST implant) the FIRST and only INTRACAMERAL biodegradable sustained-release implant to lower intraocular pressure in open-angle GLAUCOMA or ocular hypertension patients. 2020.

Anderson JM, Shive MS. Biodegradation and biocompatibility of PLA and PLGA microspheres. Adv Drug Deliv Rev. 1997;28:5–24.

Avitabile T, Marano F, Castiglione F, Bucolo C, Cro M, Ambrosio L, Ferrauto C, Reibaldi A. Biocompatibility and biodegradation of intravitreal hyaluronan implants in rabbits. Biomaterials. 2001;22:195–200.

Bausch & Lomb. Lacrisert Prescribing Information. 2016.

Bertens CJF, Gijs M, van den Biggelaar FJHM, Nuijts RMMA. Topical drug delivery devices: a review. Exp Eye Res. 2018;168:149–60.

Bethke W. New frontiers in sustained release. Rev Ophthalmol. 2015.

Bhagav P, Trivedi V, Shah D, Chandran S. Sustained release ocular inserts of brimonidine tartrate for better treatment in open-angle glaucoma. Drug Deliv Transl Res. 2011;1:161–74.

BioSpace. Lux Biosciences Announces Exclusive Worldwide License Agreement With National Institutes of Health (NIH) For Late Stage Ophthalmic Product. 2006. https://www.biospace.com/article/releases/lux-biosciences-announces-exclusive-worldwide-license-agreement-with-national-institutes-of-health-nih-for-late-stage-ophthalmic-product-.

Blizzard C, Desai A, Langh J, Buff N, Metzinger J, Goldstein M, Gelormini A, Driscoll A. Pharmacokinetics of OTX-TIC, a sustained release Travoprost Intracameral implant in rabbits. In: ARVO. Vancouver, BC: IOVS; 2019. p. 3777.

Bourges JL, Bloquel C, Thomas A, Froussart F, Bochot A, Azan F, Gurny R, BenEzra D, Behar-Cohen F. Intraocular implants for extended drug delivery: therapeutic applications. Adv Drug Deliv Rev. 2006;58:1182–202.

Boyer DS, Yoon YH, Belfort R, Bandello F, Maturi RK, Augustin AJ, Li X-Y, Cui H, Hashad Y, Whitcup SM. Three-year, randomized, sham-controlled trial of dexamethasone intravitreal implant in patients with diabetic macular edema. Ophthalmology. 2014;121:1904–14.

Brandt JD, Sall K, DuBiner H, Benza R, Alster Y, Walker G, Semba CP, Budenz D, Day D, Flowers B, Lee S, Nguyen Q, Wirta D. Six-month intraocular pressure reduction with a topical Bimatoprost ocular insert: results of a phase II randomized controlled study. Ophthalmology. 2016;123:1685–94.

Cagini C, Caricato A, Tosi G, Pascale A, Cesari C, Fiore T. Evaluation of the efficacy and safety of the ophthalmic insert Mydriasert in patients undergoing retinal angiography. Eur J Ophthalmol. 2014;24:728–34.

Callanan DG, Jaffe GJ, Martin DF, Peasron PA, Comstock TL. Treatment of posterior uveitis with a Fluocinolone Acetonide implant: three-year clinical trial results. JAMA Ophthalmol. 2008;126:1191–201.

Calles JA, Bermúdez J, Vallés E, Allemandi D, Palma S. Polymers in ophthalmology. In: Puoci F, editor. Advanced polymers in medicine. Cham: Springer International; 2015.

Campochiaro PA, Hafiz G, Shah SM, Bloom S, Brown DM, Busquets M, Ciulla T, Feiner L, Sabates N, Billman K, Kapik B, Green K, Kane F. Sustained ocular delivery of fluocinolone acetonide by an intravitreal insert. Ophthalmology. 2010;117:1393–99.e3.

Campochiaro PA, Brown DM, Pearson A, Ciulla T, Boyer D, Holz FG, Tolentino M, Gupta A, Duarte L, Madreperla S, Gonder J, Kapik B, Billman K, Kane FE. Long-term benefit of sustained-delivery Fluocinolone Acetonide vitreous inserts for diabetic macular edema. Ophthalmology. 2011;118:626–35.e2.

Campochiaro PA, Brown DM, Pearson A, Chen S, Boyer D, Ruiz-Moreno J, Garretson B, Gupta A, Hariprasad SM, Bailey C, Reichel E, Soubrane G, Kapik B, Billman K, Kane FE, Green K. Sustained delivery Fluocinolone Acetonide vitreous inserts provide benefit for at least 3 years in patients with diabetic macular edema. Ophthalmology. 2012;119:2125–32.

Campochiaro PA, Marcus DM, Awh CC, Regillo C, Adamis AP, Bantseev V, Chiang Y, Ehrlich JS, Erickson S, Hanley WD, Horvath J, Maass KF, Singh N, Tang F, Barteselli G. The port delivery system with Ranibizumab for Neovascular age-related macular degeneration: results from the randomized phase 2 ladder clinical trial. Ophthalmology. 2019;126(8):1141–15.

Carvalho R, Ko P, Murphree A, Brito R, Morenonn J, Wahjudi P, Vargas-Dougherty M, Mendes C, Matsutani G, Manders J. The preclinical safety, toxicity and pharmacokinetics (Pk) of targeted Episcleral delivery of Topotecan (TPT), Melphalan (MEL) and its combination with dexamethasone (DEX). Investig Ophthalmol Visual Sci. 2016;57.

Conway B. Recent patents on ocular drug delivery systems. Recent Pat Drug Deliv Formul. 2008;2:1–8.

Cunha-Vaz J. The blood-ocular barriers. Surv Ophthalmol. 1979;23:279–96.

De Carvalho R, Wahjudi P, Moreno J, Mendes C, Ko P, Matsutani G. Selective and sustained episcleral delivery of brimonidine: in vivo safety, toxiciy and pharmacokinetics. Investig Ophthalmol Visual Sci. 2014;55:13.

Diestelhorst M, Krieglstein GK. The ocular tolerability of a new ophthalmic drug delivery system (NODS). Int Ophthalmol. 1994;18:1–4.

Drug Delivery Sustained Release Technology. 2017. Available from: https://www.businesswire.com/news/home/20171005005490/en/Aerie-Pharmaceuticals-Announces-Drug-Delivery-Asset-Acquisition. Accessed Oct 27, 2021.

Edelhauser HF, Rowe-Rendleman CL, Robinson MR, Dawson DG, Chader GJ, Grossniklaus HE, Rittenhouse KD, Wilson CG, Weber DA, Kuppermann BD, Csaky KG, Olsen TW, Kompella UB, Holers VM, Hageman GS, Gilger BC, Campochiaro PA, Whitcup SM, Wong WT. Ophthalmic drug delivery systems for the treatment of retinal diseases: basic research to clinical applications. Invest Ophthalmol Vis Sci. 2010;51:5403–20.

Envisia Therapeutics. Envisia Therapeutics releases interim ENV515 (Travoprost XR) phase 2 data demonstrating 11-month duration-of-action after a single dose in patients with glaucoma. 2017. https://prn.to/2xECIcW. Accessed 21 May 2019.

Extension Study for the Port Delivery System With Ranibizumab (Portal) ClinicalTrials.gov. Available from: https://clinicaltrials.gov/ct2/show/NCT03683251. Accessed Oct 27, 2021.

EyePoint. YUTIQ Prescribing Information. 2021. Available from: https://yutiq.com/downloads/YUTIQ%20Prescribing%20Information_May%202021.pdf.

EyePoint Pharmaceuticals Announces U.S. Commercial Launch of DEXYCU (dexamethasone intracoular suspensio) 9%. GlobeNewswire; 2019.

Farkouh A, Frigo P, Czejka M. Systemic side effects of eye drops: a pharmacokinetic perspective. Clin Ophthalmol. 2016;10:2433–41.

FDA. FDA Approved Drug Products. 2019. https://www.accessdata.fda.gov/scripts/cder/daf/index.cfm?event=overview.process&ApplNo=210331. Accessed 12 May 2019.

Freeman WR. Intravitreal Brimonidine drug delivery system (Brimonidine DDS) in patients with geographic at-rophy: a phase 2 study. In: AAO retina subspecialty Day annual meeting; 2016.

Freeman WR, Francesco B, Souied EH, Guymer RH, Garg S, Chen FK, Rich RM, Holz FG, Li Y, Lopez FJ Kerr K. Phase 2b study of Brimonidine DDS: potential novel treatment for geographic atrophy. AVRO. 2019.

Gaudana R, Ananthula HK, Parenky A, Mitra AK. Ocular drug delivery. AAPS J. 2010;12:348–60.

GB-201, GB-202 and GB-203—Glaucoma Products. n.d. https://www.graybug.vision/our-technologies-and-pipeline/#pipeline.

Ghate D, Edelhauser HF. Ocular drug delivery. Expert Opin Drug Deliv. 2006;3:275–87.

Ghate D, Edelhauser HF. Barriers to glaucoma drug delivery. J Glaucoma. 2008;17:147–56.

Graybug Vision Presents Top Line Results of Phase 1/2a ADAGIO Study at Hawaiian Eye & REtina 2019. Business Wire. 2019.

Greaves JL, Wilson CG, Birmingham AT, Richardson MC, Bentley PH. Scintigraphic studies on the corneal residence of a new ophthalmic delivery system (NODS): rate of clearance of a soluble marker in relation to duration of pharmacological action of pilocarpine. Br J Clin Pharmacol. 1992;33:603–9.

Haghjou N, Soheilian M, Abdekhodaie MJ. Sustained release intraocular drug delivery devices for treatment of uveitis. J Ophthalmic Vis Res. 2011;6:317–29.

Haller JA, Bandello F, Belfort R, Blumenkranz MS, Gillies M, Heier J, Loewenstein A, Yoon Y-H, Jacques M-L, Jiao J, Li X-Y, Whitcup SM. Randomized, sham-controlled trial of dexamethasone intravitreal implant in patients with macular edema due to retinal vein occlusion. Ophthalmology. 2010;117:1134–46.e3.

Haller JA, Bandello F, Belfort R, Blumenkranz MS, Gillies M, Heier J, Loewenstein A, Yoon YH, Jiao J, Li X-Y, Whitcup SM. Dexamethasone intravitreal implant in patients with macular edema related to branch or central retinal vein occlusion: twelve-month study results. Ophthalmology. 2011;118:2453–60.

Hermann MM, Üstündag C, Diestelhorst M. Electronic compliance monitoring of topical treatment after ophthalmic surgery. Int Ophthalmol. 2010;30:385–90.

Hill JC. Slow-release artificial tear inserts in the treatment of dry eyes in patients with rheumatoid arthritis. Br J Ophthalmol. 1989;73:151–4.

Humayun M, Santos A, Altamirano JC, Ribeiro R, Gonzalez R, de la Rosa A, Shih J, Pang C, Jiang F, Calvillo P, Huculak J, Zimmerman J, Caffey S. Implantable MicroPump for drug delivery in patients with diabetic macular edema. Transl Vis Sci Technol. 2014;3:5–5.

Iluvien Approval History. www.drugs.com/history/iluvien.html. Accessed 12 May 2019.

Iluvien Prescribing information. Alimera Sciences. alimerasciences.com/wp-content/uploads/2016/02/Prescribing-Information.pdf. Accessed 12 May 2019.

Jaffe GJ, Martin D, David C, Andrew Pearson P, Levy B, Comstock T. Fluocinolone Acetonide implant (Retisert) for noninfectious posterior uveitis: thirty-four–week results of a multicenter randomized clinical study. Ophthalmology. 2006;113:1020–7.

Jaffe GJ, Stephen Foster C, Pavesio CE, Paggiarino DA, Riedel GE. Effect of an injectable Fluocinolone Acetonide insert on recurrence rates in chronic noninfectious uveitis affecting the posterior segment: twelve-month results. Ophthalmology. 2019;126:601–10.

Jager RD, Aiello LP, Patel SC, Cunningham ET Jr. Risks of intravitreous injection: a comprehensive review. Retina. 2004;24:676–98.

Jain RA. The manufacturing techniques of various drug loaded biodegradable poly(lactide-co-glycolide) (PLGA) devices. Biomaterials. 2000;21:2475–90.

Jervis LP. A summary of recent advances in ocular inserts and implants. J Bioequiv. 2017;9:320–3.
Joseph M, Trinh HM, Cholkar K, Pal D, Mitra AK. Recent perspectives on the delivery of biologics to back of the eye. Expert Opin Drug Deliv. 2017;14:631–45.
Kearns VR, Williams RL. Drug delivery systems for the eye. Expert Rev Med Devices. 2009;6:277–90.
Kedhar SR, Jabs DA. Cytomegalovirus retinitis in the era of highly active antiretroviral therapy. Herpes. 2007;14:66–71.
Kempen JH, Altaweel MM, Holbrook JT, Jabs DA, Louis TA, Sugar EA, Thorne JE. Randomized comparison of systemic anti-inflammatory therapy versus Fluocinolone Acetonide implant for intermediate, posterior, and Panuveitis: the multicenter uveitis steroid treatment trial. Ophthalmology. 2011;118:1916–26.
Khurana RN, Appa SN, McCannel CA, Elman MJ, Wittenberg SE, Parks DJ, Ahmad S, Yeh S. Dexamethasone implant anterior chamber migration: risk factors, complications, and management strategies. Ophthalmology. 2014;121:67–71.
Kiernan DF, Mieler WF. The use of intraocular corticosteroids. Expert Opin Pharmacother. 2009;10:2511–25.
Kimura H, Ogura Y. Biodegradable polymers for ocular drug delivery. Ophthalmologica. 2001;215:143–55.
Kumari A, Sharma P, Garg V, Garg G. Ocular inserts—advancement in therapy of eye diseases. J Adv Pharm Technol Res. 2010;1:291–6.
Kuno N, Fujii S. Biodegradable intraocular therapies for retinal disorders. Drugs Aging. 2010;27:117–34.
Kuppermann BD, Blumenkranz MS, Haller JA, Williams GA, Weinberg DV, Chou C, Whitcup SM, Dexamethasone DDS Phase II Study Group. Randomized controlled study of an Intravitreous dexamethasone drug delivery system in patients with persistent macular edema. JAMA Ophthalmol. 2007;125:309–17.
Lee DJ. Intraocular implants for the treatment of autoimmune uveitis. J Funct Biomater. 2015;6:650–66.
Lee TW, Robinson JR. Drug delivery to the posterior segment of the eye: some insights on the penetration pathways after subconjunctival injection. J Ocul Pharmacol Ther. 2001;17:565–72.
Lee SS, Robinson MR. Novel drug delivery systems for retinal diseases. Ophthalmic Res. 2009;41:124–35.
Lee Y-C, Simamora P, Pinsuwan S, Yalkowsky SH. Review on the systemic delivery of insulin via the ocular route. Int J Pharm. 2002;233:1–18.
Lee SS, Hughes P, Ross AD, Robinson MR. Biodegradable implants for sustained drug release in the eye. Pharm Res. 2010;27:2043–53.
Lee SS, Hughes P, Ross AD, Robinson MR. Advances in biodegradable ocular drug delivery systems. In: Kompella UB, Edelhauser HF, editors. Drug product development for the back of the eye. Boston, MA: Springer US; 2011.
Lewis RA, Christie WC, Day DG, Randy Craven E, Walters T, Bejanian M, Lee SS, Goodkin ML, Zhang J, Whitcup SM, Robinson MR, Aung T, Beck AD, Christie WC, Coote M, Crane CJ, Randy Craven E, Crichton A, Day DG, Jane Durcan F, Flynn WJ, Gagné S, Goldberg DF, Delan J, Starck Johnson C, Kurtz S, Lewis RA, Mansberger SL, Perera SA, Rotberg MH, Saltzmann RM, Schenker HI, Tepedino ME, Yap-Veloso MIR, Uy HS, Walters TR. Bimatoprost sustained-release implants for glaucoma therapy: 6-month results from a phase I/II clinical trial. Am J Ophthalmol. 2017;175:137–47.
Liechty WB, Kryscio DR, Slaughter BV, Peppas NA. Polymers for drug delivery systems. Annu Rev Chem Biomol Eng. 2010;1:149–73.
Lim J, Hung D, Fung A, Wieland M, Wong V. A phase 1 study evaluating the safety and evidence of efficacy of IBI-20089, a triamcinolone intravi-treal injection formulated with the Verisome ™ drug delivery technology, in patients with cystoid macular edema. Invest Ophthalmol Vis Sci. 2010;51.
Lima LH, Farah ME, Gum G, Ko P, de Carvalho RAP. Sustained and targeted episcleral delivery of celecoxib in a rabbit model of retinal and choroidal neovascularization. Int J Retina Vitreous. 2018;4:31.

Lo R, Li P-Y, Saati S, Agrawal RN, Humayun MS, Meng E. A passive MEMS drug delivery pump for treatment of ocular diseases. Biomed Microdevices. 2009;11:959.
Macha S, Mitra AK. Ocular disposition of ganciclovir and its monoester prodrugs following intravitreal administration using microdialysis. Drug Metab Dispos. 2002;30:670–5.
Mealy JE, Fedorchak MV, Little SR. In vitro characterization of a controlled-release ocular insert for delivery of brimonidine tartrate. Acta Biomater. 2014;10:87–93.
Miller RA, Brady JM, Cutright DE. Degradation rates of oral resorbable implants (polylactates and polyglycolates): rate modification with changes in PLA/PGA copolymer ratios. J Biomed Mater Res. 1977;11:711–9.
Mohammad DA, Sweet BV, Elner SG. Retisert: is the new advance in treatment of uveitis a good one? Ann Pharmacother. 2007;41:449–55.
Musch DC, Martin DF, Gordon JF, Davis MD, Kuppermann BD. Treatment of cytomegalovirus retinitis with a sustained-release ganciclovir implant. N Engl J Med. 1997;337:83–90.
Nancy M. Holekamp, Peter A. Campochiaro, Margaret Chang, Daniel Miller, Dante Pieramici, Anthony P. Adamis, Christopher Brittain, Erica Evans, Derrick Kaufman, Katie F. Maass, Shienal Patel, Shrirang Ranade, Natasha Singh, Giulio Barteselli, Carl Regillo, Archway Randomized Phase 3 Trial of the Port Delivery System With Ranibizumab for Neovascular Age-Related Macular Degeneration, Ophthalmology. 2021.
Niegvesky G, Butrus S, Abifarah H, Lee Y, Yalkowsky S. Ocular Gelfoam disc-applicator for pupillary dilation in humans. J Ocul Pharmacol Ther. 2000;16:311–5.
Nordmann J-P, Baudouin C, Renard J-P, Denis P, Lafuma A, Laurendeau C, Jeanbat V, Berdeaux G. Measurement of treatment compliance using a medical device for glaucoma patients associated with intraocular pressure control: a survey. Clin Ophthalmol. 2010;4:731.
Novel anti-VEGF for wet AMD meets safety, tolerability endpoint. Retina/vitreous. Healio; 2019.
NT-502 ECT. Neurotech Press Release; 2016.
Ocular Therapeutix Announces Dosing of First Patient in Phase 1 Clinical Trial for the Treatment of Wet AMD. Eyewire. 2019.
Ocular Therapeutix™ Announces Treatment of First Patient in Phase 1 Clinical Trial of OTX-TIC (travoprost intracameral implant) for the Treatment of Glaucoma and Ocular Hypertension. BioSpace. 2018.
Ozurdex Approval History. Development history and FDA approval process for Ozurdex. https://www.drugs.com/history/ozurdex.html.
Park JB, Lakes RS. Biomaterials : an introduction. New York: Springer; 2007.
Patel A, Cholkar K, Agrahari V, Mitra AK. Ocular drug delivery systems: an overview. World J Pharmacol. 2013;2:47–64.
Pearce W, Hsu J, Yeh S. Advances in drug delivery to the posterior segment. Curr Opin Ophthalmol. 2015;26:233–9.
Pearson PA, Comstock TL, Ip M, Callanan D, Morse LS, Ashton P, Levy B, Mann ES, Eliott D. Fluocinolone Acetonide intravitreal implant for diabetic macular edema: a 3-year multicenter, randomized, controlled clinical trial. Ophthalmology. 2011;118:1580–7.
Pijls RT, Sonderkamp T, Daube GW, Krebber R, Hanssen HH, Nuijts RM, Koole LH. Studies on a new device for drug delivery to the eye. Eur J Pharm Biopharm. 2005;59:283–8.
Pijls RT, Cruysberg LPJ, Nuijts RMMA, Dias AA, Koole LH. Capacity and tolerance of a new device for ocular drug delivery. Int J Pharm. 2007;341:152–61.
PolyActiva commences its first phase I clinical trial with potential to improve daily lives of millions of glaucoma patients. Business Wire; 2018.
Pontes de Carvalho RA, Krausse ML, Murphree AL, Schmitt EE, Campochiaro PA, Maumenee IH. Delivery from episcleral exoplants. Invest Ophthalmol Vis Sci. 2006;47:4532–9.
Psivida Corp Annual Report. 2016. http://investors.psivida.com/static-files/020cf090-42dd-467b-961c-52ded762c3d9.
Ravindran VK, Repala S, Subadhra S, Appapurapu AK. Chick chorioallantoic membrane model for in ovo evaluation of timolol maleate–brimonidine tartrate ocular inserts. Drug Deliv. 2014;21:307–14.

Robinson MR, Whitcup SM. Pharmacologic and clinical profile of dexamethasone intravitreal implant. Expert Rev Clin Pharmacol. 2012;5:629–47.
Rofagha S, Prechanond T, Stewart JM. Late spontaneous dissociation of a Fluocinolone Acetonide implant (Retisert). Ocul Immunol Inflamm. 2013;21:77–8.
Saati S, Lo R, Li P-Y, Meng E, Varma R, Humayun MS. Mini drug pump for ophthalmic use. Curr Eye Res. 2010;35:192–201.
Saenz-de-Viteri M, Fernandez-Robredo P, de Nova E, Bonet-Farriol E, Sabater AL, Zarranz-Ventura J, Caire J, Sadaba LM, Salinas-Alaman A, Garcia-Layana A. Comparative study measuring the dilatory effect of a mydriatic device (Mydriasert((R))) versus topical drops. Int J Ophthalmol. 2013;6:801–4.
Safety and preliminary efficacy study of V404 PDS in Uveitis. n.d. Available from: https://clinicaltrials.gov/ct2/show/NCT02125266.
Salyani A, Birt C. Evaluation of an eye drop guide to aid self-administration by patients experienced with topical use of glaucoma medication. Can J Ophthalmol. 2005;40:170–4.
Sandahl M, Melton D, Tully J, Yang M, Garcia A, Trevino L, Williams S. Evaluation of reproducibility and uniformity of PRINT implant manufacturing. In: Association for Research in Vision and Ophthalmology annual meeting. Honolulu, HI; 2018.
Sangwan VS, Andrew Pearson P, Paul H, Comstock TL. Use of the Fluocinolone Acetonide intravitreal implant for the treatment of noninfectious posterior uveitis: 3-year results of a randomized clinical trial in a predominantly Asian population. Ophthalmol Ther. 2015;4:1–19.
Seah SKL, Husain R, Gazzard G, Lim MCC, Hoh S-T, Oen FTS, Aung T. Use of Surodex in Phacotrabeculectomy surgery. Am J Ophthalmol. 2005;139:927–8.
Seal JR, Robinson MR, Burke J, Bejanian M, Coote M, Attar M. Intracameral sustained-release Bimatoprost implant delivers Bimatoprost to target tissues with reduced drug exposure to off-target tissues. J Ocul Pharmacol Ther. 2019;35:50–7.
Silva-Cunha A, Fialho SL, Naud M-C, Behar-Cohen F. Poly-ε-caprolactone intravitreous devices: an in vivo study. Invest Ophthalmol Vis Sci. 2009;50:2312–8.
Stephenson M. Can sustained release deliver the goods?. Rev Ophthalmol. 2018.
Streilein JW. Ocular immune privilege: the eye takes a dim but practical view of immunity and inflammation. J Leukoc Biol. 2003;74:179–85.
SurModics. Durable drug delivery coatings. 2013. http://www.surmodics.com/assets/uploads/documents/SurModics_Durable_Drug_Delivery_0513.pdf.
Swati M, Oshin M. Recent advancements in biodegradable ocular implants. Curr Drug Deliv. 2018;15:144–54.
Tan DTH, Chee S-P, Lim L, Lim ASM. Randomized clinical trial of a new dexamethasone delivery system (surodex) for treatment of post-cataract surgery inflammation. Ophthalmology. 1999;106:223–31.
Varma D. Sustained release drug delivery: closer than you think; 2018.
Weintraub A. Lux bio resurfaces with defined path for pill for rare eye disease. In: Xconomy; 2012.
Whitcup SM, Azar DT. Pharmacologic therapy of ocular disease. Berlin, Heidelberg, New York: Springer; 2017.
Williams GA, Haller JA, Kuppermann BD, Blumenkranz MS, Weinberg DV, Chou C, Whitcup SM. Dexamethasone posterior-segment drug delivery system in the treatment of macular edema resulting from uveitis or Irvine-Gass syndrome. Am J Ophthalmol. 2009;147:1048–54.e2.
Yasin MN, Svirskis D, Seyfoddin A, Rupenthal ID. Implants for drug delivery to the posterior segment of the eye: a focus on stimuli-responsive and tunable release systems. J Control Release. 2014;196:208–21.
Yasukawa T, Ogura Y, Kimura H, Sakurai E, Tabata Y. Drug delivery from ocular implants. Expert Opin Drug Deliv. 2006;3:261–73.

Development of Artificial Tears Products for Ocular Conditions

Peter A. Simmons

Abstract Artificial tears are developed for use by patients desiring relief from ocular discomfort due to various dry eye conditions, following ocular surgery, and to improve comfort of contact lenses during wear. Currently, artificial tears are over-the-counter drugs in the USA, medical devices in Europe and elsewhere, and also classified as prescription drugs in some regions. Most tear formulations include a hydrophilic polymer to provide lubrication, water retention, and viscosity on the ocular surface. In addition, other beneficial ingredients may be added, along with excipients such as buffers, tonicity agents, and preservatives. Process development may be complicated due to the differing ingredients and their sensitivity to the manufacturing conditions. Clinical validation is useful to demonstrate safety and efficacy of the product, but is not always required prior to commercial sale. Regulations governing sale of artificial tears vary considerably around the world and need to be carefully considered early in development.

Keywords Artificial tears · Lubricants · Over-the-counter · Dry eye · Post-surgical care · Contact lens rewetter

Abbreviations

BAK	Benzalkonium chloride
CE mark	Conformité Européene

P. A. Simmons (✉)
Allergan plc, Irvine, CA, USA

PCS Research Consulting, Yorba Linda, CA, USA

University of New South Wales, Sydney, Australia

S. Neervannan, U. B. Kompella (eds.), *Ophthalmic Product Development*, AAPS Advances in the Pharmaceutical Sciences Series 37,
https://doi.org/10.1007/978-3-030-76367-1_19

cPs	Centipoise
CTD	Common technical document
DED	Dry eye disease
Deg C	Degrees Celsius
DEWS	Dry Eye Workshop
FDA	Food and Drug Administration
GRAS	Generally recognized as safe
HPMC	Hydroxypropyl methylcellulose
ISO	International Standards Organization
KCl	Potassium chloride
KOL	Key opinion leader
LASIK	Laser-assisted in situ keratomileusis
LFA-1	Lymphocyte function-associated antigen 1
$MgCl_2$	Magnesium chloride
MGD	Meibomian gland dysfunction
mOsm	Milliosmolar
Na^+	Sodium ion
NaCl	Sodium chloride
NEI	National Eye Institute
OTC	Over-the-counter
USA	United States of America
USP	United States Pharmacopeia

Introduction

Artificial tears have been used for many years to replace insufficient ocular surface moisture lost due to a range of acute to chronic ocular surface conditions. Modern artificial tear formulations, in addition to replacing needed moisture, provide long-lasting lubrication for comfortable blinking and may also provide protective substances which either enhance the healing of the ocular surface from insult or provide specific nutritional support for maintenance of ocular surface health.

From a development perspective, selection of ingredients for an artificial tear formulation has some similarities to drug development in general, in that a specific target disease or condition should be identified at the onset of the project. We have found it useful to categorize target conditions into several groups, including dry eye disease (and its subtypes), post-operative care, and use with contact lenses. Each of these three conditions requires distinct physiochemical properties and clinical testing designs and has differing regulatory requirements.

Dry Eye Disease

While dry eye disease has been recognized as a distinct ocular condition, the understanding of its pathological features, etiology, and treatment paradigms has evolved markedly within the past 30 years. In 1985, an NEI consensus was published defining dry eye as "a disorder of the tear film due to tear deficiency or excessive tear evaporation which causes damage to the interpalpebral ocular surface and is associated with symptoms of ocular discomfort" (Lemp 1995).

Following that initial consensus, recognition developed regarding the important role that altered immune function has in development and pathology of chronic dry eye, leading to the expanded definition of dry eye disease developed by an international consensus dubbed the Dry Eye Workshop (DEWS): "Dry eye is a multifactorial disease of the tears and ocular surface that results in symptoms of discomfort, visual disturbance, and tear film instability with potential damage to the ocular surface. It is accompanied by increased osmolarity of the tear film and inflammation of the ocular surface" (Lemp et al. 2007).

More recently, the DEWS process reviewed further research in the field and modified the definition further: "Dry eye is a multifactorial disease of the ocular surface characterized by a loss of homeostasis of the tear film, and accompanied by ocular symptoms, in which tear film instability and hyperosmolarity, ocular surface inflammation and damage, and neurosensory abnormalities play etiological roles" (Craig et al. 2017).

A key factor in the current understanding of dry eye disease is that it may result from a number of distinct etiologies, which may exist alone or in combination in individual patients. One way to organize the various causes of dry eye is to divide them into intrinsic and extrinsic factors. Intrinsic factors include advancing age, female sex, autoimmune disease, and eyelid or blinking pathologies such as meibomian gland dysfunction and lagophthalmos (inability to completely close the eye). External factors would include ocular surgery; medication side effects; chronic environmental insult such as excess wind, heat, or air conditioning; and occupational factors including excessive computer screen use with associated low blink rate. In many patients, several etiological factors combine to reinforce a "vicious circle" that leads to activation of an inflammatory process that produces chronic dysfunction of the lacrimal secretory system, symptoms of discomfort, and progressive damage to the ocular surface (see Fig. 1).

Treatment of dry eye is staged depending upon the severity of patients' signs and symptoms, beginning with artificial tears and lifestyle modifications and progressing through pharmaceutical therapy and various surgical procedures. Artificial tears of one type or another are recommended at all stages of disease treatment.

Staged management and treatment recommendations for dry eye disease (Jones et al. 2017).

Step 1

- Education regarding the condition, its management, treatment, and prognosis.
- Modification of local environment.

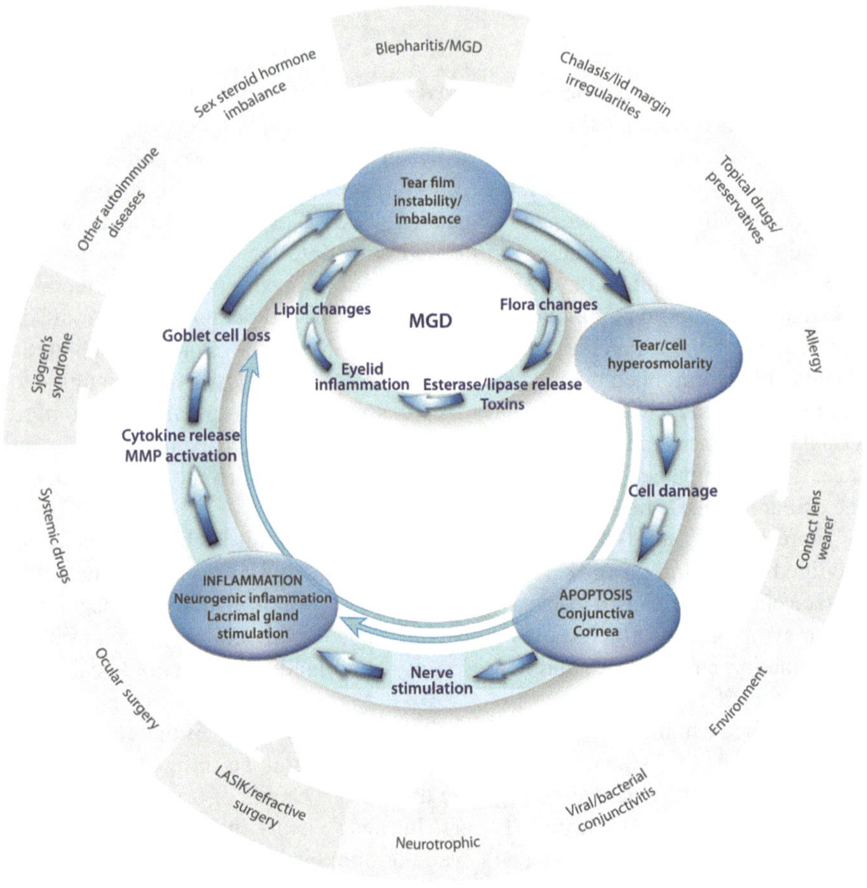

Fig. 1 The vicious circle in dry eye and meibomian gland dysfunction (Baudouin et al. 2014)

- Education regarding potential dietary modifications (including oral essential fatty acid supplementation).
- Identification and potential modification/elimination of offending systemic and topical medications.
- Ocular lubricants of various types (if meibomian gland dysfunction (MGD) is present, then consider lipid-containing supplements).
- Lid hygiene and warm compresses of various types.

Step 2

- If above options are inadequate, consider:
- Non-preserved ocular lubricants to minimize preservative-induced toxicity.
- Tea tree oil treatment for *Demodex* (if present).
- Tear conservation.
- Punctal occlusion.

- Moisture chamber spectacles/goggles.
- Overnight treatments (such as ointment or moisture chamber devices).
- In-office physical heating and expression of the meibomian glands (including device-assisted therapies, such as LipiFlow).
- In-office intense pulsed light therapy for MGD.
- Prescription drugs to manage DED.
- Topical antibiotic or antibiotic/steroid combination applied to the lid margins for anterior blepharitis (if present).
- Topical corticosteroid (limited duration).
- Topical secretagogues.
- Topical non-glucocorticoid immunomodulatory drugs (such as topical LFA-1 antagonist drugs (such as lifitegrast)).
- Oral macrolide or tetracycline antibiotics.

Step 3

- If above options are inadequate, consider:
- Oral secretagogues.
- Autologous/allogeneic serum eye drops.
- Therapeutic contact lens options.
- Soft bandage lenses.
- Rigid scleral lenses.

Step 4

- If above options are inadequate, consider:
- Topical corticosteroid for longer duration.
- Amniotic membrane grafts.
- Surgical punctal occlusion.
  Other surgical approaches (e.g., tarsorrhaphy, salivary gland transplantation).

Post-Operative Care

As mentioned above, ocular surgery may be an initiating event for development of dry eye disease. However, unlike other factors which may slowly develop and become increasingly severe over time, surgical trauma is an acute disruption of ocular surface structure and function, which follows the opposite course, being most severe immediately following the procedure, with progressive improvement as healing occurs.

The most common surgical procedure that may have effects on the lacrimal gland is cataract surgery. Since this is a procedure primarily performed upon older patients, who are already within one or more higher risk factors for dry eye-related conditions, including not only advanced age but also hormonal changes (e.g., menopause in women) and increased medication use, ocular surface discomfort is a common complaint following the surgery. Fortunately, modern cataract surgical procedures use small incisions which often heal relatively rapidly; however, treatment with artificial tears during the post-operative period is often indicated.

The advent of refractive surgery, and especially the most common procedure, LASIK, merits special attention in the area of ocular surface conditions. In this procedure, a corneal flap is created by use of either a specialized blade or a laser, with the cut going to a variable depth but through the corneal epithelium, leaving only a small "hinge" region of epithelium intact. The underlying stroma is then reshaped by the laser, and the flap is replaced onto the stromal bed. The process of flap creation and stromal reshaping severs many of the corneal nerves that normally provide sensory feedback from the corneal surface to help regulate tear secretion, thus causing a neurotrophic dry eye condition, which persists in most patients for up to 6 months and becomes permanent in some (Gomes et al. 2017). Artificial tears are therefore a standard component of LASIK post-operative care kits along with antibiotic and steroid drops and often need to be continued for an extensive time following surgery.

Contact Lens Comfort

Contact lenses can provide an excellent alternative to eyeglasses for vision correction, and modern lens designs are available for correction of astigmatism and presbyopia, making them suitable for patients with a wide variety of visual needs and of all ages. Although some "hard" contact lens materials are still used, the majority of contact lens materials in current use are a range of soft hydrogel materials that incorporate substantial quantities of water within a polymer matrix. For all lens types, but particularly for hydrogel lens wearers, dehydration, lens awareness, and lens discomfort are common issues causing patient complaints and eventual discontinuation of wear (Nichols et al. 2013).

Eye drops to be used during lens wear have been available since contact lenses were first developed and for hydrogel lenses were first designed as rewetting drops to replace lens matrix fluid lost to evaporation during the day. Further evolution of this category of product has introduced formulations with lubricating and viscosity ingredients to address discomfort. Key to this category of eye drop is the interaction with the material of the contact lens and also other solutions for storage, rinsing, and disinfection of the lenses. These interactions must be tested demonstrating lack of effect on the physical and optical properties of the lens prior to commercialization.

Concomitant Use

Many patients with symptoms of ocular discomfort that may be treated with artificial tears are also under treatment for a variety of ocular and non-ocular conditions that may have direct or indirect effects on the ocular surface. There is substantial literature, for example, on the effects of preservatives used in topical glaucoma medications and their potential for creating or exacerbating ocular surface damage and discomfort (Gomes et al. 2017). Therefore potential use of an artificial tear formulation by a patient utilizing other topical agents should be considered in the overall development plan, leading to, for example, specific inclusion or exclusion from the test population in a clinical trial, depending upon the importance of ocular comorbidities in the commercial plan for the product.

On an acute basis, an increasing niche use of ocular lubricating and comfort eye drops is in the patient population receiving injections for retinal conditions such as diabetic retinopathy, macular edema, or macular degeneration. It is common for these patients to experience pain and discomfort for some period of time following each injection, and use of an appropriate artificial tear preparation may be important for overall patient satisfaction with their therapeutic regimen.

Specialized Uses of Artificial Tears

In addition to patient use for a variety of ocular conditions, artificial tear preparations are also used by health-care providers in a number of specialized circumstances. These include use of a viscous ocular lubricant during diagnostic procedures such as gonioscopy or lubrication of the ocular surface during surgical procedures. Certain artificial tear formulations are also utilized to fill the posterior surface of scleral lenses prior to insertion (whether performed in office or in the patient's home). These uses are rarely indicated in the product labeling or tested specifically during product development; rather, they rely on case reports and peer communications as demonstrations of potential safety and efficacy (e.g., Sonsino and Schornack 2013).

Development Process

Product Requirements

At the outset of a development project for an artificial tear, certain activities common to all drug development must be completed. These include a careful examination of the target patient profile (see first section above); the physical, chemical, and pharmacological (if any) properties of the formulation to be developed; and the specific regulatory requirements for the market(s) where the product will be sold.

For all product requirements, enough specificity must be made to allow validation of the resulting formulations against each requirement; that is, do formulation properties and performance testing demonstrate (quantitatively if possible) that each requirement has been met. So for the patient profile, it is necessary to identify not only the particular indication (e.g., dry eye disease, contact lens care, etc.) but also the desired effect for that patient, such as clinically significant improvement of symptoms using a validated symptom scale in comparison to pre-treatment scores or superior performance in comparison to an existing treatment.

Physical and chemical characteristics should be specified in standard units and provided as a range suitable for the setting of specifications, which can be measured using available instrumentation both during development and also throughout commercial production. Typically for artificial tears, these include the concentrations of active or key ingredients (such as principal viscosity agent or preservative), pH, viscosity, and overall tonicity.

The actual values for physiochemical properties in commercial artificial tear formulas vary considerably. Viscosity in particular is a key determinant of clinical performance: low-viscosity tears (1–10 cPs) arc usually indicated for patients with occasional discomfort or for use with contact lenses; mid-viscosity tears (10–50 cPs) are used for more chronic dry eye conditions in patients who will tolerate some degree of blur after instillation; and high-viscosity tears (>50 cPs) are generally reserved for nighttime use and for special cases such as lagophthalmos and diagnostic procedures. pH of tears is generally controlled within the physiological range of 6.5 to 7.5; however, tonicity may also be varied in order to produce a hypotonic eyedrop, which has been shown to confer some benefit (Aragona et al. 2002), with some formulations having tonicity below 200 mOsm (see below under Excipients for further discussion). Eyedrops with high tonicity may be considered hypertonic eyedrops, which have specific applications in corneal edema and other conditions and whose ingredients and labeling are also specified in the FDA monograph (FDA 2002). Although ocular lubrication is typically listed as one of the actions for artificial tears, it is not routinely measured for tear formulations. An in vitro model for lubricity has been developed, showing that differing formulations do affect this property (Rangarajan et al. 2015).

Some artificial tear formulations may exhibit specific physiological or even pharmacological actions, for example, a desired ocular residence time or the vasoconstriction of conjunctival vessels (to reduce visible eye redness). The conditions under which these properties are demonstrated must be specified, as they may require special testing with in vitro or animal models, in addition to clinical measurements.

A key product requirement is the regulatory classification, since artificial tears are regulated in a variety of means in different regions. Conformity with the target market requirements should be specified in regard to the identity of ingredients, required labeling, and allowed indications. An overview of regulatory classifications in various global markets is provided later in this chapter.

Ingredients

Selection of ingredients in an artificial tear formulation is driven by a number of factors. In many cases, the use of an established (and therefore already approved) active ingredient such as those published in the OTC monograph for ophthalmic demulcents by FDA (2002) is essential for the commercial plan. In addition, the development project may be based upon a predicate product and thus will contain many of the same ingredients with a specific new feature, such as enhanced viscosity, removal of preservative, or addition of a novel beneficial ingredient. In all cases, ingredients of artificial tears must be available in the appropriate pharmaceutical grade and are allowed in this class of product by the regulatory authority in the target market.

Almost all artificial tear formulations are intended to lubricate and hydrate the ocular surface (although there are exceptions). Therefore, one or more hydrophilic polymers that provide viscosity at low concentration are included in most formulations. The FDA demulcent monograph lists a number of such polymers suitable as artificial tear active ingredients that are derived from cellulose, which are found in most artificial tears sold in the USA. In other regions, derivatives of polycarboxylic acid and hyaluronic acid are also common hydrophilic polymers. Each one of these polymers has specific properties, including molecular weight, presence or absence of charge in solution, ability to build viscosity at low concentration, sensitivity to pH, overall electrolyte concentration, or temperature, as well as rheological properties which may be important in determining how the polymer solution feels on the eye. Table 1 lists some of the polymers commonly used in artificial tears.

In addition to polymers, some small molecules are also considered active ingredients in the USA or elsewhere as they provide humectant or other beneficial properties to the artificial tear. Examples of these are provided in Table 2.

With the large number of possible functional ingredients (either declared as active or not), clearly the art of formulation for an artificial tear is in selecting which ingredient(s) to include, at what concentrations, and, in the case of polymers, which molecular weight or chain length. In the USA, the FDA demulcent monograph specifies the allowed concentration range for each active ingredient and limits combination of ingredients to three actives. Non-active ingredients with useful functional properties must support the function of the actives, and no specific claims may be made for them individually. For example, hydroxypropyl guar, found in the Systane® formulations, is not an active ingredient; from a regulatory viewpoint, it supports the function of the actives listed on the product label. However, the manufacturer (Alcon Labs) has made it clear in their literature that much of the therapeutic benefit of the Systane family of products derives from the hydroxypropyl guar.

Table 1 Polymers commonly used in artificial tears

Polymer	Source	Listed in US monograph?	Concentration range	Sensitivity	Other properties
Methylcellulose	Cellulose	Yes			
Carboxy-methylcellulose	Cellulose	Yes	0.1–1.5%	Viscosity affected by temperature	Available in several molecular weights which may be blended. Anionic
Hydroxypropyl-methylcellulose	Cellulose	Yes	0.2–1.5%	Viscosity affected by temperature	Neutral charge
Hyaluronic acid	Bacterial fermentation	No	0.1–0.4%	Viscosity affected by overall ionic strength	Available in varying molecular weights. Anionic; highly shear-thinning
Hydroxypropyl-guar	Guar gum	No	0.1–0.3%	Viscosity affected by pH	
Polysorbate 80	Synthetic	Yes	0.1–1.9%	Stable	Used in emulsions as a stabilizer for oils
Polyacrylic acid	Synthetic	No	0.1–1.0	Very salt-sensitive	Used as a gelling agent in gel-type tears; also stabilizes oils in emulsions
Polyvinyl alcohol	Synthetic	Yes	0.1–1.0	Stable	Common in older formulas
Polyethylene glycol	Synthetic	Yes	0.1–1.0	Stable	Low viscosity polymer typically used in combination with others

Excipients

Excipients used in artificial tear formulations have generally the same functions as excipients found in other classes of pharmaceutical products: maintaining the chemical integrity of the active agent(s) while on the shelf and delivering the active(s) to the target tissue during use. In the case of topical artificial tears, however, the selection and concentration of active ingredients may directly affect the functional

Table 2 Other ingredients commonly listed as active or primary functional ingredients

Ingredient	Source	USA monograph active?	Concentration range used	Sensitivity	Other properties
Glycerin	Synthetic or animal source	Yes	0.1–1.0	Stable	Builds osmotic strength of formulation well
Propylene glycol	Synthetic or animal source	Yes	0.1–1.0	Stable	Similar to glycerin
Trehalose	Purified from plant source	No	1.0–3.0	Stable	Disaccharide with protective properties for cells

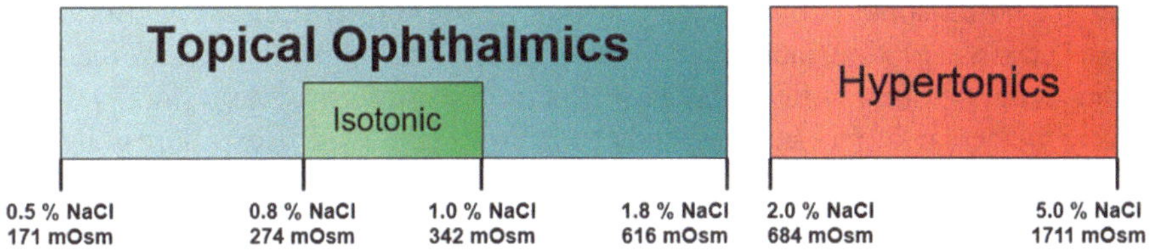

Fig. 2 FDA guidelines on tonicity (FDA 1980a)

properties of the formulation on the ocular surface – either positively or negatively – and therefore merit particular attention by the formulator.

Tonicity agents provide osmotic strength to the formulation in order to have the total osmolarity within a specific target range. The target is most commonly near the osmolarity of natural tears (300 +/– 10 mOsm); the FDA has specified that solutions between the concentrations of 0.8% and 1.0% NaCl equivalent (roughly 274 to 342 mOsm) may be considered "isotonic" (see Fig. 2). While NaCl itself is perhaps the most common tonicity agent used in artificial tear formulations, other ingredients such as mannitol or sorbitol are also commonly used. Many other excipient ingredients also provide osmotic strength to the formulation, including the buffer and beneficial electrolytes (e.g., KCl or $MgCl_2$), and some active ingredients (as defined by the FDA monograph) are also osmotically active (e.g., glycerin). A key factor in selection of osmotic agent is the potential sensitivity of the viscous polymer to the presence of specific compounds. For example, hyaluronic acid and to an even greater extent polyacrylic acid solubilize and lose viscosity in the presence of electrolytes, particularly NaCl; another example is the use of sorbitol in some of the Systane formulations to stabilize the viscosity of the hydroxypropyl guar.

Numerous artificial tear formulations are intentionally compounded at a low total ionic strength (<250 mOsm) and are usually referred to as hypotonic artificial tears. Examples include Hypotears (Novartis), first available in the 1970s, and TheraTears (Akorn). The purpose of the hypotonic formulation is an attempt to counteract the high osmolarity found in tears of dry eye patients. While there is some data to suggest that hypotonic solutions may have therapeutic benefits when compared with isotonic tears (Aragona et al. 2002), other studies have indicated that application of a hypotonic tear may influence the tear osmolarity for at most a few minutes (Holly and Lamberts 1981). In fact, since formulation of a hypotonic artificial tear typically entails reduction or elimination of added NaCl from the solution, much of the benefit of these formulas may be due to lack of interference by the salt with functionality of the lubricant polymer, as well as reduction of any pro-inflammatory stimulus provided by the excess Na^+ found in the isotonic tear formulation (Luo et al. 2005).

Some tonicity agents may be seen as having a direct beneficial effect on the therapeutic functionality of the tear formulation. As mentioned above, sorbitol in Systane formulations is described as stabilizing the function of the hydroxypropyl guar polymer matrix on the ocular surface. Addition of certain small organic osmolytes including some polyols (e.g., glycerin, erythritol, xylitol) and amino acids (e.g., betaine, carnitine, taurine) in place of added NaCl not only reduces the total sodium content of the solution but provides specific protective and restorative benefits to the ocular surface cells when they are under stress due to exposure to hypertonic solutions, which usually occurs in dry eye conditions (Corrales et al. 2008; Garrett et al. 2013; Hua et al. 2015). In this case, some of these organic osmolytes have been shown to reduce the production of pro-inflammatory markers and to prevent or reduce ocular surface damage in cell and animal models of dry eye conditions (Chen et al. 2013). Finally, by a different mechanism of action, the tonicity agent trehalose (a disaccharide) has been shown either alone or in combination with other ingredients, to provide therapeutic benefit to the ocular surface in animal models and human studies or dry eye (Matsuo 2002; Hom et al. 2017). In this case, the presumed mechanism of action is stabilization of cell surface proteins, as trehalose is not known to enter mammalian cells, unlike the smaller polyols or amino acids discussed above.

Preservatives

The most common preservative used for multi-dose ophthalmic prescription products is benzalkonium chloride (BAK). BAK has a number of advantages, including a stable shelf life, robust activity against a broad spectrum of potential contaminants, and in some cases can be shown to enhance the ocular penetration of some therapeutic agents (Okabe et al. 2005). While BAK is considered to be relatively non-toxic to the ocular tissues in comparison with earlier-generation preservatives containing mercury or other toxic materials, there is some concern with its use in chronic or frequently used medications (Noecker 2001), and almost all newer artificial tear products employ alternative preservatives or are non-preserved.

Preservatives found in current artificial tear preparations include, for example, polyquaternium-1, sodium chlorite, and sodium perborate. All of these have been shown in tissue and animal models to have lower potential for toxicity and ocular irritancy than BAK and therefore to be good choices for preserved multi-dose artificial tear formulations.

Alternatively, artificial tears may be formulated without a preservative, in which case they must be packaged either in a single-use vial or in a special multi-use container designed to maintain product sterility after initial use through the open shelf life of the product. These options are discussed further below.

Raw Material Procurement

Obviously, all raw materials utilized in manufacture of an artificial tear must meet appropriate quality standards for the market or region in which the final product will be marketed. In many cases, ingredients are readily available in an appropriate pharmaceutical grade and sufficient quantity from standard suppliers; however, when utilizing a novel ingredient, a number of factors need to be confirmed prior to moving forward. For example, a novel material not previously used in pharmaceutical products may not be available in USP- or Pharm Eur-qualified forms, but perhaps only in grades suitable for food or cosmetic use. This does not exclude use of such compounds, but additional testing and possibly purification may be required to satisfy internal or external quality standards. Additional toxicological testing may also be required, although materials used in foods are often listed as generally recognized as safe (GRAS), which reduces testing requirements (FDA 1980b).

Occasionally a desirable material may only be available in the appropriate grade from a single manufacturer. This introduces the risk of interruption of supply for the product once commercialized. Multiple suppliers, with established supply agreements, are always preferable in order to minimize this risk; exact equivalence of raw materials obtained from different suppliers, especially for specialized materials such as polymers, should however never be assumed, and appropriate testing should be carried out to establish equivalence. In addition, unique suppliers may own the license to intellectual property for use of the ingredient in specific applications or for use in specific regions; it is good practice to review all IP issues with suppliers during development of a new supply agreement and to allow sufficient time during the development process to finalize these agreements. It is not uncommon for suppliers to set one price for small quantities of a special material for research purposes and another much higher price for large quantities to be used in manufacture of a commercial product.

Process Development

In parallel with selection of ingredients, their final concentrations, and establishing reliable sources of commercial supply, process development is ongoing. Artificial tears typically contain common small-molecule ingredients such as salts, buffers, and preservatives which can easily be dissolved in water, but other ingredients such as polymers, oils, or surfactants usually require special handling. Fortunately, raw material manufacturers usually provide detailed guidance on how to solubilize their products, as well as whether solutions of their compounds may be sterilized with heat or may be filtered. For example, hydroxypropyl methylcellulose (HPMC) may be dissolved by dispersing the dry material in rapidly mixing hot water and then gradually cooling to allow dissolution (Sigma-Aldrich 2003).

The use of oils or other hydrophobic materials in artificial tear preparations merits additional discussion. Since these substances do not directly dissolve in water, additional materials that act as surfactants, emulsifiers, and stabilizers are typically added in order to produce a uniform distribution of hydrophobic materials within the overall aqueous solution. There are numerous patents covering methods and materials used in this context, which may be a helpful source, for example, while developing a new process for an emulsified tear product (e.g., Rabinovich-Guilatt et al. 2015; Gore et al. 2019). Currently available emulsion products marketed as artificial tears include Systane Balance (Alcon), containing mineral oil in an opaque white emulsion; Soothe XP (Bausch & Lomb), also containing mineral oil; and Refresh Optive Mega-3 (Allergan), containing castor and flaxseed oils, but which is virtually clear due to the lower amount of oil used and the small size of the lipid micelles within the product.

When developing a commercial process for manufacture of an artificial tear, consideration must be made to the order of addition of ingredients to the final solution. The solubility and final properties of many polymers, for example, are influenced by the presence of other ingredients in the formula. When more complex formulations are made including, for example, lipid emulsions, the formulation will often need to be made in separate parts which are then combined in a specific order under precisely controlled conditions, which will include control of temperature, exposure to air (a nitrogen overlay may be used to reduce oxidation of some materials), pressure, mixing speed, and type of mixing device.

Sterilization Methods and Strategies

Due to the considerations discussed above, sterilization of artificial tear formulations may provide a number of challenges. Small-molecule, low-viscosity materials that are water soluble may often be sterilized by filtration once they are dissolved. However, viscous polymers often cannot be sterilized in this manner, and heat sterilization is typically the first alternative to be considered. When using heat to sterilize either a full formula or part of one, careful control of the temperature and duration of heating must be employed, as the properties of many polymers are altered by heat, with often a loss of viscosity due to partial breakdown of the polymer chains. Oils and other special materials also pose challenges for sterilization and may require sterile filtration while in the pure oil state prior to emulsification under aseptic conditions. Multi-part formulations are thus often sterilized in separate parts and then must be combined aseptically to form the final product.

One alternative sterilization method that may be employed is terminal sterilization, where the final packaged product is exposed to radiation or electron beam energy (E-beam). These solutions often require off-site processing by a contract vendor and may significantly increase the total cost of goods for the product; however they may be considered as they reduce or eliminate the need for aseptic manufacture, reducing the complexity of the manufacturing process considerably. Careful testing during process development must be performed to ensure that the terminal sterilization process does not adulterate the finished product due to formation of free radicals, etc., due to interaction of the formula components or container-closure with the applied energy.

Container-Closure Issues

Until recently, most artificial tear products were packaged in multi-use plastic bottles, with a smaller number of products packaged in single-use vials to avoid the need for a preservative. Although the use of a plastic bottle may seem straightforward, it does usually require a preserved solution and in some cases is complicated by specific regulatory requirements for translucency or color. Therefore, selection of a particular bottle, tip, and cap must be made based upon a number of considerations. For example, if the formulation is sensitive to light, a colored or opaque bottle may be required. In some regions (e.g., Japan), a translucent bottle is required for licensing, in order to allow physicians and patients to see the fluid within the container.

In addition to the bottle itself, the bottle tip must be designed to deliver a consistent volume in a single drop of the artificial tear, without streaming or clogging. The actual volume delivered by a given tip will vary with the viscosity and surface tension of the solution, so drop size measurements must be conducted during development to ensure that the expected volume is being dispensed.

Single-use vials are often preferred for more vulnerable patients such as in post-surgical care or for those dry eye patients sensitive to preservatives. Production of these requires specialized form-fill-and-seal manufacturing equipment, where the container is filled with sterile product while it is being made within an aseptic environment. Alternatively, filled units may be terminally sterilized, although again care must be taken in this instance to avoid production of contaminants from reaction of the bottle plastic or the formulation itself with the externally applied energy.

Most recently, non-preserved products have been introduced in multi-use bottles which use valves and filters built into the bottle tip to prevent intake of contaminants into the bottle following dispensing of a drop (see Fig. 3). There are a number of manufacturers of these specialized bottles, which usually may be used with slight modification on the same manufacturing filling line as standard multi-dose bottles. Use of this type of container-closure may require submission of testing results to

Fig. 3 Multi-dose non-preserved eyedrop bottle showing internal mechanism (Aptar 2016)

licensing agencies and increases the cost of goods (compared to standard multi-dose packaging) considerably. Careful testing must also be performed to ensure that an existing non-preserved solution as previously made in single-use vials will have the same properties over its shelf life in a larger container, exposed to different container materials and a larger head space.

Finally, the container-closure must be able to be used by the target patient, who in many cases may be elderly and/or infirm. Ergonomic testing is therefore appropriate, particularly when a novel formulation or new bottle design is being developed. Adequate ease of use will also be confirmed during clinical evaluation.

Stability Issues

The ideal formulation will be stable under a wide range of temperatures from freezing to near boiling and have a shelf life >2 years. In practice, sufficient stability to support a shelf life of 12 to 18 months at ambient temperatures will usually be adequate for initial launch in the USA. The exact parameters required will depend upon supply chain and distribution channel requirements. For example, some retailers may require additional months of shelf life to accommodate their own warehousing and shelf stocking needs.

Regions outside of the USA may require testing under a variety of environmental conditions to model local climate and/or specific storage conditions, such as high temperature and humidity. Teams should obtain these regulatory requirements at an early stage in planning to ensure specific conditions are tested during development.

Stability testing is therefore done by storage of the final packaged product in environmentally controlled chambers usually in several sets of conditions: controlled room temperature (25 °C), accelerated temperature (40 °C), plus special conditions such as 30 °C with elevated humidity. Samples are then withdrawn at appropriate intervals (typically monthly), and key parameters determined, which should always include the concentration of any active or key ingredients such as polymers and preservatives, as well as basic physical parameters such as pH, viscosity, osmolarity, and physical appearance. Tolerance ranges for each parameter must be set in advance; these may be dictated by regulatory requirements. Several separate batches of final product should be made at commercial scale with sufficient quantities filled into the final container-closure to provide for several years of stability testing, as well as any clinical evaluation needs. Once the data is collected from a number of time points (usually a minimum of three), trends can be calculated to project product shelf life, with the accelerated temperature values used to model typically twice the projected shelf life. Figure 4 illustrates how shelf life might be

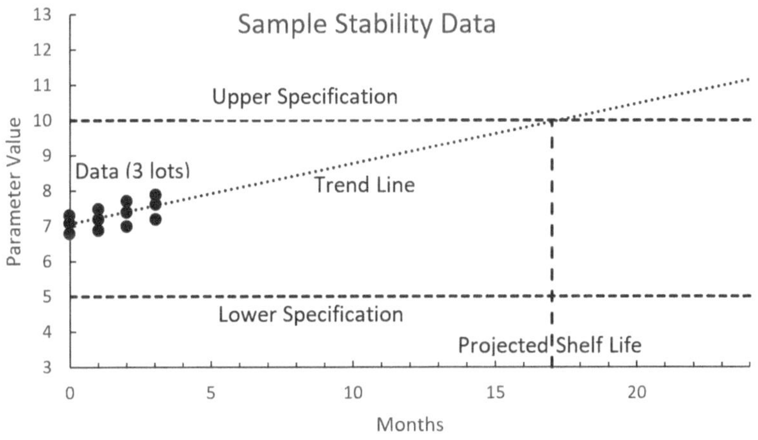

Fig. 4 Sample stability data and shelf life projection

projected based upon a parameter with specified value of 7.5 and tolerance limits of 5 to 10. The projected life is 17 months in this case, which if the data were collected at elevated temperature could justify a labeled shelf life of up to 34 months. Naturally, with multiple parameters being tested, the parameter predicting the shortest shelf life will determine what can be put on the label. Testing will continue according to a predetermined schedule to confirm projected stability with real-time data.

In addition to shelf life or expiry dating, some regions may require a discard after opening statement on the label as well, typically for 30 to 90 days. This would apply primarily to multi-use containers, as single-use vials are always labeled to be discarded after use.

Clinical Validation

In many regions, including in the USA, clinical trial data for artificial tears is not always required prior to commercial launch, provided certain minimum requirements are met for the safe manufacture (e.g., Good Manufacturing Practice) of the finished product. This fact alone may significantly reduce the overall development time and expense for a new artificial tear.

However, it is highly desirable to understand both the strengths and weaknesses of a new artificial tear formulation when in actual clinical use, which can only be truly understood by means of a randomized, masked, controlled clinical trial run according to the same standards used for pharmaceutically active drugs. Having stated this as the gold standard for clinical evidence of safety and efficacy, this type of trial is time-consuming and expensive, often taking at least a year and more than $one million US dollars to complete. Therefore, alternatives are often sought to obtain some clinical evidence without delaying commercial launch more than necessary. These alternatives may include a short open-label trial at a few clinical practices prior to commercial launch, utilization of existing data for similar products already on the market, or an "early experience" model with product being made available to clinicians for a limited time after commercial launch, with surveys of the patients and practitioners conducted following use.

In some regions, trial data may be required in order to obtain reimbursement to patients from either public health agencies or in some cases private health insurance. In France, for example, artificial tears are sold both as pharmaceutical products and as medical devices. A clinical comparison to an existing product within the same category is necessary in order to obtain reimbursement at the same level as the existing product within the French health system.

If the artificial tear is being developed within a quality system in order to support registration as a medical device using, for example, the CE mark process, clinical validation is generally required in order to confirm that the user requirements for safety, efficacy, and usability have been met. If, however, the new formulation is substantially equivalent to prior formulations by nature of similar ingredients or physical properties, prior data from these predicate products may suffice to validate the new product.

Ultimately, the need for a clinical trial may be established by the marketing strategy for the new product. As patients and doctors will certainly have numerous existing artificial tear formulations available to them already, the developers of the new product will need to distinguish it from others. A clinical trial may be able to support marketing claims such as a longer-lasting formula requiring fewer doses per day for relief of ocular symptoms, in comparison to an existing product (a desirable claim for patients) or, for eye care practitioners, a greater degree of improvement in specific ocular signs such as ocular surface damage (measured as surface staining) or tear instability (measured as tear break-up time).

Regarding the design of a clinical trial, a number of issues need to be determined, including the duration of the trial, number of patients to be enrolled, and inclusion and exclusion characteristics for those patients. For mild to moderate dry eye, relief of symptoms is usually considered paramount, as in those patients, clinical signs are often variable or absent. Symptomatic relief for lubricant eye drops is usually rapid, so a trial of a few weeks may suffice unless a specific duration is required for regulatory submission in the desired commercial region. For moderate to severe dry eye, reduction of clinical signs is usually considered more appropriate to monitor (along with symptoms), and a longer trial is typically needed to demonstrate clear improvements from baseline and to distinguish the performance of the new formulation from the control or predicate product. Other considerations for clinical trial design in dry eye disease, including different trial designs, methods of randomization, selection of subjects, placebo effects, and other confounding factors, have been discussed in the DEWS II report (Novack et al. 2017).

In addition to evaluation for use in dry eye patients, products may also be developed for use in post-surgical patients or for use as a contact lens rewetting drop. In either of these cases, separate trials may be conducted in these target populations in order to validate use for the specific indication. In the case of use with contact lenses, material compatibility testing should also be conducted along with clinical tests in order to confirm that the formulation ingredients will not impact the contact lens material itself.

Finally, some regions may require publication of trial data in order to support specific labeling claims. Presentation of results at professional conferences (with an abstract that may be referenced) is usually sufficient for some period of time (typically 6–12 months), but publication in the peer-reviewed literature is considered necessary for long-term claim support. In addition to journal publication, it is now standard practice (and in some cases legally required) to post certain details of clinical trial design and results on one or more government databases, such as clinicaltrials.gov in the USA, if the resulting data is desired for use to support marketing claims or be published in a professional journal. This requirement would therefore apply to all trials except for exploratory investigations at the earliest stage of development.

Final Steps for Commercialization

Regulatory Submissions

Regulatory requirements for artificial tear products vary considerably across the various regions of the world. Generally, one of three different regulatory paths will be present: (1) development and launch under an existing government monograph requiring no formal filing, as in the USA OTC category; (2) formal filing as a pharmaceutical agent essentially the same as for any new pharmaceutical product (exists in Africa/Middle East, Latin America, most Asian countries); and (3) registration as a medical device through the CE mark process (in Europe) or an equivalent process in some other countries (Canada for some products, some Asian countries). Medical device filings range from a simple notification to creation, submission, and review of a substantial file. For use as a contact lens rewetting drop in the USA, an application to the device division of the FDA, with review of labeling, material compatibility test results, and clinical testing, is also required prior to commercial sale. In the USA, lubricant eye drops for relief of dry eye discomfort are classified as OTC drugs (and labeled as such), while the same formulation when used with contact lenses would be classified as a medical device with labeling appropriate to that indication.

Because of these different requirements, it is essential at an early stage in development to decide the proposed indications for use, and where the final product is to be commercialized, as this will determine the extent of the regulatory file and the required data to be submitted. If commercial launch is planned in multiple regions, creation of a summary dossier in Common Technical Document (CTD) format is appropriate. CTD files contain multiple sections covering all aspects of the production and testing of the formulation, structured in an internationally agreed format; once created, required sections for each region may be excerpted and edited with local information such as the local business office, country-specific labeling, etc. added as needed. It is beyond the scope of this chapter to discuss the entire contents of the CTD dossier; however all regulatory personnel in the pharmaceutical industry are very familiar with its nature and requirements.

In some cases, such as in the USA, no file is made, but the company is subject to inspection of its manufacturing records and other files relating to the testing and ongoing safety of the product. For CE mark medical devices, maintenance of a quality system subject to periodic audit and inspection allows launch of a new product with a notification to the company's notified body testifying that the activities with accompanying documentation required for a new medical device under the quality system (e.g., adherence to various ISO standards requirements) have been completed. These two pathways to commercial launch generally require significantly less time than a conventional pharmaceutical file, which may substantially reduce the overall development time for the product.

Product Launch Activities

Product launch can occur when the various internal and external activities to manufacture and test the new artificial tear formulation have been completed and document that the product is manufactured and labeled in accordance with all applicable regulatory requirements for the region of interest, and appropriate testing has demonstrated that the product is safe for use by the end-user. Thus product development has demonstrated that a consistent formulation of known composition is being manufactured, which is labeled correctly, and sufficient data has been obtained to support the labeled shelf life.

In the USA, a standard labeling format for OTC products is followed, known as the Drug Facts. This label is directly on the package and informs the consumer of the ingredients, directions for use, and any precautions related to use of the product. An example of an artificial tear Drug Facts is shown below (Fig. 5).

Fig. 5 Drug Facts label from Systane Ultra artificial tears as sold in the USA

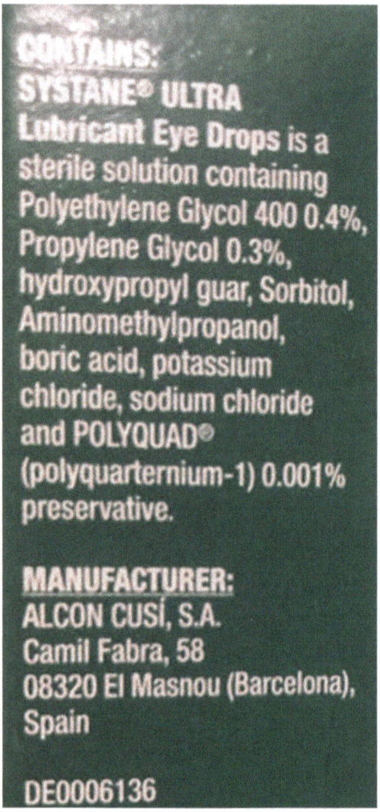

Fig. 6 List of ingredients for CE mark product label for Systane Ultra

Label format in regions outside of the USA varies, and local regulations will dictate the level of detail required; some regions may require the concentrations listed for certain key ingredients such as the polymer and preservative. If the artificial tear is marketed as a medical device, there are no listed active ingredients, and all of the ingredients are usually listed on the label in an order determined by the manufacturer. Here is the same product shown above for the USA Drug Facts label (Systane Ultra, Alcon) with the list of ingredients as manufactured in Spain (Fig. 6).

As part of the commercial launch, data collected during development is often presented at professional conferences. This data may be the results of testing in animal models of disease, results of clinical trials, and results of physical testing where the physical properties of the formation are distinct from those of other available treatments. In this process, engagement of professionals external to the developing company (often eye care practitioners) who may provide independent evaluation and support of the new product is often done. These key opinion leaders (KOLs) may be very useful in helping to explain the distinct features and benefits of the new product to their colleagues. Other than providing information that aids professionals in patient selection and management that may be specific to the new formulation, the use of artificial tears as a category is well known to eye care professionals, and special practitioner education programs are not usually needed.

The presentation of research results, developing of marketing materials, and monitoring the production quality and ongoing safety of the product are primarily the same process for artificial tear products as for other pharmaceuticals. However, the regulations around making additional claims for OTC or medical device products after initial launch may be slightly different than for prescription drugs and should be carefully reviewed for each region or country prior to implementation.

Conclusions

The development of artificial tear formulations for ocular use is complicated by the multiple indications these products are used for, the many types of ingredients that can be included, and the various regulatory processes they are licensed under. The developer must decide initially what the target patient profile will be for the proposed product and then include specific ingredients to provide hydration, viscosity, and other properties that will allow the new product to be distinguished from existing products. Clinical studies are not always required prior to commercialization, depending upon the target for licensing, and regulatory requirements vary widely from internal documentation only (as in USA) to a full pharmaceutical filing (as in some Asian and Latin American countries). Despite this, the overall development time scale for artificial tears is typically shorter than for a new pharmaceutical product, due to reduced need for lengthy clinical testing and regulatory filing.

Acknowledgments The author would like to acknowledge the many years of collaboration with talented scientists at Allergan plc, including especially Joseph Vehige, Cindy Carlisle-Wilcox, and Haixia Liu.

References

Aptar. Aptar Pharma's preservative free multi-dose ophthalmic squeeze dispenser launched in U.S. 2016. http://news.aptar.com/pharma/aptar-pharmas-preservative-free-multi-dose-ophthalmic-squeeze-dispenser-launched-in-u-s/. Accessed 23 May 2016.

Aragona P, Di Stefano G, Ferreri F. Sodium hyaluronate eye drops of different osmolarity for the treatment of dry eye in Sjögren's syndrome patients. Br J Ophthalmol. 2002;86:879–84.

Baudouin C, Aragona P, Messmer EM, et al. Role of hyperosmolarity in the pathogenesis and management of dry eye disease: proceedings of the OCEAN group meeting. Ocul Surf. 2014;11(4):246–58.

Chen W, Zhang X, Li J, et al. Efficacy of osmoprotectants on prevention and treatment of murine dry eye. Invest Ophthalmol Vis Sci. 2013;54:6287–97.

Corrales RM, Luo L, Chang EY, et al. Effects of osmoprotectants on hyperosmolar stress in cultured human corneal epithelial cells. Cornea. 2008;27:574–9.

Craig JP, Nichols KK, Akpek EK, et al. TFOS DEWS II definition and classification report. Ocul Surf. 2017;15:276–83.

Food and Drug Administration. Guidelines on tonicity for topical ophthalmic drugs. Proposed rules. Fed Regist. 1980a;45(89):30013–5. Accessed 6 May 1980.

Food and Drug Administration. GRAS substances (SCOGS) database. 1980b. https://www.fda.gov/food/generally-recognized-safe-gras/gras-substances-scogs-database.
Food and Drug Administration. Ophthalmic drug products for over-the-counter human use. Code Fed Regul. 2002;21(349):273–8.
Garrett Q, Khandekar N, Shih S, et al. Betaine stabilizes cell volume and protects against apoptosis in human corneal epithelial cells under hyperosmotic stress. Exp Eye Res. 2013;108:33–41.
Gary DN. TFOS DEWS II Clinical Trial Design Report. The Ocular Surface. 2017;15(3):635–655.
Gomes JAP, Azar DT, Baudouin C, et al. TFOS DEWS II iatrogenic report. Ocul Surf. 2017;15:516–43.
Gore AV, Giyanani J, Likitlersuang S. Stabilized omega-3 ophthalmic compositions. US patent 10,279,005. 2019. Accessed 7 May 2019.
Holly FJ, Lamberts DW. Effect of non-isotonic solutions on tear film osmolality. Invest Ophthalmol Vis Sci. 1981;20(2):236–45.
Hom MM, Berdy GJ, Downie LE, et al. Clinical evaluation of a novel lipid-containing lubricant eye drop with omega-3 oil and trehalose. In: ARVO Annual Meeting, May 9, 2017, Baltimore, Maryland, Program # 2671; 2017.
Hua X, Su Z, Deng R, et al. Effects of l-carnitine, erythritol and betaine on pro-inflammatory markers in primary human corneal epithelial cells exposed to hyperosmotic stress. Curr Eye Res. 2015;40(7):657–67.
Jones L, Downie LE, Korb D, et al. TFOS DEWS II management and therapy report. Ocul Surf. 2017;15:580–634.
Lemp MA. Report of the National eye Institute industry workshop on clinical trials in dry eyes. CLAO J. 1995;21(4):221–32.
Lemp MA, Baudouin C, Baum J, et al. The definition and classification of dry eye disease: report of the definition and classification subcommittee of the international dry eye workshop. Ocul Surf. 2007;5(2):75–92.
Luo L, Li DQ, Corrales RM, et al. Hyperosmolar saline is a proinflammatory stress on the mouse ocular surface. Eye Contact Lens. 2005;31:186–93.
Matsuo T et al. Trehalose Eye Drops in the Treatment of Dry Eye Syndrome. Ophthalmology 2002;109:2024–2029.
Nichols JJ, Jones L, Nelson JD, et al. The TFOS international workshop on contact lens discomfort. Invest Ophthalmol Vis Sci. 2013;54(11):TFOS1–TFOS203.
Noecker R. Effect of common ophthalmic preservatives on ocular health. Adv Ther. 2001;18(5):205–15.
Okabe K, Kimura H, Okabe J, et al. Effect of benzalkonium chloride on transscleral drug delivery. Invest Ophthalmol Vis Sci. 2005;46(2):703–8.
Rabinovich-Guilatt L, Lambert G, Lallemand F, et al. Emulsion compositions containing cetalkonium chloride. US patent 9,220,694. 2015. Accessed 29 Dec 2015.
Rangarajan R, Kraybill B, Ogundele A, Ketelson HA. Effects of a hyaluronic acid/hydroxypropyl guar artificial tear solution on protection, recovery, and lubricity in models of corneal epithelium. J Ocul Pharm Ther. 2015;31(8):491–597.
Sigma-Aldrich. Product information for (Hydroxypropyl) methylcellulose, Product H 9262. 2003. https://www.sigmaaldrich.com/content/dam/sigma-aldrich/docs/Sigma/Product_Information_Sheet/2/h9262pis.pdf.
Sonsino J, Schornack M. Troubleshooting scleral lenses. Contact Lens Spectr. 2013;28:26–33.

Approval of Topical Ophthalmic Generic Products in the USA: Simple to Complex Dosage Forms and Establishing Equivalency

Stephanie H. Choi, Yan Wang, and Darby Kozak

Abstract This chapter provides an overview of how generic topical ophthalmic drug products are approved in the USA. Standards for approval of generic ophthalmic drug products and the role of the Office of Generic Drugs in review of generic ophthalmic drug products are explained. Equivalence evaluation of solution and non-solution topical ophthalmic dosage forms is described, and future research needed to advance equivalence standards are presented.

Keywords Generic drugs · Complex dosage forms · Bioequivalence · Regulatory approval

Therapeutic Equivalents and Generic Products

The FDA maintains a list of all drug products approved on the basis of safety and effectiveness under the Federal Food, Drug, and Cosmetic Act (the FD&C Act). This list is presented in the publication, *Approved Drug Products with Therapeutic Equivalence Evaluations* (commonly known as the *Orange Book*), which also provides a therapeutic equivalence evaluation for all multisource products. Multisource products are pharmaceutical equivalents available from more than one manufacturer. The two basic categories of multisource products in the *Orange Book* are

S. H. Choi (✉) · Y. Wang · D. Kozak
Office of Research and Standards, Office of Generic Drugs, Center for Drug Evaluation and Research, Greenwich Biosciences plc, Food and Drug Administration,
Silver Spring, MD, USA
e-mail: stephaniechoi1122@gmail.com; Yan.Wang3@fda.hhs.gov

S. Neervannan, U. B. Kompella (eds.), *Ophthalmic Product Development*,
AAPS Advances in the Pharmaceutical Sciences Series 37,
https://doi.org/10.1007/978-3-030-76367-1_20

those FDA considers to be therapeutically equivalent (A) and those FDA has considered not be therapeutically equivalent (B).

Drug products are classified as therapeutically equivalent if they meet the following criteria:

1. They are approved as safe and effective.
2. They are pharmaceutical equivalents.[1]
3. They are bioequivalent in that they:
   (a) Do not present a known or potential bioequivalence (BE) problem and they meet an acceptable in vitro standard.
   (b) If they do present such a known or potential problem, they are shown to meet an appropriate BE standard.
4. They are adequately labeled.
5. They are manufactured in compliance with Current Good Manufacturing Practice regulations.

A multisource drug product may be approved under either Section 505(b)(2) or 505(j) of the FD&C Act. The *Orange Book* listings for ophthalmic products contain examples of both therapeutically equivalent and non-therapeutically equivalent multisource drug products that have been approved under both sections. This can lead to confusion about what is considered a generic ophthalmic product.

Section 505(j) describes the Abbreviated New Drug Application (ANDA) submissions that are reviewed by the Office of Generic Drugs and are described as generic products. Each ANDA submission identifies a Reference Listed Drug (RLD). The RLD is, in general, a name brand innovator product that contains full reports of investigations of safety and effectiveness to support approval. The ANDA does not contain its own safety or efficacy studies but relies on FDA's finding that the RLD is safe and effective. When an ANDA is approved, it generally receives an A rating in the *Orange Book*.

Section 505(b)(2) applications are New Drug Applications (NDA) that contain full reports of investigations of safety and effectiveness, where at least some of the information required for approval comes from studies not conducted by or for the applicant and for which the applicant has not obtained a right of reference or use. Section 505(b)(2) applications can result in a multisource drug product if an applicant submits a section 505(b)(2) application for a product that has the same active ingredient, dosage form, and strength as a previously approved product. This situation may occur if the section 505(b)(2) product contains some difference from the previously approved product that may not be permitted in a section 505(j) application. When a section 505(b)(2) application that is a multisource product is approved, it generally will not receive an A rating to the previously approved products. An

[1]Drug products are considered pharmaceutical equivalents if they contain the same active ingredient(s) in the same amount(s); have the same dosage form, route of administration, strength, or concentration; and meet same or compendial or other applicable standards (i.e., strength, quality, purity, and identity).

application may request an A rating, and this request will be evaluated by the FDA. Drug products submitted in a section 505(b)(2) application are reviewed and approved by the Office of New Drugs and are not considered to be generic products (even if they are therapeutic equivalents and have an A rating in the *Orange Book*).

Ophthalmic drug products submitted under either section 505(b)(2) or 505(j) can be classified as therapeutic equivalents in FDA's *Orange Book* that indicates which products are appropriate for substitution.

The Office of Generic Drugs does not regulate biosimilar products. Ophthalmic biologic products which are shown to be biosimilar to or interchangeable with an FDA-licensed biological reference product are not considered to be generic drug products and do not appear in the *Orange Book*.

Inactive Ingredient Changes

Under an ANDA, a drug product intended for ophthalmic use generally must contain the same inactive ingredients and in the same concentration as the RLD (Code of Federal Regulations 2011). However, an applicant may seek approval of an ophthalmic drug product that differs from the RLD in preservative, buffer, substance to adjust tonicity, or thickening agent provided that the applicant identifies and characterizes the differences and provides information demonstrating that the differences do not affect the safety or efficacy of the proposed drug product (Code of Federal Regulations 2011). These four excipient types are commonly referred to as "exception excipients." A formulation which includes non-exception excipient differences to the RLD (e.g., pH adjusters, humectants) may not be submitted in an ANDA, but can be submitted as an NDA (under Section 505(b)(2)) and still be classified as a therapeutic equivalent.

The ANDA Regulatory Pathway

The ANDA regulatory pathway permits approval of generic drugs on the basis of quality and BE data, without the need for new evidence of effectiveness and safety (Peters et al. 2009). An ANDA relies on the Agency's finding of safety and effectiveness for an RLD, and, as a result, that ANDA may be approved without submission of the same type and extent of information as is required for an NDA to establish the safety and efficacy of the proposed product. However, the FDA ensures that the generic product is therapeutically equivalent to the RLD by requiring applicants to demonstrate that the generic product is bioequivalent to the RLD.

Bioequivalence refers to the absence of a significant difference in the rate and extent to which the active ingredient in a pharmaceutically equivalent drug product becomes available at the site of action, when administered to subjects at the same molar dose under similar conditions (CDER/FDA n.d.-a). Bioequivalence for ophthalmic products can be demonstrated by comparing the properties of a generic product to the RLD, via comparative in vivo clinical endpoint studies, in vivo ocular pharmacokinetic studies, and/or in vitro studies. The type of study that can be used to demonstrate BE for an ophthalmic product depends on the drug product's active pharmaceutical ingredient (API), dosage form, indication, site of action, mechanism of action, and scientific understanding of drug release/drug availability and drug product characteristics (Choi and Lionberger 2016).

Offices Involved in Review of Ophthalmic Drug Applications

An ANDA comprises of three major sections: Chemistry, Manufacturing and Controls, Labeling, and Bioequivalence. Labeling and Bioequivalence modules are reviewed by the Office of Generic Drugs, while the Chemistry, Manufacturing, and Controls (CMC) module is reviewed by the Office of Pharmaceutical Quality. Specifically, the CMC module is reviewed by the Office of Lifecycle Drug Products, which is a sub-office of the Office of Pharmaceutical Quality.

Medical reviews of clinical studies submitted in NDAs are conducted by the Office of New Drugs. This includes clinical reviews for 505(b)(2) applications of ophthalmic products. The CMC module for NDAs is reviewed by the same Office of Pharmaceutical Quality but by a different sub-office, the Office of New Drug Products. The Office of Generic Drugs does not review any sections of 505(b)(2) applications.

Bioequivalence Standards for Generic Drug Approval

An ANDA applicant is responsible for developing their proposed product and associated testing and justification for BE determination. To facilitate generic drug development and to reduce the potential number of review cycles of an ANDA, FDA posts product-specific guidances (PSGs) (FDA n.d.). These PSGs provide FDA's current thinking on appropriate studies/information that an ANDA applicant may consider submitting to support BE. FDA recommends different studies to demonstrate BE for ophthalmic products based on whether the generic product's formulation contains the same inactive ingredients and in the same concentration as the reference product. Qualitative sameness (Q_1) is established if the generic product contains the same inactive ingredient(s) as the reference product. Quantitative sameness (Q_2) is established if all the inactive ingredient(s) in the generic product are within ±5% of those in the reference product. For an ophthalmic product, the FDA considers a generic product to be Q_1 and Q_2 equivalent only when all of the following three criteria are met:

1. All inactive ingredients—including the "exception excipients"—are the same as that in the RLD and in the same concentration.
2. The generic product does not contain an inactive ingredient not contained in the RLD.
3. The difference in the amount of inactive ingredients between the generic product and the RLD is less than 5%.

For any proposed ophthalmic generic product which is not Q_1 and Q_2 equivalent, a comparative pharmacokinetic or clinical endpoint BE study may be recommended. Examples of comparative clinical endpoint studies recommended for products indicated for treatment of chronic open- angle glaucoma (CDER/FDA n.d.-b, c, d, e) and treatment of pain and inflammation associated with cataract surgery (CDER/FDA n.d.-f) have been described in PSGs issued by the FDA. They are conducted in patients with the relevant indication, and the proposed generic product must show statistical equivalence to the reference product for the primary endpoint.

In general, section 505(j) is likely the most appropriate pathway for a permissible non-Q_1/Q_2 ophthalmic formulation if the ANDA (1) provides justification that the proposed formulation changes do not effect product safety; (2) demonstrates BE to the reference product via an appropriate in vivo (e.g., pharmacokinetic and/or clinical endpoint) study; and (3) meets applicable labeling requirements, which do not permit claims that differences from the RLD in formulation confer a therapeutic advantage or difference. Applications that do not meet these criteria may be more appropriate for submission under the section 505(b)(2) pathway.

Solution Products

As of 2018, there are approximately 80 approved and currently marketed ophthalmic solution RLDs (CDER/FDA 2016). These products fall into roughly seven major categories based on their indications. There are 21 products to reduce intraocular pressure, 21 to reduce ocular itching, 15 antimicrobial products, 8 for mydriasis, 8 to treat pain and inflammation, 2 to treat dry eye disease, 1 for miosis, and 11 for other indications. There are over 260 therapeutic equivalent ophthalmic solution products currently approved to be marketed. Of the 80 reference products, 60% have at least one approved therapeutic equivalent, and 40% have three or more therapeutic equivalents available.

For ophthalmic solutions which are Q_1 and Q_2 equivalent, BE is considered to be self-evident and does not need to be demonstrated (Code of Federal Regulations n.d.; Chambers 2012). Therefore, in vivo comparative clinical endpoint studies or pharmacokinetic studies are not required. Since all ingredients in a solution are dissolved, a microstructure is not present in a solution, and consequently no differences will exist between two products that are Q_1 and Q_2 when good manufacturing practices are followed. However, the generic must still demonstrate comparative pH, specific gravity, osmolality, viscosity, and buffer capacity to the reference product (CDER/FDA 2018).

These tests are recommended to ensure that any differences in excipient grade, manufacturing process, and/or container closure system do not affect the quality of the drug product formulation. In addition, these physicochemical properties can have a potential impact on therapeutic performance and/or patient perception/comfort. For example, the optimum pH for eye drops is similar to that of tear fluid which varies between individuals but is commonly cited in literature as about 7.4 (Baranowski et al. 2014). If the pH value gets outside the range of 4–8 which is tolerated by the eye, a patient may feel discomfort and irritation, potentially altering drug bioavailability due to altered tear production (Jitendra et al. 2011). In addition, the pH of the drug product may impact corneal permeability, stability, and solubility (Conroy and Maren 1995) and may have an irritating effect if it is formulated outside the typical range of the human precorneal tear film with a strong buffer capacity (Choi and Lionberger 2016; Ali and Lehmussaari 2006).

As changes in formulation may alter bioavailability at the site of action, BE of generic ophthalmic products which is not Q_1 and Q_2 equivalent is generally not considered to be self-evident. One example is that changes in the preservative or concentration of the preservative may affect the efficacy and safety of a drug product. In general, preservatives with a broad range of antimicrobial activity can result in adverse effects including tear film instability, corneal cytotoxicity, anterior chamber inflammation, and change in corneal permeability due to its nonspecificity. Most literature reports on preservative toxicity have been focused on ophthalmic products that are indicated for the control of glaucoma, as these patients with chronic eye disease use multiple drops over extended periods of time. Among all preservatives that have been used in ophthalmic drug products, benzalkonium chloride (BAK) is one of the mostly frequently evaluated preservative. On one side, the cytotoxicity of BAK to the ocular surface was documented as early as the 1970s (Gasset 1977; Gasset et al. 1974). On the other side, some recently studies showed that formulations with and without BAK did not show differences in efficacy nor adverse effects. Therefore, it is important to note that these effects are drug substance and formulation specific.

The various study results when evaluating the same preservative indicate that the potential toxic/adverse effects resulted from a preservative are drug substance and formulation specific and can influence the active ingredient's penetration through the cornea. A study using rabbits was conducted to compare the corneal permeability of latanoprost containing BAK with travoprost that did not contain BAK (McCarey and Edelhauser 2007). Corneal permeability was significantly higher for the eyes treated with latanoprost compared to those treated with travoprost. This suggests that drops containing BAK may result in increased ocular bioavailability of those drug products. However, it is important to note that these effects are drug substance and formulation specific as there are also several studies demonstrating that formulation changes do not affect BE. For example, a study comparing the safety and efficacy of latanoprost with and without BAK in patients with open angle glaucoma (Aptel et al. 2016) found that the efficacy (determined by reduction in intraocular pressure (IOP)) was similar between the two treatment groups. They also found no significant difference in adverse events, side effects, or tolerability between the two formulations. Another study (Kasai et al. 2013) investigated the safety and efficacy of preservative-free latanoprost by conducting both in vitro and clinical studies. The in vitro tests demonstrated greater cytotoxicity of preserved latanoprost to human corneal epithelial cells compared to the preservative-free formulation. The BE study, however, showed that the two formulations are similar with regard to clinical endpoint (IOP lowering).

As such, to ensure the approval of high-quality generic products that will be therapeutically equivalent, it is recommended that BE of generic products, which have permissible formulation differences (i.e., are not Q_1/Q_2 to the RLD), be demonstrated by appropriate comparative in vivo pharmacokinetic or clinical endpoint studies.

In a pharmacokinetic study in aqueous humor, samples of the aqueous humor are taken during cataract surgery and assayed for drug levels at set times after drug administration. The rate and extent of drug available in the aqueous humor must be comparable for the generic and RLD. By comparing the bioavailability of the drug in the aqueous humor between the generic and reference product, BE can be determined. However, it is important to note that large patient populations are generally required for aqueous humor PK studies due in part to the sparse sampling limitations (i.e., PK sampling is limited to one sample per patient eye) and high intra-subject variability arising, in part, from differences in cornea permeability (Harigaya et al. 2018). In addition, this type of study is limited to drugs which have a site of action adjacent to the aqueous humor, such as topically applied steroids, nonsteroidal anti-inflammatory drugs, and prostaglandin analog drug products (Cantor 1997). No other type of in vivo pharmacokinetic study has been recommended by FDA for demonstration of bioequivalence for ophthalmic products.

Although an acceptable approach to demonstrate BE, an in vivo PK study and/or comparative clinical endpoint study may not always be necessary, feasible, or the most accurate, sensitive, and reproducible approach to demonstrate BE. In vivo PK studies may not always be feasible because of limitations with respect to not knowing the site of action, sampling ease and frequency, available bioanalytical methods, or subject variability. Likewise, comparative clinical endpoint studies are often confounded by variables including, but not limited to, disease severity and variability in measuring efficacy.

For a product that is Q_1/Q_2 equivalent in composition to the RLD, the only difference it can have from the reference product is in its physicochemical properties. In vitro characterization tests are used to understand a formulation's physicochemical properties. While a direct in vitro-in vivo correlation may not be feasible to establish for many ophthalmic products, the potential ability of some physicochemical properties to impact ocular bioavailability is generally known through various studies conducted in both animals and humans. Therefore, a strict comparison of physicochemical properties will reveal any potential clinical differences between formulations that are Q1/Q2 equivalent.

In addition, alternative methods such as in vitro characterization tests are able to successfully detect formulation and manufacturing changes of ophthalmic formulations having the same composition. This was demonstrated through studies conducted by FDA's Office of Pharmaceutical Quality, in which various Q1/Q2 test products of cyclosporine ophthalmic emulsion were formulated by changing formulation and process variables (Rahman et al. 2014). Comparative physicochemical characterization tests revealed that changes in physicochemical properties were predictive of changes in manufacturing. In vitro tests can be useful in the determination of bioequivalence for products such as cyclosporine ophthalmic emulsion which possesses modest clinical efficacy. Unless extremely large sample sizes are used, clinical studies may not be sensitive enough to detect qualitative and/or quantitative differences in formulation for these types of products. In vitro tests can alternatively be used to assess differences in formulation and manufacturing to accurately predict human bioavailability of test and reference products. Accordingly, FDA has been researching in vitro BE testing methods for ophthalmics that can be expected to detect a meaningful difference between a generic non-systemically absorbed drug and its RLD in safety and therapeutic effect (see Research to Advance Equivalency Standards).

Suspension and Emulsion Products

As of 2018, there are approximately 20 approved and currently marketed ophthalmic suspension RLDs and 3 emulsion reference products (CDER/FDA 2016). These products fall into roughly six different categories based on their indication and class of active ingredient. Of those only four products have at least one therapeutic equivalent approved for marketing in the USA. Table 1 lists the approved therapeutic equivalents for ophthalmic suspensions that currently have a marketing

Table 1 Ophthalmic suspension therapeutic equivalents

Application #	Active ingredient	Strength	Proprietary name	Approval date
A064135	Dexamethasone; neomycin sulfate; polymyxin B sulfate	0.1%; EQ. 3.5 mg base/mL; 10,000 units/mL	Dexasporin	Sep 13, 1995
A062341	Dexamethasone; neomycin sulfate; polymyxin B sulfate	0.1%; EQ 3.5 mg base/mL; 10,000 units/mL	Maxitrol	May 22, 1984
A064134	Dexamethasone; tobramycin	0.1%; 0.3%	Tobramycin and dexamethasone	Oct 27, 1999
A207609	Loteprednol etabonate	0.5%	Loteprednol etabonate	Apr 17, 2019
N017469	Prednisolone acetate	1%	Omnipred	Prior to Jan 1, 1982

status of "Prescription" in the *Orange Book*. Only one suspension product (dexamethasone; neomycin sulfate; polymyxin B sulfate ophthalmic suspension 0.1%; EQ. 3.5 MG BASE/ML; 10,000 UNITS/ML) has more than one approved generic. There are currently no approved generic ophthalmic emulsion products.

Most of the products in Table 1 were approved in the 1980s and 1990s when standards for approval of generic ophthalmic products were significantly different from standards that are applied today. A waiver of in vivo bioequivalence study requirements was granted based on 21 CFR 320.22(b)(2) for ANDAs 64,135 and 62,341. For ANDA 64134, clinical endpoint studies to assess redness and itching were performed to evaluate bioequivalence (FDA 2019). For NDA 17469, the product was approved via the 505(b)(2) pathway and was determined to be therapeutically equivalent to the reference product.

For complex dosage forms such as suspension and emulsions, if a generic product is Q_1/Q_2 the same as the RLD, the only potential remaining differences from the RLD would be in its physicochemical (Q_3) properties and drug release rate. Such differences can arise from differences in the generic product's manufacturing process and formulation steps and can affect the generic product's bioavailability and dose uniformity. When a generic product's physicochemical properties and drug release rate are similar to that of the RLD, bioavailability is expected to be the same for both products. Increasingly, advanced analytical techniques and methods can provide an accurate, sensitive, and reproducible measure of differences in physicochemical properties and in vitro drug release rates.

Based upon these considerations, the FDA has recently recommended in vitro studies for demonstrating the BE of several locally acting products when the formulations of the products are Q_1/Q_2 the same as the RLD, including several ophthalmic suspension and emulsion products. The recommended in vitro BE option is product-specific and is based on testing that demonstrates comparative physicochemical characteristics and in vitro drug release to the RLD. This weight-of-evidence approach seeks to ensure the generic formulation and end-product sameness to the RLD such that the bioavailability of the proposed generic product and the RLD is expected to be the same.

The first guidance to include in vitro studies to demonstrate BE for an ophthalmic product was posted in June 2013 for cyclosporine ophthalmic emulsion (CDER/FDA n.d.-g). Table 2 below lists all the PSGs for ophthalmic suspensions and emulsions that include an in vitro study option as of July 2019.

As with any FDA guidance, PSGs provide FDA's current thinking and non-binding recommendations on a topic, and alternative approaches can be used if they satisfy the requirements of the applicable statutes and regulations. In addition, if research later uncovers additional physicochemical properties that are indicators of product differences that could influence bioavailability, the FDA would review the evidence to determine if the PSG should be revised to include those properties.

In vitro studies used to evaluate BE are aimed at comparing the physicochemical properties and microstructure of the generic and reference formulations. Even for a generic product which is Q_1 and Q_2 equivalent to the reference product in composition, the microstructure and physicochemical properties may be different because

Table 2 Product-specific bioequivalence recommendations on ophthalmic suspensions and emulsions

Active ingredient	Dosage form	Indication	Bioequivalence study type	Date recommended/ revised
Cyclosporine	Emulsion	To increase tear production in patients whose tear production is presumed to be suppressed due to ocular inflammation associated with keratoconjunctivitis sicca	Two options: (1) In vitro study for Q_1/Q_2 products, (2) clinical endpoint study	Posted 6/2013, revised 2/2016, 10/2016
Dexamethasone; neomycin sulfate; polymyxin B sulfate	Suspension	For steroid-responsive inflammatory ocular conditions for which a corticosteroid is indicated and where bacterial infection or a risk of bacterial infection exists	Two options: (1) In vitro study for Q_1/Q_2 products, (2) in vivo and in vitro studies for non-Q_1/Q_2 products: (a) PK study in aqueous humor, (b) in vitro microbial kill rate study	Posted 7/2018
Dexamethasone; tobramycin	Suspension, 0.05%/0.3%	For steroid-responsive inflammatory ocular conditions for which a corticosteroid is indicated and where superficial bacterial ocular infection or a risk of bacterial ocular infection exists	Two options: (1) In vitro study for Q_1/Q_2 products, (2) in vivo and in vitro studies for non-Q_1/Q_2 products: (a) PK study in aqueous humor, (b) in vitro microbial kill rate study	Posted 2/2010; revised 3/2012, 6/2012, 6/2013, 6/2016, 2/2019
Dexamethasone; tobramycin	Suspension, 0.1%/0.3%	For steroid-responsive inflammatory ocular conditions for which a corticosteroid is indicated and where superficial bacterial ocular infection or a risk of bacterial ocular infection exists	Two options: (1) In vitro study for Q_1/Q_2 products, (2) in vivo and in vitro studies for non-Q_1/Q_2 products: (a) PK study in aqueous humor, (b) in vitro microbial kill rate study	Posted 2/2010; revised 3/2012, 6/2012, 6/2013, 6/2016, 2/2019

(continued)

Table 2 (continued)

Active ingredient	Dosage form	Indication	Bioequivalence study type	Date recommended/ revised
Difluprednate	Emulsion	For the treatment of inflammation and pain associated with ocular surgery and for the treatment of endogenous anterior uveitis	Two options: (1) In vitro study for Q_1/Q_2 products, (2) PK study in aqueous humor	Posted 1/2016, revised 2/2017
Fluorometholone	Suspension	For the treatment of corticosteroid-responsive inflammation of the palpebral and bulbar conjunctiva, cornea, and anterior segment of the globe	Two options: (1) In vitro study for Q_1/Q_2 products, (2) PK study in aqueous humor	Posted 10/2017
Loteprednol etabonate	Suspension, 0.2%	For the temporary relief of the signs and symptoms of seasonal allergic conjunctivitis	In vitro option for Q_1/Q_2 products	Posted 2/2018
Loteprednol etabonate	Suspension, 0.5%	For the treatment of steroid-responsive inflammatory conditions of the palpebral and bulbar conjunctiva, cornea, and anterior segment of the globe such as allergic conjunctivitis, acne rosacea, superficial punctate keratitis, herpes zoster keratitis, iritis, cyclitis, selected infective conjunctivitis, when the inherent hazard of steroid use is accepted to obtain an advisable diminution in edema and inflammation	Two options: (1) In vitro study for Q_1/Q_2 products, (2) PK study in aqueous humor	Posted 3/2011; revised 4/2013, 6/2016, 2/2018
Nepafenac	Suspension, 0.1%	For the treatment of pain and inflammation associated with cataract surgery	Three options: (1) In vitro study for Q_1/Q_2 products, (2) PK study in aqueous humor, and (3) clinical endpoint study	Posted 12/2014, revised 12/2016

(continued)

Table 2 (continued)

Active ingredient	Dosage form	Indication	Bioequivalence study type	Date recommended/ revised
Nepafenac	Suspension, 0.3%	For the treatment of pain and inflammation associated with cataract surgery	Three options: (1) In vitro study for Q_1/Q_2 products, (2) PK study in aqueous humor, and (3) clinical endpoint study	Posted 12/2014, revised 12/2016
Prednisolone acetate	Suspension, 1%	For the treatment of steroid-responsive inflammation of the palpebral and bulbar conjunctiva, cornea, and anterior segment of the globe	Two options: (1) In vitro study for Q_1/Q_2 products, (2) PK study in aqueous humor	Posted 4/2014, revised 6/2016, 5/2019

they can be affected by differences in manufacturing methods used to formulate a product. Manufacturing differences can alter the arrangement of matter (microstructure) within a dosage form and can impact the physicochemical characteristics of a drug product. These differences in physicochemical characteristics can affect drug release from the vehicle, drug absorption by ocular tissues, and product stability. However, if the microstructure and physicochemical properties of a generic are equivalent to the reference product, drug release, drug absorption, and product stability will also be equivalent for both products.

Some physicochemical properties of complex emulsion and suspension products that may be affected by the manufacturing and/or or formulation process are the following:

- *Particle/globule size distribution.* Large drug particulates or globules have a different drug release rate than smaller particles, and the clearance rate may be different for differently sized particles. These factors can be important for bioavailability. Particle size can also affect the stability of the dispersion and the drug release rate. In general, particle size and size distribution are strongly dependent on the manufacturing conditions, with the ability to tune specific sizes via the manufacturing process.
- *Viscosity profile as a function of applied shear.* A high-viscosity vehicle increases the formulation contact time with ocular tissues, which can improve bioavailability by reducing the drainage rate (Zignani et al. 1995). Viscosity can also affect the rate of drug release from the vehicle (Erös et al. 1994) as well as increase the dispersion stability (i.e., slow down phase separation). In general, it is the choice, grade, and concentration of viscosity modifying excipients, and not the manufacturing process, that affect the overall formulation viscosity properties.

- *pH*. pH of the formulation can influence tear pH and drug absorption. Instilling a product into the eye that has a different pH from tears can be irritating and can cause excessive blinking and lacrimation. This, in turn, could cause the drug to be cleared away more quickly through drainage, thus reducing bioavailability. The pH can also affect stability, solubility, and permeability of the drug through ocular tissues (Rabinovich-Guilatt et al. 2004).
- *Zeta potential*. The surface charge of particles can influence interactions with ocular tissues and cell membranes. For example, positively charged particles can better adhere to negatively charged surfaces such as the cornea and conjunctiva (Lallemand et al. 2012). A high zeta potential likely will also keep particles in a state of dispersion, whereas a low zeta potential may lead to aggregation and instability.
- *Osmolality*. Osmolality is recommended to be measured because the cornea is particularly influenced by changes in the tonicity of tear fluid. A hypotonic product applied to the eye can increase the permeability of the epithelium, drawing water into the tissue (Wilson et al. 2001).
- *Surface tension*. Ophthalmic formulations with different surface tension values can affect spreading of the ophthalmic product on the ocular surface; a low surface tension relative to the tear film has been shown to improve spreading and potentially enhanced bioavailability. Surface tension also plays a role in the capillary drainage through the nasolacrimal ducts (Zhu and Chauhan 2005, 2008), which in turn will affect precorneal tear film residence time of the instilled drug product. Moreover, if the surface tension of the instilled product is significantly lower than that of the tear film, this may lead to disruption of the tear film lipid layer which may in turn cause the formation of dry spots (Siddique and Braun 2015). These dry spots will typically increase the blink rate and subsequently increase nasolacrimal drainage (Choi and Lionberger 2016; Saettone et al. 1999). All of these factors would significantly affect instilled drug product residence time.
- *Drug distribution in different phases of the formulation*. Drug distribution in the different phases of the formulation (i.e., concentration of drug in solid particles, oil globules, or micelles vs the dispersion media) may affect bioavailability and rate of absorption as drug within the dispersed phase is more readily available (Yamaguchi et al. 2005; Jensen et al. 2010). As detailed further below, this is generally a thermodynamic process such that drug distribution should be the same for Q_1/Q_2 products that also have similar Q_3 properties.

In addition, although there is currently no compendia and/or recommended method for testing the drug release rate from an ophthalmic formulation, in vitro release testing is generally recommended as part of the in vitro totality of evidence. This is because an in vitro release rate reflects the combined effect of several physical and chemical properties in both the drug substance and the drug product. Manufacturing methods and processes may change formulation attributes, thereby affecting the rate of drug release and the drug's bioavailability.

Assessing the release profiles is intended to enable a sensitive determination of any formulation and manufacturing differences. As stated by Choi and Lionberger (2016), drug release measurements are valuable for understanding the impact of Q_3 differences as a whole. Current scientific understanding may provide accurate estimates of the impact that differences in an individual physicochemical parameter may have on BE, but the interactions between parameters will be more difficult to capture. The release rate test then provides an additional layer of data to ensure that Q_3 comparisons provide an accurate assessment of the potential differences in BE between formulations. Confirmation that a proposed generic product has a comparable release rate to that of the RLD can help ensure that the proposed generic product will deliver drug to the ocular tissues for absorption in a manner comparable to that of the RLD.

Therefore, two ophthalmic products, composed of the same materials, with comparative drug release rates and physicochemical properties (e.g., they are $Q_1/Q_2/Q_3$ the same), are expected to be bioequivalent.

Ophthalmic Ointments

As of 2018, there are approximately 20 approved and currently marketed ophthalmic ointment RLDs. These products fall into roughly four different categories based on their indication and class of active ingredient. This includes eight antimicrobial products, seven antimicrobial steroid combination products, three steroids, and three products indicated for intraocular pressure reduction. Of these reference products, approximately 40% have at least three or more approved generics. This higher percentage of approved generics is due in part to these generics being approved prior to 1984 under pre-Hatch-Waxman standards.

In general, ophthalmic ointments are very similar to each other in terms of formulation composition and components, consisting of the active ingredient(s) and white petrolatum (ointment base). Some ointments also contain mineral oil and preservatives. White petrolatum is inert and is compatible with all kinds of API(s). However, it is a mixture of hydrocarbons with inherent heterogeneity. White petrolatum from different sources may vary in the microstructure and physicochemical properties (Ogita et al. 2010). Therefore, ophthalmic ointments prepared using white petrolatum from different sources may not have comparable drug release rates. In addition, different manufacturing processes may also lead to differences in physicochemical characteristics and microstructure of the product.

Accordingly, the in vitro studies for demonstration of BE of ophthalmic ointments are selected to compare critical physicochemical characteristics that can be affected by either raw material properties or manufacturing processes. Table 3 below lists all the PSGs for ophthalmic ointments that include an in vitro study option.

For in vitro studies, comparative physicochemical characterization on at least three exhibit batches of the test and reference products are requested. Tests that may be recommended include solid-state form of API, appearance, acidity and alkalinity

Table 3 Product-specific bioequivalence recommendations on ophthalmic ointments

Active ingredient	Dosage form	Indication	Bioequivalence study type	Date recommended/ revised
Bacitracin	Ointment	For the treatment of superficial ocular infections involving the conjunctiva and/or cornea caused by bacitracin-susceptible organisms	In vitro studies: Comparative physicochemical characterization	Posted 6/2012, revised 10/2016
Ciprofloxacin	Ointment	For the treatment of bacterial conjunctivitis caused by susceptible strains of various microorganisms	Two options: (1) In vitro study for Q_1/Q_2 products, (2) in vivo clinical endpoint study	Posted 9/2018
Erythromycin	Ointment	For the treatment of superficial ocular infections involving the conjunctiva and/or cornea caused by organisms susceptible to erythromycin and for prophylaxis of ophthalmia neonatorum due to *N. gonorrhoeae* or *C. trachomatis*	In vitro studies: Comparative physicochemical characterization	Posted 6/2012, 10/2016
Loteprednol etabonate	Ointment	For the treatment of post-operative inflammation and pain following ocular surgery	Two options: (1) In vitro study for Q_1/Q_2 products, (2) PK study in aqueous humor	Posted 7/2018

of the extracted ointment base, rheology, and particle size distribution. Detailed descriptions of some characterization tests are described below.

Similar to ophthalmic suspensions, the API(s) of ophthalmic ointments are also in solid state and are suspended in the ointment base. API(s) with polymorphs, hydrates, and solvents may have different physicochemical properties and, in particular, their kinetic solubility and stability. Besides the solid state of the API(s), the particle size distribution is also very critical for therapeutic efficacy of the product. Particles with a different particle size distribution have different surface areas, which may result in changes in the in vivo drug release rate. The distribution of the API particles could also affect content uniformity, which in turn affects dosing accuracy. Therefore, it is important that the API(s) used in generic product have similar solid-state form and size distribution as the API(s) of the RLD.

White petrolatum from different sources may have different appearance. In general, petrolatum with a higher degree of refinement will appear lighter in color and contain less unsaturated and polar hydrocarbons. Comparative appearance ensures that there is no significant difference in color and homogeneity macroscopically. This can also eliminate potential issues with patient acceptance. White petrolatum from different sources may vary in composition. Although acidity and alkalinity of white petrolatum are included in the current USP monograph, the acceptance criteria in the USP monograph would include white petrolatum that induces an aqueous pH anywhere in the approximate range of pH 3 to pH 8. Therefore, two ointments produced with different sources of white petrolatum may not be the same with respect to pH simply because they both are USP grade. Therefore, comparative acidity and alkalinity of the ointment (mainly the white petrolatum base) are important to ensure that patients will not experience more irritation from the generic products than the RLD.

Assessment of rheological behavior is important for determining ointment spreadability, retention time, and drug release at the site of action following administration. Generic ointments with a lower viscosity compared to that of the RLD could enhance drainage and decrease contact time with the cornea so therapeutic benefit may be reduced. However, generic ointments with a higher viscosity may prolong the duration of temporary blurring of vision which may result in issues of patience compliance. In general, ophthalmic ointments exhibit non-Newtonian shear thinning viscoelastic behavior. This is very relevant to the real use condition under which ointments will experience a range of shear rates due to blinking. Several studies have demonstrated that the rheological behaviors, particularly storage modulus of ophthalmic ointments, is resource and manufacturing processes dependent (Xu et al. 2015; Patere et al. 2018; Bao et al. 2017a). For example, Bao et al. reported that ointments prepared using the hot-melt method showed higher rheological parameters and lower drug release profiles compared to the simple mixing method. Ointments prepared using the hot melting method, but a different cooling procedure, had no significant difference in physicochemical characteristics, indicating that the quenching rate is not critical in forming the ointment matrix. Generally, ointments with higher rheological parameters have lower in vitro drug release rates. Therefore, it is important for generic products to show comparable rheological properties, and viscosity should be evaluated over a range of shear rates.

Besides comparable physicochemical characteristics, it is also important that the in vitro drug release rate of API(s) from generic and reference products is comparable. This ensures that a similar amount of API(s) will be released for absorption following administration. To date, no standardized in vitro drug release methods have been developed for ophthalmic ointments. The Franz diffusion cell is one of the most commonly used/reported apparatus for ointments, but mostly for topical applications. For ophthalmic ointments, very little information is available in the literature. Xu et al. and Ghabeish et al. have reported conducting in vitro drug release testing of ophthalmic ointments using both Franz diffusion cell and USP apparatus II with enhance cells (Xu et al. 2015; Al-Ghabeish et al. 2015; Bao et al. 2017b). Besides these two methods, Bao et al. also reported using USP Apparatus IV with semisolid adapters (Bao et al. 2017b).

Table 4 lists the physicochemical properties generally recommended for ophthalmic suspensions, emulsions, and ointments for the in vitro option. The product-specific guidance (FDA n.d.) should be consulted for specific physicochemical tests for a particular product, as recommended tests may differ from one product to

Table 4 Physicochemical tests generally recommended for ophthalmic suspensions, emulsions, and ointments

Suspensions	Emulsions	Ointments
Viscosity	Viscosity	Rheology
pH	pH	Particle size distribution
Specific gravity	Zeta potential	Appearance
Osmolality	Osmolality	Acidity and alkalinity
Surface tension	Surface tension	Solid state of API
Buffer capacity	Drug distribution in different phases within the formulation	
Appearance	Globule size distribution	
Soluble fraction of API		
Dose concentration of API		
Particle size distribution		

another. Recommendations for specific physicochemical tests may also continue to evolve as new scientific information becomes available for different dosage forms, and this will be reflected through future revisions of product-specific guidances.

Research to Advance Equivalency Standards

In July 2012, the Generic Drug User Fee Amendments (GDUFA) were passed to speed access to safe and effective generic drugs to the public. These amendments require user fees to supplement the costs of reviewing generic drug applications and to provide additional resources for regulatory science research.

OGD implements the regulatory science research program for generic drugs to support the development of new tools to evaluate drug equivalence and the development of generic drugs in all product categories. Tools include simulation tools to predict drug absorption, analytical methods for product characterization, and in vitro methods to predict in vivo performance. These tools will help address gaps in scientific knowledge that will help advance alternative methods for establishing equivalence. Research is conducted with external (academia, industry, other government agencies) and internal (FDA labs and offices) collaborators.

Since the first year of implementation of the GDUFA research program in fiscal year 2013, OGD has awarded 13 external grants and contracts and has also established a number of internal collaborations on ophthalmic research covering a wide range of dosage forms (suspensions, emulsions, ointments, implants). All projects fall under at least one of the following objectives of the ophthalmic research generic drug program:

- Develop physicochemical characterization methods to assess and compare formulation CQAs.
- Investigate key physicochemical properties (CQAs) that affect drug release and ocular bioavailability.
- Develop in vitro release testing methods which are sensitive to formulation difference and/or are predictive of in vivo release.
- Develop and better understand in vitro-in vivo correlations.
- Predictive modeling of ocular drug absorption that can assess impact of Q_3 formulation changes.

Before start of the GDUFA research program, there were no PSGs that presented an in vitro option to demonstrate BE for non-solution ophthalmic products. However, through external and internal research conducted under the ophthalmic research program, there are now two PSGs for emulsions, four PSGs for ointments, and eight PSGs for suspensions that include an in vitro option to demonstrate BE. These alternative approaches provide additional options to industry and help facilitate faster and more efficient drug development of ophthalmic generic products. FDA continues to expand the ophthalmic research program by identifying scientific gaps that preclude ANDA review and approval and also PSG development of complex ophthalmic drug products.

Conclusion

The approval of an ophthalmic generic drug product submitted as an ANDA is different from the approval of an ophthalmic therapeutic equivalent which is submitted as an NDA under Section 505(b)(2) of the Act. Standards for BE and chemistry, manufacturing, and controls ensure that ophthalmic drug products approved as generics are therapeutically equivalent to the branded product. The impact of differences in manufacturing and formulation can be detected through a rigorous comparison of excipients and physicochemical properties. The FDA reevaluates and improves standards which will enable availability of ophthalmic generics in new product categories and ensure safety and efficacy of approved products. Scientific research has contributed toward development of alternative equivalence standards for ophthalmic generic drugs that will provide additional options to industry for faster and more efficient drug development.DeclarationsThis article reflects the views of the authors and should not be construed to represent the views or policies of the US Food and Drug Administration (FDA). The authors have no declarations of interest to report.

References

Al-Ghabeish M, Xu X, Krishnaiah YS, Rahman Z, Yang Y, Khan MA. Influence of drug loading and type of ointment base on the in vitro performance of acyclovir ophthalmic ointment. Int J Pharm. 2015;495(2):783–91. https://doi.org/10.1016/j.ijpharm.2015.08.096.

Ali Y, Lehmussaari K. Industrial perspective in ocular drug delivery. Adv Drug Deliv Rev. 2006;58(11):1258–68. https://doi.org/10.1016/j.addr.2006.07.022.

Aptel F, Choudhry R, Stalmans I. Preservative-free versus preserved latanoprost eye drops in patients with open-angle glaucoma or ocular hypertension. Curr Med Res Opin. 2016;32(8):1457–63. https://doi.org/10.1080/03007995.2016.1202818.

Bao Q, Jog R, Shen J, Newman B, Wang Y, Choi S, et al. Physicochemical attributes and dissolution testing of ophthalmic ointments. Int J Pharm. 2017a;523(1):310–9. https://doi.org/10.1016/j.ijpharm.2017.03.039.

Bao Q, Shen J, Jog R, Zhang C, Newman B, Wang Y, et al. In vitro release testing method development for ophthalmic ointments. Int J Pharm. 2017b;526(1–2):145–56. https://doi.org/10.1016/j.ijpharm.2017.04.075.

Baranowski P, Karolewicz B, Gajda M, Pluta J. Ophthalmic drug dosage forms: characterisation and research methods. Scientific World Journal. 2014;2014:861904.

Cantor LB. Ophthalmic generic drug approval process: implications for efficacy and safety. J Glaucoma. 1997;6(5):344–9.

CDER/FDA. Orange Book: approved drug products with therapeutic equivalence evaluations; 2016. p. 36. http://www.accessdata.fda.gov/scripts/cder/ob/.

CDER/FDA. MAPP 5200.14, Filing review of abbreviated new drug applications. https://www.fda.gov/downloads/AboutFDA/CentersOffices/OfficeofMedicalProductsandTobacco/CDER/ManualofPoliciesProcedures/UCM574493.pdf. Accessed 18 Dec 2018.

CDER/FDA. Draft guidance on Brimonidine Tartrate. n.d.-a. http://www.fda.gov/downloads/Drugs/GuidanceComplianceRegulatoryInformation/Guidances/ucm082954.pdf.

CDER/FDA. Draft guidance on Brinzolamide. n.d.-b. http://www.fda.gov/downloads/Drugs/GuidanceComplianceRegulatoryInformation/Guidances/UCM384099.pdf.

CDER/FDA. Draft guidance on Dorzolamide Hydrochloride. n.d.-c. http://www.fda.gov/downloads/Drugs/GuidanceComplianceRegulatoryInformation/Guidances/UCM436822.pdf.

CDER/FDA. Draft guidance on Nepafenac. n.d.-d. http://www.fda.gov/downloads/Drugs/GuidanceComplianceRegulatoryInformation/Guidances/UCM428230.pdf.

CDER/FDA. Draft guidance on Cyclosporine. n.d.-e. http://www.fda.gov/downloads/Drugs/GuidanceComplianceRegulatoryInformation/Guidances/UCM358114.pdf.

CDER/FDA. Draft guidance on Difluprednate. n.d.-f. http://www.fda.gov/downloads/Drugs/GuidanceComplianceRegulatoryInformation/Guidances/UCM481813.pdf.

CDER/FDA. GDUFA regulatory science. n.d.-g. http://www.fda.gov/GDUFARegScience.

Chambers WA. Ophthalmic generics—are they really the same? Ophthalmology. 2012;119(6):1095–6. https://doi.org/10.1016/j.ophtha.2012.03.033.

Choi SH, Lionberger RA. Clinical, pharmacokinetic, and in vitro studies to support bioequivalence of ophthalmic drug products. AAPS J. 2016;18(4):1032–8. https://doi.org/10.1208/s12248-016-9932-z.

Code of Federal Regulations. Title 21 Code of Federal Regulations 314.94(a)(9)(iv); 2011.

Code of Federal Regulations. Title 21 Code of Federal Regulations 320.22(b)(1); n.d.

Erös I, Soósné-Csányi E, Selmeczi B. Influence of viscosity on drug release from ointments, creams, gels and emulsions. Acta Pharm Hung. 1994;64(2):57–61.

FDA. Drugs@FDA: FDA Approved Drug Products. 2019. https://www.accessdata.fda.gov/scripts/cder/daf/. Accesed 28 Jun 2019.

FDA. Product-specific guidances for generic drug development. n.d.. https://www.fda.gov/Drugs/GuidanceComplianceRegulatoryInformation/Guidances/ucm075207.htm.

Gasset AR. Benzalkonium chloride toxicity to the human cornea. Am J Ophthalmol. 1977;84(2):169–71.

Gasset AR, Ishii Y, Kaufman HE, Miller T. Cytotoxicity of ophthalmic preservatives. Am J Ophthalmol. 1974;78(1):98–105.

Harigaya Y, Jiang X, Zhang H, Chandaroy P, Stier EM, Pan Y. Bioequivalence study methods with pharmacokinetic endpoints for topical ophthalmic corticosteroid suspensions and effects of subject demographics. Pharm Res. 2018;36(1):13. https://doi.org/10.1007/s11095-018-2537-8.

Jensen LB, Magnussson E, Gunnarsson L, Vermehren C, Nielsen HM, Petersson K. Corticosteroid solubility and lipid polarity control release from solid lipid nanoparticles. Int J Pharm. 2010;390(1):53–60. https://doi.org/10.1016/j.ijpharm.2009.10.022.

Jitendra PK, Sharma A, Banik S, Dixit S. A new trend: ocular drug delivery system. Int J Pharmaceut Sci. 2011;2(3).

Kasai H, Aoyama Y, Kurasawa T, Imamura T, Tsuruma K, Hara H, et al. Comparison of efficacy and safety evaluation of Latanoprost formulations with and without Benzalkonium chloride. Pharmacol Pharmacy. 2013;4(4):377–84.

Lallemand F, Daull P, Benita S, Buggage R, Garrigue JS. Successfully improving ocular drug delivery using the cationic nanoemulsion. Novasorb J Drug Deliv. 2012;2012:604204. https://doi.org/10.1155/2012/604204.

McCarey B, Edelhauser H. In vivo corneal epithelial permeability following treatment with prostaglandin analogs [correction of analoges] with or without benzalkonium chloride. J Ocul Pharmacol Ther. 2007;23(5):445–51. https://doi.org/10.1089/jop.2007.0024.

Ogita Y, Takahashi Y, Iwata M, Sasatsu M, Onishi H, Hashimoto S, et al. Comparison of physical properties and drug-releasing characteristics of white petrolatums. Pharmazie. 2010;65(11):801–4.

Patere S, Newman B, Wang Y, Choi S, Vora S, Ma AWK, et al. Influence of manufacturing process variables on the properties of ophthalmic ointments of tobramycin. Pharm Res. 2018;35(9):179. https://doi.org/10.1007/s11095-018-2462-x.

Peters JR, Hixon DR, Conner DP, Davit BM, Catterson DM, Parise CM. Generic drugs—safe, effective, and affordable. Dermatol Ther. 2009;22(3):229–40.

Rabinovich-Guilatt L, Couvreur P, Lambert G, Dubernet C. Cationic vectors in ocular drug delivery. J Drug Target. 2004;12(9–10):623–33. https://doi.org/10.1080/10611860400015910.

Rahman Z, Xu X, Katragadda U, Krishnaiah YS, Yu L, Khan MA. Quality by design approach for understanding the critical quality attributes of cyclosporine ophthalmic emulsion. Mol Pharm. 2014;11(3):787–99. https://doi.org/10.1021/mp400484g.

Saettone MF, Burgalassi S, Chetoni P. Ocular bioadhesive drug delivery systems. In: Mathiowitz E, Chickering III DE, Lehr C-M, editors. Bioadhesive drug delivery systems: fundamentals, novel approaches, and development. Boca Raton: CRC Press; 1999.

Siddique JI, Braun RJ. Tear film dynamics with evaporation, osmolarity and surfactant transport. App Math Model. 2015;39(1):255–69.

Wilson CG, Zhu YP, Kurmala P, Rao LS, Dhillon B. Ophthalmic drug delivery. In: Hillery AM, Lloyd AW, Swarbrick J, editors. Drug delivery and targeting: for pharmacists and pharmaceutical scientists; 2001. p. 298–319.

Xu X, Al-Ghabeish M, Rahman Z, Krishnaiah YS, Yerlikaya F, Yang Y, et al. Formulation and process factors influencing product quality and in vitro performance of ophthalmic ointments. Int J Pharm. 2015;493(1–2):412–25. https://doi.org/10.1016/j.ijpharm.2015.07.066.

Yamaguchi M, Yasueda S, Isowaki A, Yamamoto M, Kimura M, Inada K, et al. Formulation of an ophthalmic lipid emulsion containing an anti-inflammatory steroidal drug, difluprednate. Int J Pharm. 2005;301(1–2):121–8. https://doi.org/10.1016/j.ijpharm.2005.05.036.

Zhu H, Chauhan A. A mathematical model for tear drainage through the canaliculi. Curr Eye Res. 2005;30(8):621–30. https://doi.org/10.1080/02713680590968628.

Zhu H, Chauhan A. Effect of viscosity on tear drainage and ocular residence time. Optom Vis Sci. 2008;85(8):715–25. https://doi.org/10.1097/OPX.0b013e3181824dc4.

Zignani M, Tabatabay C, Gurny R. Topical semi-solid drug delivery: kinetics and tolerance of ophthalmic hydrogels. Adv Drug Deliv Rev. 1995;16(1):51–60.

The Development and Commercialization of Sustained-Release Ocular Drug Delivery Technologies

Michael J. O'Rourke and Clive G. Wilson

Abstract There have been major advances in recent years developing new sustained-release ocular drug delivery systems; however, only a small number have achieved both global regulatory approval and commercial success. Despite the challenges, there remain significant market opportunities to enhance the delivery of currently marketed, generic, or novel therapeutics with new innovative technologies offering improved treatment options for patients suffering from major blinding diseases.

Keywords Sustained-release · Polymeric inserts · Implants · Technology platform

---

Scotia Vision is a specialized ophthalmic consulting company with expertise in global ocular drug delivery commercial and product development strategies. Founder Michael O'Rourke has over 30 years drug delivery experience across ophthalmology, periodontal, and pulmonary markets in sales, marketing, product launch, strategy development, and global commercialization. His industry career experience includes senior positions with several leading organizations and startups including 3M Pharmaceuticals, Alza, Chiron Vision, Bausch + Lomb, and GrayBug. He is currently the CEO of Re-Vana Therapeutics.

M. J. O'Rourke (✉)
Scotia Vision Consultants LLC, Tampa, FL, USA

C. G. Wilson
Strathclyde Institute of Pharmacy and Biomedical Science, Glasgow, Scotland, UK

S. Neervannan, U. B. Kompella (eds.), *Ophthalmic Product Development*,
AAPS Advances in the Pharmaceutical Sciences Series 37,
https://doi.org/10.1007/978-3-030-76367-1_21

Introduction

The estimated number of people visually impaired in the world is 285 million, of whom 39 million are blind with the remaining 246 million having greatly reduced eyesight. It is also estimated that 65% of people visually impaired and 82% of all blind are 50 years and older (World Health Organization 2010). Four major blinding diseases, age-related macular degeneration, diabetic retinopathy, diabetic macular edema, and glaucoma, due to their whole or partial impact on the posterior segment of the eye and their growing market sizes, may offer the most promising opportunities for future ocular drug delivery technologies. They affect large numbers of people and pose a significant risk of vision loss and blindness for those affected.

Current therapeutic options for these diseases may at best manage the condition, slow or halt further deterioration or disease progression. New breakthrough treatments would benefit from robust sustained delivery of the drug to the target tissues in both the posterior and anterior segments of the eye and, importantly, enhance compliance of patients with long-term treatment regimens for these chronic diseases, for example, reducing the need for frequent injections and reducing or eliminating the need for topical eye drops. Sustained-release drug delivery therapies for both small and large molecules provide significant market opportunities for the development of new therapies based on enhanced drug delivery methods and technologies.

The global pharmaceutical market continues to grow from $29.04 billion in 2019 to an estimated $42.14 billion at the end of 2024. Within this time period, retinal therapeutics are projected to grow from $13 billion to approximately $22.4 billion. In 2019 Glaucoma was the second largest segment at $4.9 billion, followed by dry eye treatment also at $4.9 billion, both projected to grow to $6.3 billion and $6.5 billion, respectively, by 2024 (Market Scope 2013) . The changing trends in revenue by specialty are shown in Fig. 1.

From the 1970s onward, eight sustained-release ophthalmic drug delivery products have been successfully brought to market. They include two short-term anterior extraocular inserts, five sustained-release intraocular delivery products targeting the posterior segment, and an intracameral implant.

Early Developments in Extraocular Sustained-Release Drug Delivery Products

The first polymeric inserts employed to release an ophthalmic drug over prolonged periods were used in the late 1800s in the UK. These gelatin inserts released cocaine for the purpose of local ocular anesthesia (Del Amo and Urtti 2008). After some brief development work in the Soviet Union on soluble ophthalmic drug inserts in the 1960s, it was the California-based Alza Corporation in 1975, with its innovative founder Alejandro Zaffaroni, who developed the first FDA-approved,

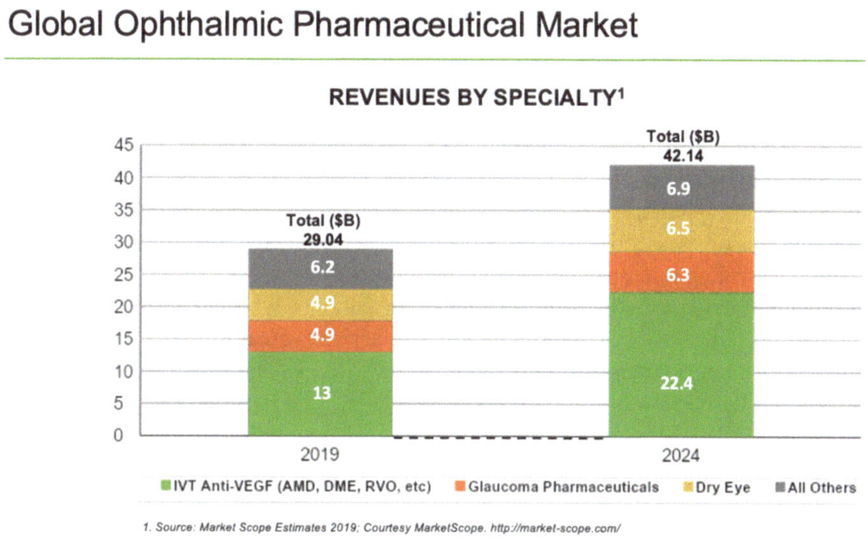

Fig. 1 Global growth by specialty, 2019–2024

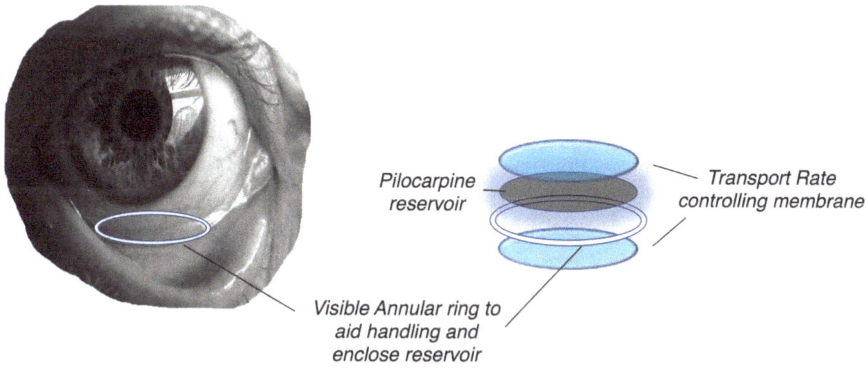

Fig. 2 Illustrating position of Ocusert in the eye and elements of the device

sustained-release ocular product. Ocusert® (pilocarpine—Pilo-20/40) was an anterior extraocular insert delivering pilocarpine to patients with glaucoma at a near constant rate, minimizing side effects by avoiding absorption peaks (Fig. 2). At a delivery rate of 50 μg h^{-1}, the effect was significantly greater than 2% w/w/ pilocarpine (Armaly and Rao 1973).

Ocusert was a breakthrough innovation from Alza, then the world's leader in drug delivery systems. Unfortunately, it was a commercial failure. Ocusert lasted only 1 week and had to be inserted in the inferior fornix by the patient. Patient compliance was poor, the unit was occasionally lost from the eye, and the product is no longer marketed. Following the failure of Ocusert, it was clear that the successful development of future drug delivery systems was not just about release rates and pharmacokinetics, but critically, it was just as much about the patient, their compliance to the new therapy, the level of comfort in the eye, and the endorsement of the physician to prescribe the product and maintain "learning curve" support for the patient. Later in 1981, Lacrisert® (hydroxypropyl cellulose ophthalmic insert 5 mg), produced by hot melt extrusion, was launched by Merck, Sharpe, and Dohme for dry eye syndrome patients with a daily dosage regime. The Lacrisert rod is placed in the lower conjunctival sac with an applicator. As it imbibes water and gels, the polymer dissolves. Lacrisert stabilizes and thickens the precorneal tear film to prolong tear breakup time in patients with moderate to severe dry eye including keratoconjunctivitis sicca. It is contra-indicated in patients who are hypersensitive to hydroxypropyl cellulose (Jones et al. 2017). Lacrisert remains on the market today (Bausch + Lomb) but with limited commercial success that may be due in part to difficulty of insertion and minority of patients experience blurriness and an oily sensation (Wander and Koffler 2009).

Later Approved Extraocular Release Products

The use of silicone punctal plugs to reversibly obstruct the canalicular duct has been well established as an effective treatment in aqueous flow-deficient dry eye syndrome, particularly that refractory to topical treatment. Additionally, punctal occlusion has been used to attempt to increase the effectiveness of topical therapy in glaucoma (Tost and Geerling 2008) and to open up the drainage in epiphora. Punctal plugs can be expelled, and thus the alternative of placing the device in the canicular canal was pursued. Intracanalicular plugs of acrylic, silicone, and collagen have been described, and an excellent review is provided by Jehinger and colleagues (Jehangir et al. 2016). In December 2018, the FDA approved Dextenza®, an intracanalicular implant for the treatment of post-operative pain. The dose, 0.4 mg of dexamethasone, is designed to be released over a 30-day period (Steinbach 2019).

Intraocular Sustained Delivery

In 1996, approximately 20 years after Ocusert, Chiron Vision and Controlled Delivery Systems (CDS), working together, launched the world's first posterior sustained-release intraocular delivery system, Vitrasert®, ganciclovir 4.5 mg intravitreal implant (Steinbach 2019; Sanborn et al. 1992). The implant was developed by Ashton and colleagues in the early 1990s. Each Vitrasert implant contained a ganciclovir tablet with the inactive ingredient, magnesium stearate (0.25%), and the tablet was coated with polyvinyl alcohol (PVA) and ethylene vinyl acetate (EVA) polymers. The EVA and PVA polymers controlled the release of the drug (Fig. 3).

Indicated for cytomegalovirus (CMV) retinitis at the height of the HIV disease epidemic, Vitrasert delivered ganciclovir for approximately 6–8 months with initial

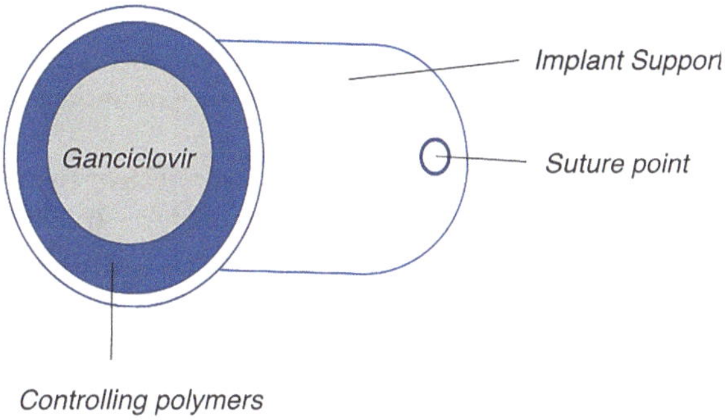

Fig. 3 Graphic of Vitrasert construction

resounding success following US and European approval. However sales rapidly declined from 1998 onward due to the increasing impact of the first protease inhibitor saquinavir which was marketed under the brand names Invirase® and then Fortovase® from Hoffman La Roche in 1995, which offered a greater degree of prevention against declining CD4 cell counts in HIV patients. Thus, there were fewer cases of CMV retinitis. Vitrasert was subsequently removed from the market in 2014.

Another 10 years passed before the world's second intraocular posterior delivery product was launched. Retisert® (fluocinolone acetonide intravitreal implant 0.59 mg), another CDS technology launched by Bausch + Lomb in 2005, was introduced to the market with an orphan drug single-indication for non-infectious posterior uveitis (NIPU), delivering a generic steroid drug for a period of about 30 months. Clinical trials were also conducted for neovascular age-related macular degeneration (w-AMD) and diabetic retinopathy (DR), but the product was never approved for these significantly more lucrative market opportunities due to the trials failing to meet their end points.

Both Vitrasert and Retisert delivered successful and innovative solutions to the new emerging pharmaceutical retinal market and no doubt prevented multiple cases of potential blindness while greatly enhancing the quality of life. However, despite significant proven efficacy, invasive insertion methods plus ocular side effects including increased intraocular pressure (IOP) and cataract formation have proven to be barriers to broader market penetration.

Vitrasert was launched at approximately $4500 and Retisert at $18,000 per implant. These premium priced products also set expectations for future pricing strategies and highlighted the need for in-depth pricing studies to assist with market acceptance and, more importantly, reimbursement. For Vitrasert and Retisert, a new era for establishing ophthalmic premium-pricing strategies tied to reimbursement methodology was emerging.

With its zero-order kinetics and efficacy rates, Vitrasert was also a landmark for demonstrating the technical viability of an intraocular drug release system. The drug-delivery technical innovation and expertise of CDS were recognized, as well as the leadership and commitment of Chiron Vision entering the retina drug delivery market at a time when few other companies had posterior-segment delivery as part of their strategic investment plan. "Big Pharma" had not perceived the ophthalmic marketplace as large enough to support a fully fledged development effort. Posterior drug delivery development was thus largely limited to smaller, "specialty" ophthalmic driven companies.

As stated by one of the early Vitrasert investigators, "Intuitively, local therapy made sense, but it hadn't been done before. It was Vitrasert that showed us how potent local therapy could be when compared with systemically administered drugs" (Martin et al. 1994). Vitrasert and Retisert also created an R&D catalyst for posterior segment product developments; they provided new treatment options for vitreo-retinal surgeons and offered the prospect of improved outcomes for their patients.

Later Intraocular Sustained-Release Drug Delivery Products

The third intraocular, sustained-release product was Ozurdex® (dexamethasone intravitreal implant 0.7 mg), launched by Allergan in 2007. Ozurdex is indicated to treat adults with swelling of the macula (macular edema) following branch retinal vein occlusion (BRVO) or central retinal vein occlusion (CRVO), noninfectious inflammation of the uvea (uveitis) affecting the back segment of the eye, and diabetic macular edema (DME).

Originally developed by Oculex Pharmaceuticals whom Allergan acquired in 2003, Ozurdex is less invasive than Vitrasert or Retisert, because it is injected via a 22-gauge needle and is bioerodible. The benefits of a single Ozurdex injection may last several months at approximately $1400 per implant. Continued development of the Ozurdex applicator led to the adoption of a coated needle manufactured by TSK Laboratory which is claimed to have better glide and penetration characteristics as measured in porcine eyes (Bakri et al. 2014).

An anterior version of Ozurdex, Surodex®, was also developed and, similar to Ozurdex, was a bioerodible implant delivering 60 μg dexamethasone that achieved a continuous aqueous steroid level for 7–10 days after which it fell to low or non-detectable levels. Intraocular placement of two Surodex implants was demonstrated as a safe and effective treatment method to reduce intraocular inflammation after cataract surgery and was superior to eye drops in reducing inflammatory symptoms (Tan et al. 2001). However clinical trials were never completed nor a product launched due most likely in part to foreseeable reimbursement challenges at that time, with limited private pay penetration. The concept however of anterior drug delivery was widely accepted as a potential breakthrough.

The fourth intraocular sustained-release product was Iluvien® (fluocinolone acetonide 0.19 mg) intravitreal implant in an applicator, from Alimera Sciences, which gained European approval in 2012 and USA approval in 2014 for the treatment of diabetic macular edema (DME) in patients who have been previously treated with a course of corticosteroids and did not have a clinically significant rise in intraocular pressure (Bailey et al. 2017).

Each non-biodegradable Iluvien implant provides a therapeutic effect of up to 36 months by delivering sustained, sub-microgram levels of fluocinolone acetonide. Now approved in multiple European countries, with further approvals and reimbursement expansion expected, Iluvien costs approx. \$8800–\$9196 per implant.

In February 2018 the fifth intraocular sustained-release product was approved, Dexycu® (dexamethasone intraocular suspension 9%), treating inflammation associated with cataract surgery. It is administered as a single intraocular dose to the posterior segment of the eye at the end of ocular surgery for the treatment of post-operative inflammation (Kiernan 2020). It is the first and only FDA-approved intraocular product with this indication. Dexycu is not an intraocular implant but is an intraocular suspension that employs Verisome® extended-release drug delivery technology, which encompasses a broad number of related, but distinct drug delivery systems capable of incorporating an extensive range of active agents, including small molecules, proteins, and monoclonal antibodies. Dexycu was launched by EyePoint Pharmaceuticals in March 2019.

The sixth intraocular sustained-release product is Yutiq™ (fluocinolone acetonide intravitreal implant) 0.18 mg from EyePoint Pharmaceuticals (Levine et al. 2020). Yutiq was approved in 2018 for the treatment of chronic non-infectious uveitis affecting the posterior segment of the eye. Yutiq utilizes the company's Durasert™ drug delivery technology and is non-bioerodible and designed to release consistently over 36 months.

Finally, Allergan's Durysta® received market approval in March 2020. The intracamerally administered implant contains 10 μg of bimatoprost in a PLGA-based matrix and is claimed to have an intraocular lowering effect of 4 to 6 months (Medeiros et al. 2020).

It is worth noting of seven intraocular product approvals, five contain steroids (Retisert, Ozurdex, Iluvien, Dexycu and Yutiq) and of those three contain the same steroid active fluocinolone acetonide. As yet all approvals have been with generic, approved therapeutics. No successful approval has yet been achieved with a novel, new chemical entity.

Intravitreal Injection (Non-Sustained Release) Drug Delivery Products

In 2006, Genentech launched Lucentis® (ranibizumab 0.5 mg or 0.05 mL of 10 mg/mL Lucentis solution), recommended for intravitreal injection (IVT) once a month (approximately every 28 days. It is now indicated for wet AMD, macular edema following retinal vein occlusion (RVO), DR, DME, and myopic choroidal neovascularization (mCNV). In November 2015 results from the Diabetic Retinopathy Clinical Research Network Protocol S provided evidence for the first major advance in the treatment of proliferative diabetic retinopathy (PDR) in more than 40 years. Lucentis treatment of PDR was at least as good as pan-retinal photocoagulation for visual acuity at 2 years and is now an effective treatment alternative to PRP. Lucentis transformed the treatment landscape for neovascular AMD, and although it offers no cure or CNV regression, it has offered renewed hope to thousands of AMD sufferers.

Another breakthrough product, Eylea® (aflibercept 2 mg, 0.05 mL IVT), from Regeneron Pharmaceuticals, was launched in 2012 for neovascular (wet) AMD and is also indicated for the treatment of macular edema following RVO, DME, and DR in patients with DME.

The recommended dose for Eylea is 2 mg (0.05 mL or 50 μL) administered by an IVT injection every 4 weeks (i.e., monthly) for the first 12 weeks (3 months), followed by 2 mg (0.05 mL) via IVT once every 8 weeks (2 months). In August 2018 the FDA approved a supplemental biologics license application for a 12-week dosing schedule of Eylea injection in wet age-related macular degeneration patients. In addition to Lucentis and Eylea, Avastin® (bevacizumab) was initially approved by the FDA as a cancer drug but is widely used off label to treat wet AMD and in some cases for macular edema. It is a full-length antibody with a molecular weight of 149 kDa, whereas ranibizumab is a 49 kDa Fab fragment. It is significantly less expensive than ranibizumab, and both bind to all isoforms of VEGF-A. Based on the Comparison of AMD Treatments Trials (CATT) study funded by the National Eye Institute (NEI), overall, at both 1 and 2 years, ranibizumab and bevacizumab had similar beneficial effects on visual acuity when the dosing regimen was the same.

As referenced previously with sales of the anti-VEGF market segment at approximately $13 billion in 2019, it demonstrates a remarkable turnaround for the posterior segment industry over the past 17–18 years. Compared to 2001, there has been a dramatic shift in revenues generated between front and back of the eye diseases (Fig. 4).

These multi-billion dollar intravitreal injection drug products have demonstrated that a scientifically defined, sustained-release implantable technology itself is not a prerequisite for commercial success but that the sustained clinical efficacy effect of the drug is critical. However, the market potential for sustained-release versions of anti-VEGF products is significant, as the onerous need for monthly or bi-monthly injections may not be ideal from a patient adherence, comfort, or safety perspective. There were approximately 24 million IVT injections in 2019 (Market Scope 2019), 6.9 million in the USA and 17.5 million ex USA, (Grzybowski et al. 2018), with the most serious but rarely occurring injection-related complications including endophthalmitis (Ta 2004), cataract, retinal detachment, and vitreous or choroidal hemorrhages (Jager et al. 2004).

Developing New Sustained-Release Technologies

With only five approved posterior-segment sustained-release products by end 2018, the challenges to successful development are clear, yet the quest to provide better treatments is unrelenting. These challenges were documented at a major drug

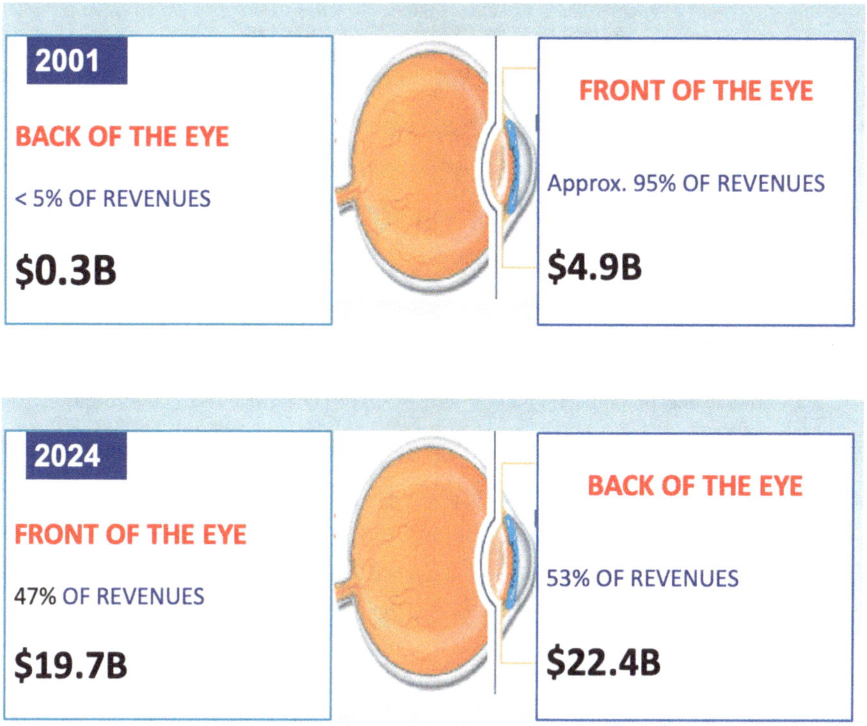

Fig. 4 How the market changed. Comparison of back of the eye vs. front of the eye diseases in 2001 and 2024 (Market Scope 2019)

delivery forum held in 2009 where five key barriers to new effective treatments were identified (SERC 2009):

- Developing a great product.
- Identifying and implementing the best method of delivery.
- Using the appropriate animal model for testing the drug's safety and efficacy.
- Identifying an adequate patient sample and developing a well-considered treatment design or plan for a clinical trial to attain a satisfactory endpoint.
- Locating a company to finance the product and guide it into the commercial market.

It is worth determining what is the optimal development strategy early in the drug research process: developing/acquiring a drug and then looking for a delivery mechanism or developing/acquiring a drug delivery technology and determining which drug(s) it can deliver (Fig. 5). As stated at the ARVO 2009 Summer Eye Research Conference, "Without a successful and convenient drug delivery system, drugs that are very effective at the bench or in preclinical testing often fail."

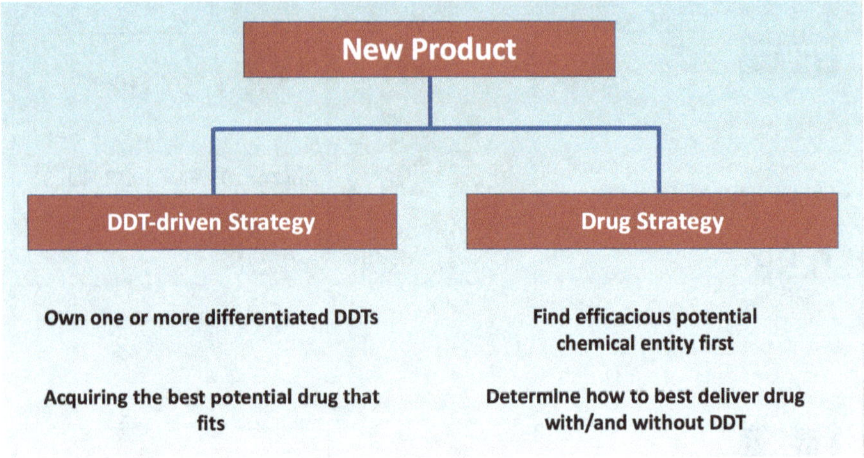

Fig. 5 New product with two distinct development pathways

Drug Delivery Development: If a Drug Delivery Technology (DDT) Platform Only Approach Is the Primary Strategy

The drug delivery systems under investigation are exemplified by two fundamental approaches and philosophies: first, longer-acting, reservoir implants with good long-term control of disease but with potential for drug or suppressive side effects and, second, shorter-acting, biodegradable or non-biodegradable inserts that potentially expose the eye to less drug or suppressive side effects but may also control disease less well. These approaches might depend on the natural course of the disease (chronic, curable, or prone to remissions and exacerbations) and the severity and reversibility of potential side effects of a drug. Drug stability and activity in the eye are also important issues to consider.

Challenges to a DDT-Focused Strategy

For a company focused on developing a DDT alone, there are commercial barriers to overcome which include a relatively low level of interest to potential API owners seeking partners since drug-device combinations may represent a higher regulatory hurdle. In most delivery systems, there is an optimum with regard to API properties due to the nature of the matrix and the required period of treatment. For example, a hydrophilic matrix will retard the release of a hydrophobic molecule to a greater extent than a hydrophilic material which might allow fast escape of API through the percolating water phase.

Finally the economics of investment may appear unattractive from return and risk perspectives, with a potentially low percentage royalties (e.g., <10%), often being the best commercial outcome expected if the technology is licensed. However upfront payments and milestones could be included.

Opportunities for DDT-Focused Strategy

Several strategic options do remain for startup companies offering a DDT without a drug. Previous models from the Alza Corporation had licensing deals or some other business development partnership that could include co-development, with the opportunity for some upfront payments on clinical milestones, and an on-going royalty stream once the product was approved and marketed. CDS with Vitrasert and Retisert technologies agreed to licensing deals with Chiron Vision and its successor, Bausch + Lomb. Many other startup companies today are in this product development life phase, and many have already secured venture capital funding or licensing deals. With carefully planned business development strategies, companies can increase their chances of funding and commercial success. Much of this will initially depend on demonstrating compelling proof of concept in pre-clinical animal models and positioning the technology with a high value proposition versus other competing technologies in development. Another strategy could include acquiring or sourcing a drug with the DDT for a specific indication, the drug being either a generic or would be classed as a generic at the estimated time of launch, for example, 10–12 years away.

Drug Delivery Development: If a New Drug Is the Primary Strategy

There are multiple development options here, and generally given a choice, assuming funding is available, this will be the preferred company positioning. With the security of a new drug in a company's portfolio or pipeline, the secondary remaining challenge will be the determination of delivery route. In the cases of Lucentis and Eylea that are new ophthalmic-indicated chemical entities, drug delivery is by the intravitreal route, with no physical sustained action drug delivery technology built into the product. However, as a second phase of development, it is feasible that intraocular drug delivery enhancements could be added and launched as a new product in the future. Companies will often develop life cycle R&D programs permitting sustained competitive advantages over the nearest competitors, thus securing greater strategic market shares.

To further explain the product development options, there are four possible combinations between the drug delivery technology and a generic drug or new chemical entity (Fig. 6).

Sustained DDT Strategies

Technology

Drug		Existing	New
	Existing	Existing Drug - Existing Delivery System e.g. Retisert. Yutiq	Existing Drug - New Delivery System e.g. Vitrasert, Ozurdex, Iluvien
	New	New Drug - Existing Delivery System None launched to date	New Drug - New Delivery System None launched to date

Fig. 6 Four possible combinations for sustained-release DDT strategies

Existing Drug-Existing Delivery System

In the case of Retisert, the delivery technology, a modified version (smaller pars planar incision and longer delivery period) of Vitrasert already existed, as did the generic drug, fluocinolone acetonide. The product development strategy envisioned was to develop a new product by combining an existing drug with a modified version of an existing delivery system and thereby create the first sustained-release, drug-delivery device for a specific indication, in this case non-infectious posterior uveitis.

Existing Drug: New Delivery System

One common strategy to date for development and approval of sustained-delivery systems is the combination of an existing drug with a new method of delivery. In these cases new products like Vitrasert, Iluvien, and Ozurdex were launched containing existing generic drugs but delivered by new drug delivery technologies, permitting new product market segments to be created.

New Drug: Existing Delivery System

This has the added advantage of a proven delivery system. However, other than proof of concept, the development pathway starts at the beginning, given that a new drug must be tested. Therefore, this strategy is high risk and expensive but with high rewards if successful.

New Drug: New Delivery System

The combination of an untested therapeutic agent with an untested delivery system is the toughest, highest-risk, and most expensive approach. Not only does the new drug have to be proven effective and approved, but it must also be compatible with a new drug-delivery technology. In addition, regulatory agencies may be more risk-averse in this situation, adding to the challenge. However, overall the rewards can be substantial, and development in the future may trend to this pathway.

Product Development

There are many factors involved in product development, and "best practice" models will vary from company to company. It is important however to have an end goal in mind and to build your plan backward from the visualized end point. This will provide a greater degree of clarity on timing and the potential level of investment required. The figure below divides the development into the three major phases, during which activities criss-cross in cycles of feasibility and optimization continues until the first definition of "product" crystallizes into product development (Fig. 7).

The pathway to developing a new product is complex, expensive, and risky. About 50% of new systemic drugs fail due to safety, toxicity, and pharmacokinetics (Pritchard et al. 2003). The National Institute of Health reported that, of 10,000 drugs at the drug discovery phase, only one is expected to achieve FDA approval after an approximately 14-year development period. By comparison, Vitrasert progressed from in vitro tests to FDA approval in 8 years and Retisert in 7 years.

Consultants with expertise in bringing new products to market, design of clinical trials, pharmacokinetics, and particularly with drug delivery systems, can be enormously helpful in smoothing out the product development pathway and shortening the time horizon for commercialization.

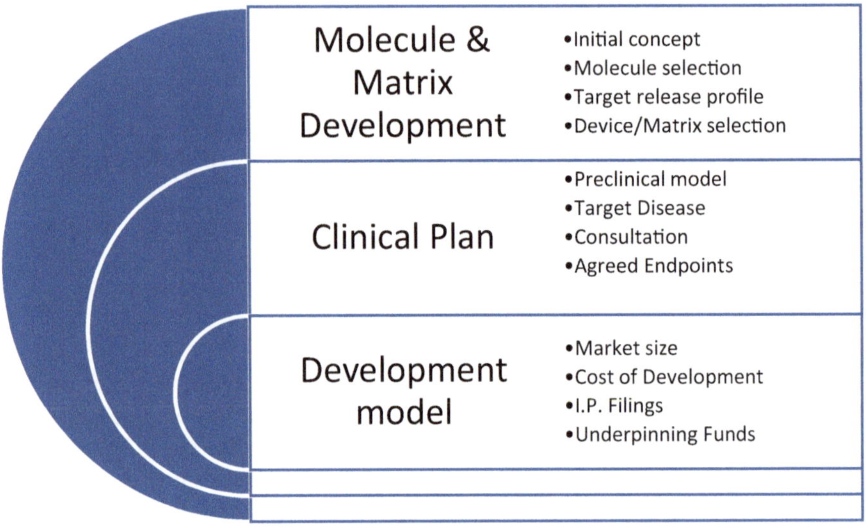

Fig. 7 Development pathway near here

The "Ideal" DDT System

Based on market research interviews conducted by Scotia Vision with industry executives in 2011, the top ten attributes illustrated in Table 1 were identified as the most desirable for developing the optimal DDT system for the posterior segment.

Target Release Profile

Four months exposure probably would not be required to treat many acute and subacute conditions. The innovator may start with a 2-month targeted release profile with perhaps less potential for side effects. It is important to keep in mind that some diseases are potentially curable (e.g., macular edema) while others are manageable, but not curable (glaucoma) which might also affect the target treatment period. Many glaucoma patients have 20–20 vision, so a longer duration, for example, 9–12 months, may be more beneficial, thus requiring less frequent treatments. Also, some diseases may require variable amount of medication to maintain a desired effect, and some drugs may have more side effects than others or if given at a constitutively high level rather than, for example, a pulsed pattern (e.g., steroids). There is always an issue of developing tolerance to a drug over time (e.g., a drug becomes ineffective if used constantly for a long time, or an increasing dose may be needed to maintain the desired effect).

Similarly developing and testing treatments for acute diseases are easier than those for chronic conditions that progress slowly over months and years (e.g.,

Table 1 Desirable DDT attributes

Item	Rationale
4–12 months delivery	Obviates frequent office visits
No adverse or minimal side effects	Avoids giving patient glaucoma and/or cataract
Ability to vary dosage—Change of posology	Customized dosing for patients—Perhaps complete withdrawal of a drug if needed
Minimal intraocular debris	Debris from drug delivery can lead to inflammation and "floaters"
Clearly developed and executed dose-ranging studies	Appropriate dose is identified in phase II or phase II/III studies to reduce risk of extended regulatory delays
High patient compliance	Better patient outcomes will trump less compliant regimens
Demonstrated safety and efficacy	This is the minimum requirement, the gatekeeper's minimum threshold
Cost-effective manufacturing	Manufacturers require acceptable gross margins to participate in this space
Continuous, controlled long-term delivery of small or large molecule therapies	Zero-order kinetics/steady-state delivery (in most cases) will meet patient/physician need for an improved treatment paradigm
Good understanding of the strategic marketing landscape, regulatory, and clinical challenges	Plan for long-term development with a competitive product, think outside the box

choroidal neovascularization (CNV) vs. geographic atrophy). It is important to stress the need for positive reimbursement without which even the greatest drug delivery technology will be a commercial failure. Successful drug delivery systems will not only have to foolproof the potential functional requirements of system but will have to find a way to address the reimbursement needs of a physician vs. the challenges of convincing society about the benefits of introduction in a choice of anti-VEGF therapy. This has been extensively modelled. Glaser and colleagues constructed a model which predicted $468 million saving to Medicare B scheme with a patient saving of $119 million if bevacizumab reimbursement was made equivalent to that for aflibercept (Glasser et al. 2020).

So what are the key success factors to consider in the product development pathway? These are generally similar to those for immediate release products, with appropriate consideration of pre-clinical proof of concept, safety and efficacy studies in cell lines and animals in preparation for the clinical phase: chemistry manufacturing and controls, engagement with IP and regulatory consultants, and then preparation for the IND submission. This could take 2 to 3 years.

Commercialization: New Sustained-Release Drug Delivery Product

The cost of development from a pre-IND to a phase IV product launch could run into several tens of millions of dollars. There are a number of external factors that can contribute to substantial costs for the development of a drug delivery system. For a completely novel approach with a new active ingredient that must be characterized from a safety and efficacy perspective combined with the novel delivery system, the costs can be substantial.

It is important to understand the approval pathway and expectation of the regulatory agency that is approached; for example, would there be a need for a progression from a phase I through to phase III pivotal studies. With a generic drug, there may be plenty of systemic information already available to allow faster progress through the process. Longer or more studies translate to greater cost, so a full understanding of the pathway for approval by the particular regulatory agency will be required from a clinical and financial perspective.

The final costs will depend on a number of factors which may include regulatory agency advice, product indication, and the final study design including number of treatment groups and sample size of treatment groups. The chosen endpoints and the length of study envisaged, together with concurrent clinical studies taking place, affect patient recruitment. Here clinical and commercial ocular drug delivery consultants can provide assistance in determining costs and timing and identifying ways to streamline the process. As given below, a two-stage approach is generally recommended to ensure costs are contained and the program meets the milestone goals in a timely manner.

Stage 1: Clinical and Regulatory Program Management

Interface and communication with the regulatory body are required to identify the timeliest, cost-effective, and efficient path to approval of the product. To that end, the existing data for the program should be reviewed and proposed clinical studies and preclinical plans generated prior to scheduling a meeting with the regulatory body. There are defined pathways to solicit advice and understanding. All too often, this step is skipped due to perceived timeline delays only to lose the time later in the development due to a lack of understanding and delay based on the unforeseen requirements imposed on the project from the agency.

Stage 2: Execution of Drug Delivery Development Program

The study plan, site selection, identification of species for pre-clinical, determination of required toxicity work, final identification of high enrolling sites, scheduling, and other matters will need to be optimized in order to project a final budget. It should be kept in mind that the product may be competing against other technologies also being developed and the site network will be very important to ensure no delays and an on-time/on-budget program.

Intellectual Property (IP)

The IP landscape is vital to understand at the outset of the development project. Generally the ocular drug delivery field can be quite complex so working with an experienced pharmaceutical and drug delivery patent attorney is important. Having a clear IP strategy that allows submission of a patent application in key geographic areas, e.g., USA, Europe, China, and Japan, is critical for progress, and the applicant will be required to demonstrate robust IP and ideally a degree of "freedom to operate." Due to the cost of IP, a carefully budgeted strategy needs to be planned. Once the leading patent is identified and submitted, as the development program proceeds, the team should continually be aware of opportunities to generate new IP to support the progress of the assets. One strategy model employed is to carve out IP that supports the value proposition and marketing strategy and maintains the competitive advantages.

Funding

The development of any new technology will require funding, most probably many millions of dollars. There are several strategies that can be employed; however going into detail is beyond the main scope of this article. However, the support from angel investors to venture capital investment funds will need to be considered. In addition, non-dilutive forms of funding, e.g., Small Business Innovation Research (SBIR) grants or grant equivalents in Europe (e.g., Innovate UK), should all be pursued.

One additional approach to consider is a strategic collaboration with a pharmaceutical partner. Typically, most major pharma companies do not have access to drug delivery technology so are willing to partner with technology companies to provide sustained release of their own novel drugs. Early feasibility deals could be negotiated to an agreed target product profile (e.g., 4–6 months delivery time). If development milestones are successful, future licensing or other business deals could be possible. In most cases the major pharma company would fund at least the early feasibility work.

Technology Pipeline

It may appear obvious but before a company embarks on their own product development program, they must fully understand the market dynamics and status of all new potentially competitive products currently in development. The acid test of determining what value proposition or competitive advantage a product will have perhaps 7–10 years from the starting point is a critical factor to discuss with the development and commercial teams.

> Ophthalmology Times in June 2011 ran the following headline – "*Drug Delivery To The Posterior Pole - In Search of The Holy Grail.*"

By inference, the "Holy Grail" has yet to be found, even with the four currently approved sustained-release products available for the posterior segment. The multi-billion dollar market for "new & innovative" ocular sustained-release products and other delivery systems, particularly to the posterior segment, therefore remains a significant market opportunity. The technology is diverse as indicated in Fig. 8.

What constitutes the "Holy Grail" is up for debate, but it could consist of the sustained-release of large molecules—proteins, peptides, or aptamers—over 4 months or more, with a high drug loading capability to enable a therapeutic dose, a sustained-release glaucoma therapy, a slow release system for geographic atrophy,

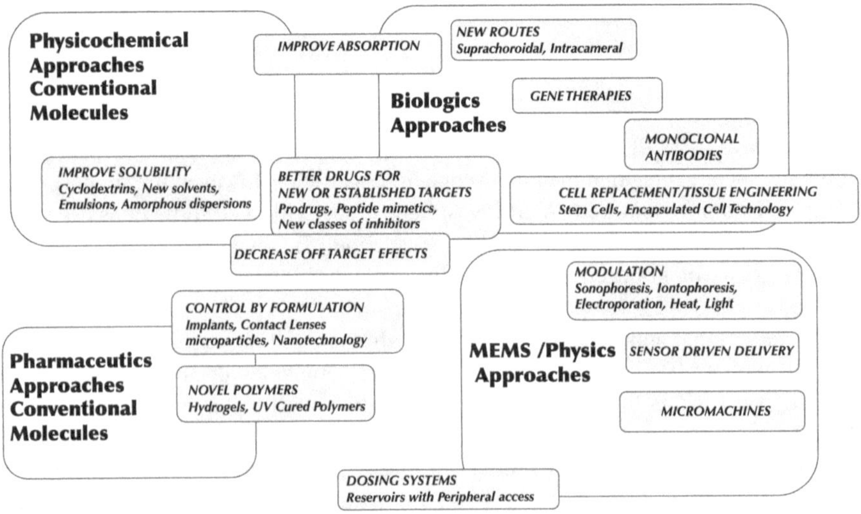

Fig. 8 Current technology horizon in ophthalmic delivery. (from Wilson and Singh)

or a microparticle-based stem cell therapy for wAMD or DR, among many others; all could be considered strong possible candidates for this honor.

There are many new products in development utilizing sustained-release technology, ranging from pre-clinical to phase III. At the end of 2016, the number of sustained-release development projects as a minimum in the various disease segments included, for example, w-AMD- DR/DME (Levine et al. 2020), glaucoma (Ta 2004), and dry eye (Del Amo and Urtti 2008).

Due to the large numbers of products in development, it should be clear the selected product and product strategy should offer "disruptive technology," that is, innovation in the market compared to what already exists or is in the development pipeline of competitors. It must offer true advantages to both patients and doctors, meet a significant market need, and be clinically feasible and potentially reimbursable. A number of protein, peptide, gene delivery, and small-molecule delivery products are in development, offering innovative strategic delivery options. Several breakthrough stem cell-based programs are now underway with some promising early successes including delivery via microparticles. Gene therapy may offer additional opportunities.

Packaging therapeutic proteins into slow-release technology of 4 months or greater has yet to be approved but is one example of a potentially disruptive development pathway. The first sustained-release approved product with a new chemical entity and not just an approved generic could be another. Any technology that can offer a several months longer duration of effect compared to current VEGF inhibitor therapies, resulting in less frequent intravitreal injection, perhaps with the regression of CNV, is a clear example of a potential disruptive technology.

Summary: The Future

From a disease perspective, we have an increased understanding of conditions that can be expected to produce new cellular targets and drug candidates. Increasing the ability to deliver an existing or new drug agent in a safe and effective way will offer multiple opportunities to tackle currently blinding diseases.

The technologies required to deliver agents specifically and effectively to the eye are rapidly evolving. These technologies will have the potential to radically alter the way many diseases are treated, especially retinal blinding diseases. The next decade promises great strides in therapy for many currently poorly treated or untreatable ocular diseases.

The future for sustained-release ocular drug delivery lies in reducing the treatment burden by innovations in delivery technology, biologics delivery, targeting gene therapy to the appropriate cell types, and combining effective small-molecule therapeutics with the appropriate drug delivery system. Patient compliance and convenience will be key driver for drug delivery; however a demonstration of improved efficacy for a new product may be essential if delivering competitive advantage is to be achieved.

Product life cycle extension strategies may include new drug delivery technologies, e.g., a new drug delivery project for Lucentis or Eylea is probably not targeting improved efficacy, but potentially fewer injections with enhanced safety and compliance. Within the next few years, several major products are coming off patent, e.g., Eylea 2020-US: 2021-Europe, Lucentis 2020-US: 2022-Europe, and Avastin 2019-US: 2022-Europe. In their place there is an emerging pipeline of biosimilar products for all three branded products. These biosimilars may be candidates for sustained-release systems.

In the future major retinal diseases such as w-AMD and DR/DME may be treated with new alternatives to the current anti-VEGF therapeutics. There are several new drugs already in late-stage clinical development with projected efficacies to approximately 12 weeks in certain patient cohorts. However it remains to be determined if these products will be successful as recent setbacks have been reported due to inflammation.

If sustained-release biologics or small-molecule retina technologies are not developed, this will create several challenges: (1) system capacity for more injections and (2) the likelihood these drugs will be used individually at different time points vs. co-formulated therapy. This further opens up an expanded and significant need for new extended-release delivery technologies. However, if new biologic therapeutics have an approved 12 weeks label from one single bolus injection, the requirement from any sustained-release technology increases to a minimum 4 months or more. For large molecule this creates a challenge but also a major opportunity. Currently there are only a few companies with the capabilities the deliver ocular biologics for 4 months or greater.

Capacity is already a major concern in Europe. PRN treatment regimens were born out of practical capacity issues and are in effect very bad for patient safety. Clearly the movement is to treat and extend, but there is still a tradeoff between maximum efficacy and capacity constraints. This is an additional key argument for what drug delivery could potentially obviate in the future. In many developing countries such as China, India, Russia, and others, practitioners have one chance to address disease morphology as they consistently lose patients to follow up. Administering a single injection of anti-VEGF does not solve a problem and indeed creates an ethical issue in initiation of a treatment for a patient that will not return. Could this ethical dilemma be addressed in a long-term delivery option?

Ideally, the drug comes first, and then delivery technology follows, but it is not an exact science. It could be ideal to have a broad drug delivery platform technology, customized to new drugs or a class of drugs. From the regulatory standpoint, it might be useful to develop a drug and its delivery system in parallel, at least when phase III is reached. Each drug compound will require a different release profile and formulation; therefore formulation work and rigor are generally the same for both existing and new compounds. Platform compatibility differs with compound solubility and molecule size, and key factors for the formulation team are the intended duration of treatment and release profile.

References

Armaly MF, Rao KR. The effect of pilocarpine Ocusert with different release rates on ocular pressure. Invest Ophthalmol Vis Sci. 1973;12(7):491–6.

Bailey C, Chakravarthy U, Lotery A, Menon G, Talks J. Real-world experience with 0.2 μg/day fluocinolone acetonide intravitreal implant (ILUVIEN) in the United Kingdom. Eye. 2017;31(12):1707–15.

Bakri SJ, Boyer DS, Albini TA, Holekamp NM, Ferrone PJ, Bhagat R, Novakovic Z. Comparison of the tissue penetration and glide force of 22-gauge thin-wall needles for intravitreal implant administration. Ophthalmic Surg Lasers Imaging Retina. 2014;45(5):430–5.

Del Amo EM, Urtti A. Current and future ophthalmic drug delivery systems: a shift to the posterior segment. Drug Discov Today. 2008;13(3–4):135–43.

Glasser DB, Parikh R, Lum F, Williams GA. Intravitreal anti-vascular endothelial growth factor cost savings achievable with increased bevacizumab reimbursement and utilization. Ophthalmology. 2020;127:1688–92. https://doi.org/10.1016/j.ophtha.2020.06.012.

Grzybowski A, Told R, Sacu S, Bandello F, Moisseiev E, Loewenstein A, Schmidt-Erfurth U. 2018 update on intravitreal injections: Euretina expert consensus recommendations. Ophthalmologica. 2018;239(4):181–93.

Jager RD, Aiello LP, Patel SC, Cunningham ET Jr. Risks of intravitreous injection: a comprehensive review. Retina. 2004;24(5):676–98.

Jehangir N, Bever G, Mahmood SM, Moshirfar M. Comprehensive review of the literature on existing punctal plugs for the management of dry eye disease. J Ophthalmol. 2016;2016:9312340.

Jones L, Downie LE, Korb D, Benitez-del-Castillo JM, Dana R, Deng SX, Dong PN, Geerling G, Hida RY, Liu Y, Seo KY. TFOS DEWS II management and therapy report. Ocul Surf. 2017;15(3):575–628.

Kiernan DF. Dexamethasone intracameral drug-delivery suspension for inflammation associated with vitreoretinal surgery. BMJ Open Ophthalmol. 2020;5(1):e000491.

Levine D, Albini TA, Fine HF, Yeh S. Emerging drug delivery systems for posterior segment disease. Ophthalmic Surg Lasers Imaging Retina. 2020;51(3):132–5.

Market Scope estimates. Ophthalmic Innovation Summit Data. Accessed 14 Nov 2013.

Market Scope estimates. Ophthalmic Innovation Summit 2019 Presentation.

Martin DF, Parks DJ, Mellow SD, Ferris FL, Walton RC, Remaley NA, Chew EY, Ashton P, Davis MD, Nussenblatt RB. Treatment of cytomegalovirus retinitis with an intraocular sustained-release ganciclovir implant: a randomized controlled clinical trial. Arch Ophthalmol. 1994;112(12):1531–9.

Medeiros FA, Walters TR, Kolko M, Coote M, Bejanian M, Goodkin ML, Guo Q, Zhang J, Robinson MR, Weinreb RN, ARTEMIS 1 Study Group. Phase 3, randomized, 20-month study of Bimatoprost implant in open-angle Glaucoma and ocular hypertension (ARTEMIS 1). Ophthalmology. 2020;127:1627–41.

Pritchard JF, Jurima-Romet M, Reimer ML, Mortimer E, Rolfe B, Cayen MN. Making better drugs: decision gates in non-clinical drug development. Nat Rev Drug Discov. 2003;2(7):542–53.

Sanborn GE, Anand R, Torti RE, Nightingale SD, Cal SX, Yates B, Ashton P, Smith T. Sustained-release ganciclovir therapy for treatment of cytomegalovirus retinitis: use of an intravitreal device. Arch Ophthalmol. 1992;110(2):188–95.

SERC. Ophthalmic drug delivery systems for the treatment of retinal disease. ARVO 2009 Summer Eye Research Conference. IOVS; 2009.

Steinbach OC. An industry update: the latest developments in therapeutic delivery covering July 2019. Ther Deliv. 2019;10(12):753–6.

Ta CN. Minimizing the risk of endophthalmitis following intravitreous injections. Retina. 2004;24:699–705.

Tan DT, Chee SP, Lim L, Theng J, Van Ede M. Randomized clinical trial of Surodex steroid drug delivery system for cataract surgery: anterior versus posterior placement of two Surodex in the eye. Ophthalmology. 2001;108(12):2172–218.

Tost FH, Geerling G. Plugs for occlusion of the lacrimal drainage system. In: Surgery for the dry eye, vol. 41. Basel: Karger; 2008. p. 193–212.
Wander AH, Koffler BH. Extending the duration of tear film protection in dry eye syndrome: review and retrospective case series study of the hydroxypropyl cellulose ophthalmic insert. Ocul Surf. 2009;7(3):154–62.
World Health Organization. Global data on visual impairments. Geneva: World Health Organization; 2010. p. 3.

Trademarks

Dexycu and Yutiq are trademarks of EyePoint Pharmaceuticals.
Eylea is a trademark of Regeneron.
Fortovase is trademark of Roche.
Iluvien is a trademark of Alimera Sciences.
Lacrisert, Vitrasert, and Retisert are trademarks of Bausch + Lomb.
Lucentis and Avastin are trademarks of Genentech.
Ozurdex is a trademark of Allergan.

Manufacturing Considerations and Challenges for AAV Ocular Gene Therapy

Gerard A. Rodrigues, Evgenyi Shalaev, Thomas K. Karami, James Cunningham, Hongwen M. Rivers, Shaoxin Feng, Dinen Shah, and Nigel K. H. Slater

Abstract Gene therapy has the potential to revolutionize the treatment of debilitating ocular diseases, and adeno-associated viral vectors (AAV) are particularly attractive. However, technical hurdles that remain in their manufacture and product formulation may hamper the development of new treatments. Such issues arise from the structural properties of AAV and differ between serotypes due to their varying capsid structure. Whereas the manufacture and quality control of AAV vectors generally exploit many techniques, materials and procedures that have been developed for other biotherapeutics, the unique characteristics of AAV capsid structure, and the AAV genome encapsidation process introduce complexities not commonly encountered hitherto. In this chapter we examine these issues and review progress with overcoming the technical challenges that arise. In particular, the development of AAVs using directed evolution or rational design techniques will result in novel

G. A. Rodrigues
Eye Care Research, Abbvie, Irvine, CA, USA

E. Shalaev · S. Feng · D. Shah
Pharmaceutical Sciences, R&D, Abbvie, Irvine, CA, USA

T. K. Karami
Pharmaceutical R&D, Allergan plc, Irvine, CA, USA

Analytical QC, Glaukos Corporation, San Clemente, CA, USA

J Cunningham
Combination Product Development, AbbVie, Irvine, CA, USA

H. M. Rivers
Allergan Aesthetics R&D, AbbVie, Irvine, CA, USA

N. K. H. Slater (✉)
Department of Chemical Engineering and Biotechnology, University of Cambridge, Cambridge, UK
e-mail: nkhs2@cam.ac.uk

S. Neervannan, U. B. Kompella (eds.), *Ophthalmic Product Development*, AAPS Advances in the Pharmaceutical Sciences Series 37,
https://doi.org/10.1007/978-3-030-76367-1_22

vectors that enable more efficient transduction of diseased tissues and which can be manufactured and stored more reliably and safely.

Keywords Ocular gene therapy · Adeno-associated virus · AAV · Biomanufacturing

Adeno-Associated Virus Vectors

AAVs are small (25-nm), non-enveloped viruses that belong to the Parvoviridae family (Schön et al. 2015). The AAV genome contains three open reading frames (ORFs) bordered by inverted terminal repeats and is packaged within an icosahedral capsid. The rep ORF codes for four proteins essential for viral genome packaging and replication. The cap ORF codes for three proteins, VP1, VP2, and VP3, which comprise the 60-subunit AAV capsid. The three capsid proteins share a common β-barrel connected by variable loops (Madigan and Asokan 2016). These loops create unique surface topologies for the different AAV serotypes that enable cell binding and entry, as well as determine immunological properties. Uptake of AAV into target cells involves binding to cell surface glycans, which is facilitated by interactions with co-receptors including fibroblast growth factor receptor, epidermal growth factor receptor, and platelet-derived growth factor receptor.

Over the past few decades, AAV has become the vector of choice for ocular gene therapy. In addition to efficiently transducing retinal cells (Schön et al. 2015), AAV offers a number of advantages over other viral vectors. Because of the need for a helper adenovirus for replication, AAV vectors are considered to be non-pathogenic (Boye et al. 2013). Additionally, once an AAV vector transduces a cell, its genetic material remains episomal, thereby reducing the potential for insertional mutagenesis. To date 12 AAV serotypes differing in their capsid sequences have been identified in primates. AAV1, 2, 4, 5, 6, 7, 8, and 9 all display tropism for retinal tissue (Lotery et al. 2003; Yang et al. 2002; Weber et al. 2003), and many of these serotypes are currently being used in ocular gene therapy studies. In recombinant AAV (rAAV) design, the rep and cap genes are substituted with the transgene and its regulatory elements. rAAV viral particles are then generated by transfecting producer cells with a plasmid containing transgene DNA flanked by the ITRs and a separate construct expressing the viral rep and cap genes. The adenovirus helper factors can then be provided either by adenovirus infection or by transfection of a third plasmid encoding these factors. rAAV2, the first recombinant AAV vector to be successfully used for gene transfer, efficiently transduces cells of the retinal pigmented epithelium (RPE) and retinal ganglion cells but is less competent at transducing photoreceptors (Auricchio et al. 2001; Gao et al. 2005; Vandenberghe et al. 2011). The tropism of rAAVs can be refined through the use of pseudotyping, the process of mixing of a capsid and genome from different viral serotypes (Auricchio 2003). Pseudotyping can confer a vector with a desired tropism as well potentially enhance transduction efficiency. Additionally, it can also help avoid issues with immunity to certain AAV serotypes due to pre-existing antibodies that could reduce the efficacy and safety of a gene therapy. Most recombinant vectors in use today are based on elements of AAV2 combined with the capsids of AAV1 (AAV2/1), AAV4 (AAV2/4), AAV5 (AAV2/5), AAV6 (AAV2/6), AAV7 (AAV2/7), AAV8 (AAV2/8), and AAV9 (AAV2/9). All AAV pseudotypes transduce RPE cells, with AAV2/1, AAV2/4, and AAV2/6 being the most efficient (Trapani et al. 2014). In contrast, only AAV2/5, AAV2/7, AAV2/8, and AAV2/9 transduce photoreceptors with AAV2/8 and AAV2/9 being most effective (Lotery et al. 2003; Auricchio et al. 2001; Vandenberghe et al. 2011; Allocca et al. 2007; Lebherz et al. 2008; Manfredi et al. 2013; Mussolino et al. 2011). AAV2/5, AAV2/8, and AAV2/9 transduce the highest percentage of cone photoreceptors (Venderbeghe et al. 2011; Manfredi et al. 2013; Alexander et al. 2007; Kamaromy et al. 2010; Mancuso et al. 2009). The tropism of naturally occurring and pseudotyped AAV vectors for retinal cells is summarized in Table 1.

An alternative method to attain cell specificity is through the use of cell-specific promoters to drive transgene expression. Most gene therapies for inherited retinal diseases (IRDs) are intended to overcome gene mutations in cells of the RPE or photoreceptors. The rhodopsin kinase 1 and interphotoreceptor retinoid binding protein (IRBP) promoters have been used to successfully express transgenes in both cone and rod photoreceptors (Young et al. 2003; Beltran et al. 2012). While the rhodopsin promoter has been used to attain rod-specific expression (Allocca et al.

Table 1 Retinal tropism of naturally occurring and pseudotyped AAV vectors

Serotype	PRs	RPE	Species	ROA	References
AAV2/1	+	–	Mse	IVT, SR	Gao et al. (2005)
AAV2	+	+	NHP	IVT, SR	Auricchio (2003)
AAV2/2	+	+	Mse	IVT, SR	Lotery et al. (2003) and Gao et al. (2005)
AAV2/3	–	–	Mse	SR	Lotery et al. (2003)
AAV2/4	–	+	Rat, dog, NHP	SR	Yang et al. (2002)
AAV5	+	+	NHP	SR	Boye et al. (2013)
AAV2/5	++	+	Mse, rat, dog, pig	IVT, SR	Lotery et al. (2003), Yang et al. (2002), Gao et al. (2005), Allocca et al. (2007), Lebherz et al. (2008), Manfredi et al. (2013) and Mussolino et al. (2011)
AAV2/6	–	+	Mse	SR	Lotery et al. (2003)
AAV2/7	+++	+	Mse	SR	Allocca et al. (2007) and Lebherz et al. (2008)
AAV8	++	+	NHP	SR	Auricchio (2003)
AAV2/8	+++	+	Mse, pig	SR	Allocca et al. (2007), Manfredi et al. (2013) and Mussolino et al. (2011)
AAV2/9	++	+	Mse, pig	SR	Allocca et al. (2007) and Manfredi et al. (2013)

PRs photoreceptors, *RPE* retinal pigmented epithelium, *ROA* route of administration, *IVT* intravitreal, *SR* subretinal, *Mse* mouse, *NHP* non-human primate

2007; Flannery et al. 1997), cone arrestin, blue opsin, and red/green opsin promoters have been used to obtain selective expression in cone photoreceptors (Alexander et al. 2007; Carvalho et al. 2011; Michalakis et al. 2010; Komáromy et al. 2008).

An area of active investigation is the development of next-generation vectors with enhanced transduction efficiency or altered tropism. Two of the more common approaches currently being used to derive novel capsid variants are rational design and directed evolution. Rational design leverages understanding of structure/function relationships to design modified virus capsids. For instance, site-directed mutagenesis of tyrosine residues in the AAV2 capsid has been conducted to create AAV2 variants that escape phosphorylation, ubiquitination and proteasomal degradation (Zhong et al. 2008). As a consequence these variants display improved nuclear transport and improved transgene expression (Zhong et al. 2008). AAV2, AAV8, and AAV9 vectors harboring these mutations have enhanced transduction relative to their wild type counterparts in vitro and in vivo (Zhong et al. 2008; Petrs-Silva et al. 2009). As an additional benefit these mutations have the potential to lower immunogenicity because less protein degradation reduces major histocompatibility complex class 1 presentation of viral antigens (Martino et al. 2013). Other examples of modifications include mutation of capsid sequences to remove antibody-binding epitopes, or the incorporation of novel ligands as a means to alter vector tropism (Kwon and Schaffer 2008). Altering AAV tropism by chemical conjugation of a targeting ligand to a capsid protein has been shown to enable transduction of normally non-permissive cell types (Kwon and Schaffer 2008). In contrast to rational design, the directed evolution approach exploits random genetic diversity together with evolutionary pressure to enrich for novel synthetic AAV capsids with desired features. Randomly generated AAV libraries with mutated capsids are screened in vivo to identify variants displaying improved tissue penetration, cell transduction or tropism. For instance, variants have been isolated that show an enhanced ability to transit through ocular barriers following intravitreal administration and thus more effectively transduce retinal cells (Kotterman and Schaffer 2014).

Manufacture and Purification of AAV

Ayuso et al. (2010) have reviewed cell culture manufacturing procedures for clinical AAV, and, in general, it follows the process scheme shown in Scheme 1. The manufacturing objectives for a clinical AAV bulk product are that it must be both safe and efficacious but produced at a price that is acceptable to healthcare providers. Clinical safety requires the reliable removal of product- and process-related impurities to a level that is judged acceptable for use in patients. Product cost control requires that the manufacturing unit operations necessary to ensure conformity of product quality with the approved specification provide high recovery and titer of infectious AAV viruses.

Commonly, transient transfection procedures have been used to express the genes required for AAV production in a cell line such as HEK293 human embryonic

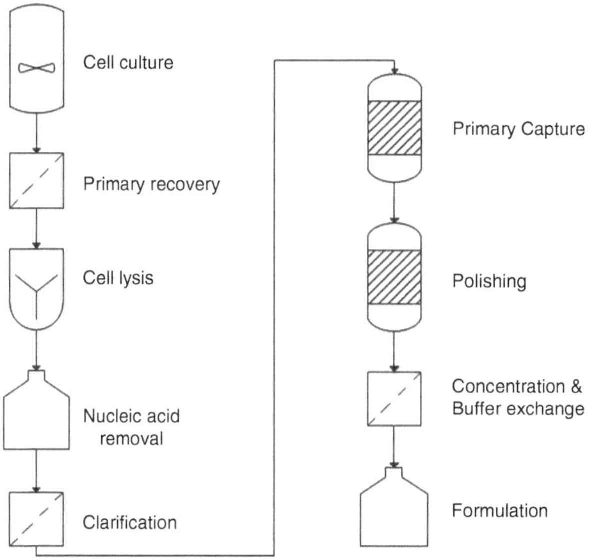

Scheme 1 Generic process flowsheet for the production of clinical grade AAV bulk product

kidney cells. This involves the transfection of three separate plasmids into the host cells: one to express the transgene, a second to encode regulatory and structural proteins, and a third that enables the viral helper functions that are essential for viral replication.

An alternative approach is to integrate the viral regulatory and structural protein encoding genes into the cell genome. Infection with a helper virus and a virus that carries the genome for the transgene are required to initiate vector production. Additional integration of the transgene into the cell genome enables the construction of a producer cell line with which AAV production commences upon infection by a helper virus. The downstream processing for each of these approaches presents different challenges.

These different approaches place specific requirements on downstream processing. For the case of transient transfection, residual amounts of the three plasmids must be separated from the AAV product, which can be achieved by virtue of the highly anionic nature of nucleic acids. AAV manufacture using a producer or a packaging cell line requires the separation of a helper and/or vector-carrying virus from the AAV product, which might be a more complex undertaking. The following discussion addresses process and product impurity issues for the transient expression process.

Contamination by process-related impurities can occur from materials involved in the production process. As with most cell-derived biological products, contamination can occur from host cell proteins (HCPs) and from any proteins present in the cell culture media, for example, serum proteins. Nucleic acid contamination can also derive from the host cells as well as from residual plasmids used for transfection (Wright 2014). Host cell-derived contamination is an unavoidable complication for AAV manufacture since it is necessary to lyse the producer cells in order to release the AAV product. Steps in the downstream processing scheme are therefore included to reduce these sources of contamination to acceptably low levels. Less clear-cut is the potential that the host cells may harbor adventitious agents and that these may co-purify with the AAV product (Wright 2014). Avoidance of this risk commonly rests upon thorough cell line characterization and validation during cell line development.

Another class of impurities derives from the physical and chemical characteristics of the AAV vector (Wright 2014; Schnodt and Buning 2017). Such product-related impurities include AAV capsids in which the viral genome is missing or in which an incomplete genome is packaged. These unpackaged virions afford no medicinal benefit and increase the burden of viral proteins that may be immunogenic in the clinical product (Schnodt and Büning 2017). Host cell nucleotides and residual plasmid material may also be encapsidated (Wright 2014), with similar consequences. Finally, malformed or degraded capsid proteins may result in non-infectious AAV. Such impurities derive from cell culture and occur to varying extents dependent upon process conditions.

Other product-related impurities may be formed at any stage of the manufacturing process or even during subsequent formulation and storage (see below). One such impurity, AAV aggregates, is concerning since they reduce product efficacy and risk potentially immunogenic viral side effects (Wright 2014). Chemical degradation of capsid proteins may occur, for example, by oxidation, deamidation, or cleavage, and could also reduce product efficacy.

A broad strategy for downstream processing of AAV that aims to control and overcome these risks may involve:

Centrifugation and cell lysis: Cells are first de-watered and washed to remove contaminants in spent cell culture media. They are then lysed to release AAV by means of either mechanical stress, hypertonic shock, or freeze-thaw procedures. Inevitably, upon lysis HCPs and nucleic acid contaminants are released that must be removed in subsequent processing, and mechanical stresses may damage AAV capsids. Optimization of this step is therefore essential to obtain the highest ratio of infectious AAV to impurities, and controls would involve determination of infectious and viral particle titer, HCPs, and host cell and total DNA.

Nucleic acid removal: Endonucleases, such as the commonly used Benzonase®, can be used to reduce nucleic acid contaminants. Reports implicate nucleic acids in the aggregation of AAV capsids (Wright et al. 2005), which may be by cross-linking between cationic regions on adjacent viruses, and this effect is also reduced by endonuclease digestion. It is assumed that encapsidated genomes within AAV particles are resistant to nuclease digestion and that infectious titer is unaffected. Process controls at this stage include assays of infectious and viral particle titer to determine the degree of encapsidation, as well as assays of HCPs, host cell, and total DNA.

Affinity chromatography: Due to the highly specific nature of the molecular interactions that lead to the affinity binding of a target protein to an immobilized ligand affinity, chromatography can efficiently remove HCPs and serum protein impurities. Clinical grade AAV2 has been produced with heparin affinity chromatography, whereby adsorption was achieved from low ionic strength buffer and the virus was eluted with a buffer of higher ionic strength (Clément and Grieger 2016; Summerford and Samulski 1999). Many AAV serotypes can be purified using AVB-Sepharose High Performance (Nass et al. 2018), an adsorbent that exploits single-domain antibody fragments as the ligand that binds a common AAV capsid epitope with high selectivity. Low pH buffers are used to elute virus, and to avoid virus degradation, the eluate pH must be adjusted to avoid lability to acidic pH. Controls would again involve the measurement of infectious and viral particle titer, HCPs, host cell, and total DNA and the degree of AAV aggregation.

Intriguingly, Wang et al. (2015) have elucidated the AVB-Sepharose binding epitope. The incorporation of this epitope in AAV8, rh.64R1, and AAV9 with the corresponding epitope of AAV3B provided increased binding affinity yet had no effect upon the vector potency. This demonstration opens the way to modify virus capsid structure to enhance flexibility in manufacturing.

Ion-exchange chromatography: Separation of infectious AAV viruses from empty, non-infectious capsids has been achieved by cesium chloride or iodixanol density gradient ultracentrifugation, but this low productivity method is tedious to conduct as part of a scaled-up manufacturing process. Recently, full and empty capsids have been separated by anion exchange chromatography, exploiting differences in capsid electrical charge arising from the anionic viral genome (Qu et al. 2007). The high ionic strength buffer used for elution may also reduce virus aggregation (see below). Infectious and viral particle titer, HCPs, host cell, and total DNA and AAV aggregates would be assayed.

Final polishing: Further reduction of HCPs and small molecular contaminants might be achieved using core-bead adsorbents. These structured matrices have a narrow pore outer shell that excludes the passage of AAV and a ligand-functionalized core that binds low-molecular-weight contaminants. For example, Nestola et al. (2015) demonstrated the use of a core-shell octylamine resin (Capto™ Core 700) to capture residual DNA and HCP while the AAV product passed in the column flow-through.

Other adsorbents and approaches have been reported. For example, heparin affinity column chromatography methods have been used for rAAV2 (Gao et al. 2000), and ceramic hydroxyapatite adsorbents have been used to purify AAV1 and AAV9 (Qua et al. 2015). However, while such chromatographic processes reduce many process- and product-related impurities to acceptable levels, some difficult challenges remain. Reliance upon anion exchange chromatography to remove empty capsids may be feasible, but removal of capsids containing host cell and helper or partial genomes is likely to be extremely challenging, if at all possible. Also, chromatographic steps do not necessarily avoid problems with AAV aggregates. Some degree of aggregate clearance can occur as aggregates may adsorb less efficiently than smaller virions, but aggregates may reform during formulation and storage. Scope remains then for further developments in processing methods.

Formulation Strategies

Ocular gene therapy products are administered by injection, and a ready-to-use (RTU) liquid formulation is the preferred dosage form. However, in order to demonstrate that a liquid biopharmaceutical product is stable during shipping and storage, extensive studies are required, and the amount of material needed is significant. Production of AAV vectors is probably the most complicated manufacturing process among pharmaceutical products, and the quantities of the material available are usually very small. As a result, the amount of the AAV material available is not sufficient to develop a stable liquid presentation, and gene therapy products are usually presented as frozen solutions. Nevertheless, limited data on stability of liquid and freeze-dried formulations are described in the literature, as discussed later in this section.

Drug product (DP) design for AAV is similar to that for other biologics, with the majority of formulations containing buffer, tonicity agent, cryoprotector, and surfactant. Although the purity of a vector is mainly controlled during drug substance (DS) manufacture (Wright 2014), production and storage of a DP can impact the potency and also generate undesirable modified species of AAV. Aggregation represents the most common degradation pathway for AAVs, although oxidation was also mentioned in the discussions of formulation studies of AAV albeit without any data present (Wright et al. 2005). Aggregates and other degradation-related impurities could represent potential immunotoxicity risks and have and also impact biodistribution and the in vivo functional activity of the AAV (Wright et al. 2005).

Formulation development of biologicals typically starts with selecting pH and ionic strength. Similar to the majority of biologics, stability of AAV depends on pH. Considering that the majority of AAV are formulated as frozen solutions, both pH in solution and in the frozen state should be taken into consideration, as pH in the frozen state can be significantly different from that in the initial solution (Wu et al. 2015). In one carefully designed study, pH of AAV formulations was measured in the frozen state and was correlated with stability of AAV during freeze-thaw. The stability improved with the pH (as measured in the frozen state) increased from 4 to 7 (Croyle et al. 2001). The sensitivity of AAV to pH could be related to the impact of pH on structure of the AAV capsid. According to Venkatakrishnan et al. (2013), decrease in pH from 7.5 to 4.0 resulted in a lost of the α-helical structure of VP1u (a unique 137–amino acid N-terminal region of VP1) in AAV1 and AAV6. The loss of α-helical structure was reversed when the pH was brought back to neutral. No changes in the VP3 common region were observed in the same pH range. Negative-stain electron microscopy data showed retention of the capsid integrity at pH of 7.5 to 4. The partial unfolding of VP1u capsid protein at acidic pH could increase propensity of the virus to hydrophobic aggregation. This would be consistent with the observed pH trend in the freeze-thaw stability of AAV, when acidic shift in pH after freezing was correlated with instability during freeze-thaw (see above). While near-neutral pH appears to be favorable for both freeze-thaw stability and structural integrity, an opposite pH trend was reported for solubility of AAV2. The decrease in pH from 10 to 4.5 resulted in a monotonous decrease in the solubility of the vector, with the higher solubility observed at basic pH. Samples with virus concentration of approximately 0.1 mg/mL were used in that study (Xie et al. 2004). Lower solubility is usually associated with a higher aggregation propensity; therefore, it is not clear at the moment if neutral pH is indeed optimal stability against aggregation in all AAVs.

Wright et al. (2005) developed an efficient and material-sparing method for formulation screening, which would facilitate selection of formulation components. The method involves dilution of concentrated AAV samples, which were initially prepared at high salt concentrations, with solutions of different excipients and monitoring aggregation of the diluted samples by dynamic light scattering. Stabilization against the dilution stress was found to be provided by charged excipients (inorganic salts and amino acids), with multi-charged salts being more effective. For example, stability of AAV2 was achieved at 180 mOsm of Mg sulfate versus 220 mOsm for Na sulfate and 300–320 mOsm for NaCl and amino acids. The inhibition of AAV aggregation by salts correlated with the ionic strength of the solution, rather than the osmolarity, with higher ionic strength preventing increase in particle size (increase in the particle size reflects aggregation). Accordingly, multivalent salts, which have higher ionic strength than monovalent ions at comparable osmolarity, required lower concentrations than NaCl to prevent aggregation of AAV. Contrary to the results obtained with the dilution stress method, no consistent trends in the impact of salt on solubility of AAVs were reported in another study. While increase in solubility by Mg^{2+} (20 mM, pH 4.5 to 7.5) was observed (Xie et al. 2004), a multi-charge anion ($citrate^{3-}$) didn't improve solubility of the virus.

The AAV purification method can also have a significant effect on aggregation. In particular, removal of DNA impurities by nuclease treatment resulted in reduced aggregation even at lower ionic strength (Wright et al. 2005). This observation is consistent with electrostatic attraction as the main driving force for the aggregation, if one assumes that the DNA impurities are sorbed on the virus capsid particle. DNA and the capsid proteins have different acid dissociation constants, resulting in the difference in charges across the capsid and between the particles. The aggregation is caused then by the electrostatic attractive interactions and ionic bridges between the sorbed DNA and the DNA-free part of another virus particle.

There are conflicting reports on the impact of non-ionic surfactants (Pluronic® F68 and polysorbate PS80) and polyhydroxy compounds (PHC) on AAV aggregation. The surfactants and PHC (including glycerol, sucrose, mannitol, trehalose, sorbitol) were found to be ineffective in the prevention of aggregation, at least at the concentrations used (1% for polysorbate 80, 10% pluronic F68, and 5% for PHC), during a dilution stress study (Wright et al. 2005). In another study, however, 25% glycerol was reported to prevent AAV aggregation (Xie et al. 2004). It should be noted that such a high concentration of glycerol is not practical in a DP formulation. Reduction of aggregation by a non-ionic surfactant (beta-octyl glucopyranoside at 0.01–0.5%) was also reported (Xie et al. 2004).

In addition to aggregation, physical loss of AAV particles due to adsorption on different surfaces of contact could also represent a significant risk. For example, up to 80% of AAV-2 were lost during centrifugal concentration with Amicon microconcentrator-100. The losses were confirmed to be the result of sorption of the virus on the membrane (Xie et al. 2004). Polyethylene glycol (12–20 kDa) at 5–20% reduced sorption-related losses to 30%, whereas a high concentration of glycerol (up to 25%) further minimized sorption.

Significant loss of virus was also observed under simulated in-use conditions, when formulations without a surfactant were exposed to different delivery devices (Bennicelli et al. 2008). Addition of the nonionic surfactant Pluronic F68 (0.001%) to the formulation allowed essentially 100% recovery of the virus. In addition, Sommer et al. (2002) reported that polysorbate 80 or Pluronic F68 at concentrations of 0.01 and 0.001%, respectively, prevented losses of the vector.

Both AAV DS and DP are usually stored as frozen solutions, and a typical AAV product undergoes several freeze-thaw cycles during manufacture and use. Freeze-thaw-induced AAV2 aggregation was shown to depend on both formulation composition and freezing temperature (−20 vs. −80 °C). Between three compositions tested, the highest level of aggregation (based on a very high DLS intensity observed) was detected in a formulation containing phosphate buffer and no cryopreservative, while the addition of a cryoprotector (sorbitol) to the phosphate buffer helped to decrease the DLS signal and therefore reduce aggregation level. Formulation with sodium citrate and Tris buffer produced the best results, with no aggregation after one freeze-thaw cycle. However, AAV aggregation was observed in the same formulation after five freeze-thaw cycles (−80 °C) (Wright et al. 2005). It should be noted that pH of a frozen solution can be significantly different from the pH of the liquid sample and also that AAV stability could be compromised in the acidic pH range. Therefore, AAV can be destabilized during freeze-thaw if pH of the frozen solution shifts to the acidic range. The freeze-induced pH changes depend on the type and concentration of buffer and the presence of other excipients; for example, a significant acidic pH shift (to up to pH 3.4) is expected if phosphate buffer is used (Wu et al. 2015), and such acidic shift can explain aggregation of AAV in phosphate buffer during freeze-thaw.

AAV products are usually stored in the frozen state. A convenient temperature condition for frozen storage is usually considered to be −15 to −25 °C (often referred as −20 °C), because of a wide availability of freezers for this temperature range, including large walk-in freezers. However, significant stability risks are associated with storage of aqueous biological solutions at −20 °C, because they are not completely frozen in this temperature range. Frozen aqueous solutions consist of at least two phases, ice and amorphous freeze-concentrate, which contains all the solutes and unfrozen portion of water; water content in the freeze-concentrate can vary between 20 wt% and 50 wt% and above, depending on the chemical composition and temperature (Levine et al. 2002). This freeze-concentrate is a viscous liquid at −20 °C, which solidifies (forming a glassy state) below its glass transition temperature. In typical biopharmaceutical systems, the glass transition temperatures of the freeze-concentrate are −35 to −50 °C or even lower. The liquid state of the freeze-concentrate, and the corresponding higher molecular mobility, would facilitate various destabilization processes, many of which depend on the rotational and translational diffusion of viruses and/or reactive species such as oxygen. Multiple destabilization pathways are associated with freeze-thaw and storage in the frozen state. The most pronounced effect of freezing is a dramatic (2 orders of magnitude or more) increase in concentrations of all components including an active ingredient (such as AAV), which could promote aggregation, and oxygen, which could trigger oxidative degradation processes. Other freeze-induced destabilization effects include pH changes; formation of extensive ice/solution interface which is often associated with destabilization of proteins; and crystallization of a cryoprotector. Many cases of chemical and physical (i.e., protein aggregation) instability in partially frozen systems have been reported (Anzo et al. 2013; Bhatnagar et al. 2007, 2008; Franks and Hatley 1991; Lund et al. 1969; Pincock 1969; Pincock and Kiovsky 1966; Schwegman et al. 2009; Singh et al. 2011; Piedmonte et al. 2007; Connolly et al. 2015; Gu et al. 2013), with the instabilities usually observed in the temperature range of −10 to −30 °C. Consequently, storage at lower temperatures, preferably below −65 °C, should be used for aqueous gene therapy products as a default.

While AAV DP can be quite stable when stored below −65 °C, shipping and storage of frozen biologics could be challenging from the practical perspectives. Therefore, AAV formulations that are stable above 0 °C are desirable. There are two options for developing refrigeration-stable biologicals, i.e., a ready-to-use liquid formulation and a lyophilized (freeze-dried) dosage form. Limited data on impact of freeze-drying on AAV and longer-term stability of liquid AAV formulations under refrigeration (2 to 8 °C) have been reported (Croyle et al. 2001; Wright et al. 2003; Howard and Harvey 2017). Croyle et al. evaluated stability of AAV during freeze-drying and subsequent storage, as well as stability of liquid formulations, using transduction activity test (Croyle et al. 2001). A noticeable loss of titer (0.3 log) was observed after lyophilization of AAV2 formulations with phosphate potassium buffer, as well as with a formulation containing 0.4% sucrose, 0.4% mannitol, and protamine. Infectivity of the latter formulation was also measured after 3 months storage at 25 °C, and no loss of titer was observed. A liquid formulation of the same virus, which was formulated with 0.4% sucrose, 0.4% mannitol, 0.001% sorbitan monolaurate [Span 20], and 0.1% protamine, was quite stable, with only 0.1 log titer loss after approx. 150 days at 4 °C. Unexpectedly, the same liquid formulation was also reported to be stable during storage for approx. 150 days at a higher temperature of 25 °C. A long-term liquid stability was also reported by Wright et al. (2003), where AAV vector in neutral phosphate buffered saline with 5% sorbitol and 0.1% polysorbate 80 showed no significant loss of transduction activity after 1 year at 2–8 °C. On the other hand, up to 40% loss in transgene expression after 7 weeks at 4 °C was observed for AAV1 virus diluted in phosphate-buffered saline containing 0.5 mM of $MgCl_2$ (Howard and Harvey 2017). Overall, while limited literature-reported studies indicate a feasibility to develop refrigeration-stable AAV formulations, additional studies are required to evaluate stability of the vectors over the prolonged time period in both liquid and lyophilized forms. A valuable commercial product usually requires a shelf life of at least 18 to 24 months and should also tolerate shipping stresses (e.g., mechanical agitation) and short-term temperature excursions.

Aggregation of Vectors

As discussed, aggregation represents a main destabilization pathway for AAV vectors. Wright et al. (2005) studied aggregation of rAAV2 by dynamic light scattering to determine the aggregation of AAV with high sensitivity. The method requires tiny amount of sample (20 μL volume) to obtain a semi-quantitative measure of AAV aggregation. AAV aggregation can also be monitored by small-angle X-ray scattering (SAXS), which can be used for samples that are not transparent, such as freeze-dried powders and frozen solutions. An example of SAXS data for AAV is shown in Fig. 1. A broad peak, which is detected in an initial sample (before performing freeze-thaw cycles), is indicative of interacting particles (i.e., viruses) present. After 10 freeze-thaw (FT) cycles, a decrease in the magnitude virus interaction peak and corresponding increase in the scattering in the low-q region occur; both observations are consistent with the reduction in monomeric viruses and increase in the aggregates as the result of freeze-thaw.

Some insight into aggregation potential, via hydrophobic interaction, in different AAV serotypes might be had by aggregation hotspot analysis of capsid proteins. This involves identifying stretches of amino acids in proteins that may be more prone to aggregation and can help in lead optimization as well as manufacturability

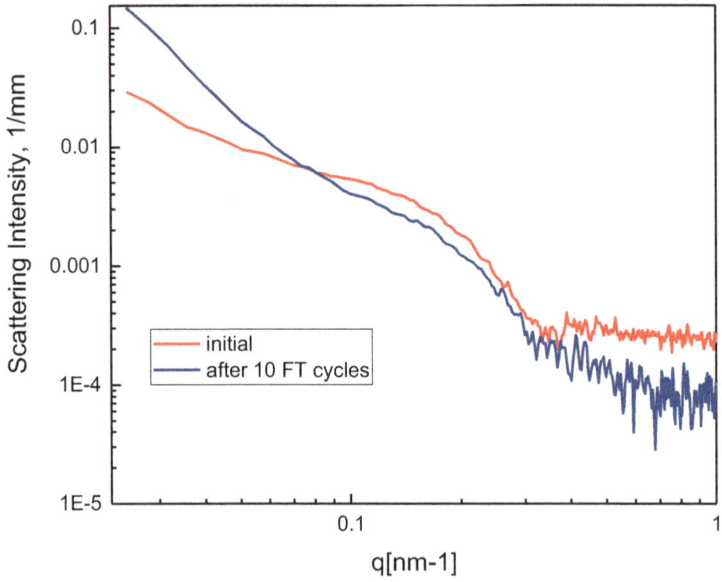

Fig. 1 Example of SAXS data for AAV

and developability assessment. Having this information can also enable the protein formulator to better design a formulation to mitigate such issues. Although this exercise is routinely used for therapeutic proteins, currently, no such information is available for gene therapy vectors like AAV. With AAVs becoming more commonly employed in therapeutic formulations, this information can be a helpful first step to better understand these unique protein carriers.

Three different serotypes of adeno-associated viruses (AAVs), AAV2, AAV5, and AAV9, were selected for aggregation hotspot analysis. The specific serotypes were chosen for the following reasons:

1. AAV2, AAV5, and AAV9 serotypes are vectors of choice for several ongoing clinical trials, and AAV2 is already being used in an FDA-approved drug (Gu et al. 2013; Wright et al. 2003; Howard and Harvey 2017).
2. Crystal structures of VP3 of these serotypes are readily available. The structure of VP3 protein of AAV2 is shown below (Fig. 2).
3. The three serotypes have different transduction efficiencies for different tissues and therefore potentially subtle structural differences that enable this (Bennett et al. 2017).

The AAV viral capsid contains a total of 60 copies of three viral proteins—VP1, VP2, and VP3 (Bennett et al. 2017). The ratio of these proteins is 1:1:10 (Bennett et al. 2017). Although structures of VP1 and VP2 are yet to be determined, the structure of VP3 for several AAV serotypes has already been determined. The entire sequence of VP3 (62 kDa) is contained within VP2, and the entire sequence of VP2 (73 kDa) is contained within VP1 (87 kDa) (https://clinicaltrials.gov/ct2/show/NCT03520712) (common C-terminal). Interestingly, VP3 has been shown to be sufficient to assemble the virus capsid and is responsible for the stability of the viral capsid. T_m (melting temperature) of the capsid has been shown to be dictated by VP3 sequence alone (Bennett et al. 2017). This is an important observation, as that allows us to study the already available structures of VP3 for the AAV serotypes to identify differences that may lead to their different behaviors in solution.

For this aggregation hotspot evaluation, VP3 structure of AAV2, AAV5, and AAV9 was obtained from Protein Data Bank (PDB). The PDB code for AAV2 is

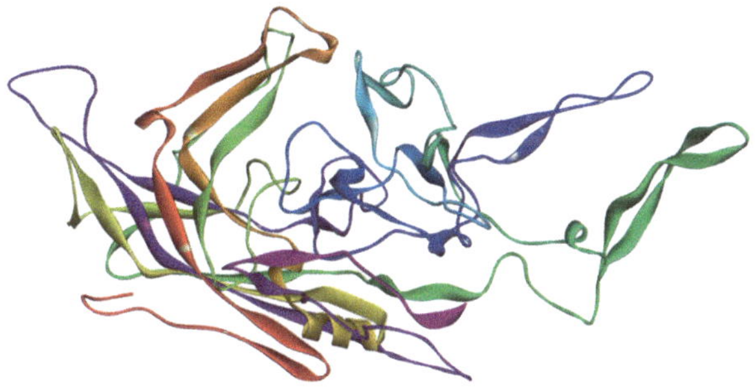

Fig. 2 Structure of VP3 of AAV2 (1LP3)

1LP3, for AAV5 is 3NTT, and for AAV9 is 3UX1. Primary sequences in FASTA format of all VP3s were also obtained from PDB.

Two separate aggregation prediction softwares that utilize different approaches to hotspot prediction were used. Aggrescan is an aggregation prediction program that takes into account potential hydrophobic interactions, structure of the protein, charge, and electrostatic interactions to determine potential aggregation-prone regions in the protein (Conchilli-Solé et al. 2007). The other program used was ZipperDB, which analyzes the structure of proteins to determine its proclivity to form cross-β structures (Zipper et al. 2010). Cross-β structures are known to be the most common structures in protein aggregates.

The identified hotspots generated for each serotype by both the prediction softwares were then compared to each other. To ensure better accuracy at identifying only the most aggregation vulnerable regions, we selected only those stretches of amino acids that were predicted by both the models as being aggregation hotspots.

The identified hotspots for the serotypes are highlighted in red in the crystal structure of their respective VP3 regions below (Figs. 3, 4, and 5). Interestingly, AAV5 can be seen as the serotype that has the least aggregation hotspots and, therefore, theoretically the lowest aggregation propensity. This explanation aligns with the observation from a previous study that identified AAV5 as being a highly stable serotype compared to AAV1, AAV2, and AAV8 (Rayaprolu et al. 2013).

AAV9 was identified as having the most aggregation vulnerable regions, via hydrophobic mechanism, followed by AAV2. Both AAV2 and AAV9 may therefore be more prone to hydrophobic aggregation. The high aggregation propensity seen in AAV9 could possibly arise from stretches of hydrophobic residues. Interestingly, it is known that AAV9 can cross the blood-brain barrier with enhanced efficiency among the different AAV serotypes, and therefore it can target the central nervous system with high efficiency (Conchilli-Solé et al. 2007). The AAV9 VP differs in three variable surface regions (VR-I, VR-II, and VR-IV) compared to AAV2 (Di Mattia et al. 2012). The residues responsible for the ability of AAV9 to cross the blood-brain barrier are still unknown, although residues in the VRs are the most likely candidates.

Additional insight into aggregation behavior of AAV can be obtained from analysis of the electrostatic interaction. The aggregation studies on AAV2 vectors by

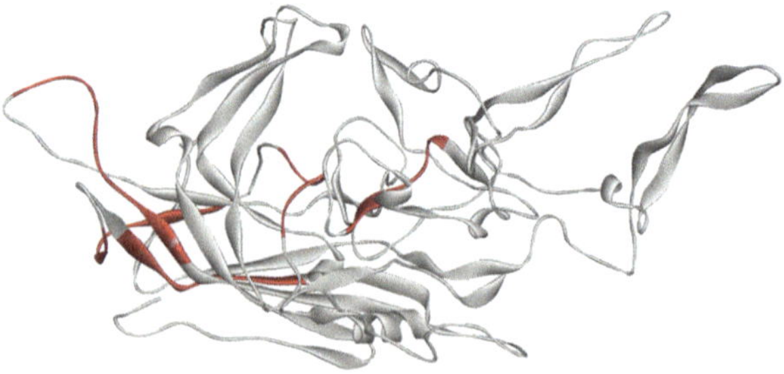

Fig. 3 Hydrophobic hotspots of VP3 of AAV2

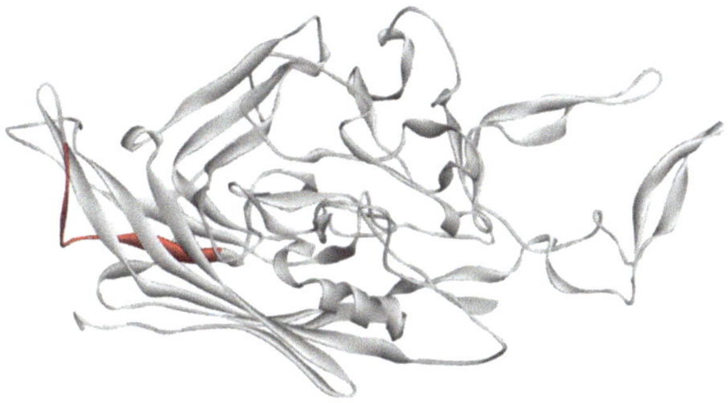

Fig. 4 Hydrophobic hotspots of VP3 of AAV5

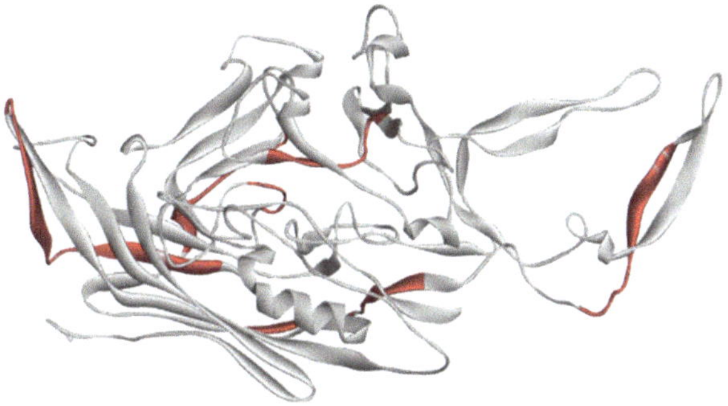

Fig. 5 Hydrophobic hotspots of VP3 of AAV9

Wright et al. (2005) showed increasing ionic strength could significantly inhibit the aggregations, but on the other hand, increasing molarity may not have the similar inhibition effect (such as adding glycerol). This experiment indicates the electrostatic interactions between AAV molecules are likely attractive. Here a simulation was run on calculating Coulombic interactions between two AAV molecules. The 3D structure of the AAV molecules is taken from Protein Data Bank (PDB) with a PDB code of 3NG9. For Coulombic interactions, only the charged residues were considered. The charges at neutral pH were assigned to all amino acid side chains. Due to charge screening in aqueous media, the Coulombic energy between two charged residues (i, j with the charges of q_i and q_j) can be modeled with a Yukawa-type potential:

$$E_{ij}\left(r_{ij}\right) = k_e \frac{q_i q_j}{r_{ij}} \exp\left(\frac{-r_{ij}}{\lambda_D}\right) \quad (1)$$

where r_{ij} is the distance between the two charges; k_e is Coulomb's constant; and λ_D is the Debye length. In an electrolyte solution at 25 °C, the Debye length can be expressed as (assuming monovalent electrolyte) (Israelachvili 1985):

$$\lambda_D(\text{nm}) = \frac{0.304}{\sqrt{I(M)}} \tag{2}$$

The above equation is consistent with Blanco et al.'s work on modeling colloidal protein interactions (B_{22}) (Blanco et al. 2013) where they got $\lambda_D(\text{nm}) \approx \frac{0.286}{\sqrt{I(M)}}$. The total electrostatic energy between two AAV molecules can be calculated as:

$$E_{\text{total}}(R) = \sum_i \sum_j E_{ij}(r_{ij}) \tag{3}$$

When two AAV molecules are separated at a fixed distance (center to center), both molecules have freedom to rotate. An energy minimization was carried out at any given distance. Due to the high icosahedral symmetry of the AAV molecules, small movements are needed to reach the energy minimum. Figure 6 shows the electrostatic energies as a function of intermolecular distance (center to center distance subtracted by the diameter of AAV, 26 nm) at different ionic strengths. Figure 7 shows the electrostatic energies as a function of ionic strength at a fixed intermolecular distance of 4 nm.

From Fig. 6, whether the Coulombic interaction is attractive (negative energy) or repulsive (positive energy) depends on the ionic strength of the solution. The electrostatic interaction is repulsive when ionic strength is below 0.02 M because the net charge of AAV molecules is highly negative (~−180). When ionic strength is higher than 0.02 M, the overall electrostatic interactions become attractive because the local attractive interactions overcome the repulsive interactions from total charges

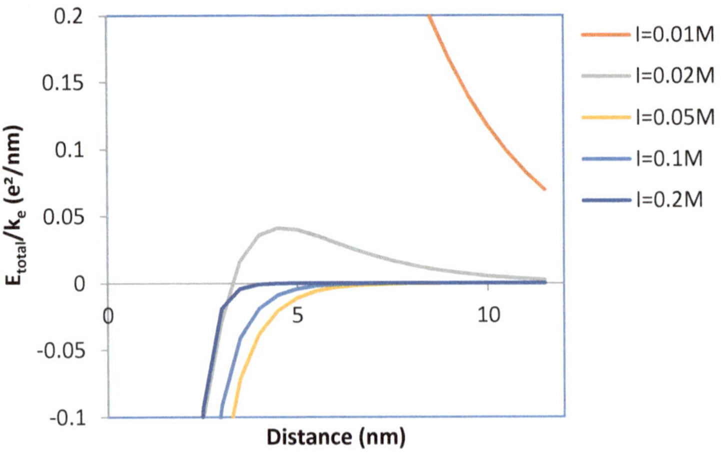

Fig. 6 The electrostatic energy between two AAV molecules as a function of intermolecular distance at different ionic strengths

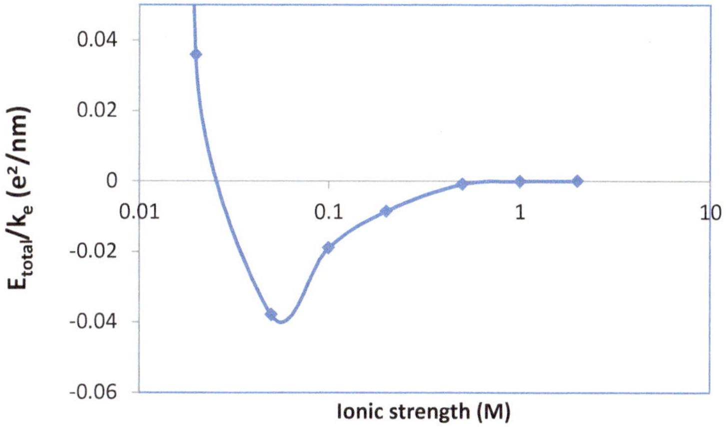

Fig. 7 The electrostatic energy between two AAV molecules as a function of ionic strength at an intermolecular distance of 4 nm

that had been screened significantly by electrolytes. At ionic strength of 0.02 M, the electrostatic interaction could change sign with changing intermolecular distance. At long distance (>3.7 nm) the interaction is repulsive, while at shorter distance the interaction becomes attractive.

Figure 7 shows another interesting property of electrostatic interactions between AAV molecules: the attractive interaction has a maximum at around 0.05 M ionic strength. At this condition, the AAV molecules have the highest tendency to aggregate because of the strong Coulombic attractions. If the ionic strength increases or decreases from the critical point, the electrostatic interactions will always be less attractive (inhibiting aggregations). This simulation results are consistent with Wright et al.'s experimental results at the ionic strength range of 0.05–0.15 M (Wright et al. 2005).

One thing that needs to be pointed out is the simulations only considered 60 VP3 proteins in the virus capsid (no X-ray structure available for VP1/VP2 proteins). The fractions of VP1 and VP2 in virus capsid are quite low: about 1/12 for each. More importantly, VP1 and VP2 have identical structure with VP3 on C-terminal region and have additional amino acid chains at the N-terminal of VP3. These additional AA chains are located inside the capsid and have no exposure to the surface. The surface charges make the dominant contributions to the Coulombic interactions, while the inside charges get screened by electrolytes (Eq. 1). Therefore, the simulations are good representative of electrostatic interactions between AAV molecules.

The development of novel and sophisticated state-of-the-art technologies with higher resolution and applications of these new technologies for characterization of biologics, including vectors, will offer a better scientific understanding of the physicochemical properties, purity, and stability of the vector-based formulations used for human gene therapy products in the future.

Conclusions and Outlook

Gene therapies for the treatment of ocular diseases have received increased attention since 2017 and the approval of voretigene neparvovec-rzyl. Further research into such therapies will continue for both IRDs and other ocular diseases. Indeed, the development pipeline is robust with many programs being evaluated. Although AAV is now firmly established as the most popular vector technology, lentivirus also remains of interest for its ability to deliver larger transgenes. Investigations of non-viral delivery also continue, for example, with the EyeCET electroporation technology that is currently being tested in a phase I/II trial. Both directed evolution and rational design techniques will likely lead to new AAV vectors with improved characteristics, including more efficient transduction of target tissues and less invasive administration (e.g., intravitreal or suprachoroidal injection).

The production, formulation, and characterization of AAV-based products commonly exploit procedures already established for other biologics, and the regulatory expectations in the United States and European Union are broadly similar. However, as seen above, certain aspects of AAV manufacturing, formulation, and quality assurance differ from those encountered with protein products, and we envisage that these will continue to develop as clinical demand for AAV increases. Regulatory expectations will likely be further clarified as products proceed through development and approval. The recent scientific advances and clinical successes have promoted gene therapy as a powerful approach with the potential to provide long-lasting therapeutic benefits. We therefore anticipate continued progress in translating gene therapy into a prominent modality for the treatment of ocular diseases.

Acknowledgments Authors thank Drs. Michael Sztucki and Theyencheri Narayanan for their expert assistance with the SAXS experiment.

References

Alexander JJ, Umino Y, Everhart D, Chang B, Min SH, Li Q, et al. Restoration of cone vision in a mouse model of achromatopsia. Nat Med. 2007;13(6):685–7.

Allocca M, Mussolino C, Garcia-Hoyos M, Sanges D, Iodice C, Petrillo M, et al. Novel adeno-associated virus serotypes efficiently transduce murine photoreceptors. J Virol. 2007;81(20):11372–80.

Anzo K, Harada M, Okada T. Enhanced kinetics of pseudo first-order hydrolysis in liquid phase coexistent with ice. J Phys Chem A. 2013;117(41):10619–25.

Auricchio A. Pseudotyped AAV vectors for constitutive and regulated gene expression in the eye. Vision Res. 2003;43(8):913–8.

Auricchio A, Kobinger G, Anand V, Hildinger M, O'Connor E, Maguire AM, et al. Exchange of surface proteins impacts on viral vector cellular specificity and transduction characteristics: the retina as a model. Hum Mol Genet. 2001;10(26):3075–81.

Ayuso E, Mingozzi F, Bosch F. Production, purification and characterization of adeno-associated vectors. Curr Gene Ther. 2010;10(6):423–36.

Beltran WA, Cideciyan AV, Lewin AS, Iwabe S, Khanna H, Sumaroka A, et al. Gene therapy rescues photoreceptor blindness in dogs and paves the way for treating human X-linked retinitis pigmentosa. Proc Natl Acad Sci U S A. 2012;109(6):2132–7.

Bennett A, et al. Thermal stability as a determinant of serotype identity. Mol Ther Methods Clin Dev. 2017;6:171–82. Accessed 15 Sep 2017. https://doi.org/10.1016/j.omtm.2017.07.003. eCollection.

Bennicelli J, Wright JF, Komaromy A, Jacobs JB, Hauck B, Zelenaia O, et al. Reversal of blindness in animal models of Leber congenital amaurosis using optimized AAV2-mediated gene transfer. Mol Ther. 2008;16(3):458–65.

Bhatnagar BS, Bogner RH, Pikal MJ. Protein stability during freezing: separation of stresses and mechanisms of protein stabilization. Pharm Dev Technol. 2007;12(5):505–23.

Bhatnagar BS, Pikal MJ, Bogner RH. Study of the individual contributions of ice formation and freeze-concentration on isothermal stability of lactate dehydrogenase during freezing. J Pharm Sci. 2008;97(2):798–814.

Blanco MA, Sahin E, Robinson AS, Roberts CJ. J Phys Chem B. 2013;117:16013–28.

Boye SE, Boye SL, Lewin AS, Hauswirth WW. A comprehensive review of retinal gene therapy. Mol Ther. 2013;21(3):509–19.

Carvalho LS, Xu J, Pearson RA, Smith AJ, Bainbridge JW, Morris LM, et al. Long-term and age-dependent restoration of visual function in a mouse model of CNGB3-associated achromatopsia following gene therapy. Hum Mol Genet. 2011;20(16):3161–75.

Clément N, Grieger JC. Manufacturing of recombinant adeno-associated viral vectors for clinical trials. Mol Ther Methods Clin Dev. 2016;3:16002.

Conchilli-Solé O, de Groot NS, Avilés FX, Vendrell J, Daura X, Ventura S. AGGRESCAN: a server for the prediction and evaluation of "hot-spots" of aggregation in polypeptides. BMC Bioinform. 2007;8:65.

Connolly BD, Le L, Patapoff TW, Cromwell MEME, Moore JMR. Protein aggregation in frozen trehalose formulations: effects of composition, cooling rate, and storage temperature. J Pharm Sci. 2015;104:4170–84.

Croyle MA, Cheng X, Wilson JM. Development of formulations that enhance physical stability of viral vectors for gene therapy. Gene Ther. 2001;8(17):1281–90.

Di Mattia MA, et al. Structural insight into the unique properties of adeno-associated virus serotype 9. J Virol. 2012;86(12):6947–58. https://doi.org/10.1128/JVI.07232-11.

Flannery JG, Zolotukhin S, Vaquero MI, LaVail MM, Muzyczka N, Hauswirth WW. Efficient photoreceptor-targeted gene expression in vivo by recombinant adeno-associated virus. Proc Natl Acad Sci U S A. 1997;94(13):6916–21.

Franks F, Hatley RHM. Stability of proteins at subzero temperatures: thermodynamics and some ecological consequences. Pure Appl Chem. 1991;63(10):1367–80.

Gao G, Qu G, Burnham MS, Huang J, Chirmule N, Joshi B, Yu QC, Marsh JA, Conceicao CM, Wilson JM. Hum Gene Ther. 2000;11:2079–91.

Gao G, Vandenberghe LH, Wilson JM. New recombinant serotypes of AAV vectors. Curr Gene Ther. 2005;5(3):285–97.

Gu JH, Beekman A, Wu T, Piedmonte DM, Baker P, Eschenberg M, Hale M, Goldenbergh M. Beyond glass transitions: studying the highly viscous and elastic behavior of frozen protein formulations using low temperature rheology and its potential implications on protein stability. Pharm Res. 2013;30:387–401.

Howard DB, Harvey BK. Assaying the stability and inactivation of AAV serotype 1 vectors. Hum Gene Ther Methods. 2017;28(1):39–48.

Israelachvili J. Intermolecular and surface forces. San Diego, CA: Academic Press; 1985.

Komáromy AM, Alexander JJ, Cooper AE, Chiodo VA, Glushakova LG, Acland GM, et al. Targeting gene expression to cones with human cone opsin promoters in recombinant AAV. Gene Ther. 2008;15(14):1049–55.

Komaromy AM, Alexander JJ, Rowlan JS, Garcia MM, Chiodo VA, Kaya A, et al. Gene therapy rescues cone function in congenital achromatopsia. Hum Mol Genet. 2010;19(13):2581–93.

Kotterman MA, Schaffer DV. Engineering adeno-associated viruses for clinical gene therapy. Nat Rev Genet. 2014;15(7):445–51.
Kwon I, Schaffer DV. Designer gene delivery vectors: molecular engineering and evolution of adeno-associated viral vectors for enhanced gene transfer. Pharm Res. 2008;25(3):489–99.
Lebherz C, Maguire A, Tang W, Bennett J, Wilson JM. Novel AAV serotypes for improved ocular gene transfer. J Gene Med. 2008;10(4):375–82.
Levine H, Shalaev E, Franks F. Solid-liquid state diagrams in pharmaceutical lyophilisation: crystallization of solutes. In: Levine H, editor. Amorphous food and pharmaceutical systems. London: The Royal Society of Chemistry; 2002. p. 200–15.
Lotery AJ, Yang GS, Mullins RF, Russell SR, Schmidt M, Stone EM, et al. Adeno-associated virus type 5: transduction efficiency and cell-type specificity in the primate retina. Hum Gene Ther. 2003;14(17):1663–71.
Lund DB, Fennema OR, Powrie WD. Enzymic and acid hydrolysis of sucrose as influenced by freezing. J Food Sci. 1969;34:378–82.
Madigan VJ, Asokan A. Engineering AAV receptor footprints for gene therapy. Curr Opin Virol. 2016;18:89–96.
Mancuso K, Hauswirth WW, Li Q, Connor TB, Kuchenbecker JA, Mauck MC, et al. Gene therapy for red-green colour blindness in adult primates. Nature. 2009;461(7265):784–7.
Manfredi A, Marrocco E, Puppo A, Cesi G, Sommella A, Della Corte M, et al. Combined rod and cone transduction by adeno-associated virus 2/8. Hum Gene Ther. 2013;24(12):982–92.
Martino AT, Basner-Tschakarjan E, Markusic DM, Finn JD, Hinderer C, Zhou S, et al. Engineered AAV vector minimizes in vivo targeting of transduced hepatocytes by capsid-specific CD8+ T cells. Blood. 2013;121(12):2224–33.
Michalakis S, Mühlfriedel R, Tanimoto N, Krishnamoorthy V, Koch S, Fischer MD, et al. Restoration of cone vision in the CNGA3−/− mouse model of congenital complete lack of cone photoreceptor function. Mol Ther. 2010;18(12):2057–63.
Mussolino C, Della Corte M, Rossi S, Viola F, Di Vicino U, Marrocco E, et al. AAV-mediated photoreceptor transduction of the pig cone-enriched retina. Gene Ther. 2011;18(7):637–45.
Nass SA, Mattingly MA, Woodcock DA, Burnham BL, Ardinger JA, Osmond SE, et al. Universal method for the purification of recombinant AAV vectors of differing serotypes. Mol Ther Methods Clin Dev. 2018;9:33–46.
Nestolaa P, Peixotoa C, Villaine L, Alvesa PM, Carrondoa MJT, Mota JPB. Rational development of two flowthrough purification strategies for adenovirus type 5 and retro virus-like particles. J Chromatogr A. 2015;1426:91–101.
Petrs-Silva H, Dinculescu A, Li Q, Min SH, Chiodo V, Pang JJ, et al. High-efficiency transduction of the mouse retina by tyrosine-mutant AAV serotype vectors. Mol Ther. 2009;17(3):463–71.
Piedmonte DM, Sumers C, McAuley A, Karamuji L, Ratnaswamy G. Sorbitol crystallization can Lead to protein aggregation in frozen protein formulations. Pharm Res. 2007;24:136–46.
Pincock RE. Reactions in frozen systems. Acc Chem Res. 1969;2:97–103.
Pincock RE, Kiovsky TE. Kinetics of reactions in frozen solutions. J Chem Educ. 1966;43:358–60.
Qu G, Bahr-Davidson J, Prado J, Tai A, Cataniag F, McDonnell J, et al. Separation of adeno-associated virus type 2 empty particles from genome containing vectors by anion-exchange column chromatography. J Virol Methods. 2007;140(1–2):183–92.
Qua W, Wanga M, Wua Y, Lva Y, Wanga Q, Xua R. Calcium-ion-modulated ceramic hydroxy-apatite resin for the scalable purification of recombinant adeno-associated virus serotype 9. J Chromatogr B. 2015;990:15–22.
Rayaprolu V, et al. Comparative analysis of adeno-associated virus capsid stability and dynamics. J Virol. 2013;87(24):13150–60. https://doi.org/10.1128/JVI.01415-13.
Schnodt M, Büning H. Improving the quality of adeno-associated viral vector preparations: the challenge of product-related impurities. Hum Gene Ther Methods. 2017;28(3):101–8.
Schön C, Biel M, Michalakis S. Retinal gene delivery by adeno-associated virus (AAV) vectors: strategies and applications. Eur J Pharm Biopharm. 2015;95(Pt B):343–52.

Schwegman JJ, Carpenter JF, Nail SL. Evidence of partial unfolding of proteins at the ice/freeze-concentrate interface by infrared microscopy. J Pharm Sci. 2009;98(9):3239–46.

Singh SK, Kolhe P, Mehta AP, Chico SC, Lary AL, Huang M. Frozen state storage instability of a monoclonal antibody: aggregation as a consequence of trehalose crystallization and protein unfolding. Pharm Res. 2011;28(4):873–85.

Sommer J, Sanftner L, Feng L, Suzuki B, Zhen Z, Powell S, et al. Optimal recovery of AAV-AADC vector from Teflon catheters using Polysorbate 80 or Pluronic F68 (abstract 768). Mol Ther. 2002;5(Suppl):S251.

Summerford C, Samulski RJ. Viral receptors and vector purification: new approaches for generating clinical-grade reagents. Nat Med. 1999;5(5):587–8.

Trapani I, Puppo A, Auricchio A. Vector platforms for gene therapy of inherited retinopathies. Prog Retin Eye Res. 2014;43:108–28.

Vandenberghe LH, Bell P, Maguire AM, Cearley CN, Xiao R, Calcedo R, et al. Dosage thresholds for AAV2 and AAV8 photoreceptor gene therapy in monkey. Sci Transl Med. 2011;3(88):88ra54.

Venkatakrishnan B, Yarbrough J, Domsic J, Bennett A, Bothner B, Kozyreva OG, et al. Structure and dynamics of adeno-associated virus serotype 1 VP1-unique N-terminal domain and its role in capsid trafficking. J Virol. 2013;87(9):4974–84.

Wang Q, Lock M, Prongay AJ, Alvira MR, Petkov B, Wilson JM. Identification of an adeno-associated virus binding epitope for AVB sepharose affinity resin. Mol Ther Methods Clin Dev. 2015;2:15040.

Weber M, Rabinowitz J, Provost N, Conrath H, Folliot S, Briot D, et al. Recombinant adeno-associated virus serotype 4 mediates unique and exclusive long-term transduction of retinal pigmented epithelium in rat, dog, and nonhuman primate after subretinal delivery. Mol Ther. 2003;7(6):774–81.

Wright JF. Product-related impurities in clinical-grade recombinant AAV vectors: characterization and risk assessment. Biomedicine. 2014;2(1):80–97.

Wright JF, Qu G, Tang C, Sommer JM. Recombinant adeno-associated virus: formulation challenges and strategies for a gene therapy vector. Curr Opin Drug Discov Devel. 2003;6(2):174–8.

Wright JF, Le T, Prado J, Bahr-Davidson J, Smith PH, Zhen Z, et al. Identification of factors that contribute to recombinant AAV2 particle aggregation and methods to prevent its occurrence during vector purification and formulation. Mol Ther. 2005;12(1):171–8.

Wu C, Shamblin S, Varshney D, Shalaev E. Advance understanding of buffer behavior during lyophilization. In: Varshney D, Singh M, editors. Lyophilized biologics and vaccines: modality-based approaches. New York: Springer; 2015. p. 25–42.

Xie Q, Hare J, Turnigan J, Chapman MS. Large-scale production, purification and crystallization of wild-type adeno-associated virus-2. J Virol Methods. 2004;122(1):17–27.

Yang GS, Schmidt M, Yan Z, Lindbloom JD, Harding TC, Donahue BA, et al. Virus-mediated transduction of murine retina with adeno-associated virus: effects of viral capsid and genome size. J Virol. 2002;76(15):7651–60.

Young JE, Vogt T, Gross KW, Khani SC. A short, highly active photoreceptor-specific enhancer/promoter region upstream of the human rhodopsin kinase gene. Invest Ophthalmol Vis Sci. 2003;44(9):4076–85.

Zhong L, Li B, Mah CS, Govindasamy L, Agbandje-McKenna M, Cooper M, et al. Next generation of adeno-associated virus 2 vectors: point mutations in tyrosines lead to high-efficiency transduction at lower doses. Proc Natl Acad Sci U S A. 2008;105(22):7827–32.

Zipper DB, Goldschmidt L, Teng PK, Riek R, Eisenberg D. Identifying the amylome, proteins capable of forming amyloid-like fibrils. Proc Natl Acad Sci U S A. 2010;107:3487–92.

Index

S. Neervannan, U. B. Kompella (eds.), *Ophthalmic Product Development*, AAPS Advances in the Pharmaceutical Sciences Series 37,
https://doi.org/10.1007/978-3-030-76367-1

T